Chickpea Breeding and Management

Chickpea Breeding and Management

Edited by

S.S. Yadav

Pulse Research Laboratory, Division of Genetics, Indian Agricultural Research Institute, New Delhi, India

R.J. Redden

Australian Temperate Field Crops Collection, Department of Primary Industries, Victorian Institute for Dryland Agriculture, Horsham, Victoria, Australia

W. Chen

United States Department of Agriculture – Agricultural Research Service (USDA-ARS), Grain Legume Genetics and Physiology Research Unit, Washington State University, Pullman, Washington State, DC, USA

B. Sharma

Division of Genetics, Indian Agricultural Research Institute, New Delhi, India

www.cabi.org

CABI is a trading name of CAB International

CABI Head Office
Nosworthy Way
Wallingford
Oxfordshire OX10 8DE
UK
Tel: +44 (0)1491 832111
Fax: +44 (0)1491 833508
E-mail: cabi@cabi.org
Website: www.cabi.org

CABI North American Office
875 Massachusetts Avenue
7th Floor
Cambridge, MA 02139
USA
Tel: +1 617 395 4056
Fax: +1 617 354 6875
E-mail: cabi-nao@cabi.org

A catalogue record for this book is available from the British Library, London, UK.

Library of Congress Cataloging-in-Publication Data

Chickpea breeding and management / edited by S.S. Yadav . . . [et al.].
p. cm.
Includes bibliographical references and index.
ISBN 1-84593-213-7 (alk. paper)
1. Chickpea. 2. Chickpea – Breeding. 3. Chickpea – Diseases and pests.
I. Yadav, S.S. (Shyam S.)
SB351.C45C45 2006
635' .657—dc22

2006027059

ISBN-13: 978 1 84593 213 8

Typeset by SPi, Pondicherry, India.
Printed and bound in the UK by Cromwell Press, Trowbridge.

Contents

Contributors

S. Abbo, *The Levi Eshkol School of Agriculture, The Hebrew University of Jerusalem, Rehovot 76100, Israel. E-mail: abbo@agri.huji.ac.il*

N. Açikgöz, *Aegean Agricultural Research Institute (AARI), PO Box 9, Menemen 35661, Izmir Turkey. E-mail: aari@egenet.com*

I.P.S. Ahlawat, *Division of Agronomy, Indian Agricultural Research Institute, New Delhi 110012, India. E-mail: ahlawat47@hotmail.com*

F. Ahmad, *Brandon University, Brandon, MB R7A 6A9, Canada. E-mail: ahmad@brandonu.ca*

S.S. Ali, *Department of Agricultural Biotechnology, Assam Agricultural University, Jorhat, Assam 785013, India. E-mail: shahin.sharif@rediffmail.com*

D.V. Amla, *National Botanical Research Institute, Rana Pratap Marg, Lucknow 226 001 India. E-mail: dvamla@rediffmail.com*

M.R. Anwar, *Primary Industries Research Victoria – Horsham 110 Natimuk Road, Horsham, VIC 3400, Australia. E-mail: muhuddin.anwar@dpi.vic.gov.au*

A. Aslam, *Biofertilizer Division, National Institute for Biotechnology and Genetic Engineering (NIBGE), PO Box 577, Jhang Road, Faisalabad, Pakistan. E-mail: asma@nibge.org*

P.N. Bahl, *A-9, Nirman Vihar, New Delhi 110092, India. E-mail: pnbahl@hotmail.com*

A. Bano, *Department of Plant Sciences, Faculty of Biological Sciences, Quaid-i-Azam University, Islamabad 45320, Pakistan. E-mail: asgharibano@yahoo.com*

G. Bejiga, *Green Focus Ethiopia, PO Box 802, Code No. 1110, Addis Ababa, Ethiopia. E-mail: geletub@hotmail.com*

J.D. Berger, *CSIRO Plant Industry, Private Bag 5, Wembley, WA 6913, Australia. E-mail: jens.berger@csiro.au; Centre for Legumes in Meditarranean Agriculture, M080, The University of Western Australia, 35 Stirling Highway, Crawley, WA 6009, Australia.*

A.F. Berrada, *Colorado State University, Arkansas Valley Research Center, Rocky Ford, Colorado, USA. E-mail: Abdel.Berrada@Colostate.Edu; aberrada@coop.ext.colostate.edu*

Z. Bishaw, *International Center for Agricultural Research in the Dry Areas (ICARDA), PO Box 5466, Aleppo, Syria. E-mail: z.bishaw@cgiar.org*

J. Brand, *Primary Industries Research Victoria – Horsham, 110 Natimuk Road, Horsham, VIC 3400, Australia. E-mail: jason.brand@dpi.vic.gov.au*

W. Chen, *USDA-ARS, Grain Legume Genetics and Physiology Research Unit, Washington State University, Pullman, WA 99164-6402, USA. E-mail: w-chen@wsu.edu*

S.L. Clement, *USDA-ARS, 59 Johnson Hall, Washington State University, Pullman, WA 99164-6402, USA. E-mail: slclement@wsu.edu*

S. Coles, *C/o School of Biological Sciences, University of Birmingham, Edgbaston, Birmingham, B15 2TT, UK. E-mail: coles@upatras.gr*

C.J. Coyne, *Plant Introduction Unit, 303 Johnson Hall, Washington State University, Pullman, WA 99164, USA. E-mail: coynec@wsu.edu*

J. Croser, *Center for Legumes in Mediterranean Agriculture, 35 Stirling Highway, Crawley, WA 6009, Australia. E-mail: jcroser@clima.uwa.edu.au*

A.M.R. Davies, *Institut Für Systematische Botanik, Menzinger Strasse 67, München, Germany. E-mail: a.m.davies@lrz.uni-muenchen.de*

C.J. Demianyk, *Cereal Research Centre, Agriculture and Agri-Food Canada, 195 Dafoe Road, Winnipeg, MB R3T 2M9, Canada. E-mail: cdemianyk@agr.ge.ca*

F. Dusunceli, *The Central Research Institute for Field Crops, PO Box 226, Ulus, Ankara, Turkey. E-mail: fazd62@yahoo.com*

M. El Bouhssini, *International Center for Agricultural Research in the Dry Areas (ICARDA), Aleppo, Syria. E-mail: m.bohssini@cgiar.org*

D. Enneking, *Australian Temperate Field Crops Collection, Department of Primary Industries, Victorian Institute for Dryland Agriculture, Horsham, Victoria 3401, Australia. E-mail: enneking@cyllene.uura.edu.au*

R. Ford, *BioMarka, School of Agriculture and Food Systems, Faculty of Land and Food Resources, The University of Melbourne, VIC 3010, Australia. E-mail: rebeccaf@unimelb.edu.au*

B.J. Furman, *International Centre for Agricultural Research in Dry Areas (ICARDA), PO Box 5466, Aleppo, Syria. E-mail: b.furman@cgiar.org*

Y. Gan, *Agriculture and Agri-Food Canada, Airport Road East, Box 1030, Swift Current, Saskatchewan, S9H 3X2, Canada. E-mail: gan@agr.gc.ca*

B. Gangaiah, *Division of Agronomy, Indian Agricultural Research Institute, New Delhi 110012, India. E-mail: bandla_gan@hotmail.com*

P.M. Gaur, *International Crops Research Institute for the Semi-Arid Tropics (ICRISAT), Hyderabad 502 324, India. E-mail: p.gaur@cgiar.org*

C.L.L. Gowda, *Genetics Resources Divisions, International Crops Research Institute for the Semi-Arid Tropics (ICRISAT), Asia Center, Hyderabad 502 324, India. E-mail: c.gowda@cgiar.org*

B.R. Gregg, *Mississippi State University (retired professor), Starkville, MS 39762, USA. E-mail: billgregg1@bellsouth.net*

M.A. Grusak, *USDA-ARS Children's Nutrition Research Center, Department of Pediatrics, Baylor College of Medicine, Houston, TX 77030-2600, USA. E-mail: mgrusak@bcm.tmc.edu*

A. Gupta, *Canny Overseas Private Limited, B-170, Priyadarshini Vihar, New Delhi 110092, India. E-mail: canny@vsnl.net*

F.Y. Hafeez, *Biofertilizer Division, National Institute for Biotechnology and Genetic Engineering (NIBGE), PO Box 577, Jhang Road, Faisalabad, Pakistan. E-mail: fauzia@nibge.org*

T.J. Higgins, *CSIRO Plant Industry, Black Mountain Laboratories, Clunies Ross Street, Black Mountain, ACT 2601, Australia. E-mail: tj.higgins@csiro.au*

F. Javadi, *Graduate School of Life and Environmental Sciences, Osaka Prefecture University, Sakai City, 599-8531, Osaka, Japan. E-mail: javadif@plant.osakafu-u.ac.jp*

D.S. Jayas, *University of Manitoba, Winnipeg, MB R3T 2N2, Canada. E-mail: jayas@ac.umanitoba.ca*

F. Kantar, *Ataturk University, Faculty of Agriculture, Department of Agronomy, 25240 Erzurum, Turkey. E-mail: fkantar@atauni.edu.tr*

E.J. Knights, *New South Wales Department of Primary Industries, Tamworth Agricultural Institute, Calala, NSW 2340, Australia. E-mail: ted.knights@dpi.nsw.gov.au*

J. Kumar, *Pulse Research Laboratory, Division of Genetics, Indian Agricultural Research Institute, New Delhi 110012, India. E-mail: jk_meher@rediffmail.com*

J. Kumar, *Hendrick Beans-for-Health Research Foundation, 11791 Sandy Row Inkerman, Ontario K0E 1J0, Canada. E-mail: jk_icrisat@yahoo.com*

C. Lluch, *Department of Plant Physiology, Faculty of Sciences, University of Granada, Campus de Fuentenueva s/n, 18071 Granada, Spain. E-mail: clluch@ugr.es*

N. Longnecker, *Centre for Learning Technology, The University of Western Australia, Perth, Australia. E-mail: longneck@cyllene.uwa.edu.au*

M. Manohar, *Pulse Research Laboratory, Division of Genetics, Indian Agricultural Research Institute, New Delhi 110012, India. E-mail: manohar.iitr@gmail.com*

M. Materne, *Victoria Institute of Dryland Agriculture, Horsham, VIC 3401, Australia. E-mail: michael.materne@dpi.vic.gov.au*

V.C. Mathur, *Division of Agricultural Economics, Indian Agricultural Research Institute, New Delhi 110012, India. E-mail: vcmathur49@yahoo.com*

N. Maxted, *C/o School of Biological Sciences, University of Birmingham, Edgbaston, Birmingham, B15 2TT, UK. E-mail: n.maxted@bham.ac.uk*

D. McNeil, *Victorian Department of Primary Industries, PMB 260, Horsham, VIC 3400, Australia. E-mail: david.mcneil@dpi.vic.gov.au*

K.E. McPhee, *USDA-ARS, Grain Legume Genetics and Physiology Research Unit, Washington State University, Pullman, WA 99164-6434, USA. E-mail: kmcphee@wsu.edu*

M. Miles, *Department of Primary Industries and Fisheries, 203 Tor Street, PO Box 102, Toowoomba, Queensland 4350, Australia. E-mail: Melina.Miles@dpi.qld.gov.au*

K. Moore, *NSW Department of Primary Industries, Tamworth Agricultural Institute, 4 Marsden Park Road, Calala, NSW 2340, Australia. E-mail: kevin.moore@dpi.nsw.gov.au*

F.J. Muehlbauer, *USDA-ARS, Grain Legume Genetics and Physiology Research Unit, Washington State University, Pullman, WA 99164-6434, USA. E-mail: muehlbauer@wsu.edu*

M. Muzquiz, *Dep. de Tecnología de Alimentos. SGIT-INIA. 28080 Madrid, Spain. E-mail: muzquiz@inia.es*

A.A. Niane, *International Center for Agricultural Research in the Dry Areas (ICARDA), PO Box 5466, Aleppo, Syria. E-mail: a.niane@cgiar.org*

G. O'Leary, *Primary Industries Research Victoria – Horsham 110 Natimuk Road, Horsham, VIC 3400, Australia. E-mail: garry.o'leary@dpi.vic.gov.au*

S. Pande, *International Crops Research Institute for the Semi-Arid Tropics (ICRISAT), Hyderabad 502324, India. E-mail: s.pande@cgiar.org*

M. Pathak, *International Crops Research Institute for the Semi-Arid Tropics (ICRISAT), Hyderabad 502324, India. E-mail: maheshpathak@rediffmail.com*

C. Pittock, *The University of Melbourne, PB 260, Horsham, VIC 3401, Australia. E-mail: Chris.Pittock@dpi.vic.gov.au*

R.P.S. Pundir, *Genetic Resources Divisions, International Crops Research Institute for the Semi-Arid Tropics (ICRISAT) Asia Center, Hyderabad 502 324, India. E-mail: s.pundir@cgiar.org*

M.M. Rahman, *Bangladesh Agricultural Research Institute, Joydebpur, Gazipur, Bangladesh. E-mail: baridres@citechco.net*

P. Raja, *Department of Agricultural Microbiology, Tamil Nadu Agricultural University, Coimbatore 641003, India. E-mail: microbiology@tnau.ac.in*

P.N. Rajesh, *Department of Crop and Soil Sciences, Washington State University, Pullman, WA 99164-6434, USA. E-mail: pnrajesh@wsu.edu*

G.V. Ranga Rao, *Genetics Resources Divisions, International Crops Research Institute for the Semi-Arid Tropics (ICRISAT), Hyderabad 502324, India. E-mail: g.rangarao@cgiar.org*

R.J. Redden, *Australian Temperate Field Crops Collection, Department of Primary Industries, Victorian Institute for Dryland Agriculture, Horsham, Victoria 3401, Australia. E-mail: bob.redden@dpi.vic.gov.au*

A.A. Reddy, *Indian Institute of Pulses Research, Kanpur 208024, Uttar Pradesh, India. E-mail: aareddy12@email.com*

T.J. Ridsdill-Smith, *CSIRO Entomology, Private Bag No 5, Wembley, WA 6913, Australia. James. E-mail: James.Ridsdill-Smith@csiro.au*

A.H. Rizvi, *Pulse Research Laboratory, Division of Genetics, Indian Agricultural Research Institute, New Delhi 110012, India. E-mail: hasan_ra@rediffmail.com*

J. Romeis, *Agroscope ART Reckenholz Tänikon, Reckenholzstr, 191, 8046 Zurich, Switzerland. E-mail: joerg.romeis@art.admin.ch*

D. Rubiales, *Institute for Sustainable Agriculture, Alameda del Obispo s/n, Apdo. 4084, 14080 Cordoba, Spain. E-mail: ge2ruozd@uco.es*

P.M. Salimath, *Department of Genetics and Plant Breeding, University of Agricultural Sciences, Dharwad, Karnataka, India. E-mail: pmsalimath@hotmail.com*

J.S. Sandhu, *Department of Plant Breeding, Genetics and Biotechnology, Punjab Agricultural University, Ludhiana 141004, India. E-mail: js_sandhuin@yahoo.com*

R.L. Sapra, *Division of Genetics, Indian Agricultural Research Institute, New Delhi 110012, India. E-mail: saprarl@gmail.com*

B. Sarmah, *Department of Agricultural Biotechnology, Assam Agricultural University, Jorhat, Assam 785013, India. E-mail: bsarmah@rediffmail.com*

A. Satyanarayana, *Nuziveedu Seeds Ltd, Hyderabad, Andhra Pradesh, India. E-mail: alapatisatyam@yahoo.com*

H.S. Sekhon, *Department of Plant Breeding, Genetics and Biotechnology, Punjab Agricultural University, Ludhiana 141004, India. E-mail: sekhonhsd@yahoo.com*

R. Serraj, *International Rice Research Institute, Dapo Box 7777, Metro Manila, Philippines. E-mail: r.serraj@cgiar.org*

H.C. Sharma, *Genetics Resources Divisions, International Crops Research Institute for the Semi-Arid Tropics (ICRISAT), Hyderabad 502324, India. E-mail: h.sharma@cgiar.org*

Y.R. Sharma, *Department of Pathology, Punjab Agricultural University, Ludhiana 141004, India. E-mail: paupulses@rediffmail.com*

B.G. Shivakumar, *Division of Agronomy, Indian Agricultural Research Institute, Pusa, New Delhi 110012, India. E-mail: bgskumar@yahoo.com*

K.H.M. Siddique, *Institute of Agriculture, The University of Western Australia, Faculty of Natural and Agricultural Sciences, 35 Stirling Highway, Crawley, WA 6009, Australia. E-mail: ksiddique@fnas.uwa.edu.au*

G. Singh, *Department of Plant Breeding, Genetics and Biotechnology, Punjab Agricultural University, Ludhiana 141004, India. E-mail: drgurdip@rediffmail.com*

U.C. Sood, *Indian Agricultural Statistics Research Institute, New Delhi 110012, India. E-mail: ucsud@iasri.res.in*

J. Speijers, *Western Australia Department of Agriculture, 3 Baron-Hay Court, South Perth, WA 6151, Australia. E-mail: jspeijers@agric.wa.gov.as*

P.C. Stevenson, *Natural Resources Institute, University of Greenwich, Chatham, ME4 4TB, UK and Royal Botanic Gardens, Kew, Surrey, TW9 3AB, UK. E-mail: p.c.stevenson@gre.ac.uk*

B. Suma, *Department of Genetics and Plant Breeding, University of Agricultural Sciences, Dharwad, Karnataka, India. E-mail: suma_bgpb@yahoo.com*

S.P. Sundaram, *Department of Agricultural Microbiology, Tamil Nadu Agricultural University, Coimbatore 641003, India. E-mail: profsundaram@yahoo.com*

N.A. Tejera, *Department of Plant Physiology, Faculty of Sciences, University of Granada, Campus de Fuentenueva s/n, 18071 Granada, Spain. E-mail: natejera@ugr.es*

C. Toker, *Department of Field Crops, Faculty of Agriculture, Akdeniz M080, The University, TR-07059 Antalya, Turkey. E-mail: toker@akdeniz.edu.tr*

N.C. Turner, *Centre for Legumes in Mediterranean Agriculture, M080, The University of Western Australia, 35 Stirling Highway, Crawley, WA 6009, Australia. E-mail: ncturner@clima.uwa.edu.au*

H.D. Upadhyaya, *Genetics Resources Divisions, International Crops Research Institute for the Semi-Arid Tropics (ICRISAT) Asia Center, Hyderabad 502 324, India. E-mail: h.upadhyaya@cgiar.org*

L.J.G. van der Maesen, *Biosystematics Group and National Herbarium of the Netherlands, Wageningen University, Gen. Foulkesweg 37, 6703 BL, Wageningen, The Netherlands. E-mail: josvandermaesen@wur.nl*

A.J.G. van Gastel, *International Center for Agricultural Research in the Dry Areas (ICARDA), PO Box 5466, Aleppo, Syria (current address: Harspit 10, 8493KB, Terherne, The Netherlands). E-mail: tvangastel@hetnet.nl*

T. Warkentin, *Crop Development Centre, University of Saskatchewan, Saskatoon, SK S7N 5A8, Canada. E-mail: warkentin@sask.usask.ca*

N.D.G. White, *Cereal Research Centre, Agriculture and Agri-Food Canada, 195 Dafoe Road, Winnipeg, MB R3T 2M9, Canada. E-mail: nwhite@agr.ge.ca*

J.A. Wood, *Tamworth Agricultural Institute, NSW Department of Primary Industries, Tamworth, NSW 2340, Australia. E-mail: jenny.wood@dpi.nsw.gov.au*

Z. Xaxiao, *Chinese Academy of Agricultural Sciences, Beijing, China. E-mail: zongxx@mail.caas.net.cn*

M. Yadav, *College of Business Administration, Tarleton State University (Texas A&M University System), Stephenville, TX 76402, USA. E-mail: manav.yadav@gmail.com*

S.S. Yadav, *Pulse Research Laboratory, Division of Genetics, Indian Agricultural Research Institute, New Delhi 110 012, India. E-mail: shyamsinghyadav@yahoo.com*

N.T. Yaduraju, *National Agricultural Technology Programme, KAB II, Pusa, New Delhi 110012, India. E-mail: yaduraju@yahoo.co.uk*

J.P. Yenish, *Extension Weed Scientist, Washington State University, Pullman, Washington, USA. E-mail: yenish@wsu.edu*

M.A. Zahid, *Pulses Program, National Agricultural Research Centre, PO NIH, Park Road, Islamabad 45500, Pakistan. E-mail: ashrafzahid@hotmail.com*

H.B. Zhang, *Department of Soil and Crop Sciences, Texas A&M University, College Station, 2474 TAMU, TX 77843-2474, USA. E-mail: hbz7049@tamu.edu*

Preface

Chickpea, an ancient crop of modern times, was first cultivated at least 9500 years ago in the Fertile Crescent, from Turkey to Iran, at the beginning of agriculture. Chickpea cultivation in the Indian subcontinent dates back at leat 4000 years. Chickpea is cultivated in nearly 50 countries around the world. Due to its high nutritional value, it is an integral part of the daily dietary system for millions of people. Chickpea dominates international markets over other legume crops and its trading is more than $8 billion annually. The consumers' preferences for extra-large seed size have provided an excellent opportunity for its premium price and higher profitability. Therefore, this ancient crop has been accepted as the crop of modern management.

Plant breeding and the production of new cultivars is widely regarded as underpinning agriculture and the development of society. Yet crop failures and risks associated with genetic uniformity on large cultivated areas, yield stagnation (below potentially attainable levels) and persistent failures to achieve sustainable production increases in important local ecologies are wide spread problems. What is frequently not recognized is that continuing success requires a long-term, sustainable commitment to the effective utilization of plant genetic resources by enhancing and expanding the genetic base from which future cultivars will be generated.

Chickpea is traditionally a low-input crop and is grown extensively in the moisture stress environments. The global chickpea production has increased only marginally, unlike the manifold increase in cereal production over the last 40 years. There are many constraints to production from diseases, insects-pests, soil problems, environmental stresses and non-adoption of modern management techniques.

This book on chickpea comprises 30 chapters. Chapters 1–7 present the history, origin, taxonomy, ecology, consumption pattern, nutritional and antinutritional factors along with geographic distribution. Chapters 8–14 explain production management, cropping systems, nitrogen fixation, nutrient, weed and irrigation management, integrated crop production technologies and profitability in cultivation. Chapters 15–20 highlight genetics and cytogenetics, wild relatives, management of biodiversity, breeding approaches and achievements as well as quality seed production. Chapters 21 and 22 describe advancement in genomics, recent molecular techniques and development of transgenic chickpea. Chapters 23–26 discuss various biotic and abiotic stress management techniques and seed storage. Chapters 27–30 cover world trade, crop modelling, yield stability, parameters and chickpea growers.

An interdisciplinary modern approach has involved more than 5–6 diverse international experts in the writing of various chapters, to provide a global perspective of the interaction of technologies for international breeding approaches, production, protection, trade, domestic uses and various key economic issues arising internationally. This book provides latest reviews of chickpea publications as well as new findings for scientists, students, technologists, economists, traders, consumers and farmers.

Foreword

This comprehensive survey of chickpea management and improvement offers a wealth of information, which should prove highly useful for agricultural research and development.

Numerous features of chickpea point to an increasingly important role for the crop during the coming decades as farmers, researchers, development specialists and others strive to achieve sustainable agricultural development. The third most important food legume globally, chickpea has high nutritional value, is more important in international markets than other food legumes, contributes importantly to soil fertility management by serving as a generous source of nitrogen and is especially well suited to environments characterized by moisture stress.

These traits should prove especially valuable as small farmers come to grips with the changes that are sweeping global agriculture. Prominent among these is economic globalization, which makes it more necessary and possible for farmers to strengthen their ties with markets. Another trend is climate variability, which is already translating into more frequent drought and flooding in tropical and subtropical agriculture. Chickpea will be a useful resource for helping developing countries adapt to those impacts, particularly in India, where they are projected to be severe and where the burden of hunger and poverty is still great.

During recent years, important advances in chickpea varietal improvement have been registered. As a result, it has proved possible to raise yields and thus intensify production without expanding the area in production.

On the strength of those gains, further efforts are needed to overcome the diverse constraints of chickpea production. For that purpose, the knowledge contained in this book should prove valuable.

KATHERINE SIERRA
Chair, CGIAR, and Vice President,
Sustainable Development Network, World Bank

Foreword

On behalf of the United States Department of Agriculture (USDA) I am pleased to reflect on the significance of this world reference book on chickpea.

Cicer arietinum, or chickpea, is one of the oldest grain legumes farmed by modern humans; in Turkey and Syria, archaeological evidence of chickpeas – possibly domesticated – has been dated to around 8000 BC. It became widely cultivated in the Mediterranean, North Africa, the Middle East and India, where today about two-thirds of the global chickpea crop is grown. In 2004, over 45 countries produced approximately 8.6 million tonnes of chickpea; among edible pulses, its position in the global market for human consumption is second only to dry beans.

Approximately 6.5 billion people currently live on our planet, and by 2050, that number is projected to rise by almost 50% to over 9 billion. We will be increasingly faced with the demand to produce more food for more people with fewer resources, and we will need to draw more heavily on high-quality crops to provide for this growing demand. Chickpea has a head start in this agricultural race. The seeds are packed with protein, fibre, calcium, potassium, phosphorous, iron and magnesium. Like other legumes, it is a key crop in cereal-based production systems because it supports weed management and enriches the soil through nitrogen fixation. Their wild ancestors have passed on drought-tolerant characteristics that allow domestic chickpea varieties to thrive in semi-arid regions, reducing the need for irrigation. Hair-like structures on its stems, leaves and seed pods secrete acids that provide a first-line defence against insect pests, saving farmers time and money in pesticide applications in certain parts of the world, as well as helping to protect fragile environments.

USDA is pleased to be part of the global community contributing to chickpea research. In Colorado, USDA scientists found that during times of water stress, chickpea surpassed soybeans and field peas in the growth of fine root structures to capture moisture lingering deep in the soil layers. Working with colleagues in Idaho, North Dakota and California, USDA researchers developed new chickpea varieties that exhibit genetic resistance to Ascochyta blight, and they are studying ways to improve chickpea's genetic resistance to Fusarium wilt. USDA nutritionists have found that the leaves of the chickpea contain more calcium, phosphorus and potassium than spinach or cabbage, as well as containing iron, zinc, magnesium, and other minerals. Our colleagues in US Land Grant Universities (LGUs) have also made numerous contributions to the body of knowledge about chickpea. The National Plant Germplasm System, a cooperative programme by USDA, the LGUs and private organizations to preserve the genetic diversity of plants, contains over 4000 accessions of chickpea that originated from almost 50 countries. These germplasm resources are made available to scientists worldwide to support plant breeding and other research programme on chickpea.

The collection of research in this volume, painstakingly edited by Shyam Yadav, Bob Redden, Weidong Chen and Balram Sharma, represents outstanding work from some of the finest scientists around the world; over 80 contributors leave no stone unturned in the study of the chickpea. I am grateful that they have chosen to share their expertise, and am confident that their hard work and teamwork will continue to benefit agricultural producers and consumers in the years to come.

EDWARD B. KNIPLING
Administrator, Agricultural Research Service
U.S. Department of Agriculture

Acknowledgements

The editorial team would like to acknowledge their sincere thanks to all the contributors for their efficiency and patience. Editing multi-author texts is not always an easy process. In this case, however, it has been both enjoyable and relatively painless. All the contributors responded efficiently and speedily to the collective hectoring and pressure exerted by the editors, with the consequences that the script was delivered on time and without problems. This made the job of the editors immeasurably easier, and our job in collecting the script and preparing the final text for the publisher relatively straightforward.

Several people have rendered invaluable assistance in making this publication possible and we thank them all.

We would like to express our sincere thanks to Manav Yadav, who worked as a Project Manager/Communications Coordinator/Administrator for this book. He brought the various editors and authors from around the world together, and helped in making *Chickpea Breeding and Management* a truly global reference book. He was also instrumental in creating and maintaining a special e-mail id specifically for this book (chickpea.book@gmail.com) which was an effective medium of communication for the contributors from the initial to the final stages of publishing.

1 History and Origin of Chickpea

R.J. Redden[1] and J.D. Berger[2,3]

[1]Australian Temperate Field Crops Collection, Department of Primary Industries, Victorian Institute for Dryland Agriculture, Horsham, Victoria 3401, Australia; [2]CSIRO Plant Industry, Private Bag 5, Wembley, WA 6913, Australia; [3]Centre for Legumes in Mediterranean Agriculture, M080, The University of Western Australia, 35 Stirling Highway, Crawley WA 6009, Australia

Domestication

Chickpea was domesticated in association with other crops of wheat, barley, rye, peas, lentil, flax and vetch (Harlan, 1971; Abbo *et al.*, 2003a), and with sheep, goats, pigs and cattle (Diamond, 1997), as part of the evolution of agriculture in the Fertile Crescent 12,000–10,000 years ago (Zohary and Hopf, 1973; Bar-Yosef, 1998). In this broad arc extending from western Iran through Iraq, Jordan and Israel to south-east Turkey, there developed a 'balanced package of domesticates meeting all of humanity's basic needs: carbohydrate, protein, oil, animal transport and traction, and vegetable and animal fibre for rope and clothing' (Diamond, 1997).

Initially in this region wild plants were harvested in the ancient hunter-gatherer cultures, with evidence of cultivated crops appearing in archeological records from 7500 BC (Harlan, 1971) and earlier dates are plausible (Hillman *et al.*, 1989). Possibly the onset of an extended cool dry period triggered interest in securing food supply with domestication and the beginning of agriculture (Wright, 1968; Smith, 1995). The time frame for domestication of einkorn wheat from the wild appears to be within a few centuries (Heun *et al.*, 1997); however, the present distribution of wild relatives may not reflect distribution at the period of domestication (Diamond, 1997), and further crop diversity may have evolved outside of the centre of origin (Harlan, 1971).

The earliest records of chickpea used as food are: 8th millennium BC at Tell el-Kerkh (Tanno and Willcox, 2006) and Tell Abu Hureyra Syria (Hillman, 1975); 7500–6800 BC at Cayonu Turkey (van Zeist and Bottema, 1972); and 5450 BC at Hacilar Turkey (van der Maesen, 1984). These remains often cannot distinguish between wild *Cicer* and the domesticated *Cicer arietinum* (Ladizinsky, 1998), but at Tell el-Kerkh both *Cicer arietinum* and the progenitor *Cicer reticulatum* were clearly identified, this being the earliest date for cultivation of chickpea (Tanno and Willcox, 2006). Archeological records of domestic chickpea are sparse because the distinguishing seed beak has often broken off in carbonized seed; however, its cultivation is well documented from 3300 BC onwards in Egypt and the Middle East (van der Maesen, 1972).

With the present-day distribution of *C. reticulatum* Ladz., the wild progenitor (Ladizinsky and Adler, 1976), chickpea is likely to have been domesticated in south-east Anatolia (Fig. 1.1) although the current distribution may be of relic populations since the Tell el-Kerkh occurrence is a further 260 km away (Tanno and Willcox, 2006). This is supported by the distribution of early Neolithic chickpea which was confined to the Fertile Crescent, particularly in modern Anatolia and the eastern Mediterranean (Fig. 1.1, Table 1.1). In the late Neolithic era chickpea spread westwards to modern Greece (Fig. 1.1, Table 1.1). By the Bronze Age chickpea had been disseminated widely to Crete in the west, upper Egypt in the south, eastwards through present-day Iraq to the Indian subcontinent, where remains have been found in Harrapan settlements in Pakistan (Vishnu-Mittre and Savithri, 1982) and a variety of sites in Uttar Pradesh and Maharashtra (Fig. 1.1, Table 1.1). By the Iron Age, chickpea consolidated its distribution in South and West Asia, and appeared in Ethiopia for the first time (Fig. 1.1, Table 1.1).

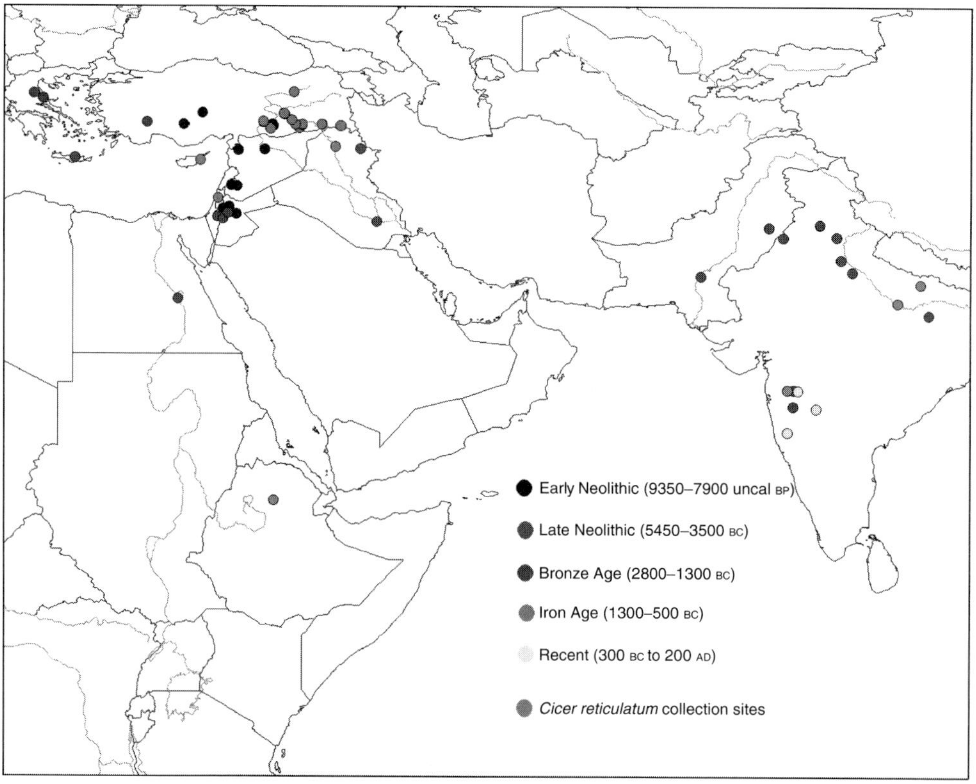

Fig. 1.1. Excavation sites containing ancient chickpea or *Cicer* spp. (for details see Table 1.1). Present-day collection sites of *Cicer reticulatum*, the wild progenitor of cultivated chickpea, are included as a reference area for possible domestication.

Table 1.1. Prehistoric sites containing chickpea or unidentified *Cicer* spp. listed in chronological order.

Remains	Site	Country	Period	Age	Reference
Cicer arietinum	Tell el-Kerkh	Syria	Early Neolithic (9350–7900 uncal BP)	7260 BC	Tanno and Willcox (2006)
Cicer sp.	Nevali Cori	Turkey	Early Neolithic (9350–7900 uncal BP)	7250 BC	Pasternak (1998)
Cicer sp.	Çayönü	Turkey	Early Neolithic (9350–7900 uncal BP)	7150 BC	van Zeist and de Roller (1991, 1992)
Cicer sp.	Nevali Cori	Turkey	Early Neolithic (9350–7900 uncal BP)	7150 BC	Pasternak (1998)
Cicer sp.	Asikli Höyük	Turkey	Early Neolithic (9350–7900 uncal BP)	6700 BC	van Zeist and de Roller (1995)
Cicer sp.	Wadi el-Jilat	Jordan	Early Neolithic (9350–7900 uncal BP)	6600 BC	Garrad and Colledge (1996)
C. arietinum	Jericho	Palestine	Early Neolithic (9350–7900 uncal BP)	6500 BC	Hopf (1983)
C. arietinum	Ain Ghazal	Jordan	Early Neolithic (9350–7900 uncal BP)	6350 BC	Rollefson *et al.* (1985)
C. arietinum	Ghoraife	Syria	Early Neolithic (9350–7900 uncal BP)	6275 BC	van Zeist and Bakker-Heeres (1982)
Cicer sp.	Ramad	Syria	Early Neolithic (9350–7900 uncal BP)	6100 BC	van Zeist and Bakker-Heeres (1982)
C. arietinum	Abu Hureyra	Syria	Early Neolithic (9350–7900 uncal BP)	6020 BC	de Moulins (2000)
C. arietinum	Çatalhöyük	Turkey	Early Neolithic (9350–7900 uncal BP)	6000 BC	Fairbairn *et al.* (2002)
C. arietinum	Otzaki	Greece	Late Neolithic (5450–3500 BC)	5500 BC	Kroll (1981)
C. arietinum	Hacilar	Turkey	Late Neolithic (5450–3500 BC)	5450 BC	Helbaek (1970)
C. arietinum	Dimini	Greece	Late Neolithic (5450–3500 BC)	3500 BC	Kroll (1979)
C. arietinum	Tel Arad	Israel	Bronze Age (2800–1300 BC)	2800 BC	Hopf (1978)
C. arietinum	Bab edh-Dhra	Jordan	Bronze Age (2800–1300 BC)	2800 BC	McCreery (1979)

Continued

Table 1.1. *Continued*

Remains	Site	Country	Period	Age	Reference
C. arietinum	Ur	Iraq	Bronze Age (2800–1300 BC)	2500 BC	Ellison *et al.* (1978)
C. arietinum	Lachish	Israel	Bronze Age (2800–1300 BC)	2300 BC	van der Maesen (1987)
C. arietinum	Harrapa	Pakistan	Bronze Age (2800–1300 BC)	2000 BC	Fuller (2000)
C. arietinum	Mohenjo Daro	Pakistan	Bronze Age (2800–1300 BC)	2000 BC	Fuller (2000)
C. arietinum	Atranji-khera	India	Bronze Age (2800–1300 BC)	2000 BC	Chowdury *et al.* (1971)
C. arietinum	Kalibangan	India	Bronze Age (2800–1300 BC)	2000 BC	Vishnu-Mittre and Savithri (1982)
C. arietinum	Tell Baz-mosian	Iraq	Bronze Age (2800–1300 BC)	1900 BC	Helbaek (1965)
C. arietinum	Lal Quila	India	Bronze Age (2800–1300 BC)	1750 BC	Saraswat (1992)
C. arietinum	Sanghol	India	Bronze Age (2800–1300 BC)	1700 BC	Saraswat (1992)
C. arietinum	Inamgaon	India	Bronze Age (2800–1300 BC)	1650 BC	Saraswat (1992)
C. arietinum	Nevasa	India	Bronze Age (2800–1300 BC)	1650 BC	Saraswat (1992)
C. arietinum	Hulas	India	Bronze Age (2800–1300 BC)	1600 BC	Vishnu-Mittre *et al.* (1985)
C. arietinum	Senuwar	India	Bronze Age (2800–1300 BC)	1500 BC	Saraswat (1992)
C. arietinum	Daimabad	India	Bronze Age (2800–1300 BC)	1500 BC	Saraswat (1992)
C. arietinum	Crete	Crete	Bronze Age (2800–1300 BC)		van der Maesen (1972)
C. arietinum	Deir el Medineh	Egypt	Bronze Age (2800–1300 BC)	1400 BC	Darby *et al.* (1977)
C. arietinum	Jorwe	India	Iron Age (1300–500 BC)	1150 BC	Kajale (1991)
C. arietinum	Narhan	India	Iron Age (1300–500 BC)	1150 BC	Saraswat (1992)
C. arietinum	Tell Megiddo	Israel	Iron Age (1300–500 BC)	1000 BC	van der Maesen (1987)
C. arietinum	Sringavera-pura	India	Iron Age (1300–500 BC)	825 BC	Saraswat (1992)
C. arietinum	Nimrud	Iraq	Iron Age (1300–500 BC)	700 BC	Helbaek (1966)
C. arietinum	Salamis	Cyprus	Iron Age (1300–500 BC)	550 BC	Renfrew (1967)
C. arietinum	Lalibela cave	Ethiopia	Iron Age (1300–500 BC)	290 BC	Dombrowski (1970)
C. arietinum	Nevasa	India	Recent (300 BC–200 AD)	200 BC	van der Maesen (1987)

Table 1.1. *Continued*

Remains	Site	Country	Period	Age	Reference
C. arietinum	Kolhapur	India	Recent (300 BC–200 AD)	50 BC	Kajale (1991)
C. arietinum	Tripuri	India	Recent (300 BC–200 AD)	0 AD	Kajale (1991)
C. arietinum	Bokardan	India	Recent (300 BC–200 AD)	0 AD	Kajale (1991)
C. arietinum	Ter	India	Recent (300 BC–200 AD)	25 AD	Kajale (1991)

Most Neolithic dates are presented in the literature as uncalibrated years before present (bp), so the conversion to bc is very approximate.

Chickpea is used for both food and medicinal or herbal purposes by Homer in the *Iliad* (1000–800 BC), in Roman, Indian and medieval European literature (van der Maesen, 1972). The crop spread with the cluster of founder crops from the Fertile Crescent into Europe and west-central Asia from the *c.*5500 BC onwards (Harlan, 1992; Damania, 1998; Harris,1998).

Chickpea was introduced to the New World by the Spanish and Portuguese in the 16th century AD, and kabuli types moved to India from the Mediterranean via the Silk Road in the 18th century (van der Maesen, 1972). Desi chickpea was probably imported to Kenya by Indian immigrants during the later 19th century (van der Maesen, 1972). Recently chickpea breeding programmes began in the USA, Australia (first cultivar, Tyson, released in 1979; Knights and Siddique, 2002) and Canada.

The progenitor wild relative for domestic chickpea is *C. reticulatum*, the only relative in the primary gene pool, with restricted distribution in south-east Turkey where domestication may have occurred (Ladizinsky and Adler, 1975; Ladizinsky, 1998). The two species are interfertile, with similar seed proteins, and are morphologically similar, but some domestic accessions may differ from *C. reticulatum* by a reciprocal inversion, a paracentric inversion or by location of chromosomal satellites (Ladizinsky, 1998). The secondary gene pool contains only one species, *C. echinospernum*, which differs from the domestic by a single reciprocal translocation, and hybrids tend to be sterile (Ladizinsky, 1998). All other annual and perennial *Cicer* spp. are genetically isolated in the tertiary gene pool and equidistant from the domestic species as per amplified fragment length polymorphism (AFLP) diversity analyses (Nguyen *et al.*, 2004).

Changes with domestication initially included the loss of dormancy, followed by reduced pod dehiscence, larger seed size, larger plant size and variants with more erect habit and reduced anthocynin pigmentation (Smartt, 1984; Ladizinzky, 1987). However, the key to chickpea domestication was the change from a winter habit with an autumn sowing to a spring habit, which avoided or reduced the threat of lethal infestation of the endemic ascochyta pathogen complex (Abbo *et al.*, 2003a).

Domestication of chickpea (Fig. 1.2) may well have followed a different evolutionary sequence from that of other founder crops domesticated first in

Fig. 1.2. Seed of ATFCC accession of desi chickpea CA0805 originating from India, with pod shell.

the Fertile Crescent (Abbo *et al.*, 2003b). The archeological appearance of large seed type is usually associated with domestication, and the records for large seeded chickpea have a period of no fossil remains between 9000 and 6000 BP, in contrast to the continuous records for the other founder crops. However, both annual and perennial wild relatives in the region are adapted to winter cropping, whereas domestic chickpea is spring-sown for summer cropping, and the main differentiation between the wild and domestic species is the loss of responsiveness to vernalization, a polygenetic trait. Domestic chickpea is also characterized by larger plant and seed size than the *C. reticulatum* wild progenitor (Fig. 1.3).

In the Fertile Crescent winter rainfall is predominant and conducive to devastating ascochyta blight, which may cause complete yield loss in winter-grown susceptible chickpea varieties. Summer cropping provides an escape from this disease, with a relatively dry spring and summer. This disease is much more severe for chickpea than for winter-cropped pea and lentil. Hence the loss of vernalization and the advent of summer cropping, coincident with introduction of summer crops sorghum, millet and sesame around 6000 BP, enabled chickpea to reappear as a significant crop (Abbo *et al.*, 2003a).

In contrast to other grain legume crops, loss of seed dormancy and reduction of pod shattering do not appear to be the key traits for domestication in chickpea (Ladizinsky, 1987). The most widespread cultivation of chickpea is in summer crop regions of the Middle East and North America, and dry winter areas of India, whereas winter cropping in Australia collapsed due to the appearance of ascochyta blight, and is only now re-emerging with the release of disease-resistant varieties (Pande *et al.*, 2005). Domestic chickpea has been

Fig. 1.3. *Cicer reticulatum*, ATC # 42268, synonym ILWC 108, originating from Turkey.

selected for adaptation to residual stored soil moisture (Abbo *et al.*, 2003b). The evolution of summer cropping has resulted in a temporal separation of the reproductive phases between the cultivated and the wild chickpea progenitor. This isolation of wild relatives, restricted distribution of the progenitor, 'founder' effect with domestication of this progenitor and change from a winter to a summer crop have all contributed to the bottlenecks in genetic diversity of the domestic species (Abbo *et al.*, 2003b). The 'founder' effect is dealt with in more detail by Abbo *et al.* (Chapter 16, this volume).

Evolution of Crop Types

Domestic chickpea and its wild relatives are in the genus *Cicer*, which is in the family *Fabaceae* within the tribe *Cicerae* (USDA, 2005). The diploid chromosome number is 16, though 14 has been reported for some landraces, the species *C. songaricum* and for some accessions of *C. anatolicum* (van der Maesen, 1972). The domestic crop is self-pollinating. Pollination is completed in the flower bud stage, before bees visit open flowers in the field (van der Maesen, 1972). Usually only one seed per pod is set.

The very large seeded kabuli domestic types appear to have evolved from the smaller seeded desi types (Moreno and Cubero, 1978). Molecular diversity analyses clearly separate the two groups, which have further subgroups (Iruela *et al.*, 2002). The kabuli types, also known as 'macrosperma', are characterized

Fig. 1.4. Kabuli type seed, ATC # 41017, originating from Turkey.

by large pods, seeds, leaves and a taller stature (van der Maesen, 1972; Moreno and Cubero, 1978). Flowers and seed are mainly white though other colours occur with a low frequency. Most kabuli have a 'ram's head' seed shape (Fig. 1.4).

The distribution of genetic diversity in the kabuli is much narrower than in the desi predominant chickpea type. The desi have small pods, seeds, leaflets and a small stature. However, wide genetic diversity occurs in the desi types for flower, pod, seed and vegetative colour, variation in seed surface and shape (Moreno and Cubero, 1978).

The geographic distribution differs for these two types, with the kabuli tending to be restricted to the western Mediterranean where the desi are mainly absent. The desi range more widely from the eastern Mediterranean to central Asia and the Indian subcontinent (Moreno and Cubero, 1978). The evidence points to recent evolution of the kabuli types from the desi, in association with selection for white seed colour and for better food quality (Moreno and Cubero, 1978).

This is consistent with a very limited distribution of the *C. reticulatum* progenitor in south-eastern Turkey, with an ancient reproductive separation from the summer crop domesticates, and full cross-compatibility with each of the kabuli and desi types. The desi and kabuli types are also near uniform in cytoplasm. This indicates no evolution of hybridization barriers (Moreno and Cubero, 1978).

The desi and kabuli groups tend to have maintained distinct morphological types, with only rare appearance of intermediate forms (Iruela *et al.*, 2002). These differences have become blurred in modern plant breeding programmes that attempt to combine large seed size with the local adaptation and vigour of desi types (Yadav *et al.*, 2004). Other geographic distributions of traits in the desi include a concentration of black seed in Ethiopia, and certain parts of Turkey, Iran and India, but not central Asia (van der Maesen, 1972). Tormentose leaves are particularly associated with small-stature desi types from India (S.S.Yadav, IARI, New Delhi, 2006 personal communication).

Centres of Diversity

The primary centre of diversity is in the Fertile Crescent where the crop was domesticated, and with the geographic spread of chickpea secondary centres of diversity developed, some older than 2000 years in Mediterranean Europe, the Indian subcontinent and north-east Africa, and some more recently in Mexico and Chile with post-Colombus introduction (van der Maesen, 1972).

The distribution of old landraces and wild relatives of chickpea occurs in three main regions from 8° to 52°N latitude and 8°W to 85°E longitude: (i) western Mediterranean, Ethiopia, Crete and Greece; (ii) Asia-minor, Iran and Caucasus; and (iii) Central Asia, Afghanistan and the Himalayan region (van der Maesen, 1972). The distribution of wild relatives is covered by Abbo *et al.* (Chapter 16, this volume). Domestic chickpea is now widely cultivated in southern South America, Australia, both the European and African Mediterranean regions, the Balkans, Ethiopia and East Africa, southern Asia from Myanmar to Iran, and the Middle East encompassing Iraq, Israel and Turkey (van der Maesen, 1972).

In the largest genebank for chickpea landraces at the International Crops Research Institute for the Semi-Arid Tropics (ICRISAT), Hyderabad, India, with 17,250 accessions, the greatest diversity is from India (6930), followed by Iran (4850), Ethiopia (930), Afghanistan (700), Pakistan (480), Turkey (470), Mexico (390), Syria (220), Chile (139), former Soviet Union (133) and various other regions from southern Europe, northern Africa, East Africa, South America and North America.

At the International Centre for Agricultural Research in the Dry Areas (ICARDA), Aleppo, Syria, with a regional mandate for kabuli chickpea, the genebank has 12,070 landrace accessions mainly sourced from Iran (1780), Turkey (970), India (410), Chile (340), Uzbekistan (300), Spain (280), Tunisia (270), Morocco (230), Bulgaria (210), Portugal (170), Russian federation (160), Mexico (160), Jordan (150), USA (120), Bangladesh (110), Tajikistan (110), Azerbaijan (110) and various other regions with less than 100 (Ethiopia, Italy, Palestine, Algeria, Egypt, through to northern Europe, and tropical Africa and South America).

Both genebanks broadly reflect the available sources of genetic diversity in landraces, with a clear preponderance in the Indian subcontinent, Iran, the broad stretch of the Middle East and Ethiopia, indicating that important secondary centres of diversification followed the spread of the crop out of the Fertile Crescent.

North and south of the Himalayas, from India to Iran, from the Central Asian republics to Georgia, and from Turkey to Ethiopia, there is a broad overlap between landrace diversity and the distribution of wild *Cicer* spp. (van der Maesen, 1972). However, the primary and secondary gene pools currently, and possibly historically, have a very limited distribution in south-eastern Turkey (Ladizinsky, 1998). This limited geographic distribution and the separation of reproductive phases with the conversion of the crop to a spring habit points to an absence of introgressive gene flow from wild to domestic gene pools post-domestication (Abbo *et al.*, 2003b). Thus the domestic gene pool has developed from a narrow genetic base, albeit with adaptation selection for a wide range of agri-ecological niches: mainly spring–summer-cropped except in the Indian subcontinent, with winter cropping on residual moisture (or irrigated), or rainfed-winter-cropped in Australia in association with development of resistance to ascochyta as a major pathogen (Pande *et al.*, 2005).

Domestication as well as distribution and use of wild relatives are described by Berger and Turner (Chapter 3, this volume) and Abbo *et al.* (Chapter 16, this volume).

Crop Production

Chickpea has many local names: *hamaz* (Arab world), *shimbra* (Ethiopia), *nohud* or *lablabi* (Turkey), *chana* (India) and *garbanzo* (Latin America) Muehlbauer and Tullu, 1997. World production of chickpea as a grain crop has varied from 6.6 million tonnes in 2000–2001 to 9.5 million tonnes in 1998–1999 (Agriculture and Agri-Food Canada, 2005).

Details of world production are given by Knights *et al.* (Chapter 7, this volume).

Prospects

There is a wide variety of plant types, seed characteristics and end uses for chickpea as a food crop. It is widely cultivated from the tropics to high latitudes and in a range of management packages from dryland on residual moisture, rainfed with spring sowing, rainfed with winter sowing and spring-summer-irrigated. This book presents opportunities for more strategic use of both domestic and wild germplasm, improved crop management and integrated packages of development of pest and disease management. There are ways of managing overall farm plans for the generally better economic returns with cereals, in rotation with chickpea, which has increased production risks. The final determination of the crop's success lies with the consumer and market competition based on price and quality of food products. Opportunities in the major cropping regions depend on both reducing the crop risk and improving productivity.

References

Abbo, S., Shtienberg, D., Lichtenzveig, J., Lev-Yadun, S. and Gopher, A. (2003a) The chickpea, summer cropping, and a new model for pulse domestication in the ancient near east. *The Quarterly Review of Biology* 78(4), 435–448.

Abbo, S., Berger, J. and Turner, N. (2003b) Evolution of cultivated chickpea: four bottlenecks limit diversity and constrain adaptation. *Functional Plant Biology* 30, 1081–1087.

Agriculture and Agri-Food Canada (2005) Chickpeas: situation and outlook. Bi-weekly bulletin. Available at: www.agr.gc.ca/mad-dam/e/bulletine/v14e/v14n03e.htm

Bar-Yosef, O. (1998) The Natufian culture in the Levant, threshold to the origins of agriculture. *Evolutionary Anthropology* 16(5), 159–177.

Chowdury, K.A., Saraswar, K.S., Hasam, S.M. and Gaur, R.G. (1971) 4000–3500 year old barley, rice and pulses from Atranjikhera. *Science and Culture* 37, 531–533.

Damania, A.B. (1998) Diversity of major cultivated plants domesticated in the Near East. In: Damania, A.B., Valkoun, J., Willcox, G. and Qualset, C.O. (eds) *The Origins of Agriculture and Crop Domestication*. ICARDA, Aleppo, Syria, pp. 51–64.

Darby, W.J., Ghaloungui, P. and Grivetti, L. (1977) *Food: The Gift of Osiris*. Academic Press, London.

de Moulins, D. (2000) Abu Hureyra 2: plant remains from the Neolithic. In: Moore, A.M.T., Hillman, G.C. and Legge, A.J. (eds) *Village on the Euphrates: From Foraging to Farming at Abu Hureyra*. Oxford University Press, Oxford, pp. 399–416.

Diamond, J. (1997) Location, location, location: the first farmers. The American Association for the Advancement of Science. *Science* 278, 1243–1244.

Dombrowski, J. (1970) Preliminary report on excavation in Lalibela and Natchabiet caves, Begemeder. *Annales d' Ethiopie* 8, 21–29.

Ellison, R., Renfrew, J., Brothwell, D. and Seelwy, N. (1978) Some food offerings from Ur, excavated by Sir Leonard Wooley, and previously unpublished. *Journal of Archeological Science* 5, 167–177.

Fairbairn, A., Asouti, E., Near, J. and Martinoli, D. (2002) Macro-botanical evidence for plant use at Neolithic Çatalhöyük, south-central Anatolia, Turkey. *Vegetation History and Archaeobotany* 11, 41–54.

Fuller, D.Q. (2000) Fifty years of archeobotanical studies in India: laying a solid foundation. In: Settar, S. and Korisettar, R. (eds) *Indian Archeology in Retrospect: Archeology and Interactive Disciplines*. Manohar, New Delhi, pp. 247–363.

Garrad, A. and Colledge, S.L.M. (1996) The emergence of crop cultivation and caprine herding in the "Marginal Zone" of the Southern Levant. In: Harris, D.R. (ed.) *Origins and Spread of Agriculture and Pastoralism in Eurasia*. Smithsonian Institution Press, Washington, DC, pp. 204–226.

Harlan, J. (1971) Agricultural origins: centers and noncenters. *Science* 174, 468–474.

Harlan, J. (1992) Crops and Man. American Society of Agronomy, Crop Science Society of America, Madison, Wisconsin, pp. 63–262.

Harris, D.R. (1998) The spread of neolithic agriculture from the levant to western Central Asia. In: Damania, A.B., Valkoun, J., Willcox, G. and Qualset, C.O. (eds) *The Origins of Agriculture and Crop Domestication*. ICARDA, Aleppo, Syria, pp. 54–64.

Helbaek, H. (1965) Isin-Larson and Horian food remains at Tell Bazmosian in the Dokan Valley. *Sumer* (Baghdad) 19, 27–35.

Helbaek, H. (1966) The Plant Remains. In: Mallowan, M.E.L. (ed.) *Nimrud and Its Remains*. Collins, London, pp. 613–620.

Helbaek, H. (1970) The plant husbandry at Hacilar. In: Mellaart, J. (ed.) *Excavation at Hacilar*. Edinburgh University Press, Edinburgh, UK, pp. 189–244.

Heun, M., Schafer-pregl, R., Klawan, D., Castagna, R., Accerbi, M., Borghi, B., and Salamini, F. (1997) Site of einkorn wheat domestication identified by DNA fingerprinting. *Science* 278, 1312–1314.

Hillman, G.C. (1975) The plant remains from Tell Abu Hureya in Syria: a preliminary report. In: Moore, A.M.T. (ed.) *The Excavation of Tell Abu Hureya in Syria: A Preliminary Report. Proceedings of the Prehistory Society* 41, 70–73.

Hillman, G.C., Colledge, S.M. and Harris, D.M. (1989) Plant-food economy during the epipaleolithic period at Tell Abu Hureya, Syria: dietary diversity, seasonality, and modes of exploitation. In: Hillman, G.C. and Harris, D.R. (eds.) *Forages and Farming: The Evolution of Plant Exploitation*. Unwin-Hyman, London, pp. 240–266.

Hopf, M. (1978) Plant remains, strata V-I. In: Amiran, R. (ed.) *Early Arad: The Chalcothic Settlement and the Early Bronze Age City*. Israel Exploration Society, Jerusalem, pp. 64–82.

Hopf, M. (1983) The pottery phases of the Tell and other finds. In: Kenyon, K.M. and Holland, T.A. (eds) *Excavation at Jericho*. British School of Archeology in Jerusalem, London, pp. 576–621.

Iruela, M., Rubio, J., Cubero, J.I., Gil, J. and Millan, T. (2002) Phylogenetic analysis in the genus *Cicer* and cultivated chickpea using RAPD and ISSR markers. *Theoretical and Applied Genetics* 104, 643–651.

Kajale, M.D. (1991) Current status of Indian palaeoethnobotany: introduced and indigenous food plants with a discussion of the historical and evolutionary development of Indian agriculture and agricultural systems in general. In: Renfrew, J.M. (ed.) *New Light on Early Farming: Recent Developments in Palaeoethnobotany*. Edinburgh University Press, Edinburgh, UK, pp. 155–189.

Knights, E.J. and Siddique, K.H.M. (2002) Chickpea status and production constraints in Australia. In: Abu Bakr, M., Siddique, K.H.M. and Johansen, C. (eds) *Integrated Management of Botrytis Grey Mould of Chickpea in Bangladesh and Australia*. Bangladesh Agricultural Research Institute (BARI), Joydepur, Gazipur, Bangladesh, pp. 33–41.

Kroll, H. (1979) Kulturpflanzen aus Dimini. *Archeo-Physika* 8, 173–189.

Kroll, H. (1981) Thessalische Kulturpflanzen. *Zeitschrift Archeol* 15, 97–103.

Ladizinsky, G. (1987) Pulse domestication before cultivation. *Economic Botany* 41(1), 60–65.

Ladizinsky, G. (1998) *Plant Evolution Under Domestication*. Kluwer Academic, Dordrecht, The Netherlands, pp.174–176.

Ladizinsky, G. and Adler, A. (1976) The origin of chickpea *Cicer arietinum* L. *Euphytica* 25, 211–217.

McCreery, D. (1979) Flotation of the Bab edh-Dhra and Numeira plant remains. *Annals of the American School of Orient Research* 46, 165–169.

Moreno, M. and Cubero, J.I. (1978) Variation in *Cicer arietinum* L. *Euphytica* 27, 465–485.

Muehlbauer, F.J. and Tullu, A. (1997) New crop fact sheet: chickpea *Cicer arietinum* L. *New York Times* 18 November. Available at: http://hort.purdue.edu/newcrop/crops/cropfactsheets/chickpea.html

Nguyen, T.T., Taylor, P.W.J., Redden, R.J. and Ford, R. (2004) Genetic diversity estimates in *Cicer* using AFLP analysis. *Plant Breeding* 123, 173–179.

Pande, S., Siddique, K.H.M., Kishore, G.K., Bayaa, B., Gaur, P.M., Gowda, C.L.L., Bretag, T.W. and Crouch, J.H. (2005) Ascochyta blight of chickpea (*Cicer arietinum L*) : a review of biology, pathogenicity, and disease management. *Australian Journal of Agricultural Research* 56, 317–322.

Pasternak, R. (1998) Investigations of botanical remains from Nevali CoriPPNB, Turkey: short interim report. In: Damania, A.B., Valkoun, J., Wilcox, G. and Qualset, C.O. (eds) *Origins of Agriculture and Crop Domestication*. ICARDA, Aleppo, Syria, pp. 170–177.

Renfrew, J.M. (1967) Carbonized seeds and fruits from the funeral pyres of Salamis, 6th–5th century BC. In: Kargeorghis, V. (ed.) *Excavations in the Necropolis of Salamis*. Harrison & Sons, UK.

Rollefson, G.O., Simmonds, A., Donaldson, M.L., Gillespie, W., Kafafi, Z., Kohler-Rollefson, I.U., McAdam, E., Rolston, S.L. and Tubb, M.K.

(1985) Excavation at the Pre-Pottery Neolithic B village of Ain Ghazal (Jordan). *Mitteilungen der Deutschen Orient-Gesselschaft zu Berlin* 117, 69–116.

Saraswat, K.S. (1992) Archaeobotanical remains in ancient cultural and socio-economical dynamics of the Indian sub-continent. *Palaeobotanist* 40, 514–545.

Smith, B.D. (1995) *Emergence of Agriculture.* Scientific American Library, Series 54, p. 231.

Smartt, J. (1984) evolution of pulse legumes. 1. Mediterranean pulses. *Experimental Agriculture* 20, 275–296.

Tanno, K.-i. and Willcox, G. (2006) The origins of cultivation of *Cicer arietinum* L. and *Vicia faba* L.: early finds from Tell el-Kerkh, north–west Syria, late 10th millennium B.P. *Vegetation History and Archaeobotany* 15(3), 197–204.

USDA (2005) GRIN species of *Cicer*. Available at: www.ars-grin.gov/cgi-bin/npgs/html/splist.pl?2600

van der Maesen, L.J.G. (1972) *Cicer* L., a monograph of the genus, with special reference to the chickpea (*Cicer arietinum* L.), its ecology and distribution. Mendelingen Landbouwhogeschool Wageningen, Wageningen, The Netherlands, 1–341.

van der Maesen, L.J.G. (1984) Taxonomy, distribution and evolution of chickpea. In: Witcombe, J.R. and Erskine, W. (eds) *Genetic Resources and Their Exploitation-Chickpeas, Faba Beans and Lentils*. Martinus Nijhoff/Junk, The Hague, The Netherlands, pp. 95–104.

van der Maesen, L.J.G. (1987) Origin, history and taxonomy of chickpea. In: Saxena, M.C. and Singh, K.B. (eds) *The Chickpea*. CAB International, Wallingford, UK, pp. 11–34.

van Zeist, W. and Bakker-Heeres, J. (1982) Archeobotanical studies in the Levant 1. Neolithic sites in the Damascus basin: Aswad, Ghoraife, and Ramad. *Palaeohistoria* 24, 165–256.

van Zeist, W. and Bottema, S. (1972) Vegetation history of the eastern Mediterranean and the near east during the last 20,000 years. In Bintliff, J.L. and van Zeist, W. (eds) *Paleoclimates, Palaenvironments and Human Communities in the Eastern Mediterranean Region in Later Prehistory*. British Archeological Reports, International Series 133.

van Zeist, W. and de Roller, G. (1991, 1992) The plant husbandry of aceramic Çayönü, SE Turkey. *Palaeohistoria* 33/34, 65–96.

van Zeist, W. and de Roller, G. (1995) Plant remains from Asikli Höyük, a pre-pottery Neolithic site in central Anatolia. *Vegetation History and Archaeobotany* 4, 179–185.

Vishnu-Mittre and Savithri, R. (1982) Food economy of the Harappans. In: Possehl, G.L. (ed.) *Harappan Civilization: A Contemporary Perspective*. Oxford & IBH, New Delhi, pp. 205–221.

Vishnu-Mittre, Saraswat, K.S., Sharma, A. and Chanchala. (1985) *Indian Archeology: A Review*. Archeological Survey of India, New Delhi, pp. 146–15.

Wright, H.E. (1968) Geological aspects of the archeology in Iraq. *Science* 161, 334.

Yadav, S.S., Turner, N.C., Meuhlbauer, F.J., Knights, E.J., Redden, B., McNeil, D., Berger, J. and Kumar, J. (2004) Enhancing adaptation of large seeded kabuli chickpea to drought prone environments. 4th International Crop Science Congress, Brisbane. In: Fischer, T., Turner, N., Angus, J., McIntyre, L., Robertson, M., Borrell, A. and Lloyd, D. (eds) *Australian Society of Agronomy*. Brisbane, Australia, p. 117.

Zohary, D. and Hopf, M. (1973) Domestication of pulses in the old world. *Science* 182, 887–894.

2 Taxonomy of the Genus *Cicer* Revisited

L.J.G. VAN DER MAESEN,[1] N. MAXTED,[2] F. JAVADI,[3]
S. COLES[2] AND A.M.R. DAVIES[4]

[1]Biosystematics Group and National Herbarium of the Netherlands, Wageningen University, Gen. Foulkesweg 37, 6703 BL, Wageningen, The Netherlands; [2]C/o School of Biological Sciences, University of Birmingham, Edgbaston, Birmingham, B15 2TT, UK; [3]Graduate School of Life and Environmental Sciences, Osaka Prefecture University, Sakai City, 599-8531, Osaka, Japan; [4]Institut Für Systematische Botanik, Menzinger Strasse 67, München, Germany

Introduction

The genus *Cicer* L. (Leguminosae, *Cicereae*) comprises 9 annual and 35 perennial species that have a centre of diversity in south-western Asia, with remote, endemic species found in Morocco and the Canary Islands (van der Maesen, 1987, Map 2.1). The genus is the member of the monogeneric tribe *Cicereae* Alef., subfamily *Papilionoideae*, family *Leguminosae*. It was historically included in the legume tribe *Vicieae* Alef., but Kupicha (1977) presented detailed taxonomic evidence to support the tribal distinction of the genus from the other *Vicieae* genera, *Vicia* L., *Pisum* L., *Lens* Adans., *Lathyrus* L. and *Vavilovia* A.Fed. To this end Kupicha (1977) reinstated the monogeneric tribe *Cicereae* Alef. originally proposed by Alefeld (1859) and provided a detailed generic description (Kupicha, 1981). The genus was revised by van der Maesen (1972), who largely adhered to the classification of Popov (1929). The latter divided the species into 2 subgenera – *Pseudononis* Popov and *Viciastrum* Popov; 4 sections – *Monocicer* Popov, *Chamaecicer* Popov, *Polycicer* Popov and *Acanthocicer* Popov; and 14 series. The major taxonomic divisions were made on the basis of flower size, life span, growth habit, whether the plants are woody or herbaceous, and form of leaf apex, terminating in a tendril, spine or

This chapter reviews the taxonomy of the genus *Cicer* (Leguminosae, *Cicereae*), taking into account the 1972 revision and the 1987 update by van der Maesen, and new data available since then. Brief descriptions enumerate the 44 species known at present. Improved identification aids for *Cicer* taxa are presented: a dichotomous key and a tabular key. Infrageneric classification is discussed, on the basis of results of molecular investigations.

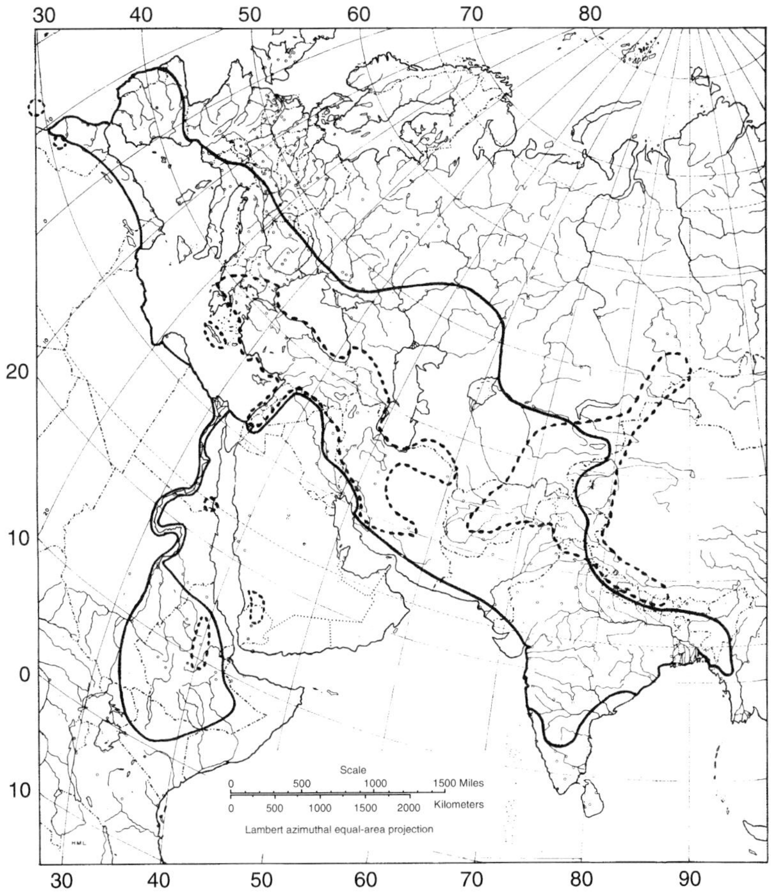

Map 2.1. ——Main areas of chickpea cultivation; ------Areas of occurrence of wild *Cicer* species.

leaflet. The *Cicer* taxa present within the former Soviet Union (Popov, 1929; Linczevski, 1948; Czrepanov, 1981) were also reviewed by Seferova (1995), who rearranged the subgeneric classification. Recently, Javadi and Yamaguchi (2004a, 2005) conducted a molecular study on 29 species of the genus, suggesting further adaptation of the classification above species level.

Only one of the 44 species currently recognized is cultivated on a large scale, the chickpea (*Cicer arietinum* L.). Today it is cultivated in 49 countries (FAO, 2006) from the Mediterranean basin to India, Ethiopia, East Africa and Mexico. It is a major pulse crop in Asia and Africa, which, combined, account for 96% of total world production. In 2004, the world production of chickpea was 8.6 million tonnes, from 11.2 million hectares of land. These figures show that the chickpea accounts for almost 15% of the total land area used for cultivation of pulses (Singh, 1990).

There has been an increasing demand for higher-yielding, as well as disease-, insect-, wilt- and drought-resistant cultivars. Chickpea breeders increasingly look to the wild relatives to supply genes to meet this demand, which has resulted in definite improvement (van der Maesen *et al.*, 2006). This interest has focused primarily on those species in the gene pools closest to the crop. Using the Harlan and de Wet (1971) gene pool concept, the chickpea gene pool may be characterized as follows:

Crop = GP1a	GP1b	GP2	GP3
Cicer arietinum	*C. echinospermum*	*C. bijugum*	Other *Cicer* species
		C. judaicum	
	C. reticulatum	*C. pinnatifidum*	

Thus, those in GP1b are most closely related to chickpea, with those in GP2 and GP3 more remotely related.

The threat status for *Cicer* species in the wild has not been systematically reviewed but six *Cicer* species were included in the 1997 World Conservation Union (IUCN) List of Threatened Plants (Walter and Gillett, 1998) using the pre-1994 categories: the six species, *C. atlanticum*, *C. echinospermum*, *C. floribundum*, *C. graecum*, *C. isauricum* and *C. reticulatum*, each being categorized as rare (R). The ecogeography and conservation status of *Cicer* species was recently reviewed by Hannon *et al.* (2001). Major international *Cicer ex situ* collections (see also Berger *et al.*, 2003) are maintained by the International Crops Research Institute for the Semi-Arid Tropics (ICRISAT), Hyderabad, India (17,244 accessions); the International Centre for Agricultural Research in the Dry Areas (ICARDA), Aleppo, Syria (9682 accessions, 292 wild); the US Department of Agriculture, Agricultural Research Service (USDA-ARS), Western Regional Plant Introduction Station (WRPIS), Pullman, Washington (4602 accessions) (Hannon *et al.*, 2001). There is little information available in the literature regarding *in situ* conservation of any wild *Cicer* species; although species will obviously occur in protected areas, their maintenance will not be the focus of the reserve management plan. However, Hannon *et al.* (2001) in their review of *in situ* conservation of *Cicer* spp. found one report of *in situ* conservation of one rare species in Bulgaria (Gass *et al.*, 1996), but they report widespread actual loss of wild species habitat in the areas considered to be the centre of diversity for *Cicer*.

Since 1972 (only) four species have been described as new to science: *C. heterophyllum* Contandriopoulos, Pamukçuoglu and Quézel (Contandriopoulos *et al.*, 1972), *C. reticulatum* Ladizinsky (Ladizinsky, 1975), *C. canariense* A.Santos and G.P.Lewis (Santos Guerra and Lewis, 1985), which was sufficiently distinct to warrant the erection of the monospecific subgenus *Stenophylloma* A.Santos and G.P.Lewis, and *C. luteum* Rassul. and Sharip. (Rassulova and Sharipova, 1992). In recent years collection of wild *Cicer* spp. has become more difficult, as uncertainty about the execution of new biodiversity access and benefit sharing slowed down exploration activities (van der Maesen *et al.*, 2006). The current interest in wild relatives and addition of new taxa has resulted in improved identification aids to enable accurate identification of the currently described *Cicer* taxa.

Keys

For the preparation of keys, refer to Coles (1993) and Coles *et al.* (1998). Descriptive Language for Taxonomy (DELTA) format (Dallwitz, 1980; Dallwitz *et al.*, 1996) and KEY software (Dallwitz, 1974) were used to obtain the keys. A total of 103 characters were used (60 vegetative, 20 inflorescence, 6 pod, 13 seed, 3 phytogeographical, 1 cytological); for each species each character was scored from herbarium specimens and, wherever not available, from the literature. For the eight species whose specimens could not be traced or loaned, the character states that could be found in the literature were included in the data-set. It is difficult to find 'good' diagnostic characters to consistently and accurately identify the *Cicer* taxa. This was especially true for section *Polycicer*. The taxa of this section are obviously closely related and their specific relationships warrant further study.

Out of 103 characters, 48 were used to produce the dichotomous and tabular keys (see Appendices 1 and 2, respectively). One of the advantages of multiaccess keys (see Coles *et al.*, 1998) over traditional dichotomous keys is that the user is not forced to obtain the identification by passing through the character set in a specific preordained sequence. It allows the user to ignore particular characters or sets of characters and still obtain identification. For example, if the specimen lacks fruits or seed, the identification can be based on vegetative and flower characters alone. Multiaccess keys are now more commonly presented as interactive computer programs and the *Cicer* data-set is also available in LUCID format and can be obtained on demand from the second author. LUCID is a powerful and highly flexible knowledge management software application designed to help users with identification and diagnostic tasks developed by the Centre for Biological Information Technology, The University of Queensland, Brisbane, Australia (www.lucidcentral.com).

Collection and Conservation of *Cicer*

Even though there has been much recent conservation activity focused on *Cicer* taxa, there remains a shortage of available herbarium specimens for many species, especially for the rarer taxa from Central Asia. Several species are known only from their type collection (*C. stapfianum* Rech.f. and *C. subaphyllum* Boiss.), and very few specimens exist for others (such as *C. grande* (Popov) Korotkova, *C. incanum* Korotkova, *C. balcaricum* Galushko, *C. baldshuanicum* (Popov) Lincz.) (see Yakovlev *et al.*, 1996). Confirmation of the continued existence of these species is urgently required; collection trips to the type localities should be given priority. Recently, *C. subaphyllum* was found again near the locus classicus (J. Javadi, Kuh Ayub, Iran, 1999 personal communication), as well as *C. stapfianum*. The existence of other taxa, such as *C. rassuloviae* Lincz. (= *C. multijugum* Rassul. and Sharip.) (Czrepanov, 1981) requires verification. These species are poorly known in the taxonomic and ecogeographic sense, and therefore are difficult to conserve and impossible to utilize at present.

Following Maxted's attempts to systematically collect *Cicer* taxa from Central Asia (N. Maxted, Rome, 1990, unpublished data; N. Maxted and C. Sperling, Rome, 1991, unpublished data), it seems likely that each of the recognized species recorded from the region has a much broader geographical distribution than is currently recognized and that there are several species yet to be described. More collection of specimens and accompanying distribution data is urgently required from this region. Political inaccessibility is a recurring problem in sensitive areas such as Iraq, Iran, Afghanistan and India (Kashmir) (see Townsend and Guest, 1974; van der Maesen, 1979).

The general need for a more detailed study of *Cicer* taxa may be illustrated by the case of *C. cuneatum* Hochst. ex Rich., which was thought to occur only in Ethiopia, and south-east Egypt (Jebel Elba) until collected from Saudi Arabia in 1982 by Collenette (*Collenette* 3559, 19.04.82, Herb. K). Likewise, *C. montbretii* Jaub. and Spach had a known distribution in south Albania, south-east Bulgaria (Kusmanov, 1976; Strid, 1986), and European and Aegean Turkey (van der Maesen, 1972); it therefore seemed likely that it would also occur in the adjacent, but relatively unexplored, areas of northern Greece. A specimen from the region was indeed determined recently by Coles and Per Lasson (*Phitos et al.*, 25.06.73, Herb. UPA).

It can be concluded, therefore, that Central Asia should be given priority for further collection activities. A shortage of material from the former Soviet Union, particularly within the new Central Asian republics, was especially noted during investigations by Maxted and others. Furthermore, it is unlikely that all the species endemic to this region have been discovered and described.

Short Species Descriptions with Ecogeographic Notes

1. *C. acanthophyllum* Borissova, Novit. Syst. Pl. Vasc., Leningrad 6: 167 (1970). Type: USSR-Tadzhik SSR: Pamir, Schach-Darja river, Korshinky s.n. (Holo: LE). Also considered as *C. macracanthum* subsp. *acanthophyllum* (Borissova) Seferova (1995).

Erect or shrubby perennial, (15–)20–35(–50) cm high. Leaves paripinnate, (20–)32–55(–65) mm long, terminating in a spine, with 8–20 obovate leaflets, 2–6(–8) × (1–)3–5 mm. Stipules single spines, 2–8(–15) mm long, or perules on lower leaves. Racemes 1(–2)-flowered, corolla purple. Seeds obovoid, circumference ~10 mm. Flowering July–August. Afghanistan, India (Kashmir), N Pakistan, Tadzhikistan; 2500–4000 m.

2. *C. anatolicum* Alefeld, Bonplandia 9: 349 (1861). Type: Turkey, in shrubs on the Boz dagh, Boissier s.n. (Holo: G). Syn. *C. glutinosum* Alef., l.c.; *C. songaricum* Jaub. *et Spach* 1842.

Erect perennial, 15–45(–60) cm tall. Leaves imparipinnate or paripinnate, (20–)50–70(–110) mm long, terminating in a simple or branched tendril or leaflet with a tendrilous midrib, with (3–)8–14(–18), obovate or elliptic leaflets, 6–14(–20) × 4–13 mm. Stipules triangular, 1–6(–14) × (1–)3–7(–12) mm. Racemes 1–2-flowered, corolla purple, pubescent. Seeds obovoid to subglobular, circumference 12–13 mm. Flowering May–August(–September). 2*n* = 14, 16. Armenia, NW and W Iran, N Iraq, Turkey; 500–4200 m.

3. ***C. arietinum*** Linnaeus, Sp. pl. ed. 1–2: 738 (1753). Type: Spain or Italy, Hortus Cliffortianus 370 (Lecto: BM). Syn. *C. grossum* Salisb., Prodr. 340 (1796); *C. sativum* Schkuhr, Handb. 2: 367 (1805); *C. physodes* Reichb., Fl. Germ. Excurs 532 (1830; *C. rotundum* Jord. ex Alef., Oest. Bot. Zeitschr. 9, 11: 356 (1859).

Erect annual, 12–50(–100) cm tall. Leaves imparipinnate, 25–75 mm long, with (7–)11–17 oblong-obovate to elliptic leaflets, (5–)10–16(–21) × (3–)7–14 mm. Stipules triangular to ovate, 3–6(–11) × (1–)2–4(–7) mm. Racemes 1(–2)-flowered, corolla white, pink, purple or light blue. Seeds obovoid or globular, smooth or rough but never reticulate or echinate, large, very variable in colour. Whole plant very variable. Cultivated species (Chickpea). 2*n* = 14, 16, 24, 32, 33. Widespread in cultivation in the semiarid tropics and warm temperate zones at (0–)110–2400 m. Not found naturally outside cultivation, but escapes occur.

4. ***C. atlanticum*** Cosson ex Maire, Bull. Soc. Hist. Nat. Afr. Nord 19: 42 (1928). Type: Morocco, Mt. Gourza, Goundafa, Humbert and Maire (Lecto: P). Syn. *C. maroccanum* Popov., Trudy Prikladnoi Botanike Genetike i Selekcii 21: 188 (1929).

Prostrate perennial, 4–10 cm. Leaves imparipinnate, (5–)15–30 mm long, with (3–)9–15 obovate or flabellate leaflets, 5–10 × 3–5 mm. Stipules triangular or ovate, 1–3 × 2 mm. Racemes 1-flowered, corolla pink or purple. Seeds ovoid. Flowering June–August. Morocco; 2700–2900 m.

5. ***C. balcaricum*** Galushko, Novit. Syst. Plant. Vasc. 6: 174–176 (1970). Type: USSR-Balkarskaya ASSR, Balcaria, Caucasus, source of Baksan river, nr. Elbrus village, Galushko and Kurdjashova 25–8–1964 (Holo: LE). Closely related to *C. anatolicum* and *C. songaricum*.

Erect or shrubby perennial, 30–60 cm tall. Stems sparsely pubescent. Leaves imparipinnate or paripinnate, 60–100 mm long, terminating in a simple tendril or leaflet, with 12–16(–18), oblong-obovate, obovate, elliptic or flabellate leaflets, 10–15 mm long. Stipules ovate-lanceolate, 5–7 mm long. Racemes 2-flowered, corolla purple, pubescent or glabrous. Seeds obovoid. Flowering July–August. Armenia, Azerbaijan, Georgia; 2000 m.

6. ***C. baldshuanicum*** (M.G. Popov) Linczevski, Not. Syst. Herb. Inst. Bot. Acad. Sci. USSR 9: 112 (1949). Type: USSR: Tadzhik SSR, lower Darvaz Mts between Chovaling and Yakh-su rivers, from Zagara pass to Jakh-su river, Michelson 1428 (Holo: LE). Basionym: *C. flexuosum* Lipsky subsp. *baldshuanicum* Popov, Trudy Prikladnoi Botanike Genetike i Selekcii 21: 211 (1929).

Erect perennial, 30–60 cm high. Stems mostly eglandular pubescent. Leaves imparipinnate or paripinnate, 50–100(–150) mm long, terminating in a simple or branched tendril, or a leaflet with a tendrilous midrib, with 8–16, rounded, obovate or flabellate leaflets, (5–)10–20(–22) × 4–15 mm. Stipules flabellate, 4–8 mm long. Racemes 1–2-flowered, corolla purple. Flowering April–July. Tadzhikistan; 1600–2000 m.

7. ***C. bijugum*** K.H. Rechinger, Arkiv. Bot., Stockholm, andra ser. 1: 510 (1952). Type: Syria, Azaz, Haradjian 4442 (Holo: G).

Prostrate to erect annual, 10–30 cm tall. Leaves imparipinnate, (13–)18–25(–44) mm long, with (3–)5(–7) oblong-obovate or obovate leaflets, (5–)7–12(–18) × 3–8 mm. Stipules ovate-lanceolate or flabellate, 2–5 × 1–2 mm. Racemes 1-flowered, corolla pink or purple. Seeds subglobular, circumference 7–12 mm, echinate. Flowering (April–)May–June. $2n = 16$. N, W Iran, N Iraq, N Syria, SE Turkey; 500–1300 m.

8. ***C. canariense*** A. Santos and G.P. Lewis, Kew Bull. 41: 459–462 (1986). Type: Canary Islands, La Palma, Caldera de Taburiente, alluvial soils in pine forest near Lomo de Las Chozas. 1200 m. Santos and Fernandez s.n. (Holo: ORT; iso: K).

Scrambling perennial, height 50–200 cm. Stems sparsely pubescent. Leaves paripinnate, (40–)70–110 mm long, terminating in a simple tendril, with 32–63 linear leaflets, 15–32 × 0.5–1 mm. Stipules triangular, 4–7 × 1.5–3 mm. Racemes 1–4(–7)-flowered, arista absent, corolla pink or violet. Seeds subglobular or globular, circumference 9–12 mm. Flowering August–September. $2n = 16$ and 24. Canary Islands, La Palma and Tenerife; 900–1400 m.

9. ***C. chorassanicum*** (Bunge) M.G. Popov, Bull. Appl. Bot. Genet. Pl. Breed. 21(1): 180 (1929). Type: Iran, Khorassan Mts near Sabzevar, Bunge s.n. (Holo: G). Basionym: *Ononis chorassanica* Bunge, in Boissier, Fl. Orient. 2: 63 (1872). Syn. *C. trifoliatum* Bornm., Bull. Herb. Boiss. Sér. 2, 5: 849 (1905). Type: Persia, Khorassan Mts near Ssabzewar, Bunge s.n. (Holo: G).

Prostrate to erect annual, 5–12(–15) cm tall. Leaves imparipinnate, 12–18(–22) mm long, with 3 flabellate leaflets, 5–8(–10) × 3–6(–9) mm. Stipules perular, 0.5–1 × 0.5(–1) mm. Racemes 1(–2)-flowered, corolla white or pink. Seeds obovoid, circumference 9–12 mm. Flowering (April–)May–July. $2n = 16$. N and C Afghanistan, N and NE Iran; 1400–3300 m.

10. ***C. cuneatum*** Hochstetter ex Richard, Tent. Fl. Abyss. 1: 195 (1847). Type: Ethiopia, nr. Gapdiam in Tigre, Schimper Sect. 2: 810 (Lecto: P; isolecto: BM, G, K, L, M, MPU).

Erect or climbing annual, 20–40(–60) cm tall. Leaves paripinnate, 30–70(–90) mm long, terminating in a ramified tendril (the only annual species with tendrils), with (8–)14–22 oblong-obovate or elliptic leaflets, 5–9(–12) × 1–5 mm. Stipules flabellate, 2–7 × 1–4 mm. Racemes 1-flowered, corolla pink or purple. Seeds globular, circumference 9–10 mm. Flowering October–November: Ethiopia, January–February: Egypt. $2n = 16$. Egypt, Ethiopia, Saudi Arabia; 1000–2200 m.

11. ***C. echinospermum*** P.H. Davis, Notes RBG Edinb. 29: 312 (1969). Type: Turkey, Urfa, Tel Pinar, Sintenis 747 (Holo: K). Syn. *C. edessanum* Stapf ex Bornm., Beih. Bot. Centr.bl 19, 2: 248 (1906).

Prostrate or erect annual, 20–35 cm tall. Leaves imparipinnate, (20–)30–40(–46) mm long, with 7–11(–14) oblong-obovate leaflets, 4–11(–14) × 2–5 mm. Stipules perular, 2–4 × 1–3 mm. Racemes 1-flowered, corolla purple. Seeds obovoid, echinate. Flowering May. $2n = 16$. N Iraq, Turkey; 600–1100 m.

12. ***C. fedtschenkoi*** Linczevski, Not. Syst. Herb. Inst. Bot. Acad. Sci. USSR 9: 108 (1949). Type: USSR-Tadzhik SSR, Schugnan, Badzhan-kutal, Fedtschenko 27–07–1904 (Holo: LE).

Syn. *C. fedchenkoi* Lincz. (orthogr. variant); *C. songaricum* var. *schugnanicum* Popov, *Trudy Prikladnoi Botanike Genetike i Selekcii* 21: 219 (1929); *C. songaricum* var. *pamiricum* Lipsky ex Paulsen, Bot. Tidskr. 19: 162 (1909).

Erect perennial, 14–25(–35) cm tall. Leaves imparipinnate or paripinnate, 30–80(–100) mm long, terminating in a simple tendril or a leaflet with a tendrilous midrib, with (7–)11–19(–30), obovate leaflets, 5–10(–15) × 3–7(–10) mm. Stipules triangular or ovate, 7–12(–15) × 5–8(–13) mm, equal to or larger than the leaflets. Racemes 1-flowered; arista with a small flabellate leaflet at the tip; corolla purple. Seeds deltoid. Flowering June–August. N and NE Afghanistan, Iran, S Kirghizistan, Tadzhikistan; 2500–4200 m.

13. ***C. flexuosum*** Lipsky, Acta Hort. Petrop. 23: 102 (1904). Type: USSR-Kazakh SSR, Kuletschek, Karatau, Regel 200 (Lecto: LE). Syn. *C. flexuosum* Lipsky subsp. *tianschanicum* var. *robustum* Popov, Bull. Univ. As. Centr. 15 suppl. 14 (1927); *C. flexuosum* Lipsky subsp. *robustum* Popov, Trudy Prikladnoi Botanike Genetike i Selekcii 21: 210 (1929).

Erect or shrubby perennial, 30–70 cm in height. Leaves imparipinnate or paripinnate, 50–100(–150) mm long, terminating in a simple or branched tendril, or rarely a leaflet, with 6–20 obovate, flabellate, or elliptic leaflets, 4–11(–15) × (2–)4–8(–12) mm. Stipules ovate or flabellate, 2–5(–7) × 2–4(–8) mm. Racemes 1–3-flowered, corolla purple, pubescent. Seeds obovoid, circumference 12–13 mm. Flowering May–July. S Kirgizistan, Tadzhikistan, Uzbekistan; 500–2400 m.

14. ***C. floribundum*** Fenzl, Pugillus Plant. Nov. Syr. Taur. occ. 1:4 (1842). Type: Turkey, Taurus Mts, Gülek, Kotschy 167 (Holo: W).

Erect perennial, 10–35 cm high. Leaves imparipinnate or paripinnate, (24–)50–80(–105) mm long, terminating in a simple or branched tendril, or leaflet, with (8–)10–16(–19) oblong to elliptic leaflets, (6–)10–16(–20) × 4–8(–10) mm. Stipules triangular or flabellate, (3–)5–9(–12) × 2–6(–10) mm. Racemes 1–5-flowered; arista with clavate leaflet at the tip; bracts flabellate, large ~4mm and dentate; corolla purple or violet. Seeds subglobular, circumference 3–4 mm. Flowering June–July. $2n = 14$. S Turkey; 800–1700 m.

15. ***C. graecum*** Orphanides, in Boiss., Diagn. Ser. 2(2): 43 (1856). Type: Greece above Trikkala, Mt. Kyllene, Orphanides 578 (Holo: P; iso: BR, K, W).

Erect perennial, 19–60 cm tall. Leaves imparipinnate or paripinnate, (30–)50–100 mm long, terminating in a simple or branched tendril, or a leaflet with a tendrilous midrib, with (6–)11–19 oblong or elliptic leaflets, 6–16(–22) × 3–7(–10) mm; margin entirely serrated except extreme base. Stipules triangular or ovate, 4–8(–12) × 2–8 mm. Racemes 1–6-flowered, peduncle ending in an arista or bract; arista with a clavate leaflet at the tip; bracts flabellate, ~4 mm, dentate; corolla purple. Seeds globular, circumference 4 mm. Flowering June–July. Greece; 1000–1400 m.

16. ***C. grande*** (M.G. Popov) Korotkova, Bot. Mat. Herb. Inst. Bot. Acad. Sci. Uzbek. 10: 18 (1948). Type: USSR-Uzbek SSR, Pamir-alai, Kugitang, Popov and Vvedensky 494 (Lecto: TAK). Basionym: *C. flexuosum* Lipsky ssp. *grande* Popov, Bull. Univ. As. Centr. 15, suppl. 15 (1927).

Erect perennial, 20–50 cm tall. Leaves paripinnate, 60–110 mm long, terminating in a simple or branched tendril, or a leaflet with a tendrilous midrib, sparsely pubescent, with 8–12 oblong-obovate or elliptic leaflets, 10–25(–27) × 6–12 mm. Stipules flabellate or perular, 5–10 × 5–9 mm. Racemes 1–2-flowered, corolla pubescent. Seeds obovoid. Flowering June. Uzbekistan; 1000–2000 m.

17. ***C. heterophyllum*** Contandriopoulos, Pamukçuoglu and Quézel, Biol. Gallo-Hellen. 4(1): 12–15 (1972). Type: Turkey, Antalya prov., forests of *Pinus brutia* and *Quercus cerris* between Manavgat and Akseki, 3 km S of Didere, N exposition, 1100 m, sample 6-J-3 (Holo: MARSSJ; iso: WAG).

Erect perennial, 40–70 cm tall. Leaves imparipinnate, (50–)80–120 mm long, with (9–)11–17 oblong, obovate or elliptic leaflets, of two sizes, 25–35 × 6–12 mm or 15–25 × 12–18 mm; leaflets entirely serrated except extreme base. Stipules triangular, 3–6 × 2–3 mm. Racemes 2–6-flowered; arista with a clavate leaflet at the tip; corolla pale yellow. Flowering July. $2n = 16$. Turkey; 1100 m.

18. ***C. incanum*** Korotkova, Not. Syst. Herb. Inst. Biol. Zool. Acad. Sci. Uzbek. 10: 17 (1948). Type: USSR-Tadzhik SSR, W Pamir-alai, Jakkabag-darja, Botshantshev and Butkov 720 (Holo: TAK). Considered syn. of *C. macracanthum* Popov by Seferova (1995).

Erect perennial, 20–30 cm tall. Leaves paripinnate, 25–40 mm long, terminating in a spine, with 8–12 obovate leaflets, 2–6 × 4–5 mm; teeth of leaflet midribs not pronounced. Stipules 3–7 mm long, equal to or larger than the leaflets. Racemes 1–2-flowered, corolla pubescent. Flowering August. Tadzhikistan; 2000–3000 m.

19. ***C. incisum*** (Willdenow) K. Maly, in Ascherson and Graebner, Syn. Mittel-Europ. Fl. 6(2): 900 (1909). Type: Crete, herb. Willdenow (Holo: B). Basionym: *Anthyllis incisa* Willd., Sp. pl. ed. 3, 2: 1017 (1802); *C. ervoides* (Sieb.) Fenzl, Ill. Syr. Taur. in Russegger, Reise 1: 892 (1841), basionym *Ononis ervoides* Sieb., Riese Kreta 1817, 1: 325 (1823); *C. pimpinellifolium* Jaub. *et Spach*, Ann. Sci. Nat. Sér. 2, 17: 228 (1842); *C. minutum* Boiss. and Hoh., Diagn. Sér. 1, 9: 130 (1849); *C. adonis* Orph. ex Nyman, Consp. Fl. Eur. 200 (1878); *C. caucasicum* Bornm., Fedde Rep. 50: 139 (1941).

Prostrate perennial, 5–16(–25) cm tall. Leaves imparipinnate, 6–16 mm long, with (3–)5–7(–9) obovate or flabellate leaflets, (2–)4–10 × 1–6 mm. Stipules ovate to flabellate, 2–5 × 1–3 mm. Racemes 1(–2)-flowered, corolla pink, purple or light blue. Seeds obovoid or subglobular, circumference ~10 mm. Flowering (May–)June–August(–September). $2n = 16$. Armenia, Greece (including Crete), Georgia, Iran, Lebanon, Syria, Turkey; 1400–2740 m.

20. ***C. isauricum*** P.H. Davis, Notes RBG Edinb. 29: 311 (1969). Type: Turkey, Antalya prov., Akseki, Huber-Morath 17174 (Holo: Hb Huber-Morath).

Erect perennial, 20–40 cm tall. Stems sparsely pubescent. Leaves imparipinnate, 30–80 mm long, with 7–13 oblong-obovate leaflets, 7–25 × (5–)9–12(–16) mm; leaflets entirely serrated except extreme base. Stipules triangular or perular, 2–5 × 1–2(–3) mm. Racemes 1–3(–4)-flowered; arista with a clavate leaflet at the tip; corolla white. Seeds subglobular. Flowering June–July(–August). $2n = 16$. S Turkey; 1000–1750 m.

21. *C. judaicum* Boissier, Diagn. Sér. 2(9): 130 (1849): Zohary, Fl. Palaest. 2: 193 (1992). Type: Palestine, Jerusalem, Boissier s.n. (Holo: G).

Prostrate to erect annual, 10–40 cm. Leaves imparipinnate, (14–)20–35(–43) mm long, with 7–14 doubly incised, obovate or oblong-obovate leaflets, 4–9 × 1.5–5(–8) mm. Stipules triangular, ovate or ovate-lanceolate, 1–3 × 1–2(–3) mm. Racemes 1(–2)-flowered, corolla purple or light blue. Seeds deltoid, circumference 7–10 mm, weakly bilobed. Flowering March–May. $2n = 16$. Israel, Lebanon; 0–500 m.

22. *C. kermanense* Bornmüller, Bull. Herb. Boiss. Ser. 2(5): 969 (1905). Type: Iran, Kerman prov., Kuh-i-Dschupar, Bornmüller 3678 (Lecto: JE). Syn. *C. spiroceras* Jaub. *et Spach* subsp. *kermanense* (Bornm.) Popov, Trudy Prikladnoi Botanike Genetike i Selekcii 21: 196 (1929).

Erect or shrubby perennial, 30–50 cm tall. Leaves paripinnate, 70–100(–130) mm long, terminating in a simple tendril, with (6–)16–24, remote, flabellate leaflets, 2–9 × 5–8(–15) mm. Stipules perular, 1–2 × 0.5–1 mm. Racemes 1–2-flowered, corolla white or pink. Seeds obovoid, circumference 9–11 mm. Flowering May–June(–July). SE Iran; 2300–3300 m.

23. *C. korshinskyi* Linczevski, Not. Syst. Herb. Inst. Bot. Acad. Sci. USSR 9: 110 (1949). Type: USSR-Tadzhik SSR, Darvaz Mts, Imam-Askara Mt., Linczevski 1179 (Holo: LE).

Erect perennial, 50–80 cm tall. Stems sparsely pubescent. Leaves imparipinnate or paripinnate, 40–80 mm long, terminating in a spiny curl, simple tendril or a leaflet, with 10–13, obovate or elliptic leaflets, (5–)10–17(–20) × (4–)6–10(–13) mm. Stipules ovate or flabellate, 5–10 × 5–12 mm. Racemes 1-flowered. Flowering June–August. Iran, Tadzhikistan; 2500–2600 m.

24. *C. laetum* Rassulova and Sharipova, Fl. Tadzhiskoj SSR 5: 634 (1978). Type: Tadzhikistan, southern slope of Alai Mts, near Pitautkul, above Mt. Czonkirkiz, T.F.Koczkareva, L.S.Rjabkova and E.P.Zhogoleva 4528 (Holo: DU).

Erect or ± ascendent perennial, ca. 50 cm tall. Leaves imparipinnate, 4.5–8 cm long, with 5–10 pairs of obovate leaflets, 2–4 × 1.5–2 mm, apex toothed. Stipules foliaceous, 2–4 mm long, toothed, hairy. Racemes 1-flowered, arista 6–10 mm, corolla yellow-bluish, 2–2.5 cm long. Pods and seeds unknown. Flowering June. Tadzhikistan, East Pamir-Alai, 2500 m.

25. *C. luteum* Rassulova and Sharipova, Izv. Akad. Nauk Respubl. Tadzhik. Otd. Biol. Nauk. 1: 51–52 (1992). Type: USSR-Tadzhik SSR, Vachanici,

Jschkaschim, between Kalik-Dara and Charvik, forest at 3900 m, Kurbanbekov 539, 4.08.1962 (Holo: TAD).

Erect perennial, 24–29 cm tall. Leaves imparipinnate, 45–110 mm long, with 6–18, obovate leaflets, 5–11 mm. Stipules equal to or longer than leaflets. Racemes 1-flowered; arista with a small flabellate leaflet at the tip; corolla yellow. Flowering July–August. Tadzhikistan; 3900–4000 m.

26. ***C. macracanthum*** M.G. Popov, Bull. Univ. As. Centr. 15, suppl.: 16 (1927). Type: USSR-Tadzhik SSR, Pamir-alai, Guralash, Popov in Herb. As. Med. 205 (Holo: TAK). Syn.

C. songaricum Steph. var. *spinosum* Aitch., Bot. J. Linn. Soc. 18: 49 (1881); *C. garanicum* Boriss. is considered a synonym by Seferova (1995)

Erect perennial, (10–)20–50(–60) cm tall. Leaves paripinnate, (13–)25–40(–70) mm, terminating in a spine. Tendrils absent, with (6–)10–18(–22) obovate or flabellate leaflets, 1.5–6(–8) × 1.5–5 mm. Stipules double spines, (1–)6–25(–33) mm long, (one long, one short). Racemes 1(–3)-flowered, corolla purple, pubescent or glabrous. Seeds obovoid, circumference 11–12 mm. Flowering June–August. Afghanistan, India (Kashmir), Iran, N Pakistan, Tadzhikistan, Turkestan, Uzbekistan; (1200–) 2743–4250 m.

27. ***C. microphyllum*** Bentham, in Royle, Ill. Bot. Himal. 200 (1839). Type: India, Himachal Pradesh, Shalkur (Hungarung), Royle s.n. (Holo: K). Syn.: *C. jacquemontii* Jaub. et Spach, Ann. Sci. Nat. Sér. 2, 18: 231 (1842).

Erect perennial, (20–)40–60 cm in height. Leaves paripinnate, (38–)50–140 mm long, terminating in a simple tendril or a leaflet with a tendrilous midrib, with (8–)14–28(–37) obovate or elliptic leaflets, 4–10(–14) × 3–7(–10) mm. Stipules triangular, ovate, ovate-lanceolate or flabellate, 3–10(–16) × (2–)4–10(–15) mm, equal to or larger than the leaflets. Racemes 1–3-flowered, corolla purple, pubescent or glabrous. Seeds obovoid, circumference 10–12 mm. Flowering June–July(–September). $2n$ = 14, 16. E Afghanistan, China, India (Kashmir), Nepal, Pakistan, Tadzhikistan; (2000–)2500–4600(–5600) m.

28. ***C. mogoltavicum*** (M.G. Popov) Koroleva, Fl. Tadzhik. 5: 600 (1937). Type: USSR-Tadzhik SSR, Mogol-Tau Mts., Katar-Bulak, Popov and Vvedensky, Herb. As. Med. 264 (Holo: TAK). Basionyms: *C. flexuosum* Lipsky subsp. *tianschanicum* var. *mogoltavicum* Popov, Bull. Univ. As. Centr. 15, suppl.: 15 (1927); *C. flexuosum* Lipsky subsp. *mogoltavicum* Popov, Trudy Prikladnoi Botanike Genetike i Selekcii 21: 211 (1929).

Shrubby perennial, 60–80 cm tall. Stems sparsely pubescent. Leaves paripinnate, 50–160 mm long terminating in a simple or branched tendril, with 16–22, remote, flabellate leaflets, (2–)3–7 × 2–8 mm. Stipules flabellate or perular, 2–4 × 1–4 mm. Racemes (1–)2-flowered, corolla violet, pubescent. Seeds obovoid. Flowering April–June. Tadzhikistan; 1500 m.

29. ***C. montbretii*** Jaubert *et Spach*, Ann. Sci. Nat. Sér. 2(17): 229 (1842). Type: Turkey, Kaz Dagh, (Mt. Gargaro, Gassadagh). Aucher-Eloy 1146 (Lecto: P; iso: OXF).

Erect perennial, 25–45 cm tall. Leaves imparipinnate, (42–)60–90(–100) mm long, with (8–)11–19 oblong to oblong-obovate or elliptic leaflets, (8–)13–22(–27) × (4–)7–10(–13) mm; leaflets entirely serrated except extreme base. Stipules triangular or ovate, (3–)5–8(–10) × 3–6(–8) mm. Racemes (1–)2–5-flowered; peduncle ending in an arista with a clavate leaflet at the tip; corolla white with a purple blotch. Seeds obovoid, circumference 11–12 mm. Flowering March–June(–August). $2n = 16, 24$. S Albania, SE Bulgaria, N Greece, European and Aegean Turkey; 0–1200 m.

30. ***C. multijugum*** Maesen, Meded. Landbouwhogeschool Wageningen 72(10): 91 (1972). Type: Afghanistan, Koh-i-Baba, Köie 2630 (Holo: C). Considered by Czerepanov (1995) a synonym of *C. microphyllum* Benth.

Erect perennial, 10–30 cm tall. Leaves imparipinnate, (30–)60–130 mm long, with 19–35(–41) obovate leaflets, (3–)5–13 × (2–)4–7 mm. Stipules triangular or ovate, 5–13 × 3–9 mm, equal to or larger than leaflets. Racemes 1-flowered, corolla purple or violet. Seeds deltoid. Flowering June–July(–August). Afghanistan; 3000–4200 m.

31. ***C. nuristanicum*** Kitamura, Acta Phytotax. Geobot. Kyoto 16: 136 (1956). Type: Afghanistan, Nuristan, Voma-Chatrass, Kitamura s.n. (Holo: KY).

Erect perennial, 20–45 cm tall. Leaves paripinnate or imparipinnate, (30–)72–100(–140) mm long, terminating in a simple or branched tendril, or a leaflet with a tendrilous midrib, with (9–)11–20(–28), obovate leaflets, (5–)8–14(–18) × (4–)6–10(–13) mm. Stipules triangular, (1–)2–5(–7) × 2–5 mm. Racemes 1–2-flowered, corolla purple, pubescent. Seeds obovoid. Flowering June–July(–August). E Afghanistan, India (Kashmir), N Pakistan; 2134–4600 m.

32. ***C. oxyodon*** Boissier and Hohenacker, Diagn. Ser. 1(9): 129 (1849). Type: Iran, Elburz Mt., Uston Bag nr. Passgala, Kotschy 287 (Holo: P).

Erect perennial, 20–55 cm high. Leaves paripinnate, 38–100(–140) mm long, terminating in a simple or branched tendril, with (4–)10–14(–16) flabellate leaflets, 5–10(–17) × 4–10(–17) mm. Stipules triangular, ovate or flabellate,(1–)2–5 × 1–3(–5) mm. Racemes 1–2-flowered, corolla cream or pale yellow. Seeds obovoid to globular. Flowering (May-) June–July. Afghanistan, Iran, N Iraq; 500–2980 m.

33. ***C. paucijugum*** (M.G. Popov) Nevski, Acta Inst. Bot. Acad. Sci. USSR sér. 1(4): 260 (1937). Type: USSR-Servshan above Chodsha-i-fil village, Kugitang-tau, Nevski 485 (Holo: LE).

Erect perennial, 18–35 cm tall. Leaves imparipinnate, (19–)22–40 mm long, terminating in a foliate spine or leaflet, with 7–10 obovate leaflets, 4–11 × 3–6 mm. Stipules flabellate, 2–7 × 2–7 mm. Racemes 1-flowered, corolla purple. Flowering June–July. E Kazakhstan, Tadzhikistan; 2900 m.

34. ***C. pinnatifidum*** Jaubert *et Spach*, Ann. Sci. Nat. Sér. 2(18): 227 (1842). Type: Asia Minor, Montbret s.n. cultiv. at Paris (Holo: P).

Prostrate to erect annual, 10–20(–40) cm tall. Leaves imparipinnate, (12–)20–40 mm long, with 5–15 oblong to obovate leaflets, 3–8(–12) × 1.5–5(–7) mm.

Stipules ovate-lanceolate or flabellate, 2–5(–7) × 1–2 mm. Racemes 1-flowered, corolla pink or purple. Seeds deltoid, circumference 8–12 mm, strongly bilobed. Flowering March–June. $2n = 16$. Armenia, N Iraq, N Syria, Turkey; 250–1500 m.

35. ***C. pungens*** Boissier, Diagn. Sér. 2(2): 44 (1856). Type: Afghanistan, Yomutt nr. Kabul, Griffith 1608 (Holo: K). Syn.: *C. spinosum* Popov, Bull. Univ. As. Centr. 15, suppl. 15 (1927).

Flexuous, erect, spiny, perennial, 15–40 cm tall. Leaves paripinnate, (7–)12–30(–55) mm long, terminating in a spine, with 4–6(–10) sub-sessile, obovate leaflets, (3–)6–11(–13) × 3–7(–9) mm. Stipules ovate or perular, 2–6(–8) × (0.5–)2–5 mm. Racemes 1(–2)-flowered, corolla purple, pubescent. Seeds obovoid, circumference 11–13 mm. Flowering (May–)June–July(–August). $2n = 14$. Afghanistan, W Tadzhikistan; (1800–)2300–4200 m.

36. ***C. rassuloviae*** Linczevski (Czrepanov, 1981); Syn. *C. multijugum* Rassulova and Sharipova non Maesen, Fl Tadzhiskoj SSR 5: 633 (1978). Type: Tadzhikistan, Hissar Mts, Varzob river above Obi-Odshuk, S. Junussov 6187 (Holo: DU).

Perennial, adscendent from the base, 30–40 cm tall. Leaves imparipinnate, 7–16 pairs of flabellate leaflets, 3–6 × 3–6 mm, apex 7–12-toothed. Stipules flabellate-cuneiform, 3–6 mm long. Racemes 1-flowered, corolla yellow-bluish, 19 mm long. Seeds ovoid, 5 mm long, brown-black. Flowering July. Tadzhikistan, W Pamir-Alai, endemic; 1700 m.

37. ***C. rechingeri*** Podlech, Mitt. Bot. Staatssamml. München 6: 587 (1967). Type: Afghanistan, Baghlan, Middle Andarab Valley, NE of Deh-Salah in the Upper Kasan Valley, Podlech 11700 (Holo: M; iso: E, W).

Perennial, branched, up to 40 cm tall. Leaves paripinnate, 4–6 cm long, ending in a spine, with (8)10–20 rotundate-ovate leaflets with truncate-emarginate incised apex, 2–5 mm long, 3–5 mm wide. Stipules triangular-lanceolate, 1–3 teeth, 2–6 mm long, 1–2.5 mm wide, largest ones at the base of the plant. Racemes 1–2-flowered, arista a spine of 1–1.5 cm, corolla (pale) violet, 12–15 mm long. Seeds not known. Flowering July–August. Afghanistan, endemic; 2400–3600 m.

38. ***C. reticulatum*** Ladizinsky, Notes RBG Edinb. 34: 201–202 (1975). Type: Turkey, Mardin prov., nr. Dereici, ca. 9 km E of Savur on gulley, edge of vineyard, Ladizinsky s.n. (Holo: HUJ).

Prostrate to erect annual, 20–35 cm tall. Leaves imparipinnate, 15–28(–40) mm, with 7–11 oblong-obovate sometimes doubly incised leaflets, 4–11 × 2–4 mm. Stipules ovate, 1–3 × 1–4 mm. Racemes 1-flowered, arista absent, corolla purple. Seeds obovoid, large, circumference 15–19 mm, reticulate. Flowering May–June. $2n = 16$. Turkey; 650–1100 m.

39. ***C. songaricum*** Stephan ex de Candolle, Mem. Lég. 8: 349 (1825). Type: USSR-Songaria (Dzhungarskyi Alatau), Stephan ex Herb. Prescott (Holo: OXF). For synonymy see van der Maesen (1972).

Erect or shrubby perennial, 15–60 cm tall. Leaves paripinnate, (28–)35–60(–75) mm long, terminating in a simple tendril or a leaflet with a tendrilous midrib, with (8–)12–18, obovate leaflets, (2–)6–10(–15) × (2–)4–8 mm. Stipules triangular, ovate or flabellate,(3–)5–11 × (2–)6–12 mm, equal to or larger than leaflets. Racemes 1–2-flowered, arista without or rarely with a clavate leaflet at the tip, corolla purple, pubescent. Seeds obovoid. Flowering (May–) June–July (–August). E Kazakhistan, Kirghizistan, Tadzhikistan, Uzbekistan; 1000–4000 m.

40. ***C. spiroceras*** Jaubert *et Spach*, Ann. Sci. Nat. Sér. 2(18): 233 (1842). Type: Iran, Isfahan, Aucher-Eloy 1126 (Holo: P, iso: G, K, P). Paratype: Iran, s.l., Aucher-Eloy 4357 (BM, G, K, OXF, P, WU).

Erect or shrubby perennial, 25–75 cm tall. Stems sparsely pubescent. Leaves paripinnate, 30–90(–120) mm long, terminating in a simple tendril, with (5–)8–14(–22), remote, flabellate leaflets, (2–)4–7(–11) × (2–)4–9(–15) mm. Stipules flabellate or perular, 1–5 × 0.5–1 mm. Racemes 1–2(–4)-flowered, corolla violet, pubescent or glabrous. Seeds obovoid, or subglobular, circumference ~12 mm. Flowering May–July(–August). Iran; 1600–3500 m.

41. ***C. stapfianum*** K.H. Rechinger, Bot. Jahrb. Syst. 75: 339 (1951). Type: Iran, Fars prov., Kuh-e-Bul NNE of Shiraz, Stapf 625 (Holo: W; iso: K).

Shrubby perennial, ~25 cm tall. Stems sparsely pubescent. Leaves paripinnate, (20–)30–55 mm long, terminating in a spine, with (4–)6–10 spine-bearing leaflets, 3–9 mm long, sometimes with 1 or 2 pairs of glabrous flabellate leaflets, (3–)5–10 × 3–7 mm on basal leaves. Stipules perular, 2–5 × 1–3 mm. Racemes 1–2-flowered, corolla glabrous. Seeds obovoid. Flowering August. Iran; 4000 m.

42. ***C. subaphyllum*** Boissier, Diagn. Sér. 1(6): 44 (1845). Type: Iran, Fars prov., Kuh-Ajub Mts, Mt. Jobi near Persepolis, Kotschy 403 (Holo: P; iso: BM, C, G, JE, K, L, M, OXF, W, WAG).

Erect to shrubby perennial, 30–40 cm high. Plants glabrous except pedicels and calyx. Leaves paripinnate, 40–100(–140)mm long, terminating in a spiny curl. Tendrils absent, with 4–14(–20), spine-bearing remote leaflets, 2–8 mm long. Stipules perular, 1–4 × 0.5–1(–2) mm. Racemes 1–2-flowered, arista spiny, pedicels and calyx pubescent. Seeds obovoid. Flowering May. Iran; 2000 m.

43. ***C. tragacanthoides*** Jaubert et Spach, Ann. Sci. Nat. Bot. Sér. 2(18): 234 (1842). Type: Iran, Elamout Mts, Aucher-Eloy 4337 (Holo: P; iso: BM, G, OXF, P, W).

Erect perennial, 11–35 cm tall. Stems sparsely pubescent. Leaves paripinnate, (25–)40–60(–90) mm long, not grooved above terminating in a spine, with (8–)10–16(–20), flabellate or obovate leaflets, 1–5 × 1–5 mm. Stipules perular, 1–3(–4) × 0.5–2 mm. Racemes 1–2-flowered, corolla purple or violet, pubescent. Seeds subglobular, circumference ~11 mm. Flowering June–July (–August). Afghanistan, Iran, S Turkmenia; 1500–3800 m.

43a. ***C. tragacanthoides*** var. ***tragacanthoides*** Type: see species.

Erect perennial, 15–26 cm tall. Stems sparsely pubescent. Leaves paripinnate, 40–50(–60) mm long, not grooved above terminating in a spine, with (8–) 14–16, remote, flabellate or obovate leaflets, 1–3 × 1–3 mm. Stipules perular,

1–3 × 1–2 mm. Racemes 1-flowered, corolla purple, pubescent. Seeds subglobular. Flowering June–July(–August). Afghanistan, Iran, S Turkmenia; 1500–3800 m.

43b. *C. tragacanthoides* var. ***turcomanicum*** Popov, 230 (1928/9). Type: USSR-Turkman SSR, Kopet-dagh Mts., Karanka gorge nr. Ashkhabad, Litwinow 243 (Lecto: LE). Syn.: *C. straussii* Bornm., Beih. Bot. Centralbl. 27, 2: 344 (1910). Type: Iran, Mt. Schuturunkuh, Strauss s.n. 8/1903 (Holo: JE); *C. kopetdaghense* Lincz., Not. Syst. Herb. Bot. Acad. Sci. USSR 9: 111 (1949). Type: Ashkhabad, Litwinow 243 (LE).

Erect perennial, 11–35 cm tall. Stems sparsely pubescent. Leaves paripinnate, (25–)30–60(–90) mm long, not grooved above, terminating in a spine, with (8–)10–16(–20), flabellate or obovate leaflets, (1–)2–5 × (1–)2–5 mm. Stipules perular, 1–3(–4) × 0.5–2 mm. Racemes 1–2-flowered, corolla purple or violet, pubescent. Seeds subglobular, circumference ~11 mm. Flowering June–July(–August). Iran, S Turkmenia; 1500–3000 m.

44. *C. yamashitae* Kitamura, Acta Phytotax. Geobot. Kyoto 16: 135 (1956). Type: Afghanistan, between Sarobi and Kabul, Yamashita and Kitamura s.n. (Holo: KY). Syn.: *C. longearistatum* K.H. Rechinger, Biol. Skrift. Kong. Dansk. Vidensk Selsk 9, 3: 201 (1957). Type: Afghanistan, nr Sarobi, Volk 1887 (Holo: W).

Prostrate to erect annual, (10–)21–30 cm tall. Leaves imparipinnate, 10–30 mm long, with 3–7 oblong, oblong-obovate or oblong-elliptic leaflets, 5–17 × 2–5(–6) mm. Stipules perular, 1–2.5 × 0.5–1(–2) mm. Racemes 1-flowered, arista very long, (5–)10–20 mm, corolla pink. Seeds deltoid, circumference 8–12 mm, colour very variable. Flowering May–June. $2n = 16$. Afghanistan; 900–2800 m.

Molecular Phylogeny of the Genus *Cicer*

The genus *Cicer* has been traditionally classified into two subgenera (*Pseudononis* and *Viciastrum*) and four sections (*Cicer*, *Chamaecicer*, *Polycicer* and *Acanthocicer*) based on morphological traits and geographical distribution (Popov, 1929; van der Maesen, 1972, 1987). Morphological homoplasy (i.e. life cycle and flower size) and lack of diagnostic synapomorphies hinder sectional monophyly based on morphology in the infrageneric classification of van der Maesen (1972, 1987). Therefore, molecular sequence data were the obvious choice to try to find a robust phylogeny on which to base future taxonomic classifications.

Despite recent intensive molecular studies on the genus *Cicer* (i.e. Kazan and Muehlbauer, 1991; Ladbi *et al.*, 1996; Ahmad, 1999; Iruela *et al.*, 2002; Sudupak *et al.*, 2002, 2003; Javadi and Yamaguchi, 2004b), little attention has been given to the aspect of infrageneric classification of the genus at molecular level.

The chloroplast and nuclear genomes have been extensively surveyed to reconstruct plant phylogeny (e.g. Olmstead and Palmer, 1994; Sang, 2002).

In legumes, the chloroplast *trnK/matK* and *trnS–trnG* and also the internal transcribed spacer (ITS) region of nuclear DNA markers provided very useful information at lower taxonomic levels (e.g. Steele and Wojciechowski, 2003; Frediani and Caputo, 2005; Kenicer *et al.*, 2005). A combination of both chloroplast and nuclear DNA sequences can provide complementary information on the evolution of the genus. Analysis of the chloroplast genome will provide information about the maternal evolutionary relationships between the species, and the biparentally inherited nuclear DNA will provide independent data from which to infer evolutionary relationships.

In the present study the nuclear ITS and the chloroplast *trnK/matK* and *trnS–trnG* regions were used to investigate phylogeny of *Cicer*, which provides a frame for the reclassification of the genus. Twenty-nine *Cicer* species representing two subgenera and all four traditionally recognized sections were sampled. We selected *Lens ervoides* (Brign.) Grande, *Pisum sativum* L. and *Vicia sativa* L. (tribe *Vicieae*) as outgroups.

Total genomic DNA was extracted from freshly collected leaf from plants grown in pots in Osaka Prefecture University greenhouse or leaf material from herbarium specimens using a cetyltrimethyl ammonium bromide (CTAB) protocol with minor modification (Doyle and Doyle, 1987).

Polymerase chain reaction (PCR) amplification and sequencing reactions of the plastid and nuclear regions were performed using the following universal primers: the *trnK/matK* region was amplified using a primer pair of trnK685F/trnK2R* and trnK1L/matK1932R, and the region was sequenced using trnK685F, trnK2R, matK4La, matK1100L and trnK1L primers (Hu *et al.*, 2000). ITS4 and ITS5 primers (White *et al.*, 1990) were used both for amplification and sequencing of the ITS1 and ITS2 spacer along with the 5.8S gene. The *trnS–trnG* region amplified and sequenced using trnS-F and trnG-R primers (Xu *et al.*, 2000). DNA amplifications, purification and cycle sequencing were performed following Javadi and Yamaguchi (2004a).

Phylogenetic analyses were performed with phylogenetic analysis using parsimony (PAUP*) 4.0b10 (Swofford, 2002) using the parsimony analysis. A heuristic tree search was conducted using MULTREES, STEEPEST DESCENT off, and ACCTRAN optimizations and tree-bisection-reconnection (TBR) branch swapping. Searches were replicated 1000 times. Clade support was assessed using bootstrap value (Felsenstein, 1985) implemented in PAUP*4.0b10 (Swofford, 2002) using 1000 replicates. An incongruence length difference (ILD; Farris *et al.*, 1995) test was conducted using PAUP* version 4.0b10 (Swofford, 2002) to determine the congruence between plastid and nr ITS data-sets. The test was performed with 100 replicates, using a heuristic search option with random taxon addition and TBR branch swapping.

The ILD test result ($P = 0.1$) indicates an acceptable degree of congruence among the three components of the combined data-set (Farris *et al.*, 1995). The combined nuclear and plastid data matrix consisted of 3850 characters, of which 420 were potentially informative. The combined parsimony analyses, with gaps included, produced 100 equally parsimonious trees of 1085 steps with a consistency index of 0.877 (0.792 excluding uninformative characters) and retention index of 0.878. The strict consensus tree is shown in Fig. 2.1.

The maximum parsimony analyses of combined nrDNA and cpDNA data indicate that three major clades correspond to biogeographic relationships in the genus *Cicer* and contradict the monophyly of each section (Fig. 2.1). Clade A (African group, bootstrap support 100%) includes African species, *C. cuneatum* (section *Cicer*) and *C. canariense* (section *Polycicer*). Clade B (west-central Asian group, bootstrap support 99%) contains species belonging to four sections mainly distributed at high altitudes of western and Central Asia and sister group species of eastern Europe that are growing at low altitudes (Aegean-Mediterranean group) (Fig. 2.1). Clade C (Mediterranean group, bootstrap support 99%) consists of six species in section *Cicer* (*C. arietinum*, *C. echinospermum*, *C. reticulatum*, *C. bijugum*, *C. judaicum* and *C. pinnatifidum*) and one in section *Chamaecicer* (*C. incisum*) which are grouped into two subclades (C1 and C2) and distributed in the eastern part of the Mediterranean region (Fig. 2.1).

What can we tell about infrageneric classification of the genus *Cicer* from the present study? In section *Cicer* (subgenus *Pseudononis*), the monophyly of the six species in clade C supports the cpDNA *trn*T-F sequence results (Javadi and Yamaguchi, 2004a, 2005). Section *Cicer* is characterized by small flowers, imparipinnate leaves or rachis ending in a tendril, and erect or prostrate stems (van der Maesen, 1972, 1987). All these characters appear elsewhere in the genus. According to our results, section *Cicer* is polyphyletic and its traits will need to be reanalysed or weighted differently to define monophyletic subgroups around the type species of section *Cicer*.

The small section *Chamaecicer* (subgenus *Pseudononis*) with two species, *C. chorassanicum* (annual) and *C. incisum* (perennial), shows conflicting pattern in molecular data (clades C and B, Fig. 2.1). Non-monophyly of section *Chamaecicer* is also congruent with geographical distribution pattern of these taxa. The trifoliolate species *C. chorassanicum* is a taxon of north and central Afghanistan, and north and north-east Persia, and grows under dry conditions at high altitude where members of sections *Polycicer* and *Acanthocicer* (van der Maesen, 1972) are quite common. The placement of *C. incisum* (section *Chamaecicer*) in clade C is supported by its morphological traits and geographical pattern. The appearance of section *Chamaecicer* in two distinct clades based on molecular data calls into question the integrity of the section.

Section *Polycicer* (subgenus *Viciastrum*) has a widely distributed pattern that includes members of section *Acanthocicer* (subgenus *Viciastrum*) at high altitude in Persia and Central Asia. Morphologically, species assigned in section *Acanthocicer* share the tragacanthoid and spiny plant shape (van der Maesen, 1972, 1987), which is related to their dry habit. However, our results suggest that section *Acanthocicer* is not monophyletic; four species belong to section *Acanthocicer* (*C. tragacanthoides*, *C. subaphyllum*, *C. macracanthum* and *C. pungens*) and are grouped with members of section *Polycicer* (clade B, Fig. 2.1).

C. montbretii, *C. floribundum*, *C. isuaricum* and *C. graecum* (section *Polycicer*) form a well-supported monophyletic group at the base of clade B. They also share morphological features such as erect habit, entirely dentate margins of leaflet (except near the base), inflorescence with 2–5 flowers and arista with clavate leaflet at the tip (van der Maesen, 1972). Biogeographically, these four

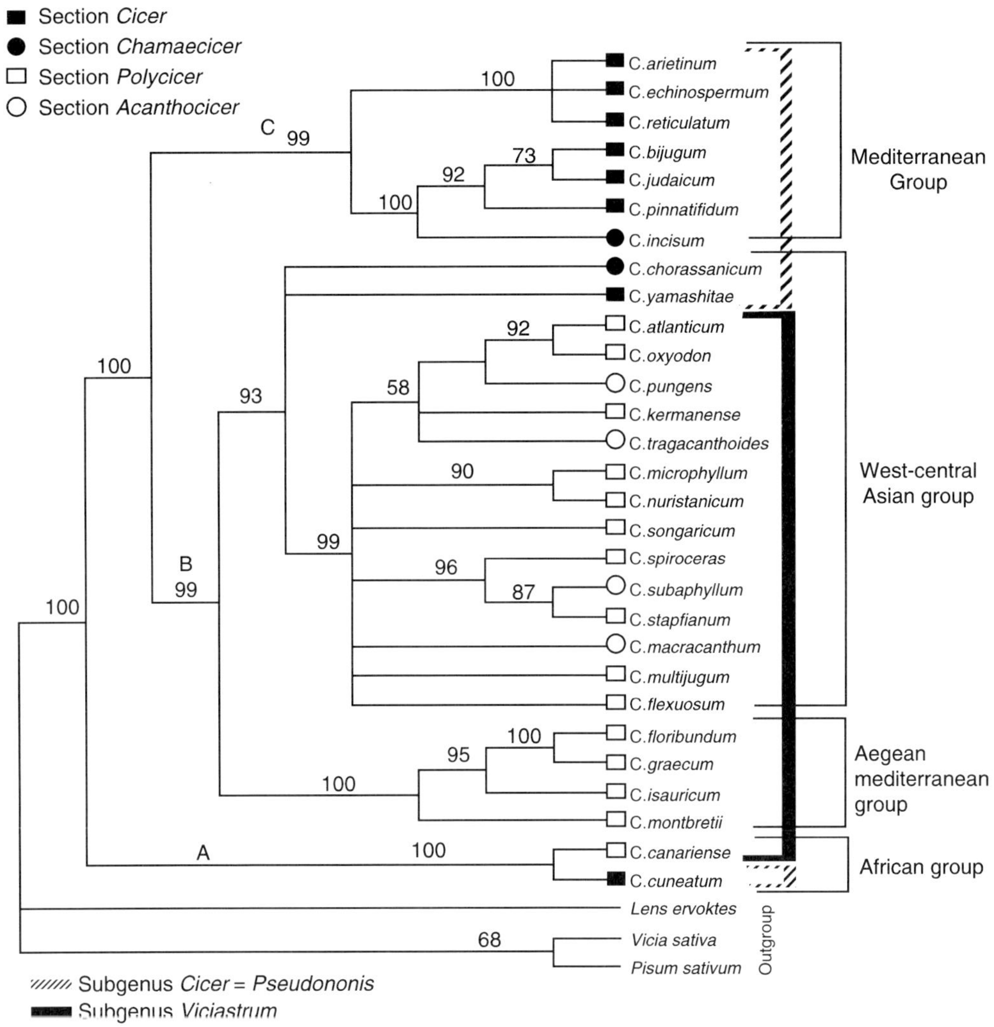

Fig. 2.1. Strict consensus tree of 100 maximum parsimonious trees of combined nrITS, cp *trnK/matK* and *trnS–trnG* data (Tree length = 1085, CI = 0.877 and RI = 0.878).

species are distinct within the section in having relatively humid forest distributions in eastern Europe and southern Turkey at lower altitudes (0–1700 m). This group was treated as a series, Europaeo-Anatolica by van der Maesen (1972); now we distinguish it informally as the Aegean-Mediterranean group.

The two African species *C. canariense* (section *Polycicer*) and *C. cuneatum* (section *Cicer*) form a highly supported basal clade (bootstrap support 100%) in the phylogenetic tree (Fig. 2.1), which is in agreement with the previous studies (Iruela *et al.*, 2002; Javadi and Yamaguchi, 2004a,b; Frediani and Caputo, 2005). These two species are also morphologically distinct and share the synapomorphic trait of a climbing habit. Geographically, they are growing in isolated areas in Africa. *C. canariense*, endemic to the Canary Islands, was first described by Santos Guerra and Lewis (1985) and was placed in a new monospecific subgenus *stenophyllum* (Santos Guerra and Lewis, 1985). *C. cuneatum* is distributed in Ethiopia, north-east Sudan, south-east Egypt and Saudi Arabia (van der Maesen, 1972, 1987). Our analyses suggest the exclusion of *C. cuneatum* and *C. canariense* from sections *Cicer* and *Polycicer*, respectively. It appears that this small group of species is well differentiated, with a number of molecular and morphological (vetch-like) synapomorphies making them appear quite different. Sampling another African species, *C. atlanticum*, may also provide better insight into the differentiation of the African species in the genus *Cicer*.

The current molecular study from both chloroplast and nuclear regions of *Cicer* has demonstrated that the intuitive classification systems devised for the genus in the past (van der Maesen, 1972, 1987) inadequately reflect the natural groupings within the genus. In general, morphological features for delimiting sections show high homoplasy, but ecology and geographical habitat are the most important features to express relationships and are highly congruent with molecular data.

Acknowledgements

This chapter is mainly adapted from both earlier and recent papers by the authors, from Coles *et al.* (1998), from a paper in preparation by Javadi, Wojciechowski and Yamaguchi, and from another in preparation by Davies, Maxted and van der Maesen. We thank the curators of the many Herbaria who provided specimens on loan and all those who helped with information, as well as those of the Gene banks that provided seeds of *Cicer* spp. ANK, IZMIR and W provided leaf samples for the molecular analyses.

References

Ahmad, F. (1999) Random amplified polymorphic DNA (RAPD) analysis reveals genetic relationships among the annual *Cicer* species. *Theoretical and Applied Genetics* 98, 657–663.

Alefeld, F. (1859) Ueber die Vicieen. *Oesterreichische Botanische Zeitschrift* 9, 352–366.

Berger, J., Shahal, A. and Turner, N.C. (2003) Ecogeography of annual wild *Cicer* species: the poor state of the world collection. *Crop Science* 43, 1076–1090.

Coles, S. (1993) The creation of identification aids for *Cicer* L. species. MSc thesis, Department of Biology, University of Birmingham, Birmingham, UK, pp. 1–152.

Coles, S., Maxted, N. and van der Maesen, L.J.G. (1998) Identification aids for *Cicer* (Leguminosae, *Cicereae*) taxa. *Edinburgh Journal of Botany* 55, 243–265.

Contandriopoulos, J., Pamukçuoglu, A. and Quézel, P. (1972) A propos des *Cicer* vivaces du pourtour Mediterranéen oriental. *Biologia Gallo-Hellenica* 4(1), 1–18.

Czrepanov, S.K. (1981) Plantae vasculares URSS: 230. Nauka, Leningrad, Russia.

Dallwitz, M.J. (1974) A flexible computer program for generating diagnostic keys. *Systematic Zoology* 23(1), 50–57.

Dallwitz, M.J. (1980) A general system for coding taxonomic descriptions. *Taxon* 29(1), 41–46.

Dallwitz, M.J., Paine, T.A. and Zurcher, C. (1996) User's guide to the DELTA system: a general system for coding taxonomic descriptions.

CSIRO Australian Division of Entomology, Report No. 13, Canberra, Australia.
Davies, A.M.R., Maxted, N. and van der Maesen, L.J.G. (in prep.) A taxonomic study of the genus *Cicer* L. (Leguminosae, Cicereae). *Blumea.*
Doyle, J.J. and Doyle, J.L. (1987) A rapid DNA isolation procedure for small quantities of fresh leaf tissue. *Phytochemical Bulletin* 19, 11–15.
FAO (2006) FAOSTAT Database faostat.fao.org/.
Frediani, M. and Caputo, P. (2005) Phylogenetic relationships among annual and perennial species of the genus *Cicer* as inferred from ITS sequences of nuclear ribosomal DNA. *Biologia Plantarum* 49(1), 47–52.
Farris, J.S., Källersjio, M., Kluge, A.G. and Bult, C. (1995) Testing significance of incongruence. *Cladistics* 10, 315–319.
Felsenstein, J. (1985) Confidence limits on phylogenies: an approach using the bootstrap. *Evolution* 39, 783–791.
Gass, T., Ambrose, M., Leguen, J., Hadjichristodoulou, A. and Blixt, S. (1996) Review of national grain legume collections. In: Gass, T., Ambrose, M., Leguen, J., Hadjichristodoulou, A. and Blixt, S. (eds) *Report of First Working Group on Grain Legumes*. IPGRI, Rome, pp. 3–11.
Hannon, R., Açikgöz, N. and Robertson, L.D. (2001) Chickpeas (*Cicer* L.) In Maxted, N. and Bennett, S.J. (eds) *Plant Genetic Resources of Legumes in the Mediterranean*. Kluwer Academic, Dordrecht, The Netherlands, pp. 115–124.
Harlan, J.R. and de Wet, J.M.J. (1971) Towards a rational classification of cultivated plants. *Taxon* 20, 509–517.
Hu, J.M., Lavin, M., Wojciechowski, M.F. and Sanderson, M.J. (2000) Phylogenetic systematics of the tribe Millettieae (Leguminosae) based on chloroplast *trn*K/*mat*K sequences and its implications for evolutionary patterns in Papilionoideae. *American Journal of Botany* 87(3), 418–430.
Iruela, M., Rubio, J., Cubero, J.I., Gil, J. and Milián, T. (2002) Phylogenetic analyses in the genus *Cicer* and cultivated chickpea using RAPD and ISSR markers. *Theoretical and Applied Genetics* 104, 643–651.
Javadi, F. and Yamaguchi, H. (2004a) Interspecific relationships of the genus *Cicer* L. (Fabaceae) based on *trn*T-F sequences. *Theoretical and Applied Genetics* 109(2), 317–322.
Javadi, F. and Yamaguchi, H. (2004b) RAPD and seed coat morphology variation in annual and perennial species of the genus *Cicer* L. *Genetic Resources and Crop Evolution* 51, 783–794.
Javadi, F. and Yamaguchi, H. (2005) Is there congruency between traditional classification and molecular relationships in the genus *Cicer* L.? Insights from nuclear and chloroplast sequence data. Abstract P 1163, XVII International Botanical Congress, Vienna, p. 422.
Kazan, K. and Muehlbauer, F.J. (1991) Allozyme variation and phylogeny in annual species of *Cicer* (Leguminosae). *Plant Systematics and Evolution* 175, 11–21.
Kenicer, G.J., Tadashi, K., Pennington, R.T. and Murata, J. (2005) Systematics and biogeography of *Lathyrus* (Leguminosae) based on internal transcribed spacer and cpDNA sequence data. *American Journal of Botany* 92, 1199–1209.
Kupicha, F.K. (1977) The delimitation of the tribe Vicieae and the relationships of *Cicer* L. *Botanical Journal of the Linnean Society* 74, 131–162.
Kupicha, F.K. (1981) *Cicereae* Alef. In: Polhill, R.M. and Raven, P.H. (eds) *Advances in Legume Systematics*, Part 1. Royal Botanic Gardens, Kew, UK, pp. 381–382.
Kusmanov, R. (1976) *Cicer* L. In: Jordanov, D. (ed.) *Flora Republicae Popularis Bulgaricae* VI. Serdicae, In Aedibus Academiae Scientiarum Bulgaricae, pp. 441–442.
Ladbi, M., Robertson, L.D., Singh, K.B. and Charrier, A. (1996) Genetic diversity and phylogenetic relationships among the annual *Cicer* species as revealed by isozyme polymorphism. *Euphytica* 88, 181–188.
Ladizinsky, G. (1975) A new *Cicer* from Turkey. *Notes from the Royal Botanic Garden Edinburgh* 34, 201–202.
Linczevski, I.A. (1948) *Cicer* L. In: Komorov, V.L. (ed.) *Flora of the U.S.S.R.* 13. Moskova-Leningrad, Izdatel'stvo Akademii Nauk SSSR, pp. 294–309.
Olmstead, R.G. and Palmer, J.D. (1994) Chloroplast DNA systematics: a review

of methods and data analysis. *American Journal of Botany* 81, 1205–1224.

Popov, M.G. (1929) The genus *Cicer* and its species. *Trudy Prikladnoi Botanike Genetike i Selekcii* 21, 1–254.

Rassulova, M.A. and Sharipova, B. (1992) New species of the genus *Cicer* from Tajikistan. *Izvestya Akademia Nauk Respubliki Tadzhikistan, Biologiceskii Nauk* 1(125), 51–52.

Sang, T. 2002. Utility of low-copy nuclear gene sequences in plant phylogenetics. *Critical Reviews in Biochemistry and Molecular Biology* 37(3), 121–147.

Santos Guerra, A. and Lewis, G.P. (1985) A new species of *Cicer* (Leguminosae, Papilionoideae) from the Canary Islands. *Kew Bulletin* 41, 459–462.

Seferova, I.V. (1995) Konspekt sistemy roda *Cicer* (Fabaceae). Synopsis of the system of genus *Cicer* (Fabaceae). *Botaniceskii Zhurnal (Leningrad)* 80(8), 96–104.

Singh, K.B. (1990) Status of chickpea in the world. *International Chickpea Newsletter* 22, 10–16.

Steele, K.P. and Wojciechowski, M.F. 2003. Phylogenetic analyses of tribes Trifolieae and Vicieae based on sequences of the plastid gene *matK* (Papilionaceae: Leguminosae). In: Klitgaard, B.B. and Bruneau, A. (eds) *Advances in Legume Systematics* 10. Royal Botanic Garden, Kew, UK, pp. 355–370.

Strid, A. (1986) *Cicer* L. In: Strid, A. (ed.) *Mountain Flora of Greece* 1. Cambridge University Press, Cambridge, pp. 480–481.

Sudupak, M.A., Akkaya, S. and Kence, A. (2002) Analysis of genetic relationships among perennial and annual *Cicer* species growing in Turkey using RAPD makers. *Theoretical and Applied Genetics* 108(5), 937–944.

Sudupak, M.A., Akkaya, S. and Kence, A. (2003) Genetic relationships among perennial and annual *Cicer* species growing in Turkey assessed by AFLP. *Theoretical and Applied Genetics* 108(5), 937–944.

Swofford, D.L. (2002) PAUP*: Phylogenetic Analysis Using Parsimony (*and other methods). Version 4.0610. Sunderland, Massachusetts, Sinauer Associates.

Townsend, C.C. and Guest, E. (1974) *Cicer* L. In: *Flora of Iraq* 3. Ministry of Agriculture and Agrarian Reform of the Republic of Iraq, Baghdad, pp. 505–512.

van der Maesen, L.J.G. (1972) *Cicer* L., a monograph of the genus, with special reference to the chickpea (*Cicer arietinum* L.), its ecology and cultivation. *Mededelingen Landbouwhogeschool Wageningen* 72–10, 1–342.

van der Maesen, L.J.G. (1979) *Cicer* L. In: Rechinger, K.H. (ed.) *Flora Iranica* 140, 1–15, t. 1–22. Graz, Akademische Druck- und Verlagsanstalt.

van der Maesen, L.J.G. (1987) Origin, history and taxonomy of chickpea. In: Saxena, M.C. and Singh, K.B. (eds) *The Chickpea*. CAB International/ICARDA, Wallingford, UK, pp. 11–34.

van der Maesen, L.J.G., Pundir, R.P.S. and Upadhaya, H.D. (2006) Use made of wild legume relatives in breeding. In: *Abstracts Proceedings, International Food Legume Research Conference*. 18–22 October 2005, New Delhi, India, pp. 29–30.

Walter, K.S. and Gillett, H.J. (1998) *1997 IUCN Red List of Threatened Plants*. Compiled by the World Conservation Monitoring Centre. IUCN – The World Conservation Union, Gland, Switzerland and Cambridge, pp. 1–862.

White, T.J., Bruns, T., Lee, S. and Taylor, J. (1990) Amplification and direct sequencing of fungal ribosomal RNA genes for phylogenetics. In: Innis, M., Gelfand, D., Sninsky, J. and White, T. (eds) *PCR Protocols: A Guide to Methods and Applications*, Academic Press, San Diego, pp. 315–322.

Xu, D.H., Abe, J., Sakai, M., Kanazawa, A. and Shimamoto, Y. (2000) Sequence variation of non-coding regions of chloroplast DNA of soybean and related wild species and its implications for the evolution of different chloroplast haplotypes. *Theoretical and Applied Genetics* 101, 724–732.

Yakovlev, G.P., Sytin, A.K. and Roskov, Y.R. (1996) Legumes of Northern Eurasia: a checklist. Royal Botanic Gardens, Kew, UK. Cicereae, pp. 70–75.

APPENDIX 1
A DICHOTOMOUS KEY TO *CICER* SPECIES

1. Annual 2
– Perennial 10
2. Leaves imparipinnate; tendrils absent 3
– Leaves paripinnate; rachis terminating in a branched tendril .... 10. ***C. cuneatum***
3. Leaf with 3 flabellate leaflets, with only the apex serrate .... 9. ***C. chorassanicum***
– Leaf with (3–)5–17, oblong-obovate, obovate or elliptic leaflets, with a margin two thirds to one half serrate 4
4. Seeds echinate 5
– Seeds not echinate 6
5. Leaf with 7–11(–14) leaflets; stipules perular; peduncle 6–11(–15) mm, occasionally ending in an arista, 0.1–1(–2) mm long; seeds obovate .... 11. ***C. echinospermum***
– Leaf with (3–)5(–7) leaflets; stipules ovate-lanceolate or flabellate; peduncle (1–)3–7 mm, ending in an arista 1–3 mm long; seeds subglobular ..... 7. ***C. bijugum***
6. Corolla length 7–13 mm; seeds obovoid or globular, not bilobed, with a circumference of 15–19 mm 7
– Corolla length 4–9 mm; seeds deltoid, bilobed, with a circumference of 7–12 mm 8
7. Plant height 12–50(–100) cm; leaf with (7–)11–17 leaflets; stipules 3–6 (–11) × (1–)2–4(–7); arista 0.2–3(–5) mm; seeds smooth or scabrous, not reticulate, colour very variable .... 3. ***C. arietinum***
– Plant height 20–35 cm; leaf with 7–11 leaflets; stipules 1–3 × 1–4 mm; arista absent; seeds scabrous, always reticulate, brown to grey .... 38. ***C. reticulatum***
8. Peduncular arista (5–)10–20 mm long; stipules perular .... 44. ***C. yamashitae***
– Peduncular arista 0.5–3(–5) mm long; stipules triangular, ovate-lanceolate or ovate 9
9. Leaf petiole length 5–12 mm; leaflets doubly incised; calyx 4–8 mm long; seeds weakly bilobed, 3–4mm long, circumference 7–10 mm .... 21. ***C. judaicum***
– Leaf petiole length (5–)10–17 mm; leaflets not doubly incised; calyx 3–6 mm long; seeds strongly bilobed, 3–6 mm long, circumference 8–12 mm .... 34. ***C. pinnatifidum***
10. Leaflets entirely laminate 11
– Leaflets mostly spiniform 42
11. Tendrils present 12
– Tendrils absent 29
12. Leaflets linear, 15–32 × 0.5–1 mm; peduncular arista absent (Canary Islands) .... 8. ***C. canariense***

– Leaflets flabellate, obovate, oblong-obovate, elliptic, oblong, or rounded, up to 27 × 17 mm; peduncular arista present 13

13. Stipules equal to or larger than leaflets 14
– Stipules smaller than leaflets... 16

14. Flowers solitary; corolla glabrous; peduncular arista with a small flabellate leaflet at the tip 15
– Flowers 1–3 in raceme, corolla pubescent or glabrous; peduncular arista without flabellate leaflet, or very rarely with a clavate leaflet at the tip 16

15. Flowers purple; peduncle 30–70 mm long; pedicel 3–7 mm. .... 12. ***C. fedtschenkoi***
Flowers yellow; peduncle 26–50 mm long; pedicel 2–3 mm. .... 25. ***C. luteum***

16. Leaf (28–)35–60(–75) mm long, with (8–)12–18 leaflets; peduncular arista without leaflet or very rarely with a clavate leaflet at the tip; racemes 1–2-flowered, corolla pubescent .... 39. ***C. songaricum***
– Leaf (38–)50–140 mm long, with (8–)14–28(–37) leaflets; peduncular arista always lacking a leaflet at the tip; racemes 1–3-flowered, corolla pubescent or glabrous .... 27. ***C. microphyllum***

17. Plants with minute bracts usually c. 1 mm; peduncular arista spinose or not, never with a leaflet at the tip; leaflet margins two thirds to one half serrate 18
– Plants with large flabellate bracts, c. 4mm; peduncular arista with a clavate leaflet at the tip, leaflet margins entirely serrate, except at the extreme base 28

18. Corolla small, 10–15 mm long 19
– Corolla larger, (10–)15–27 mm long 20

19. Stems sparsely pubescent; leaflets remote; racemes 1–2(4)-flowered, corolla violet, pubescent or glabrous .... 40. ***C. spiroceras***
– Stems densely pubescent; leaflets close; racemes 1–2-flowered, corolla cream, glabrous ....32. ***C. oxyodon***

20. Stems densely pubescent.. 21
– Stems sparsely pubescent 25

21. Leaf with (3–)8–24(–28) leaflets, each (2–)6–14(–20) mm long............ 22
– Leaf with 8–12 leaflets, each 10–25(–27) mm long........... 16. ***C. grande***

22. Leaf rachis ending in a simple or branched tendril, leaflet, or tendril-like leaflet; leaflets crowded; stipules of medium size (1–)2–7(–14) × (1–)2–7(–12) mm; corolla purple 23
– Leaf rachis always ending in a simple tendril; leaflets remote; stipules perular 1–2 × 0.5–1 mm; corolla white or pink............ 22. ***C. kermanense***

23. Plants with glandular and eglandular hairs; (6–)11–20(–28) leaflets per leaf; leaflets not crowded; seeds obovoid 24
– Plants with glandular hairs only; (3–)8–14(–18) leaflets per leaf; leaflets crowded; seeds subglobular or obovoid ....2. ***C. anatolicum***

24. Stems flexuous; leaf ending in a simple or branched tendril (occasionally a leaflet); stipules ovate or flabellate; racemes 1–3-flowered .... 13. ***C. flexuosum***

– Stems straight; leaf ending in a simple or branched tendril or tendrilous leaflet; stipules triangular; racemes 1–2-flowered........31. ***C. nuristanicum***

25. Pubescence eglandular and glandular.. 26

– Pubescence glandular only... 27

26. Plant 30–60 cm high; hairs mostly eglandular; leaf ending in a simple or branched tendril, or a leaflet with a tendrilous midrib; 8–16 leaflets per leaf; leaflets rotundate, obovate or flabellate, (5–)10–20(–22) × 4–15 mm; calyx 7–9 mm .... 6. ***C. baldshuanicum***

– Plant 60–80 cm high; hairs eglandular and glandular; leaf ending in a simple or branched tendril; 16–22 leaflets per leaf; leaflets flabellate, (2–)3–7 × 2–8 mm; calyx 10–12 mm ....28. ***C. mogoltavicum***

27. Stems flexuous; leaf ending in a simple tendril or leaflet; number of leaflets 12–16(–18); racemes 2-flowered......................................5. ***C. balcaricum***

– Stems straight or slightly flexuous; leaf ending in a simple tendril (upper leaflets), leaflet (lower leaves), or a spiny curl; number of leaflets 10–13; racemes 1-flowered...23. ***C. korshinskyi***

28. Plants 19–60 cm high; pedicels 2–10 mm long; racemes 1–6-flowered; seeds globular (Greece)...15. ***C. graecum***

– Plants 10–35 cm high; pedicels (4–)7–10(–12) mm long; racemes 1–5-flowered; seeds subglobular (Turkey)............................14. ***C. floribundum***

29. Leaf terminating in a leaflet.. 30

– Leaf terminating in a sturdy spine 36

30. Leaflets large (7–)13–35 × (4–)7–18 mm, margin serrate except extreme base; peduncular arista with a clavate leaflet at the tip; racemes 1–6-flowered; corolla white or pale yellow.. 31

– Leaflets small (2–)4–13 × 1–7 mm, two thirds to one half of margin serrate; peduncular arista without clavate leaflet at tip; racemes 1(–2)-flowered; corolla purple, pink or light blue... 33

31. Stems sparsely pubescent; leaf with 7–13 leaflets; racemes 1–3(–4)-flowered...20. ***C. isauricum***

– Stems densely pubescent; leaf with (8–)11–19 leaflets; racemes (1–)2–6-flowered.. 32

32. Plant 25–45 cm high; racemes (1–)2–5-flowered; corolla white with a purple blotch ....29. ***C. montbretii***

– Plant 40–70 cm high; racemes 2–6-flowered; corolla pale yellow .... 17. ***C. heterophyllum***

33. Plant 10–35 cm high; leaflets obovate; calyx strongly gibbous; corolla length (13–)16–22 mm 34

– Plant 4–16(–25) cm high; leaflets obovate or flabellate; calyx weakly dorsally gibbous; corolla length 5–11(–15) 35

34. Leaf (19–)22–40 mm long, with 7–10 leaflets, ending in a foliate spine or leaflet; stipules flabellate and smaller than leaflets; peduncle 13–20 mm long ....33. ***C. paucijugum***

– Leaf (30–)60–130 mm long, with 19–35(–41) leaflets, ending in a leaflet; stipules triangular or ovate, equal to or larger than leaflets; peduncle 27–40(–70) mm long .... 30. ***C. multijugum***

35. Leaf 6–16 mm long, with (3–)5–7(–9) leaflets; corolla 5–8(–12) mm long (Greece, Turkey, Syria, Lebanon, Iran, former USSR)............ 19. ***C. incisum***
– Leaf (5–)15–30 mm long, with (3–)9–15 leaflets; corolla 11–15 mm long (Morocco) .... 4. ***C. atlanticum***

36. Stipules mostly spiniform 37
– Stipules entirely laminate.. 38

37. Stipules with double spines, one long, one short, (1–)6–25(–33) mm; leaflets obovate or flabellate; peduncular arista 3–10(–25) mm long; corolla pubescent or glabrous ....40. ***C. macracanthum***
– Stipules with single spines 2–8(–15) mm long, or perular at plant base; leaflets obovate; peduncular arista (4–)10–20(–28) mm long; corolla glabrous .... 1. ***C. acanthophyllum***

38. Stems sparsely pubescent; leaf rachis grooved or flattened above; leaflets (8–)10–18(–20), rotundate, flabellate, or obovate.................................. 39
– Stems densely pubescent, leaf rachis grooved above; leaflets 4–12 obovate.. 41

39. Leaf rachis grooved above or not; leaflets rounded; leaflets crowded; corolla 9–15 mm long, pubescent or not .... 37. ***C. rechingeri***
– Leaf rachis not grooved above; leaflets flabellate or obovate; leaflets not crowded; corolla 10–25 mm long, pubescent.. 40

40. Plant height 15–26 cm; branches 6–16(–20) mm long; leaflets 1–3 × 1–3 mm, remotely spaced on a 40–50(–60) mm long leaf; peduncular arista 2–7(–9) mm; racemes 1-flowered.................. 43a. ***C. tragacanthoides*** var. ***tragacanthoides***
– Plant height 11–35 cm; branches 15–30(–40) mm; leaflets (1–)2–5 × (1–)2–5 mm, not remote, on a (25–)30–60(–90) mm long leaf; peduncular arista (2–)7–15 mm; racemes 1–2-flowered .... 43b. ***C. tragacanthoides*** var. ***turcomanicum***

41. Leaf with 4–6(–10) leaflets; tooth of leaflet midrib pronounced; stipules smaller than leaflets .... 35. ***C. pungens***
–Leaf with 8–12 leaflets; tooth of leaflet midrib not pronounced; stipules equal to or larger than leaflets...18. ***C. incanum***

42. Stems glabrous; leaf terminating in a spiny curl; leaflets remote, all spiny .... 42. ***C. subaphyllum***
–Stems sparsely pubescent; leaf terminating in a spine; leaflets crowded, mostly spiny, occasionally with 1–2 pairs of flabellate (3–)5–10 × 3–7 mm leaflets on the basal leaves...41. ***C. stapfianum***

APPENDIX 2
A TABULAR KEY TO *CICER* SPECIES

Characters Species	1	2	3	4	5	6	7	8	9	10	11	12	13	14	15	16	17	18	19	20	21
arietinum	A	12–50(–100)	A/C	A	C	A	25–75	A	C	B	N/A	(7–)11–17	C	A	B/D	(5–)10–16(–21)	(3–)7–14	B	A	A	A/B
reticulatum	A	20–35	A	A	C	A	15–28(–40)	A	C	B	N/A		B/C	A	B			B	A	A	B
bijugum	A		A/B/C	A	C	A	(13–)18–25 (–44)	A	C	B	N/A	(3–)5(–7)	C	A	B/C	(5–) 7–12 (–18)		B	A/B	A	C/D
echino-spermum	A	20–35	A	A	C	A	(20–)30–40 (–46)	A	C	B	N/A	7–11(–14)	C	A	B	4–11 (–14)		B	A/B	A	E
judaicum	A	10–40	A	A	C	A	(14–)20–35 (–43)	A	C	B	N/A		C	A	B/C		1.5–5 (–8)	B	B	A	A/B /C
pinnatifidum	A	10–20(–40)	B	A	C	A	(12–)20–40	A	C	B	N/A		B	A	B	3–8 (–12)	1.5–5 (–7)	B	B	A	C/D
cuneatum	A	20–40(–60)	C	A	A	B	30–70 (–90)	A	B	A	B	(8–)14–22	C	A	B/D	5–9 (–12)		B	A	A	D
yamashitae	A	(10–)21–30	B	A	A	A		A	C	B	N/A		C	A	A/B/ D		2–5(–6)	B	B	A	E
choras-sanicum	A	5–12(–15)	B/C	A	C	A		A	C	B	N/A	3	B	A	E	5–8 (–10)	3–6(–9)	C	B	A	E
incisum	B	5–16(–25)	A	A	C	A		A	C	B	N/A	(3–5)5–7 (–9)	C	A	C/E	(2–) 4–10		C	B	A	B/D
atlanticum	B		C	B	A	A	(5–)15–30	A	C	B	N/A	(3–)9–15	C	A	C/E			C	B	A	A/B
canariense	B	50–200	C	B	A	B	(40–) 70–110	A	B	A	A	32–63	C	A	G		0.5–1	N/A	N/A	A	A
kermanense	B	30–50	B/C	A	C	B	70–100 (130)	A	B	A	A	(6–) 16–24	A	A	E		5–8 (–15)	B/C	A/B	A	E
oxyodon	B	20–55	B/C	A	C	B	38–100 (–140)	A	B	A	A/B	(4–) 10–14 (–16)	B	A	E	5–10 (–17)	4–10 (–17)	B	A	A	A/B/D
spiroceras	B	25–75	C	B	A	B	30–90 (120)	A	B	A	A	(5–) 8–14 (–22)	A	A	E	(2–) 4–7 (–11)	(2–) 4–9 (–15)	C	A	A	D/E
subaphyllum	B	30–40	B	C	A	B	40–100 (–140)	A	A	B	N/A	4–14 (20)	A	B	N/A		U	N/A	N/A	A	E

Continued

APPENDIX 2 *Continued*

Characters Species	1	2	3	4	5	6	7	8	9	10	11	12	13	14	15	16	17	18	19	20	21
anatolicum	B	15–45 (–60)	A/B	A	N/A	B	(20–) 50–70 (–110)	A	B/C	A	A/B	(3–) 8–14 (–18)	C	A	C/D	6–14 (–20)		B	A	A	A
balcaricum	B	30–60	C	B	C	A/B	60–100	U	B/C	A	A	12–16 (–18)	U	A	B/C/E		U	B	U	A	C
floribundum	B		B/C	A	A	A/B	(24–) 50–80 (–105)	A	B/C	A	A/B	(8–) 10–16 (–19)	C	A	B/D	(6–) 10–16 (–20)	4–8 (–10)	A	A	A	A/D
graecum	B	19–60	C	A	C	A/B	(30–) 50–100	A	B/C	A	A/B	(6–) 11–19	B/C	A	A/C/D	6–16 (–22)	3–7 (–10)	A	A/B	A	A/B
heterophy-llum	B	40–70	C	A	C	A/B	(50–) 80–120	A	C	B	N/A	(9–) 11–17	B	A	A/C/D	16–35		A	A/B	A	A
isauricum	B	20–40	B/C	B	A	A	30–80	A	C	B	N/A		C	A	A/B		(5–) 9–12 (–16)	A	A/B	A	A/E
montbretii	B	25–45	A/B/C	A	A	A	(42–) 60–90 (–100)	A	C	B	N/A	(8–) 11–19	C	A	A/B/D	(8–) 13–22 (–27)	(4–) 7–10 (–13)	A	A/B	A	A/B
baldshuani-cum	B	30–60	A/B	B	A	A	50–100 (–150)	A	B/C	A	A/B		B	A	C/E/F	(5–) 10–20 (–22)		B/C	B	A	D
flexuosum	B	30–70	C	A	C	A/B	50–100 (–150)	A	B/C	A	A/B		B	A	C/D/E	4–11 (–15)	(2–) 4–8 (–12)	B	A/B	A	B/D
grande	B	20–50	A/B	A	A	A/B	60–110	A	B/C	A	A/B		C	A	B/D	10–25 (–27)		B	U	A	D/E
incanum	B	20–30	C	A	C	B	25–40	U	A	B	N/A		U	A	C			U	B	A	U
korshinskyi	B	50–80	A/B	B	A	B	40–80	A	A/B/C	A	A		C	A	C/D	(5–) 10–17 (–20)	(4–) 6–10 (–13)	B	U	A	B/D
mogoltavicum	B	60–80	A/C	B	C	A/B	50–160	A	B	A	A/B	16–22	A	A	E	(2–) 3–7		C	A	A	D/E

nuristanicum	B	20–45	A	A	A	B	(30) –72–100 (140)	A	B/C	A	A/B	(9–) 11–20 (–20)	B	A	C	(5–) 8–14 (–18)	(4–) 6–10 (–13)	B	A/B	A	A
fedtschenkoi	B	14–25 (–35)	A	A	C	A/B	30–80 (100)	A	B/C	A/B	A	(7–) 11–19 (–30)	C	A	C	5–10 (–15)	3–7 (–10)	B	A	A	A/B
luteum	B	24–29	U	A	A	A/B	45–110	U	C	B	N/A		U	A	C		U	B	U	A	A
multijugum	B		B/C	A	A	B	(30–) 60–130	A	C	B	N/A	19–35 (–41)	C	A	C	(3–) 5–13	(2–) 4–7	B	A	A	A/B
paucijugum	B	18–35	A	A	A	B	(19–) 22–40	A	A/C	B	N/A		C	A	C			B	A	A	D
songaricum	B	15–60	A/B/C	A	A	B	(28–) 35–60 (75)	A	B/C	A	A	(8–) 12–18	B/C	A	C	(2–) 6–10 (–15)	(2–) 4–8	B/C	A/B	A	A/B/D
microphyllum	B	20– (40–60)	A/B	A	A	B	(38–) 50–140	A	B/C	A	A	(8–) 14–28 (–37)	C	A	C/D	4–10 (–14)	3–7 (–10)	B	A/B	A	B/C
pungens	B	15–40	C	A	C	B	(7–) 12–30 (–55)	A	A	B	N/A	4–6 (–10)	B/C	A	C	(3–) 6–11 (13)	3–7 (–9)	C	A	A	B/E
rechingeri	B	20–40	B/C	B	C	B	(22–) 30–50 (90)	A/B	A	B	N/A	(8–) 10–18 (–20)	B/C	A	F			C	A	A	E
stapfianum	B	25	C	B	A	B	(20–) 30–55	A	A	B	N/A	(4) 6–10	C	A/B	E		(3–)5–10	B	U	A	E
macracan-thum	B	(10–) 20–50 (–60)	C	A	A	B	(13–) 25–40 (–70)	A	A	B	N/A	(6–) 10–18 (–22)	B/C	A	C/E	1.5–6 (–8)	1.5–5	C	A	B	N/A
acantho-phyllum	B	(15–) 20–35 (–50)	A/C	A	C	B	(20–) 32–55 (–65)	A	A	B	N/A		B	A	C	2–6 (–8)	(1–) 3–5	B/C	A	B	N/A
tragacan-thoides var. tragacan-thoides	B	15–26	C	B	C	B	40–50 (–60)	B	A	B	N/A	(8–) 14–16	A	A	C/E			C	A/B	A	E
tragacan-thoides var. turcom-anicum	B		C	B	C	B	(25–) 30–60 (–90)	B	A	B	N/A	(8–) 10–16 (–20)	B	A	C/E	(1–) 2–5	(1–) 2–5	C	A/B	A	E

Continued

APPENDIX 2 *Continued*

Characters Species	22	23	24	25	26	27	28	29	30	31	32	33	34	35	36	37	38	39	40	41	42
arietinum	N/A	3–6 (–11)	(1–) 2–4 (–7)	1 (–2)	(5–) 13–18 (–30)	A	0.2 –3(–5)	B	B	A/B	(5–) 8–11 (–14)	B	A/C/D/F	7–9 (–13)	B	B/D	A/B		17	B	B
reticulatum	N/A			1	(3–) 6–11	B	N/A	N/A	N/A	A		B	D		B	B	B		15–19	B	A
bijugum	N/A			1	(1–) 3–7	A		B	B	A		B	C/D		B	C	B			A	B
echinos-permum	N/A			1	6–11 (–15)	A/B	0–1 –1(–2)	B	B	A	(5–) 7–12	B	D	(5–) 7–12	B	B	B	7	U	A	B
judaicum	N/A		1–2 (–3)	1 (–2)	(3–) 7–20	A	0.5–3	B	B	A	(3–) 5–9	B/C	D/F		B	E	B			B	B
pinnatifidum	N/A	2–5 (–7)		1	(3–) 5–20 (–30)	A	0.5–3 (–5)	B	B	A		B	C/D		B	E	B			B	B
cuneatum	N/A			1	(6–) 10–20 (–30)	A	1–6 (–12)	B	B	A		B	C/D		B	D	B			B	B
yarnashitae	N/A	1–2.5	0.5–1 (–2)	1	4–9 (–15)	A	(5–) 10–20	B	B	A		B/C	C	7	B	E	B			B	B
chorassanicum	N/A	0.5–1	0.5 (–1)	1 (–2)	(1.5–) 2–6	A		B	B	A		B	A/C		B	B	B			B	B
incisum	N/A			1 (–2)	7–20 (–40)	A/B	0.5–2 (–4)	B	B	A	8–16 (–27)	B	C/D/F	5–8 (–12)	B	B/C	B	3–4	10	B	B
atlanticum	N/A		2	1		A		B	B	A		B	C/D		B	A	B	U	U	B	B
canariense	N/A		1.5–3	1–4 (–7)	35–100	B	N/A	N/A	N/A	B		B/C	C/E		B	C/D	B			B	B
kermanense	N/A		0.5–1		20–30 (–60)	A	4–10 (–20)	A	B	A		A	A/C	18	U	B	B			B	B
oxyodon	N/A	(1–) 2–5	1–3 (–5)		(10–) 15–40 (–70)	A	(2–) 6–10 (–20)	B	B	A/B	(2–) 4–8 (–10)	A	B/G		B	B/C/D	B		B	U	B
spiroceras	N/A		0.5–1	1–2 (–4)	(9–) 15–25 (–40)	A	(2–) 6–13	A	B	A	4–7 (–10)	B	E	10–13 (–15)	A/B	B/C	B		12	B	B

subaphyllum	N/A		0.5–1 (–2)	1 Feb	(10–) 20–36 (–46)	A		A	B	A		B	U		B	B	B	5	U	B	U
anatolicum	N/A	1–6 (–14)	(1–) 3–7 (–12)	1 Feb	(12–) 25–50	A	3–10 (–14)	A/B	B	A		A	D	(10–) 16–25	A	B/C	B			B	B
balcaricum	N/A		U	2	35–45	A		A	B	U	3–10 (–12)	B	D	20–25	A/B	B	B		U	U	U
floribundum	N/A	(3–) 5–9 (–12)	2–6 (–10)		(17–) 45–67	A		B	C	C	(4–) 7–10 (–12)	A	D/E		B	C	B			B	B
graecum	N/A	4–8 (–12)			30–68	A	4–8 (–13)	B	C	C		A	D		B	D	U	4	4	B	U
heterophyllum	N/A				36–100	A		B	C	A		A	G	18–20	B	U	U	U	U	U	U
isauricum	N/A		1–2 (–3)	1–3 (–4)	25–30 (–38)	A		A/B	C	A/B		A	A		B	C	B	U	U	U	U
montbretii	N/A	(3–) 5–8 (–10)	3–6 (–8)	(1–) 2–5	30–52 (–60)	A	(7–) 11–16	B	C	A	(4–) 7–12	A	A	16–20 (–28)	B	B	B			B	B
baldshuanicum	N/A		U		30–60	A		B	B	A		A/B	D	20	U	U	U	U	U	U	U
flexuosum	N/A	2–5 (–7)	2–4 (–8)		11–30 (–80)	A	5–15 (–18)	A	B	A/B		A	D	15–27	A	B	B			B	B
grande	N/A				30–40	A	3	A	B	A		A	U	23	A	B	B	7	U	B	B
incanum	N/A		U		15–25	A	U	A	B	U	U	A/B	U	20	A	U	U	U	U	U	U
korshinskyi	N/A			1	25–60	A		A	U	U	U	A/B	U	20	U	U	U	U	U	U	U
mogoltavicum	N/A			(1–) 2	25–55	A		B	B	A		A	E	15–20 (–22)	A	B	B	5	U	B	B
nuristanicum	N/A	(1–) 2–5 (–7)			30–50	A		A/B	B	A	(2–) 6–12	A	D	15–25	A	B	B	5.5	U	B	B
fedtschenkoi	N/A	7–12 (–15)	5–8 (–13)	1	30–70	A	(2–) 8–11	B	A	A		A	D	(180) 20–25 (–30)	B	E	B	4	U	B	B
luteum	N/A		U	1	26–50	A	U	B	A	B		A	G	20–28	B	U	U	U	U	U	U
multijugum	N/A			1	27–40 (–70)	A	(2–)6–9	B	B	A	(3–) 5–9	A	D/E	16–22	B	E	B	3	U	B	B
paucijugum	N/A			1	13–20	A		B	B	A		A	D	(13–) 18–23	B	U	U	U	U	U	U

Continued

APPENDIX 2 *Continued*

Characters Species	22	23	24	25	26	27	28	29	30	31	32	33	34	35	36	37	38	39	40	41	42
songaricum	N/A	(3–) 5–11	(2–) 6–12		16–25 (–45)	A	(1–) 7–16	B	B/C	A/B		A	D	19–25 (–31)	A	B	B		U	B	B
microphyllum	N/A	3–10 (–16)	(2–) 4–10 (–15)		(14–) 19–45 (–66)	A	(1–) 5–14	A/B	B	A	(4–) 7–12	A	D	(17–) 20–25	A/B	B	B			B	B
pungens	N/A	2–6 (–8)	(0.5) 2–5	1(–2)	(12–) 15–30 (–50)	A	7–15 (–20)	A	B	B	(2–) 6–10 (–14)	A	D	(11–) 16–25	A	B	B			B	B
rechingeri	N/A		0.5–1 (2.5)	1 Feb	(14–) 24–30 (–40)	A	(5–) 9–15	A	B	A	3–7 (–10)	A	D		A/B	U	U	U	U	U	U
stapflanum	N/A				20–30	A		A	B	U	7	B	U	18	B	B	B	5	U	U	U
macracanthum	B	(1–) 6–25 (–33)	0.5	1(–3)	12–23 (–50)	A	3–10 (–25)	A	B	A		A	D		A/B	B	B			B	B
acanthophyllum	A	2–8 (–15)	0.5–1 (–2)	1(–2)	(15–) 26–42 (–70)	A	(4–) 10–20 (–28)	A	B	A	6–10 (–15)	A	D		B	B	B		10	B	B
tragacanthoides var. tragacanthoides	N/A			1	15–25	A	2–7 (–9)	A	B	A		A	D		A	C	B		U	B	B
tragacanthoides var. turcomanicum	N/A	1–3 (–4)	0.5–2		(12–) 20–25	A	(2–) 7–15	A	B	A	6–11 (–15)	A	D/E	14–25	A	C	B		11	B	B

Characters/Species	43	44	45	46	47
arietinum	A/B	4–8(–15)	A		B
reticulatum	A/B	U	A		B
bijugum	A/B		A		B
echinospermum	A	U	A		B
judaicum	A		A		A
pinnatifidum	B	(5–)10–17	A		A
cuneatum	B	U	A		B
yamashitae	B	U	A		A

chorassanicum	B		A	B
incisum	B	U	A	B
atlanticum	B	U	A	B
canariense	B	U	A	B
kermanense	B	U	A	U
oxyodon	B		A	B
spiroceras	B	U	A	B
subaphyllum	U	U	A	B
anatolicum	B		A	B
balcaricum	B	U	A	B
floribundum	B	U	A	B
graecum	B	U	A	B
heterophyllum	B	U	A	U
isauricum	B	U	A	U
montbretii	A/B		A	B
baldshuanicum	B	U	A	U
flexuosum	B	U	A	B
grande	B	U	A	B
incanum	B	U	B	U
korshinskyi	A/B	U	A	U
mogoltavicum	B	U	A	B
nuristanicum	B	U	A	B
fedtschenkoi	B	U	B	A
luteum	B	U	B	U
multijugum	B	U	B	B
paucijugum	B	U	A	U
songaricum	B	U	B	B
microphyllum	B	U	B	B
pungens	B	U	A	B
rechingeri	B	U	A	U
stapfianum	B	U	A	B
macracanthum	B	U	N/A	B
acanthophyllum	B	U	N/A	B
tragacanthoides var. *tragacanthoides*	B	U	A	B
tragacanthoides var. *turcomanicum*	B	U	A	B

The characters used in the tabular key are:
1. Life cycle: A, annual; B, perennial.
2. Plant height (cm).
3. Stem orientation: A, straight; B, slightly flexuous; C, flexuous.
4. Stem pubescence: A, pubescent; B, sparsely pubescent; C, glabrous.
5. Plant pubescence: A, glandular; B, eglandular; C, glandular and eglandular.
6. Leaves: A, imparipinnate (with a terminal leaflet); B, paripinnate (without a terminal leaflet).

Continued

APPENDIX 2 *Continued*

7. Rachis length (mm).
8. Rachis: A, grooved above; B, not grooved above.
9. Rachis terminating in a: A, spine; B, tendril; C, leaflet.
10. Tendrils: A, present; B, absent.
11. Tendril form: A, simple; B, branched.
12. Number of leaflets on rachis.
13. Leaflets: A, remote; B, neither remote nor crowded; C, crowded.
14. Leaflets: A, laminate; B, spine bearing.
15. Lamina shape: A, oblong; B, oblong-obovate; C, obovate; D, elliptic; E, flabellate; F, rounded; G, linear.
16. Leaflet length (mm).
17. Leaflet width (mm).
18. Leaflet margin: A, entirely serrate; B, top two thirds serrate; C, apex serrate only.
19. Tooth of leaflet midrib: A, pronounced; B, not pronounced.
20. Stipules: A, laminate; B, spiny.
21. Stipule shape: A, triangular; B, ovate; C, ovate-lanceolate; D, flabellate; E, perular (scale-like).
22. Spiny stipule form: A, single spine; B, double spine.
23. Stipule length (mm).
24. Stipule width (mm).
25. Flower raceme number or range.
26. Peduncle length (mm).
27. Peduncle type: A, ending in arista; B, not ending in arista.
28. Arista length (mm).
29. Arista form: A, spinose; B, not spinose.
30. Arista form: A, with tiny flabellate leaflet at tip; B, without leaflet at tip; C, with clavate leaflet at tip.
31. Bract shape: A, triangular, less than or equal to 1 mm; B, triangular, more than one mm; C, flabellate, large, c. 4 mm, dentate.
32. Pedicel length (mm).
33. Calyx base: A, strongly dorsally gibbous; B, weakly dorsally gibbous; C, not dorsally gibbous.
34. Corolla colour: A, white; B, cream; C, pink; D, purple; E, violet; F, light blue; G, pale yellow.
35. Corolla length (mm).
36. Corolla pubescence: A, pubescent; B, glabrous.
37. Seed shape: A, ovoid; B, obovoid; C, subglobular; D, globular; E, deltoid.
38. Seed coat texture: A, smooth; B, rough.
39. Seed length (mm).
40. Seed circumference (mm).
41. Seed surface: A, echinate; B, not echinate.
42. Seed surface: A, reticulate; B, not reticulate.
43. Leaflet dentation: A, doubly incised; B, not doubly incised.
44. Rachis petiole length (mm).
45. Laminate stipules: A, smaller than leaflets; B, equal to or larger than leaflets.
46. Total calyx length (mm).
47. Seeds: A, bilobed; B, not bilobed.

3 The Ecology of Chickpea

J.D. Berger[1,2] and N.C. Turner[2]

[1]CSIRO Plant Industry, Private Bag 5, WA 6913, Australia; [2]Centre for Legumes in Mediterranean Agriculture, M080, The University of Western Australia, 35 Stirling Highway, Crawley, WA 6009, Australia

Introduction

Ecology is the scientific study of interrelationships among organisms, and between them and all aspects of their environment (Allaby, 1998). Toker *et al.* (Chapter 23, this volume) discuss the abiotic stresses encountered by chickpea (*Cicer arietinum* L.) in the field. This chapter focuses largely on the physiological ecology of chickpea by examining chickpea functioning in relation to its environment. As the evolution of chickpea is uniquely different from other members of the West Asian Neolithic crop assemblage, and plays an important role in determining the habitat range of the crop (Abbo *et al.*, 2003a), we begin by reviewing the history of chickpea. The production environments of chickpea are defined by mapping its worldwide distribution and conducting a climate analysis, as it plays a dominant role in determining where and when chickpea can be grown. Abiotic and biotic stresses encountered by chickpea are categorized by environment and cropping systems in relation to these are discussed. Finally, the physiology of chickpea is briefly reviewed to add perspective to the evidence for ecotype formation in the crop.

Chickpea Evolutionary Bottlenecks and Current Distribution

The domestication, archaeology and early history of chickpea are discussed in Chapter 1. Currently, chickpea is grown in 60 countries across the world except in Antarctica. Five centres of diversity are recognized: (i) the Mediterranean basin (for kabuli – white seed coat – types); (ii) Central Asia; (iii) West Asia (a secondary centre for pea-shaped forms); (iv) the Indian subcontinent (desi – coloured seed coat – types); and (v) Ethiopia (a secondary centre for desi types) (van der Maesen, 1984). Despite this wide distribution, production is skewed.

Average production data from 2000 to 2005 indicate that 73% of chickpea came from South Asia, 13% from West Asia and North Africa (WANA), 6% from North America, 4% from East Africa and surrounding areas and 2% from Australia (FAO, 2006). Europe, South America, East and Central Asia contributed <1% of global production, respectively.

This skewed distribution, where much of the crop is produced in the semi-arid tropics, despite originating in Mediterranean south-east Anatolia, is likely to be an outcome of chickpea's unique evolution. Abbo *et al.* (2003a) document a series of bottlenecks in the development of chickpea, which had the effect of limiting diversity and adaptive potential in the crop. In contrast to cereals, where the wild progenitors are spread over a wide area and domestication is likely to have occurred independently at multiple locations (Willcox, 2005), *Cicer reticulatum* (wild variety) has an extremely restricted habitat range in south-east Anatolia (Fig. 1.1). A narrow distribution implies limited adaptive potential and exacerbates the founder effect as the opportunity for multiple domestication events is limited. Therefore, it is suggested that the genetic base of chickpea has always been relatively narrow compared to other crops of the WANA (Abbo *et al.*, 2003a). Given that *C. reticulatum* is a Mediterranean winter annual, it is assumed that early chickpea was an autumn-sown winter crop like the other members of WANA. Subsequently, it became a spring-sown crop, presumably to minimize the probability of ascochyta blight (Abbo *et al.*, 2003a,b), which remains the major biotic constraint to chickpea in the Mediterranean to this day (Singh and Reddy, 1996). Indeed, chickpea is traditionally grown as a spring-sown crop throughout the WANA region for this reason (see discussion in Abbo *et al.*, 2003b). The shift from autumn to spring sowing was an additional bottleneck, which further narrowed the genetic diversity of chickpea, and had important adaptive implications. Winter-adaptive traits such as vernalization responsiveness and low temperature tolerance during the vegetative and reproductive phases were lost from the cultigen, but still remain in the annual wild relatives (Singh *et al.*, 1998; Abbo *et al.*, 2002; Berger *et al.*, 2005). Abbo *et al.* (2003b) suggest that the shift from autumn to spring sowing in West Asia coincided with the introduction of summer crops such as sorghum (Sorghum bicolor (L.) Moench) and sesame (Sesamum indicum L.) in the early Bronze Age. If this is correct, it is likely that this shift, which effectively turned chickpea into a post-rainy season, warm season crop, facilitated its subsequent distribution to the semiarid tropics to the east (Indian subcontinent) and south (Ethiopia). This is borne out by the climate analysis which follows.

Global Chickpea Distribution and Climate Analysis

To date, the approach taken in the literature to describe chickpea environments has been somewhat fragmented. Chickpea habitats in WANA have been well described, with soil types, rainfall and temperature ranges mapped for each country individually, and more detailed information graphed for key locations (Saxena *et al.*, 1996). Outside the WANA region there is far less information available. Broad environmental categories (Khanna-Chopra and Sinha, 1987) or key centres of cultivation (van der Maesen, 1972) have been represented by single locations and graphs of temperature, rainfall and photoperiod over the year presented. Although this provides a clear snapshot of gross differences, it tends to oversimplify reality and makes comparisons within and between regions difficult. Importantly in the context of this chapter, it does not encourage an ecophysiological approach to investigating adaptation, because the finer points of habitat characterization are not captured. With the advent of high-resolution interpolated climate surfaces and user-friendly geographical information system (GIS) software freely available in the public domain (Hijmans *et al.*, 2001, 2005; New *et al.*, 2002), the task of characterizing chickpea habitats has become much simpler (Figs 3.1 and 3.2). Moreover, multivariate techniques such as principal components analysis can effectively integrate related variables and present a more holistic habitat characterization than mapping key descriptors individually.

Global chickpea distribution was defined by plotting all accessions with passport data from the International Crops Research Institute for the Semi-Arid Tropics (ICRISAT), Hyderabad, India; International Centre for Agricultural Research in the

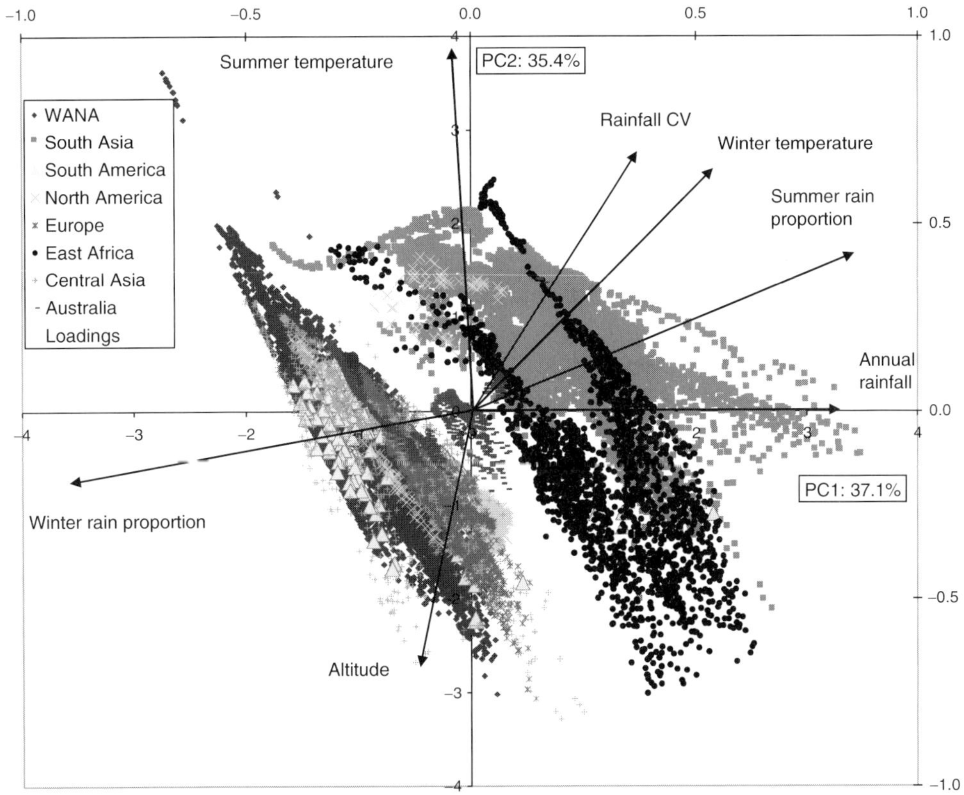

Fig. 3.1. Principal components analysis of climate of global chickpea growing regions. Factors were quartimax rotated with Kaiser normalization to maximize the information summarized within single dimensions for PC1 and PC2, so that these scores could be mapped effectively (Fig. 3.2a and b).

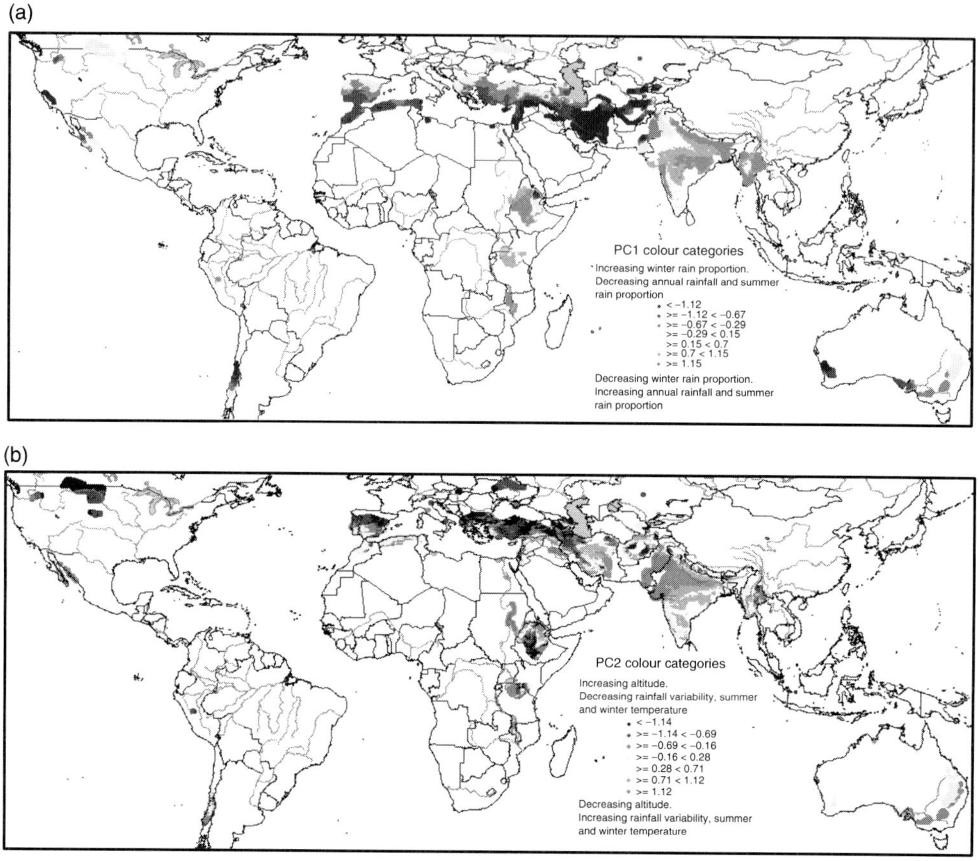

Fig. 3.2. (a) Global chickpea distribution classified by PC1 scores capturing the 'Mediterranean-ness' of the climate (see Fig. 3.1 and Table 3.1). Navy blue zones are particularly Mediterranean, while red is the subtropical extreme. (b) Global chickpea distribution classified by PC2 scores capturing altitude (negative direction), rainfall variability, summer and winter temperatures (positive direction) (see Fig. 3.1 and Table 3.1). Altitude tends to decrease, while rainfall variability, summer and to a lesser extent winter temperature tend to increase as the chickpea growing region moves from navy blue to red. Note: while the Americas have been moved to the east to save space in the figure, latitudes have not been modified.

Dry Areas (ICARDA), Aleppo, Syria; and United States Department of Agriculture (USDA) collections ($n = 7759$) to establish approximate boundaries within regions. These were modified using pre-existing maps wherever possible (Cubero, 1980; van der Maesen, 1984; Saxena *et al.*, 1996). This approach captures chickpea distribution at a single time point, and may not be strictly accurate, as cropped areas fluctuate annually. However, it has the advantage of defining habitats for germplasm that has been collected, and is available for use by breeders and scientists. For regions without extensive germplasm collections or pre-existing maps (i.e. North America and Australia), growing areas were defined with the assistance of local chickpea breeders. Distribution shapefiles were converted into 10′ grids (~16 km resolution) using DIVA-GIS (Hijmans *et al.*, 2001) and climate data extracted (Hijmans *et al.*, 2005) for each grid point. Subsequently all areas at an altitude >3000m were excluded because a search of germplasm passport data suggested that chickpea was unlikely to grow at these elevations. In the northern hemisphere summer is from June to August and winter from December to February. While this summer period does not define the hottest annual quarter for Ethiopia and parts of South Asia, it does provide consistency across the globe, which facilitates interpretation of the results. In the southern hemisphere the seasons were reversed: winter is from June to August and summer from December to February. Multivariate trends in the data were analysed with SPSS v. 14.

Principal components analysis (PCA) demonstrated regional climatic similarities and differences very effectively, explaining almost 73% of total variance in two components (Fig. 3.1). PC1 captured the 'Mediterranean-ness' of the climate by positively loading annual precipitation and the proportion falling in summer, and negatively loading the proportion of rain falling in winter (Table 3.1). Thus, total rainfall and the proportion of summer rain tend to increase, while winter rainfall decreases from left to right in Fig. 3.1. In PC2 summer temperature was contrasted with altitude (Table 3.1). Rainfall variability and winter temperatures were also positively loaded on PC2, but their respective vectors run at almost 45° through the upper right quadrant of Fig. 3.1 because of their comparatively high PC1 coefficients. The chickpea regions largely clustered into two separate climate types:

1. Central Asia, WANA, Europe, much of South and North America and parts of Australia form a 'Mediterranean-type' group on the left side of Fig. 3.1, characterized by consistent, low annual rainfall, low winter temperatures and winter-dominant precipitation. Locations in the upper left quadrant of Fig. 3.1 represent stressful environments with a very narrow window for growth. These areas in the Egyptian Nile valley, central Iran, Iraq and Pakistan receive the lowest annual precipitation, predominantly in winter, and combine the highest summer temperatures with moderate winter temperatures.
2. South Asia, East Africa, parts of Australia and North and South America form a cluster largely on the right side of Fig. 3.1, and are characterized by high winter temperatures, high and variable rainfall, largely falling in summer. Note

Table 3.1. Factor loading for PCA of chickpea climate across the distribution range.

Trait	PC1	PC2
Summer rain proportion (%)	0.84	0.42
Annual rainfall (mm)	0.82	0.01
Mean winter temperature (°C)	0.54	0.65
Monthly rainfall CV (%)	0.37	0.69
Mean summer temperature (°C)	−0.04	0.96
Altitude (m)	−0.12	−0.67
Winter rain proportion (%)	−0.89	−0.19
Variance explained (%)	37.1	35.4

that both clusters have a wide spread of similar PC2 values, indicative of a wide range of summer temperatures and altitudes. Clustered around the origin is a large group of Australian locations characterized by an intermediate climate.

Mediterranean-type environments (low–medium PC1 scores)

Figure 3.2 gives a geographical context to the PCA. Chickpea throughout WANA, Central Asia, southern Europe, western USA, Chile and southern Australia is grown in typical Mediterranean climates with relatively low annual precipitation, which largely occurs in winter. Within these regions there are degrees of Mediterranean-ness, which are clearly highlighted in Fig. 3.2a and categorized in Table 3.2a. The navy blue zone in Fig. 3.2a is the most arid winter-dominant rainfall category, receiving an average of 219 mm rainfall/year, 51% of which falls in winter (Table 3.2a). These chickpea habitats are found in central Iran, central Pakistan, parts of Afghanistan, inner eastern Mediterranean, parts of North Africa, California, northern Chile and Western Australia. The navy blue zones in the southern hemisphere receive more annual rainfall than those in the north (Table 3.2a). In the Mediterranean basin chickpea is grown in the dark to medium blue zones, where annual rainfall is much higher (488–544 mm) and not as winter-dominant (31–43%, Table 3.2a). Moving north into Europe, chickpea is found in the light blue to yellow zones with 640–700 mm annual rainfall (Fig. 3.2a), occurring equally in winter and summer (Table 3.2a). Similar trends are evident in other chickpea-growing regions. In the USA, California is considerably more Mediterranean than the Pacific North-west. In Chile there is a northerly trend of increasing Mediterranean-ness (Fig. 3.2a), where winter rainfall proportion decreases from 66% to 50%, as annual rainfall increases from 330 to 1360 mm. In southern Australia there is a westerly trend of increasing Mediterranean-ness and a northerly trend towards summer-dominant rainfall along the east coast.

Summer-dominant rainfall environments (medium–high PC1 scores)

Most of the chickpea-growing regions in South Asia, East Africa and Peru are high-precipitation, summer-dominant rainfall environments (Fig. 3.2a). Within the Indian subcontinent there is a strongly decreasing rainfall cline from the south-east to the north-west (Table 3.2a), reflected in the colour gradient in Fig. 3.2a. Almost 50% of the chickpea production in India comes from Madhya Pradesh (Ali and Kumar, 2003), a central state enclosing much of the high rain and/or summer-dominant rainfall area in the subcontinent (dark red zone). The light to dark blue areas in the north-west of the Indian subcontinent are not Mediterranean-type climates. These arid zones (<385 mm/year), largely in Pakistan, receive some winter rain (9–22%), but the bulk of annual rain falls in summer (48–61%, Table 3.2a). The blue zones in East Africa also largely reflect differences in annual rainfall (Table 3.2a). Only the navy blue area receives

Table 3.2a. Descriptive statistics for the principal chickpea-growing regions categorized by PC1 scores capturing the 'Mediterranean-ness' of the climate (see Fig. 3.1 and Table 3.1).

Region	Navy blue	Dark blue	Medium blue	Light blue	Yellow	Orange	Red
			Annual rainfall				
Australia	387 ± 6.4	370 ± 3.7	469 ± 3.8	604 ± 3	694 ± 4.8	–	–
Central Asia	245 ± 2.9	346 ± 4.4	433 ± 10.5	535 ± 15.5	552 ± 12.8	547 ± 5.8	–
East Africa	197 ± 14.1	190 ± 7.4	326 ± 17.6	341 ± 16.5	542 ± 11.4	826 ± 5.3	1114 ± 8
Europe	–	544 ± 3.9	553 ± 4.7	642 ± 6.5	764 ± 17.7	–	–
North America	241 ± 8	279 ± 19.1	349 ± 17.5	362 ± 2.1	405 ± 2.1	–	–
South America	330 ± 15.8	613 ± 23.1	975 ± 11.3	1212 ± 14.2	1361 ± 28.2	1167	2103 ± 207.9
South Asia	176 ± 10.4	147 ± 17.5	223 ± 5.8	385 ± 7.1	675 ± 3.7	988 ± 3.5	1635 ± 10.2
WANA	206 ± 3.1	488 ± 4.4	507 ± 4.4	558 ± 6.6	609 ± 88.5	–	–
Total	219 ± 2.6	475 ± 3.2	480 ± 3.4	523 ± 3.7	622 ± 3.4	959 ± 3.2	1468 ± 8.5
			Winter rain (%)				
Australia	50 ± 0.4	44 ± 0.3	27 ± 0.3	18 ± 0.1	12 ± 0.4	–	
Central Asia	47 ± 0.5	36 ± 0.5	27 ± 0.9	19 ± 0.7	14 ± 0.2	11 ± 0.9	–
East Africa	38 ± 0.9	30 ± 0.7	24 ± 1.5	11 ± 0.7	6 ± 0.3	4 ± 0.2	3 ± 0.1
Europe	–	37 ± 0.2	31 ± 0.2	25 ± 0.2	24 ± 0.5	–	–
North America	51 ± 0.2	33 ± 0.7	27 ± 1	11 ± 0.2	8 ± 0.1	–	
South America	66 ± 0.4	61 ± 0.3	56 ± 0.4	51 ± 0.3	50 ± 0.6	1013 ± 0.5	
South Asia	39 ± 1.3	22 ± 0.6	14 ± 0.1	9 ± 0.2	3 ± 0.1	2 ± 0.03	2 ± 0.04
WANA	51 ± 0.2	43 ± 0.1	35 ± 0.2	30 ± 0.2	27 ± 2.9	–	
Total	51 ± 0.2	42 ± 0.1	30 ± 0.2	19 ± 0.2	6 ± 0.1	3 ± 0.04	3 ± 0.04
			Summer rain (%)				
Australia	10 ± 0.3	12 ± 0.1	22 ± 0.3	36 ± 0.1	45 ± 0.7	–	–
Central Asia	1 ± 0.1	5 ± 0.4	12 ± 0.8	21 ± 0.9	27 ± 0.5	32 ± 1.7	–
East Africa	26 ± 0.7	33 ± 0.6	38 ± 1.2	52 ± 1.5	42 ± 1.2	43 ± 0.7	60 ± 0.4
Europe	–6 ± 0.1	13 ± 0.2	24 ± 0.3	24 ± 0.7	–	–	
North America	1 ± 0.1	19 ± 2	28 ± 1.9	43 ± 0.1	44 ± 0.1	–	–
South America	1 ± 0.1	2 ± 0.1	4 ± 0.1	6 ± 0.1	7 ± 0.3	38	36 ± 0.2
South Asia	23 ± 2	48 ± 1.4	58 ± 0.4	61 ± 0.5	57 ± 0.4	68 ± 0.1	64 ± 0.2
WANA	1 ± 0	4 ± 0.1	10 ± 0.1	13 ± 0.2	15 ± 1.3	–	–
Total	1 ± 0.1	6 ± 0.1	18 ± 0.3	34 ± 0.3	51 ± 0.3	64 ± 0.2	63 ± 0.2

more rain in winter than in summer. In the dark blue zone rainfall is equally distributed, and thereafter becomes increasingly summer-dominant (Table 3.2a).

The Mexican chickpea regions are arid (119–284 mm/year), summer-dominant rainfall areas, where the proportion of both summer and winter rainfall increases (36–43% and 33–46%) with annual rainfall. In North America, moving eastwards from the Pacific North-west to the Northern Plains and Canada, rainfall increases to some extent (~360–410 mm/year), and becomes summer-dominant (43–44%).

Temperature and altitude (PC2)

Figure 3.2b puts PC2 into the geographical perspective. Altitude tends to decrease, while rainfall variability, summer and to a lesser extent winter temperature tend to increase as the chickpea-growing region moves from navy blue to red. Because winter temperatures and rainfall variability are not modelled by PC2 alone (Table 3.1), there are discrepancies in these variables between regions which are addressed individually below. From Europe to the Mediterranean there is a predominant southerly trend of increasing PC2 scores punctuated by higher elevations in Spain, the Balkans, central Anatolia and eastwards into the Caucasus. From south-eastern Anatolia to northern Syria and Iraq, and from western to central Iran, the PC2 gradient is particularly sharp (Fig. 3.2b), culminating in average summer temperatures of 32.5°C (Table 3.2b). Within PC2 categories mean winter temperatures in WANA and Europe are similar, ranging from −1.7°C to 12.9°C in WANA, and from 0.6°C to 10.5°C in Europe (Table 3.2b). However, altitudes and rainfall variability tend to be lower in any given category in Europe than in WANA (Table 3.2b).

South Asia, the Nile Valley, Mexico and subtropical Australia are dominated by high PC2 scores. In the Indian subcontinent there is a tight gradient in the north between the Terai and the Himalayan foothills, whereas in the plains PC2 scores tend to increase in a north-westerly direction in northern latitudes, and westerly in the south (Fig. 3.2b). These trends reflect the monsoonal pattern, which commences in June in the south, and a month later in the north and decreases average temperatures throughout. Throughout South Asia, peak annual temperatures are reached just prior to the onset of the monsoon. In fact, there is little difference in the mean temperature of the warmest quarter between the southern and northern halves of the subcontinental chickpea distribution (30.8°C vs 31.9°C). However, mean winter temperatures differ markedly: 22.1°C in the south, compared with 16.8°C in the north. PC2 scores in Myanmar are generally the lowest in South Asia, but increase towards the warmer central zone. Winter temperatures and rainfall variability are underestimated by PC2 categories in South Asia, ranging from a relatively mild 12.5–20.6°C for the former to a comparatively high 90.7–138.4% for the latter (Table 3.2b). Chickpea in Central Asia is grown in areas with a wide range of temperatures and rainfall variability, testament to the rapid changes in altitude found in the region. For Central Asia, winter temperatures are lower (−4.6–7.9°C) and altitudes higher (502–2101 m) than predicted by the average values for the PC2 categories (Table 3.2b).

East Africa is dominated by low PC2 scores, indicative of high altitude and low mean summer temperature (Table 3.2b). Similar to South Asia, winter

Table 3.2b. Descriptive statistics for the principal chickpea-growing regions categorized by PC2 scores capturing altitude (negative direction), rainfall variability, summer and winter temperatures (positive direction) (see Fig. 3.1 and Table 3.1).

Region	Navy blue	Dark blue	Medium blue	Light blue	Yellow	Orange	Red
				Altitude (m)			
Australia	–	612 ± 47	253 ± 7.3	213 ± 3.9	272 ± 3.8	126 ± 19.1	50 ± 4.7
Central Asia	2101 ± 39	1424 ± 55	923 ± 46	617 ± 30	615 ± 26	502 ± 34	924 ± 32
East Africa	2194 ± 14	1650 ± 11	1287 ± 10	1004 ± 17	792 ± 11	601 ± 10	299 ± 14
Europe	1013 ± 16	494 ± 12	452 ± 10	250 ± 7	96 ± 12	–	–
North America	810 ± 7	718 ± 7	322 ± 19	–	62 ± 8	246 ± 27	32 ± 6
South America	1222 ± 99	635 ± 67	414 ± 31	358 ± 21	189 ± 19	–	–
South Asia	1529 ± 43	1204 ± 12	826 ± 15	517 ± 11	409 ± 5	303 ± 3	181 ± 2
WANA	1710 ± 13	1465 ± 15	1274 ± 17	960 ± 18	755 ± 15	787 ± 11	482 ± 14
Total	1541 ± 11	1085 ± 10	910 ± 10	611 ± 9	524 ± 6	412 ± 5	218 ± 3
				Mean summer temperature (°C)			
Australia	–	22.2 ± 0.1	23.2 ± 0.1	25.3 ± 0.1	26 ± 0.1	26.1 ± 0.1	31.2 ± 0.1
Central Asia	17 ± 0.2	21.7 ± 0.1	23.9 ± 0.1	25.4 ± 0.1	27.4 ± 0.1	29.3 ± 0.1	30.7 ± 0.1
East Africa	16.9 ± 0.1	20.3 ± 0.1	22.5 ± 0	24.4 ± 0.1	26.7 ± 0.1	28.3 ± 0.1	32.1 ± 0.1
Europe	17.9 ± 0.1	20.2 ± 0	22.7 ± 0	24 ± 0.1	25.8 ± 0.1	–	–
North America	18.1 ± 0.05	20 ± 0.02	21 ± 0.09	–	24.8 ± 0.13	26.7 ± 0.14	30.1 ± 0.13
South America	14.7 ± 0.5	17.5 ± 0.3	18.4 ± 0.1	19.7 ± 0.1	21.4 ± 0.3	–	–
South Asia	20.8 ± 0.18	22.5 ± 0.07	24.5 ± 0.07	25.9 ± 0.05	27.4 ± 0.03	29 ± 0.02	31.5 ± 0.03
WANA	18.7 ± 0.04	21.6 ± 0.03	24 ± 0.04	25.9 ± 0.05	27.6 ± 0.06	30.5 ± 0.06	32.5 ± 0.08
Total	18 ± 0.03	20.8 ± 0.02	23.2 ± 0.03	25.3 ± 0.03	27.3 ± 0.03	29.3 ± 0.02	31.6 ± 0.03
				Mean winter temperature (°C)			
Australia	–	8.4 ± 0.3	9.7 ± 0.1	11.9 ± 0	12.8 ± 0.1	14.3 ± 0.2	24.3 ± 0.1
Central Asia	–4.6 ± 0.2	–1.1 ± 0.3	1.3 ± 0.2	2.8 ± 0.1	3.9 ± 0.1	5.5 ± 0.1	7.9 ± 0.1
East Africa	16.5 ± 0.1	19.1 ± 0.1	20.7 ± 0.1	22.5 ± 0.2	24.2 ± 0.2	24.9 ± 0.2	24.6 ± 0.2

Continued

Table 3.2b. *Continued*

Region	Navy blue	Dark blue	Medium blue	Light blue	Yellow	Orange	Red
	Mean winter temperature (°C)						
Europe	0.6 ± 0.1	1.3 ± 0.2	6.5 ± 0.1	9.5 ± 0.1	10.5 ± 0.1	–	–
North America	–10.3 ± 0.1	–8.4 ± 0.12	1.8 ± 0.08	–	8.5 ± 0.14	13.9 ± 0.67	18.8 ± 0.14
South America	5.4 ± 0.9	8.6 ± 0.7	8.4 ± 0.1	9.7 ± 0.1	11.5 ± 0.1	–	–
South Asia	12.5 ± 0.42	14.6 ± 0.17	17 ± 0.15	20.4 ± 0.12	20.6 ± 0.07	19.5 ± 0.06	17 ± 0.05
WANA	–1.7 ± 0.06	1 ± 0.05	4 ± 0.07	7 ± 0.08	8.1 ± 0.08	8.6 ± 0.06	12.9 ± 0.11
Total	0.6 ± 0.16	2.3 ± 0.15	8.9 ± 0.12	11.7 ± 0.12	15.5 ± 0.12	17.1 ± 0.1	16.9 ± 0.05
	Monthly rainfall CV (%)						
Australia	–	27.9 ± 3.4	26.1 ± 0.6	41.2 ± 0.5	68 ± 0.6	87.3 ± 1.5	119.8 ± 0.2
Central Asia	69.1 ± 1.8	73 ± 3	70.7 ± 2.5	75.2 ± 1.7	93.9 ± 1.2	103.9 ± 1.2	127.6 ± 2.7
East Africa	88 ± 0.9	95.2 ± 1.3	95.3 ± 1	108.8 ± 1.4	105.4 ± 1.7	107.2 ± 2.3	139.3 ± 5
Europe	29.1 ± 0.3	32.6 ± 0.4	44.2 ± 0.5	58.4 ± 0.4	66.3 ± 0.4	–	–
North America	61.9 ± 0.38	69.1 ± 0.32	42.1 ± 0.31	–	87.8 ± 0.52	103.1 ± 2.54	116.3 ± 1.12
South America	8 3 ± 3	82.2 ± 2.8	99.3 ± 1	110.7 ± 0.9	113.3 ± 1.4	–	–
South Asia	90.7 ± 1.3	89.6 ± 0.89	95.4 ± 0.79	103.7 ± 0.62	110.2 ± 0.48	128.1 ± 0.4	138.4 ± 0.42
WANA	56.8 ± 0.43	67 ± 0.55	75.8 ± 0.57	83.1 ± 0.51	88 ± 0.45	96.1 ± 0.45	114 ± 2.32
Total	60.7 ± 0.42	63.2 ± 0.46	69.8 ± 0.51	79.6 ± 0.48	99.5 ± 0.38	120.2 ± 0.4	135.2 ± 0.5

temperatures and rainfall variability are relatively high, ranging from 16.5°C to 24.9°C, and from 88.0% to 139.3%, respectively. As was the case with PC1, Ethiopia has a wide range of PC2 scores in close proximity, indicative of a wide habitat diversity, which is reflected in the diversity of the germplasm found there (Harlan, 1969).

In North America PC2 scores increase with latitude along the west coast, and from the Pacific North-west to the Northern Plains (Fig. 3.2b). Medium to navy blue zones in the north are particularly cold in winter, ranging from −10.3°C to −1.8°C, despite their relative low altitude (322–810 m, Table 3.2a). South America is characterized largely by low to medium PC2 scores, with relatively high rainfall variability (82.2–116.3%), and low summer temperatures (14.7–21.4°C) and altitudes (189–1222 m, Table 3.2b).

In Australia, the area for cultivating chickpea is characterized by intermediate PC2 scores, increasing from east to west in the southern, Mediterranean-type region, and from south to north along the east coast. In the summer-dominant rainfall regions of the northern east coast, PC2 scores tend to increase from east to west. Altitudes and rainfall variability tend to be relatively low (50–612 m, 27.9–119.8%) within their respective PC2 categories.

This climate analysis demonstrates that while the world's chickpea-growing regions can be grouped into four coarse rainfall and temperature categories (Mediterranean rainfall distribution, cool or warm climate; summer-dominant rainfall, cool or warm climate), there is a climatic diversity within and between regions captured by the sliding scale of PC1 and PC2 scores (Fig. 3.2a and b). The rest of this chapter will discuss chickpea adaptation from an agroclimatic perspective.

Chickpea Adaptation: Stresses, Cropping Systems and Traits from Different Habitats

Stresses

Biotic and abiotic stresses encountered by chickpea can be classified into coarse agroclimatic categories (Table 3.3). Drought is almost ubiquitous and is exacerbated by heat stress in the warmer Mediterranean and summer-dominant rainfall chickpea-growing areas. Cold and frost stresses are most prevalent in the cool regions characterized by low PC2 scores, but also in selected warm regions, such as northern South Asia (Table 3.3), and are expected to increase in WANA as winter-sowing and higher altitude cultivation are increasingly adopted (Saxena, 1993). Among the diseases, ascochyta blight is regarded as the principal constraint to productivity. Evaluation of cold tolerance and *Ascochyta* resistance at ICARDA in the 1980s graphically illustrated the significance of these stresses in Mediterranean climates (Singh, 1990). Of the 15,000 lines screened against ascochyta blight, and 4500 against winter cold, only 18 and 15 accessions, respectively, were resistant and there was no evidence of combined resistance (Singh, 1990). Ascochyta blight is ubiquitous throughout Mediterranean climates, but is also found in cool intermediate summer-dominant rainfall regions of North America, and to a lesser extent, in subtropical Australia (Table 3.3). Only in extremely warm or isolated Mediterranean environments such as the Nile Valley or Chile, respectively, does ascochyta blight present less of a threat to chickpea production. Given the capacity of the disease to spread rapidly once it has established a foothold (Galdames and Mera, 2003), isolation is unlikely to remain an effective barrier for long. The fusarium wilt–root rot complex is the primary biotic stress of most of the summer-dominant rainfall chickpea-growing regions (Table 3.3), and is estimated to be responsible for 10% annual yield loss in India (Singh and Dahiya, 1973). These diseases also occur in Mediterranean-type climates (Table 3.3), but are generally ranked behind ascochyta blight in terms of breeding priorities. Botrytis grey mould is another widely distributed disease of chickpea (Table 3.3), which causes significant damage in north-western South Asia, particularly from the northern Indian states to Nepal and Bangladesh (Gurha *et al.*, 2003). Insect pests of chickpea can also be classified by region and climate (Table 3.3). Pod borer and bruchids are more significant in summer-dominant rainfall zones, while leaf miner is more common in the Mediterranean.

Cropping Systems

Chickpea-sowing strategies vary with environment so as to fit the crop into the farming system and minimize exposure to the prevalent stresses. The widest range of sowing strategies is found in the Mediterranean climates, depending on the relative intensity of the principal stresses (ascochyta blight, cold/frost and terminal drought; Table 3.3).

Autumn-sown rainy season crop maturing in late spring or early summer

This is the system of choice for regions with relatively warm winters and low *Ascochyta* pressure (low–medium PC1 and medium–high PC2) because it capitalizes on within-season rainfall and minimizes exposure to terminal drought. In WANA this system is traditionally used in the Nile Valley (Masadeh *et al.*, 1996) and the warmer areas of Iran (entezari system; Sadri and Banai, 1996). It is interesting to note that these areas fall into the high-stress zone on the upper left quadrant of Fig. 3.2, which is characterized by the lowest, most winter-dominant annual rainfall, and the highest summer temperatures of the world's chickpea-growing regions. In both countries chickpea is grown with supplemental irrigation in these areas to ameliorate drought stress. Recently winter sowing and drip-irrigation has been adopted by ~90% of Israeli chickpea farmers (S. Abbo, The Hebrew University of Jerusalem, 2006, personal communication). Outside the WANA region autumn sowing is used in the relatively warm Mediterranean climates of California and Australia. Australia was relatively *Ascochyta*-free until the mid-1990s, whereupon chickpea production declined sharply in the Mediterranean, and is recovering only now following the release of more resistant varieties and the adoption of prophylactic management practices (Knights and Siddique, 2002). While winter temperatures are moderate in Mediterranean Australia (Table 3.2b), autumn sowing exposes chickpea to suboptimal temperatures during flowering

Table 3.3. Biotic and abiotic stresses categorized by region and climate type.

Region	Abiotic stresses	Biotic stresses	Reference
Mediterranean rainfall distribution (low–medium PC1 scores), cool climate (low–medium PC2 scores)			
Europe: Spain	–	FW, DRR, AB	Cubero *et al.* (1990)
WANA: Iraq	Drought, frost, salinity	AB, DRR, BRR, PB, LM, B, W	Abbas *et al.* (1996)
WANA: Turkey	Drought, heat, frost	AB, LM, PB, B, W	Kusmenoglu and Meyveci (1996)
North America: USA, Pacific North-west	–	AB, FW, WRR	F. Muehlbauer, Washington State University, 2006 (personal communication); Acosta-Gallegos *et al.* (1990)
South America: Chile	Drought, salinity	FW, DRR, AB	M. Mera, INIA-Carillanca, 2006 (personal communication)
Mediterranean rainfall distribution (low–medium PC1 scores), warm climate (medium–high PC2 scores)			
North America: USA, California	–	AB, FW, WRR	F. Muehlbauer, Washington State University, 2006 (personal communication); Acosta-Gallegos *et al.* (1990)
Australia, Mediterranean zone	Drought, cold, waterlogging	AB, BGM, W	Knights and Siddique (2002)
WANA: Algeria	Drought, heat, cold/frost	FW, AB, PB, LM, W	Maatougui *et al.* (1996)
WANA: Morocco	Drought, heat	AB, FW, WRR, LM, B	Amine *et al.* (1996)
WANA: Tunisia	Drought	AB, DRR, LM, PB	Haddad *et al.* (1996)
WANA: Syria	Drought, heat, frost	AB, FW, DRR, PB, LM, W	El-Ahmed *et al.* (1996)
WANA: Jordan	Drought, frost	AB, LM, PB, W	Masadeh *et al.* (1996)
WANA: Israel	Heat, drought, cold/frost	AB, FW, PB, LM	S. Abbo, The Hebrew University of Jerusalem, 2006 (personal communication)
WANA: Iran	Drought, frost, salinity	FW, AB, PB, CW, W	Sadri and Banai (1996)
WANA: Egypt	Heat, salinity, alkalinity	WRR, DRR, AB, B, W	Johansen *et al.* (1996); Khattab and El-Sherbeeny (1996)
Summer-dominant rainfall (medium–high PC1 scores), cool climate (low–medium PC2 scores)			
East Africa: Ethiopia	Drought, waterlogging, frost	FW, DRR, CR, PB, B	Bejiga and Eshete (1996)
East Africa	Drought, heat	FW, DRR, CR, PB, B	Bejiga (1990); Saxena (1993)
Europe: Poland	–	FW, WRR, BGM	Mazur *et al.* (2002)

Continued

Table 3.3. *Continued*

Region	Abiotic stresses	Biotic stresses	Reference
	Mediterranean rainfall distribution (low–medium PC1 scores), cool climate (low–medium PC2 scores)		
Australia, summer rainfall zone	Drought, cold/frost	AB, PRR, BGM, PB, W	Knights and Siddique (2002)
North America: Canada	Late frost–cold, drought	AB, FW, WRR	T. Warkentin, University of Saskatchewan, 2006 (personal communication)
North America: USA, Northern Plains	–	AB, FW, WRR	F. Muehlbauer, Washington State University, 2006 (personal communication); Acosta-Gallegos *et al*. (1990)
South Asia: Nepal	Cold, early drought	BGM, FW, DRR, PB, B	Asthana *et al*. (1990); Johansen *et al*. (1996); Pande *et al*. (2003); Stevenson *et al*. (2005)
	Summer-dominant rainfall (medium–high PC1 scores), warm climate (medium–high PC2 scores)		
North America: Mexico	Drought, cold, heat	FW, DRR, CR, BGM, PB	P. Manjarrez, INIFAP, 2006 (personal communication); Acosta-Gallegos *et al*. (1990); Singh (1993)
East Africa: Sudan	Drought, heat	FW, DRR, WRR, PB, B, W	Faki *et al*. (1996)
South Asia: Bangladesh	Drought, heat, waterlogging	BGM, FW, CR, PB, CW	Islam *et al*. (1991); Johansen *et al*. (1996); Abu Bakr *et al*. (2002)
South Asia: Myanmar	Drought	FW, WRR, DRR, PB,	Virmani (1996)
South Asia: India, northern	Drought, heat, cold	FW, DRR, AB, BGM, PB, B	van Rheenen (1991); Saxena (1993); Nayyar and Chander (2004)
South Asia: India, southern	Drought, heat, salinity	FW, DRR, BGM, PB, B	van Rheenen (1991); Saxena (1993)

FW, Fusarium wilt (*Fusarium oxysporum*); DRR, dry root rot (*Rhizoctonia bataticola*); WRR, wet root rot (*R. solani*); BRR, black root rot (*F. solani*); AB, ascochyta blight (*Ascochyta rabiei*); CR, collar rot (*Sclerotinium rolfsii*); BGM, botrytis grey mould (*Botrytis cinerea*); PRR, phytophtora root rot (*Phytophtora medicaginis*); PB, pod borer (*Helicoverpa* spp.); LM, leaf miner (*Liriomyza cicerina*); CW, black cutworm (*Agrotis ipsilon*); B, bruchids (*Callosobruchus* spp.); W, weeds.

and can delay podset by more than 30 days (Berger *et al.*, 2004, 2005). Early flowering increases yield stability and specific adaptation to terminal drought, but increases the risk of encountering low temperatures. To optimize chickpea adaptation to Australia, reproductive chilling tolerance is an important breeding priority (Knights and Siddique, 2002).

Spring-sown post-rainy season crop maturing in summer

This is the traditional chickpea-cropping system of Mediterranean climates in WANA and beyond (low–medium PC1 and PC2), which minimizes the risk of winter frosts, chilling and disease stresses, and allows farmers to make planting decisions based on stored soil moisture profiles (Walker, 1996). In general, sowing date is negatively correlated to winter temperature, and the season ranges from February–June in North Africa to May–August/September in the Pacific North-west of the USA (F. Muehlbauer, Washington State University, 2006, personal communication) and higher altitudes in Turkey (Kusmenoglu and Meyveci, 1996). As a result, warm regions have relatively longer growing seasons with a shorter average day length (Walker, 1996). Chile is a special case, where chickpea is treated as a bona fide summer crop (August–March), not because of excessively cold winters, but because summer temperatures are relatively low (Table 3.2b), and chickpea production in winter is not as profitable as the other options (M. Mera, INIA-Carillanca, 2006, personal communication). The stress-avoidance strategy is applied differently throughout WANA. For example, in Tunisia winter stresses are minimized by geography, as the crop is grown in low-elevation (<600 m) deep clay loams in semiarid areas, avoiding heavy rainfall (>1000 mm/year), as well as cold and frost-prone areas (Haddad *et al.*, 1996). Conversely, in Turkey, where chickpea is grown across a wide elevation range, time of sowing is used to avoid winter stresses. In central and eastern Anatolia (900–2000 m), where there are 80–120 frosty days/year, chickpea is a summer crop, whereas in western Anatolia, which is much warmer in winter, the season is more typically Mediterranean, running from February to June (Kusmenoglu and Meyveci, 1996).

Although spring sowing in Mediterranean climates avoids the twin stresses of cold and ascochyta blight, it delays crop phenology and increases the exposure to terminal drought. This stress severely limits the yield potential of the traditional Mediterranean chickpea-cropping system, and therefore researchers at ICARDA and elsewhere became strong proponents of winter sowing once breeding programmes had delivered combined ascochyta blight and cold resistance (Singh, 1990). Subsequent research in Syria and Lebanon has demonstrated that winter sowing can double dry matter production (Hughes *et al.*, 1987) and water use efficiency (Brown *et al.*, 1989), as well as produce taller plants with 70% (692 kg/ha) more seed yield than the spring-sown crop averaged over 10 years (Singh *et al.*, 1997). Nevertheless, except in Israel, winter sowing has not been widely adopted by farmers in WANA to date, and therefore there is little feedback on which habitats are most suitable (Walker, 1996). However, it is clear that winter sowing will intensify *Ascochyta* and frost stresses, and reduce the need for drought tolerance (Walker, 1996). Therefore, high-altitude and cool climate of the Mediterranean represented by low PC2 scores in Fig. 3.2b are expected to be particularly stressful for winter-sown chickpea.

In summer-dominant rainfall regions two different sowing strategies are used for chickpea production depending on temperature. Dryland spring-sown within rainy season cropping is used in cool areas (medium–high PC1, low–medium PC2) such as the Northern Plains in the USA and Canada (F. Muehlbauer, personal communication). This region experiences very cold winters (Fig. 3.2b, Table 3.2b). The chickpea season (late April/May–August/September) coincides with the narrow frost-free window suitable for cropping and maximizes the use of the relatively low annual precipitation (Fig. 3.2a, Table 3.2a). Early sowing is preferred, despite delayed germination at soil temperatures of 8–12°C, to prolong the growing season and start flowering early, around the peak of summer in July (Gan *et al.*, 2002). This cropping system exposes chickpea to considerable *Ascochyta* risk during the wet, cool spring and low temperatures at the season end, which can delay maturity and damage seeds (McKay *et al.*, 2002). The low temperature finish is considered to be the primary abiotic stress in western Canada, and therefore there is a need for semi-determinate varieties that will mature regardless of soil moisture status (T. Warkentin, University of Saskatchewan, 2006, personal communication).

The autumn-sown post-rainy season cropping system is responsible for the bulk of the world's chickpea production, and is used throughout the warm summer-dominant rainfall (medium–high PC1 and PC2) regions of South Asia, East Africa, Mexico and north-eastern Australia. Here chickpea is generally grown as a dryland crop on stored soil moisture, on neutral to alkaline fine-textured soils with good water-holding capacity (Johansen *et al.*, 1996). Extremely low annual rainfall areas such as Sudan (Faki *et al.*, 1996) and north-western Mexico (P. Manjarrez, INIFAP, 2006, personal communication) are exceptions, where chickpea is grown with pre-sowing irrigation, and occasional supplementation within the growing season. In South Asia chickpea sowing is determined by the end of the preceding rainy (kharif) cropping season, ranging from early October in the south and central India to mid-November in Nepal (Chaurasia, 2001; Berger *et al.*, 2006). Early sowing is preferred to reduce exposure to terminal drought (Yadav *et al.*, 1998), but must be balanced against high temperature stress at germination because the temperature gradient is very steep from September to November, particularly in north India. In the south, crop duration is short, typically around 100 days in Hyderabad (Saxena, 1984), and the growing season finishes in late January or early February. Minimum and maximum temperatures vary between 15°C and 30°C, and there is little change after flowering commences in December (Berger *et al.*, 2006). Consistent high temperatures cause high rates of evapotranspiration, leading to severe terminal drought. Crop duration in the north is far longer, between 150 and 160 days at Hisar (Saxena, 1984), and the season finishes in March or April. Temperatures in the vegetative phase are 5–10°C lower in the north, but increase very rapidly after flowering in January to early February, ending up with maxima only marginally lower than in the south (Berger *et al.*, 2006). Although podset in the northern regions is often delayed because of low temperatures during flowering (Saxena, 1984), the rapid subsequent temperature increase also imposes considerable terminal drought stress.

In north-eastern Australia chickpea is usually sown in mid-May to limit the exposure to high temperatures after October. The crop can be dry-sown if there is sufficient stored soil moisture; otherwise farmers wait for an opening rain, which can delay sowing until August in some seasons (E.J. Knights, Tamworth Agricultural Institute, 2006, personal communication). The crop can face high temperature stress at maturity, as average November maxima range from 27°C to 31°C. However, because rainfall tends to increase from October onwards through much of the region, chickpea does not encounter the same terminal degree of drought stress as in South Asian environments.

Chickpea adaptation traits across the habitat range

Given that chickpea is grown in distinctly different habitats characterized by different climates, stresses and cropping systems, it follows that chickpea must have differentiated into distinct ecotypes, reflecting local selection pressures, particularly in those regions where the crop has been grown for millennia (Fig. 3.1). There has been considerable germplasm evaluation over the last 30 years, ranging from resistance screening and characterization of thousands of lines by the international centres (Singh, 1990; Upadhyaya, 2003) to multi-environment trials investigating genotype–environment (G × E) interaction in a variety of countries (Berger *et al.*, Chapter 30, this volume), to detailed physiological studies based on small numbers of accessions. However, this research has not paid much attention to germplasm provenance or provided insight into the interaction between germplasm origin and behaviour in different environments. Therefore our understanding of chickpea physiology is largely a general one, and does not explain why or how chickpea from different habitats differ. The picture of chickpea emerges as a typical stress-avoiding competitive ruderal (Grime, 1979), in which the major stresses are avoided by a combination of sowing strategies and appropriate phenology. The timing of flowering is determined by additive positive responses to day length and temperature (Roberts *et al.*, 1985). Carbon (C) and nitrogen (N) fixation rates peak immediately prior to podset, and decline rapidly during pod fill (Hooda *et al.*, 1986; Kurdali, 1996), in line with falling leaf water potential and stomatal conductance (Leport *et al.*, 1998, 1999). The bulk of seed N is remobilized from vegetative tissue formed prior to flowering, whereas C mobilization from pre-flowering assimilates is low (Hooda *et al.*, 1989, 1990; Kurdali, 1996; Davies *et al.*, 2000). This implies that seeds are filled as rates of photosynthesis decline during pod fill, augmented by recycling of respired C within the pods (Furbank *et al.*, 2004; Turner *et al.*, 2005). Drought stress speeds up chickpea phenology by decreasing the thermal time taken from emergence to flowering, pod fill and maturity (Singh, 1991), and decreases leaf water potential, photosynthesis, pod number and yield (Davies *et al.*, 1999; Leport *et al.*, 2006). The primary adaptive strategy to drought stress in chickpea appears to be escape through early phenology (Silim and Saxena, 1993a,b; Siddique *et al.*, 2001; Berger *et al.*, 2004, 2006). However, chickpea also has a number of characteristics consistent with dehydration postponement and tolerance, such as deep rooting (Saxena *et al.*, 1994), high soil water extraction (Zhang *et al.*, 2000) and osmotic adjustment (Morgan *et al.*, 1991; Lecoeur *et al.*, 1992; Leport *et al.*, 1999; Moinuddin and Khanna-Chopra, 2004; Basu *et al.*, 2006). Although temperatures >30°C reduce germination, duration of flowering, pod fill, podset, proportion of fertile seeds and yield, chickpea is more tolerant of high temperature stress than cool season grain legumes such as faba bean, lentil and field pea (Summerfield *et al.*, 1984; Saxena *et al.*, 1988; van Rheenen *et al.*, 1997; Gan *et al.*, 2004). Under non-lethal cold stress chickpea delays germination (Auld *et al.*, 1988; Gan *et al.*, 2002) and podset (Berger *et al.*, 2005) more than well-adapted species like pea and the annual wild *Cicer* sp. Similarly, leaf expansion rates are relatively low under cool conditions, and this, combined with small leaf size, means that winter chickpea absorbs less incident radiation (<50% photosynthetically available radiation (PAR)) than other grain legumes, in the early part of the Mediterranean growing season (Mwanamwenge *et al.*, 1997).

To what extent do these characteristics change across different habitats? There is a paucity of such information in the chickpea literature. Studies on geographical patterns of morphological and agronomic variation in global chickpea germplasm collections have not explained how these traits interact with the environment to play a role in adaptation (Jana and Singh, 1993; Upadhyaya, 2003). Upadhyaya (2003) demonstrated a preponderance of kabuli-type characters (white flower and light seed colour, large, smooth-textured seeds) in East Asia, the Mediterranean and Europe, while desi characters (pink flowers, yellow to brown angular seeds with a rough texture) were more common in Africa, and South as well as South-east Asia. These findings reflect the Mediterranean preference for kabuli types, and the predominance of desi types in India and East Africa. Agronomic traits describing phenology, plant architecture and productivity varied widely within regions (Upadhyaya, 2003), which is to be expected given the range of habitats found within any given region (Fig. 3.2).

The best evidence for ecotype formation in chickpea has come from studies that focus on germplasm provenance. Evaluation of a recent *C. judaicum* collection, in which habitats were described in great detail, demonstrated that this wild species of chickpea is a typical stress-avoider (Ben-David, The Hebrew University of Jerusalem, 2006, personal communication). As habitats become more stressful, in terms of rainfall (amounts, distribution, variance), temperature (minima, maxima, frost day distribution), wind speed, radiation, relative humidity and soil type, reproductive strategies became increasingly conservative, with accessions flowering earlier and accumulating less biomass (Ben-David, The Hebrew University of Jerusalem, 2006, personal Communication). G × E studies across Australia and India have demonstrated similar tendencies in chickpea (Berger *et al.*, 2004, 2006). The germplasm with the earliest phenology came from short-season habitats in Gujarat, southern and central India, followed by northern India and Australia, and finally the Mediterranean. Germplasm with early phenology was widely adapted throughout Australia (Berger *et al.*, 2004), and was particularly effective in stressful terminal drought environments, such as Merredin in Western Australia, and Sehore, Jabalpur and Gulbarga in central and southern India (Berger *et al.*, 2006). In addition to early phenology, genotypes specifically adapted to low yielding sites in India were characterized by low yield responsiveness, low biomass and high harvest index (Berger *et al.*, 2006). Conversely, genotypes that were well adapted to higher-yielding northern Indian sites were almost exclusively sourced from Punjab, Haryana, New Delhi and Uttar

Pradesh, and characterized by later, highly responsive phenology, whereby flowering was delayed significantly at late flowering northern sites. Extending the vegetative phase under long-season conditions increased biomass accumulation prior to reproduction, and delayed flowering until temperatures became sufficiently warm to support podset. As chickpea is indeterminate, the delayed phenology increased both source and sink potential in material specifically adapted to productive northern sites (Berger *et al.*, 2006). Regional differences have also been observed in the evaluation of Ethiopian drought-tolerant germplasm. The highest drought response index was collected from drought-prone areas in Gonda, Gojam and Shiva (Anbessa and Bejiga, 2002).

Conclusion

It is likely that chickpea's poor performance under cold and chilling stress, as well as its relative tolerance of high temperatures, are outcomes of the crop's unique evolution from a Mediterranean winter-annual life cycle, to post-rainy season spring sowing, and subsequent dissemination to warmer, summer-dominant rainfall regions. Nevertheless, through its evolutionary path chickpea has spread to a wide range of habitats characterized by different climates, stresses and cropping systems, all of which place different selection pressures on the crop. Our understanding of the interaction between these selection pressures and ecotypic adaptation is only rudimentary at this stage. Given the explosion of data available for habitat characterization, the potential exists to make great strides in the understanding of adaptation in chickpea, by exploiting this information whenever the crop is evaluated, be it in resistance screening, multi-environment trials or detailed physiological investigations. This ecophysiological approach can widen the impact of research from the specific (i.e. pertaining only to a particular group of genotypes) to the general (i.e. pertaining to genotypes representing different habitat types), and should be encouraged.

Acknowledgements

The authors would like to acknowledge generous research funding support from the Commonwealth Scientific and Industrial Research Organization (CSIRO), the Centre for Legumes in Mediterranean Agriculture (CLIMA) at the University of Western Australia, the Australian Centre for International Agricultural Research (ACIAR) and the Australian Grains Research and Development Corporation (GRDC). This chapter, particularly the definition of crop distribution, stresses and cropping systems, would not have been possible without generous feedback and advice from the global community of chickpea breeders and scientists. The authors would especially like to thank Fred Muehlbauer, Mario Mera, Tom Warkentin, Renuka Shrestha, Shahal Abbo, Pedro Manjarrez, Kadambot Siddique, Robert Hijmans, Dirk Enneking, Col Douglas, Kharaiti Mehra, Jason Brand, William Martin, Michael Materne, Larn McMurray, Eric Armstrong, Tanveer Khan and Ted Knights. We also thank Steve Milroy for his incisive feedback on the manuscript.

References

Abbas, A.I., Ali, A.H. and Ibrahim, K.S. (1996) Chickpea in Iraq. In: Saxena, N.P., Saxena, M.C., Johansen, C., Virmani, S.M. and Harris, H. (eds) *Adaptation of Chickpea in the West Asia and North African Region*. ICRISAT, Hyderabad, India; ICARDA, Aleppo, Syria, pp. 35–46.

Abbo, S., Lev-Yadun, S. and Galwey, N. (2002) Vernalization response of wild chickpea. *New Phytologist* 154, 695–701.

Abbo, S., Berger, J.D. and Turner, N.C. (2003a) Evolution of cultivated chickpea: four genetic bottlenecks limit diversity and constrain crop adaptation. *Functional Plant Biology* 30, 1081–1087.

Abbo, S., Shtienberg, D., Lichtenzveig, J., Lev-Yadun, S. and Gopher, A. (2003b) The chickpea, summer cropping, and a new model for pulse domestication in the ancient near east. *The Quarterly Review of Biology* 78, 435–438.

Abu Bakr, M., Ali Hussein, S., Ali Afzal, M. and Rahman, M.A. (2002) Chickpea status and production constraints in Bangladesh. In: Bakr, M.A., Siddique, K.H.M. and Johansen, C. (eds) '*Integrated Management of Botrytis Grey Mould of Chickpea in Bangladesh and Australia*', Summary proceedings of a project inception workshop, Bangladesh Agricultural Research Institute (BARI), Joydebpur, Gazipur, Bangladesh, pp. 63–69.

Acosta-Gallegos, J.A., Gomez-Garza, R.M., Morales-Gomez, J.A. and Andrade-Arias, E. (1990) Recent advances in chickpea improvement and prospects for the nineties: Americas. In: van Rheenen, H.A. and Saxena, M.C. (eds) *Chickpea in the Nineties: Proceedings of the Second International Workshop on Chickpea Improvement*. ICRISAT, Hyderabad, India, pp. 267–268.

Ali, M. and Kumar, S. (2003) Chickpea research in India: an overview. In: Ali, M., Kumar, S. and Singh, N.B. (eds) *Chickpea Research in India*. Indian Institute of Pulses Research, Kanpur, India, pp. 1–14.

Allaby, M. (1998) *A Dictionary of Ecology*. Oxford University Press, New York.

Amine, M., Boulif, M. and Beniwal, S.P.S. (1996) Chickpea in Morocco. In: Saxena, N.P., Saxena, M.C., Johansen, C., Virmani, S.M. and Harris, H. (eds) *Adaptation of Chickpea in the West Asia and North African Region*. ICRISAT, Hyderabad, India; ICARDA, Aleppo, Syria, pp. 101–109.

Anbessa, Y. and Bejiga, G (2002) Evaluation of Ethiopian chickpea landraces for tolerance to drought. *Genetic Resources and Crop Evolution* 49, 557–564.

Asthana, A.N., Bharati, M.P., Haqqani, A.M., Moe, K. and Rahman, M.M. (1990) Recent advances in chickpea improvement and prospects for the nineties: South Asia. In: van Rheenen, H.A. and Saxena, M.C. (eds) *Chickpea in the Nineties: Proceedings of the Second International Workshop on Chickpea Improvement*. ICRISAT, Hyderabad, India, pp. 261–262.

Auld, D.L., Bettis, B.L., Crock, J.E. and Kephart, K.D. (1988) Planting date and temperature effects on germination emergence and seed yield of chickpea. *Agronomy Journal* 80, 909–914.

Basu, P.S., Turner, N.C., Berger, J.D., Ali, M., Siddique, K.H.M. and Chaturvedi, S.K. (2006) Osmotic adjustment of chickpea (*Cicer arietinum* L.) is not associated with changes in carbohydrate composition or leaf gas exchange under drought. *Annals of Applied Biology* (in press).

Bejiga, G. (1990) Recent advances in chickpea improvement and prospects for the nineties: Eastern Africa. In: van Rheenen, H.A. and Saxena, M.C. (eds) *Chickpea in the Nineties: Proceedings of the Second International Workshop on Chickpea Improvement*. ICRISAT, Hyderabad, India, pp. 265–266.

Bejiga, G. and Eshete, M. (1996) Chickpea in Ethiopia. In: Saxena, N.P., Saxena, M.C., Johansen, C., Virmani, S.M. and Harris, H. (eds) *Adaptation of Chickpea in the West Asia and North African Region*. ICRISAT, Hyderabad, India; ICARDA, Aleppo, Syria, pp. 137–153.

Ben-David, R., Lev-Yadun, S., Can, C. and Abbo, S. (2006) Ecogeography and demography of *Cicer judaicum* Boiss., a wild annual relative of domesticated chickpea. *Crop Science* 46, 1360–1370.

Berger, J.D., Turner, N.C., Siddique, K.H.M., Knights, E.J., Brinsmead, R.B., Mock, I., Edmondson, C. and Khan, T.N. (2004) Genotype by environment studies across Australia reveal the importance of phenology for chickpea (*Cicer arietinum* L.) improvement. *Australian Journal of Agricultural Research* 55, 1–14.

Berger, J.D., Buck, R.P., Henzell, J.M. and Turner N.C. (2005) Evolution in the genus *Cicer* – vernalization response and low temperature pod set in chickpea (*C. arietinum* L.) and its annual wild relatives. *Australian Journal of Agricultural Research* 56, 1191–1200.

Berger, J.D., Ali, M., Basu, P.S., Chaudhary, B.D., Chaturvedi, S.K., Deshmukh, P.S., Dharmaraj, P.S., Dwivedi, S.K., Gangadhar, G.C., Gaur, P.M., Kumar, J., Pannu, R.K., Siddique, K.H.M., Singh, D.N., Singh, D.P., Singh, S.J., Turner, N.C., Yadava, H.S. and Yadav, S.S. (2006) Genotype by environment studies demonstrate the critical role of phenology in adaptation of chickpea (*Cicer arietinum* L.) to high and low yielding environments of India. *Field Crops Research* 98, 230–244.

Brown, S.C., Gregory, P.J., Cooper, P.J.M. and Keatinge, J.D. (1989) Root and shoot growth and water use of chickpea (*Cicer arietinum* L.) grown in dryland conditions effects of sowing date and genotype. *Journal of Agricultural Science* 113, 41–50.

Chaurasia, P.C.P. (2001) Integrated management of Botrytis gray mold (*Botrytis cinerea*) of chickpea at Taraharapp In: Neupane, R.K., Yadav, N.K. and Darai, R. (eds) *Proceedings of National Winter Crops Research Workshop: Grain Legumes*. Regional Agricultural Research Station, Parwanipur, Nepal, pp. 80–84.

Cubero, J.I. (1980) Research on chickpea in Spain. In: Green, J.M., Nene, Y.L. and Smithson, J.B. (eds) *Proceedings of the International Workshop on Chickpea Improvement*. ICRISAT, Hyderabad, India pp. 268–269.

Cubero, J.I., Moreno, M.T. and Gil, J. (1990) Research on chickpea in Spain. *Options Méditerranéennes* 9, 157–161.

Davies, S.L., Turner, N.C., Siddique, K.H.M., Plummer, J.A. and Leport, L. (1999) Seed growth of desi and kabuli chickpea (*Cicer arietinum* L.) in a short season Mediterranean type environment. *Australian Journal of Experimental Agriculture* 39, 181–188.

Davies, S.L., Turner, N.C., Palta, J.A., Siddique, K.H.M. and Plummer, J.A. (2000) Remobilisation of carbon and nitrogen supports seed filling in chickpea subjected to water deficit. *Australian Journal of Agricultural Research* 51, 855–866.

El-Ahmed, H., Tawil, W., El-Ahmed, M., Al-Seoud, A. and Nassif, A.M. (1996) Chickpea in Syria. In: Saxena, N.P., Saxena, M.C., Johansen, C., Virmani, S.M. and Harris, H. (eds) *Adaptation of Chickpea in the West Asia and North African Region*. ICRISAT, Hyderabad, India; ICARDA, Aleppo, Syria, pp. 55–66.

Faki, H., Mohamed, A.I.S. and Ali, M.E.K. (1996) Chickpea in Sudan. In: Saxena, N.P., Saxena, M.C., Johansen, C., Virmani, S.M. and Harris, H. (eds) *Adaptation of Chickpea in the West Asia and North African Region*. ICRISAT, Hyderabad, India; ICARDA, Aleppo, Syria, pp. 155–169.

FAO (2006) FAOSTAT. Food and Agriculture Organization of the United Nations.

Furbank, R.T., White, R., Palta, J.A. and Turner N.C. (2004) Internal recycling of respiratory CO_2 in pods of chickpea (*Cicer arietinum* L.): the role of pod wall, seed coat, and embryo. *Journal of Experimental Botany* 55, 1687–1696.

Galdames, R. and Mera, M. (2003) First report of Ascochyta blight of chickpea caused by *Ascochyta rabiei* in Chile. *Plant Disease* 87, 603.

Gan, Y.T., Miller, P.R., Liu, P.H., Stevenson, F.C. and McDonald, C.L. (2002) Seedling emergence, pod development, and seed yields of chickpea and dry pea in a semi-arid environment. *Canadian Journal of Plant Science* 82, 531–537.

Gan, Y.T., Wang, J., Angadi, S.V. and McDonald, C.L. (2004) Response of chickpea to short periods of high temperature and water stress at different developmental stages. In: *New Directions for a Diverse Planet: Proceedings of the 4th International Crop Science Congress*. Brisbane, Australia.

Grime, J.P. (1979) *Plant Strategies and Vegetation Processes*. Wiley, Chichester, UK.

Gurha, S.N., Singh, G. and Sharma, Y.R. (2003) Diseases of chickpea and their management. In: Ali, M., Kumar, S. and Singh, N.B. (eds) *Chickpea Research in India*. Indian Institute of Pulses Research, Kanpur, India, pp. 195–227.

Haddad, A., Halila, M.H. and Jendoubi, M. (1996) Chickpea in Tunisia. In: Saxena, N.P., Saxena, M.C., Johansen, C., Virmani, S.M. and Harris, H. (eds) *Adaptation of Chickpea in the West Asia and North African Region*. ICRISAT, Hyderabad, India; ICARDA, Aleppo, Syria, pp. 111–122.

Harlan, J.R. (1969) Ethiopia: a centre of diversity. *Economic Botany* 23, 309–314.

Hijmans, R.J., Guarino, L., Cruz, M. and Rojas, E. (2001) Computer tools for spatial analysis of plant genetic resources data. 1. DIVA-GIS. *Plant Genetic Resources Newsletter* 127, 15–19.

Hijmans, R.J., Cameron, S.E., Parra, J.L., Jones, P.G. and Jarvis, A. (2005) Very high resolution interpolated climate surfaces for global land areas. *International Journal of Climatology* 25, 1965–1978.

Hooda, R.S., Rao, A.S., Luthra, Y.P., Sheoran, I.S. and Singh, R. (1986) Partitioning and utilization of carbon and nitrogen for dry matter and protein production in chickpea (*Cicer arietinum*). *Journal of Experimental Botany* 37, 1492–1502.

Hooda, R.S., Sheoran, I.S. and Singh, R. (1989) Distribution of carbon 14 carbon dioxide during ontogeny of chickpea *Cicer arietinum* grown at two moisture levels. *Annals of Applied Biology* 114, 367–376.

Hooda, R.S., Sheoran, I.S. and Singh, R. (1990) Partitioning and utilization of carbon and nitrogen for dry matter and protein production in chickpea *Cicer arietinum* L. under control and stress conditions. *Indian Journal of Experimental Biology* 28, 280–283.

Hughes, G., Keatinge, J.D.H., Cooper, P.J.M. and Dee, N.F. (1987) Solar-radiation interception and utilization by chickpea (*Cicer arietinum* L.) crops in Northern Syria. *Journal of Agricultural Science* 108, 419–424.

Islam, M.O., Matiur Rahman, M., Rahman, A., Sarker, A. and Zaman, W. (1991) Status and future breeding strategies for chickpea. In: *Advances in Pulses Research in Bangladesh*. ICRISAT, Joydebpur, Gazipur, Bangladesh, pp. 11–17.

Jana, S. and Singh, K.B. (1993) Evidence of geographical divergence in kabuli chickpea from germplasm evaluation data. *Crop Science* 33, 626–632.

Johansen, C., Saxena, N.P. and Saxena, M.C. (1996) Comparisons of abiotic constraints to chickpea production in WANA and SAT. In: Saxena, N.P., Saxena, M.C., Johansen, C., Virmani, S.M. and Harris, H. (eds) *Adaptation of Chickpea in the West Asia and North African Region*. ICRISAT, Hyderabad, India; ICARDA, Aleppo, Syria, pp. 181–188.

Khanna-Chopra, R. and Sinha, S.K. (1987) Chickpea: physiological aspects of growth and yield. In: Saxena, M.C. and Singh, K.B. (eds) *The Chickpea*. CAB International, Wallingford, UK, pp. 163–190.

Khattab, A.M.A. and El-Sherbeeny, M.H. (1996) Chickpea in Egypt. In: Saxena, N.P., Saxena, M.C., Johansen, C., Virmani, S.M. and Harris, H. (eds) *Adaptation of Chickpea in the West Asia and North African Region*. ICRISAT, Hyderabad, India; ICARDA, Aleppo, Syria, pp. 127–136.

Knights, E.J. and Siddique, K.H.M. (2002) Chickpea status and production constraints in Australia. In: Abu Bakr, M., Siddique, K.H.M. and Johansen, C. (eds) *Integrated Management of Botrytis Grey Mould of Chickpea in Bangladesh and Australia*. Bangladesh Agricultural Research Institute, Joydebpur, Gazipur, Bangladesh, pp. 33–41.

Kurdali, F. (1996) Nitrogen and phosphorus assimilation, mobilization and partitioning in rainfed chickpea (*Cicer arietinum* L.). *Field Crops Research* 47, 81–92.

Kusmenoglu, I. and Meyveci, K. (1996) Chickpea in Turkey. In: Saxena, N.P., Saxena, M.C., Johansen, C., Virmani, S.M. and Harris, H. (eds) *Adaptation of Chickpea in the West*

Asia and North African Region. ICRISAT, Hyderabad, India; ICARDA, Aleppo, Syria, pp. 67–84.

Lecoeur, J., Wery, J. and Turc, O. (1992) Osmotic adjustment as a mechanism of dehydration postponement in chickpea *Cicer arietinum* L. leaves. *Plant and Soil* 144, 177–189.

Leport, L., Turner, N.C., French, R.J., Tennant, D., Thomson, B.D. and Siddique, K.H.M. (1998) Water relations, gas exchange and growth of cool-season grain legumes in a Mediterranean-type environment. *European Journal of Agronomy* 9, 295–303.

Leport, L., Turner, N.C., French, R.J., Barr, M.D., Duda, R., Davies, S.L., Tennant, D. and Siddique, K.H.M. (1999) Physiological responses of chickpea genotypes to terminal drought in a Mediterranean type environment. *European Journal of Agronomy* 11, 279–291.

Leport, L., Turner, N.C., Davies, S.L. and Siddique, K.H.M. (2006) Variation in pod production and abortion among chickpea cultivars under terminal drought. *European Journal of Agronomy* 24, 236–246.

Maatougui, E.H., Bouznad, Z. and Labdi, M. (1996) Chickpea in Algeria. In: Saxena, N.P., Saxena, M.C., Johansen, C., Virmani, S.M. and Harris, H. (eds) *Adaptation of Chickpea in the West Asia and North African Region*. ICRISAT, Hyderabad, India; ICARDA, Aleppo, Syria, pp. 89–99.

Masadeh, A., Hawash, I. and Al Majali, N. (1996) Chickpea in Jordan. In: Saxena, N.P., Saxena, M.C., Johansen, C., Virmani, S.M. and Harris, H. (eds) *Adaptation of Chickpea in the West Asia and North African Region*. ICRISAT, Hyderabad, India; ICARDA, Aleppo, Syria, pp. 47–54.

Mazur, S., Nawrocki, J. and Kucmierz, J. (2002) Fungal diseases of chickpea (*Cicer arietinum* L.) cultivated in the south region of Poland. *Plant Protection Science* 38, 332–335.

McKay, K., Miller, P., Jenks, B., Riesselman, J., Neill, K., Buschena, D. and Bussan, A.J. (2002) Growing chickpea in the Northern Great Plains. North Dakota State University A-1236, Fargo.

Moinuddin, and Khanna-Chopra, R. (2004) Osmotic adjustment in chickpea in relation to seed yield and yield parameters. *Crop Science* 44, 449–455.

Morgan, J.M., Rodriguez-Maribona, B. and Knights, E.J. (1991) Adaptation to water deficit in chickpea breeding lines by osmoregulation: relationship to grain yields in the field. *Field Crops Research* 27, 61–70.

Mwanamwenge, J., Siddique, K.H.M., Sedgley, R.H. (1997) Canopy development and light absorption of grain legume species in a short season Mediterranean-type environment. *Journal of Agronomy and Crop Science* 179, 1–7.

Nayyar, H. and Chander, S. (2004) Protective effects of polyamines against oxidative stress induced by water and cold stress in chickpea. *Journal of Agronomy and Crop Science* 190, 355–365.

New, M., Lister, D., Hulme, M., and Makin, I. (2002) A high-resolution data set of surface climate over global land areas. *Climate Research* 21, 1–25.

Pande, S., Bourai, V.A., Neupane, R.K. and Joshi, P.K. (2003) Chickpea production: constraints and promotion of integrated pest management in Nepal. *ICRISAT Information Bulletin* No. 66, Hyderabad, India.

Roberts, E.H., Hadley, P. and Summerfield, R.J. (1985) Effects of temperature and photoperiod on flowering in chick peas (*Cicer arietinum*). *Annals of Botany* 55, 881–892.

Sadri, B. and Banai, T. (1996) Chickpea in Iran. In: Saxena, N.P., Saxena, M.C., Johansen, C., Virmani, S.M. and Harris, H. *Adaptation of Chickpea in the West Asia and North African Region*. ICRISAT, Hyderabad, India; ICARDA, Aleppo, Syria, pp. 23–33.

Saxena, M.C. (1993) The challenge of developing biotic and abiotic streess resistance in cool-season food legumes. In: Singh, K.B. and Saxena, M. (eds) *Breeding for Stress Tolerance in Cool-Season Food Legumes*. Wiley, Chichester, UK, pp. 3–14.

Saxena, M.C., Saxena, N.P. and Mohamed, A.K. (1988) High temperature stress. In: Summerfield, R.J. (ed.) *World Crops: Cool Season Food Legumes*. Kluwer Academic, Dordrecht, The Netherlands, pp. 845–856.

Saxena, N.P. (1984) Chickpea. In: Goldsworthy, P.R. and Fisher, N.M. (eds) *The Physiology of Tropical Field Crops*. Wiley, Chichester, UK, pp. 419–452.

Saxena, N.P., Krishnamurthy, L. and Johansen, C. (1994) Registration of a drought resistant chickpea germplasm. *Crop Science* 33, 1424.

Saxena, N.P., Saxena, M.C., Johansen, C., Virmani, S.M. and Harris, H. (1996) *Adaptation of Chickpea in the West Asia and North African Region*. ICRISAT, Hyderabad, India; ICARDA, Aleppo, Syria.

Siddique, K.H.M., Regan, K.L., Tennant, D. and Thomson, B.D. (2001) Water use and water use efficiency of cool season grain legumes in low rainfall Mediterranean-type environments. *European Journal of Agronomy* 15, 267–280.

Silim, S.N. and Saxena, M.C. (1993a) Adaptation of spring-sown chickpea to the Mediterranean basin. 2. Factors influencing yield under drought. *Field Crops Research* 34, 137–146.

Silim, S.N. and Saxena, M.C. (1993b) Adaptation of spring-sown chickpea to the Mediterranean basin. 1. Response to moisture supply. *Field Crops Research* 34, 121–136.

Singh, K.B. (1990) Winter chickpea: problems and potential in the Mediterranean region. *Options Méditerranéennes. Série A, Séminaires Méditerranéens*. Zaragoza, Spain, pp. 25–34.

Singh, K.B. (1993) Problems and prospects of stress resistance breeding in chickpea. In: Singh, K.B. and Saxena, M. (eds) *Breeding for Stress Tolerance in Cool-Season Food Legumes*, Wiley, Chichester, UK, pp. 17–35.

Singh, K.B. and Dahiya, B.S. (1973) Breeding for wilt resistance in chickpea. *Symposium on Wilt Problems and Breeding for Wilt Resistance in Bengal Gram*. IARI, New Delhi, India, pp. 13–14.

Singh, K.B. and Reddy, M.V. (1996) Improving chickpea yield by incorporating resistance to ascochyta blight. *Theoretical and Applied Genetics* 92, 509–515.

Singh, K.B., Malhotra, R.S., Saxena, M.C. and Bejiga, G. (1997) Superiority of winter sowing over traditional spring sowing of chickpea in the Mediterranean region. *Agronomy Journal* 89, 112–118.

Singh, K.B., Robertson, L.D. and Ocampo, B. (1998) Diversity for abiotic and biotic stress resistance in the wild annual *Cicer* species. *Genetic Resources and Crop Evolution* 45, 9–17.

Singh, P. (1991) Influence of water deficits on phenology growth and dry matter allocation in chickpea (*Cicer arietinum*). *Field Crops Research* 28, 1–16.

Stevenson, P.C., Pande, S., Pound, B. and Neupane, R.K. (2005) A strategy for wealth generation through chickpea production. *ICRISAT Information Bulletin* No. 70, Hyderabad, India.

Summerfield, R.J., Hadley, P., Roberts, E.H., Minchin, F.R. and Rawsthorne, S. (1984) Sensitivity of chickpeas (*Cicer arietinum* L.) to hot temperatures during the reproductive period. *Experimental Agriculture* 20, 77–93.

Turner, N.C., Davies, S.L., Plummer, J.A. and Siddique, K.H.M. (2005) Seed filling in grain legumes under water deficits, with emphasis on chickpeas. *Advances in Agronomy* 87, 211–250.

Upadhyaya, H.D. (2003) Geographical patterns of variation for morphological and agronomic characteristics in the chickpea germplasm collection. *Euphytica* 132, 343–352.

van der Maesen, L.J.G. (1972) Cicer *L., a Monograph of the Genus, with Special Reference to the Chickpea (*Cicer arietinum *L.)*. Veenman & Sons, Wageningen, The Netherlands, pp. 9–133.

van der Maesen, L.J.G. (1984) Taxonomy, distribution and evolution of chickpea. In: Witcombe, J.R. and Erskine, W. (eds) *Genetic Resources and their Exploitation-Chickpeas, Faba beans and Lentils*. Martinus Nijhoff/Junk, The Hague, The Netherlands, pp. 95–104.

van Rheenen, H.A. (1991) Chickpea breeding: progress and prospects. *Plant Breeding Abstracts* 61, 998–1009.

van Rheenen, H.A., Singh, O. and Saxena, N.P. (1997) Using evaluation techniques for photoperiod and thermo-insensitivity in pulses improvement. In: Asthana, A.N. and Ali, M. (eds) *Recent Advantages in Pulses Research*. Indian Society of Pulses Research and Development, IIPR, Kanpur, India, pp. 443–458.

Virmani, S.M. (1996) Geographical information system: its use and relevance in adaptation of crops in varied agroenvironments. In: Saxena, N.P., Saxena, M.C., Johansen, C., Virmani, S.M. and Harris, H. (eds) *Adaptation of Chickpea in the West Asia and North African Region*. ICRISAT, Hyderabad, India; ICARDA, Aleppo, Syria, pp. 13–19.

Walker, G.K. (1996) Climatic adaptation of chickpea in WANA. In: Saxena, N.P., Saxena, M.C., Johansen, C., Virmani, S.M. and Harris, H. (eds) *Adaptation of Chickpea in the West Asia and North African Region*. ICRISAT, Hyderabad, India; ICARDA, Aleppo, Syria, pp. 175–180.

Willcox, G. (2005) The distribution, natural habitats and availability of wild cereals in relation to their domestication in the Near East: multiple events, multiple centres. *Vegetation History and Archaeobotany* 14, 534–541.

Yadav, V.S., Yadav, S.S. and Panwar, J.D.S. (1998) Effect of planting time on physiological attributes and grain yield in desi and kabuli chickpea genotypes. *Indian Journal of Plant Physiology* 3, 292–295.

Zhang, H., Pala, M., Oweis, T. and Harris, H. (2000) Water use and water use efficiency of chickpea and lentil in a Mediterranean environment. *Australian Journal of Agricultural Research* 51, 295–304.

4 Uses, Consumption and Utilization

S.S. Yadav,[1] N. Longnecker,[2] F. Dusunceli,[3] G. Bejiga,[4] M. Yadav,[5] A.H. Rizvi,[1] M. Manohar,[1] A.A. Reddy,[6] Z. Xaxiao[7] and W. Chen[8]

[1]*Pulse Research Laboratory, Division of Genetics, Indian Agricultural Research Institute, New Delhi 110012, India;* [2]*Centre for Learning Technology, The University of Western Australia, Perth, Australia;* [3]*The Central Research Institute for Field Crops, PO Box 226, Ulus, Ankara, Turkey;* [4]*Green Focus Ethiopia, PO Box 802, Code No. 1110, Addis Ababa, Ethiopia;* [5]*College of Business Administration, Tarleton State University (Texas A & M University System), Stephenville, TX, 76402, USA;* [6]*Indian Institute of Pulses Research, Kanpur 208024, Uttar Pradesh, India;* [7]*Chinese Academy of Agricultural Sciences, Beijing, China;* [8]*USDA-ARS, Grain Legume Genetics and Physiology Research Unit, Washington State University, Pullman, WA 99164, USA*

Introduction

Since the beginning of agriculture, grain legumes have had multiple uses depending on the utilization of different parts of the plant. The seeds are used dry or green, the legumes green and the plant dry as straw for animal feed and green as fodder or as organic manure. The dried grain is used as animal feed or for human consumption. In the latter case, it is used whole or hulled, as flour, boiled or roasted. The flour is used on its own or mixed with other flours, generally made from cereals. Legumes are served as a main dish, either alone or accompanying meat or fish, as a snack, green or dried (Hernándo Bermejo and León, 1994).

The same diversity exists in the cultivation systems (extensively dry or irrigated, purely horticultural and winter or spring) and in postharvest handling (for fresh consumption, dry storage or immediate use). Packaging can be simple and the product can be frozen, canned or pre-cooked. Grain legumes are a source of oil with which protein-rich cake is made and they also have other interesting substances for industry and pharmacology.

In addition to everything that the plant offers directly, legumes have constantly been accompanied by species that produce carbohydrates, i.e. cereals in temperate zones or roots and tubers in tropical zones. This is due not only to their great nutritional value, but also to their ability to fix atmospheric nitrogen, which is then released into the soil, a fact that has been perceived intuitively by farmers throughout the ages because of its effects on crops and one that necessitated the inclusion of legumes in all farming.

The essential scientific needs of the first-order crops were met since the mid 19th century, and in particular with genetic improvement in the 20th century. Grain legumes were not among them, unless they had a horticultural use, in which case they became part of a refined human diet as a complement and not as a basic food. The legumes of the poor, which had served as a source of protein for thousands of years and which still perform this function through subsistence farming in developing countries, were relegated to marginal regions away from trade routes and scientific interest, and away from new agricultural techniques and improvement work.

In sum, the situation of the chickpea dealt with here applies also to numerous other legumes, and is valued by the farmer, the consumer and the industry. Nevertheless, their cultivation is decreasing from year to year. The ultimate cause of their marginalization is their late arrival in the marketing world.

Prehistorically, the Indian vegetarian dietary system was evolved on a very sound footing considering the basis of health benefits. The vegetarian dietary system is mentioned in ancient books like the Vedas and Puranas, which are estimated to have been written down between 1500 BCE and 300 AD. Now it is well known that non-vegetarian dietary systems are creating health hazards like mad cow disease and bird flu in different parts of the world. High cholesterol level, cardiological and diabetic problems are associated with non-vegetarian dietary systems, whereas the vegetarian dietary system helps in maintaining good health. This was the reason to introduce the concept of vegetarian dietary system in the Indian society during ancient times.

The majority of legume crops are directly associated with Indian civilization and constitute an important component of the vegetarian dietary system. India is the largest producer and consumer of pulses in the world, accounting for 33% of the world area and 20% of the world production. Most of the legume crops like chickpea (*Cicer arietinum* L.), pigeon pea (*Cajanus cajan* (L.) Mill sp.), green gram (*Vigna radiata* (L.) Wilczek), black gram (*Vigna mungo* (L.) Hepper), lentils (*Lens culinaris* Medik), field pea, cowpea, moth bean, horse gram, *Lathyrus* and Rajmash are predominantly cultivated in different parts of India that produce 14.94 million tonnes of pulses from 23.44 million hectares. Most of these pulses or legumes are dried and preserved for consumption throughout the year. However, a significant proportion of chickpea, field pea and beans are consumed as green vegetables. In India, these pulses are the major sources of dietary protein and are essential in cereal-based diets due to their complementary character to cereal proteins.

The World Health Organization (WHO) recommends a per capita consumption of pulses of 80 g/day, and the Indian Council of Medical Research (ICMR) has recommended a minimum consumption of 47 g/day. The actual consumption in India, however, is much less, ~30–34 g/day/person due to non-availability and high prices of pulses. Due to a change in dietary habit in view of the growing health consciousness and preference for vegetarian proteins, there is an increasing demand of grain legumes in the developed world not only as food but also as feed, and people are shifting to the vegetarian system from non-vegetarian one in the developed world.

Although India produces a wide variety of pulses, chickpea alone accounts for >25% of the total area and 40% of the total production nationally. Internationally, India accounts for >60% in both area and production. Like for India, chickpea is also important for Pakistan and Bangladesh; therefore, it is an important food item in the diets of all income groups in the Indian subcontinent, which has a much higher usage of chickpea than any other region of the world.

Per Capita Consumption

Production plus import minus exports of a particular country can be considered as per capita consumption in the absence of actual consumption data. World per capita consumption was 1.34 kg/year in 2004. It fluctuates between 1.29 and 1.61 kg/capita/year. The highest per capita consumption of chickpea is in Turkey (6.65 kg/year) followed by India (5.37 kg/year), Myanmar (4.54 kg/year), Jordan (4.27 kg/year) and Pakistan (4.11 kg/year). There has been a drastic reduction in chickpea consumption in Turkey since 1990. In India, per capita consumption fluctuates between 5 and 6 kg/year. A slight decline has been observed in Pakistan, while Myanmar, Jordan and Iran showed an increase in consumption. Overall, per capita consumption is high in developing countries due to various preparations for daily consumption Fig 4.1

Fig. 4.1. Various Chickpea recipies.

and low or negligible in developed countries. Some countries that have less production of chickpea but significant consumption are Jordan (4.27 kg/capita/year), Algeria (1.94 kg/capita/year) and Spain (2.68 kg/capita/year) (Table 4.1).

Chickpea Consumption in the Indian Subcontinent

In India and surrounding countries, the desi types are used whole, shelled and split to produce dhal, or ground into fine flour called *besan*. *Besan* is used in many ways for cooking, for example, mixed with wheat flour to make *roti* or *chapatti*, and for making sweets and snacks. It is also used as a vegetable.

Traditionally, chickpea is one of the most favoured of all pulses in Indian society. It is consumed in north, west, east and south of India, in tribal areas, villages and cities – everywhere chickpea has found a place in the daily diet. During cropping season green leaves are used as a vegetable, fully developed green pods are used in vegetable dishes, rice and pulav, and some are roasted with salt. After harvesting and threshing, dried seeds are used for the preparation of dhal, which has an attractive yellow colour, and is used in various preparations. From dhal, a yellow flour known as *besan* is prepared, and both dhal and flour are used in various preparations. Chickpea dhal blends well with vegetables, meat and sauces, and can be used to make a main course and snack items. Considering its wide uses in every home as well as on a commercial scale, its acceptability in Indian society is universal. The major and direct methods of processing are detailed below.

Table 4.1. Consumption of chickpea (kg/capita/year). (From FAOSTAT, 2006.)

Country	1988–1990	1993–1995	1998–2000	2002–2004
Developing countries				
Turkey	7.89	9.10	7.24	6.84
India	5.38	5.85	6.08	5.03
Myanmar	2.84	1.80	1.71	4.54
Jordan	3.81	3.87	3.06	4.48
Pakistan	4.46	4.39	5.34	4.38
Iran	1.90	5.13	2.76	2.96
Algeria	2.17	1.32	1.68	1.93
Ethiopia	2.00	1.54	2.30	1.84
Mexico	1.06	1.13	0.36	1.13
Syria	2.15	0.89	2.77	3.67
Bangladesh	0.68	0.86	0.48	0.64
Tanzania	0.99	0.64	0.60	0.19
Developed countries				
Spain	2.31	2.07	2.30	2.73
Italy	0.44	0.43	0.39	0.50
Canada	0.04	0.10	5.14	0.27
Australia	3.63	1.19	−0.33	1.13
World	1.27	1.37	1.46	1.27

Consumption = production + import – export.

Methods of utilization

Consumption of *besan* is most common in India, followed by consumption as dhal, and the least preferred preparation is as whole grain. In other Asian countries, dhal is the most preferred form followed by whole grain. In Europe and North America, most of the consumption of chickpea is as whole grain.

Several traditional processing practices are still used to convert chickpea into a consumable form. These processes include soaking, sprouting, fermenting, boiling, steaming, roasting, parching and frying. Some important food products based on these methods of preparation are listed in Table 4.2.

Green leaves as vegetable

In Indian villages and rural areas where chickpea is cultivated, people collect the top portion of vegetative plants for cooking. These green leaves are washed and cut into small pieces, mixed with green leaves of *Brassica*, *Chenopodium* and *Melilotus alba* and cooked for 30–40 min at high temperature. The preparation is then mixed with maize flour, salt, chillies and spices and eaten as green vegetable known as *saag* in Hindi (J. Kumar, New Delhi, 2006, personal communication). It is very tasty and healthy. Farmers also chop the top portion of a standing crop and sell it in nearby markets. Early planting of chickpea is a common practice in dry areas; when the crop is about 35 days old, they allow sheep and goats into chickpea fields and get good money out of it.

Green pods and seeds

Green immature chickpea pods harvested a week or two before they mature are consumed as snacks. Sometimes the whole plant is roasted, the pods then shelled and consumed. Some people separate pods from plants and green seeds from pods, and then roast and consume them. Green seeds are also used as vegetable and for pulav. The green seeds separated from pods have less starch and protein, and more sugars than the mature form. They are easily digested, even when raw. This is predominant mostly in the areas of chickpea cultivation. However, green harvested crop is also marketed in big cities by farmers and traders. Curries are also made of fresh green seeds, dried whole seeds and dhal, and are eaten with bread (Pushpamma and Geervani, 1987).

Uses of sprouted whole seed

In the Indian subcontinent, use of sprouted chickpea as breakfast is a very old practice due to the belief that it controls diabetic and cardiological problems. Sprouted chickpea is used in small quantities along with salt, coriander leaves, tomato, onion, garlic and lemon juice. It is slightly sweet due to hydrolysis of the starch. It

Table 4.2. Some important preparations of chickpea around the world. (From Janbunathan and Umaid Singh, 1990.)

Food	Component	Method	Country
Dhal	Decorticated dry split cotyledons	Boiled in water to a soft consistency, fried with spices and consumed with cereals	Bangladesh, India, Nepal and Pakistan
Chhole	Whole seed	Prepared and consumed similar to above	Afghanistan, Bangladesh, India, Iran and Pakistan
Pakora	*Besan* (dhal flour)	*Besan* is fried in oil and consumed as a snack	Egypt, India, Iran, Pakistan and Sudan
Kadi	*Besan*	*Besan* is boiled with butter milk and used as curry	Indian subcontinent
Unleavened bread	Whole seed/*besan*	Chickpea flour is mixed with wheat flour and roti is prepared	Ethiopia, India, Pakistan and Syria
Kiyit injera	Whole seed	Fermented	Ethiopia
Roasted	Whole seed	Grains are heated at 245–250°C for 2 min	Afghanistan, Ethiopia, India, Iraq, Iran and Nepal
Homos-Bi-tehineh	Whole seed	Soaked, boiled and mixed with other ingredients	Egypt, Jordan, Lebanon, Syria, Tunisia and Turkey
Tempeh	Decorticated split seed	Fermented product	Canada and the USA
Leblebi	Whole seed	Boiled in water with salt and pepper	Jordan, Tunisia and Turkey
Dhokla	*Besan*	Fermented with green gram flour	India
Salad	Whole seed	Boiled in water and served with other vegetables	Australia, Canada, Mexico, Spain and the USA
Green immature seeds	Whole green seed	Raw, salted or roasted	Ethiopia, India, Iran, Nepal, Pakistan and Sudan

may be boiled and used directly or dried in the shed, roasted and powdered before use, which is known as malting. It is used preferentially in infant feeding mixtures.

Generally, in this process chickpea seeds are washed and soaked in water for 5–6 h at room temperature. After washing, all the seeds are kept in a fine cotton cloth for 24–48 h at room temperature for sprouting. During this time healthy seeds will start to germinate. Germinated and sprouted seeds are washed and mixed with other vegetables and consumed as breakfast.

Uses for parching purposes

In the Indian subcontinent, the dry whole seeds are frequently used for parching purposes. In this process, large-seeded cultivars are prepared. Seeds are soaked in water for 4–6 h for imbibition and sun-dried for 1 day. After proper drying, the seeds are mixed with pure hot sand in a big pot over fire for a few minutes. The testa of the seeds cracks and cotyledons become very soft. Sometimes seeds are coloured with turmeric and salted for taste before being put on the fire. Approximately 5–10% of total chickpea production in India is utilized for this process. The parched seeds are used throughout India.

Soaked and boiled

Mostly in villages, the dried whole seeds are soaked in water for 8–10 h and then boiled for 25–30 min. Sometimes they are fried in oil with salt and lemon juice, and onion is added before eating. In this process, seeds become soft and tasty. This preparation can be performed at any time.

Soaked and puffed

Another popular method of processing chickpea is puffing. This is largely a commercial process and requires high temperature. For puffing, the seeds are sprinkled heavily with water or salt water, allowed to remain damp for about 5 min, and roasted with hot sand at 240–250°C for 1–2 min. In puffing, the husk loosens rapidly and is removed by winnowing. Puffed chickpea is light and ready to eat. Sometimes puffed chickpea is powdered and used in vegetable dishes as a garnish or made into brittles with sugar (Table 4.3) (Pushpamma and Geervani, 1987).

Vegetables

In the Indian subcontinent, chickpea is frequently used as a special vegetable with rice, fried purées (a kind of bread) and *samosas*. This usage peaks at different times with religious occasions and weddings. In this process the whole seed is soaked in water for 6–8 h and then boiled for 30–40 min.Various spices, chillies, onions and tomatoes are also added to this preparation, which has a very special flavour and an excellent taste.

Table 4.3. Form of consumption of chickpea in different countries. (From Pushpamma and Geervani, 1987.)

Legume form/process description	Types of product	Countries
Fresh immature green pods		
Boiled (5–10 min)	Salted	India, Iran, Nepal, Pakistan and Sudan
Roasted (5–10 min)	Pod walls removed	Ethiopia, India, Pakistan and Peru
Whole seed		
Soaked (4–6 h) and boiled (20–30 min)	Seasoned and served alone or with vegetable purée or served as salad with salt and lemon juice	Greece, India, Iran and Italy
Soaked (2–3 h) and steamed (10 min) 15 lb		Italy, North Africa and Spain
Soaked (4–6 h) and germinated (24 h)	Seasoned (salted) and boiled, or seasoned, dried and roasted	Nepal and India
Soaked (10–15 min) and puffed (240–250°C) (2–3 min)	Dehusked	Indian subcontinent
Soaked in salt solution and puffed (240–250°C) (2–3 min)	Consumed with husk	Indian subcontinent, West Asia and North Africa
Soaked (10–15 min) and puffed (240–250°C) (2–3 min)	Coated with hot sugar syrup and consumed as snack	Ethiopia, West Asia and North Africa
Dhal		
Boiled (25–30 min) (until very soft) or	Mashed and used as plain dhal	Ethiopia and Indian subcontinent
Steamed (10 min)	Mashed and used as purée, soup combined with jaggery and used as filling for sweets	India
Boiled with vegetable/ green leafy vegetable/ meat of soft consistency (15–20 min)	Curries	Indian subcontinent and Turkey
Steamed with vegetable/ meat (soft) (15–20 min)	Curries	Indian subcontinent, turkey, West Asia and North Africa
Boiled and mashed, seasoned with olive oil and lime	Homos-Bi-tehineh consumed with Arabic bread	West Asia
Boiled, mashed and cooked with milk	Consumed as beverage with sugar	India
Roasted (130–150°C) (5 min)	Powdered dry, and wet-ground with spices and used	India

Continued

Table 4.3. *Continued*

Legume form/process description	Types of product	Countries
Soaked (4 h)	Ground coarsely, dough flattened and deep-fried	Indian subcontinent, West Asia and North Africa
	Fried and salted	Indian subcontinent
	Cooked in sugar syrup and set into crystalline product	Indian subcontinent
Flour		
Batter (thin)	Pancakes	Nepal, Afghanistan and Indian subcontinent
	Covering/binding for cutlets	Indian subcontinent
	Used for thickening gravies	Indian subcontinent
Batter (thick)	Fermented, steamed and garnished	Indian subcontinent
Dough	Extruded into different shapes, pakoda or noodle shape, and fried	Indian subcontinent and Iran
	Fried in irregular shapes, or into circles	Indian subcontinent and Iran

Dhal

The marketing and consumers' demand for dhal has created a niche for big business in the Indian subcontinent. In this process, whole seeds are soaked in water to increase their water content to ~25–30%. After this, they are dried to lower the moisture percent to ~15–17%. Then they are dehusked and split. After splitting, the cotyledons are coated with water or oil for shine and coloration. This entire process is essential for making dhal.

Dhal has a special advantage over the whole seed in that it needs no soaking before boiling, as it has no husk, and can be cooked in a few minutes. It is cooked until tender, soft or very soft, depending on the desired texture of the finishing product. If dhal is to be served as a vegetable with green leafy vegetables and meat, it is cooked until tender. If it is required to be mashed, it is cooked until soft. Dhals are sometimes pressure-cooked to reduce the cooking time even further (Pushpamma and Geervani, 1987).

Mashed dhal is used in soups and thin sauces. Boiled dhal may be ground with jaggery (unrefined sugar) and used as filling of snacks. If a sweet beverage is desired, boiled dhal can be mashed and cooked with milk and sugar. Chickpea is the only pulse that blends well with jaggery and is the most popular pulse for use in preparing sweets and desserts (Table 4.3).

Roasted dhal is popular in the Indian subcontinent. It can be powdered or ground with water and combined with ground spices. Such a ground mixture can be made either with a single pulse or with the combination of others. Because of its strong aroma, roasted dhal is more frequently used than other legumes in preparing such mixtures (*chutneys*). These *chutneys* are used as adjuncts to boiled rice. Dhal is used to make snacks. The soaked dhal is ground coarsely and made into patties and deep-fried. Plain dhal is also soaked and deep-fried, salted and consumed. These two fried products are very popular snacks in both rural and urban communities (Pushpamma and Geervani, 1987).

Chickpea flour as *besan*

In the Indian subcontinent, chickpea flour or *besan* is very popular. It has a big market and there is more consumer choice. Traditionally, dhal is converted into flour or *besan* in flour mills. This flour is of different categories like very fine flour, medium fine and rough or coarse, and is used for different purposes. Flour products can be made quickly. Flour is used in preparations of various kinds of fast food, snacks (*pakoras*), sweets (*ladoos* and *burfis*) and driads like *bhoojya*. *Satoo* is a favourite drink used in summer. It is made by mixing roasted chickpea flour with roasted barley and wheat flour. *Dhokla*, another famous item, is also made of chickpea flour with rice, fermented overnight and steamed. It is very popular and consumed in different parts of India mostly as breakfast. Chickpea flour can be made into dough, extruded into different shapes and deep-fried. These fried products are served as snacks in canteens, restaurants and fast-food centres (Pushpamma and Geervani, 1987).

Pakoras

Chickpea flour is mixed with water and a semi-liquid paste is prepared. Various vegetables such as spinach, onion, potato, chilli, brinjal and cabbage can be mixed with this paste and fried in vegetable oil to prepare different *pakoras*. These *pakoras* are tasty and popular in the Indian subcontinent. *Pakora* preparations and eating habits are now spreading to European countries, the USA, Australia and China.

Chickpea *chapatti* or bread

Chickpea-based *chapatti* or bread is very popular in Indian villages and cities. Throughout India, Pakistan, Bangladesh, Nepal and Sri Lanka, these are made of chickpea flours mixed with wheat in various ratios as per the taste of the individual family, e.g. 1:1, 1:3 and 1:5. Sometimes pure chickpea *chapattis* are served in hotels and restaurants in the cities, mostly as a special item during lunch and dinner.

Chickpea as animal feed

In rural areas in the Indian subcontinent, the farming community keeps draft animals for agricultural work like land preparation, planting, loading and unloading, irrigation work, threshing, sugarcane crushing and land levelling. Animals are also reared for the dairy and for meat. For all these animals, green fodder, straw and grain concentrate are essential. People in rural areas are well aware that the animals are essential for their productive agriculture and family support. Thus, the developing community increasingly recognizes the many links between human health and the practice and products of agriculture. The educated progressive farmers are now pursuing opportunities for using these links to achieve more productive agriculture and better health, as they understand that healthy agriculture leads to healthy people. Therefore, the farmers provide healthy feeds to their valuable animals. Legume concentrate in general and chickpea in particular are excellent feeds for such animals.

Animal feed includes various legumes, but chickpea concentrate is most popular among villagers. There are different kinds of feed like chickpea by-products and whole-chickpea products.

Chickpea by-products

These by-products are the remaining materials when a main fine product like dhal or flour is prepared. During such preparations whole chickpea seeds are dehusked in processing plants, thus removing the seed coat. While dehusking, some small portion of seed is also broken. Dehusking the large-seeded variety is better than for small-seeded ones, as the broken material is less in percentage when compared with small-seeded varieties. These small broken pieces are mixed with the separated seed coat and used as animal feed in many villages. The cost of this mixed feed is approximately $25–30/q.

This mixed feed is rich in protein, calcium, iron, zinc and fibre. It improves animal health, productivity and vitality. In milking animals, daily milk-producing capacity increases significantly.

Whole-seed product

Whole-seed chickpea products are also very popular, and flour or split chickpea after soaking in water are used as animal feed. The percentage of such feeding is currently lower due to high prices in the market. However, in both the products ~10% of total chickpea production is used as animal feed.

Direct consumption of whole grain (kabuli type)

The whole kabuli-type seeds, which are preferred for direct human consumption, come from Mexico followed by Australia, Iran and Turkey. The most preferred quality attributes are seed size, colour, good taste and thin seed coat in that order. Prices range from $500 to $950/q depending upon seed size.

Direct consumption of whole grain (desi type)

Desi chickpea consumers prefer golden yellow, soft and light seeds, which taste good. Locally produced desi chickpea is preferred, but imports from Myanmar are also accepted as they are of similar quality. There is wide variation in prices in terminal markets, which is mainly influenced by short-term demand and supply. But in general, price ranges from $350 to $500/q.

Consumption

There is a well-known proverb that healthy agriculture produces healthy people and healthy people belong to healthy nations. The nutritional richness of legumes was recognized prehistorically, and people converted from the non-vegetarian to the vegetarian dietary system in the Indian subcontinent. Various pulses, which are excellent sources of energy to humans, animals and other living organisms, have been reported and documented in the earliest literature of Indian origin for their unique and nutritional values. Therefore, legume cultivation in general and chickpea in particular was established prehistorically in this region.

The average pulse consumption is about 27 g/person/day in rural India. The major pulse-consuming states are Uttar Pradesh (35 g/person/day), Maharashtra (32.7 g/person/day) and Karnataka (31.7 g/person/day), whereas less consumption was reported in Orissa (15 g/person/day), Kerala (15.3 g/person/day) and West Bengal (15.3 g/person/day). On the other hand, the major chickpea-consuming states are Punjab, Harayana, Rajasthan and western Uttar Pradesh (Reddy, 2004). This shows a wide diversity in the consumption of pulse crops in terms of quantity and variety among different states.

A case study of Maharashtra during 1993–1994 showed that pulse consumption was less among the poor (32 g/person/day) compared to the rich (54 g/person/day); less among schedule casts (38 g/person/day) compared to others (43 g/person/day); and also less among landless and marginal farmers

Table 4.4. Status of nutrient intake and population deficient in intake in rural Maharashtra during 1993–1994. (From Richard and Kumar, 2002.)

Social group	Consumption of pulses (g/capita/day)	Protein intake (g/capita/day)	Percent of population deficient
Income group			
Very poor	32	57	32
Moderately poor	40	60	14
Non-poor-lower	45	73	9
Non-poor-higher	54	87	3
Social group			
Scheduled tribe	43	66	21
Scheduled caste	38	68	18
Others	43	71	15
Landholding class			
Landless	41	65	59

Continued

Table 4.4. *Continued*

Social group	Consumption of pulses (g/capita/day)	Protein intake (g/capita/day)	Percent of population deficient
Submarginal	39	65	50
Marginal	41	69	48
Small	46	75	40
Medium	51	80	26
Large	55	88	19
Educational status of head of households			
Illiterate	42	67	19
Below primary	41	71	15
Above primary	43	71	15
Technical	46	77	8

Cut-off point for estimating protein deficiency is 60 g protein/day:
1. Very poor <Rs 190; moderately poor Rs 190–265; non-poor-lower Rs 265–355; non-poor-higher ≥Rs 355.
2. Landless 0 acres; submarginal 1–2.5 acres; small 2.5–5 acres; medium 5–10 acres; large >10 acres.

(40 g/person/day) than bigger landholders (55 g/person/day). Along the same lines, the uneducated consumed less pulses (41 g/person/day) than the educated (46 g/person/day). The detailed status of nutrient intake and population deficient in intake in rural Maharashtra during 1993–1994 is presented in Table 4.4 (Richard and Kumar, 2002).

In India, numerous uses of chickpea are found throughout all segments of the society. The preferences and choices of people and traders for different varieties are similar. A summary of preferred quality attributes of the most commonly used chickpea varieties in India is presented in Table 4.5.

New diversified preparations from chickpea

During the last two decades, new types of food products such as a garbanzo bean chips, bean bread and bean soups have been developed and tested for nutritional quality. Some food supplements and infant feeding mixtures have also been tested for acceptability and nutritional quality in different parts of the world. These mixtures are basically intended to supplement protein in the cereal-based low-protein diets of people in developing countries. Low-cost supplementary foods have been developed, using chickpea, at different levels and in combination with cereals and millets.

Effect of processing on the nutritional quality of chickpea-based products

Information on the evaluation of the nutrient composition of chickpea-based food is limited. Much interest has been shown in studying the effect of a single process such as boiling, roasting, germination and fermentation on nutritional quality. However, in developing a product, the food is often subjected to more

Table 4.5. Quality characteristics and common uses of some chickpea varieties in India. (From Agbola *et al.*, 2002a,b.)

Variety	Preferred quality characteristics	Common use(s)
Kabuli	Cream or white Large and uniform seed Good taste and cooks fast	Direct food use
Desi	High recovery rate (for dhal) Light brown or yellowish Large and uniform seed Thin seed coat Low moisture content High recovery rate	Split chickpea for making dhal Direct food use Flour (*besan*)
Mosambi	Light brown Uniform seed Good puffing quality Good taste and cooks faster	Puffed chickpea (phutana) Direct food use
Kantewala	Large and uniform seed Light brown Low moisture content High recovery rate	Split chickpea for making dhal Flour (*besan*)
Annigeri	Large and uniform seed Medium brown Good puffing quality High recovery rate (roasted dhal) Good taste and cooks faster	Puffed chickpea (phutana) Roasted dhal
G5	Large and uniform seed Good puffing quality Good taste and cooks faster	Puffed chickpea (phutana)
Green gram	Dark green Large and uniform seed	Direct food use
Gulabi	Light brown Thin seed coat Low moisture content	Puffed chickpea (phutana)

than one process. For example, germination may be followed by boiling, roasting and further boiling, or by steaming and so on. In addition, other ingredients such as oil, acidic vegetables or alkaline substances such as bicarbonates may be used in cooking. The interaction of all the ingredients used in the development of the food will have an effect on the sensory and nutritional quality of the product.

Many of the processes involved in food preparation have beneficial effects. They improve not only taste, aroma, digestibility and acceptance from consumers but also nutritional quality, and reduce unwanted material. For example, soaking has been shown to reduce trypsin inhibitor activity and haemagglutinating activity. Part of the inhibitors leach out during soaking, which will have beneficial effects on health in addition to the reduction in cooking time. Boiling softens the husk, due to the reaction of phytate with insoluble calcium and magnesium pectates in the cell walls, to produce soluble pectate.

During the roasting of dhal, a brown colour develops and the aroma of seeds improves, imparting a highly acceptable quality to the roasted dhal. In puffing, the seed becomes light (bulk density is reduced) due to shrinkage of the endosperm and loss of water. The starch is dextrinized. The changes that take place during puffing also impart a special texture and taste to chickpea and enable them to be used in several food preparations. Most importantly germination and fermentation will increase nutritional qualities and digestibility significantly.

Indian Food Preparations from Chickpea

India, the major consumer of chickpea in the world, has diversified uses. A list of food preparations in different Indian states has been given in Table 4.6. Most of the Indian preparations are from *besan* and dhal. Preparations from whole seeds are also common. Mostly desi types are used for preparations that include dhal and flour. Kabuli type is preferred for preparations from whole seed. For chhole and roasted preparations seed size is an important quality characteristic. For flour and dhal, the recovery percent is the important quality characteristic for which traders and consumers pay premium price.

Recently, research stations of agricultural universities developed new products from chickpea by using germinated seed for preparation of chhole and *chapatti*. The chhole preparation from the germinated chickpea is more acceptable compared with non-germinated chickpea because nutritional quality and digestibility are increased after germination. *Chapattis* are rich in protein and fat, and are easily digestible.

Table 4.6. List of food preparations in different states of India.

Name of state	Product	Highlight
Uttar Pradesh	Curries	Prepared from chickpea flour
Gujarat	*Dhokla*	Prepared from desi chickpea flour
Uttar Pradesh and Punjab	Chana sag	Prepared from green leaves of chickpea
Madhya Pradesh	*Besan* burfi	Prepared from flour of desi chickpea
Uttar Pradesh, Madya Pradesh, Punjab and Bihar	Chilla	Prepared from chickpea flour with light seasoning
Uttar Pradesh, Madhya Pradesh and Bihar	Puffed chickpea	Roasted whole chickpea grain (desi)
Uttar Pradesh, Madhya Pradesh, Bihar, Punjab and Maharastra	Chhole	Prepared from whole seed chickpea (kabuli type is preferred)
Punjab, Uttar Pradesh and Maharastra	Mixed *chapatti*	Prepared from mixture of wheat flour, and flour of chickpea and other millets

Table 4.7. Quality characteristics and common uses of some chickpea varieties in India. (From a market survey.)

Use	Preferred quality characteristics
Kabuli type (direct use)	Cream or white, large and uniform seed, good taste and cooks fast
Desi type (direct use)	Light brown or yellowish, large and uniform seed, thin seed coat, low moisture content
Flour and split dhal	Large and uniform seed, light brown, low moisture content, high recovery rate
Puffing and roasting	Large and uniform seed, medium to light brown, good puffing quality, high recovery rate (roasted dhal), good taste and cooks faster

Product Segmentation in Indian Chickpea Markets

The demand for chickpea is segmented depending largely on end use. Four main segments are identified: (i) direct food use of whole grain (Kabuli type); (ii) whole-grain market (desi type); (iii) split and flour market; and (iv) roasted and puffed chickpea market (Table 4.7).

Chickpea Consumption in Australia

Consumption of chickpea in Australia is much lower than that in the Indian subcontinent, the Mediterranean or the Middle East. This is largely cultural. Australia has no history of consuming chickpea and no traditional cuisine that incorporates this seed. With a growing multiculturalism and consequent effect on cuisine, curry, *hummus* and *felafel* are popular and are relatively easy to find. Whole chickpea is most often used in soups, curries and salads in homes, but chickpea remains uncommon in the diet of most Australians. There have been concerted efforts in the last 10 years to increase Australian domestic consumption.

Non-profit support for promotion of domestic consumption

Grains Research and Development Corporation (GRDC) is an Australian grower and federal government-supported research body that has funded initiatives to increase domestic grain and pulse consumption since the mid-1990s. The largest and longest running support group is GoGrains – run since 1998 with support from the food industry by Trish Griffiths, an accredited practising dietician with BRI Australia. GoGrains provides information to health professionals, consumers and the media.

GoGrains predominantly aims to increase awareness of the nutritional content and health benefits of grains. The primary focus of GoGrains is on wheat, the main grain produced in Australia. GoGrains also provides information about other grains and pulses.

Since 2004, GoGrains has provided a monthly e-mail-delivered news service with updates about current research and health news. GoGrains has also commissioned reviews about the effects of grains and pulses on health (e.g. Venn and Mann, 2004).

With GRDC support, both GoGrains and the Centre for Legumes in Mediterranean Agriculture (CLIMA) have produced teaching resources for primary schools to educate children about grains and pulses. Increased awareness and familiarity could encourage consumption.

In 1998, GRDC and CLIMA cosponsored the production of a cookbook on pulses, *Passion for Pulses* (Longnecker, 1999), which heavily features chickpea. A survey of Australian consumers has found that low pulse consumption is due to a perception that pulses are too hard to prepare (Yann Campbell Hoare Wheeler, 1998). Many people revealed that they did not know what to do with pulses such as chickpea. The main objective of the cookbook was to demonstrate that pulses are delicious when prepared well and easy to incorporate into a modern, busy lifestyle.

To support the launch of the cookbook, a media campaign was funded by GRDC. Dr Longnecker was the spokesperson and was supported by a professional publicist, WolfeWords. The 1999 campaign succeeded in getting significant coverage for pulses, a topic not renowned for gaining media attention.

As a follow-up for promoting domestic chickpea consumption, GRDC supported a 3-year study (2000–2003) by food scientists and medical professionals across four states (Longnecker, 2004). The health researchers examined effects of consumption of chickpea on the risk factors for heart disease. There was a lower need for insulin secretion after a meal containing chickpea than after a wheat-based control meal (Nestel *et al.*, 2004). Johnson *et al.* (2005) found that eating whole chickpea resulted in lower rapidly available glucose in blood than eating bread products with chickpea flour, implying that whole chickpea has a lower glycemic index. These results are significant for people with diabetes. Whole chickpea in a meal may help diabetics manage their disease by having lower 'spikes' of insulin after single meals (Fig. 4.2).

In 2004, release of the health research results was used as the basis of another media campaign coordinated by the same team that worked on the cookbook campaign. In conjunction with release of the research results, a web site (www.passionforpulses.com) was launched with information about health benefits of pulses and some recipes.

Food industry promotion

Many businesses with chickpea products have made significant marketing efforts with displays at field days and food shows, as well as point-of-sale demonstrations. Smaller businesses include Quickpulse, Chic Nuts® and TLC (The Lentil Company).

Fig. 4.2. Photo of chickpea. (From www.passionforpulses.com.)

Quickpulse was a small, family-owned business that provided vacuum-sealed, pre-cooked chickpea as well as prepared chickpea curry. Because of its small size, Quickpulse had difficulty securing shelf space in supermarkets, but was available in some grocery stores and health food shops. At one point, Quickpulse's prepared meal was available on Ansett, one of Australia's major airlines. However, when the carrier went bust, so did the contract. Quickpulse was bought by a larger company, Sanitarium. It remains to be seen whether Sanitarium, with its larger clout in supermarkets, will do more with the pulse lines.

Chic Nuts® provides roasted chickpea as a snack. The proprietor has aggressively marketed his product, and it is available in grocery stores in some states and in health food shops.

Up to the middle of 2000, TLC provided recipes and testing at field days to support its products that included vacuum-sealed bags of raw chickpea and other pulses. They targeted chefs and other high-end-users, as well as consumers.

Larger food companies that provide raw or tinned chickpea also have advertising campaigns that promote domestic consumption. Edgell is one of the largest suppliers of tinned chickpea in Australia. It runs regular advertising campaigns to increase domestic awareness about how to use chickpea and other pulses. Their campaigns target major women's magazines and other opportunities. McKenzies markets packaged dry pulses and provides recipes at point of sale.

Evaluation of chickpea promotion

No formal evaluation of promotion campaigns is publicly available. Anecdotally, it appears that Australians are eating more chickpea and other pulses. Chickpea seems to be a common item on restaurant menus. *Hummus* is now widely available in mainstream supermarkets.

The authors do not know of any comprehensive survey of domestic sales for the Australian pulse food industry. Such research would provide more

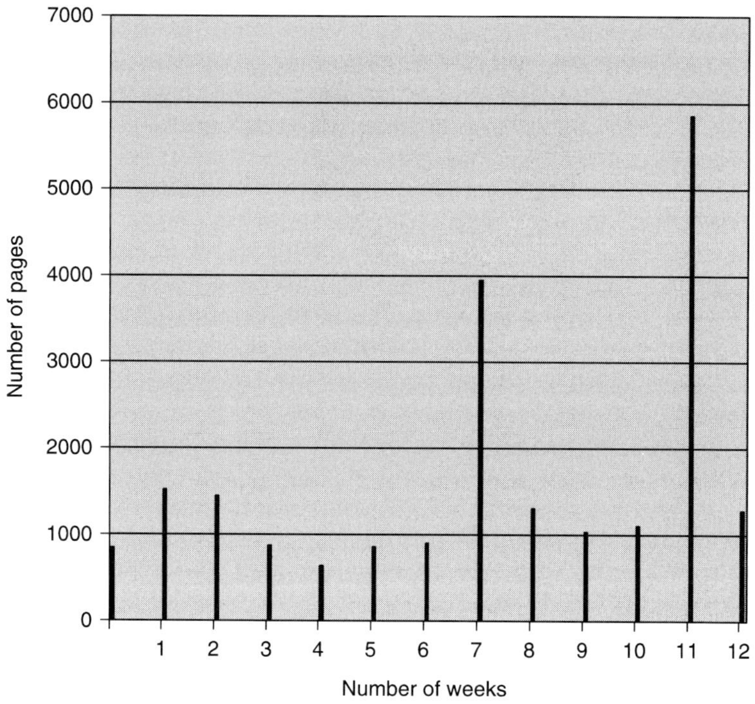

Fig. 4.3. Number of pages at www.passionforpulses.com that were visited during the 2004 media campaign that reported health benefits of chickpea consumption. Two main media reports occurred on 7th and 11th week. Week first started on 25 January, 2004 and ended on 18 April, 2004.

accurate information about trends in chickpea consumption than what is available from FAOSTATS (Table 4.1). This would allow better evaluation of effectiveness of chickpea promotion to date. Such knowledge could improve decisions to increase domestic chickpea consumption in Australia. A number of countries with traditionally low domestic consumption (e.g. Australia and Canada) might benefit from a collaborative approach to such research and promotion.

Chickpea Consumption in the USA and Mexico

The variety of chickpea under production in the USA is mainly the kabuli type, and the majority of the US production is exported mostly to European countries. According to the USA Dry Pea and Lentil Council, ~45,000 t of chickpea is produced in the USA and 56% is exported for about $15 million, 22% consumed domestically, 13% saved as seeds and ~2% used as feed or waste. In the domestic chickpea consumption, 50% is sold as whole chickpea either in dry packages or in cans, 30% as dry soup mixes, 17% in canned or jarred soups, and ~3% as unique value-added products like snacks. Likewise, almost all the chickpea produced in Mexico is the large-seeded kabuli type and the majority is for export. Domestic consumption is mainly in the form of whole chickpea either in dry packages or in cans. The most common use of chickpea in the USA and Mexico is whole chickpea consumed with salads or cooked in stews (Weidong Chen, USA, 2006, personal communication).

Peru

In Peru, chickpea is used in both green and dry forms. Dry chickpea is consumed boiled and roasted. These forms are always mixed with rice or vegetables.

Chile

The most common method of consumption of chickpea in Chile is as a whole grain using the decorticated seeds. Less frequently, the grain is ground into flour for soups or creams.

Chickpea Consumption in Canada

Domestic utilization

More than 90% of chickpea is consumed in the countries where they are produced. Chickpea is used almost exclusively for human consumption. The desi-type seed must be dehulled and is used whole or split or milled. Domestic use consists of food, feed, seed, dockage and waste. Only small volumes of low-quality chickpea are used for livestock feed; however, nutritional analysis indicates that they make an excellent feed for hogs, cattle and poultry.

Healthy diet

Pulses, including chickpea, are increasingly used in health-conscious diets to promote general well-being and reduce the risk of illness. They are low in sodium and fat, high in protein and are an excellent source of both soluble and insoluble fibre, complex carbohydrates, vitamins (especially B vitamins) and minerals (especially potassium, phosphorus, calcium, magnesium, copper, iron and zinc). Chickpea is an inexpensive, high-quality source of protein.

Since chickpea is high in fibre, low in sodium and fat, and also cholesterol-free, it is a healthy food that is beneficial to the prevention of coronary and cardiovascular disease. It may also lower blood cholesterol levels due to their high content of soluble fibre and vegetable protein.

Chickpea consumption can be beneficial in the management of type-2 diabetes because it has a low glycemic index of 55 or less, indicating that their effect on blood glucose is less than that of many other carbohydrate-containing foods. It also has other health effects, such as reducing blood lipids, which may help to reduce some serious complications of diabetes.

Flour made from chickpea is gluten-free and is a very nutritious option for people with celiac disease. Chickpea fits well in vegetarian diets as it is a good source of iron and protein, and complements the amino acid profile of cereal grains and nuts.

Insoluble dietary fibre consumption can be beneficial to a healthy colon and has been associated with reducing the risk of colon cancer. In addition, diets high in fibre have beneficial effects on weight loss because they deliver more bulk and less energy. Chickpea is an excellent source of the vitamin B folate, which is an essential nutrient. In addition, folate consumption during pregnancy has been shown to reduce the risk of neural tube defects (Agriculture and Agri-food Canada, 2004).

Chickpea Consumption in Spain

The names *Cicero*, pods chiche and chickpea derive from the latin word *Cicer* and the Spanish word *chicharo*, which seems to be a common name for dry seeds of legumes, including some *Lathyrus* species. The Spanish word *garbanzo* seems to be a pre-Roman indigenous name, since it has no connection with either Greek or Arabic. The antiquity of the crop in Spain seems evident (Mercedes, Madrid, 2006, personal communication).

The average diet of the Spaniards is a model that is followed by other countries, mainly because of the great variety of products in the regions that are consistent with the standard recommended nutrition. Today, the Spanish gastronomy and nutrition can be included in the 'Mediterranean diet', which means something more than a healthy and well-balanced alimentation. It is admitted by food experts from all over the world that natural products having different flavours play a very important role. Legumes are one of the most relevant components of this diet and their consumption has always had a very important socio-economical role in Spain.

The agricultural, economical and social developments that took place in Spain in the early 1960s led to the preference of crops other than legumes. Although these crops are still of great importance in Spain, the planting area of legumes has been considerably reduced since the 1960s. This decrease, which could be somehow justified with the Spanish agriculture moving towards other crops, does not correspond with the parallel evolution of the consumption. This fact has forced a significant increase in the volume of imports; however, the quality of these imported legumes often does not correspond to the quality of the indigenous legumes. Since the mid-1980s, legumes have ceased to be a prime objective of the Agricultural and Nutritional Policy in Spain.

In spite of this, large quantities of legumes are still being consumed in Spain. The nutritional consumption habits are very different from north to south, east to west, small rural villages to big city centres and from young people to older citizens. The statistics refer to a yearly consumption of 6 kg pulses/person, which amounts to a total of ~250 million kg, with an almost equal share of the three most common types of legumes: chickpea (37%), lentil (33%) and beans

(30%). Most (90%) are consumed at home, 6% in restaurants and 4% in institutions such as schools, hospitals, catering centres.

Regarding chickpea, the national average consumption is 2.3 kg/person/year. In Castilla-Leon and Andalucía regions, the consumption is higher than 3 kg/person/year. The metropolitan areas have an annual consumption of ~1.8 kg/person, a figure that is quite low. According to official statistics, the consumption of many legumes sold in bulk still continues in Spain (20–30% of the total), similar to the main distribution channels typical of the past.

The image of chickpea has started to change. Supermarkets and hypermarkets in towns monopolize the distribution, absorbing 50% of the whole national sales. In the rural areas, there are very high percentages of self-consumption, bulk purchasing in little markets or direct purchases from the packing firms. However, the market of the big cities offers all sorts of possibilities to meet the challenges of modern times. Furthermore, the chickpea possesses the necessary conditions to be packaged in tins or pre-cooked, in order to meet some demands of a segment of the population in modern times. In Spain, traditionally in the market of tinned products, the *cocido madrileño* (traditional Spanish dish) is known for its very high quality. Cooked chickpea is normally sold in glass bottles, which reduces cooking time as it is ready for consumption after receiving a suitable dressing. Traditional dishes containing chickpea as the main food have the necessary ingredients to guarantee a complete, tasty meal and this fact has now been confirmed by the science of nutrition.

In general, the chickpea crop is planted over an area of ~81,300 ha, which produces 70,400 t. Spain imports ~59,000 t of chickpea and also exported ~4600 t/year during 1999–2003. The total utilization within Spain is about 1,24,900 t; of this 6600 t is used as seed, 5000 t in animal nutrition and 1,13,300 t for human consumption.

Chickpea Consumption in Italy

Chickpea has also been cultivated in Italy for domestic consumption for many years. The most famous preparation of chickpea is *pasta cche sardi*. Among the major specialities are *pasta cche sardi* (pasta with sardines), *panelle* – a pancake made with chickpea flour, seafood, fish and vegetable specialities. Sweets such as *sfince, torrone* and ice creams are also included in this delicious menu. The low-quality chickpea is also used as livestock feed for cattle and poultry.

Chickpea Consumption in Turkey

Chickpea has a long history in Turkey, where it is known as national food crop. It is also used as *leblebi*, a kind of snack. Turkish people believe that *leblebi* reduces acidity in the stomach, helping to cure ulcers. Cultivars that have dark-coloured seeds are used as animal feeds.

Use as meal ingredient

In Turkey, chickpea is used as a raw food source without much industrial processing. It is one of the main ingredients of the human diet. It is mostly used as a main meal or as ingredients for other meals. Approximately 90% of the product is used as ingredients in meals. As main meal, it is cooked with meat cubes (red meat) and various vegetables like potatoes and carrots, and it usually accompanies some kind of pulav (or pilaf) of cereals such as rice pulav or bulgur pulav (cracked wheat). It can be used as ingredients in these pulav in a proportion of ~5%. Chickpea is usually soaked in water overnight before it is cooked. It is then boiled with other ingredients as whole grains. In general, kabuli types and large grains are preferred for meals.

Use as *leblebi*

Leblebi is a well-known snack. Brown-red coloured grains are preferred for this. Grains are soaked in water and the seed coat is removed. Then the grains are soaked and brewed several times before they are roasted in large ovens. The roasted grains can be covered with many different ingredients, such as soybean flavour, chocolate and various spices. Different types of *leblebi* are produced using different combinations of ingredients to give different flavours. Finally these are packed in different quantities (100–1000 g).

Green snack

Fresh green chickpea is also consumed as snacks in the early season. For this, plants are pulled out during the late grain-filling stage and sold in bunches. The pods are opened by hand and seeds are eaten green. Although this type of consumption is not well studied, it is likely to amount to ~2–3% of production.

Other uses

Chickpea has other minor uses include grinding of grains and use of flour in pastries (Kusmenoglu and Meyveci, 1996), making of bread in some parts of rural areas, use of small grains in feeds and use of plant debris in hay mixtures. As in many African countries chickpea is also used as green manure in southern regions to improve soil fertility (Özdemir, 2002).

Chickpea Consumption in Africa

Chickpea is traditionally grown in East and North African countries. It is consumed in different forms. In Ethiopia the green seed is utilized as vegetable or cooked and salted to be served as snacks. It is consumed mainly in the form of sauces made of *kik* (dehulled split and spiced) and *shiro* (slightly roasted, powdered and spiced). It is also consumed as *kollo* (soaked and roasted), *nifro* (boiled), *shimbra* (unleavened bread of chickpea with spice) and *dabo* or *kitta* (leavened bread). *Shimbra asa* (chickpea fish) is mainly used by the followers of the Ethiopian Coptic church during the two fasting months. It is used to simulate fish where it is not available as a source of protein during the fasting period when animal products are not consumed. *Butucha* (a thick paste) is most popular in the northern parts of Ethiopia, particularly in Gonder. Chickpea is also used to make *mitad shiro* (thick, relatively drier paste made on a clay pan, and is also used to make *injera*). Approximately 10% of chickpea is mixed with soybean and wheat to make *faffa*, which is an infant and child food. It is consumed as *hommos*, chickpea *yani* and chickpea cookies in North African countries. It is coated with sugar and sold as sweets in Egypt. Chickpea consumption is generally high during Ramadan period in countries like Sudan and Egypt. In some countries it is used as a substitute for coffee (G. Bejiga, Addis Ababa, 2006, personal communication).

Sudan and Egypt

In Sudan and Egypt, chickpea (kabuli type) is boiled in water with salt and sesame oil to produce *Balilah*, a popular energy-giving food, eaten especially during the fasting period of Ramadan. It is mixed with onions, chillies, garlic and baking powder, which are ground together to form a dough. The dough is divided into small round balls and fried in vegetable oil to make *Tammia*, which is consumed for breakfast or supper. Sometimes immature pods are boiled for use as green vegetable and eaten whole.

Tunisia

Chickpea (*homos*) is mainly consumed as *leblebi*, which is prepared by boiling water with salt and pepper and is a famous luncheon dish. It is also used to make many other dishes in Tunisia. Recently, a considerable amount of chickpea has been substituted for coffee in this country.

Chickpea Consumption in Arabic Countries

In the Middle East, consumption is based on a popular dish known as *hummus*, which is produced from mashed chickpea mixed with oil and spices. The kabuli types are used mainly in salads and vegetable mixes. They are also used in preparing a wide variety of snacks, soups, sweets and as condiments. Smaller-sized kabuli types are also milled for flour. Kabuli types are substituted for desi types if the price is competitive.

Israel

In Israel, like most of the Middle East, the vast majority of chickpea consumption is of the kabuli type. It is used for *falafel, hummus* as cooked grain in dishes with rice and meat, as roasted snack (*edamame*, in Arabic) like sunflower and watermelon seeds, and also as roasted green pods (*hamle malane* in Arabic). In general, the chickpea consumption pattern in Israel is similar to that in the neighbouring countries (Syria, Lebanon and Jordan) (S. Abbo, Rehorot, 2006, personal communication).

Iraq, Syria, Lebanon and Jordan

In Iraq, chickpea is boiled or parched. It is also eaten raw, roasted or cooked, or in the form of soup. *Tashrib* is a delicious variety of Baghdad soup or stew including chickpea with pieces of meat and unleavened bread.

More than 75% of chickpea is consumed in the form of *hummus bi tahineh, falafel* and *tisquieh* in Syria. The other popular preparations are *hummus moudames, hummus mashwi* (roasted chickpea), *mlabes al kedameh* (sugar-coated roasted chickpea) and *kaak* dough (hard type of cake). The seed may be eaten green as *malianeh* and also as *baleelah* (soaked, boiled and salted seed). Chickpea is also made into a hot drink, similar to coffee.

Iran

In Iran, the predominant variety of chickpea is a fairly large-seeded kabuli type (25–35 g/100 seeds), which is used almost exclusively for cooking as a vegetable and with rice. A smaller seed type is primarily used for roasting and as a snack. The seed has a reddish brown or black seed coat, which is normally removed. The yellow green splits are then used as a vegetable. The black seed varieties are often used.

Afghanistan

Chickpea is grown in the Takhar, Kunduz, Herat, Badakhshan, Mazar-Sharif, Smangan, Ghazni and Zabol provinces of Afghanistan. Both kabuli and desi types are grown mostly under rainfed conditions in alternate cropping with wheat. Average crop duration is 90–100 days depending upon the variety used. Farmers mostly use the local varieties; however, some improved varieties have also been introduced. Average yield is 1.4–1.8 million tonnes/ha.

Chickpea is mainly used as vegetable (*chhole*), mixed with some meat preparations. Roasted chickpea with or without salt is popular and eaten along with dry fruits like resins and almonds. Chickpea flour (*besan*) is used for making *pakoras*, sweets and mixed with minced meat while preparing meat balls.

Generally, chickpea is consumed as a whole grain. The kabuli types are preferred to the desi types. Commonly, chickpea is also roasted, boiled, roasted with salt or roasted with sugar, and eaten in rural and urban areas (Javed Rizvi, Kabul, 2006, personal communication).

Chickpea Consumption in China

Distribution and production

Chickpea is cultivated in Xinjiang, Gansu, Qinghai, Inner Mongolia and Yunnan on an economic scale, and in Sha'anxi, Shanxi, Ningxia, Hebei and Heilongjiang provinces sporadically. Xinjiang and Gansu provinces are the most stable and main chickpea-producing areas in China. In Xinjiang, chickpea is cultivated in Uigur ethnic area, including the Mulei, Qitai, Wushi and Baicheng counties around the Tianshan mountain range (Fig. 4.4). Wild species of chickpea are scattered within Qinghai county near the Altai mountain in Xinjiang province. According to ancient documents, chickpea has been used for Uigur ethnic medicine 2500 years ago.

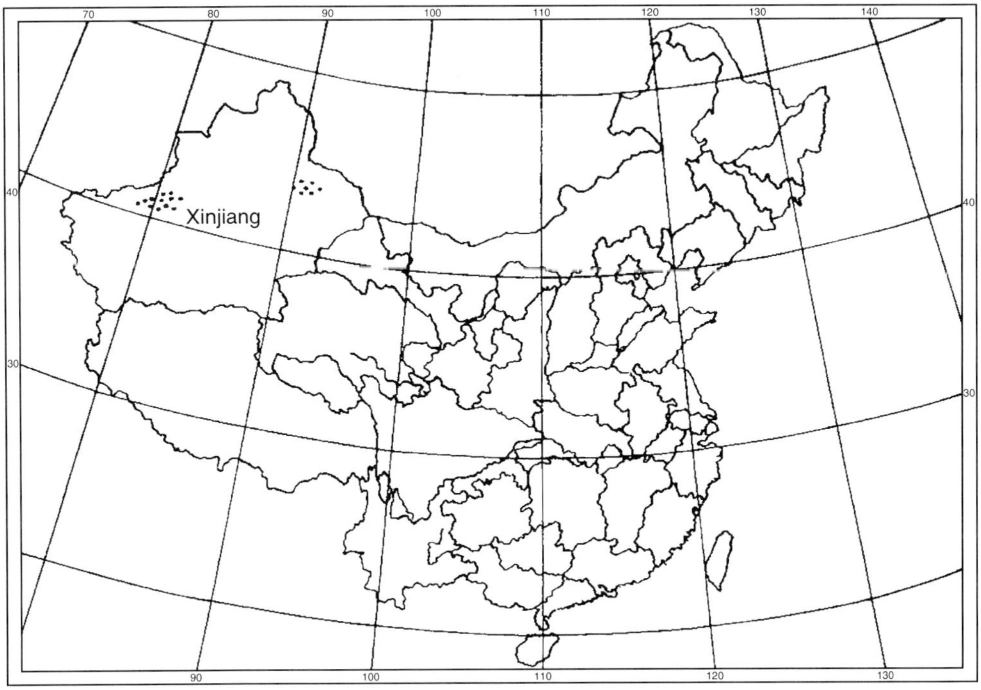

Fig. 4.4 . Chickpea distribution in Xinjiang.

Processing and usage

1. *Snacks* – Fried chickpea with salt, fried and coated chickpea, baked chickpea, boiled chickpea, fast chickpea porridge, chickpea cake;
2. *Meal* – Wheat flour mixed with chickpea flour for noodle and bread, chickpea flour mixed with edible oil and spices for various kinds of bread and salad souse, chickpea paste, congee with various grains, fresh chickpea items, 50% rice and 50% chickpea boiled together for main dish;
3. *Other products* – Milk powder mixed with chickpea flour for chickpea milk, canned chickpea, frozen green chickpea for vegetable.

Introduction and adaptation test of chickpea lines from International Crops Research Institute for the Semi-Arid Tropics (ICRISAT), Hyderabad, India, and International Centre for Agricultural Research in the Dry Areas (ICARDA), Aleppo, Syria, are the major activities for chickpea research in China till now (Zong Xaxiao, 2006, personal communication).

Chickpea Consumption in Myanmar

Tofu, sometimes also called *doufu* (often in Chinese recipes) or *bean curd* (literal translation), is a food of Chinese origin, made by coagulating soy milk, and then pressing the resulting curds into blocks. The making of tofu from soy milk is similar to the technique of making cheese from milk. Wheat gluten, or seitan, in its steamed and fried forms, is often mistakenly called *tofu* in Asian or vegetarian dishes.

Tofu made from chickpea

Burmese tofu (*to hpu* in Burmese) is a type of tofu made from chickpea (*chana dhal*) flour instead of soybeans; the Shan variety uses yellow split peas instead. Both types are yellow in colour and generally found only in Myanmar, though the Burman variety is also available in some overseas restaurants serving Burmese cuisine. Burmese tofu has very little flavour or smell of its own. Tofu can be prepared either as savouries or sweet dishes, which act as a canvas for presenting the flavours of the other ingredients used.

Chickpea Consumption in Nepal

Chickpea is principally consumed as a pulse in *Terai*, Inner *Terai* and other places in the hills. Leaves are frequently used as vegetables. Seeds are eaten raw, boiled as vegetables, spiced or cooked. Flour is commonly used for *satoo* (flour of roasted chickpea and barley mixed with salt or sugar in water). It is especially recommended for patients suffering from acidity or other gastric problems. Flour is also used for making sweets and dhal for making snacks.

Summary

In general, per capita consumption of chickpea is higher in Asia and the Middle East, and low or negligible in developed countries. There is a case for increasing chickpea consumption in developing Asian and African countries due to the widespread protein malnutrition in these regions. There is also a case for increasing chickpea consumption in countries with plentiful protein supply because of the apparent health benefits related to the consumption of chickpea and other pulses.

Numerous traditional and popular dishes are prepared from chickpea, including dhal in the Indian subcontinent, Turkish *leblabi* and *hummus* in the Middle East. This legume is consumed in various forms, like fresh immature green seeds, whole dry seed, dhal and flour. The preparation is mostly made by boiling, roasting, germination and fermentation. Different forms of preparation often result in improved quality and taste. Soaking is known to reduce trypsin inhibitor activity and haemagglutinating activity. Boiling dhal softens the husk, and germination reduces cooking time. When dhal is roasted, the aroma of seeds improves. In puffing, the starch in seed is dextrinized. Fermentation makes more nutrients available by increasing the digestibility. Because of these advantages chickpea is used in the preparation of a large number of dishes.

Per capita consumption is highest in Turkey at ~7–9 kg/annum. In Turkey, chickpea is used as whole grain in pulav, roasted as the snack *leblebi* and, to a lesser extent, fresh green seeds are eaten as snacks.

Total consumption is highest in India, which accounts for ~5–6 kg/capita/annum. Consumption in the form of dhal and *besan* is most common, followed by whole grain. Consumption demand of chickpea is segmented depending on end use, i.e. whole grain (kabuli and desi types), dhal and flour, and puffed and roasted products. In the case of kabuli type, consumers are willing to pay premium for seed size, whereas in the case of desi type, they are willing to pay premiums for seed shape, colour and size. Considering the above preferences, there is large scope for increasing quality traits like seed size and decreasing the antinutritional factors of desi type through appropriate breeding methods.

References

Agbola, F.W., Kelley, T.G., Bent, M.J. and P-Rao, P. (2002a) Chickpea marketing in India: challenges and opportunities. *Agribusiness Perspectives*, 6 June. Available at: http://www.agrifood.info/perspectives/2002/Agbola.html

Agbola, F.W., Kelley, T.G., Bent, M.J. and P-Rao, P. (2002b) Eliciting and evaluating market preferences with traditional food crops: the case of chickpea in India. *International Food and Agribusiness Management review* 5, 5–21.

Agriculture and Agri-food Canada (2004) Chickpeas: situation and outlook. 2004-09-14 | Volume 17 Number 15 | ISSN 1494-1805 | AAFC No. 2081/E.

FAOSTAT Database (2006) Food and Agriculture Organization, Rome. Available at: www.fao.org.

Hernándo Bermejo, J.E. and León, J. (1994) Neglected crops: 1492 from a different perspective. *Plant Production and Protection,* Series No. 26. FAO, Rome, Italy, pp. 289–301.

Jambunathan and Umaid Singh (1990) Present status and prospects for utilisation of chickpea in ICRISAT. In: *Chickpea in the Nineties: Proceedings of the Second International Workshop on Chickpea Improvement.* 4–8 December 1989. ICRISAT, Hyderabad, India, pp. 41–46.

Johnson, S.K, Thomas, S.J. and Hall, R.S. (2005) Palatability and glucose, insulin and satiety responses of chickpea flour and extruded chickpea flour bread eaten as part of a breakfast. *European Journal of Clinical Nutrition* 59, 169–176.

Kusmenoglu, I. and Meyveci, K. (1996) Chickpea in Turkey. In: Saxena, N.P., Saxena, M.C., Johansen, C., Virmani, S.M. and Haris, H. (eds) *Adaptation of Chickpea in the West Asia and North Africa Region.* ICRISAT, Hyderabad, India; ICARDA, Aleppo, Syria.

Longnecker, N. (1999) *Passion for Pulses.* UWA Press, Perth, Western Australia.

Longnecker, N. (2004) Determination and promotion of health benefits of pulses with special emphasis on chickpeas. Final report to Grains Research and Development Committee Canberra, Australia.

Nestel, P., Cehun, M. and Chronopoulos, A. (2004) Long-term consumption and single meal effects of chickpeas on plasma glucose, insulin and triacylglycerol concentrations. *American Journal of Clinical Nutrition* 79, 390–395.

Özdemir, S. (2002) *Grain Legumes.* Hasad Publications, Turkey, p. 142.

Pushpamma, P. and Geervani, P. (1987) Utilization of chickpea. In: Saxena, M.C. and Singh, K.B. (eds) *The Chickpea.* CAB International, Wallingford, UK, pp. 357–368.

Reddy, A.A. (2004) Consumption pattern, trade and production potential of pulses. *Economic and Political Weekly* 39(44), 4854–4860.

Richard, O.M. and Kumar, P. (2002) Dietary pattern and nutritional status of rural house hold in Maharashtra. *Agricultural Economics Research Review* 15(2), 111–122.

Venn, B.J. and Mann, J.I. (2004) Cereals, grains and diabetes. *European Journal of Clinical Nutrition* 58(11), 1443–1461.

Yann Campbell Hoare Wheeler (1998) GoGrains Consumer Research. BRI, North Ryde, Australia.

5 Nutritional Value of Chickpea

J.A. Wood[1] and M.A. Grusak[2]

[1]Tamworth Agricultural Institute, NSW Department of Primary Industries, Tamworth, NSW 2340, Australia; [2]USDA-ARS Children's Nutrition Research Center, Department of Pediatrics, Baylor College of Medicine, Houston, TX 77030–2600, USA

Introduction

Adequate nutrition via food is a necessity of human life. Food provides energy and essential macro- and micronutrients required for growth, tissue maintenance and the regulation of metabolism and normal physiological functions. Besides these essential nutrients, foods of plant origin supply various non-nutritive phytochemicals that promote good health and reduce the incidence of many chronic diseases. The World Health Organization (WHO) recognizes the importance of plant foods in the diet, recommending >400 g/day consumption of fruits and vegetables, not including tubers (WHO/FAO, 2003).

Chickpea (*Cicer arietinum* L.) and other pulse crops are staple foods in many countries and play an enhanced role in the diets of vegetarians around the world. Pulses are a primary source of nourishment and, when combined with cereals, provide a nutritionally balanced amino acid composition with a ratio nearing the ideal for humans. Frequent consumption of pulses is now recommended by most health organizations (Leterme, 2002). Chickpea is a good source of energy, protein, minerals, vitamins, fibre, and also contains potentially health-beneficial phytochemicals.

The nutritional value of chickpea has been documented in numerous publications; however, there are few reviews that compare the nutrition of desi (coloured seed coat) and kabuli (white seed coat) types, their dhal or flour, and the use of chickpea as a green vegetable. In addition to these topics, this chapter will also cover health benefits and the effect of common processing techniques on the nutritional value of chickpea.

Nutritional Composition

The nutritional quality of seeds can vary depending on the environment, climate, soil nutrition, soil biology, agronomic practices and stress factors (biotic and abiotic). The literature covering the proximate and chemical composition of desi and kabuli seeds is summarized in Tables 5.1–5.4. In general, the cotyledon and embryo make up most of the nutritionally beneficial part of the seed whilst the seed coat contains many of the antinutritional factors (ANFs). Desi types have a thicker seed coat than the kabuli types, reflected most obviously in the greater fibre content of desi seeds (Knights and Mailer, 1989).

Energy

Energy is often expressed as gross energy (MJ/kg) or as a caloric value (Kcal/100g) and refers to the amount of energy contained in a food. Energy values for chickpea have been reported at 14–18MJ/kg (334–437Kcal/100g) for desi types and 15–19MJ/kg (357–446Kcal/100g) for kabuli types. The kabuli types generally have slightly higher energy values than desi types grown under identical conditions due to a smaller seed coat component. The WHO recommends a high consumption of energy-dilute foods rich in non-starch polysaccharides (NSPs) such as chickpea, other vegetables and fruits (WHO/FAO, 2003).

Protein and amino acid

Most legumes have high nitrogen contents, due to their ability to fix atmospheric nitrogen through a symbiotic association with soil microbes. The protein concentration of chickpea seed ranges from 16.7% to 30.6% and 12.6% to 29.0% for desi and kabuli types, respectively, and is commonly 2–3 times higher than cereal grains. Chickpea has been specifically used to treat protein malnutrition and kwashiorkor in children (Krishna Murti, 1975).

Protein in the diet is essential in providing the body with amino acids to build new proteins for tissue repair and replacement, and to synthesize enzymes, antibodies and hormones. The amino acid composition of chickpea is well balanced, apart from the limited sulphur amino acids (methionine and cysteine), and is high in lysine. Hence, chickpea is an ideal companion to cereals, which are known to be higher in sulphur amino acids but limited in lysine.

Lipid and fatty acid

Chickpea exhibits higher lipid content than other pulses, with wide genotypic variation. The total lipid concentration of desi and kabuli types ranges from 2.9% to 7.4% and 3.4% to 8.8%, respectively (Table 5.3), which although high for pulses is low for some grain legumes (e.g. soybean and groundnut). The total lipid content of chickpea mainly comprises polyunsaturated (62–67%), mono-unsaturated (19–26%) and saturated (12–14%) fatty acids (Table 5.3). Hence, the small amount of lipid in chickpea is mostly of the beneficial kind (mono-unsaturated and polyunsaturated) rather than saturated fats that have been linked to heart and circulatory diseases.

Table 5.1. General composition and carbohydrate composition of desi and kabuli chickpea seeds.

		Desi whole seed				Kabuli whole seed			
Parameter	Unit	Minimum	Maximum	Number of cultivars	References	Minimum	Maximum	Number of cultivars	References
General									
Seed coat	%	10.1	21.89	111	1–9	4.5	9.5	31	1, 3, 6–10
Gross energy	MJ/kg	16.2	18.3	6	11–13	14.958	18.7	3	11–13
Calorific value	Cal/100 g	334	387	15	2, 11, 14	357.5	391	26	11, 15, 16
Protein	%	16.7	30.57	1898	1–6, 8, 11–14, 17–30	12.6	29	1749	1, 3, 6, 8, 10–13, 15, 16, 19–23, 26, 27, 30–34
Ash	%	2.04	4.2	180	1, 2, 5, 11–14, 19, 21, 22, 24–30	2	3.9	156	1, 10–13, 15, 16, 19, 21, 22, 26, 27, 30–34
Fat	%	2.9	7.42	1954	1, 2, 5, 11–14, 19, 21, 22, 24–30	3.4	8.83	167	1, 11–13, 15, 16, 19, 21, 22, 26, 27, 30–34
Crude fibre	%	3.7	13	107	1, 2, 5, 7, 12–14, 19, 20, 22, 25–27, 29	1.17	4.95	92	1, 7, 12, 13, 15, 16, 19, 20, 22, 26, 31–34
CHO, total	%	50.64	64.9	48	2, 5, 14, 22, 23, 25, 26, 29	54.27	70.9	38	11, 15, 22, 26, 31
Carbohydrates									
NSP, total	%	–	12.2	2	35	–	8.78	1	36
Total soluble sugars	%	5.33	11.8	17	3, 6, 27	6.65	7.5	2	31
Total reducing sugars	%	2.61	4.77	10	17	2.25	2.42	2	31
Total non-reducing sugars	%	1.12	1.89	10	17	4.4	5.08	2	31
Sugars, total	%	2.2	10.7	26	1, 20, 37	5.5	10.85	21	1, 11, 20

Continued

Table 5.1. *Continued*

		Desi whole seed				Kabuli whole seed			
Parameter	Unit	Minimum	Maximum	Number of cultivars	References	Minimum	Maximum	Number of cultivars	References
Starch	%	32	56.3	44	1, 3, 6, 17, 19, 20, 38	30	57.2	18	1, 3, 19, 20, 27, 36, 38, 39
Amylose	% of starch	20	42	7	3, 6, 33, 38, 40–42	20.7	46.5	6	3, 38, 42–46
Resistant starch	%	–	3.39	1	25	0.31	16.43	2	36, 47, 48
ADF	%	9.4	16.7	37	3, 6, 7, 13, 20, 27, 30	3.24	12.2	33	3, 6, 7, 13, 20, 27, 30
NDF	%	15.6	30.2	20	7, 13, 20, 27	5.16	16	21	7, 13, 20, 27
Dietary fibre	%	18.4	22.7	10	7, 11, 30	10.6	16.63	11	7, 11, 30
Soluble dietary fibre	%	–	0.0	1	25	–	–	–	–
Insoluble dietary fibre	%	–	13.9	1	25	–	–	–	–
Hemicellulose	%	3.5	8.8	8	7, 27	4	7.3	8	7, 27
Pectic substances	%	1.5	3.8	7	7	2.4	4.1	8	7, 27
Lignin	%	1.3	17.0	14	7, 13, 20, 27	0.01	2.1	19	7, 13, 20, 27

1. Jambunathan and Singh (1980); 2. Agrawal and Singh (2003); 3. Saini and Knights (1984); 4. Awasthi *et al.* (1991); 5. Suryawanshi *et al.* (1998); 6. Saini (1989); 7. Singh (1984); 8. Singh and Jambunathan (1981); 9. Knights and Mailer (1989); 10. Lopez Bellido and Fuentes (1990); 11. Nutrient Data Laboratory (USDA Agricultural Research Service) (2005); 12. Batterham (1989); 13. Petterson *et al.* (1997); 14. Agrawal and Bhattacharya (1980); 15. el Hardallou *et al.* (1980); 16. FAO/USDA (1987); 17. Mittal *et al.* (1999); 18. Vijayalakshmi *et al.* (2001); 19. Viveros *et al.* (2001); 20. Ramalho Ribeiro and Portugal Melo (1990); 21. Sharma and Goswami (1971); 22. Shobhana *et al.* (1976); 23. Amirshahi and Tavakoli (1970); 24. Khanvilkar and Desai (1981); 25. de Almeida Costa *et al.* (2006); 26. Rossi *et al.* (1984); 27. Salgado *et al.* (2001); 28. Iqbal *et al.* (2006); 29. Patel and Rajput (2003); 30. Avancini *et al.* (1992); 31. Zaki *et al.* (1996); 32. Sotelo *et al.* (1987); 33. Madhusudhan and Tharanathan (1995); 34. Dodok *et al.* (1993); 35. Perez-Maldonado *et al.* (1999); 36. Bravo (1999); 37. Labavitch *et al.* (1976); 38. Jarvis and Iyer (2000); 39. Dalgetty and Baik (2003); 40. Madhusudhan and Tharanathan (1996); 41. Lineback and Ke (1975); 42. Hoover and Ratnayake (2002); 43. Biliaderis and Tonogai (1991); 44. Biliaderis *et al.* (1981); 45. Biliaderis *et al.* (1980); 46. Yanezfarias *et al.* (1997); 47. Garcia-Alonso *et al.* (1998); 48. Saura-Calixto *et al.* (1993).

Table 5.2. Amino acid and mineral composition of desi and kabuli chickpea seeds.

		Desi whole seed				Kabuli whole seed			
Parameter	Unit	Minimum	Maximum	Number of cultivars	References	Minimum	Maximum	Number of cultivars	References
Amino acids									
Alanine	g/16 g N	2.64	6.06	85	1–6	2.81	5.66	44	1–4, 6–10
Arginine	g/16 g N	5.1	12.79	117	1–6, 11, 12	5.14	13.2	44	1, 2, 4, 6–10, 12
Aspartic acid	g/16 g N	8.36	14.19	83	1–6	8.49	14.26	44	1–4, 6–10
Cystine	g/16 g N	0.5	1.93	121	1–5, 11, 13	0.53	1.77	39	1, 2, 4, 8
Glutamic acid	g/16 g N	12.94	21.09	84	1–6	13.14	18.9	44	1–4, 6–10
Glycine	g/16 g N	2.44	5.1	85	1–6	2.49	4.43	42	1, 2, 4, 6, 7, 9, 10
Histadine	g/16 g N	1.48	5.71	115	1–6, 11–13	1.4	4.2	519	1, 2, 4, 6–10, 12, 13
Isoleucine	g/16 g N	2.15	5.82	114	1–6, 11, 13	3.2	6.03	47	1, 2, 4, 6–10, 13
Leucine	g/16 g N	3.12	9.95	114	1–6, 11, 13	5.73	9.09	51	1, 2, 4, 6–10, 13
Lysine	g/16 g N	4.86	9.62	120	1–6, 11–13	4.85	8.71	53	1, 2, 4, 6–10, 12, 13
Methionine	g/16 g N	0.64	2.29	175	1–6, 11–14	0.61	3.63	54	1, 2, 4, 6–10, 12–14
Phenylalanine	g/16 g N	2.76	6.83	114	1–6, 11, 13	4.06	8.41	51	1, 2, 4, 6–10, 13
Proline	g/16 g N	2.82	4.98	83	1–3, 5, 6	2.88	5.28	41	1, 2, 6, 7, 9, 10
Serine	g/16 g N	3.26	6.08	86	1–6	3.13	5.89	43	1, 2, 4, 6, 7, 9, 10
Threonine	g/16 g N	1.18	5.58	119	1–6, 11, 13, 14	2.59	4.93	48	1, 2, 4, 6–10, 13, 14
Tryptophan	g/16 g N	0.39	1.30	80	1, 4–6, 11, 13, 14	0.31	2.36	15	1, 6, 8–10, 13, 14
Tyrosine	g/16 g N	0.71	3.81	102	1–6, 11, 13	1.1	4.21	46	1, 2, 4, 6–10, 13
Valine	g/16 g N	2.03	5.55	114	1–6, 11, 13	3.11	6.22	47	1, 2, 4, 6–10, 13
Minerals									
Calcium	mg/100 g	68	269	185	1, 3, 5, 6, 11, 14–19	40	267	73	1, 3, 6–10, 12, 14, 16, 17, 20
Magnesium	mg/100 g	107.4	168	95	1, 3, 5, 6, 16, 17, 19	10	239	29	1, 3, 6, 9, 10, 16, 17
Phosphorus	mg/100 g	169	860	180	1, 3, 5, 6, 11, 12, 14–17, 19	159	930	78	1, 3, 6–10, 12, 14, 16, 17, 20

Continued

Table 5.2. *Continued*

		Desi whole seed				Kabuli whole seed			
Parameter	Unit	Minimum	Maximum	Number of cultivars	References	Minimum	Maximum	Number of cultivars	References
Potassium	mg/100 g	230	1272	142	1, 3, 5, 6, 16	220	1333	43	1, 3, 6, 8–10, 16
Sodium	mg/100 g	1.0	101	74	1, 3, 5, 6, 18	2.06	64	28	1, 3, 6, 8–10, 18
Sulphur	mg/100 g	160	220	84	1	160	200	23	1
Copper	mg/100 g	0.31	11.6	114	1, 3, 5, 6, 18	0.27	1.42	27	1, 3, 6, 9, 10, 16, 17
Iron	mg/100 g	3	12	113	1, 3, 5, 6, 18, 19	3.2	14.3	62	1, 3, 6–10, 12, 14, 16, 17, 20
Manganese	mg/100 g	1.78	5.16	80	1, 3, 5, 6	0.09	9.4	31	1, 3, 6, 9, 10, 16, 17
Molybdenum	mg/100 g	0.03	0.13	20	1	0.07	0.13	9	1
Zinc	mg/100 g	2.2	20	101	1, 3, 5, 6, 18	2	5.4	31	1, 3, 6, 9, 10, 16, 17
Cobalt	μg/100 g	11.0	35.0	8	1	6.0	41.0	5	1
Selenium	μg/100 g	5.0	100.0	15	1, 3	0.5	10.0	8	1, 3

1. Petterson *et al.* (1997); 2. Rossi *et al.* (1984); 3. Nutrient Data Laboratory (USDA Agricultural Research Service) (2005); 4. Salgado *et al.* (2001); 5. Iqbal *et al.* (2006); 6. Avancini *et al.* (1992); 7. el-Hardallou *et al.* (1980); 8. FAO/USDA (1987); 9. Madhusudhan and Tharanathan (1995); 10. Dodok *et al.* (1993); 11. Khanvilkar and Desai (1981); 12. Sharma and Goswami (1971); 13. Batterham (1989); 14. Shobhana *et al.* (1976); 15. Agrawal and Bhattacharya (1980); 16. Jambunathan and Singh (1981); 17. Viveros *et al.* (2001); 18. Awasthi *et al.* (1991); 19. Patel and Rajput (2003); 20. Zaki *et al.* (1996).

Table 5.3. Fatty acid composition and vitamins of desi and kabuli chickpea seeds.

		Desi whole seed				Kabuli whole seed			
Parameter	Unit	Minimum	Maximum	Number of cultivars	References	Minimum	Maximum	Number of cultivars	References
Fatty acids									
Myristic (14:0)	% in oil (of total FAs)	0.2	0.2	2	1, 2	0.01	0.2	11	1–5
Palmitic (16:0)	% in oil	9.1	11.5	6	1, 2	3.3	10.8	12	1–5
Palmitoleic (16:1)	% in oil	–	0.3	1	2	0.002	0.2	4	1, 2, 5
Stearic (18:0)	% in oil	0.3	1.8	6	1, 2	0.05	5.3	12	1–5
Vaccenic (18:1, 7)	% in oil	1.2	1.4	4	1	0.7	1.3	9	1
Elaidic (18:1, 9t)	% in oil	1	1	1	1	0.7	1.0	2	6
Oleic (18:1, 9c)	% in oil	17.6	23.3	5	1	17.37	32	11	5, 6
Linoleic (18:2)	% in oil	45.95	61.5	13	1, 2	16.4	70.4	14	1–5
Linolenic (18:3)	% in oil	2.16	4.8	6	1, 2	0.3	3.3	12	1–5
Total 18:1		0.6	0.8	5	1	0.7	0.9	6	6
Arachidic (20:0)	% in oil	0.5	0.6	5	1	0.5	0.7	4	6
Eicosenoic (20:1)	% in oil	0.4	0.8	5	1	0.5	0.6	3	6
Behenic (22:0)	% in oil	0.1	0.1	1	1	0.3	0.5	4	6
Lignoceric (24:0)	% in oil	–	13.4	1	2	12	14	2	1–3
Total saturated	% in oil	–	28.8	1	2	5	29	2	1–3
Total mono-unsaturated	% in oil	–	29.0	1	2	19	29.0	2	1–3
Total polyunsaturated	% in oil	–	57.6	1	2	62	67	2	1–3
Vitamins									
Vit A (β-carotene equivalent)	μg/100 g	–	40	1	2	9.6	49	9	2, 7, 8
Vit C (ascorbic acid)	mg/100 g	–	4.0	3	2	–	–	–	
Vit E (tocopherols and tocotrienols)	mg/100 g	–	0.82	1	2	0.82	13.7	8	2, 8
Thiamine (B_1)	mg/100 g	–	0.48	42	2	0.41	0.58	11	2, 3, 5
Riboflavin (B_2)	mg/100 g	–	0.21	48	2	0.106	1.58	7	2, 3, 5
Niacin (B_3)	mg/100 g	–	1.54	43	2	1.5	1.76	5	2, 3
Pantothenic acid (B_5)	mg/100 g	–	1.59	20	2	–	0.83	1	2
Vit B_6	mg/100 g	–	0.54	26	2	–	0.49	1	2
Folate (B_9)	μg/100 g	150	557	16	2, 9, 10	347	437	1	2, 11
Vit K	μg/100 g	–	9.0	1	2	–	9.1	1	2

1. Petterson *et al.* (1997); 2. Nutrient Data Laboratory (USDA Agricultural Research Service) (2005); 3. FAO/USDA (1987); 4. Madhusudhan and Tharanathan (1995); 5. Dodok *et al.* (1993); 6. Biliaderis *et al.* (1980); 7. Abbo *et al.* (2005); 8. Atienza *et al.* (1998); 9. Dang *et al.* (2000); 10. Babu and Srikantia (1976); 11. Lin *et al.* (1975).

Table 5.4. Antinutritional factors, phytoestrogens and phytosterol content of desi and kabuli chickpea seeds.

		Desi whole seed				Kabuli whole seed			
Parameter	Unit	Minimum	Maximum	Number of cultivars	References	Minimum	Maximum	Number of cultivars	References
Antinutritional factors									
Oligosaccharides	%	1.2	3.9	82	1, 2	0.4	2.8	35	2, 3
Phytate	%	0.37	1.13	63	2	0.25	0.99	14	2, 3
Tannins (total)	%	0.36	0.72	53	2, 4	0.12	0.51	39	2, 3, 5
Tannins (condensed)	%	0.01	0.09	47	2, 4	0	0.04	37	2, 4
Trypsin inhibitor (TIA)	mg/g	1.16	15.7	129	1, 2, 4, 6, 7	1.39	12.1	62	2, 4–8
Chymotrypsin inhibitor (CTIA)	mg/g	2.40	13.19	77	1, 2, 7	3	10.74	24	2, 5–7
Polyphenols	mg/g	0.84	6.00	19	4, 7, 9, 10	0.02	2.2	10	4, 7, 10
Saponins	mg/g	17.7	56.0	3	4, 11, 12	0.7	21.9	1	4, 13
Phytoestrogens (isoflavones)									
Formononetin	μg/100 g	–	215.0	1	14	94.3	126.0	2	14
Biochanin A	μg/100 g	–	838.0	1	14	1420	3080	2	14
Daidzein	μg/100 g	–	11.4	1	14	34.2	192.0	2	14
Genistein	μg/100 g	–	76.3	1	14	69.3	214.0	2	14
Coumestrol	μg/100 g	–	5.0	1	14	0.00	0.0	2	14
Phytoestrogens (lignans)									
Secoisolariciresinol	μg/100 g	–	8.4	1	14	6.7	8.1	2	14
Matairesinol	μg/100 g	–	0.0	1	14	0.0	0.0	2	14
Phytosterols	mg/100 g	–	39	1	15	35	40	2	15, 16

1. Saini (1989); 2. Petterson *et al.* (1997); 3. Viveros *et al.* (2001); 4. Salgado *et al.* (2001); 5. Zaki *et al.* (1996); 6. Batterham (1989); 7. Singh and Jambunathan (1981); 8. Sotelo *et al.* (1987); 9. Agrawal and Bhattacharya (1980); 10. Avancini *et al.* (1992); 11. Fenwick and Oakenfull (1983); 12. Kerem *et al.* (2005); 13. Ruiz *et al.* (1996); 14. Mazur (1998); 15. Nutrient Data Laboratory (USDA Agricultural Research Service) (2005); 16. Sánchez-Vioque *et al.* (1998).

Essential fatty acids are those that cannot be synthesized by the human body and must be supplied through the diet. The two most important essential fatty acids are the omega-6 (linoleic) and omega-3 (linolenic) fatty acids. They are necessary for normal growth, physiological functions and cell maintenance. The major fatty acid in chickpea is linoleic acid, with desi oil containing 46–62% of the acid and kabuli oil containing 16–56%. Oleic acid, a monounsaturated fatty acid, is the next most common type with 18–23% found in desi oil and 19–32% found in kabuli oil. Linoleic acid has been shown to be hypocholesterolemic and can reduce the likelihood of atherosclerosis and coronary heart disease. The high linoleic acid levels in chickpea can explain lowered serum cholesterol levels in chickpea feeding trials (Mathur *et al.*, 1968; Jaya *et al.*, 1979). Conversely, palmitic acid, a saturated fatty acid, that is hypercholesterolemic and adverse to health, is found in comparatively small amounts in chickpea.

The lipid content of foods is often responsible for its flavour, which in the case of chickpea may contribute to its 'nutty' taste. On the other hand, deterioration of food quality and formation of objectionable flavours can sometimes occur during storage and processing due to enzymatic oxidation of unsaturated fatty acids. This can lead to rejection of poorly stored legumes for human consumption, although the problem is more common in legumes with higher lipid concentrations such as soybean and groundnut.

Carbohydrates

Carbohydrates are the major nutritional component in chickpea, with 51–65% in desi type and 54–71% in kabuli type. The major classes of carbohydrates are monosaccharides, disaccharides, oligosaccharides and polysaccharides.

Monosaccharides and disaccharides

Monosaccharides are single sugar units, whereas disaccharides consist of 2 monosaccharides joined by a glycosidic bond. Free monosaccharides and disaccharides are the most readily available sources of energy; however, only small amounts are present in dried, mature pulse seeds. Most sugars are found as components of glycosides, oligosaccharides and polysaccharides. The most common monosaccharides in chickpea seeds are glucose (0.7%), fructose (0.25%), ribose (0.1%) and galactose (0.05%) (Sanchez-Mata *et al.*, 1998). The most abundant freely occurring disaccharides in chickpea are sucrose (1–2%) and maltose (0.6%).

Oligosaccharides

Oligosaccharides are generally defined as polymeric sugars made up of 2–4 monosaccharides. Legume seeds are rich sources of the raffinose family of oligosaccharides (RFO) in which galactose is present in α-D-1$\rightarrow$6-linkage. They are often considered ANFs because they are neither hydrolysed nor absorbed by the human digestive system, passing into the large intestine where they undergo fermentation by colonic bacteria with the release of gases causing flatulence. Oligosaccharides, therefore, may be classified as part of dietary fibre.

Chickpea contains 0.4–2.8% and 1.2–3.9% oligosaccharides in desi and kabuli seeds, respectively (Table 5.4). Similar amounts are also found in beans (*Phaseolus vulgaris*), which are less than the content in peas (*Pisum sativus*), but more than that in faba beans (*Vicia faba*) and lentils (*Lens culinaris*) (Kozlowska *et al.*, 2001).

The most important oligosaccharides in chickpea are raffinose, stachyose, ciceritol and verbascose, reported in amounts of up to 2.2%, 6.5%, 3.1% and 0.4%, respectively (Table 5.4). Interestingly, ciceritol, which was discovered only in 1983, does not belong to the raffinose family (Quemener and Brillouet, 1983). It is most abundant in chickpea – hence the name – and is present in only a few other pulses such as lentil, lupin and kidney bean. Quemener and Brillouet (1983) also showed that, unlike the RFO, ciceritol does not contribute significantly to flatulence. However, the ingestion of large quantities of chickpea has been known to cause flatulence.

Polysaccharides

Polysaccharides are high-molecular-weight polymers of monosaccharides. They are generally classified by their function in plants, either as an energy reserve or by providing structural support, although there is some overlap. The two main storage polysaccharides in legumes are galactomannans and starch, of which chickpea contains only starch.

STARCH Starch is the principal carbohydrate constituent in chickpea seeds (30–57%) and is the main dietary energy source derived from chickpea. A high proportion of the human requirement for energy is met by starch from seeds and tubers.

Starch is composed of two types of glucose polymer (amylose and amylopectin), with differing properties, occurring together in a starch 'granule'. Amylose is an essentially linear molecule and comprises 20–41% and 23–47% in desi and kabuli types, respectively, depending on the analytical method used (Table 5.1). The remaining 59–80% and 53–77%, respectively, is amylopectin, a highly branched molecule. Comparatively, cereal starches contain less amylose (20–25%) and more amylopectin (75–80%) than pulses. Legume starches also differ from cereal and tuber starches in terms of starch granule structure and crystallinity, referred to as C-type starch. The type of starch (A, B or C) and the proportion of amylose to amylopectin and their respective characteristics are responsible for many of the differences in seed, flour and dough behaviour during processing.

On ingestion, starch is cleaved by pancreatic α-amylase in the duodenum and is hydrolysed by enzymes in the small intestine to produce glucose, an energy source (Gray, 1991; Kozlowska *et al.*, 2001). Legume starch has been reported to be less digestible than cereal starch, possibly due to less amylopectin, higher levels of ANFs, starch granule structure differences and/or cell wall components blocking enzyme access to the granules (Thorne *et al.*, 1983; Jenkins *et al.*, 1987; Tovar *et al.*, 1992; Carre *et al.*, 1998).

Starch can also be divided according to digestibility as soluble, insoluble or resistant. Digestibility varies according to plant species and genotype, chemical and physical granule structure and method of food preparation. Soluble and insoluble starches are completely digested and absorbed, differing only in the speed at which these processes occur. Starch was thought to be fully digested until Englyst *et al.* (1992) recognized a portion of undigested starch (usually from wholegrain or processed foods) that passed into the large intestine. It was named *resistant starch* and defined as 'the sum of starch and products of starch degradation not absorbed in the small intestine of healthy humans' (Asp and Bjoerck, 1992). As such, resistant starch is now classified as a component of dietary fibre. Kabuli seeds have been reported to contain 16.43% resistant starch and desi seeds 3.4–10.3% (Table 5.1).

NON-STARCH POLYSACCHARIDES NSPs are polysaccharides that are not part of starch and can be classified as soluble or insoluble (Choct, 1997). Soluble NSP includes hemicellulose and pectic substances. Hemicelluloses and pectic substances are usually present in amorphous matrixes around cellulose or can act as binding agents between cells in the middle lamella. Hemicellulose contents range from 3.5% to 9% and pectic substances from 1.5% to 4% in chickpea (Table 5.1). Soluble NSP is digested slowly due to its hygroscopic and gummy nature. Insoluble NSP includes cellulose, some insoluble hemicellulose and other polysaccharides, and is indigestible. Cellulose is present in the cell walls of plants and contributes to most of the rigidity and strength of plant structures (often referred to as crude fibre). Desi types contain 4–13% cellulose whilst kabuli types have less (1–5%), primarily due to a thinner seed coat.

Bravo (1999) measured 3.41% soluble NSP (mainly uronic acid and arabinose), 5.37% insoluble NSP (mainly arabinose and glucose) and 8.78% total NSP in kabuli seeds. In chickpea the total NSP fraction is dominated by the monosaccharide residues of glucose and arabinose followed by xylose and mannose, mainly in the form of insoluble NSP from the seed coat (Perez-Maldonado *et al.*, 1999).

DIETARY FIBRE Monosaccharides, disaccharides and the majority of starch are generally classified as available carbohydrates. These are enzymically digested into smaller sugar units within the small intestine and are a rapidly available energy source. Resistant starch and NSP (including oligosaccharides) are classified as unavailable carbohydrates and constitute dietary fibre. Kamath and Belavady (1980) reported chickpea as having more non-available carbohydrates (sum of non-cellulosic polysaccharides, cellulose and lignin) compared to pigeon pea (*Cajanus cajan*), black gram (*P. mungo*) and mung bean (*P. aureus*).

Dietary fibre includes unavailable carbohydrates and lignin (not a carbohydrate, but a major constituent of plant cell walls). It is often difficult to compare dietary fibre contents as numerous methods of calculation are reported in the literature. Legume seeds are generally characterized by their relatively high content of dietary fibre compared to other grains. Much of this fraction comes from the seed coat; hence kabuli types have lower dietary fibre content (11–16%) than desi types (19–23%). A high content of dietary fibre in diets has been shown to have a negative effect on nutrient digestibility in animals, especially monogastrics (Choct *et al.*, 1999; Hughes and Choct, 1999). NSP is generally the largest fraction of dietary fibre in grain legume seeds, the rest comprising resistant starch and lignin. NSP and resistant starch are fermented to varying degrees in the colon.

Lignin is not a carbohydrate but is a polyphenolic constituent of plant cell walls. Lignin's role in plant cells is to cement and anchor NSP and strengthen cell walls to prevent biochemical degradation and physical damage. Lignin is classified as a constituent of dietary fibre as it is indigestible and not fermented in the colon.

Dietary fibre, as a proportion of total carbohydrate, constitutes ~25% in peas and 27% in faba beans (Gdala, 1998). In chickpea, kabuli types have 20–21% and desi types 35–37% (Table 5.1). This large difference between desi and kabuli types is mainly due to differing seed coat contents, with desi types having seed coat with 2–3 times higher levels of lignin than kabuli types (Singh, 1984).

Fibre can also be calculated as acid detergent fibre (ADF) and neutral detergent fibre (NDF). This terminology is usually seen in reference to stock-feeds, more so than for human nutrition. The ADF consists of cellulose, lignin, bound protein and acid-insoluble ash portions, which are all indigestible. The NDF consists of the ADF and hemicellulose. The NDF is used to predict dry matter intake whilst the ADF is inversely correlated with digestible energy (Table 5.1).

Fibre was traditionally considered an antinutritive factor that diluted the energy content and decreased the nutrient availability of food. This is still considered a problem for intensive monogastric livestock production where rapid growth is preferred, but is also a problem in malnourished populations where a high-fibre diet can inhibit the absorption of other nutrients leading to deficiency-related diseases. On the other hand, a high-fibre diet has many health benefits in affluent societies, especially where obesity, heart disease and diabetes are a growing problem. Dietary fibre is fermented slowly by microflora in the large intestine to provide energy, producing beneficial short-chain fatty acids and gaseous by-products (Southgate, 1992).

Minerals

Chickpea plants obtain minerals from their soil environment and partition these to their seeds. Roots utilize specific and/or selective transport proteins to obtain all the minerals (usually as ions) that are essential for plant growth and development (i.e. Ca, Mg, K, P, S, Cl, B, Fe, Mn, Zn, Cu, Ni and Mo) (Grusak and DellaPenna, 1999). Together these minerals constitute what is referred to as the ash fraction of the seed. For the plant, these minerals are important for many metabolic activities including photosynthesis, respiration, chlorophyll synthesis, cell division and various responses to biotic and abiotic stress. Although most of the essential plant minerals are also essential for humans, there are certain minerals required by humans (e.g. Na, I, Se and Cr) that plants do not require. Fortunately, these additional minerals can be absorbed by plants and accumulated in seeds and other edible tissues. Chickpea seeds can thus supply several minerals essential for humans. Although the concentrations of any given mineral vary depending on genotype and environmental constraints (Table 5.1), chickpea contributes significantly to several minerals in the human diet.

Dietary intake recommendations for essential minerals differ slightly from country to country. However, if the US Recommended Dietary Allowance (RDA) is used as an example, a single 100g serving of cooked chickpea can provide as much as 40% of the adult RDA for some minerals (Table 5.5).

With respect to the macronutrient minerals, it is worth noting that chickpea is a rich source of phosphorus and magnesium (Table 5.5). Phosphorus is a major element in hydroxyapatite, a key inorganic constituent of bone, and is also critical in several cellular compounds such as phospholipids, phosphoproteins, nucleic acids and adenosine triphosphate (ATP) (Arnaud and Sanchez, 1996). Thus, it serves basic roles both in human structure and metabolism. Magnesium is crucial in a wide range of metabolic reactions, and additionally contributes to bone health and bone density (Shils, 1996). Approximately 60–65% of total body magnesium in an adult is present in bone. Magnesium is involved as a cofactor in at least 300 enzymatic steps, many of which are linked to energy metabolism; thus, magnesium adequacy is believed to be critical for proper maintenance of body weight and the prevention of syndromes related to cardiovascular disease (Grundy *et al.*, 2006).

Table 5.5. Contribution of a single 100g serving of chickpea to the US Recommended Dietary Allowance (RDA) or adequate intake (AI) for adults.

Mineral	Maximum adult US RDA or AI[a] (mg)	Amount in 100g of cooked chickpea[b] (mg)	Percent contribution to adult RDA or AI
Potassium (K)	4700	291	6
Calcium (Ca)	1200	49	4
Phosphorus (P)	700	168	24
Magnesium (Mg)	420	48	11
Iron (Fe)	18	2.9	16
Zinc (Zn)	11	1.5	14
Manganese (Mn)	2.3	1.0	43
Copper (Cu)	0.9	0.35	39
Selenium (Se)	0.055	0.004	7

[a]RDAs are the daily levels of intake of essential nutrients judged to be adequate to meet the known nutrient needs of practically all healthy persons. Values presented are the highest RDAs either for male or female adults, excluding pregnant or lactating women. The value for potassium is the AI value, which is the amount believed to cover the needs of all individuals in a group, but lack of data prevents specification of the percentage of individuals covered by this intake. All values are from the dietary reference intake reports of the US Institute of Medicine (*Dietary Reference Intakes for Calcium, Phosphorus, Magnesium, Vitamin D, and Fluoride*, 1997; *Dietary Reference Intakes for Vitamin C, Vitamin E, Selenium, and Carotenoids*, 2000; *Dietary Reference Intakes for Vitamin A, Vitamin K, Arsenic, Boron, Chromium, Copper, Iodine, Iron, Manganese, Molybdenum, Nickel, Silicon, Vanadium, and Zinc*, 2001; and *Dietary Reference Intakes for Water, Potassium, Sodium, Chloride, and Sulfate*, 2004). Available at: www.nap.edu

[b]Values are from US Department of Agriculture, Agricultural Research Service (2005) for mature, cooked, boiled without salt chickpea seeds (NDB No. 16057). Note that these are only a subset of the minerals that can be found in chickpea seeds (see Table 5.2).

Although contributing only 4–6% of the adult RDA for calcium and potassium (Table 5.5), the contribution of chickpea seeds to the daily intake of these minerals is none the less important. Calcium plays a critical role in building bone density, with calcium deficiency leading to reductions in bone density and the potential development of rickets in children or osteoporosis in older adults (Arnaud and Sanchez, 1996). In the Middle East, where chickpea use is high and the consumption of dairy products (a better source of calcium) is minimal to non-existent, chickpea can account for a large proportion of the calcium consumed. This potential for chickpea has not gone unnoticed as researchers have been working to increase calcium concentration in the seeds for the purpose of human consumption (Abbo *et al.*, 2000). Potassium is an important intracellular cation in the body that plays a crucial role in the maintenance of cell membrane potential, energy metabolism and membrane cotransport of other ions (Luft, 1996). Because of its role in these processes, adequate potassium intake is important for the contraction of muscle groups such as the heart.

For several of the essential micronutrient minerals, a single serving of chickpea can provide even higher amounts of the RDA (Table 5.5). Although chickpea has the potential to supply ~40% of the adult RDA for manganese and copper, or ~15% for iron and zinc, it must be remembered that seed concentrations can vary across genotypes. However, as these minerals are required in such low daily amounts, most chickpea varieties would in fact be able to supply sizeable amounts of the RDA. Of these micronutrient minerals, iron and zinc are needed in the highest daily amounts (18 mg or 11 mg, respectively). Iron is a major constituent of blood, being a component of haemoglobin, and thus plays a crucial role in oxygen delivery throughout the body (Yip and Dallman, 1996). Due to its redox potential, iron is also involved in many haeme-containing compounds or iron sulphur enzymes that are essential for electron transportation, respiration and energy metabolism. Zinc is another metal that plays a crucial role in energy metabolism, as it is involved in numerous catalytic, structural or regulatory processes (Cousins, 1996). Human deficiencies in both iron and zinc are unfortunately quite widespread (Bouis, 2005), especially throughout countries of the developing world; thus, chickpea has the potential to contribute to daily iron and zinc intake, and help alleviate these problems of micronutrient malnutrition.

Manganese, copper and selenium can be found in chickpea seeds, along with several other metals such as molybdenum and cobalt (Table 5.2). Chickpea can provide ~40% of the adult RDA for manganese and copper in a single serving, and ~7% of the selenium RDA (Table 5.5). Plant selenium concentration, however, is quite variable depending on soil availability of selenium (Mikkelsen *et al.*, 1989). Because soil selenium concentrations can differ quite dramatically from location to location, chickpea's contribution to the selenium RDA varies regionally. Minerals like iron and zinc have important functions in metalloproteins, or as cofactors or activators of specific enzymes. Copper plays a role in antioxidant defence and is also a constituent of several oxidase enzymes (Linder, 1996). Selenium, although required in lesser quaantities, is also quite important. It occurs as selenomethionine or selenocysteine, which take the place of methionine or cysteine in various enzymes (Levander and Burk, 1996). Selenium-dependent glutathione peroxidases contain selenocysteine; these enzymes work to reduce hydroxyperoxides, thereby disrupting their potential to damage membranes, proteins, etc.

Vitamins

Vitamins are essential for normal bodily growth and many metabolic processes. They can be obtained from food, and are consumed as the vitamin themselves or as precursors that require conversion into vitamin-active compounds in the body. Chickpea contains several water-soluble vitamins such as the B-complex vitamins (including folate) and vitamin C, as well as several lipid-soluble vitamins such as vitamin A (found as provitamin A carotenoids), vitamin E (tocopherols and tocotrienols) and vitamin K.

Water-soluble vitamins

The B-complex vitamins include vitamin B_1 (thiamine), vitamin B_2 (riboflavin), vitamin B_3 (niacin), vitamin B_6 and pantothenic acid (Table 5.3). Thiamine (vitamin B_1) is present in small amounts in a wide variety of plant and animal foods; however, the process of refining cereal and pulse products removes much of the thiamine content. Riboflavin (vitamin B_2) is also present in small amounts in chickpea, but is not active until conversion after absorption in the small intestine. Niacin (vitamin B_3) is associated with protein; hence, high-protein foods such as chickpea are rich sources of niacin. Vitamin B_6 exists in 3 major chemical forms (pyridoxine, pyridoxal and pyridoxamine). Grain legumes, including chickpea, are among the richest sources of pyridoxine. Pantothenic acid (vitamin B_5) is widely distributed in food and is important in the metabolism and breakdown of carbohydrates, proteins and fats in the body. Vitamin B_6 performs a wide variety of functions in the body. It is essential for synthesizing enzymes and neurotransmitters required for protein metabolism and normal nerve communication, respectively. Vitamin B_6 is also important for immune system function, blood glucose control, red blood cell metabolism and the conversion of tryptophan to niacin. Folic acid (or folate – the anion form) is also known as vitamin B_9. Folacin is a provitamin that is converted to the active form (folic acid) for use by the body. Folic acid is required for RNA synthesis and DNA replication, and is very important for new cell growth and maintenance. Chickpea contains 150–557 μg/g folate (Table 5.3). There is evidence that folic acid may help prevent cancers induced by DNA damage (Christensen, 1996; Giovannucci *et al.*, 1998; Coppen and Bolander-Gouaille, 2005).

Ascorbic acid (vitamin C) is readily available from fruits and vegetables. Chickpea contains 4 mg/100 g vitamin C, although it is easily destroyed by processes such as cooking and prolonged storage. Vitamin C is a strong antioxidant and is required for the production of collagen in connective tissues (including skin, mucous membranes, teeth and bones), synthesis of neurotransmitters and the synthesis of carnitine (important for fatty acid metabolism).

Lipid-soluble vitamins

Vitamin A (retinol) helps maintain healthy skin and mucous membranes, immune system regulation, cell division and differentiation, bone growth, reproduction and vision. Whilst chickpea does not contain vitamin A per se (only found in animal foods), they are rich sources of carotenoids. Carotenoids are primarily responsible for the yellow colour of chickpea cotyledons and some carotenoids can be readily converted to vitamin A after consumption. Abbo *et al.* (2005) found that chickpea seeds contain a higher concentration of carotenoids than the engineered carotene-boosted, proof-of-concept 'golden rice'. There are more that 600 carotenoids identified from plants, of which only 10% are precursors for vitamin A. Of these, β-carotene has been reported to be the most efficient precursor of vitamin A (Olson and Kobayashi, 1992; Olson, 1996; Paiva and Russell, 1999). Chickpea contains up to 49 μg/100 g β-carotene (Table 5.3) existing in both the cotyledon and seed coat (Atienza *et al.*, 1998). In addition to serving as sources of vitamin A, some carotenoids have been shown to act as antioxidants in laboratory studies. Antioxidants help eliminate free radicals in the body, thereby protecting cells from potential damage. Free radicals are by-products of oxygen metabolism in the body that have been suggested as contributors to the development of many chronic diseases (Olson, 1996; Paiva and Russell, 1999). β-carotene has been shown to be protective against coronary heart disease (Gey *et al.*, 1993). Chickpea contains other carotenoids that do not have vitamin A activity, such as lutein and zeaxanthin, but which are believed to have other health-promoting properties such as the prevention of atherosclerosis and age-related macular degeneration (Institute of Medicine and Food and Nutrition Board, 2001).

Vitamin E is fat-soluble and exists in at least eight different forms of either tocopherol or tocotrienol. Of these eight, only α-, β- and γ-tocopherols and α- and β-tocotrienols are widespread in nature. The most important sources of vitamin E in the diet are the plant seed oils (Coultate, 1996). Chickpea is a reasonable source as the seed contains ~3–9% lipids (Table 5.1) and up to 13.7 mg/100 g vitamin E (Table 5.2). The most important characteristic of vitamin E is as an antioxidant, more powerful than vitamins A or C. It has been reported to aid in the prevention of atherosclerosis and cardiovascular disease (by limiting oxidation of low-density lipoprotein (LDL) cholesterol and preventing formation of blood clots), cancer (by protecting cell membranes from free radical damage and blocking the formation of carcinogens in the stomach) and cataracts. Vitamin E also plays a role in immune function, DNA repair and other metabolic processes. Much of the literature has focused on α-tocopherol, the most active form and a powerful antioxidant in humans. γ-tocopherol has also been shown to be potentially protective against chronic conditions such as inflammation. In addition, the natural admixture of tocopherol forms in natural vegetable oils has an additive and synergistic relationship that provides a higher antioxidant effect than any single form on its own and a broader range of health benefits (Saldeen and Saldeen, 2005).

Vitamin K (phylloquinone) is required for the modification of certain proteins required for blood coagulation, bone metabolism and vascular health. Chickpea contains a low concentration of vitamin K compared to leafy vegetables, but higher than fruits and most animal products (Nutrient Data Laboratory, USDA Agricultural Research Service, 2005).

Bioavailability

It is important to mention that not all nutrients in food are fully utilized. A nutrient's bioavailability is a concept defined as the percent of a specific nutrient, in a given food source or meal, that is digested, absorbed and ultimately utilized by the individual (Grusak, 2002). Bioavailabilities for a given nutrient will vary from food to food, depending on the level of promoters or inhibitors in that food or in the accompanying diet. Individual nutrients can also have different percent bioavailabilities in a common food or meal. It should be noted that average bioavailability factors are already taken into account when expert panels develop dietary guidelines, such as the US RDA or WHO nutrient intake recommendations.

Non-nutritive Health-beneficial Components

More than 2000 years ago Hippocrates said: 'Let food be your medicine and medicine be your food.' Sears (2000) stated that food is the most potent medicine of all. At no time have these statements been more relevant than at the dawn of the 21st century. For the first time in 10,000 years, since agriculture first fuelled human development, what we eat is being associated with deaths due to chronic diseases worldwide, especially in high-income countries. Of all the diseases recorded in 2001, 46% were chronic (including cardiovascular disease, diabetes, cancers and obesity). These chronic diseases accounted for 60% of all deaths in the world, but are largely preventable by improvements in diet and exercise (WHO/FAO, 2003). The proportion of chronic diseases is expected to increase from 46% to 57% by 2020 and have already started to occur at younger ages (WHO/FAO, 2003). As a result of industrialization, urbanization and globalization of food markets, we are now facing an epidemic of obesity and a danger that some children will die at an earlier age than their parents (Jadad, 2005).

Plant foods contain many constituents that are non-nutritive in nature, yet may promote good health. The presence of many different types of proteins and other smaller molecules, including alkaloids, isoflavones, polyphenolics and a variety of oligosaccharides, make pulse seeds unique. Experimental evidence has demonstrated the beneficial activity of pulse components in the prevention and treatment of various diseases. This has prompted a reappraisal of pulses in the diet recently. By and large, these results strongly support the claim that a diet that includes a regular intake of pulses, including chickpea, is one of the ways to maintain and improve health.

Carbohydrates

All carbohydrates are not created equal. Some break down quickly during digestion and can raise blood glucose to dangerous levels. These are the foods that have high glycaemic indexes (GIs). Other carbohydrates break down more slowly, releasing glucose gradually into our bloodstream and are said to have low GIs. GI is defined as a measure of the blood glucose response to carbohydrates within a food as a percentage of the response to an equivalent carbohydrate dose of glucose (Monro and Williams, 2000).

Diabetes is a chronic disease where the body either does not produce insulin or does not respond to insulin properly. Insulin is usually released into our bloodstream after a meal to convert glucose into energy, thereby reducing our blood glucose levels. The inability of diabetics to produce or use insulin causes their blood glucose levels to rise dangerously after eating; the higher the food's GI, the more dangerous the potential increase in blood glucose levels.

Legumes generally have a GI of less than half that of white and wholemeal bread (Kozlowska *et al.*, 2001) and chickpea probably has the lowest GI among the food grains. FAO/WHO listed chickpea with the GIs of 44 (raw) and 47 (canned) compared to white bread at 100 (FAO/WHO, 1998). Mendosa (2005) reported raw chickpea as having a GI of 7–11 compared to more than 100 for white bread and potatoes. In addition, Nalwade *et al.* (2003) found that boiled chickpea had a low GI of 21.45, compared to boiled lentil (27.63), boiled rice (61.18) and bread (76.55). It would therefore be beneficial to include chickpea in the diets of diabetics.

Other health-beneficial carbohydrates include oligosaccharides and resistant starch, which can serve as prebiotics (Guillon and Champ, 2002). Prebiotics stimulate growth and activity of beneficial bacteria in the gastrointestinal tract (e.g. *Bifidobacterium* and *Lactobacillus*) over harmful bacteria (e.g. *Salmonella* spp., *Helicobacter pylori*, *Clostridium perfringens*). More than half of the 'functional foods' on the Japanese market contain prebiotic oligosaccharides as the active component, with the aim of promoting favourable gut microflora to improve human health (FAO/WHO, 1998).

Fibre is known to decrease the bioavailability of many mineral components of the diet, but it also has many advantages. High-fibre diets decrease disorders of the bowel such as constipation, diverticulitis and cancer. This is achieved through increased faecal bulk (diluting potential carcinogens, improving bowel musculature and decreasing the time potential carcinogens spend in the bowel) and through bacterial fermentation (producing short-chain fatty acids such as acetic, propionic and butyric acids), which stimulate growth of healthy colonic epithelial cells and promote death of bowel tumour cells (Gurr and Asp, 1994). The by-products of this reaction are gases (carbon dioxide, hydrogen and methane). High-fibre diets also lower the faecal pH, thereby aiding in the excretion of protein metabolites, which are potent carcinogenic substances (Binghams, 1990).

Fibre in the diet increases bile acid excretion in the bowel, which can reduce blood lipid levels (useful in the prevention and treatment of cardiovascular disease) and beneficially influence cholesterol metabolism (Vahouny *et al.*, 1988; Wolever and Miller, 1995; Vanhoof and Schrijver, 1997). Studies have shown that replacing animal products with legumes (such as in vegetarian diets) can reduce the incidence of cancers of the digestive tract through decreased saturated fat and increased dietary fibre in the diet. Cassidy *et al.* (1994) examined food consumption in 12 countries and found an inverse correlation between colorectal cancer incidence and starch or NSP intake, and a positive association with protein and fat.

A high dietary intake of NSP or dietary fibre has been recognized by the WHO as being protective against obesity (WHO/FAO, 2003). Low GI foods may also play a role, although more studies are needed to prove this. In addition, a high intake of NSP has been shown to reduce blood glucose and insulin levels, and is likely to be protective against diabetes (WHO/FAO, 2003). A minimum daily intake of 20 g NSP is recommended. Chickpea is rich in NSP with a low GI. Approximately 40 g of chickpea is sufficient to satisfy the daily NSP recommended by the WHO (Table 5.1).

Phenolics

Phenolics are one of the most numerous and diverse of all the plant metabolites. They are an integral part of both human and animal diets and can range from simple molecules, such as phenolic acids, to highly polymerized compounds, such as tannins. The basic phenolic backbone consists of an aromatic ring substituted with one or more hydroxyl units. The most common phenolics (found in all vascular plants) are the polyphenolics and lignins (Shahidi and Naczk, 1995). Hundreds of new polyphenolic structures are being discovered every year (Williams and Grayer, 2004; Martens and Mithöfer, 2005). The main classes of polyphenolic compounds are classified by their chemical structures. They are simple phenols, benzoquinones, phenolic acids, acetophenones, phenylacetic acids, hydroxycinnamic acids, phenylpropenes, coumarins, chromones, naphthoquinones, xanthones, stilbenes, anthraquinones, flavonoids, lignans and lignins (Bravo, 1998). In legumes, the main phenolics in the seed are phenolic acids, flavonoids and lignans.

Historically, plant polyphenolics interested scientists for their contribution to plant physiology, including their role in growth and reproduction, pigmentation and provision of resistance to some pathogens, predators and environmental conditions. Polyphenolics also proved to be valuable in several industrial applications, such as the production of tanning solutions, paints, paper, cosmetics and rubber coagulation. From a nutritional point of view, polyphenolics were traditionally considered ANFs due to the adverse effects of tannins binding with macromolecules (such as dietary protein, carbohydrate and digestive enzymes), thereby reducing food digestibility and bioavailability. Polyphenolics have also been used in the food industry as natural food colorants, preservatives and in the clarification of wine, beer and fruit juices. Polyphenolics are also responsible for the taste sensations of astringency and bitterness, thus influencing the sensory and nutritional qualities of food. Oxidation of polyphenolics during storage or processing often results in changes in the organoleptic properties of food. For example, the desirable taste of tea develops from polyphenolic oxidation during processing, and in red wine, modification of astringency with ageing. Conversely, the enzymatic and non-enzymatic browning reactions of phenolic compounds are responsible for the formation of undesirable colour and flavour in fruits and vegetables. In the case of chickpea, prolonged storage under high temperature and humidity will cause browning reactions of the phenolic compounds in the seed coat.

Only recently has interest in polyphenolic compounds, mostly the flavonoids, surfaced due to their antioxidant capacity and potential benefits to human health, such as in the treatment and prevention of cancer, cardiovascular disease, hypertension, hypercholesterolemia, atherosclerosis, bacterial and viral infection, diarrhoea, ulcers, inflammation and allergies (Bravo, 1998; Martens and Mithöfer, 2005).

Polyphenolics are common in most plant foods (vegetables, cereals, legumes, fruits, nuts, etc.), with wide variations in content influenced by genetic and environmental factors, agronomic practices and storage conditions. Legumes generally have higher polyphenolic contents than cereals, which contain <1% of polyphenolic compounds. In pulses, the seed coat usually contains the majority of polyphenolic compounds. Hence, desi types will have a higher concentration than kabuli types due to the larger seed coat content. Singh and Jambunathan (1981) found that ~75% of the polyphenolic content was present in the seed coat of desi types. In addition, the darker varieties of legumes usually contain higher amounts of polyphenolics, such as red kidney beans, black beans (*P. vulgaris*) and black gram (*Vigna mungo*). This also holds true within legume species, so that desi chickpea will generally have higher polyphenolic contents than the kabuli variety, and darker desi seeds will generally contain more than lighter-coloured seeds. Desi and kabuli chickpea contain 0.84–6.00 and 0.02–2.20 mg/g polyphenols, respectively (Table 5.4).

Phenolic acids

Kaushik *et al.* (1996) and Sosulski and Dabrowski (1984) analysed the phenolic acids in cotyledon and seed coat fractions of desi and kabuli types, respectively. They used different methods and came up with different answers (Table 5.6). Large variations between desi varieties were also observed.

Flavonoids

Flavonoids are the most common and widely distributed group of plant phenolics, with more than 9000 compounds described (Martens and Mithöfer, 2005). There are 13 classes of flavonoids (Bravo, 1998): chalcones, dihydrochalcones, aurones, flavones, flavonols, dihydroflavonols, flavanones, flavanols, flavandiols (also called leucoanthocyanidin), anthocyanidins, isoflavonoids, biflavonoids and proanthocyanidins (also called condensed tannins). Their basic structure is derived from diphenylpropanes. Flavonoids commonly occur as glycoside derivatives in plants but can also exist as free monomers.

Anthocyanins, isoflavoids (isoflavones, pterocarpans), flavones (in aerial parts), flavondiols and tannins have been detected in chickpea seeds (Harborne, 1994; Bravo, 1998). The flavone 3,7,4′-trihydroxyflavanone was named 'garbanzol' after its discovery in chickpea (Kühnau, 1976). In addition, it is plausible that chickpea seeds may contain flavonols, as they are precursors to anthocyanins. Most flavonoids are of relatively low molecular weight and are soluble to some degree. However, others can be essentially unextractable, linked to cell wall components such as polysaccharides and lignin. There is

Table 5.6. Phenolic acid composition of desi and kabuli chickpea cotyledons and seed coats.

Cotyledons		Desi whole seed				Kabuli whole seed			
Parameter	Unit	Minimum	Maximum	Number of cultivars	References	Minimum	Maximum	Number of cultivars	References
Gallic acid	µg/g	0.0	22.9	6	1	–	–	–	–
Protocatechuic acid	µg/g	0.0	22.5	6	1	–	–	–	–
p-hydroxybenzoic acid	µg/g	16.3	39.4	6	1	–	0	1	2
Chlorogenic acid	µg/g	0.0	31.3	6	1	–	–	–	–
Vanillic acid	µg/g	0.0	0.0	6	1	–	–	–	–
Caffeic acid	µg/g	0.0	0.0	6	1	–	–	–	–
p-coumaric acid	µg/g	0.0	0.0	6	1	–	0		
Ferulic acid	µg/g	0.0	0.0	6	1	–	43	1	2
Syringic acid	µg/g	–	–	–	–	–	52	1	2
Total phenolic acids	µg/g	16.3	98.1	6	1	–	95	1	2
Seed coat									
Gallic acid	µg/g	0.0	25.9	6	1	–	0.0	1	2
Protocatechuic acid	µg/g	0.0	15.0	6	1	–	0.0	1	2
p-hydroxybenzoic acid	µg/g	0.0	21.3	6	1	–	8.0	1	2
Chlorogenic acid	µg/g	0.0	0.0	6	1	–	–	–	–
Vanillic acid	µg/g	0.0	–	6	1	–	–	–	–
Caffeic acid	µg/g	0.0	–	6	1	–	–	–	–
p-coumaric acid	µg/g	0.0	22.8	6	1	–	3.0	1	2
Ferulic acid	µg/g	0.0	0.0	6	1	–	0.0	1	2
Total phenolic acids	µg/g	16.3	74.4	6	1	–	11.0	1	2
Total phenolic acids in seed	µg/g	42.2	172.5	6	1	–	106	1	2

1. Kaushik *et al.* (1996); 2. Sosulski and Dabrowski (1984).

very little literature on the polyphenolic composition of chickpea. Most of the literature has focused on tannins in tea, anthocyanidins in wine and isoflavonoids in soybean.

ANTHOCYANINS Anthocyanins are water-soluble plant pigments often responsible for the orange to red (sometimes blue, violet or magenta) colour of flowers, fruits and seeds of higher plants. Anthocyanins are the glycosides of anthocyanidins (e.g. pelargonidin, malvidin, cyanidin) and play an important role in pollinator attraction and seed dispersal. Most of the literature focuses on anthocyanins of flowers and more recently on red wine. Relatively little work has been done on anthocyanins as a dietary component; however, Kong *et al.* (2003) recently reviewed the literature on the health-promoting benefits of anthocyanins outlining their antioxidant, anti-inflammatory, anti-oedema, anti-ulcer and antitumour activities. Hence, anthocyanins may play a role in the prevention of coronary heart disease, inflammatory diseases and some cancers.

PHYTOESTROGENS Phytoestrogens are able to mimic the hormone oestrogen and activate or block oestrogen receptors in the body. Several different classes of molecules have been identified as phytoestrogens (Mazur, 1998; Rubio, 2003). Phytoestrogens are converted by microflora into biologically active compounds that are structurally similar, but not identical, to human oestrogens (Setchell *et al.*, 1981). These hormone-like structures allow phytoestrogens to have oestrogenic activity in animals. Thus phytoestrogen is essentially a functional classification (Ganora, 2003–2005). The plant family most abundant in phytoestrogens is the *Leguminosae*.

Dietary phytoestrogens have been found to possess many pharmacologic attributes, as they are oestrogenic/anti-oestrogenic, hypocolesterolemic, anti-atherogenic, antioxidative, chemoprotective, antiviral, antibacterial and anti-insecticidal, and have been found to stimulate endothelial cell proliferation and platelet activation (Setchell *et al.*, 1981; Adlercreutz *et al.*, 1991, 2000a,b; Mazur and Adlercreutz, 2000). The perceived beneficial health effects of phytoestrogens include possible prevention or delay of hormone-related cancers (breast, prostate and colon), cardiovascular disease, diabetes, osteoporosis, inflammation and menopausal symptoms. In addition, they may be beneficial to the immune system and to brain function. Breast cancer is around six times lower in eastern Asia than in the USA, possibly due to higher consumption of phytoestrogens through soybeans and other legumes in the former. The isoflavone intake has been estimated to range from 15 to 200 mg/day in the traditional Japanese diet compared with <1 mg/day in western diets (de Kleijn *et al.*, 2001). In addition, Japanese women excrete 100–1000 times more urinary oestrogens than western women (Adlercreutz *et al.*, 1991). Isoflavones, coumestans (coumestrol) and lignans are the most powerful phytoestrogens found in plants. However, not all isoflavones are oestrogenic.

Lignans, although not in the flavonoid class of polyphenolics, are also powerful phytoestrogens. Mazur *et al.* (1998) examined lignans in chickpea and found that both desi and kabuli types contained significant amounts of the lignan secoisolariciresinol, but no matairesinol (Table 5.4).

ISOFLAVONOIDS Isoflavones are a group of diphenolic secondary metabolites. Although 22 families of plants produce and accumulate isoflavones, only certain species contain medicinally significant amounts (Dixon, 2001). The best-studied dietary phytoestrogens are the soy isoflavones and the flaxseed lignans. Chickpea is one of the richest dietary sources of isoflavones, second only to soybean (Mazur *et al.*, 1998). The four most common active isoflavones are diadzein, genistein, formononetin and biochanin A. Biochanin A and formononetin are metabolized to genistein and diadzein, respectively, after ingestion. Diadzein and genistein exhibit strong activity in oestrogen receptor assays. Soybean has high glycetein, diadzein and genistein contents, whereas chickpea has high biochanin A content. Wall *et al.* (1988) found that chickpea seeds contain small amounts of biochanin A, formononetin and diadzein. In addition to diadzein and formononetin, Jaques *et al.* (1987) previously extracted three other isoflavones from chickpea cotyledons (calycosin 3′-hydroxy-formonetin, pseudobaptigenin and pratensein) and two isoflavone glycosides (diadzein 7-O-glucoside and formononetin 7-O-glucoside). Mazur *et al.* (1998) examined phytoestrogens in chickpea seeds recently and found that the major isoflavone type was biochanin A, containing 838 μg/100 g and 1420–3080 μg/100 g in desi and kabuli types, respectively. Chickpea contains much more biochanin A than any other pulse, containing 7–300 times more than the second highest pulse, pigeon pea (*C. cajan*). The next most abundant polyphenolic was formononetin, containing 215 μg/100 g and 94–126 μg/100 g (desi and kabuli, respectively). Chickpea seeds also contained smaller concentrations of genistein and diadzein.

Coumestans are another type of isoflavonoid and include the plant steroid coumestrol. Coumestans are thought to have the greatest oestrogenic activity of all the isoflavones. In the study by Mazur *et al.* (1998), the desi type contained a very small amount of coumestrol, whereas the kabuli type did not.

Isoflavones have been shown to have anticancer properties also (Mathers, 2002). Girón-Calle *et al.* (2004) showed that an extract containing isoflavones from chickpea had an inhibitory effect on the growth of epithelial tumours with no detrimental effect on healthy epithelial cells.

Tannins

Tannins are compounds of intermediate to high molecular weight and their name originated from their tanning ability. They are also able to form insoluble complexes with carbohydrates and proteins, and the precipitation of salivary proteins is responsible for astringency in tannin-rich foods. In addition, polyphenolics often form complexes with minerals, reducing intestinal absorption. For these reasons, polyphenolics (particularly tannins) have traditionally been considered antinutrients and will be described in more detail by Muzquiz and Wood (Chapter 6, this volume). Chickpea seed contains less than 0.04% and 0.09% condensed tannins in kabuli and desi types, respectively (Petterson *et al.*, 1997; Salgado *et al.*, 2001). The total tannin content is also low at 0.12–0.51% (kabuli) and 0.36–0.72% (desi).

Tannins have recently received some positive attention for the roles they play in health. Many tannins have been found to exhibit chemoprotective, antiviral and antibacterial activity, and may be beneficial in the prevention and/or treatment of acquired immune deficiency syndrome (AIDS), cardiovascular disease and various cancers (Hertog *et al.*, 1993, 1995; Khanbabaee and van Ree, 2001). The ability of polyphenolics to chelate with metals can inhibit some reactions that produce active oxygen radicals in the body. In addition, polyphenolics are very effective antioxidants (chelating with free radicals in the same way as they do with minerals), and can retain their free radical scavenging capacity after forming complexes with metal ions (Afanas'ev *et al.*, 1989). This antioxidant activity can continue right through the digestive tract because tannins are not absorbed. Polyphenolics' protective effect against cardiovascular disease is due to their antioxidant ability inhibiting LDL oxidation. Inhibition of starch-degrading enzymes and polyphenolic binding with starch itself can beneficially decrease the glycaemic and insulinemic response to a meal. Similarly, it has been shown that polyphenolic compounds may have beneficial hypocholesterolemic effects by binding with lipids and cholesterol.

Plant sterols

Plant sterols, or phytosterols, are present in small quantities in legume seeds. They are essential components of plant cell membranes and are most abundant as sterol glucosides and esterified sterol glucosides. Their chemical structure resembles cholesterol; hence they are able to inhibit the absorption of dietary cholesterol in the small intestine. This action can help lower LDL blood cholesterol in humans and may help minimize the risk of coronary heart disease. Plant sterols have also been shown to have anticancer properties (Champ, 2002; Mathers, 2002). The sterol content of chickpea seeds and flour (*besan*) is 35 and 39 mg/100 g, respectively (Nutrient Data Laboratory, USDA Agricultural Research Service, 2005). Similarly, Sánchez-Vioque *et al.* (1998) reported the total sterol content to be 0.04% in defatted chickpea flour, comprising β-sitosterol (83%), campesterol (9%), stigmasterol (6%) and δ-5-avenasterol (2%).

Saponins

Chickpea has a higher content of saponins (25–56 mg/g) than soybean and other pulses, with molecular structures that consist of sugars linked to triterpenes (Fenwick and Oakenfull, 1983; Kerem *et al.*, 2005). In plants, saponins appear to help fight infections and microbial invasions. As a dietary component, saponins have traditionally been classified as ANFs (refer to Muzquiz and Wood, Chapter 6, this volume). However, there are many benefits in consuming saponins as there is some evidence that they may have hypocholesteremic and anticancer properties, stimulate the immune system, ward off microbial and fungal infections, protect against viruses (including human immunodeficiency virus – HIV) and may even act as a spermicide (Thompson, 1993; Ruiz, 1996; Anderson and Major, 2002; Champ, 2002; Madar and Stark, 2002; Mathers, 2002). Some saponins are 50% sweeter than sucrose, and can therefore be used as healthier, non-caloric sweeteners (Kinghorn and NamCheol, 1997).

Enzyme inhibitors

Chickpea contains two main enzyme inhibitors: protease and α-amylase. Both types of inhibitors have been classified as ANFs due to their ability to inhibit protein and starch digestion (refer to Muzquiz and Wood, Chapter 6, this volume). Chickpea generally contains protease inhibitors of 1–16 mg/g trypsin inhibitor and 2–13 mg/g chymotrypsin inhibitor. Protease inhibitors have been shown to be anticarcinogenic (Champ, 2002; Mathers, 2002). Chickpea generally contains 5–11 units/g of α-amylase inhibitor (Singh *et al.*, 1982). α-Amylase inhibitors have the ability to reduce starch digestion, thereby lowering blood glucose levels, and may be useful in the treatment and prevention of obesity and diabetes mellitus (Lajolo and Genovese, 2002; Oneda *et al.*, 2004).

Effects of Processing

Chickpea seeds are rarely eaten in a raw state. Instead, they are traditionally processed prior to consumption by various physical, biochemical or heating methods, which will change dietary composition. Physical methods include dehulling, milling and soaking, whilst biochemical methods include sprouting and fermentation. Heating methods can be wet (boiling, pressure-cooking, canning or extrusion) or dry (roasting, puffing or frying).

Removal of the seed coat

Chickpea seed coats are important for seed protection against mechanical injury, pests, diseases and premature germination. The seed coat is composed primarily of insoluble NSP in the form of cellulose, lignin, polyphenolics and minerals (Singh, 1984). As desi chickpea has a thicker seed coat, changes in composition due to seed coat removal are more pronounced in desi types than in kabuli seeds (10–22% compared to 5–10%, respectively) (Table 5.1).

Removal of the seed coat by dehulling produces dhal or flour with a lower content of dietary fibre and a higher nutritional value than the whole seed. The protein content of desi chickpea increased by 116–118% after dehulling (from 17.7–25.9% protein in whole seed to 20.6–30.5% in dhal), whereas the crude fibre content decreased markedly from 7.1–10.8% in whole seed to 0.7–1.3% in dhal (Jambunathan and Singh, 1980). The same trend for kabuli types is evident but is much less dramatic due to their smaller seed coat content. In addition, there are slight differences between the nutritional value of desi and kabuli dhal. Singh (1984) reported higher NDF, dietary fibre and hemicellulose in desi compared to kabuli cotyledons.

The majority of tannins (79–86%) are located in the seed coat of chickpea (Shahidi and Naczk, 1995). Similarly, Rao and Deosthale (1982) found that dehulling resulted in 75–93% loss of tannin content. Hence, dehulling removes much of the beneficial and antinutritional effects of tannins.

Dehulling does not remove the effects of oligosaccharides, resistant starch or protease inhibitors as these compounds are located in the cotyledons. Removal of the seed coat may actually increase the content of these compounds as dehulling lowers seed weight. Deshpande *et al.* (1982) examined the effect of dehulling on other ANFs and digestibility of beans (*P. vulgaris* L.). Dehulling increased the phytic acid content and inhibitor activities of trypsin, chymotrypsin and α-amylase, which increased the *in vitro* digestibility by 2–4%.

Chickpea seed coat digestion produced unexpectedly high gas release, suggesting the tannins were binding to carbohydrates, protecting them from digestion and passing into the large intestine (Sreerangaraju *et al.*, 2000). Dehulling, therefore, may reduce the amount of flatus caused by chickpea consumption.

Grinding to flour

Grinding whole seeds or dhal into flour will result in damage to starch granules, which are more susceptible to enzymic degradation on consumption. In addition, mineral contents like calcium, iron and zinc significantly decreased in flour products. Tuan and Phillips (1991) reported improved starch digestibility from grinding, but lower protein digestibility. Bhama and Sadana (2004) found that both the protein and starch digestibility of chickpea flour products (*halwa* and *burfi)* decreased, whereas products of chickpea dhal (boiled and fried dhal) had increased protein and starch digestibility.

Soaking

Soaking is often performed before, or in conjunction with, other processing steps such as sprouting, fermenting, cooking and canning. Soaking allows the seeds to imbibe water into their cells, which swell as they hydrate. Leaching of soluble molecules, such as monosaccharides and disaccharides, oligosaccharides, soluble polyphenolics and phytic acid, occurs during this process. Soaking overnight causes a 53% loss in the tannin content of chickpea (Rao and Deosthale, 1982). Spaeth (1987) reported starch granules and protein bodies leaching from *P. vulgaris* cotyledons during the first 30 min of soaking. Soaking has also been found to increase *in vitro* starch digestibility (Bishnoi and Khetarpaul, 1993). This may be due to leaching of polyphenolics and phytic acid that inhibit the activity of α-amylase. Saxena *et al.* (2003) found that soaking of chickpea reduced the contents of proteins, phenols, trypsin and amylase inhibitors.

The soaking fluid can also influence the types and amounts of compounds leached. A dilute sodium bicarbonate solution was shown to enhance the leaching of oligosaccharides, relative to water (Jood *et al.*, 1985). Saxena *et al.* (2003) found that soaking of chickpea in salt solutions (NaCl and $NaHCO_3$) resulted in larger protein content reductions compared to soaking in water, and when combined with pressure-cooking, exacerbated protease inhibitor losses.

Sprouting

Sprouts are produced by germinating seeds, whereby hydrolytic enzymes are released and cause changes in the physical properties and functionality of seed components (Czukor *et al.*, 2001). Isoflavones are not present in sprouts, probably mobilized during germination; hence, many of the health-beneficial properties owing to phytoestrogens are lost (Dziedzic and Dick, 1982). Germination causes similar mobilization of other ANF compounds, resulting in sprouts with improved nutritional value compared to the raw seed (Luz Fernandez, 1988).

Jaya and Venkataraman (1981) reported that germination for 96 h decreased the carbohydrate content of chickpea. They found that the total carbohydrate content decreased from 61% to 53% and starch content from 41% to 25%, whilst total sugar increased from 15% to 22%. Jood *et al.* (1985) found that germination for 24 h significantly reduced the contents of raffinose, stachyose and verbascose in chickpea, and were totally eliminated after 48 h germination. Veena *et al.* (1995) found that sprouting reduced the resistant starch content from 10.3% to 6.8% in chickpea seeds, but increased the dietary fibre and total NSP. Mahadevamma and Tharanathan (2004) also found that the content of dietary fibre increased after germination. In addition, the *in vitro* starch digestibility of chickpea has been shown to increase from 40% to 50% following germination (Shekib, 1994). Kelkar *et al.* (1996) also found that sprouting enhanced the *in vitro* carbohydrate digestibility of chickpea. These changes would result in a significantly increased carbohydrate digestibility and flatus reduction.

Soaking and germination resulted in losses of 59% and 64% tannins after 24 and 48 h, respectively (Rao and Deosthale, 1982). Finney *et al.* (1982) reported an increase in the protein efficiency ratio (PER), after germination, without any alteration in the amino acid profile. Germination for 24 h increased the content of the vitamins riboflavin, thiamine and niacin in chickpea (Geervani and Theophilus, 1980). Hence, sprouting is a nutritious way to consume chickpea, but some of the health-beneficial effects of non-nutritive compounds may be lost or impaired by this method of preparation.

Fermentation

Reyes-Moreno *et al.* (2004) recently optimized the fermentation process to produce *tempeh* flour. The resulting flour had higher protein, lysine contents and *in vitro* protein digestibility, and lower lipid, tannin and phytic acid contents than raw chickpea flour. Other common Indian recipes (*idli* and *dhokla*) are also prepared from fermented pulse batters, including chickpea. Fermentation of *idli* and *dhokla* chickpea batters has been reported to decrease phytic acid content, increase tannin content, increase iron content and bioavailability, and increase *in vitro* carbohydrate digestibility (Kelkar *et al.*, 1996; Reddy *et al.*, 2000). Aliya and Geervani (1981) reported that fermentation of *dhokla* batter increased the biological value (BV) and net protein utilization (NPU), whilst PER and true digestibility (TD) were not affected, and protein content decreased by 6–8%. Bhama and Sadana (2004) found fermentation of chickpea flour to produce *wara* and *dhokla* increased both protein and starch digestibility. Veena *et al.* (1995) found that fermentation increased the dietary fibre and total NSP contents, and reduced the resistant starch content in chickpea seeds.

Availability of amino acids was increased by fermentation, while trypsin inhibitor and oligosaccharide concentrations decreased (Zamora, 1979).

Fermentation increased the vitamin content of thiamine and riboflavin (Aliya and Geervani, 1981), but decreased the mineral contents of calcium, iron and zinc in fermented *wara* and *dhokla* batters (Bhama and Sadana, 2004). Fermentation was also the most efficient processing method for decreasing the saponin content of chickpea (Fenwick and Oakenfull, 1983).

Boiling/Pressure-cooking/Microwaving

Cooking of pulses improves nutritional value due to the decrease or destruction of most ANFs and increased solubility of many nutrients. Soaking of pulses before cooking is a common practice. However, changes in nutritional value will depend on the intensity and duration of heat whilst cooking, which is influenced by the method used.

General

Cooking is reported to reduce the concentration of proteins, minerals, soluble sugars, starch, ANF, NDF, cellulose, hemicellulose, lignin, phytic acid and tannins in pulses (including chickpea), while protein and starch digestibility are improved (ur-Rehman *et al.*, 2004).

Pressure-cooking, boiling and microwaving have been shown to increase the *in vitro* protein, carbohydrate and starch digestibilities of chickpea (Kelkar *et al.*, 1996; Marconi *et al.*, 2000; Bhama and Sadana, 2004). The influence of different cooking methods (boiling, pressure-cooking and microwaving) on the nutritional value of chickpea was investigated by ur-Rehman and Shah (1998). They found that the *in vitro* digestibility increased from 40% (raw) to 68%, 85% and 82% for boiling, pressure-cooking and microwaving, respectively. Khatoon and Prakash (2004) also found that pressure-cooking resulted in higher increases in the *in vitro* starch and protein digestibilities than boiling.

ur-Rehman and Shah (1998) found that the tannin content was reduced by 30%, 74% and 61% for boiling, pressure-cooking and microwaving, respectively. Protein, minerals and fat contents were also reduced by cooking, and this effect was much larger after pressure-cooking than after microwaving. A minimum amount of protein and fat was lost from chickpea during boiling. de Almeida Costa *et al.* (2006) compared the chemical composition of raw and cooked pea, bean, chickpea and lentil seeds. There was a slight increase in protein content and a slight reduction in resistant starch for all the cooked legumes in comparison to their raw form. Cooked chickpea had the highest lipid content, lowest insoluble fibre content and no detectable soluble dietary fibre. Sotelo *et al.* (1987) reported that boiling kabuli seeds reduced the mineral content and increased the fat content, with no change in protein or fibre contents, digestibility or PER. Bhama and Sadana (2004) found that the calcium, iron and zinc contents significantly decreased after boiling chickpea dhal.

Heat causes denaturation of enzymes (proteases, α-amylases) and degradation of vitamins. Steaming, boiling and pressure-cooking decreased contents of riboflavin, thiamin and niacin in chickpea (Geervani and Theophilus, 1980; Aliya and Geervani, 1981). Khatoon and Prakash (2004) found decreased levels of thiamin in chickpea after pressure-cooking and microwaving. Excessive heating also reduced the nutritive value of protein (Singh, 1985). Boiling chickpea improved body weight gain and plasma cholesterol-lowering effect in rates compared to raw seeds, without changing the PER (Wang and McIntosh, 1996). Bhatty *et al.* (2000) found that boiling chickpea halved PER and resulted in decreased total protein and amino acid contents. However, TD and NPU increased due to changes in carbohydrates as a result of cooking. Parihar *et al.* (1999) found that protein losses were not as great after pressure-cooking compared to boiling. Fat losses occurred both during boiling and pressure-cooking to the same degree (Parihar *et al.*, 1999).

Effect of cooking on carbohydrates

Cooking using wet heat causes leaching of monosaccharides, disaccharides and other low-molecular-weight nutrients into the processing water. Cooking at high temperatures can also lead to solubilization and/or depolymerization of dietary fibre, which can alter physiological effects in the gastrointestinal tract. Maillard reaction products may also form due to cooking, thereby adding to the lignin content (FAO/WHO, 1998).

By definition, cooking involves the gelatinization of starch. Gelatinization involves the swelling of starch granules (in the presence of heat and moisture) resulting in the irreversible loss of granule structure and crystallinity. During this process, amylose leaches out of the starch granules. On cooling, some starch molecules begin to reassociate and recrystallize by a process called retrogradation. Retrogradation is a common result of food processing (e.g. it is the cause of bread staling and gravy thickening). Partially retrograded starch can be fermented in the colon to produce significant amounts of butyrate and may help in colon cancer prevention (Guillon and Champ, 2002). The degree of gelatinized and retrograded starch in a food will affect its rate of digestion. Gelatinized starch is much more susceptible to amylolytic enzymes during digestion than raw starch, thereby increasing the GI. Gelatinized starch is easily digested, whereas retrograded starch is difficult to digest and is therefore classified as a type of resistant starch. In agreement, Marconi *et al.* (2000) found that cooking chickpea seeds (by boiling or microwaving) increased the content of rapidly digestible starch (from 36% to 80%) and decreased the level of resistant starch (from 27% to 10%). Similarly, boiled, hot-preserved, canned and a *cocido* preparation all reduced the amount of resistant starch considerably (from 31.60% to 0.65%, 1.46%, 0.84% and 0.85%, respectively) and the rapidly digestible starch increased eight- to tenfold (Bravo, 1999). Veena *et al.* (1995) also found that pressure-cooking reduced the resistant starch content from 10.3% to 3.4% in chickpea seeds.

However, other studies have shown the resistant starch content to increase during cooking. Boiling and pressure-cooking increased resistant starch content, although boiling had the greatest effect (Mahadevamma *et al.*, 2004; Mahadevamma and Tharanathan, 2004). This increase in resistant starch content caused beneficial faecal bulking, short-chain fatty acid production and anaerobic bacterial counts *in vivo* when fed to rats (Mahadevamma *et al.*, 2004).

Total NSP contents are generally unaffected by cooking (boiling, microwaving, *cocido* and preserving); however, a redistribution from insoluble to soluble NSP has been shown (Bravo, 1999; Marconi *et al.*, 2000). In contrast, Veena *et al.* (1995) found an increase in dietary fibre and total NSP content of chickpea seeds after pressure-cooking, whilst Mahadevamma and Tharanathan (2004) found the dietary fibre content of chickpea seeds to increase after boiling and decrease after pressure-cooking. Scanning electron microscopy (SEM) studies showed that cooking destroyed cotyledon cell walls and modified parenchyma cells, which was consistent with observed NSP and starch digestibility increases (Marconi *et al.*, 2000). The increased carbohydrate and starch digestibilities of boiled chickpea resulted in a higher, but still acceptable, GI compared to raw chickpea. Boiled chickpea had a much lower GI of 21.45, compared to boiled lentil (27.63), boiled rice (61.18) and bread (76.55), and would be an appropriate food for diabetics (Nalwade *et al.*, 2003).

Roasting or Puffing

Roasting is generally a dry-heat process, and as such nutrient loss is generally not as great as in wet-heat cooking methods (Geervani and Theophilus, 1980). For example, losses in lysine content were less after roasting than after boiling or pressure-cooking. Roasting has been reported to reduce *in vitro* digestibility by 15–28% (Kowsalya Murty and Kantharaj Urs, 1980). Roasting also significantly reduced riboflavin, thiamine and niacin contents (Rao, 1974; Geervani and Theophilus, 1980). Roasting at high temperatures eliminated all oligosaccharides and sucrose from hyacinth bean after 2 min and should have a similar effect on chickpea (Revilleza *et al.*, 1990). In addition, roasting resulted in a 15–18% decrease in free lipids and an increase in bound lipids (Kowsalya Murty and Kantharaj Urs, 1979). Kelkar *et al.* (1996) reported that roasting decreased the *in vitro* carbohydrate digestibility of chickpea, probably due to the formation of resistant starch. However, Mahadevamma *et al.* (2004) and Veena *et al.* (1995) both found no significant effect of roasting on the resistant starch content of chickpea dhal.

Extrusion

The results of extrusion experiments can vary depending on the temperature, moisture content and screw speeds applied. Mahadevamma and Tharanathan (2004) found that extrusion had no significant effect on the resistant starch content of chickpea dhal. However, extrusion generally leads to a loss of total sugars and oligosaccharides, and increases starch digestibility (Tuan and Phillips,

1991; Borejszo and Khan, 1992). Extrusion following soaking significantly decreased antinutrient contents (phytic acid, tannins, phenols, α-amylase, trypsin inhibitors) and improved the *in vitro* protein digestibility (el-Hady and Habiba, 2003). In contrast, Tuan and Phillips (1991) found no effect of extrusion on protein digestibility. Wang and McIntosh (1996) found significant increases in body weight gain (but no change in PER) and plasma cholesterol-lowering effects in rats similar to those in boiled chickpea. This suggests that extrusion and boiling improves the nutritional value of chickpea by similar amounts.

Alonso *et al.* (2001) reported a lowering in phytic acid, tannin and lectin contents following extrusion of kidney bean and pea. Extrusion reduced the content of stachyose in pea, reduced oligosaccharides in kidney bean and increased solubility of dietary fibre (FAO/WHO, 1998). Iron, calcium and phosphorus contents of pea increased after extrusion, whilst kidney bean had additional increases in magnesium, zinc and copper (Alonso *et al.*, 2001). In contrast, Poltronieri *et al.* (2000) reported that extruded kabuli flour had similar iron bioavailability to that of raw flour.

Frying

A variety of sweet and savoury dishes are commonly prepared from chickpea dhal or flour, alone or in combination with cereals. Dry ground flours, or wet ground doughs and batters are used to prepare many dishes. Frying reduces sucrose content and starch digestibility of legume seeds (Jood *et al.*, 1985; Kelkar *et al.*, 1996). Kelkar *et al.* (1996) also found frying to decrease the *in vitro* carbohydrate digestibility of chickpea. In contrast, Bhama and Sadana (2004) found increased protein and starch digestibility in fried dhal and deep-fried *pakoras*, but decreased calcium, iron and zinc contents. Mahadevamma and Tharanathan (2004) found that frying chickpea dhal had no significant effect on the resistant starch content. Annapure *et al.* (1998) compared the oil content of *sev* (an extruded and deep-fried Indian product) produced using various cereal and pulse flours, including chickpea. There was no difference in total oil content when different flours were used, but oil content was minimized when fried in cottonseed oil.

Canning

Canning reduced thiamine and riboflavin contents in chickpea (Rao, 1974). Canning reduced the amount of resistant starch considerably (31.60–0.84%) and the rapidly digestible starch increased eight- to tenfold (Bravo, 1999). Canned chickpea also has higher levels of rapidly available glucose than boiled chickpea (Hawkins and Johnson, 2005).

Vacuum-packaging

Commercially pre-cooked or vacuum-packaged chickpea has higher levels of readily digestible starch than canned and boiled chickpea. It also has higher levels of rapidly available glucose than boiled chickpea, suggesting that the commercially pre-cooked or vacuum-packaged chickpea may have higher and less beneficial GIs than boiled chickpea (Hawkins and Johnson, 2005).

Non-seed Edible Parts

Immature green pods and young tender leaves are cooked and eaten as vegetables, especially in India and Nepal (Awasthi and Tandan, 1987; Agte *et al.*, 2000). The leaves, in particular, are cooked (steamed or boiled) as spinach, and used in salads and sauces. As they are photosynthetic tissues, these green preparations are rich in several vitamins (e.g. folate, vitamin A, vitamin C, vitamin K and β-carotene) and various health-beneficial phytochemicals (e.g. other carotenoids). Singh *et al.* (2001) found chickpea leaves to have higher contents of iron, vitamin C and α-carotene than spinach, mint, carrots and coriander. Tender chickpea leaves are also replete with several human-essential mineral nutrients (Ibrikci *et al.*, 2003), and for most of these minerals, the concentrations in chickpea leaves far exceed those previously reported for spinach or cabbage.

Conclusion

Chickpea is an important part of human nutrition and should be consumed more often by western societies. It has good nutritional value with few ANFs and may play a role in the prevention and treatment of many chronic diseases. Genetic variation has been shown for many of the nutrients, although little research has been done on others. There is potential to breed new varieties to further optimize the nutritional value of chickpea, either through traditional breeding or via genetic modification.

Acknowledgement

The authors wish to thank the Information and Library Services Staff of the New South Wales Department of Primary Industries (particularly F. Drum and S. Giles), and Dr J. Kumar for help with the preparation of this chapter. This work was funded to M. Grusak, in part by the US Department of Agriculture, Agricultural Research Service under Cooperative Agreement Number 58-6250-1-001, and in part by the HarvestPlus Project under Agreement Number 58-6250-4-F029.

References

Abbo, S., Grusak, M.A., Tzuk, T. and Reifen, R. (2000) Genetic control of seed weight and calcium concentration in chickpea seed. *Plant Breeding* 119, 427–431.

Abbo, S., Molina, C., Jungmann, R., Grusak, M.A., Berkovitch, Z., Reifen, R., Kahl, G. and Winter, P. (2005) Quantitative trait loci governing carotenoid concentration and

weight in seeds of chickpea (*Cicer arietinum* L.). *Theoretical and Applied Genetics* 111, 185–195.
Adlercreutz, H., Honjo, H., Higashi, A., Fotsis, T., Hamalainen, E., Hasegawa, T. and Okada, H. (1991) Urinary excretion of lignans and isoflavonoid phytoestrogens in Japanese men and women consuming a traditional Japanese diet. *American Journal of Clinical Nutrition* 54, 1093–1100.
Adlercreutz, H., Mazur, W., Bartels, P., Elomaa, V.V., Watanabe, S., Wahala, K., Landstrom, M., Lundin, E., Bergh, A., Damber, J.E., Aman, P., Widmark, A., Johansson, A., Zhang, J.X. and Hallmans, G. (2000a) Phytoestrogens and prostate disease. *Journal of Nutrition* 130, 658S–659S.
Adlercreutz, H., Mazur, W., Stumpf, K., Kilkkinen, A., Pietinen, P., Hulten, K. and Hallmans, G. (2000b) Food containing phytoestrogens, and breast cancer. *BioFactors* 12(1–4), 89–93.
Afanas'ev, I.B., Dorozhko, A.L., Brodskii, A.V., Kostyuk, V.A. and Potapovitch, A.L. (1989) Chelating and free radical scavenging mechanisms of inhibitory action of rutin and quercetin in lipid peroxidation. *Biochemical Pharmacology* 38, 1763–1769.
Agrawal, P. and Bhattacharya, L. (1980) Proximate composition of seeds of improved varieties of bengal gram (*Cicer arietinum* Linn.). *Seed Research* 8, 5–14.
Agrawal, K. and Singh, G. (2003) Physicochemical and milling quality of some improved varieties of chickpea (*Cicer arietinum*). *Journal of Food Science and Technology* 40, 439–442.
Agte, V.V., Tarwadi, K.V., Mengale, S. and Chiplonkar, S.A. (2000) Potential of traditionally cooked green leafy vegetables as natural sources for supplementation of eight micronutrients in vegetarian diets. *Journal of Food Composition and Analysis* 13, 885–891.
Aliya, S. and Geervani, P. (1981) An assessment of the protein quality and vitamin B content of commonly used fermented products of legumes and millets. *Journal of the Science of Food and Agriculture* 32, 837–842.
Alonso, R., Rubio, L.A., Muzquiz, M. and Marzo, F. (2001) The effect of extrusion cooking on mineral bioavailability in pea and kidney bean seed meals. *Animal Feed Science and Technology* 94, 1–13.
Amirshahi, M.C. and Tavakoli, M. (1970) Protein content of different varieties of five species of pulse crops. In: *Proceedings of a Symposium: Improving Plant Protein by Nuclear Techniques*. IAEA and FAO, Vienna, pp. 331–336.
Anderson, J.W. and Major, A.W. (2002) Pulses and lipaemia, short- and long-term effect: potential in the prevention of cardiovascular disease. *British Journal of Nutrition* 88, S263–S271.
Annapure, U.S., Singhal, R.S. and Kulkarni, P.R. (1998) Studies on deep-fat fried snacks from some cereals and legumes. *Journal of the Science of Food and Agriculture* 76, 377–382.
Arnaud, C.D. and Sanchez, S.D. (1996) Calcium and phosphorus. In: Ziegler, E.E. and Filer, L.J. Jr (eds) *Present Knowledge in Nutrition*, 7th edn. International Life Sciences Institute, Washington, DC, pp. 245–255.
Asp, N.G. and Bjoerck, I. (1992) Resistant starch. *Trends in Food Science and Technology* 3, 111–114.
Atienza, J., Sanz, M., Herguedas, A., Alejos, J.A. and Jimenez, J.J. (1998) β-Carotene, α-tocopherol and δ-tocopherol contents in dry legumes: influence of cooking. *Food Science and Technology International* 4, 437–441.
Avancini, S.R.P., Sales, A.M., de Aguirre, J.M. and Mantovani, D.M.B. (1992) Chemical composition and nutritional value of chick peas produced in São Paulo State. *Coletanea do Instituto de Tecnologia de Alimentos* 22, 145–153.
Awasthi, C.P. and Tandan, P.K. (1987) Quality studies on some common green leafy vegetables grown in India. *Progressive Horticulture* 19, 207–212.
Awasthi, C.P., Abidi, A.B. and Chowdhury, A.R. (1991) Studies on the nutritional quality of different varieties of chickpea. *Indian Journal of Agricultural Research* 25, 21–26.
Babu, S. and Srikantia, S.G. (1976) Availability of folates from some foods. *American Journal of Clinical Nutrition* 29, 376–379.

Batterham, E.S. (1989) Use of chickpeas (*Cicer arietinum* L.) in animal breeding. *Nutrition Reviews* 56, 317–333.

Bhama, S. and Sadana, B.K. (2004) In vitro digestibility and minerals availability from bengal gram dhal and flour products. *Journal of Food Science and Technology* 41(4), 459–461.

Bhatty, N., Gilani, A.H. and Nagra, S.A. (2000) Effect of cooking and supplementation on nutritional value of gram (*Cicer arietinum*). *Nutrition Research* 20, 297–307.

Biliaderis, C.G. and Tonogai, J.R. (1991) Influence of lipids on the thermal and mechanical properties of concentrated starch gels. *Journal of Agricultural and Food Chemistry* 39, 833–840.

Biliaderis, C.G., Maurice, T.J. and Vose, J.R. (1980) Starch gelatinization phenomena studied by differential scanning calorimetry. *Journal of Food Science* 45, 1669–1680.

Biliaderis, C.G., Grant, D.R. and Vose, J.R. (1981) Structural characterization of legume starches. I. Studies on amylose, amylopectin, and beta-limit dextrins. *Cereal Chemistry* 58, 496–502.

Binghams, S.A. (1990) Mechanisms and experimental and epidemiological evidence relating dietary fiber (non-starch polysaccharides) and starch to protection against large bowel cancer. *Proceedings of the Nutrition Society* 49, 153–171.

Bishnoi, S. and Khetarpaul, N. (1993) Effect of domestic processing and cooking methods on *in vitro* starch digestibility of different pea cultivars (*Pisum sativum*). *Food Chemistry* 47, 177–182.

Borejszo, Z. and Khan, K. (1992) Reduction of flatulence-causing sugars by high temperature extrusion of pinto bean high starch fractions. *Journal of Food Science* 57, 771–777.

Bouis, H.E. (2005) Micronutrient fortification of plants through plant breeding: can it improve nutrition in man at low cost? *Proceedings of the Nutrition Society* 62, 403–411.

Bravo, L. (1998) Polyphenols: chemistry, dietary sources, metabolism, and nutritional significance. *Nutrition Reviews* 56, 317–333.

Bravo, L. (1999) Effect of processing on the non-starch polysaccharides and in vitro digestibility of legumes. *Food Science and Technology International* 5, 415–423.

Carre, B., Melcion, J.P., Widiez, J.L. and Biot, P. (1998) Effects of various processes of fractionation, grinding and storage of peas on the digestibility of pea starch in chickens. *Animal Feed Science and Technology* 71, 19–33.

Cassidy, A., Bingham, S.A. and Cummings, J.H. (1994) Starch intake and colorectal cancer risk: an international comparison. *British Journal of Cancer* 69, 937–942.

Champ, M.M.-J. (2002) Non-nutrient bioactive substances of pulses. *British Journal of Nutrition* 88, S307–S318.

Choct, M. (1997) Feed non-starch polysaccharides: chemical structures and nutritional significance. *Feed Milling International* June, 13–26.

Choct, M., Hughes, R.J. and Annison, G. (1999) Apparent metabolisable energy and chemical composition of Australian wheat in relation to environmental factors. *Australian Journal of Agricultural Research* 50, 447–451.

Christensen, B. (1996) Folate deficiency, cancer and congenital abnormalities: is there a connection? *Tidsskrift for den Norske Laegeforening* 116, 250–254.

Coppen, A. and Bolander-Gouaille, C. (2005) Treatment of depression: time to consider folic acid and vitamin B12. *Journal of Psychopharmacology* 19, 59–65.

Coultate, T.P. (ed.) (1996) *Food: The Chemistry of Its Components*. The Royal Society of Chemistry, Cambridge, UK.

Cousins, R.J. (1996) Zinc. In: Ziegler, E.E. and Filer, L.J. Jr (eds) *Present Knowledge in Nutrition*, 7th edn. International Life Sciences Institute, Washington, DC, pp. 293–306.

Czukor, B., Bogracheva, T., Cserhalmi, Z., Fornal, J., Schuster-Gajzágó, I., Kovács, E.T., Lewandowicz, G. and Soral-Smietana, M. (2001) Processing. In: Hedley, C.L. (ed.) *Carbohydrates in Grain Legume Seeds: Improving Nutritional Quality and Agronomic Characteristics*. CAB International, Wallingford, UK.

Dalgetty, D.D. and Baik, B.K. (2003) Isolation and characterization of cotyledon fibres from peas, lentils and chickpeas. *Cereal Chemistry* 80, 310–315.

Dang, J., Arcot, J. and Shrestha, A. (2000) Folate retention in selected processed legumes. *Food Chemistry* 68, 295–298.

de Almeida Costa, G.E., da Silva Queiroz-Monici, K., Reis, S.M.P.M. and de Oliveira, A.C. (2006) Chemical composition, dietary fibre and resistant starch contents of raw and cooked pea, common bean, chickpea and lentil legumes. *Food Chemistry* 94, 327–330.

de Kleijn, M.J.J., van der Schouw, Y.T., Wilson, P.W.F., Adlercreutz, H., Mazur, W., Grobbee, D.E. and Jacques, P.F. (2001) Intake of dietary phytoestrogens is low in postmenopausal women in the United States: the Framingham Study. *Journal of Nutrition* 131, 1826–1832.

Deshpande, S.S., Sathe, S.K., Salunkhe, D.K. and Cornforth, D.P. (1982) Effects of dehulling on phytic acid, polyphenols, and enzyme inhibitors of dry beans (*Phaseolus vulgaris* L.). *Journal of Food Science* 47, 1846–1850.

Dixon, R.A. (2001) Natural products and plant disease resistance. *Nature* 411, 843–847.

Dodok, L., Ali, M.A., Hozova, B., Halasova, G. and Polacek, I. (1993) Importance and utilization of chickpea in cereal technology. *Acta Alimentaria* 22, 119–129.

Dziedzic, S.Z. and Dick, J. (1982) Analysis of isoflavones in bengal gram by high-performance liquid chromatography. *Journal of Chromatography A* 234, 497–499.

el-Hady, E.A.A. and Habiba, R.A. (2003) Effect of soaking and extrusion conditions on antinutrients and protein digestibility of legume seeds. *Lebensmittel Wissenschaft und Technologie* 36, 285–293.

El Hardollou, S.B., El Tinay, A.H. and Nour, A.A.M. (1980) Chemical characteristics of some legumes grown in the Sudan. *Sudan Journal of Food Science and Technology* 12, 35–42.

Englyst, H.N., Kingman, S.M. and Cummings, J.H. (1992) Classification and measurement of nutritionally important starch fractions. *European Journal of Clinical Nutrition* 46, S33–S50.

FAO/USDA (1987) *Food Composition Tables for the Near East*. FAO/USDA, Rome.

FAO/WHO (1998) Carbohydrates in human nutrition: report of a Joint FAO/WHO Expert Consultation. In: *FAO Food and Nutrition Paper*. Food and Agriculture Organization and the World Health Organization, Rome, pp. 1–95.

Fenwick, D.E. and Oakenfull, D. (1983) Saponin content of food plants and some prepared foods. *Journal of the Science of Food and Agriculture* 34, 186–191.

Finney, P.L., Beguin, D. and Hubbard, J.D. (1982) Effects of germination on bread-baking properties of mung bean (*Phaseolus aureus*) and garbanzo bean (*Cicer arietinum* L.). *Cereal Chemistry* 59, 520–524.

Ganora, L. (2003–2005) *The Phytochemistry of Herbs: Phytoestrogens*. Boulder, Colorado. www.herbalchem.net

Garcia-Alonso, A., Goni, I. and Saura-Calixto, F. (1998) Resistant starch and potential glycaemic index of raw and cooked legumes (lentils, chickpeas and beans). *Zeitschrift für Lebensmittel-Untersuchung und -Forschung* 206, 284–287.

Gdala, J. (1998) Composition, properties, and nutritive value of dietary fibre of legume seeds: a review. *Journal of Animal and Feed Sciences* 7, 131–149.

Geervani, P. and Theophilus, F. (1980) Effect of home processing on the nutrient composition of certain high yielding legume varieties. *Indian Journal of Nutrition and Dietetics* 17, 443–446.

Gey, K.F., Moser, U.K., Jordan, P., Staehelin, H.B., Eichholzer, M. and Ludin, E. (1993) Increased risk of cardiovascular disease at suboptimal plasma concentrations of essential antioxidants: an epidemiological update with special attention to carotene and vitamin C. *American Journal of Clinical Nutrition* 57, 787S–797S.

Giovannucci, E., Stampfer, M.J., Colditz, G.A., Hunter, D.J., Fuchs, C., Rosner, B.A., Speizer, F.E. and Willett, W.C. (1998) Multivitamin use, folate, and colon cancer in women in the Nurses' Health Study. *Annals of Internal Medicine* 129, 517–524.

Girón-Calle, J., Vioque, J., del Mar Yust, M., Pedroche, J., Alaiz, M. and Millán, F.

(2004) Effect of chickpea aqueous extracts, organic extracts, and protein concentrates on cell proliferation. *Journal of Medicinal Food* 7, 122–129.

Gray, J. (1991) *Starches and Sugars: A Comparison of Their Metabolism in Man*. International Life Sciences Institute (ILSI), Springer, London.

Grundy, S.M., Cleeman, J.I., Daniels, S.R., Donato, K.A., Eckel, R.H., Franklin, B.A., Gordon, D.J., Krauss, R.M., Savage, P.J., Smith, S.C. Jr, Spertus, J.A. and Costa, F. (2006) Diagnosis and management of the metabolic syndrome: an American Heart Association/National Heart, Lung, and Blood Institute scientific statement: executive summary. *Circulation* 112, e285–e290.

Grusak, M.A. (2002) Enhancing mineral content and bioavailability in plant food products. *Journal of the American College of Nutrition* 21, 178S–183S.

Grusak, M.A. and DellaPenna, D. (1999) Improving the nutrient composition of plants to enhance human nutrition and health. *Annual Review of Plant Physiology and Plant Molecular Biology* 50, 133–161.

Guillon, F. and Champ, M.M.-J. (2002) Carbohydrate fractions of legumes: uses in human nutrition and potential for health. *British Journal of Nutrition* 88, S293–S306.

Gurr, M.I. and Asp, N.-G. (1994) Dietary fibre. *ILSI Europe Concise Monograph Series*. International Life Sciences Institute, Washington, DC.

Harborne, J.B. (1994) *The Flavonoids: Advances in Research Since 1986*. CRC Press, Boca Raton, Florida.

Hawkins, A. and Johnson, S.K. (2005) In vitro carbohydrate digestibility of whole-chickpea and chickpea bread products. *International Journal of Food Sciences and Nutrition* 56, 147–155.

Hertog, M.G.L., Feskens, E.J.M., Hollman, P.C. (1993) Dietary antioxidant flavonoids and risk of coronary heart disease: the Zutphen Elderly Study. *The Lancet* 342, 1007–1011.

Hertog, M.G.L., Kromhout, D., Aravanis, C. *et al.* (1995) Flavonoid intake and long-term risk of coronary heart disease and cancer in the seven countries study. *Archives of International Medicine* 155, 381–386.

Hoover, R. and Ratnayake, W.S. (2002) Starch characteristics of black bean, chick pea, lentil, navy bean and pinto bean cultivars grown in Canada. *Food-Chemistry* 78, 489–498.

Hughes, R.J. and Choct, M. (1999) Chemical and physical characteristics of grains related to variability in energy and amino acid availability in poultry. *Australian Journal of Agricultural Research* 50, 689–701.

Ibrikci, H., Knewtson, S.J.B. and Grusak, M.A. (2003) Chickpea leaves as a vegetable green for humans: evaluation of mineral composition. *Journal of the Science of Food and Agriculture* 83, 945–950.

Institute of Medicine and Food and Nutrition Board (2001) *Dietary Reference Intakes for Vitamin A, Vitamin K, Arsenic, Boron, Chromium, Copper, Iodine, Iron, Manganese, Molybdenum, Nickel, Silicon, Vanadium, and Zinc*. National Academy Press, Washington, DC.

Iqbal, A., Khalil, I.A., Ateeq, N. and Khan, M.S. (2006) Nutritional quality of important food legumes. *Food Chemistry* 97, 331–335.

Jadad, A. (2005) Agriculture and its role in human health. In: *Food and a Healthy Value Proposition* (no formal proceedings published, only presentation abstracts published). Toronto, Canada, p. 17.

Jambunathan, R. and Singh, U. (1980) Studies on desi and kabuli chickpea (*Cicer arietinum* L.) cultivars. 1. Chemical composition. International Workshop on Chickpea Improvement, ICRISAT, Hyderabad, India, pp. 61–66.

Jambunathan, R. and Singh, U. (1981) Studies on desi and kabuli chickpea (*Cicer arietinum* L.) cultivars. 3. Mineral and trace element composition. *Journal of Agricultural and Food Chemistry* 29, 1091–1093.

Jaques, U., Kessman, H. and Barz, W. (1987) Accumulation of phenolic compounds and phytoalexins in sliced and elicitor-treated cotyledons of *Cicer arietinum* L. *Zeitschrift für Naturforschung C* 42c, 1171.

Jarvis, C. and Iyer, L. (2000) Understanding pulse starch behaviour during cooking. In: Wootton, M., Batey, I.L. and Wrigley, C.W. (eds) *Cereals 2000: Proceedings of the 11th ICC Cereal and Bread Congress and of the 50th Australian Cereal Chemistry Conference*. Surfers Paradise, Australia, pp. 349–355.

Jaya, T.V. and Venkataraman, L.V. (1981) Changes in the carbohydrate constituents of chickpea and greengram during germination. *Food Chemistry* 7, 95–104.

Jaya, T.V., Venkataraman, L.V. and Krishnamurthy, K.S. (1979) Influence of germinated legumes on the levels of tissue cholesterol and liver enzymes of hypercholesterolemic rats: influence of ungerminated and germinated chickpea and green-gram (whole). *Nutrition Reports International* 20, 371–381.

Jenkins, D.J.A., Cuff, D., Wolever, T.M.S., Knowland, D., Thompson, L., Cohen, Z. and Prokipchuk, E. (1987) Digestibility of carbohydrate foods in an ileostomate: relationship to dietary fiber, in vitro digestibility, and glycaemic response. *American Journal of Gastroenterology* 82, 709–717.

Jood, S., Mehta, U., Randhir, S. and Bhat, C.M. (1985) Effect of processing on flatus-producing factors in legumes. *Journal of Agricultural and Food Chemistry* 33, 268–271.

Kamath, M.V. and Belavady, B. (1980) Unavailable carbohydrates of commonly consumed Indian foods. *Journal of the Science of Food and Agriculture* 31, 194–202.

Kaushik, S.B., Deepak, M., Sand, N.K. and Pant, R.C. (1996) Chickpea seed phenolics in relation to *in situ* germination. *Plant Physiology and Biochemistry* 23, 169–172.

Kelkar, M., Shastri, P. and Rao, B.Y. (1996) Effect of processing on *in vitro* carbohydrate digestibility of cereals and legumes. *Journal of Food Science and Technology* 33, 493–497.

Kerem, Z., German-Shashoua, H. and Yarden, O. (2005) Microwave-assisted extraction of bioactive saponins from chickpea (*Cicer arietinum* L.). *Journal of the Science of Food and Agriculture* 85, 406–412.

Khanbabaee, K. and van Ree, T. (2001) Tannins: classification and definition. *Natural Product Reports* 18, 641–649.

Khanvilkar, S.G. and Desai, B.B. (1981) Genotypic variations in protein quality and nutritional composition of chick pea (*Cicer arietinum* L.). *Journal of Maharashtra Agricultural Universities* 6, 226–228.

Khatoon, N. and Prakash, J. (2004) Nutritional quality of microwave-cooked and pressure-cooked legumes. *International Journal of Food Sciences and Nutrition* 55, 441–448.

Kinghorn, A.D. and NamCheol, K. (1997) Discovery of highly sweet substances from plants. *Revista de Farmacia e Bioquimica da Universidade de São Paulo* 33, 63–75.

Knights, E.J. and Mailer, R.J. (1989) Association of seed type and colour with establishment, yield and seed quality in chickpea (*Cicer arietinum*). *Journal of Agricultural Science* 113, 325–330.

Kong, J.-M., Chia, L.-S., Goh, N.-K., Chia, T.-F. and Brouillard, R. (2003) Analysis and biological activities of anthocyanins. *Phytochemistry* 64, 923–933.

Kowsalya Murty, S. and Kantharaj Urs, M. (1979) Changes in bengal gram (*Cicer arietinum* L.) lipids during heat processing. *Journal of Food Science and Technology* 11, 87–89.

Kowsalya Murty, S. and Kantharaj Urs, M. (1980) Effect of toasting bengal gram (*Cicer arietinum* L.) on lysine availability and *in vitro* digestibility of proteins. *Journal of Food Science and Technology* 17, 200–201.

Kozlowska, H., Aranda, P., Dostalova, J., Frias, J., Lopez-Jurado, M., Pokorny, J., Urbano, G., Vidal-Valverde, C. and Zdyunczyk, Z. (2001) Nutrition. In: Hedley, C.L. (ed.) *Carbohydrates in Grain Legume Seeds: Improving Nutritional Quality and Agronomic Characters*. CAB International, Oxon, UK.

Krishna Murti, C.R. (1975) Biochemical studies on bengal gram. *Journal of Scientific and Industrial Research* 34, 266–281.

Kühnau, J. (1976) The flavonoids. A class of semi-essential food components: their role on human nutrition. *World Review on Nutrition and Diet* 24, 117–191.

Labavitch, J.M., Freeman, L.E. and Albersheim, P. (1976) Structure of cell walls. *The Journal of Biological Chemistry* 251, 5904–5910.

Lajolo, F.M. and Genovese, M.I. (2002) Nutritional significance of lectins and enzyme inhibitors from legumes. *Journal of Agricultural and Food Chemistry* 50, 6592–6598.

Leterme, P. (2002) Recommendations by health organizations for pulse consumption. *British Journal of Nutrition* 88, S239–S242.

Levander, O.A. and Burk, R.F. (1996) Selenium. In: Ziegler, E.E. and Filer, L.J. Jr (eds) *Present Knowledge in Nutrition*, 7th edn. International Life Sciences Institute, Washington, DC, pp. 320–328.

Lin, K.C., Luh, B.S. and Schweigert, B.S. (1975) Folic acid content of canned garbanzo beans. *Journal of Food Science* 40, 562–565.

Linder, M.C. (1996) Copper. In: Ziegler, E.E. and Filer, L.J. Jr (eds) *Present Knowledge in Nutrition*, 7th edn. International Life Sciences Institute, Washington, DC, pp. 307–319.

Lineback, D.R. and Ke, C.H. (1975) Starches and low-molecular-weight carbohydrates from chick pea and horse bean flours. *Cereal-Chemistry* 52, 334–347.

Lopez Bellido, L. and Fuentes, M. (1990) Cooking quality of chickpea. In: Saxena, M.C., Cubero, J.I. and Wery, J. (eds) *Present Status and Future Prospects of Chickpea Crop Production and Improvement in the Mediterranean Countries (CIHEAM – Options Méditerranéennes: Série A. Séminaires Méditerranéens, No. 9)*. Mediterranean Agronomic Institute of Zaragoza, Spain, pp. 113–125.

Luft, F.C. (1996) Potassium and its regulation. In: Ziegler, E.E. and Filer, L.J. Jr (eds) *Present Knowledge in Nutrition*, 7th edn. International Life Sciences Institute, Washington, DC, pp. 272–276.

Luz Fernandez, M. (1988) The modification of nutritional and functional properties of chickpea (*Cicer arietinum*) by germination. *Dissertation Abstracts International. Section B: The Sciences and Engineering* 49.

Madar, Z. and Stark, A.H. (2002) New legume sources as therapeutic agents. *British Journal of Nutrition* 88, S287–S292.

Madhusudhan, B. and Tharanathan, R.N. (1995) Legume and cereal starches – why differences in digestibility? I. Isolation and composition of legume (greengram and bengalgram) starches. *Starch* 47, 165–171.

Madhusudhan, B. and Tharanathan, R.N. (1996) Legume and cereal starches: why differences in digestibility? III. Structural studies of linear and branched fractions of chickpea and finger millet starches. *Carbohydrate Research* 284, 101–109.

Mahadevamma, S. and Tharanathan, R.N. (2004) Processing of legumes: resistant starch and dietary fiber contents. *Journal of Food Quality* 27, 289–303.

Mahadevamma, S., Shamala, T.R. and Tharanathan, R.N. (2004) Resistant starch derived from processed legumes: *in vitro* and *in vivo* fermentation characteristics. *International Journal of Food Sciences and Nutrition* 55, 399–405.

Marconi, E., Ruggeri, S., Cappelloni, M., Leonardi, D. and Carnovale, E. (2000) Physicochemical, nutritional, and microstructural characteristics of chickpeas (*Cicer arietinum* L.) and common beans (*Phaseolus vulgaris* L.) following microwave cooking. *Journal of Agricultural and Food Chemistry* 48, 5986–5994.

Martens, S. and Mithöfer, A. (2005) Flavones and flavone synthases. *Phytochemistry* 66, 2399–2407.

Mathers, J.C. (2002) Pulses and carcinogenesis: potential for the prevention of colon, breast and other cancers. *British Journal of Nutrition* 8, S273–S279.

Mathur, K.S., Khan, M.A. and Sharma, R.D. (1968) Hypocholesterolemic effect of Bengal gram: a long-term study in man. *British Medical Journal* 1, 30–31.

Mazur, W. (1998) Phytoestrogen content in foods. *Bailliere's Clinical Endocrinology and Metabolism* 12, 729–742.

Mazur, W. and Adlercreutz, H. (2000) Overview of naturally occurring endocrine-active substances in the human diet in relation to human health. *Nutrition* 16, 654–658.

Mazur, W.M., Duke, J.A., Wahala, K., Rasku, S. and Adlercreutz, H. (1998) Isoflavonoids and lignans in legumes: nutritional and health aspects in humans. *Journal of Nutritional Biochemistry* 9, 193–200.

Mendosa, D. (2005) Chana Dal. *Mendosa.com: Your On-line Diabetes Resource*. Boulder, Colorado. Available at: http://www.mendosa.com/chanadal.html

Mikkelsen, R.L., Page, A.L. and Bingham, F.T. (1989) Factors affecting selenium accumulation by agricultural crops. In: Jacobs, L.J. (ed.) *Selenium in Agriculture and the Environment*. American Society of Agronomy, Soil Science Society of America, Madison, Wisconsin, pp. 65–94.

Mittal, M., Nigam, P.K., Rao, S.K. and Rao, S. (1999) Effect of environment on seed quality composition in chickpea (*Cicer arietinum* L.). *Legume Research* 22, 211–221.

Monro, J.A. and Williams, M. (2000) Concurrent management of postprandial glycaemia and nutrient intake using glycaemic glucose equivalents, food composition data and computer-assisted meal design. *Asia Pacific Journal of Clinical Nutrition* 9, 67–73.

Nalwade, V.M., Snehalatha Reddy, N., Bhosale, B.S. and Swami, V.D. (2003) Glycaemic index of commonly consumed Indian foods. *Indian Journal of Nutrition and Dietetics* 40, 341–343.

Nutrient Data Laboratory, USDA Agricultural Research Service (2005) *USDA National Nutrient Database for Standard Reference*. Beltsville, Maryland. Available at: www.nal.usda.gov/fnic/foodcomp/search/

Olson, J.A. (1996) Benefits and liabilities of vitamin A and carotenoids. *Journal of Nutrition* 126, 1208S–1212S.

Olson, J.A. and Kobayashi, S. (1992) Antioxidants in health and disease: overview. *Proceedings of the Society for Experimental Biology and Medicine* 200, 245–247.

Oneda, H., SeungJae, L. and Inouye, K. (2004) Inhibitory effect of 0.19 alpha-amylase inhibitor from wheat kernel on the activity of porcine pancreas alpha-amylase and its thermal stability. *Journal of Biochemistry* 135, 421–427.

Paiva, S.A.R. and Russell, R.M. (1999) β-Carotene and other carotenoids as antioxidants. *Journal of the American College of Nutrition* 18, 426–433.

Parihar, P., Mishra, A., Gupta, O.P., Rajput, L.P.S. and Singh, A. (1999) Effect of cooking processes on nutritional quality of some common pulses. *Advances in Plant Sciences* 12, 15–20.

Patel, P. and Rajput, L.P.S. (2003) Studies on nutritional value and functional properties of 'dhokla' blended with defatted soyflour. *Jawaharlal Nehru Krishi Vishwa Vidyalaya (JNKVV) Research Journal* 37, 28–34.

Perez-Maldonado, R.A., Mannion, P.F. and Farrell, D.J. (1999) Optimum inclusion of field peas, faba beans, chick peas and sweet lupins in poultry diets. I. Chemical composition and layer experiments. *British Poultry Science* 40, 667–673.

Petterson, D.S., Sipsas, S. and Mackintosh, J.B. (1997) *The Chemical Composition and Nutritive Value of Australian Pulses*, 2nd edn. Grains Research and Development Corporation, Kingston, Australia.

Poltronieri, F., Areas, J.A.G. and Colli, C. (2000) Extrusion and iron bioavailability in chickpea (*Cicer arietinum* L.). *Food Chemistry* 70, 175–180.

Quemener, B. and Brillouet, J.M. (1983) Ciceritol, a pinitol digalactoside from seeds of chickpea, lentil and white lupin. *Phytochemistry* 22, 1745–1751.

Ramalho Ribeiro, J.M.C. and Portugal Melo, I.M. (1990) Composition and nutritive value of chickpea. In: Saxena, M.C., Cubero, J.I. and Wery, J. (eds) *Present Status and Future Prospects of Chickpea Crop Production and Improvement in the Mediterranean Countries (CIHEAM – Options Méditerranéennes: Série A. Séminaires Méditerranéens, No. 9)*. Mediterranean Agronomic Institute of Zaragoza, Spain, pp. 107–111.

Rao, G.R. (1974) Effect of heat on the proteins of groundnut and bengal gram. *Indian Journal of Nutrition and Dietetics* 11, 268–372.

Rao, P.U. and Deosthale, Y.G. (1982) Tannin content of pulses: varietal differences and effects of germination and cooking. *Journal of the Science of Food and Agriculture* 33, 1013–1016.

Reddy, S.N., Ardhapurkar, S. and Vijaya, N. (2000) Effect of fermentation on phytate phosphorus and tannin contents and bio-availability of iron of the selected foods. *Indian Journal of Nutrition and Dietetics* 37, 181–187.

Revilleza, M.J.R., Mendoza, E.M.T. and Raymundo, L.C. (1990) Oligosaccharides in several Philippine indigenous food legumes: determination, localization and removal. *Plant Foods for Human Nutrition* 40, 83–93.

Reyes-Moreno, C., Cuevas-Rodriguez, E.O., Milan-Carrillo, J., Cardenas-Valenzuela, O.G. and Barron-Hoyos, J. (2004) Solid state fermentation process for producing chickpea (*Cicer arietinum* L.) tempeh flour: physicochemical and nutritional characteristics of the product. *Journal of the Science of Food and Agriculture* 84, 271–278.

Rossi, M., Germondari, I. and Casini, P. (1984) Comparison of chickpea cultivars: chemical composition, nutritional evaluation and oligosaccharide content. *Journal of Agricultural and Food Chemistry* 32, 811–814.

Rubio, L.A. (2003) Portal vein, hepatic vein and circulating plasma amino acids in rats given diets based on lactalbumin, faba bean (*Vicia faba*) and chickpea (*Cicer arietinum*). *Animal Science (Pencaitland)* 77, 3–10.

Ruiz, R.G. (1996) A study of saponins in legumes of importance to both human and animal nutrition. Thesis. Open University, Milton Keynes, UK.

Ruiz, R.G., Price, K., Rose, M., Rhodes, M. and Fenwick, R. (1996) A preliminary study on the effect of germination on saponin content and composition of lentils and chickpeas. *Zeitschrift für Lebensmittel-Untersuchung und-Forschung* 203, 366–369.

Saini, H.S. (1989) Chickpea quality. In: Brinsmead, R.B. and Knights, E.J. (eds) *Australian Chickpea Workshop*. Australian Institute of Agricultural Science, Brisbane, Australia, pp. 169–173.

Saini, H.S. and Knights, E.J. (1984) Chemical constitution of starch and oligosaccharide components of 'desi' and 'kabuli' chickpea (*Cicer arietinum*) seed types. *Journal of Agricultural and Food Chemistry* 32, 940–944.

Saldeen, K. and Saldeen, T. (2005) Importance of tocopherols beyond a-tocopherol: evidence from animal and human studies. *Nutrition Research* 25, 877–889.

Salgado, P., Lallèsa, J.P., Toulleca, R., Mouratob, M., Cabralb, F. and Freire, J.P.B. (2001) Nutrient digestibility of chickpea (*Cicer arietinum* L.) seeds and effects on the small intestine of weaned piglets. *Animal Feed Science and Technology* 91, 197–212.

Sanchez-Mata, M.C., Penuela-Teruel, M.J., Camara-Hurtado, M., Diez-Marques, C. and Torija-Isasa, M.E. (1998) Determination of mono-, di-, and oligosaccharides in legumes by high-performance liquid chromatography using an amino-bonded silica column. *Journal of Agricultural and Food Chemistry* 46, 3648–3652.

Sánchez-Vioque, R., Clemente, A., Vioque, J., Bautista, J. and Millán, F. (1998) Polar lipids of defatted chickpea (*Cicer arietinum* L.) flour and protein isolates. *Food Chemistry* 63, 357–361.

Saura-Calixto, F., Goñi, I., Bravo, L. and Mañas, E. (1993) Resistant starch in foods: modified method for dietary fiber residues. *Journal of Food Science* 58, 642–643.

Saxena, A.K., Chadha, M. and Sharma, S. (2003) Nutrients and antinutrients in chickpea (*Cicer arietinum* L.) cultivars after soaking and pressure cooking. *Journal of Food Science and Technology* 40, 493–497.

Sears, B. (2000) *The Soy Zone*. Regan Books, New York.

Setchell, K.D.R., Lawson, A.M., Borriello, S.P., Harkness, R., Gordon, H. and Morgan, D.M.L. (1981) Lignin formation in man: microbial involvement and possible roles in relation to cancer. *Lancet* 2, 4–7.

Shahidi, F. and Naczk, M. (1995) *Food Phenolics*. Technomic Publishing Company, Lancaster, Pennsylvania.

Sharma, K.P. and Goswami, A.K. (1971) Chemical evaluation of some high yielding varieties of bengal gram (*Cicer arietinum*). *Indian Journal of Agricultural Research* 5, 109–111.

Shekib, L.A. (1994) In-vitro digestibility and microscopic appearance of germinated

legume starches and their effect on dietary protein utilization. *Food Chemistry* 18, 163–168.
Shils, M.E. (1996) Magnesium. In: Ziegler, E.E. and Filer, L.J. Jr (eds) *Present Knowledge in Nutrition*, 7th edn. International Life Sciences Institute, Washington, DC, pp. 256–264.
Shobhana, S.P.S., Nainawatee, H.S. and Lal, B.M. (1976) Chemical composition of some improved varieties of pulses. *Journal of Food Science and Technology* 13, 49–51.
Singh, U. (1984) Dietary fiber and its constituents in desi and kabuli chickpea (*Cicer arietinum* L.) cultivars. *Nutrition Reports International* 29, 419–426.
Singh, U. (1985) Nutritional quality of chickpea (*Cicer arietinum* L.): current status and future research needs. *Qualitas Plantarum Plant Foods for Human Nutrition* 35, 339–351.
Singh, U. and Jambunathan, R. (1981) Studies on desi and kabuli chickpea (*Cicer arietinum* L.) cultivars: levels of protease inhibitors, levels of polyphenolic compounds and *in vitro* protein digestibility. *Journal of Food Science* 46, 1364–1367.
Singh, U., Kherdekar, M.S. and Jambunathan, R. (1982) Studies on desi and kabuli chickpea (*Cicer arietinum* L.) cultivars. The levels of amylase inhibitors, levels of oligosaccharides and *in vitro* starch digestibility. *Journal of Food Science* 47, 510–512.
Singh, G., Asha, K. and Sehgal, S. (2001) Nutritional composition of selected green leafy vegetables, herbs and carrots. *Plant Foods for Human Nutrition* 56, 359–364.
Sosulski, F.W. and Dabrowski, K.J. (1984) Composition of free and hydrolyzable phenolic acids in the flours and hulls of ten legume species. *Journal of Agricultural and Food Chemistry* 32, 131–133.
Sotelo, A., Flores, F. and Hernandez, M. (1987) Chemical composition and nutritional value of Mexican varieties of chickpea (*Cicer arietinum* L.). *Plant Foods for Human Nutrition* 37, 299–306.
Southgate, D.A.T. (1992) *Starch and Dietary Fibre*. Health Education Authority, London.
Spaeth, S.C. (1987) Pressure-driven extrusion of intracellular substances from bean and pea cotyledons during imbibition. *Plant Physiology* 85, 217–223.
Sreerangaraju, G., Krishnamoorthy, U. and Kailas, M.M. (2000) Evaluation of bengal gram (*Cicer arietinum*) husk as a source of tannin and its interference in rumen and post-rumen nutrient digestion in sheep. *Animal Feed Science and Technology* 85, 131–138.
Suryawanshi, R.P., Reddy, N.S. and Sawate, A.R. (1998) Physico-chemical characteristics and acceptability of snacks of different varieties of chickpea (*Cicer arietinum*). *Journal of Food Science and Technology* 35, 179–182.
Thompson, L.U. (1993) Potential health benefits and problems associated with antinutrients in foods. *Food Research International* 26, 131–149.
Thorne, M.J., Thompson, L.U. and Jenkins, D.J.A. (1983) Factors affecting starch digestibility and the glycaemic response with special reference to legumes. *American Journal of Clinical Nutrition* 38, 481–488.
Tovar, J., Björck, I.M. and Asp, N.G. (1992) Incomplete digestion of legume starches in rats: a study of pre-cooked flours containing retrograded and physically inaccessible starch fractions. *Journal of Nutrition* 122, 1500–1507.
Tuan, Y.H. and Phillips, R.D. (1991) Effect of the hard-to-cook defect and processing on protein and starch digestibility of cowpeas. *Cereal Chemistry* 68, 413–418.
ur-Rehman, Z. and Shah, W.H. (1998) Effect of microwave and conventional cooking on some nutrients, antinutrient and protein digestibility of chick-peas (*Cicer arietinum*). *Bangladesh Journal of Scientific and Industrial Research* 33, 200–205.
ur-Rehman, Z., Salariya, A.M. and Samreen, S. (2004) Physico-chemical characteristics of commonly consumed legumes after domestic processing. *Pakistan Journal of Scientific and Industrial Research* 47, 66–68.
Vahouny, G.V., Satchithanandam, S., Chen, I., Tepper, S.A., Kritchevsky, D., Lightfoot, F.G. and Cassidy, M.M. (1988) Dietary

fiber and intestinal adaptation: effects on lipid absorption and lymphatic transport in the rat. *American Journal of Clinical Nutrition* 47, 201–206.

Vanhoof, K. and de Schrijver, R. (1997) Consumption of enzyme resistant starch and cholesterol metabolism in normo- and hypercholesterolemic rats. *Nutrition Research* 17, 1331–1340.

Veena, A., Urooj, A. and Puttaraj, S. (1995) Effect of processing on the composition of dietary fibre and starch in some legumes. *Nahrung* 39, 132–138.

Vijayalakshmi, N.V.S., Kumar, J. and Rao, T.N. (2001) Inheritance of protein content in chickpea. *Legume Research* 24, 28–31.

Viveros, A., Brenes, A., Elices, R., Arija, I. and Canales, R. (2001) Nutritional value of raw and autoclaved kabuli and desi chickpeas (*Cicer arietinum* L.) for growing chickens. *British Poultry Science* 42, 242–251.

Wall, M.E., Wani, M.C., Manikumar, G., Abraham, P., Taylor, H., Hughes, T.J., Warner, J. and McGivney, R. (1988) Plant antimutagenic agents. 2. Flavonoids. *Journal of Natural Products* 51, 1084–1091.

Wang, Y.H.A. and McIntosh, G.H. (1996) Extrusion and boiling improve rat body weight gain and plasma cholesterol lowering ability of peas and chickpeas. *Journal of Nutrition* 126, 3054–3062.

WHO/FAO (2003) Diet, nutrition and the prevention of chronic diseases: report of a joint WHO/FAO expert consultation. In: *WHO Technical Report Series 916*. World Health Organization (WHO) and the Food and Agriculture Organization (FAO) of the United Nations, Geneva, p. 160.

Williams, C.A. and Grayer, R.J. (2004) Anthocyanins and other flavonoids. *Natural Product Reports* 21, 539–573.

Wolever, T.M.S. and Miller, J.B. (1995) Sugars and blood glucose control. *American Journal of Clinical Nutrition* 62(Suppl. 1), 212S–221S.

Yanezfarias, G.A., Morenovalencia, J.G., Falconvilla, M.D. and Barronhoyos, J.M. (1997) Isolation and partial characterization of starches from dry beans (*Phaseolus vulgaris*) and chickpeas (*Cicer arietinum*), grown in Sonora, Mexico. *Starch-Starke* 49, 341–345.

Yip, R. and Dallman, P.R. (1996) Iron. In: Ziegler, E.E. and Filer, L.J. Jr (eds) *Present Knowledge in Nutrition*, 7th edn. International Life Sciences Institute, Washington, DC, pp. 277–292.

Zaki, A.M., Fetouh, S.A.A., Mohany, H. and El-Morsi, E.A. (1996) Biochemical changes of some chemical constituents in chickpea seeds during germination and processing. *Annals of Agricultural Science, Moshtohor* 34, 995–1022.

Zamora, A.F. (1979) Nutritional evaluation of fermented cowpeas (*Vigna sinensis*) and chickpeas (*Cicer arietinum*). *Dissertation Abstracts International B* 39, 4799.

6 Antinutritional Factors

M. MUZQUIZ[1] AND J.A. WOOD[2]

[1]*Dep. de Tecnología de Alimentos. SGIT-INIA. 28080 Madrid, Spain;*
[2]*Tamworth Agricultural Institute, NSW Department of Primary Industries, Tamworth, NSW 2340, Australia*

Introduction

Recently, there has been an increased interest regarding nutrition, not only in the scientific community, but also in the wider society. Nowadays, the focus is not simply on eating for survival, but to be informed of what we eat to enjoy the best quality of life for longer. Diet is now recognized as important, not only for availing nutrition, but also for the prevention and cure of diseases, especially when these diseases are caused by insufficient, excessive or unbalanced dietary intake.

At present, one of the most controversial subjects of discussion is the establishment of an optimum human diet. Dietary habits differ widely in each part of the world; as are the types of food consumed. A balanced diet must contain all the necessary foods to maintain optimum nutrition. The diet must supply:

1. Sufficient nutrient energy (calories) to carry out metabolic and physical processes.
2. Nutrients with rheological properties and regulatory functions (Muzquiz *et al.*, 2003).

It is very important that all nutrients must be balanced in the diet. The Food and Agriculture Organization–World Health Organization (FAO–WHO) Group of Experts (Nishida *et al.*, 2004) established the following dietary recommendations: protein must not exceed 15% of total calories, carbohydrates 60% (3% as fibre) and not more than 25% of lipids.

The latest WHO–FAO (2003) recommendation in relation to human nutrition and health is a varied and balanced diet comprising 75% of plant-based foods and 25% of animal-based foods.

Numerous epidemiological studies have shown that a plant-based diet can reduce the risk of many chronic diseases. The plant kingdom satisfies an important part of human diets and, depending on plant species, different parts of the plant are used: roots, tubers, stems, leaves, flowers, fruits and seeds. The legume family is one of the largest in terms of number of species. The chickpea (*Cicer arietinum* L.) is the fifth most important legume in the world, on the basis of total production, after soybeans, groundnuts, beans and peas. On the basis of seed characteristics and geographic origin, chickpea is grouped into two types: (i) desi (Indian origin) and (ii) kabuli (Mediterranean and Middle East). Kabuli seeds are traditionally large, rounded, white to cream coloured, and are used almost exclusively as a vegetable by cooking whole seeds. The seeds of desi cultivars are smaller, more angular with a prominent beak and colour normally ranging from yellow-brown to dark brown (and rarely black or green). These cultivars are traditionally dehulled and directly cooked as 'dhal' or milled to flour, 'besan' (Chavan *et al.*, 1989).

Chickpea is high in protein, carbohydrates and fibre, low in fat and provides many vitamins and minerals (Chapter 5, this volume). Nevertheless, when plant foods are consumed, they are often associated with a series of compounds known as antinutrients, which generally interfere with the assimilation of some nutrients. In some cases, these can be toxic or cause undesirable physiological effects (e.g. flatulence). However, recent epidemiological studies have demonstrated that many antinutrients may be beneficial, in small quantities, in the prevention of diseases like cancer and coronary diseases. For this reason they are now often called non-nutritive or bioactive compounds; although they may lack nutritive value, they are not always harmful (Muzquiz, 2000a). The term phytochemicals has also been used and emphasizes the plant source of the majority of these disease-preventing non-nutritive compounds. There is no doubt that many antinutritional compounds can be considered as phytochemicals (Bloch and Thomson, 1995; Hasler *et al.*, 2004).

Legume seeds contain large number of compounds that are qualified as phytochemicals with significant potential benefits to human health (e.g. as anticarcinogenic, hypocholesterolemic or hypoglycemic agents). Non-nutritive compounds vary considerably in their biochemistry. They can be proteins (protease inhibitors, α-amylases and lectins), glycosides (α-galactosides, vicine and convicine), tannins, saponins and alkaloids. Hence, their methods of extraction, determination and quantification are very specific. They do not appear in all plants, and their physiological effects are diverse.

Some of these compounds are important in plant defence mechanisms against predators or environmental conditions. Others are reserve compounds, accumulated in seeds as energy stores in readiness for germination (Roberts and Wink, 1998).

The importance of plant foods is now recognized all over the world. This is especially so in the European Union, where the impact of plant foods on human health, an increased appreciation of the Mediterranean diet and the current problem of bovine spongiform encephalopathy reinforce the importance of these non-nutritive compounds.

Chickpea seed, compared with other legumes, is relatively devoid of protein antinutrients, such as lectins, but it contains protein inhibitors, α-galactosides, phytates, saponins and tannins (Table 6.1). Chickpea, although

Table 6.1. Bioactive compounds and their presence in chickpea seeds.

Class	Relative abundance, in comparison with other legume seeds	Biological activity
Protease inhibitors	++++	Impair protein digestion, antitumoral
Oligosaccharides	+++	Prebiotics, flatulence
Phytates	+++	Metal overloading, glycemic index lowering
Saponins	+++	Growth inhibitors, lower plasma cholesterol
Polyphenols	++	Antioxidant
Isoflavones	±	Phytoestrogens, metabolic control
Lectins	±	Severe growth depression, antitumoral

++++: very abundant; +++: abundant; ++: low; and ±: very low.

lower in antinutritional factors than many other legumes, is rarely used in feed formulations because price is usually a strong deterrent. Chickpea is almost exclusively consumed by humans in various dishes or as snacks.

Available data shows that the balance between deleterious and beneficial effects of these compounds depends on their chemical structure, concentration, time of exposure and interaction with other dietary components. Thus, they can be considered as antinutrients and/or pronutrients with negative and/or positive effects on health, respectively. Hence, it is important to know not only the quantities but also the types of compounds in food and how they affect the human body.

The scientific understanding of the mechanism of action of these non-nutritional compounds or phytochemicals on an organism is an important challenge for the future. This was recently emphasized at the 4th International Workshop on Antinutritional Factors in Legume Seeds and Oilseeds: 'Recent advances of research in antinutritional factors in legume seeds and oilseeds' (2004). Special attention was paid to the beneficial and harmful effects of these compounds and to their mechanism of action in human nutrition. Thus, scientific interest in antinutritional factors is now focusing on potential beneficial applications of these compounds.

Bioactive Compounds

Oligosaccharides

The term oligosaccharide is derived from the Greek word *olio-*, which means *few*. They are composed of 3–10 monosaccharides joined together by glycosidic bonds. The most common oligosaccharides in the plant kingdom are α-galactosides (Kadlec, 2001). Oligosaccharides are not of great importance in foods except for a series of galactosylsucroses (often termed α-galactosides) that occur in legumes and some vegetables, and fructo-oligosaccharides found in cereals, onions and some other plants. The most ubiquitous group within the α-galactosides is the raffinose family of oligosaccharides. The family includes raffinose (a trisaccharide), stachyose (a tetrasaccharide) and verbascose (a pentasaccharide).

Raffinose is present in all parts of the chickpea plant but accumulates in the seeds and roots during development. The concentration of raffinose in the seeds increases to 0.09–3.0 g/100 g (desi) and 0.01–2.8 g/100 g (kabuli) as the seeds mature and dry (Table 6.2). Most of the drought-resistant leguminous crops contain higher raffinose contents and could contribute, in part, to their tolerance (Arora, 1983). Both raffinose and stachyose occur in chickpea seeds and leaves, and have been shown to provide frost tolerance in plants (Castonguay *et al.*, 1995). Stachyose contents range from 0.48 to 5.35 g/100 g for desi type and 0.35 to 6.48 g/100 g for kabuli types. Chickpea has only small amounts of verbascose, up to 0.41 g/100 g reported in desi (Saini and Knights, 1984) and is absent in most of kabuli and some desi varieties.

Ciceritol is also an α-galactoside, but does not belong to the raffinose family of oligosaccharides. It is an α-D-digalactoside of pinitol and was first discovered in chickpea (hence the name ciceritol) by Quemener and Brillouet (1983). It is thought that ciceritol was mistakenly identified as manninotriose in previous literature. Chickpea seeds contain 2.11–3.10 g/100 g and 1.24–2.79 g/100 g ciceritol in desi and kabuli types, respectively (Table 6.2).

The α-galactosides are well known as antinutritional factors for causing flatulence. The first information on the antinutritive effects of α-galactosides was reported back in 1917 by Kuriyama and Mendel (1917). Flatulence occurs because mammals (including humans) have no α-galactosidase present in their intestinal mucosa, which is required to hydrolyse these compounds. Hence, ingestion results in α-galactosides passing into the large intestine where anaerobic fermentation of bacteria occurs producing gaseous by-products (e.g. hydrogen, CO_2 and traces of methane gas). Western populations have poor tolerance to flatulence, and are sometimes accompanied by diarrhoea and abdominal pain if consumed in large quantities. Raffinose and stachyose ingestion causes flatulence (Fleming, 1981), but not ciceritol, perhaps due to easier hydrolysis of this compound because to its different structure (Quemener and Brillouet, 1983). This may explain the reason of less flatulence of chickpea than lentil and beans (Sanchez-Mata *et al.*, 1998).

The lack of digestion is also a constraint as far as intensive monogastric livestock producers are concerned, as oligosaccharides in the diet can reduce growth performance. In addition, a higher content of α-galactosides (higher than that found in normal diets) has been shown to reduce the absorption capacity of the small intestine by changing its osmotic pressure (Wiggins, 1984; Zdunczyk *et al.*, 1998). In any case, intestinal digestion of α-galactosides can be improved by providing supplementary diets with exogenous α-galactosidase (Kozlowska *et al.*, 2001). This is not an issue in the case of chickpea as only severely damaged seed would be economically competitive for animal feed compared to other grains.

Contrary to previous reports, research by Sandberg *et al.* (1993) recently showed that ~30% of dietary raffinose and stachyose was degraded and digested in the stomach and small intestine of humans.

Table 6.2. Oligosaccharide contents of desi and kabuli seeds.

Oligosaccharides	Desi content (g/100 g)	Number of cultivars	References	Kabuli content (g/100 g)	Number of cultivars	References
Raffinose	0.09–3.0	29	Rao and Belavady (1978); Rossi *et al*. (1984); Saini and Knights (1984); Mulimani and Ramalingam (1997); Mansour and Khalil (1998); Salgado *et al*. (2001)	0.01–2.8	22	Rossi *et al*., (1984); Saini and Knights (1984); Mulimani and Ramalingam (1997); Sanchez-Mata *et al*. (1998); Salgado *et al*., (2001)
Stachyose	0.48–5.35	19	Rossi *et al*. (1984); Saini and Knights (1984); Mansour and Khalil (1998); Salgado *et al*. (2001)	0.35–6.48	21	Rossi *et al*., (1984); Saini and Knights (1984); Sanchez-Mata *et al*., (1998); Salgado *et al*., (2001)
Verbascose	0.01–0.41	8	Saini and Knights (1984); Salgado *et al*. (2001)	0.00–0.37	2	Salgado *et al*., 2001
Ciceritol	2.11–3.10	10	Rossi *et al*. (1984)	1.24–2.79	12	Rossi *et al*., (1984); Sanchez-Mata *et al*., (1998)

Although α-galactosides have little food value (are partially digested, if at all), this does not imply the absence of health benefits. In fact, α-galactosides convey many benefits to both humans and monogastric animals. Since they pass mostly undigested into the lower gut, they are a constituent of dietary fibre and can act as a prebiotic. Dietary fibre also is highly beneficial (refer to Chapter 5, this volume). In addition, the prebiotic effect is derived from metabolism of α-galactosides by gas-producing bacteria, which increase the colonic population of bifidobacteria. Bifidobacteria are advantageous to human health by suppressing intestinal putrefaction, reducing both constipation and diarrhoea, stimulating the immune system and increasing resistance to infection (Mitsuoka, 1996).

Phytic acid

Phytic acid, myo-inositol-(1,2,3,4,5,6) hexakis-phosphate and its salts are the major sources of phosphorus in legume seeds (Urbano *et al.*, 2000). Ravindran *et al.* (1994) found that the phytate phosphorus in chickpea was 51.2% of the total phosphorus content. The total phytic acid content in chickpea has been reported to vary from 0.3% to 1.8% (Duhan *et al.*, 1989; Ravindran *et al.*, 1994; Burbano *et al.*, 1995; Rincón *et al.*, 1998; Martinez *et al.*, 2002). Phytic acid content can vary with genotype, climate, type of soil and year.

Phytic acid has been considered an antinutrient as it binds with other nutrients and makes them indigestible. Excessive phytic acid in the diet can have a negative effect on mineral balance because of the insoluble complexes it forms with essential minerals (Cu^{2+}, Zn^{2+}, Fe^{3+} and Ca^{2+}), which causes poor mineral bioavailability (Zhou and Erdman, 1995; Urbano *et al.*, 2000). Phytic acid is able to make complex with proteins also, decreasing protein solubility. Therefore, phytates have negative impact on enzyme activity and there is evidence of its negative effects on key digestive enzymes like lipase, α-amylase, pepsin, trypsin and chymotripsin (Thompson, 1993; Greiner and Konietzny, 1996b; Urbano *et al.*, 2000). The binding of phytic acid to these enzymes reduces nutrient digestibility. Phytic acid also binds with starch through phosphate linkages (Lajolo *et al.*, 2004).

The ability of phytic acid to bind with minerals, proteins or starch, directly or indirectly, may alter solubility, functionality, digestibility and absorption of these nutrients. In addition, monogastric animals have a limited ability to hydrolyse phytates and release phosphate for absorption due to a lack of intestinal phytases (Zhou and Erdman, 1995; Greiner and Konietzny, 1996a; Urbano *et al.*, 2000).

However, there are some beneficial effects of phytic acid, such as reduced bioavailability, and therefore toxicity, of heavy metals (e.g. cadmium and lead) present in the diet (Rimbach *et al.*, 1996; Rimbach and Pallauf, 1997). Several studies have provided evidence of antioxidant properties of phytic acid *in vitro* (Lajolo *et al.*, 2004). These effects are mainly mediated through its iron and copper chelating properties, although the molecular mechanisms are not fully understood. Contrary to this was research by Minihane and Rimbach (2002) who demonstrated that under *in vivo* conditions phytic acid did not always have a significant effect on oxidant or antioxidant status.

Phytases are enzymes that are widely distributed throughout nature: in plants, certain animal tissues and microorganisms. They have been studied intensively in the last few years because of their ability to reduce the phytate content in monogastric and human foods (Greiner and Konietzny, 1996b). This is done by hydrolysing phytic acid to a series of lower phosphate esters of myo-inositol and phosphate. Two types of phytases are known: 3-phytases (EC 3.1.3.8) and 6-phytases (EC 3.1.3.26), indicating the susceptible phosphoester bond that is predominantly attacked by the enzyme. The pathway of dephosphorylation of myo-inositol hexakis-phosphate by phytases purified from different legume seeds has been established by Maiti *et al.* (1974) and modified by Greiner *et al.* (2002).

Different structural isomers of myo-inositol phosphates can be generated during enzymatic degradation. They can have different physiological functions; hence, their identification is of great importance to exploit the full potential of naturally occurring phytases (Greiner *et al.*, 2002). The myo-inositol phosphates, IP_6 and IP_5, have the worst antinutritional effects, as the smaller molecules (IP_4, IP_3, IP_2 and IP_1) have a lower capacity to complex with inorganic cations. The major inositol phosphate in chickpea is IP_6 (Burbano *et al.*, 1995).

Some myo-inositol phosphates, including IP_6 from soybean, have been suggested to have beneficial health effects, such as amelioration of heart disease by controlling hypercholesterolemia and atherosclerosis, prevention of kidney stone formation and a reduced risk of colon cancer (Greiner *et al.*, 2002).

Tannins

The term 'tannin' originated from the ability of these compounds to tan animal skins to produce leather, and dates back to ancient times. Most, but not all, tannins have this ability and can also form insoluble complexes with carbohydrates and protein. In fact, the precipitation of salivary proteins is responsible for the distinct astringency sensation of tannin-rich foods. Tannins are believed to play a role in plant protection (against infection, insect and animal damage) as these types of stresses cause increased plant tannin levels (Haslam, 1998).

Tannins are now defined as polyphenolic secondary metabolites of higher plants, which fall into four specific groups depending on their basic chemical structure: (i) gallotannins; (ii) ellagitannins; (iii) complex tannins; and (iv) condensed tannins (also known as proanthocyanidins). Figure 6.1 shows this classification published by Khanbabaee and van Ree (2001). Currently, more than 1000 naturally occurring tannins are structurally identified (Quideau and Feldman, 1996).

Previously, tannins were divided into two groups based on their ability to fractionate hydrolytically (with acid, alkali, hot water or enzymatic action) as: (i) hydrolysable tannins (including gallotannins and ellagitannins) or (ii) condensed tannins. This definition was first elucidated by Freudenberg (1919). This simple categorization means that these definitions are still most commonly used in nutritional analyses today, despite ignoring tannins with molecular weights below 1000 Da (Khanbabaee and van Ree, 2001).

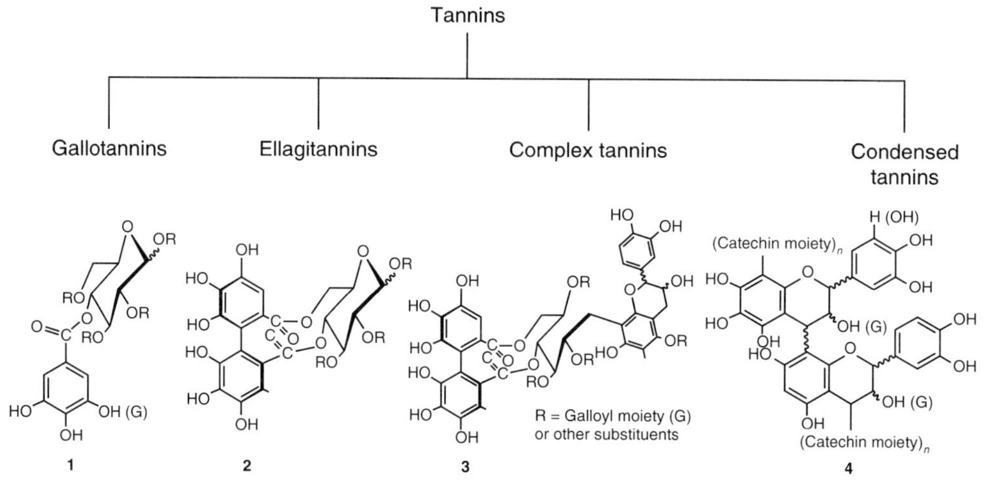

Fig. 6.1. Tannin classification. (From Khanbabaee and van Ree, 2001.)

The tannin contents of chickpea from the literature are listed in Table 6.3. Chickpea has been reported to contain approximately 0.36–0.72 g/100 g and 0.12–0.51 g/100 g total tannin (desi and kabuli types, respectively), of which 0–13% is condensed tannin. Chickpea contains low tannin contents, with concentrations (both total and condensed) much lower than faba bean (*Vicia faba*), pigeon pea (*Cajanus cajan*) and lentil (*Lens culinaris*).

The hydrolysable tannins are not considered to have more antinutritional effects, but are responsible (with some other polyphenolic compounds) for the colour of chickpea seed coats. The condensed tannins (proanthocyanidins) are high-molecular-weight polymers (commonly ~5000 Da) and are known to bind dietary protein, thereby reducing protein digestion. This antinutritional effect can be significant in some other legumes, but only very small amounts of condensed tannins are found in chickpea (<0.1% in desi and <0.04% in kabuli; Table 6.3). Garcia-Lopez *et al.* (1990) found that tannins from chickpea caused no significant reduction in iron absorption, in contrast to tannins from tea, coffee and wine. In addition, tannins are located primarily in the seed coat; hence, the removal of the seed coat during or before food preparation will practically eliminate tannins from the diet.

Table 6.3. Tannin contents of desi and kabuli seeds.

Chickpea type	Tannins	Content (g/100 g)	Number of cultivars	References
Desi	Total tannin	0.36–0.72	53	Petterson *et al.* (1997); Salgado *et al.* (2001)
	Condensed tannin	0.01–0.09	47	Petterson *et al.* (1997); Salgado *et al.* (2001)
Kabuli	Total tannin	0.12–0.51	39+	Zaki *et al.* (1996); Petterson *et al.* (1997); Viveros *et al.* (2001)
	Condensed tannin	0.00–0.04	37	Petterson *et al.* (1997); Salgado *et al.* (2001)

Protein antinutrients

The most widely studied antinutrient proteins in legumes are the enzyme inhibitors (pancreatic proteases and α-amylases) and the lectins (Lajolo *et al.*, 2004; Pusztai *et al.*, 2004).

Protease inhibitors

Legume seed protease inhibitors can have a major impact on seed nutritional value as they inhibit the function of digestive enzymes, such as trypsin and chymotrypsin, by competitive binding. These protease inhibitors contain no carbohydrates and belong to two different families: the Kunitz family and the Bowman-Birk family. Protease inhibitors from both families have been found in chickpea seeds (Domoney, 1999; Lajolo and Genovese, 2002; Lajolo *et al.*, 2004; Srinivasan *et al.*, 2005). Both families are capable of inhibiting trypsin and chymotrypsin. A large number of isoforms of the Bowman-Birk inhibitor (BBI) have been described in soybean, and have differing properties depending on their chemical structure. Saini *et al.* (1992) found 6–8 isoinhibitors in chickpea capable of inhibiting trypsin (EC 3.4.21.4) and chymotrypsin (EC 3.4.21.1). Smirnoff *et al.* (1976) found two active fragments of protease inhibitor in chickpea seeds, designated as A and B. Fragment A inhibited trypsin but not chymotrypsin, and fragment B inhibited chymotrypsin but not trypsin. The inhibition of human trypsin and chymotrypsin by chickpea seed extracts was reported by Gertler *et al.* (1982) and Belitz *et al.* (1982), and that of bovine enzymes by Borchers and Ackerson (1947), Abramova and Chernikov (1964) and Belitz *et al.* (1982). Although the porcine pancreatic proteinases have been subjected to inhibition studies using extracts from a range of leguminous species (Rascon *et al.*, 1985), there is no report on similar inhibitory studies with chickpea. The protease inhibitor contents of desi and kabuli seeds from the literature are reported in Table 6.4. Chickpea contains considerably higher amounts of trypsin inhibitor than the other commonly consumed Indian pulses, but contains much less than soybean (Sumathi and Patabhiraman, 1976). However, this is unlikely to be a problem as protease inhibitors are heat-labile and chickpea is usually cooked before consumption.

Antinutritional effects and health benefits have been ascribed to the presence of certain amounts and types of protease inhibitors. The effect of trypsin inhibitors on animal growth is a consequence of inhibition of intestinal protein digestion, since the presence of inhibitors in diets consisting of free amino acids also led to decreased growth (Lajolo *et al.*, 2004).

The Kunitz inhibitor and BBIs have been found to cause an enlargement of the pancreas (hypertrophy and hyperplasia) and hypersecretion of digestive enzymes (sulphur-rich proteins) in rodents and birds. This results in a loss of the sulphur-rich endogenous proteins that would cause growth depression, as legume seed proteins are generally deficient in sulphur amino acids (Lajolo and Genovese, 2002).

On the other hand, protease inhibitors have been linked, over the last two decades, to health-promoting properties (Champ, 2002) and are considered as

Table 6.4. Protease inhibitor contents of desi and kabuli seeds.

Chickpea type	Inhibitor type	Content (mg)	Number of cultivars	References
Desi	Trypsin inhibitor	1.16–15.7	129+	Singh and Jambunathan (1981); Batterham (1989); Petterson *et al.* (1997); Saini (1997); Salgado *et al.* (2001)
	Chymotrypsin inhibitor	2.40–13.19	77+	Singh and Jambunathan (1981); Batterham (1989); Petterson *et al.* (1997); Saini (1992)
Kabuli	Trypsin inhibitor	1.39–12.1	55+	Singh and Jambunathan (1981); Batterham (1989); Zaki *et al.* (1996); Petterson *et al.* (1997); Salgado *et al.* (2001)
	Chymotrypsin inhibitor	3.0–10.74	25+	Singh and Jambunathan (1981); Batterham (1989); Petterson *et al.* (1997); Salgado *et al.* (2001)

natural bioactive substances (Hill, 2004). Protease inhibitors may act as anticarcinogenic agents (Clemente *et al.*, 2004). BBIs have been shown to be effective in preventing or suppressing carcinogen-induced transformation *in vitro* and carcinogenesis in animal assays. The BBI achieved investigational new drug (IND) status from the Food and Drug Administration (FDA) in 1992 for this purpose (Kennedy, 1995), and studies with humans showed no toxic effects of BBI (Amstrong *et al.*, 2000). Use of BBI inhibited the growth and survival of human prostate cancer cells (Kennedy and Wan, 2002). It also reduced the incidence and frequency of colon tumours in dimethylhydrazine-treated rats. However, this effect was not observed with autoclaved BBI, suggesting that protease inhibitor activity was necessary for anticarcinogenic activity (Kennedy *et al.*, 2002).

Dietary protease inhibitors have to survive digestive process in the gastrointestinal tract in order to exert an effect, either local or systemic. There is not much information in the literature regarding the intestinal recovery of non-nutritional factors, probably due to methodological difficulties (Rubio *et al.*, 2005). BBIs were effective in preventing or suppressing carcinogens when fed to animals (Pusztai *et al.*, 2004). According to Clawson (1996), the effect of dietary protease inhibitors would be indirect. Clawson hypothesized that dietary protease inhibitors may act by inducing the synthesis and distribution of endogenous protease inhibitors, which would have widespread effects on cell growth and behaviour.

α-Amylase inhibitors

α-Amylases (α-1,4-glucan-4-glucanohydrolases) are endoamylases that catalyse the hydrolysis of α-D-(1,4) glycosidic linkages, which occur in starch and related compounds. They play a major role in the carbohydrate metabolism of animals and humans by providing them glucose as an energy source and as a building block for synthesis of other sugars. Among α-amylase inhibitors (αAIs) found in plants, legume αAIs, especially αAIs from beans, have received considerable attention (Whitaker, 1988). Jaffe *et al.* (1973) screened 95 legume cultivars for αAI levels and found that lima beans (*Phaseolus lunatus*), mung beans (*Phaseolus aureus*) and horse gram (*Dolichos biflorus* L.) had the highest levels of inhibitory activity. Although little is known about the screening of αAIs in chickpea, Mulimani *et al.* (1994) determined the αAI activity of 28 varieties of chickpea and found variations ranging from 11.6 to 51.4 inhibitory units/g. Singh *et al.* (1982) observed that the amylase inhibitor activity on pancreatic amylase of chickpea cultivars ranged from 7.8 to 10.5 units/g (desi) and 5.6 to 10.0 units/g (kabuli), with considerable variations among these cultivars. A similar variation but of lower magnitude was observed using salivary amylase. A comparison under similar assay conditions indicated that the amylase inhibitor activity had a stronger influence on pancreatic amylase than salivary amylase for both desi and kabuli cultivars. Jaffe *et al.* (1973) reported that the partially purified kidney bean inhibited the salivary amylase more than the pancreatic amylase. This shows that amylase inhibitors from different legume seeds may exhibit unequal activity against different enzymes.

The most probable function of αAI in the plant is protection from predatory insects by inhibiting their digestive amylase. These αAIs have been shown to inhibit pancreatic amylases (human and porcine), human salivary amylase and insect amylase (Le Berre-Anton *et al.*, 1997). αAIs are deleterious or antimetabolic for many predators of crop plants. The genes for both of these factors have been introduced into a number of plants and the resultant transgenic crops have a significantly increased resistance to predators (Gatehouse *et al.*, 1994).

αAIs reduce amylase activity and starch digestion in the gut when given orally to humans (Singh *et al.*, 1982). As a result, they lower postprandial increases in circulating glucose and insulin. These inhibitors may therefore prove to be useful in the treatments of obesity or diabetes mellitus.

The stability of αAI to proteolytic degradation *in vivo* does not seem to have been evaluated. However, the findings that orally administered αAI could greatly reduce intraluminal amylase levels and starch digestion in rats and humans suggests that nutritionally significant quantities of αAI survive passage through the small intestine in a fully functional form (Pusztai *et al.*, 1995).

High dietary intakes of αAIs can however cause a number of potentially deleterious alterations in the body metabolism of experimental animals. Significant enlargement of the small intestine and of the pancreas was evident in rats given αAI (Grant *et al.*, 1998).

Lectins

Lectins (haemagglutinins) are glycoproteins, which are able to reversibly bind to specific sugars and glycoproteins on the surface of cells in the gut wall, thereby interfering with nutrient breakdown and absorption. This reaction is manifested, *in vitro*, by agglutination of red blood cells from various animal species. Lectins are extremely specific; different types having different interactions on toxicity, blood groups, mitogenesis, digestion and agglutination.

Traditionally, lectins have been measured by their haemagglutinating activity (Grant, 1991). However, when possible, enzyme-linked immunosorbent assay (ELISA)-based immunological methods that recognize specific antibodies are being increasingly used due to their higher specificity (Hajós *et al.*, 1996; Muzquiz *et al.*, 2001).

Kolberg *et al.* (1983) isolated a lectin from seed extracts of *C. arietinum* and N-terminal amino acid sequence analyses indicated only one type of chain, suggesting that the lectin is probably dimeric. Papainized human erythrocyte of the different ABO groups were agglutinated equally well by the *Cicer* lectin, whereas untreated cells reacted weakly and only in the presence of bovine serum albumin.

Cuadrado *et al.* (Madrid, 2005, personal communication) analysed lectins of desi and kabuli types (Table 6.5). Two haemagglutination assay systems were used: native rat blood cells and blood cells treated with trypsin to increase the sensitivity of the assay. *Phaseolus vulgaris* cvs. Processor and Pinto were included in each assay as standards and 1 HU (haemagglutination unit) was defined as the amount of material in the last dilution at which 50% of the cells were agglutinated. These results showed that the chickpea had low reactivity with all the cells, with values similar to that of *Lupinus* ssp. (Pedrosa *et al.*, 1997). Hettiarochchy and Kantha (1982) found haemagglutinin levels of 400 HU/g in chickpea; half the activity measured in soybean (*Glycine max*). Grant *et al.* (1983) found that among the 15 legumes they studied, chickpea seeds had the lowest haemagglutinin activities, and were non-toxic.

Nutritionally, dietary lectins vary considerably in the nature and extent of their antinutritional effects. Lectins can be toxic; they can interfere with hormone balance and deplete nutrient reserves leading to severe growth depression and a high incidence of deaths. However, lectins may be beneficial by stimulating gut function, limiting tumour growth and ameliorating obesity (Pusztai *et al.*, 2004).

Saponins

Saponins derive their name from their ability to form stable, soap-like foams in aqueous solutions. They are a complex and chemically diverse group of compounds. Chemically saponins are composed of a steroidal or triterpene aglycone linked to one, two or three saccharide chains of varying size and complexity by

Table 6.5. Initial screening of lectin content of two varieties of chickpeas by determination of haemagglutination activity of sample extracts.

	Native rat cells		Trypsin-treated rat cells	
	HA (μg)	HU/100 g E+6	HA (μg)	HU/100 g E+6
Cicer arietinum Athenas	12,500	0.008	390	0.26
C. arietinum Tizón	12,500	0.008	1,560	0.06
Std (Processor)	49	2.04	0.05	2,000
Std (Pinto)	12,500	0.008	49	2.04

HA: haemagglutination activity (amount of material [μg] containing 1 HU); HU: haemagglutination unit.

ester and ether linkages. The complexity of the saponin structure (and thereby their diversity of biological activities) depends on the variability of the aglycone structure, the attachment position of the glycosidic moieties and the nature of these glycosides. Aglycones are generally linked to D-galactose, L-rhamnose, D-glucose, D-xylose, D-mannose and D-glucuronic acids, some of which may be acetylated (Fenwick *et al.*, 1991). All legumes have triterpene-type saponins.

The chemical composition of soybean (*G. max*) saponins has been extensively investigated. They are triterpene saponins (known as soyasaponins) of which more than 10 types have been isolated (Shibuya *et al.*, 2006). Soyasapogenol B was identified as the aglycone in saponins from different legume species (Price *et al.*, 1987, 1988; Ayet *et al.*, 1996). Tava *et al.* (1993) found 17.7 and 21.9 g/kg of sapogenol-B in desi and kabuli types, respectively. Ruiz *et al.* (1996, 1997) showed the presence of soyasaponin VI, a conjugated form of soyasaponin I, in many legumes, and was the only saponin detected in the seeds of chickpea. Soyasaponin VI may have an important physiological role in preventing lipid peroxidation of DNA and proteins by free radical attack.

A range of techniques has been used for saponin analysis based on their chemical structure and biological activity (Fenwick *et al.*, 1991; Oleszek, 2002). Kerem *et al.* (2005) observed two distinct saponins in various chickpea extracts using thin layer chromatography. Identification of the major saponin as a 2,3-dihydro-2,5-dihydroxy-6-methyl-4H-pyran-4-one (DDMP)-conjugated saponin was verified using ^{1}H and ^{13}C nuclear magnetic resonance (NMR).

Ingestion of saponin-containing plant foods by humans and animals has been associated with both deleterious and beneficial effects (Oakenfull and Sidhu, 1989). There are two main reasons for their physiological activity: (i) their powerful physical interactions, as surface-active agents, with other components of the digesta and (ii) their ability to interact with the membranes of mucosal cells. Sometimes, the latter causes profound changes in the associated membrane biochemistry. These physiological interactions can inhibit nutrient uptake.

Potential benefits from the consumption of food saponins include a reduced risk of cardiovascular disease and some cancers. Dietary saponins have been repeatedly shown to lower plasma cholesterol in animals. However, their hypocholestoremic effect in humans is more speculative. They may also have anticarcinogenic properties, as suggested by a recent rodent study in which feeding a saponin-containing diet inhibited the development of preneoplastic lesions in the colon (Koratkar and Rao, 1997). However, these results may be irrelevant to humans (Ridout *et al.*, 1988). Kerem *et al.* (2005) found that the major saponin in chickpea seeds possesses antifungal properties. In addition, some saponins have been reported to inhibit *in vivo* human immunodeficiency virus (HIV) infectivity (Thompson, 1993).

Allergenicity

Vioque *et al.* (1999) found that chickpea PA2 albumin proteins have the ability to agglutinate papainized human erythrocytes and can act as allergens in susceptible individuals.

Aflatoxin (and other mycotoxins) has been reported in chickpea, from trace to 220 ppb, and the contamination increases with time in storage (Nahdi *et al.*, 1982). Aflatoxin is a known allergen commonly found in groundnuts.

Processing

Processing generally improves the nutrient profile of legume seed by increasing *in vitro* digestibility of proteins and carbohydrates from approximately 40% up to 98% (ur-Rehman and Shah, 1998; el-Adawy, 2002; Naveeda and Jamuna, 2004). At the same time there are reductions in some antinutritional compounds (Muzquiz *et al.*, 1996; Hajós and Osagie, 2004; Jiménez-Martinez *et al.*, 2004).

The following are the traditional processing techniques: (i) non-heat processing such as imbibition, germination, dehulling or fermentation and (ii) heat processing, such as cooking at atmospheric pressure, autoclaving or roasting. Most antinutritional factors are heat-labile, such as α-galactosides, protease inhibitors and lectins, so cooking would eliminate any potential ill effects before consumption. On the other hand, tannins, saponins and phytic acids are heat-stable, but can be reduced by dehulling, soaking, germination and/or fermentation.

Dehulling is a common processing method of producing dhal in desi type, which is then cooked or milled to flour. Dehulling removes the seed coat, which contains many of the antinutritional factors, such as tannins ~90% (Rao and Deosthale, 1982). Therefore, the resulting dhal contains higher concentrations of protein, carbohydrates and vitamins.

Imbibition is a common practice before cooking whole grains, mainly to reduce cooking time (Wood and Harden, 2006). However, soaking also reduces certain antinutritional factors, which leach into the soaking medium, such as oligosaccharides, protease inhibitors and some tannins (Saxena *et al.*, 2003). The amount of leaching will vary depending on the soaking medium (water, salt solution or bicarbonate solution) and soaking time. Rao and Deosthale (1982) found that the tannin content of chickpeas had reduced by almost 50% after soaking overnight in water.

The effects of germination and cooking treatments on the nutritional composition and antinutritional factors of chickpeas were studied by el-Adawy (2002). He found that germination was more effective in reducing phytic acid, stachyose and raffinose in chickpea than cooking. In contrast, Greiner *et al.* (1998) reported a large reduction in phytate due to germination in lentils, but not in chickpeas. Rao and Deosthale (1982) found germination to decrease tannin content of chickpeas by a further 10% after soaking overnight. In contrast, Khaleque *et al.* (1985) found no significant decrease in chickpea tannin content during germination.

Cooking (boiling, autoclaving and microwave cooking) was found to be more effective in reducing trypsin inhibitors, haemagglutinin activity, tannins and saponins than germination (el-Adawy, 2002). Rao and Deosthale (1982) found that cooking, without prior soaking, decreased tannin contents by 70%. Cooking has been reported to decrease the oligosaccharide content (Rao and Belavady,

1978; Savitri and Desikachar, 1985), whereas Kantha *et al.* (1973) found that cooking made no significant reduction to the flatus production of dhal.

Fermentation significantly decreased both phytic acid and tannin contents in chickpea tempeh flour (Reyes-Moreno *et al.*, 2004). Saponin contents of soybean were halved in fermented tempeh compared with the raw seed (Fenwick and Oakenfull, 1983). Fermentation also resulted in a significant reduction of the raffinose content of chickpeas (Zamora and Fields, 1979).

Increased demand for food products, whose nutritional and sensory quality is retained or improved, has seen the introduction of new food processing technologies like extrusion/cooking (high temperature/short time [HTST] extrusion) and microwaving. The advantages of hot extrusion against other heat-processing methods are improved nutritive value of the raw material. It is a versatile processing technique that can produce a variety of products of different form, texture and organoleptic characters, tailored to market demand, from similar raw materials. It can also reduce thermolabile factors such as lectins and trypsin inhibitors and increase starch and protein digestibility (Urdaneta *et al.*, 2003). Alonso *et al.* (2001) extruded pea (*Pisum sativum*) and kidney bean (*P. vulgaris*) meals, which reduced tannin, oligosaccharide, starch, non-starch polysaccharides (NSP) and lectin contents, and increased mineral availability. el-Hady and Habiba (2003) found that soaking of chickpea meals followed by extrusion led to significant decreases in antinutrients, such as phytic acid, tannins, phenols, α-amylase and trypsin inhibitors. The *in vitro* protein digestibility of legume extrudates was also improved by this processing. In contrast, Poltronieri *et al.* (2000) found no significant difference in the chemical composition of chickpeas after extrusion and pressure-cooking. Wang and McIntosh (1996) found that both boiling and extrusion of chickpea significantly increased the body weight gain (BWG) and lowered plasma cholesterol levels in rats, compared with a casein control diet, but did not improve the protein efficiency ratio (PER). Raw and extruded chickpea flours may be considered for fortification of widely consumed cereal-based food products (e.g. tortillas, bread, cookies and pasta) to improve their nutritional composition (Wood, 2000; Reyes-Moreno *et al.*, 2002).

There are contradictory results concerning the effect of microwaving on nutrients; some suggest that increased nutritive value is only obtained when low power is used for a short exposure time. It seems that power is the most important parameter affecting the results, but few studies on the optimization of the physical parameters are also involved. Studies on soybean, using dry or soaked seed, and using different time and temperature conditions, show a reduction in trypsin inhibitor and lectin content with improved starch digestibility, without any significant effect on nutritional quality (Rajkó *et al.*, 1997; Mahungu *et al.*, 1999). Autoclaving was found to be more effective in reducing lupin allergenicity than extrusion, boiling or microwaving (Alvarez-Alvarez *et al.*, 2005).

Genetic Improvement

Genetic improvement of nutritive value can modify the content of non-nutritive compounds. Plant breeding genetically eliminates the alkaloids in different species of *Lupinus*, and decreases vicine and convicine in *V. faba*.

In classical plant breeding, genetic modifications may occur spontaneously in nature, and humans have consciously tried to exploit these in agriculture for crop improvement and animal husbandry (Duc *et al.*, 1989). Classical plant breeding is more often performed under a controlled environment, making deliberate and specific crosses to maximize the benefits of each parent into a new genotype. This method is only possible where significant genotypic variation exists within the species or within wild species of the same family.

At present, there are many possibilities for the use of genetically engineered plants. Genetic engineering allows selective modifications by the insertion or elimination of known single genes. Modern genetic manipulation explores the possibility of acting on the information contained in hereditary material more quickly, by adding genes from different species, or modifying gene expression to create genetically modified organisms (GMO) (Monsanto, 2001).

There is no doubt that GMOs and particularly genetically modified plants open an important area that is one of the most promising research areas from the agriculture, food and health perspective. Chickpea, like other grain legumes, contains seed protein that is relatively deficient in the sulphur amino acids, cysteine and methionine. In order to increase the concentration of the nutritionally essential sulphur amino acids in seed protein, a transgene encoding a methionine- and cysteine-rich protein, sunflower seed albumin (SSA), was transferred to chickpea. The results suggested that free methionine and *O*-acetylserine acted as signals that modulated chickpea seed protein composition in response to the variation in sulphur demand, as well as in response to variation in the nitrogen and sulphur status of the plant (Chiaiese *et al.*, 2004). Such transgenic chickpea seeds have the potential to reduce the amount of synthetic methionine required in chickpea-based diets (Brenes *et al.*, 2004).

Nevertheless, considerable protest has been raised regarding the safety of genetically engineered foods. Ewen and Pusztai (1999) investigated the deleterious effects of a GMO potato containing a foreign lectin on rats. The possibility of allergen transfer by genetic engineering was discovered when a methionine-producing gene from Brazil nut (*Bertholletia excelsa*) was incorporated into soybean to increase its protein quality. The 2S albumin of Brazil nut (allergen Ber e 1) expressed in transgenic soybean is recognized by immunoglobulin (IgE) antibodies of patients who are allergic to Brazil nut (Nordlee *et al.*, 1996). These examples highlight the importance of knowing, in advance, the characteristics of dietary components (including allergenic properties), whose genes are being transferred to ensure that there will be no side effects.

Conclusion

This chapter clearly demonstrates the complexity of these non-nutritive compounds in chickpea seed both from biochemical and physiological points of view. Taking into account the comparison between beneficial and deleterious effects of these bioactive compounds, the most important challenge to be faced in the future is to know their *in vivo* effects and how to modify their concentrations to our advantage.

References

Abramova, E.P. and Chernikov, M.P. (1964) Contents of proteinase inhibitor in seeds of some leguminous plants. *Voprosy Pitaniya* 23, 13–16.

Alonso, R., Rubio, L.A., Muzquiz, M. and Marzo, F. (2001) The effect of extrusion cooking on mineral bioavailability in pea and kidney bean seed meals. *Animal Feed Science Technology* 94, 1–13.

Alvarez-Alvarez, J., Guillamón, E., Crespo, J.F., Cuadrado, C., Burbano, C., Rodriguez, J., Fernandez, C. and Muzquiz, M. (2005) Effects of extrusion, boiling, autoclaving and microwave heating on lupin allergenicity. *Journal of Agricultural and Food Chemistry* 23, 1294–1298.

Amstrong, W.B., Kennedy, A.R., Wan, X.S., Atiba, J., McLaren, E. and Meyskens, F.L. (2000) Single-dose administration of Bowman-Birk inhibitor concentrate in patients with oral leukoplakia. *Cancer Epidemiology Biomarkers and Preventions* 9, 43–47.

Arora, S.K. (1983) *Chemistry and Biochemistry of Legumes*. IFIS Publishing, New Delhi, India.

Ayet, G., Muzquiz, M., Burbano, C., Robredo, L.M., Cuadrado, C. and Price, K. (1996) Determinación de saponinas en las principales leguminosas cultivadas en España. (Determination of saponins as the main legumes cultivated in Spain.) *Food Science and Technology International* 2, 96–100.

Batterham, E.S. (1989) Use of chickpeas (*Cicer arietinum* L.) in animal breeding. In: Brinsmead, R.B. and Knights, E.J. (eds) *Australian Chickpea Workshop*. Australian Institute of Agricultural Science, Brisbane, Australia, pp. 21–27.

Belitz, H.D., Lynen, F. and Weder, J.K.P. (1982) Comparative studies on the inhibitory action of some legume seeds, potato tubers, and bran against human and bovine proteinases. *Zeitschrift fur Lebensmittel-Untersuchung und–Forschung* 174, 442–446.

Bloch, A.S. and Thomson, C.A. (1995) Position of the American Dietetic Association: phytochemicals and functional foods. *Journal of American Dietetic Association* 95, 493–496.

Borchers, R. and Ackerson, C.U. (1947) Trypsin inhibitor IV. Occurrence in seeds of the leguminosa and other seeds. *Archives of Biochemistry* 13, 291–293.

Brenes, A., Jansman, A.J.M. and Marquardt, R.R. (2004) Recent progress on research on the effects of antinutritional factors in legume and oil seeds in monogastric animals. In: Muzquiz, M., Hill, G.D., Pedrosa, M.M. and Burbano, C. (eds) *Proceedings of the Fourth International Workshop on Antinutritional Factors in Legume Seeds and Oilseeds*. EAAP Publication No. 110, Wageningen, The Netherlands, pp. 195–217.

Burbano, C., Muzquiz, M., Osagie, A., Ayet, G. and Cuadrado, C. (1995) Determination of phytate and lower inositol phosphates in Spanish legumes by HPLC methodology. *Food Chemistry* 62, 321–326.

Castonguay, Y., Nadeau, P., Lechasseur, P. and Chounard, L. (1995) Differential accumulation of carbohydrates in alfalfa cultivars of contrasting winter hardiness. *Crop Science* 35, 509–516.

Champ, M.M. (2002) Non-nutrient bioactive substances of pulses. *British Journal of Nutrition* 88, 307–319.

Chavan, J.K., Kadam, S.S. and Salunkkhe, D.K. (1989) Chickpea. In: Salunkhe, D.K. and Kadam, S.S. (eds) *Handbook of World Food Legumes: Nutritional Chemistry, Processing Technology, and Utilization*, Vol 1. CRC Press, Boca Raton, Florida, pp. 247–299.

Chiaiese, P., Ohkama-Ohtsu, N., Molving, L., Godfree, R., Dove, D., Hocart, C., Fujiwara, T., Higgins, T.J.V. and Tabe, L.M. (2004) Sulphur and nitrogen nutrition influence the response of chickpea seeds to an added, transgenic sink for organic sulphur. *Journal of Experimental Botany* 55, 1889–1901.

Clawson, G.A. (1996) Protease inhibitors and carcinogenesis: a review. *Cancer Investigation* 14, 597–608.

Clemente, A., Mackenzie, D.A., Johnson, I.T. and Domoney, C. (2004) Investigation of legume seed protease inhibitors as potential anti-carcinogenic proteins. In: Muzquiz, M., Hill, G.D., Pedrosa, M.M. and Burbano, C. (eds) *Proceedings of the Fourth International Workshop on Antinutritional Factors in Legume Seeds and Oilseeds*. EAAP Publication No. 110, Wageningen, The Netherlands, pp. 137–142.

Domoney, C. (1999) Inhibitors of legume seeds. In: Sewry, P.R. and Casey, R. (eds) *Seed Proteins*. Kluwer Academic, Dordrecht, The Netherlands, pp. 697–719.

Duc, G., Sixdenier, G., Lila, M. and Furstoss, V. (1989) Search of genetic variability for vicine and convicine content in *Vicia faba* L. A first report of a gene which codes for nearly zero-vicine and zero-convicine contents. In: Huisman, J., van der Poel, A.F.B. and Liener, I.E. (eds) *Recent Advances of Research in Antinutritional Factors in Legume Seeds*. Pudoc, Wageningen, The Netherlands, pp. 305–313.

Duhan, A., Chauhan, B.M., Punia, D. and Kapoor, A. (1989) Phytic acid content of chickpea (*Cicer arietinum*) and black gram (*Vigna mungo*): varietal differences and effect of domestic processing and cooking methods. *Journal of the Science of Food and Agriculture* 49, 449–455.

el-Adawy, T.A. (2002) Nutritional composition and antinutritional factors of chickpeas (*Cicer arietinum* L.) undergoing different cooking methods and germination. *Plant Foods for Human Nutrition* 57, 83–97.

el-Hady, E.A.A. and Habiba, R.A. (2003) Effect of soaking and extrusion conditions on antinutrients and protein digestibility of legume seeds. *Lebensmittel Wissenschaft und Technologie* 36, 285–293.

Ewen, S.W.E. and Pusztai, A. (1999) Effect of diets containing genetically modified potatoes expressing *Galanthus nivalis* lectin on rat small intestine. *Lancet* 354, 1353–1354.

Fenwick, D.E. and Oakenfull, D. (1983) Saponin content of food plants and some prepared foods. *Journal of the Science of Food and Agriculture* 34, 186–191.

Fenwick, G.R., Price, K.R., Tsukamoto, C. and Okubo, K. (1991) Saponins. In: D'Mello, J.P.F., Duffus, C.M. and Duffus, J.H. (eds) *Toxic Substances in Crop Plants*. Royal Society of Chemistry, Cambridge, pp. 285–327.

Fleming, S.E. (1981) A study of relationships between flatus potential and carbohydrate distribution in legume seeds. *Journal of Food Science* 46, 794–798.

Freudenberg, K. (1919) Über Gerbstoffe. II.: Chebulinsauri. *Berichte der Deutschen Chemischen Gesellschaft* 52, 1238.

Garcia-Lopez, S., Erdman, J.W. and Sherman, A.R. (1990) Iron retention by rats from casein-legume test meals: effect of tannin level and previous diet. *Journal of Nutrition* 120, 760–766.

Gatehouse, A.M.R., Hilder, V.A., Powell, K.S., Wang, M., Davidson, G.M., Gatehouse, L.N., Down, R.E., Edmons, H.S., Boulter, D., Newell, C.A., Merryweather, A., Hamilton, W.D.O. and Gatehouse, J.A. (1994) Insect resistant transgenic plants: choosing the gene to do the job. *Biochemical Society Transaction* 4, 944–949.

Gertler, A., Posaz, I., Cohen, N. and Gottesfeld, F. (1982) Inhibition of human and dog serine pancreatic proteinases by naturally occurring high-molecular-weight inhibitors from plant or animal origin. *Biochemical Medicine* 27, 143–153.

Grant, G. (1991) Legumes. In: D'Mello, J.P.F., Duffus, C.M. and Duffus, J.H. (eds) *Toxic Substances in Crop Plants*. Royal Society of Chemistry, Cambridge, pp. 49–67.

Grant, G., More, L.J., McKenzie, N.H., Stewart, J.C. and Pusztai, A. (1983) A survey of the nutritional and haemagglutination properties of legume seeds generally available in the UK. *British Journal of Nutrition* 50, 207–214.

Grant, G., Murray, S., Gravells, E., Duguid, T.J., Bardocz, S. and Pusztai, A. (1998) The body metabolism of mammals can be adversely affected by enzyme inhibitors from legume seeds. In: Jansman, A.J.M., Hill, G.D., Huisman, J. and van der Poel, A.F.B. (eds) *Proceedings of the Third International Workshop on Antinutritional Factors in Legume Seeds and Rapeseed*. Wageningen, The Netherlands, pp. 233–237.

Greiner, R. and Konietzny, U. (1996a) Phytate hydrolysis in black beans by endogenous and exogenous enzymes. In: Bardocz, S., Muzquiz, M. and Pusztai, A. (eds) Effects of antinutrients on the nutritional value of legume diets. *Proceedings of the Fourth Scientific Workshop.* Office for Official Publications of the European Communities, Luxembourg, pp. 19–26.

Greiner, R. and Konietzny, U. (1996b) Construction of a bioreactor to produce special breakdown products of phytate. *Journal of Biotechnology* 48, 153–159.

Greiner, R., Pedrosa, M.M., Muzquiz, M., Ayet, G., Cuadrado, C. and Burbano, C. (1998) Effect of germination on phytate content and phytase activity in legumes. In: *Third European Conference on Grain Legumes.* Valladolid, Spain, pp. 82–83.

Greiner, R., Larsson-Alminger, M., Carlsson, N-G., Muzquiz, M., Burbano, C., Cuadrado, C., Pedrosa, M.M. and Goyoaga, C. (2002) Pathway of dephosphorylation of myo-inositol hexakisphosphate by phytases of legume seeds. *Journal of Agricultural and Food Chemistry* 50, 6865–6870.

Hajós, G. and Osagie, A.U. (2004) Technical and biotechnological modifications of antinutritional factors in legume and oilseeds. In: Muzquiz, M., Hill, G.D., Pedrosa, M.M. and Burbano, C. (eds) *Proceedings of the Fourth International Workshop on Antinutritional Factors in Legume Seeds and Oilseeds.* EAAP Publication No. 110, Toledo, Spain, pp. 293–305.

Hajós, G., Gelencsér, E., Pusztai, A. and Bardocz, S. (1996) Measurements of the antinutrient contents of legume samples. In: Bardocz, S., Gelencser, E. and Pusztai, A. (eds) *COST98-Effects of Antinutrients on the Nutritional Value of Legume Diets,* Vol. 1. Office for Official Publications of the European Communities, Luxembourg, pp. 130–134.

Haslam, E. (1998) *Practical Polyphenolics: From Structure to Molecular Recognition and Physiological Action.* Cambridge University Press, Cambridge.

Hasler, C.M., Bloch, A.S., Thomson, C.A., Enrione, E. and Manning, C. (2004) Position of the American Dietetic Association: functional foods. *Journal of American Dietetic Association* 104, 814–826.

Hettiarochchy, N.S. and Kantha, S.S. (1982) Trypsin inhibitor and phytohaemagglutinin contents in the seeds of six legumes commonly consumed in Sri Lanka. *Journal of the National Science Council of Sri Lanka* 9, 269–272.

Hill, G.D. (2004) Grain legumes and oilseeds: the way ahead. In: Muzquiz, M., Hill, G.D., Pedrosa, M.M. and Burbano, C. (eds) *Proceedings of the Fourth International Workshop on Antinutritional Factors in Legume Seeds and Oilseeds.* EAAP Publication No. 110, Wageningen, The Netherlands, pp. 353–364.

Jaffe, W.G., Moreneo, R. and Wallis, V. (1973) Amylase inhibitors in legume seeds. *Nutrition Reports International* 7, 169–174.

Jiménez-Martinez, C., Pedrosa, M.M., Muzquiz, M. and Dávila-Ortiz, G. (2004) Elimination of quinolizidine alkaloids, α-galactosides and phenolic compounds from *Lupinus campestris* seed by aqueous, acid and alkaline thermal treatment. In: Muzquiz, M., Hill, G.D., Pedrosa, M.M. and Burbano, C. (eds) *Proceedings of the Fourth International Workshop on Antinutritional Factors in Legume Seeds and Oilseeds.* EAAP Publication No. 110, Toledo, Spain, pp. 343–346.

Kadlec, P.E. (2001) Carbohydrate chemistry. In: Hedley, C.L. (ed.) *Carbohydrates in Grain Legume Seeds: Improving Nutritional Quality and Agronomic Characters.* CAB International, Wallingford, UK, p. 352.

Kantha, S.S., Sundravalli, O.E. and Desai, B.L.M. (1973) Effect of cooking and germination on the flatus inducing capacity of some legumes. In: Jaffe, W.G. (ed.) *Nutritional Aspects of Common Beans and Other Legumes.* Archives of Latinoamerican Nutrition, Venezuela, pp. 133–138.

Kennedy, A.R. (1995) The evidence for soybean products as cancer preventive agents. *Journal of Nutrition* 125, 733S–743S.

Kennedy, A.R. and Wan, X.S. (2002) Effects of the Bowman-Birk inhibitor on growth, invasion, and clonogenic survival of human prostate epithelial cells and prostate cancer cells. *Prostate* 50, 125–133.

Kennedy, A.R., Billings, P.C., Wan, X.S. and Newberne, P.M. (2002) Effects of Bowman-Birk inhibitor on rat colon carcinogenesis. *Nutrition and Cancer* 43, 174–186.

Kerem, Z., German-Shashoua, H. and Yarden, O. (2005) Microwave-assisted extraction of bioactive saponins from chickpea (*Cicer arietinum* L.). *Journal of the Science of Food and Agriculture* 85, 406–412.

Khaleque, A., Elias, L.G., Braham, J.E. and Bressani, R. (1985) Studies on the development of infant foods from plant protein sources: I. Effect of germination of chickpea on the nutritive value and digestibility of proteins. *Archives of Latinoamerican Nutrition* 35, 315–325.

Khanbabaee, K. and van Ree, T. (2001) Tannins: classification and definition. *Natural Product Reports* 18, 641–649.

Kolberg, J., Elsen, T.E. and Sletten, K. (1983) Properties of a lectin purified from the seeds of *Cicer arietinum*. *Hoppe Seyler's Zeitschrift fur Physiologische Chemie* 364, 655–664.

Koratkar, R. and Rao, A.V. (1997) Effect of soya bean saponins on azoxymethane-induced preneoplastic lesions in the colon of mice. *Nutrition and Cancer* 27, 206–209.

Kozlowska, H., Aranda, P., Dostalova, J., Frias, J., Lopez-Jurado, M., Kozlowska, H., Pokorny, J., Urbano, G., Vidal-Valverde, C. and Zdyunczyk, Z. (2001) Nutrition. In: Hedley, C.L. (ed.) *Carbohydrates in Grain Legume Seeds: Improving Nutritional Quality and Agronomic Characters*. CAB International, Wallingford, UK, p. 352.

Kuriyama, S. and Mendel, L.B. (1917) The physiological behaviour of raffinose. *Journal of Biological Chemistry* 31, 125–147.

Lajolo, F.M. and Genovese, M.I. (2002) Nutritional significance of lectins and enzyme inhibitors from legumes. *Journal of Agricultural and Food Chemistry* 50, 6592–6598.

Lajolo, F.M., Genovese, M.I., Pryme, I.F. and Dale, M. (2004) Beneficial (antiproliferative) effects of different substances. In: Muzquiz, M., Hill, G.D., Pedrosa, M.M. and Burbano, C. (eds) *Proceedings of the Fourth International Workshop on Antinutritional Factors in Legume Seeds and Oilseeds*. EAAP Publication No. 110, Wageningen, The Netherlands, pp.123–135.

Le Berre-Anton, V., Bompard-Gilles, C., Paya, F. and Rouge, P. (1997) Characterization and functional properties of the α-amylase inhibitor (α-AI) from kidney bean (*Phaseolus vulgaris*) seeds. *Biochimica et Biophysica Acta* 1343, 31–40.

Mahungu, S.M., Diaz-Mercado, S., Li, J., Schwenk, M., Singletary, K. and Faller, J. (1999) Stability of isoflavones during extrusion processing of corn/soy mixture. *Journal of Agricultural and Food Chemistry* 47, 279–284.

Maiti, I.B., Majumber, A.L. and Biswas, B.B. (1974) Purification and mode of action of phytase from *Phaseolus aureus*. *Phytochemistry* 13, 1047–1051.

Mansour, E.H. and Khalil, A.H. (1998) Reduction of raffinose oligosaccharides in chickpea (*Cicer arietinum*) flour by crude extracellular fungal alpha-galactosidase. *Journal of the Science of Food and Agriculture* 78, 175–181.

Martinez, B., Ibañez, M.V. and Rincón, F. (2002) Acido fítico: aspectos nutricionales e implicaciones analíticas. *Archivos Latinoamericanos de Nutrición* 52, 3.

Minihane, A.M. and Rimbach, G. (2002) Iron absorption and the iron binding and anti-oxidant properties of phytic acid. *International Journal of Food Science and Technology* 37, 741–748.

Mitsuoka, T. (1996) Intestinal flora and human health. *Asia Pacific Journal of Clinical Nutrition* 5, 2–9.

Monsanto, S.L. (2001) Evaluación de la seguridad de la soja roundup ready, even to 4–3–2. In: Monsanto Agricultura España, S.L. (ed.) Cuaderno Tecnico 1.

Mulimani, V.H. and Ramalingam, R. (1997) Enzymatic degradation of raffinose family sugars in chickpea flour. *World Journal of Microbiology and Biotechnology* 13, 583–585.

Mulimani, V.H., Rudrappa, G. and Supriya, D. (1994) α-Amylase inhibitors in chick pea (*Cicer arietinum* L.). *Journal of the Science of Food and Agriculture* 64, 413–415.

Muzquiz, M. (2000a) *Factores antinutricionales en fuentes proteicas*. In: Vioque, J., Clemente, A., Bautista, J. and Millán. F.

(eds) *Jornada Internacional sobre Proteínas Alimentarías*. Sevilla, Spain.

Muzquiz, M., Cuadrado, C., Ayet, G., Robredo, L.M., Pedrosa, M.M. and Burbano, C. (1996) Changes in non-nutrient compounds during germination. In: Bardocz, S., Gelencser, E. and Pusztai, A. (eds) *Effects of Antinutrients on the Nutritional Value of Legume Diets: Proceedings of the Second Scientific Workshop in Budapest*. Budapest, Hungary, pp. 124–129.

Muzquiz, M., Burbano, C., Cuadrado, C. and Martin, M. (2001) Analytical methods for determination of compounds with no nutritive value. In: Jacobsen, H.J., Muzquiz, M. and Hassa, A. (eds) *Handbook on Common Bean Related Laboratory Methods*. Galicia, Spain, pp. 11–26.

Muzquiz, M., Cuadrado, C., Guillamón, E., Goyoaga, C., Altares, P., Varela, A., Pedrosa, M.M. and Burbano, C. (2003) *Implicación en nutrición y salud de compuestos tóxicos y no-nutritivos de leguminosas*. In: Junta de Andalucía (eds) *1as Jornadas de la Asociación Española de Leguminosas*. Córdoba, Spain, pp. 80–82.

Nahdi, S., Nusrath, M., Batool, H. and Nagamani, V. (1982) Aflatoxin contamination and seed mycoflora of gram (*Cicer arietinum* L.) during storage. *Indian Journal of Botany* 5, 196–198.

Naveeda, K. and Jamuna, P. (2004) Nutritional quality of microwave-cooked and pressure-cooked legumes. *International Journal of Food Sciences and Nutrition* 55, 441–448.

Nishida, C., Uauy, R., Kumanyika, S. and Shetty, P. (2004) The Joint WHO/FAO expert consultation on diet, nutrition and the prevention of chronic diseases: process, product and policy implications. *Public Health Nutrition* 7, 245–250.

Nordlee, J.A., Taylor, S.L., Townsend, J.A., Thomas, L.A. and Bush, R.K. (1996) Identification of a Brazil-nut allergen in transgenic soybeans. *New England Journal of Medicine* 14, 688–692.

Oakenfull, D.G. and Sidhu, G. (1989) Saponins. In: Cheeke, P.R. (ed.) *Toxicants of Plant Origin*, Vol. 2. CRC Press, Boca Raton, Florida, p. 131.

Oleszek, W.A. (2002). Chromatographic determination of plant saponins. *Journal of Chromatography A* 967, 147–162.

Pedrosa, M.M., Burbano, C., Cuadrado, C. and Muzquiz, M. (1997) Levels of non-nutrient compounds in dietary legumes. In: Bardocz, S., Muzquiz, M. and Pusztai, A. (eds) *Effects of antinutrients on the nutritional value of legume diets: Proceedings of the Fourth Scientific Workshop*. Office for Official Publications of the European Communities, Luxembourg, pp. 75–81.

Petterson, D.S., Sipsas, S. and Mackintosh, J.B. (1997) *The Chemical Composition and Nutritive Value of Australian Pulses*, 2nd edn. Grains Research and Development Corporation, Kingston, Australia.

Poltronieri, F., Areas, J.A.G. and Colli, C. (2000) Extrusion and iron bioavailability in chickpea (*Cicer arietinum* L.). *Food Chemistry* 70, 175–180.

Price, K.R., Johnson, I.T. and Fenwick, R.G. (1987) The chemistry and biological significance of saponins in foods and feeding stuff. *CRC Critical Reviews in Food Science and Nutrition* 26, 27–131.

Price, K.R., Eagles, J. and Fenwick, G.R. (1988) Saponin composition of 13 varieties of legume seed using fast atom bombardment mass-spectrometry. *Journal of the Science of Food and Agriculture* 42, 183–193.

Pusztai, A., Grant, G., Duguid, T.J., Brown, D.S., Peumans, W.J., van Damme, E.J.M. and Bardocz, S. (1995) Inhibition of starch digestion by an α-amylase inhibitor reduces the efficiency of utilisation of dietary proteins and lipids and retards the growth of rats. *Journal of Nutrition* 125, 1554–1562.

Pusztai, A., Bardocz, S. and Martín-Cabrejas, M. (2004) The mode of action of ANFs on the gastrointestinal tract and its microflora. In: Muzquiz, M., Hill, G.D., Pedrosa, M.M. and Burbano, C. (eds) *Proceedings of the Fourth International Workshop on Antinutritional Factors in Legume Seeds and Oilseeds*. EAAP Publication No. 110, Wageningen, The Netherlands, pp. 87–100.

Quemener, B. and Brillouet, J.M. (1983) Ciceritol, a pinitol digalactoside from seeds of chickpea, lentil and white lupin. *Phytochemistry* 22, 1745–1751.

Quideau, S. and Feldman, K.S. (1996) Ellagitannin chemistry. *Chemical Reviews* 96, 475–504.

Rajkó, R., Szabó, G., Vidal-Valverde, C., and Kovács, E. (1997) Designed experiments for reducing antinutritive agents in soybean by microwave energy. *Journal of Agricultural and Food Chemistry* 45, 3565–3569.

Rao, P.U. and Belavady, B. (1978) Oligosaccharides in pulses: varietal differences and effects of cooking and germination. *Journal of Agricultural and Food Chemistry* 26, 316–319.

Rao, P.U. and Deosthale, Y.G. (1982) Tannin contents of pulses: varietal differences and effect of cooking and germination. *Journal of the Science of Food and Agriculture* 33, 1013–1016.

Rascon, A., Seidl, D.S., Jaffé, W.G. and Aizman, A. (1985) Inhibition of trypsins and chymotrypsins from different animal species: a comparative study. *Comparative Biochemistry Physiology* 82B, 375–378.

Ravindran, V., Ravindran, G. and Sivalogan, S. (1994) Total and phytate phosphorus contents of various foods and feed stuffs of plant origin. *Food Chemistry* 50, 133–136.

Reyes-Moreno, C., Milán-Carrillo, J., Rouzaud-Sandez, O., Garzón-Tiznado, J.A. and Mora-Escobedo, R. (2002) Dehulling/softening/extrusion (DSE): technological alternative to improve nutritional quality of chickpea (*Cicer arietinum* L.). *Agrociencia* 36, 181–189.

Reyes-Moreno, C., Cuevas-Rodriguez, E.O., Milan-Carrillo, J., Cardenas-Valenzuela, O.G. and Barron-Hoyos, J. (2004) Solid state fermentation process for producing chickpea (*Cicer arietinum* L.) tempeh flour: physicochemical and nutritional characteristics of the product. *Journal of the Science of Food and Agriculture* 84, 271–278.

Ridout, C.L., Wharf, G., Price, L.R., Johnson, L.T. and Fenwick, G.R. (1988) UK mean daily intakes of saponins-intestine-permeabilising factors in legumes. *Food Science and Nutrition* 42, 111–116.

Rimbach, G. and Pallauf, J. (1997) Cadmium accumulation, zinc status, and mineral bioavailability of growing rats fed diets high in zinc with increasing amounts of phytic acid. *Biological Trace Element Research* 57, 59–70.

Rimbach, G., Pallauf, J. and Walz, O.P. (1996) Effect of microbial phytase on cadmium accumulation in pigs. *Archives fur Tierernahrung* 49, 279–286.

Rincón, F., Martínez, B. and Ibáñez, M.V. (1998) Proximate composition and antinutritive substances in chickpea (*Cicer arietinum* L.) as affected by the biotype factor. *Journal of the Science of Food and Agriculture* 78, 382–388.

Roberts, M.F. and Wink, M. (1998) Introduction. In: Roberts, M.F. and Wink, M. (eds) *Alkaloids. Biochemistry, Ecology, and Medicinal Applications.* Plenum Press, New York, pp. 1–7.

Rossi, M., Germondari, I. and Casini, P. (1984) Comparison of chickpea cultivars: chemical composition, nutritional evaluation and oligosaccharide content. *Journal of Agricultural and Food Chemistry* 32, 811–814.

Rubio, L.A., Pérez, A., Muzquiz, M., Burbano, C., Cuadrado, C. and Pedrosa, M.M. (2005) Ileal digestibility of deffated soybean, lupin and chickpea meal in cannulated Iberian pigs. 2. Fatty acids and carbohydrates. *Journal of the Science of Food and Agriculture* 85, 1322–1328.

Ruiz, R.G., Price, K., Rose, M., Rhodes, M. and Fenwick, R. (1996) A preliminary study on the effect of germination on saponin content and composition of lentils and chick peas. *Zeitschrift fur Lebensmittel-Untersuchung und -Forschung* 203(4), 366–369.

Ruiz, R.G., Price, K.R., Fenwick, R.G. and Rhodes, M.J.C. (1997) Changes in saponin content and composition of chickpeas and lentils during germination and cooking. In: Bardocz, S., Muzquiz, M. and Pusztai, A. (eds) *Effects of antinutrients on the nutritional value of legume diets: Proceedings of the Fourth Scientific Workshop.* Office for Official Publications of the European Communities, Luxembourg, pp. 15–18.

Saini, H.S. and Knights, E.J. (1984) Chemical constitution of starch and oligosaccharide components of 'desi' and 'kabuli' chickpea (*Cicer arietinum*) seed types. *Journal of Agricultural and Food Chemistry* 32, 940–944.

Saini, H.S., Weder, J.K.P. and Knights, E.J. (1992) Inhibitor activities of chickpeas (*Cicer arietinum* L.) against bovine, porcine and human trypsin and chymotrypsin. *Journal of the Science of Food and Agriculture* 60, 287–295.

Salgado, P., Lallésa, J.P., Toulleca, R., Mouratob, M., Cabralb, F. and Freire, J.P.B. (2001) Nutrient digestibility of chickpea (*Cicer arietinum* L.) seeds and effects on the small intestine of weaned piglets. *Animal Feed Science and Technology* 91, 197–212.

Sanchez-Mata, M.C., Penuela-Teruel, M.J., Camara-Hurtado, M., Diez-Marques, C. and Torija-Isasa, M.E. (1998) Determination of mono-, di-, and oligosaccharides in legumes by high-performance liquid chromatography using an amino-bonded silica column. *Journal of Agricultural and Food Chemistry* 46, 3648–3652.

Sandberg, A.S., Andersson, H., Carlsson, N.G. and Sandström, B. (1993) Degradation of soybean oligosaccharides in the stomach and small intestine of human ileostomy subjects. In: Schlemmer, U. (ed.) *Proceedings of the International Conference on Bioavailability 93: Nutritional, Chemical and Food Processing Implications of Nutrient Availability*. Karlsruhe, Germany, pp. 197–201.

Savitri, A. and Desikachar, H.S.R. (1985) A comparative study of flatus production in relation to the oligosaccharide content of some legumes. *Nutrition Reports International* 31, 337–344.

Saxena, A.K., Chadha, M. and Sharma, S. (2003) Nutrients and antinutrients in chickpea (*Cicer arietinum* L.) cultivars after soaking and pressure cooking. *Journal of Food Science and Technology* 40, 493–497.

Shibuya, M., Hoshino, M., Katsube, Y., Hayashi, H., Kushiro, T. and Ebizuka, Y. (2006) Identification of β-amyrin and sophoradiol 24-hydroxylase by expressed sequence tag mining and functional expression assay. *FEBS Journal* 273, 948–959.

Singh, U. and Jambunathan, R. (1981) Studies on desi and kabuli chickpea (*Cicer arietinum* L.) cultivars: levels of protease inhibitors, levels of polyphenolic compounds and in vitro protein digestibility. *Journal of Food Science* 46, 1364–1367.

Singh, U., Kherdekar, M.S. and Jambunathan, R. (1982) Studies on desi and kabuli chickpea (*Cicer arietinum* L.) cultivars. The levels of amylase inhibitors, levels of oligosaccharides and in vitro starch digestibility. *Journal of Food Science* 47, 510–512.

Smirnoff, P., Khalef, S., Birk, Y. and Applebaum, S.W. (1976) A trypsin and chymotrypsin inhibitor from chickpeas (*Cicer arietinum*). *Biochemical Journal* 157, 745–751.

Srinivasan, A., Giri, A.P., Harsulkar, A.M., Gatehouse, J.A. and Gupta, V.S. (2005) A Kunitz trypsin inhibitor from chickpea (*Cicer arietinum* L.) that exerts anti-metabolic effect on podborer (*Helicoverpa armigera*) larvae. *Plant Molecular Biology* 57, 359–374.

Sumathi, S. and Patabhiraman, T.N. (1976) Natural plant enzyme inhibitors. Part II. Protease inhibitors of seeds. *Indian Journal of Biochemistry and Biophysics* 13, 52–56.

Tava, A., Oleszek, W., Jurzysta, M., Berardo, N. and Odoardi, M. (1993) Alfalfa saponins and sapogenins: isolation and quantification in two different cultivars. *Phytochemical Analysis* 4, 269–274.

Thompson, L.U. (1993) Potential health benefits and problems associated with antinutrients in foods. *Food Research International* 26, 131–149.

Urbano, G., López-Jurado, M., Aranda, P., Vidal-Valverde, C., Tenorio, E.J. and Porres, E. (2000) The role of phytic acid in legumes: antinutrient or beneficial function? *Journal of Physiology and Biochemistry* 56, 283–294.

Urdaneta, E., Alberdi, J., Aranguren, P., Barrenetxe, J., Pedrosa, M.M. and Marzo, F. (2003) Efecto del extrusionado de Lupinus (*Lupinus albus* L. var. Multolupa) en ratas en crecimiento. In: Junta de Andalucía (eds) *1as Jornadas de la Asociación Española de Leguminosas*. Córdoba, Spain, pp. 115–116.

ur-Rehman, Z. and Shah, W.H. (1998) Effect of microwave and conventional cooking on some nutrients, antinutrient and protein digestibility of chick-peas (*Cicer arietinum*). *Bangladesh Journal of Scientific and Industrial Research* 33, 200–205.

Vioque, J., Sanchez-Vioque, R., Clemente, A., Pedroche, J., Bautista, J. and Millan, F. (1999) Purification and partial characterization of chickpea 2S albumin. *Journal of Agricultural and Food Chemistry* 47, 1405–1409.

Viveros, A., Brenes, A., Elices, R., Arija, I. and Canales, R. (2001) Nutritional value of raw and autoclaved kabuli and desi chickpeas (*Cicer arietinum* L.) for growing chickens. *British Poultry Science* 42, 242–251.

Wang, Y.H.A. and McIntosh, G.H. (1996) Extrusion and boiling improve rat body weight gain and plasma cholesterol lowering ability of peas and chickpeas. *Journal of Nutrition* 126, 3054–3062.

Whitaker, J.R. (1988) α-Amylase inhibitors of higher plants and microorganisms. In: Kinsella, J.E. and Soucle, W.G. (eds) *Food Proteins: Proceedings of the Protein and Co-products Symposium*. American Oil Chemists' Society, Champaign Il, New York, pp. 354–380.

Wiggins, H.S. (1984) Nutritional value of sugars and related compounds undigested in the small intestine. *Proceedings of the Nutrition Society* 43, 69–75.

Wood, J.A. (2000) Chickpea fortified pasta – a preliminary study. In: Wootton, W., Batey, I.L. and Wrigley, C.W. (eds) *Cereals Health and Life – Technology for the New Millennium: Proceedings of the 11th International Cereal and Bread Congress Incorporating the 50th Australian Cereal Chemistry Conference*. Surfers Paradise, Australia, pp. 212–215.

Wood, J.A. and Harden, S. (2006) A method to estimate the hydration and swelling properties of chickpeas (*Cicer arietinum* L.). *Journal of Food Science* 71, E190–E195.

Zaki, A.M., Fetouh, S.A.A., Mohany, H. and El-Morsi, E.A. (1996) Biochemical changes of some chemical constituents in chickpea seeds during germination and processing. *Annals of Agricultural Science, Moshtohor* 34, 995–1022.

Zamora, A.F. and Fields, M.L. (1979) Nutritive quality of fermented cowpeas and chickpeas. *Journal of Food Science* 44, 234–236.

Zdunczyk, Z., Juskiewicz, J., Frejnagel, S. and Gulewicz, K. (1998) Influence of alkaloid and oligosaccharides from white lupin seeds on utilization of diets by rats and absorption of nutrients in the small intestine. *Journal of Animal and Feed Science* 72, 143–154.

Zhou, J.R. and Erdman, J.W. (1995) Phytic acid in health and disease. *Critical* Reviews *in Food Science and Nutrition* 35, 495–508.

7 Area, Production and Distribution

E.J. Knights,[1] N. Açıkgöz,[2] T. Warkentin,[3] G. Bejiga,[4] S.S. Yadav[5] and J.S. Sandhu[6]

[1]*New South Wales Department of Primary Industries, Tamworth Agricultural Institute, Calala, NSW 2340, Australia;* [2]*Aegean Agricultural Research Institute (AARI), PO Box 9, Menemen 35661, Izmir Turkey;* [3]*Crop Development Centre, University of Saskatchewan, Saskatoon, SK S7N 5A8, Canada;* [4]*Green Focus Ethiopia, PO Box 802, Code No. 1110, Addis Ababa, Ethiopia;* [5]*Pulse Research Laboratory, Division of Genetics, Indian Agricultural Research Institute, New Delhi 110012, India;* [6]*Department of Plant Breeding, Genetics and Biotechnology, Punjab Agricultural University, Ludhiana 141004, India*

Global Importance

Chickpea has an important role in feeding humanity. Amongst the annual seed crops it ranks 14th in terms of area and 16th in production. According to Food and Agriculture Organization (FAO) statistics for 1996–2005, the chickpea area averaged nearly 11 million hectares worldwide (1.2% of total crop area) and production was slightly more than 8 million tonnes (0.34%). However, amongst pulse crops chickpea has consistently maintained a much more significant status, ranking second in area (15.3% of total) and third in production (14.6%) (Fig. 7.1).

Any analysis of global chickpea production relies heavily on FAO statistics. These are generally the best source of information available, but in developing countries, where most of the production is consumed on-farm or locally, the published figures may underestimate supply (Oram and Agcaoili, 1994). The area may also be underestimated in these countries where chickpea is a component of intercropping, mixed cropping or relay cropping. The FAO has listed 51 countries producing chickpea during the last 10 years (Fig. 7.2) (FAO, 2006). East Asia was the dominant region (74.9% of total production) and India the dominant nation (65.1%). Eight other countries averaged more than 100,000 t during this period: Pakistan (7.5% of world production), Turkey (7.5%), Iran (3.4%), Mexico (2.8%), Australia (2.4%), Canada (2.0%), Ethiopia (1.8%) and Myanmar (1.7%) (Table 7.1). A further 14 countries averaged between 10,000 t and 100,000 t, 19 averaged between 1000 t and 10,000 t and 11 averaged less than 1000 t. In Colombia and the Dominican Republic, collection of statistics, or production itself, ceased prior to 1996. The FAO did not collect statistics on

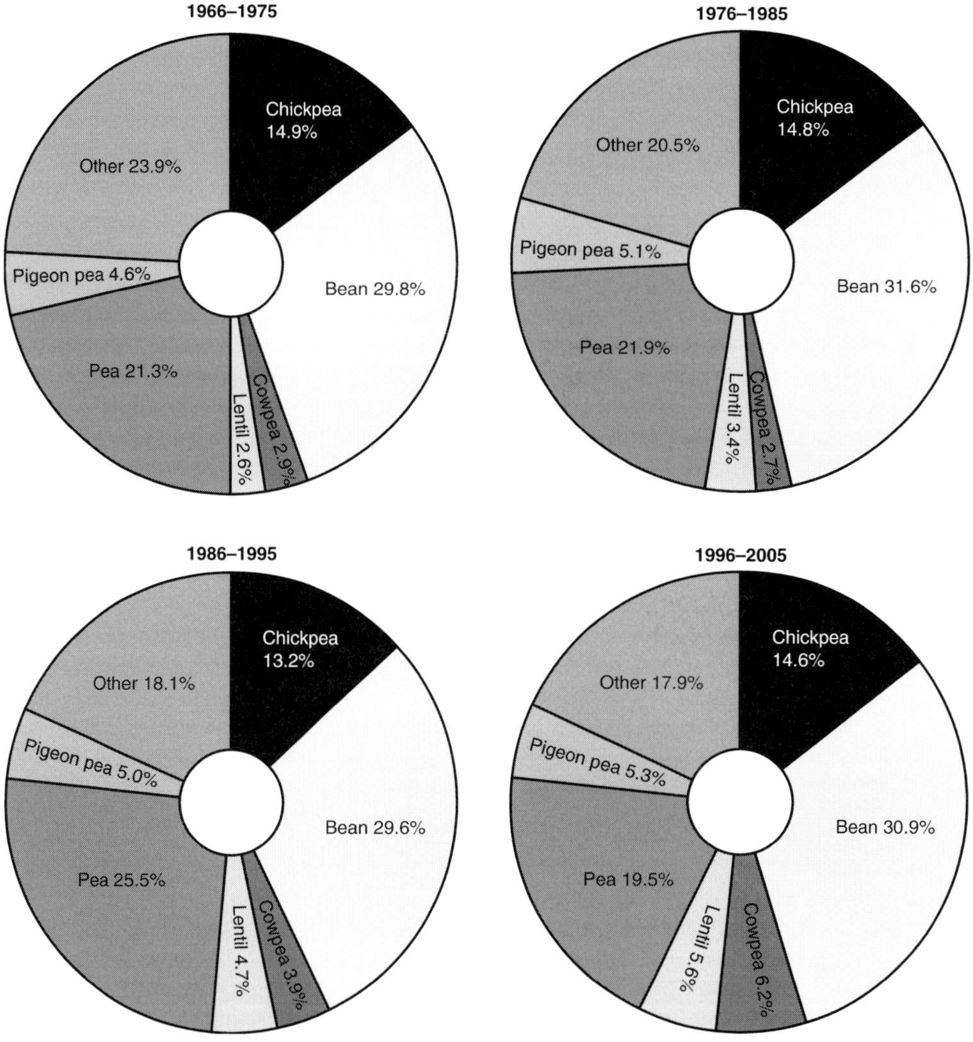

Fig. 7.1. Average world production of chickpea and other pulse crops for the 10-year periods between 1966 and 2005 (soybean and groundnut are excluded). (From FAO, 2006.)

the remaining producing countries. Of these, the USA has a significant industry: a peak area of 60,000 ha was sown in 2003 (Smith and Jimmerson, 2005). Some sources (e.g. Malhotra *et al.*, 2000) also refer to minor chickpea industries in those former Soviet Central Asian Republics (e.g. Azerbaijan, Georgia, Kyrgyzstan, Tajikistan and Turkmenistan) that are not included in the FAO statistics, and small areas are still likely to be grown in war-torn Afghanistan.

Chickpea is predominantly a rainfed crop, grown mainly on residual moisture and in drought-prone environments; only a small area (<10%) is irrigated. Unsurprisingly, world average yields are low (768 kg/ha during 1996–2005). There is much regional variation: the highest mean yield is in North and Central

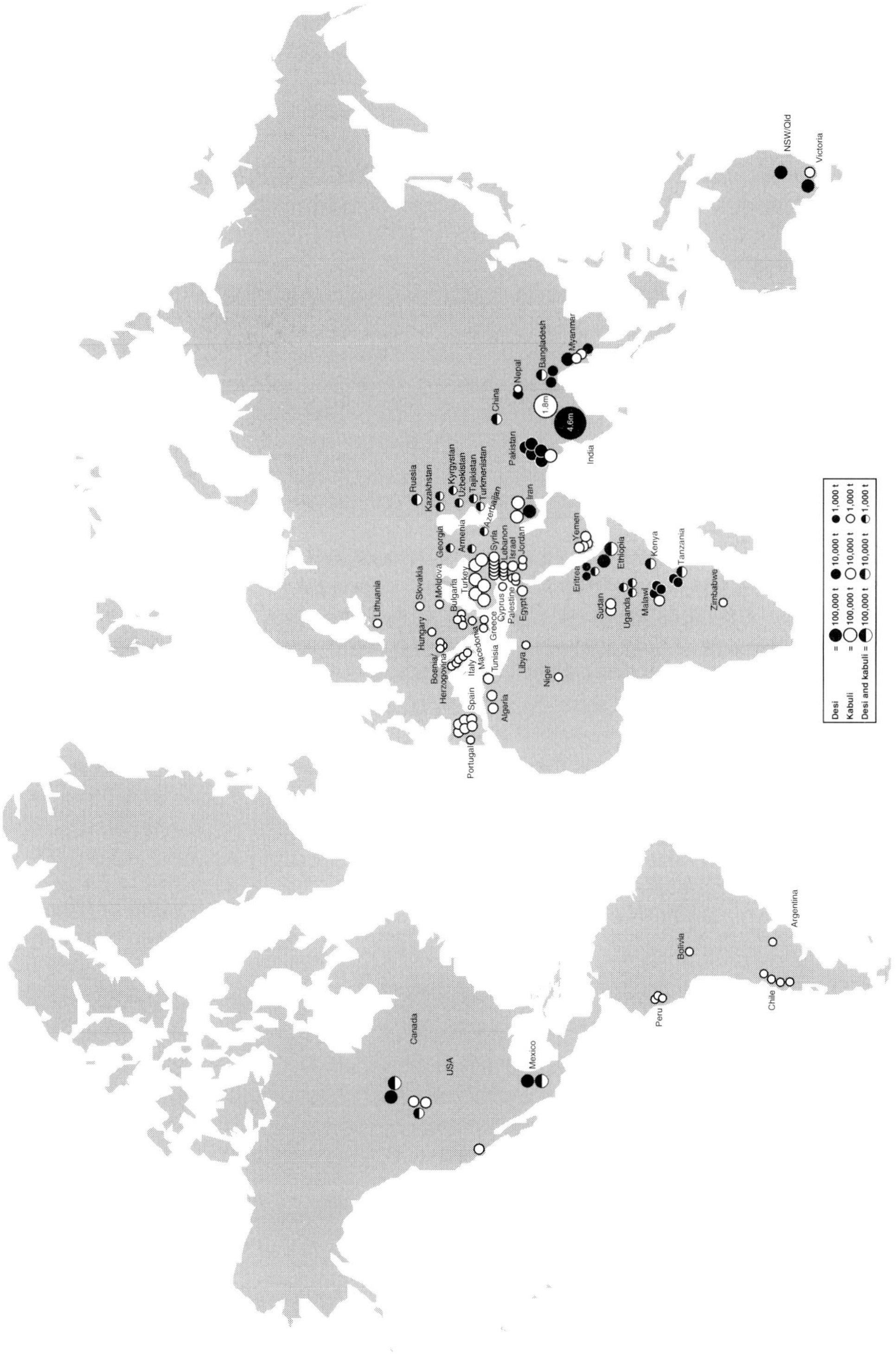

Fig. 7.2. World distribution of desi and kabuli chickpea production, 1996–2005. (From FAO, 2006.)

Table 7.1. Mean area, production and yield of chickpea for selected countries for the 10-year periods between 1966 and 2005. (From FAO, 2006.)

	1966–1975			1976–1985			1986–1995			1996–2005		
Country	Area (ha)	Production (t)	Yield (t/ha)	Area (ha)	Production (t)	Yield (t/ha)	Area (ha)	Production (t)	Yield (t/ha)	Area (ha)	Production (t)	Yield (t/ha)
India	7,665,310	4,660,310	0.608	7,487,720	4,937,980	0.658	6,729,190	4,860,200	0.720	6,790,370	5,382,610	0.791
Pakistan	1,032,163	540,860	0.526	1,018,565	488,266	0.479	1,015,630	482,890	0.475	1,024,810	621,710	0.602
Turkey	121,980	137,600	1.136	240,487	259,500	1.120	762,298	742,050	0.982	665,200	620,000	0.931
Iran	127,380	71,684	0.563	154,330	96,740	0.634	447,879	210,295	0.474	716,841	277,147	0.385
Mexico	208,995	193,050	0.923	169,965	184,791	1.070	119,555	161,147	1.357	144,957	233,354	1.599
Australia	–	–	–	11,967	15,300	1.154	143,319	144,837	1.045	215,200	196,806	0.967
Canada	–	–	–	–	–	–	1,068	1,500	1.396	125,325	163,111	1.271
Ethiopia	172,680	119,837	0.693	154,729	111,939	0.731	127,109	93,848	0.741	169,905	144,868	0.863
Myanmar	127,491	63,327	0.496	154,228	109,194	0.690	146,203	110,841	0.744	163,506	144,030	0.844
Syria	49,652	35,531	0.725	63,177	45,252	0.695	66,229	42,141	0.623	88,842	61,914	0.687
Spain	166,531	99,420	0.589	94,495	55,612	0.587	69,540	45,864	0.682	89,890	56,396	0.616
Morocco	116,910	81,591	0.693	63,140	39,422	0.599	75,380	46,358	0.615	65,428	38,020	0.583
Yemen	–	–	–	–	–	–	25,822	37,173	1.433	30,103	37,007	1.230
Malawi	21,250	13,800	0.649	27,610	18,000	0.655	76,917	29,806	0.392	88,100	35,000	0.397
Tanzania	57,097	13,056	0.250	34,700	10,850	0.300	62,600	21,200	0.337	66,200	27,300	0.412
Bangladesh	78,847	59,862	0.774	133,020	93,600	0.707	96,572	69,719	0.722	40,925	26,739	0.749
Sudan	2,233	1,927	0.863	1,618	1,572	1.007	1,660	2,037	1.175	11,130	22,700	1.983
Israel	2,113	2,883	1.426	2,688	3,300	1.311	4,834	7,066	1.464	6,548	12,257	1.863
Egypt	3,460	5,713	1.667	7,032	10,981	1.575	7,088	12,428	1.763	6,909	12,250	1.775
Italy	35,389	29,997	0.915	13,427	15,350	1.154	5,322	5,833	1.099	4,373	5,109	1.166
Chile	12,761	5,997	0.494	12,115	6,408	0.522	11,181	10,116	0.903	4,578	4,004	0.834
Portugal	53,927	19,597	0.365	33,363	12,165	0.369	5,093	2,947	0.593	2,289	1,340	0.585

America, reflecting significant irrigated production in Mexico, comparatively favourable rainfed environments in the USA and Canada, and a high-input, mechanized agriculture; the lowest yield (648 kg/ha) is in West Asia where the principal constraints are short growing seasons, diseases (especially ascochyta blight), low inputs and sometimes poor husbandry.

Global Trends in Area, Production and Yield

An array of factors produce large year to year variations in global chickpea area and yield. They include unreliable rains, terminal drought, diseases, insects, market access, on-farm profitability and government policy. During 1961–2005 chickpea area ranged from 8.7 to 12.2 million hectares and the average yield from 0.49 to 0.82 t/ha, while the largest variation was for total production (4.9–9.4 million tonnes). Overall, there was only a modest increase in production (0.6% per year), which was entirely due to higher yields (0.7% per year); crop area actually declined slightly during this period (−0.1% per year) (Fig. 7.3a). Globally, per capita consumption has declined because the increase in production has not kept pace with population growth. For the triennium 1961–1963 world per capita consumption was 2.3 kg; this had declined to 1.2 kg by 2000–2002.

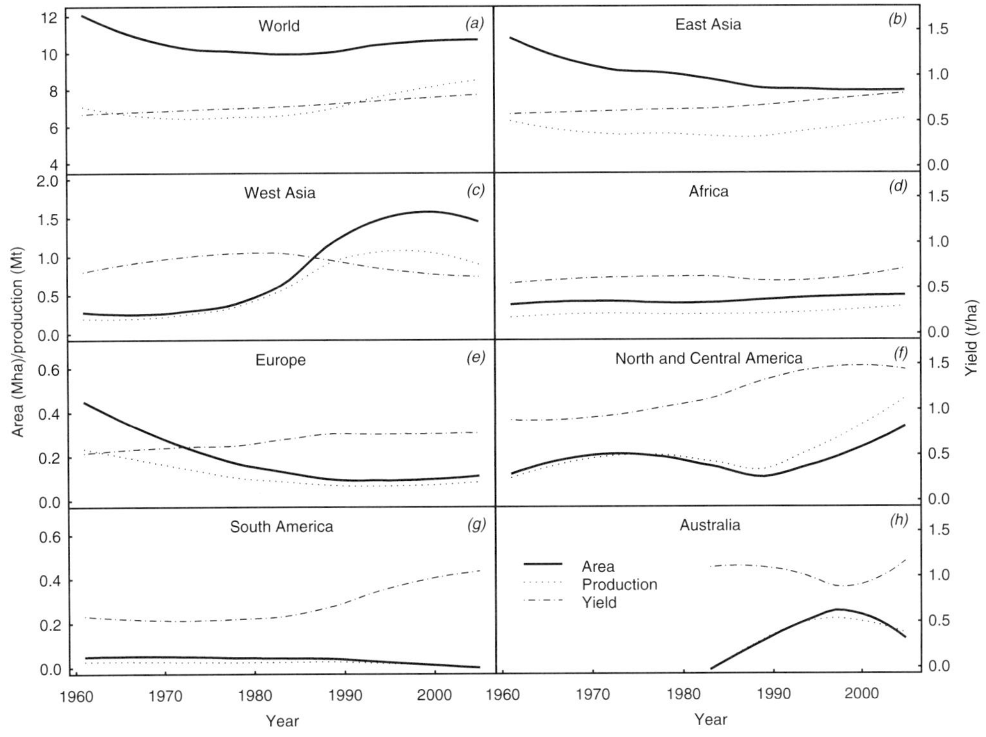

Fig. 7.3. World and regional trends for chickpea area, production and yield, 1961–2005. (From FAO, 2006.)

Regional Trends in Area, Production and Yield

The relative stability of world chickpea area over the last four decades has tended to obscure significant and sometimes dramatic changes at the regional and national levels. Important chickpea industries have collapsed in some countries (e.g. Portugal) but prospered in others (e.g. Turkey). Some countries have seen a substantial relocation (e.g. India) and, recently, major new industries have emerged in some developed countries (e.g. Australia and Canada) having a comparative production advantage.

There are no official statistics for the breakdown of global production into the two seed types, desi (coloured seed coat) and kabuli (white seed coat). This can be inferred only from production statistics and estimates of the desi/kabuli balance in each country. The chickpea crop is exclusively kabuli throughout Europe, North Africa, South America and much of West Asia. However, either desi or kabuli type accounts for a significant minority of production in many of the main producing countries, especially in East Asia, East Africa, North America and Australia. The relative proportions of desis and kabulis appear to have changed over time, both regionally and globally. Singh *et al.* (1985) estimated that kabulis comprised 14.2% of total production in the early 1980s; in countries where both types were grown, the percentage of kabulis varied from 1% for Ethiopia to 5% for India and Pakistan, and 30% for Mexico and Iran. These percentages have generally increased over time. Presently, kabulis are the dominant type (covering 76% of area) in Iran (Sabaghpour *et al.*, 2003) and were estimated to comprise 15% of India's production in 2005. The contribution of kabulis to world production was estimated to be 30% during 1996–2005. The change in the kabuli/desi balance has occurred almost exclusively as a result of the growth in kabuli production: over the period 1966–2005, mean kabuli production was estimated to have risen from 0.8 to 2.6 million tonnes compared to an almost static desi production of 5.4–5.6 million tonnes.

East Asia (India, Pakistan, Myanmar, Bangladesh, Nepal and China)

East Asia dominates the world's chickpea production, accounting for 75% during 1996–2005; India alone produced 65%. The majority (85%) were desi types. During 1961–2005 production remained almost static. A significant decline in crop area (–0.7% per year) was offset by a mean yield increase of 0.7% per year (Fig. 7.3b). The overall picture of a comparatively stable production (mean 0.1% per year increase in production) obscures a number of significant trends at the national level.

The most profound change has been a substantial migration of the Indian chickpea industry away from the far north to the centre and south of the country. Up to the late 1960s chickpea was concentrated in the Indo-Gangetic Plain, principally in the states of Uttar Pradesh, Punjab, Haryana, Bihar and West Bengal. Collectively, they contributed about 65% of the total Indian production. Then the

Green Revolution began to impact on the chickpea industry: expansion of the rice industry and development of wheat varieties suitable for late sowing made rice/wheat rotations highly profitable, and gradually chickpea was displaced to marginal rainfed areas and afforded minimal or no inputs. In the last 50 years the chickpea area in these states has declined by almost 5 million hectares, although this has been more pronounced in some states than others: during 1960–2002 the chickpea area fell by 68% in Uttar Pradesh, the leading producer, 77% in West Bengal and 88% in Bihar and Punjab (including Haryana).

Approximately half of this 'lost' area was taken up by states in central and southern India. By 2001/02 these states had accounted for 80% of the national chickpea area and production (Fig. 7.4). Madhya Pradesh became the largest producer (37% of national chickpea area) followed by Rajasthan (15%), Maharashtra (13%), Karnataka (8%) and Andhra Pradesh (4%).

The increase in chickpea area in the central and southern zones was underpinned by the availability of short-duration varieties that were also resistant to diseases, especially fusarium wilt. This led to a rapid increase in annual yield in some states, e.g. 3% in Maharashtra and Andhra Pradesh. Overall, a sustained increase in yield, despite the general displacement of chickpea to short-season, low-rainfall environments, has enabled the Indian chickpea industry to offset a 30% reduction in crop area over the period 1966–2005 (Table 7.1).

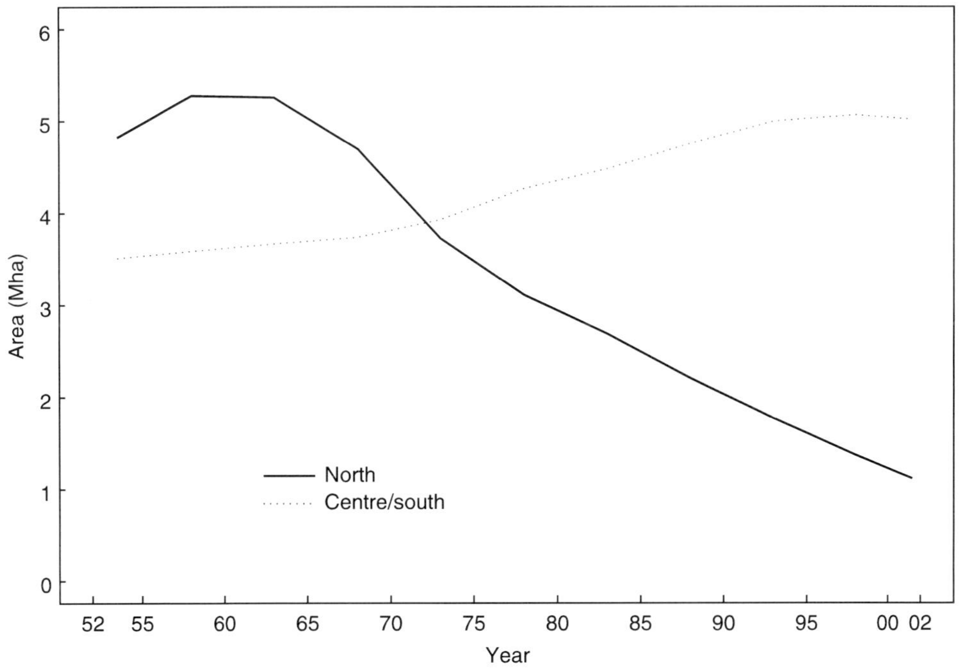

Fig. 7.4. Area of chickpea production in northern (Uttar Pradesh, Punjab, Haryana, Bihar and West Bengal) and central/southern states of India, 1952–2002. (Data are 5-year averages except for the periods 1952–1955 and 2001–2002.) (From van der Maesen, 1972; and Yearbook on Agriculture, Ministry of Agriculture, Government of India.)

More significant declines in chickpea areas have occurred in Bangladesh and Nepal. During 1966–2005, the 10-year average fell by 48% in Bangladesh and 59% in Nepal. Amongst the multiplicity of suggested causes, the foliar disease botrytis grey mould was identified as the most important (Musa *et al.*, 2001; Pande *et al.*, 2005). However, in both countries the trend is slowly reversing, as integrated disease and pest management packages become more widely adopted, particularly in the vast areas left fallow after rice harvest (Musa *et al.*, 2001). In Pakistan, the world's second largest producer of chickpea, the crop area has remained stable at just over 1 million hectares. Yields are generally lower than for India, reflecting the very marginal environments in which the crop is grown, but like India, yields have increased significantly over the last decade (Table 7.1). Myanmar is the only East Asian country to show an increase in crop area (33% over the period 1966–2005). During the same period, mean production rose by 127%, mainly due to a 70% increase in yield (Table 7.1). Sustained productivity increases have been underpinned by improved varieties: adoption surveys during the 2003/04 season showed that 93% of Myanmar's chickpea area was sown with varieties recently released by the International Centre for Research in the Semi-Arid Tropics (ICRISAT), Hyderabad, India.

West and Central Asia (Turkey, Syria, Israel, Lebanon, Jordan, Saudi Arabia, Yemen, Iraq, Iran, Armenia, Azerbaijan, Georgia, Kazakhstan, Kyrgyzstan, Uzbekistan, Tajikistan, Turkmenistan and Afghanistan)

West and Central Asia accounted for 14% and 12% of world chickpea area and production, respectively, during 1996–2005. Kabulis comprised an estimated 93% of production and 37% of world kabuli production (the same as that for East Asia). Growth in crop area averaged 4.7% per year during 1961–2005 (Fig. 7.3c), with most of the increase occurring in Turkey (trebling to 630,000 ha from the early 1980s) and Iran (trebling to 750,000 ha from the late 1980s). Increase in overall production (averaging 4.2% per year) did not keep pace with the increase in area; yield decreased by an average 0.4% per year during that period.

The rapid industry expansion in Turkey resulted from a major change in the cropping system. This was the replacement of large areas of fallow with alternative crops, a transition driven largely by government incentives (Acikgoz *et al.*, 1994). Fallowing had been a traditional practice on the Anatolian Plateau, where most of Turkey's chickpea is produced, because of low and variable winter and/or spring rainfall. A year of fallow typically preceded and followed a cereal crop (wheat or barley). The first attempt to encourage the use of fallows was the pilot Corum-Cankiri Rural Development Project, established in the northern transitional area in the late 1970s. Its success in lifting production of alternative crops led to the more extensive, longer-term Utilization of Fallow Areas Project in 1982. Government funding provided incentives such as credit for certified seed and fertilizers. The first phase of the project (1982–1986) targeted the transfer of 1.4 million hectares to alternative crops; an additional 1.7 million hectares were targeted in the project's second phase (1987–1991). A 670,000ha expansion of the chickpea area over the project's duration attested to its effectiveness. Significantly, chickpea area fell by 28% in the 14 years following the withdrawal of incentives, although some of this decline may be attributed to substitution by more profitable crops in the recently developed irrigation areas.

Africa (Ethiopia, Eritrea, Kenya, Uganda, Malawi, Tanzania, Zimbabwe, Morocco, Algeria, Tunisia, Libya, Niger, Egypt and Sudan)

Africa contributed less than 5% of the global chickpea area and less than 4% of production during 1996–2005. Most production (69%), a mixture of desis and kabulis, was in near-equatorial East Africa. The rest was split between rainfed kabulis in countries along the Mediterranean coast (20%) and those under irrigation in Egypt and Sudan (11%). Overall, desis comprised about 60% of production. Historically, the chickpea industry in Africa has been stable compared to other regions. During 1961–2005 regional production rose by an average 0.7% per year, resulting mainly from an increase in area (0.5% per year) and to a lesser extent in yield (0.2% per year) (Fig. 7.3d).

Ethiopia has always been the largest African producer, averaging more than 40% since the 1960s. Production did, however, decline sharply during the late 1980s and early 1990s, mainly as a consequence of land nationalization imposed by the communist government since the late 1970s and low, fixed prices. The industry gradually recovered to its previous level following the government's overthrow in 1991, and subsequent market liberalization and land redistribution. Industry recovery was also assisted by the release of many improved varieties and a government supported scheme to produce and distribute high-quality, treated seed.

Europe (Spain, Portugal, Italy, Greece, Cyprus, Bosnia/Herzogovina, Macedonia, Slovakia, Hungary, Bulgaria, Moldova, Russia and Lithuania)

Europe accounts for only 1% of the world's chickpea area and production, virtually all of which is the kabuli type. Spain is the dominant country, having 82% and 69% of the region's area and production, respectively, during 1996–2005. There has been a consistent decline in the industry, characterized by an average 4% per year fall in area and 3.2% in production (Fig. 7.3e). The decline has been more pronounced in some countries than in others: mean output fell by 59% in Spain, 85% in Italy and Greece, and 95% in Portugal for the 40-year period 1961–2005.

The significant decline in Europe's chickpea industry has a number of causes, although the combination and relative importance varies from one country to another. These include demographic changes (chickpea is traditionally a food for rural communities, but there has been significant migration to urban areas), increased affluence (a substitution of animal for legume protein), production costs (especially manual harvest), cheaper imports of Mexican kabulis, declining use of chickpea as a stockfeed, diseases (mainly ascochyta blight and fusarium wilt), and poor public investment in research, breeding and technology transfer (Barradas, 1990; Cubero *et al.*, 1990).

North and Central America (Canada, USA and Mexico)

North and Central America accounted for ~5% of the world's chickpea production during 1996–2005, split evenly between desis and kabulis. Historically Mexico has been the leading producer, growing a mix of high-quality, large-seeded kabulis under irrigation and a greater volume of rainfed desis, mainly used for animal feed. Since the 1930s there has also been a small, comparatively stable kabuli industry (<5000 ha) based in California. Emerging chickpea industries in the Pacific North-west of the USA in the mid-1980s, and to a greater extent in the Canadian province of Saskatchewan in the mid-1990s, have recently challenged Mexico's position as the region's leading chickpea producer and exporter (Table 7.1).

The chickpea area in north-western USA quickly rose to a peak of 5000 ha in 1986, but the inadvertent introduction of ascochyta blight, and its subsequent rapid spread on susceptible varieties, caused a decline in production. A concerted programme of germplasm introduction and breeding, supported by government contributions and grower levies, quickly led to the release of locally adapted, *Ascochyta*-resistant varieties. A restoration of grower confidence saw a rebound in the industry, with production peaking at 51,000 ha in 2001.

The introduction of chickpea to the Canadian prairies was a response to the need for crop diversification. Profitability of cereal-dominated farming had fallen significantly by the late 1960s, and in the following decade there was a shift to alternative crops. A phase of experimentation during the 1970s and 1980s led to the release of Canada's first chickpea variety in 1993. Ascochyta blight quickly emerged as a serious problem, but the fortuitous release of resistant varieties in north-western USA, and their broad suitability for the Saskatchewan environment, facilitated a rapid growth in the Canadian chickpea industry. This growth was further assisted by a timely period of unusually high export prices, the strategic choice of pioneer farmers with close links to seed cleaning and export chains, and funding of research and development by government and grower levies. By 2001 the area of chickpea had reached 467,000 ha, concluding a period of growth unprecedented anywhere else in the world. However, the industry shrank almost as rapidly, and by 2004 had declined by 90%. Foremost among the reasons for this sudden reversal were inadequate varietal resistance to ascochyta blight (coupled with increasing aggressiveness of the pathogen), onset of unseasonal frosts that severely affected seed quality of crops maturing in the field and reduced export prices.

South America (Chile, Peru, Argentina, Columbia and Bolivia)

Chickpea has always been a minor crop in South America. The combined area of less than 10,000 ha (mean 1996–2005) is less than 0.1% of the world area.

According to FAO statistics Chile is the leading producer with just over half of the region's production. (Although there are no available statistics for Columbia beyond 1994, earlier substantial production suggests that the crop can still be grown there.) Production peaked at about 25,000t during the 1970s, most likely because of favourable export prices, but declined during 1961–2005 by an average 2.1% per year (Fig. 7.3g). Only Kabuli types are grown in South America.

Australia

Chickpea is a comparatively new crop in Australia: the first commercial crops were grown in 1979. Crop areas, sown mainly (>90%) with desi types, rose steadily until a peak of 309,000ha was reached in 1998. During 1996–2005 chickpea area and production in Australia were 2.0% and 2.4%, respectively, of the world total.

The rapid farmer adoption of chickpea was largely due to improved profitability relative to that for two of the major agricultural enterprises, sheep and wheat (Rees *et al.*, 1994). In particular, decades of exploitative cereal monoculture had increased the incidence of cereal diseases and weed problems, and seriously reduced soil nitrogen levels. Chickpea helped alleviate these problems, especially in the north-eastern cropping belt where there were limited options for alternative winter pulses. An increased demand for desi chickpeas in the Indian subcontinent from the 1980s, coupled with generally low import tariffs, provided the essential export markets for a country in which local consumption was negligible. Major epidemics of the exotic disease ascochyta blight in 1998 and 1999 were a turning point for the Australian industry. A lack of resistant varieties and the high cost of fungicidal protection led to a 50% decline in crop area by 2005 (Fig. 7.3h).

References

Acikgoz, N., Karaca, M., Er, C. and Meyveci, K. (1994) Chickpea and lentil production in Turkey. In: Muehlbauer, F.J. and Kaiser, W.J. (eds) *Expanding the Production and Use of Cool Season Food Legumes: Proceedings of the Second International Food Legume Research Conference*. 12–16 April 1992. Kluwer Academic, Cairo, Egypt, pp. 388–398.

Barradas, M.I. (1990) Chickpea breeding in Portugal. In: Saxena, M.C., Cubero, J.I. and Wery, J. (eds) *Present Status and Future Prospects of Chickpea Crop Production and Improvement in the Mediterranean Countries: Proceedings of the Zaragoza/Spain Seminar*. 11–13 July 1988. INO-Reproducciones, Zarogoza, Spain, pp. 157–161.

Cubero, J.I., Moreno, M.T. and Gil, J. (1990) Chickpea breeding in Spain. In: Saxena, M.C., Cubero, J.I. and Wery, J. (eds) *Present Status and Future Prospects of Chickpea Crop Production and Improvement in the Mediterranean Countries: Proceedings of the Zaragoza/Spain Seminar*. 11–13 July 1988. INO-Reproducciones, Zarogoza, Spain, pp. 151–155.

FAO (2006) FAOSTAT.data, 2006.

Malhotra, R.S., Erskine, W., Ergashev, N., Beniwal, S.P.S. and ShevtShov, V. (2000)

Chickpea: present status and future prospects in Central Asia and the Caucasus. *Grain Legumes* 29, 25–26.

Musa, A.M., Harris, D., Johansen, C. and Kumar, J. (2001) Short duration chickpea to replace fallow after *aman* rice: the role of on-farm seed priming in the High Barind Tract of Bangladesh. *Experimental Agriculture* 37, 509–521.

Oram, P.A. and Agcaoili, M. (1994) Current status and future trends in supply and demand of cool season food legumes. In: Muehlbauer, F.J. and Kaiser, W.J. (eds) *Expanding the Production and Use of Cool Season Food Legumes: Proceedings of the Second International Food Legume Research Conference*. 12–16 April 1992. Kluwer Academic, Cairo, Egypt, pp. 3–42.

Pande, S., Stevenson, P., Narayana Rao, J., Neupane, R.K., Chaudhary, R.N., Grzywacz, D., Bo, V.A. and Krishna Kishore, G. (2005) Reviving chickpea production in Nepal through integrated crop management, with emphasis on Botrytis grey mould. *Plant Disease* 89, 1252–1262.

Rees, R.O., Brouwer, J., Mahoney, J.E., Walton, G.H., Brinsmead, R.B., Knights, E. and Beech, D.F. (1994) Pea and chickpea production in Australia. In: Muehlbauer, F.J. and Kaiser, W.J. (eds) *Expanding the Production and Use of Cool Season Food Legumes: Proceedings of the Second International Food Legume Research Conference*. 12–16 April 1992. Kluwer Academic, Cairo, Egypt, pp. 412–425.

Sabaghpour, S.H., Sadeghi, E. and Malhotra, R.S. (2003) Present status and future prospects of chickpea cultivation in Iran. In: Sharma, R.N., Shrivastava, G.K., Rathore, A.L., Sharma, M.L. and Khan, M.A. (eds) *Chickpea Research for the Millennium: Proceedings of International Chickpea Conference*. 20–22 January 2003. Indira Gandhi Agricultural University, Raipur, India, pp. 436–443.

Singh, K.B., Reddy, M.V. and Malhotra, R.S. (1985) Breeding kabuli chickpeas for high yield, stability, and adaptation. In: Saxena, M.C. and Varma, S. (eds) *Proceedings of International Workshop on Faba Beans, Kabuli Chickpeas, and Lentils in the 1980s*. ICARDA, Aleppo, Syria, pp. 71–90.

Smith, V.H. and Jimmerson, J. (2005) Chickpeas (garbanzo beans). Briefing No. 55, Agricultural Marketing Policy Centre, Montana State University.

van der Maesen, L.J.G. (1972) Cicer *L., A Monograph of the Genus, with Special Reference to the Chickpea (*Cicer arietinum *L.), Its Ecology and Cultivation*. Mededelingen Landbouwhogeschool, Wageningen, The Netherlands.

8 Chickpea: *Rhizobium* Management and Nitrogen Fixation

F. KANTAR,[1] F.Y. HAFEEZ,[2] B.G. SHIVAKUMAR,[3] S.P. SUNDARAM,[4] N.A. TEJERA,[5] A. ASLAM,[2] A. BANO[6] AND P. RAJA[4]

[1]Ataturk University, Faculty of Agriculture, Department of Agronomy, 25240 Erzurum, Turkey; [2]Biofertilizer Division, National Institute for Biotechnology and Genetic Engineering (NIBGE), PO Box 577, Jhang Road, Faisalabad, Pakistan; [3]Division of Agronomy, Indian Agricultural Research Institute, Pusa, New Delhi 110012, India; [4]Department of Agricultural Microbiology, Tamil Nadu Agricultural University, Coimbatore 641003, India; [5]Department of Plant Physiology, Faculty of Sciences, University of Granada, Campus de Fuentenueva s/n, 18071 Granada, Spain; [6]Department of Plant Sciences, Faculty of Biological Sciences, Quaid-i-Azam University, Islamabad 45320, Pakistan

Introduction

Chickpea (*Cicer arietinum* L.) is a major food legume in many countries of Asia, the Middle East and South and Central America, and is cultivated worldwide on an area of more than 11 million hectares. Not only is chickpea an important source of protein in human diets, but it also plays a significant role in maintaining soil fertility, particularly in dry lands by fixing atmospheric nitrogen (N).

Chickpea obtains a significant proportion of its N requirement through symbiotic N fixation. Inoculation of chickpea with adequate number of rhizobia results in a significant increase in the number of nodules, nodule dry weight and N fixation (Sattar *et al.*, 1993). Growth of chickpea is dependent on nodulation by effective rhizobial strains. The International Centre for Research in the Semi-Arid Tropics (ICRISAT) Asia Centre, Hyderabad, India, launched a programme in 1983 to select high-nodulating (HN) varieties of chickpea as a strategy to improve both stability and potential of yields (Rupela, 1992, 1994). HN and normal-nodulating (NN) varieties were reported in chickpea (Rupela, 1994). Mixed-culture inoculation of *Rhizobium* with associative N-fixing *Azospirillum* sp. and *Bacillus subtilis* and phosphorus (P)-solubilizing *B. megaterium* significantly increased the total number and dry weight of nodules and consequently the grain yield of chickpea (Kantar, Turkey, 2006, personal communication).

Management practices to produce larger biomass yields under lower soil nitrate conditions may result in increased N fixation by chickpea (Horn *et al.*, 1996). Rotation is a key strategy for improving production and maintaining fertility of the soil in chickpea-based cropping systems. Of particular relevance to the overall grain yield responses is the contribution of chickpea to soil N status through its ability to fix atmospheric N and effects on root pathogens of antecedent crops (Felton *et al.*, 1995), thus improving the N nutrition and yield of subsequent cereals by as much as 70% (Aslam *et al.*, 2003).

Host Specificity and Interaction at Molecular Level

Cross-inoculation group specificity

Although legume–*Rhizobium* symbioses are nuptials between two immensely different genomes, both partners contribute equally to this relationship in a strictly controlled and coordinated manner. Many signals and codes are exchanged by both contenders to find the attuned partner in the rhizosphere (Broughton *et al.*, 2000). Entry into the plants is restricted only to those bacteria that have specific and right 'keys' to a series of legume 'doors' or 'locks'. It is well known that the legume–*Rhizobium* interaction falls into the cross-inoculation groups, where certain rhizobial strains nodulate only certain legumes (Hirsch *et al.*, 2001).

Recent studies elucidate the mechanism of symbiogenesis signal transduction, morphogenesis of the symbiosome, metabolic integration and molecular cross-talk between the partners (Provorov and Vorobyov, 2000). The genetic requirements for nodulation are divided between the partners, with both containing genes specifically expressed in the presence of the other (Djordjevic *et al.*, 1987). Symbiotic control is implemented at many checkpoints in which the three symbiotic signals of flavonoids, nod factors and cell surface carbohydrates play the most important role in determining the specificity during the symbiosis process.

Regarding physiological characteristics of the N fixation process in chickpea, this plant should be classified both as an amide and ureide exporter, on the basis of the concentration of both types of N products found in the shoot (Thavarajah *et al.*, 2005). Specific activities of the enzymes involved in sucrose breakdown increase in parallel with the nitrogenase activity of the nodules and the total N content of the plant until flowering stage (Copeland *et al.*, 1995). In addition, the glutamine synthetase, glutamate synthase, phosphoenolpyruvate carboxylase and malate dehydrogenase activities also increase in the nodules with culture age, whereas rubisco activity and chlorophyll content of leaves decrease (Soussi *et al.*, 1998).

Infection, nodule development and nitrogen fixation

The formation of infection thread and its development into a functional nodule is a highly complicated and controlled process. Nodule development starts with the infection of root hair, and root hair curling is the first sign of infection. The infection thread penetrates into the root hair cell and the bacteria are embedded in a translucent matrix. The root hair cell remains vacuolated, having thick cytoplasm. The bacteria trapped within the infection thread proliferate and after 3 days the infection thread enters the cortical cells, which are converted to meristematic state with a small vacuole, large nucleus and clear nucleotide. Cell division starts and nodule primordia develop into a nodule that appears as an outgrowth on the root. Bacteria are released from the infection thread by endocytosis and differentiate into bacteroids surrounded by peribacteroid membrane. Mature nodules are fully occupied with rhizobia. Developing and senescing of chickpea nodules have general morphological and ultrastructural features that are characteristic of indeterminate nodules. These features include the presence of persistent meristematic tissue at the distal ends of the multilobed nodules, and a gradient of cells at different stages of development towards the proximal point of attachment of the nodules to the parent root. The cytoplasm of infected cells in the N-fixing region of the nodules is densely packed with symbiosomes, most of which contain a single bacteroid. Electron-dense inclusions in the intercellular spaces in the N-fixing region, and plasmodesmata that connect the infected cells with other uninfected cells, are important features of a chickpea nodule. Uninfected cells in the central region are smaller than the infected cells and are highly vacuolated. Poly-β-hydroxybutyrate granules have not been reported in bacteroids. In the later stages of development, the infected cells become enlarged and highly vacuolated, and eventually lose their contents (for details see Perret *et al.*, 2000; Long, 2001; Daniel, 2004).

Taxonomic Status of Chickpea *Rhizobium*

History of chickpea rhizobia

Nodulating rhizobia in chickpea were studied occasionally and only a few strains belonging to this group were described in the rhizobial taxonomic studies (Nour *et al.*, 1994). Thus, contradictory conclusions were made regarding the taxonomic status of chickpea rhizobia.

Vincent (1974) placed the chickpea-infecting rhizobia in the pea inoculation group (*Rhizobium leguminosarum*). Gaur and Sen (1979) tested the nodulation capacity of a large number of strains of *Cicer* rhizobia in different legume species and a multitude of other rhizobia in *C. arietinum* and recommended the placement of *C. arietinum* and its root nodule bacteria under a separate cross-inoculation group. Moreover, chickpea rhizobia were reported to form a unique group on the basis of host specificity (Dadarwal, 1980). On the other hand, it was proposed that chickpea strains should be classified as *R. loti* or *Bradyrhizobium* strains, depending on their generation time (Jarvis *et al.*, 1982). Chickpea-nodulating rhizobia were broadly divided into fast- and slow-growing forms based on their growth characteristics, but the level of DNA homology for both slow- and fast-growing strains is very low (Chakrabarti *et al.*, 1986). In *Bergey's Manual of Systematic Bacteriology*, chickpea rhizobia are classified as members of either *R. loti* or an undetermined group of *Bradyrhizobium* sp. (Jordan, 1984). Serological characterization showed the antigenic uniqueness of chickpea rhizobia and the lack of serological cross-reaction with other rhizobia. Cadahia *et al.* (1986) stated that the culture characteristics and polymorphism of *nifH* and *nifD* genes are not the proper criteria to define groups within the chickpea rhizobia.

On the basis of these contradictory conclusions, Nour *et al.* (1994, 1995) re-examined the diversity of the geographically different chickpea rhizobia at species level using a polyphasic approach and reported five genomic species in 30 isolates forming two phylogenetic lineages distinct from previously described species. The DNA homology values of both phylogenetic groups were <38%, indicating that the group organisms belong to different species. They suggested that the chickpea rhizobia cannot be assigned to a previously described species.

Current status

Current taxonomy reveals a wider diversity of legume-nodulating bacteria at the genus, species and intraspecies levels. Martinez-Romero *et al.* (2000) split the genus *Rhizobium* into six genera: *Bradyrhizobium*, *Rhizobium*, *Azorhizobium*, *Sinorhizobium*, *Mesorhizobium* and *Allorhizobium*, and identified 29 species. The chickpea-infecting rhizobia have been transferred to genus *Mesorhizobium* (Jarvis *et al.*, 1997). Hence *R. ciceri* and *R. mediterraneum* have been renamed as *Mesorhizobium ciceri* and *M. mediterraneum* (Nour *et al.*, 1994, 1995; Jarvis *et al.*, 1997). Some strains of *M. loti*, *M. trianshanese* and *M. amorphae* have been reported to nodulate chickpea in Portugal and Spain (Peix *et al.*, 2001; Laranjo *et al.*, 2004).

Ecology of Chickpea *Rhizobium*

Natural occurrence and diversity

Distribution and nodulation of chickpea vary with the site and conditions (Aouni, 1998; Laranjo *et al.*, 2002). However, chickpea-specific mesorhizobia have been reported to be present in the soils where chickpea is grown (Aouni *et al.*, 2001; Laranjo *et al.*, 2002). Some of the chickpea rhizobia have wide distribution, while other genomic species appear to be confined to one geographical area. *M. ciceri* is found in the Mediterranean, North Africa, North America, the Indian subcontinent and Russia (Nour *et al.*, 1995). The population varies from 1.5×10^3 to 4.3×10^6 nfu/g soil in Portugal (Laranjo *et al.*, 2002). Sandy soils or hydromorphic soils do not contain any chickpea rhizobial population (Aouni *et al.*, 2001). The number of nodules varies from 10 to 38/plant with 60–500 mg nodule dry weight/plant, their length is 3–4 mm and they turn red after 2 weeks due to the presence of haemoglobin (Aouni *et al.*, 2001). Some non-specific bacterial species like *Sinorhizobium medicas*, *Agrobacterium* sp. and *Ochrobactrum* sp. have also been isolated from chickpea nodules (Naseem *et al.*, 2005). However, *S. medicas* symbiosis was ineffective (Aouni *et al.*, 2001). Chickpea nodules also harbour non-symbiotic bacteria along with the symbiotic rhizobia. An *Agrobacterium* strain has been isolated from the chickpea root nodules and also has the ability to solubilize phosphate *in vitro* (Hameed *et al.*, 2005). The role of these non-symbiotic bacteria is unclear, but might be helping in the process of nodulation (Hameed *et al.*, 2004). *Paenibacillus rhizosphaerae* have been isolated from chickpea rhizosphere (Rivas *et al.*, 2005).

Survival of chickpea rhizobia

Rhizobia can better survive in the presence of the compatible host, but it can also lead a non-symbiotic life in soils where the host has never been grown or grown a long time ago (Aouni *et al.*, 2001). In the vicinity of plant roots, root exudates and root debris are the most likely substrates that must be available to rhizobial population to survive in non-symbiotic conditions and maintain the population in fallow soils (Roberts and Schmidt, 1983). The most problematic environments for rhizobia are marginal lands with low rainfall, extreme temperatures, acidic soils of low nutrient status and poor water-holding capacity (Bottomley, 1991). Other environmental stresses may include photosynthate deprivation, salinity, heavy metals, soil nitrates and biocides (Walsh, 1995) that affect the survival of the rhizobia and/or nodule metabolism directly. They can tolerate up to 2% salt and can grow at pH up to 9.5 and temperature 40°C (Maatallah *et al.*, 2002). A positive correlation has been observed between genetic variation and pH of the soil for chickpea rhizobia (Laranjo *et al.*, 2002). Nodulation is poor in acid soils because of sensitivity of the nodulation events, such as attachment, root hair curling and formation of infection thread.

A wide physiological diversity exists among chickpea rhizobia. They are mostly gram-negative, aerobic, non-spore-forming rods. Colonies on yeast extract–mannitol agar are convex and opaque at 28°C (Nour *et al.*, 1994, 1995; Naseem *et al.*, 2005). Rhizobial cultures are usually maintained in agar slants or lyophilized and stored at −70°C. There is, however, well-documented evidence of phenotypic and genotypic instability in cultures maintained on agar medium (Thies *et al.*, 2001). Such variations may originate from spontaneous mutation (Lochner *et al.*, 1991) or clonal dimorphism (Sylvester-Bradley *et al.*, 1988) and sometimes result in altered symbiotic properties (Naseem *et al.*, 2005). Phenotypic changes in rhizobia are of concern and stringent monitoring and attention to possible causal factors are required (Thies *et al.*, 2001; Naseem *et al.*, 2005).

Genetic diversity among chickpea rhizobial population is high as determined by random amplification of polymorphic DNA (RAPD), direct amplified polymorphic DNA (DAPD), polymerase chain reaction restriction fragment length polymorphism (PCR-RFLP) and 16 SrDNA sequencing (Laranjo *et al.*, 2002; Karim *et al.*, 2003). The biggest source of variation in bacteria is the horizontal transfer of the genes mainly found in the plasmids (Levin and Bergstrom, 2000). Transfer of these genetic elements or even ordinary gene is facilitated by hitch-hiking on conjugative plasmids or phages or by being picked up as free DNA by hosts with transformation, transduction and conjugation mechanisms

(Levin and Bergstrom, 2000). Acquisition and loss of these genes or whole plasmids, which is a continuous and dynamic process, seems to play a major role in the evolution of rhizobia. Plasmid exchange or genome rearrangements have been reported in rhizobia and agrobacteria (Truchet *et al.*, 1985) that result in loss or gain of symbiotic performance (Zhang *et al.*, 2001). There is evidence of transfer of large chromosomal segments (Sullivan and Ronson, 1998) and plasmids (Naseem *et al.*, 2005) in the chickpea-specific mesorhizobia.

Symbiotic effectiveness and number of nodules per plant in chickpea are not related with the number of plasmids. Up to four plasmids have been reported in chickpea rhizobia (Laranjo *et al.*, 2004), but strains having no plasmid may also have symbiotic ability in chickpea (Laranjo *et al.*, 2002, 2004). It is possible that in isolates without plasmids, the symbiotic genes are located on the chromosomes (Sullivan *et al.*, 1995; Laranjo *et al.*, 2002) as a single copy, while the isolates with one symbiotic plasmid have several copies, which could lead to a higher symbiotic efficiency.

Better N_2 fixation can be achieved by selecting superior rhizobia. However, selection of these rhizobia needs to take into consideration not only their N_2-fixing capacity, but also competitive ability against native rhizobia that are frequently ineffective in N_2 fixation. In order to achieve this, rhizobia have to be selected under natural conditions in competition with the native rhizobia (Rengel, 2002). Symbiotic effectiveness also depends upon the chickpea cultivars (Laranjo *et al.*, 2002). The symbiotic characteristics can be considered in breeding for better symbiosis as chickpea genotypes differ in the total number and weight of nodules per plant, as well as the weight of individual nodules and status of N fixation (Hafeez *et al.*, 1987, 1998; Dangaria *et al.*, 1994).

Rhizobium Inoculation and Nitrogen Fixation

Response to *Rhizobium* inoculation

Rhizobium inoculation is the most effective and sustainable N management strategy in chickpea. Due to symbiotic association between *Rhizobium* and chickpea roots, a good but variable quantity of N is made available to host chickpea plants. The extent of N fixation in chickpea due to *Rhizobium* inoculation often varies from place to place, season to season and cultivation practices. Further, the native *Rhizobium* population already present in the soil may nullify the advantage of *Rhizobium* inoculation. However, the *Rhizobium* inoculation may result in good response in terms of higher productivity of chickpea in areas with lower population of effective strains in the soil and new areas of cultivation. The residual effect of chickpea–*Rhizobium* symbiosis can often be seen in the succeeding crops grown in sequence (Sharma *et al.*, 1995).

Factors affecting nitrogen fixation

The effect of *Rhizobium* inoculation and the quantity of fixed N may vary with diverse agroclimatic conditions, cropping systems and cultivation practices

(Gupta *et al.*, 1988). Nodulation and, consequently, total impact will be quite dissimilar under contrasting environmental conditions and agronomic practices. They are broadly categorized into the following types.

Strain variation and competitive ability of rhizobia

The ability of inoculated rhizobia to nodulate and fix N more efficiently than naturally occurring rhizobial strains determines the efficiency of the microsymbiont. If the naturally occurring rhizobia or indigenous strains of rhizobia are in large quantities, the inoculated rhizobial strains cannot sufficiently colonize the rhizosphere and nodulate crop plants. Likewise, the number of nodules formed in chickpea varies with the competitive ability of rhizobial strains (Kantar *et al.*, 2003). Effectiveness of chickpea–*Rhizobium* symbioses varies either due to variations in the N-fixing potential of *Rhizobium* strains or due to host–genotypic compatibility. Efficient and competitive *Rhizobium* inoculant is the basis for establishing chickpea N fixation under field conditions. Great care has to be taken while choosing a competitive strain of *Rhizobium* for bio-inoculant development.

A large number of strains have been screened in different chickpea-growing areas for their efficiency in terms of nodulation and their influence on the productivity of crops with consistently better results over the non-inoculated control, but large variations have been observed in their effectiveness due to growing conditions or genetic make-up of the strains.

Physiological state and specificity of host plants

Inadequate photosynthesis due to plant diseases and the physiological state of the host plant have strong influence on the *Rhizobium*–legume symbiosis (Peoples *et al.*, 1995). As a result of physiological stress, photosynthate deprivation may occur, which, in turn, will deprive the carbon source of the nodule bacteria. The growth stage of the plants becomes more significant in determining the N fixation under moisture stress conditions. Under moisture stress, nodulation and N fixation is greatly affected at vegetative stage than at reproductive stage (Pena-Cabriales and Castellanos, 1993). Similarly, nitrogenase activity and N fixation also varies in response to plant growth stage. The highest activity of nitrogenase and N fixation was detected during the reproductive stage (Dorosinskii *et al.*, 1979). The host specificity is the key factor in determining the colonization potential of rhizobia. The chickpea–*Rhizobium* symbiosis is highly specific and extensive studies of Gaur and Sen (1979) demonstrated the uniqueness of the chickpea rhizobia. About 99% of the chickpea isolates studied were able to nodulate only the original host plant. As strains vary in their effectiveness in nodulation and N fixation, the chickpea cultivars also differ and play a more important role in regulating N fixation (Pareek and Chandra, 2003).

Environmental and soil conditions

Environmental and soil conditions have profound effect on the performance of *Rhizobium* strains and N fixation. Soil salinity (Zurayk *et al.*, 1998) or acidity (Chandra and Pareek, 1991), presence of air pollutants, heavy elements (Pal, 1996), nematodes (Chalal and Chalal, 1991) and root rots (Rajasekhar *et al.*,

1991) adversely affect the performance of inoculated *Rhizobium* culture. The chickpea is more salt-sensitive than *Rhizobium*, roots are more sensitive than shoots and N fixation is more sensitive than plant growth (Elsheikh and Wood, 1990). But the increasing organic matter content was complementary to the nodulation. Among the environmental conditions, factors like rainfall, water-logging and extremely high or low temperatures have a serious deleterious influence on the normal activity of the chickpea *Rhizobium*.

Quality of Rhizobium *inoculants*

Quality control is an essentially important factor for the establishment of any inoculant in field conditions. Worldwide minimum standards range from 5 × 107 to 1 × 109cells/g of freshly prepared inoculants. Increasing the number of rhizobia always exhibits positive effect on crop improvement. In a study conducted to find out the effect of increasing rhizobia population on crop performance, increasing the *Rhizobium* population applied to the seed from 1.9 × 104 to 1.9 × 106 increased nodule weight from 65 to 393 mg/plant and, most importantly, grain yield from 1.9 to 2.1 t/ha (Roughley *et al.*, 1993). Use of sterile carrier material for inoculant preparation prevents contamination and enhances shelf-life of the bio-inoculant. Rhizobial numbers decline rapidly in non-sterile carriers during storage because of competition with contaminating organisms. Steam sterilization at 1240°C for 3.5 h has been used in many countries to produce high-quality inoculants (Strijdom and Deschodt, 1976). Excessive heat renders peat unfavourable for subsequent growth and survival of rhizobia. Gamma irradiation at 50 kGy was equally effective for the production of high-quality inoculants with peat as carrier material. Generally, a loss of efficiency of the rhizobia with increasing inoculant age is observed (Catroux *et al.*, 2001).

Agronomic practices

The performance of even the best *Rhizobium* inoculants often gets masked due to poor agronomic practices. Selection of appropriate *Rhizobium* strains, chickpea cultivars, time of sowing, method of inoculation, method of sowing, growing conditions like irrigated or rainfed farming, nutrient management and use of agrochemicals have greater influence on the performance of *Rhizobium* inoculation leading to higher productivity of chickpea. Pendimethalin, a very common pre-emergence herbicide used for weed control, was highly toxic to nodulation and soil bacteria.

Management of Nitrogen Fixation in Chickpea

Microbiological and plant-related aspects

The successful production of chickpea depends on many factors, including matching of rhizobial strain and host cultivar. It is necessary to identify chickpea cultivars and *Mesorhizobium* strains for superior N fixation capacity. N fixation and, consequently, productivity of chickpea is strongly related to several ever-changing and interacting environmental, genetic and agronomic parameters. *M. ciceri* inoculated to chickpea seeds is often exposed to adverse environmental conditions, which affect its survival and subsequent effectiveness, and hence soil-applied granular inoculants have been developed. The comparison of seed inoculation (liquid or peat-based powder) with soil inoculation (granular inoculant) revealed that nodule formation in the seed inoculation treatments was restricted to the crown region of the root system, whereas soil inoculation enhanced nodulation on the lateral roots (Kyei-Boahen *et al.*, 2002a). The amount of N_2 fixed was typically lower for the liquid inoculant than peat and granular inoculants, and lateral root nodules contributed significantly to N_2 fixation and yield in chickpea (Kyei-Boahen *et al.*, 2002a). Ounane *et al.* (2003) showed a significant interaction between the strains and water stress and observed that the highest N fixation was obtained under moderate water stress, whereas severe stresses affected almost all strains at flowering stage.

Estimations using ^{15}N on several chickpea cultivars and native rhizobia from Algeria revealed a high N-fixing efficiency of symbiosis, being able to obtain between 55% and 72% of total plant N derived from air (Mefti, 2003). A comparison of desi and kabuli cultivars inoculated with different *M. ciceri* strains showed that isotopic (^{15}N) effect was influenced by the infecting rhizobial strain in kabuli type, but not in the desi type (Kyei-Boahen *et al.*, 2002b). The nitrogenase activity should be determined along with nodule number and nodule dry weight as chickpea ILC1919 from Syria inoculated with *M. ciceri* ch-191 strain achieved higher nitrogenase activity but lower nodule number and dry weight (Tejera *et al.*, 2006).

With the development of new cultivars for winter sowing, chickpea production has expanded into drier areas of the Mediterranean where low or less effective populations or indigenous rhizobia can limit N_2 fixation. Field studies using several cultivars and both native rhizobial populations and introduced rhizobial strains have indicated that cultivar–strain interactions are significant for seed yield, N yield and N_2 fixation; thus, the use of the best strain increased the amount of N derived from fixation from 52% to 72% (Beck, 1992). N harvest index is suggested as a criterion to improve seed yield in chickpea (Ayaz *et al.*, 2004).

The assessment of chickpea germplasm maintained in ICRISAT and in the International Centre for Agricultural Research in the Dry Areas (ICARDA, Aleppo, Syria) has resulted in the identification of sources of resistance to various biotic and abiotic stresses as well as high-yielding and adapted lines (Malhotra *et al.*, 1999). In addition, exotic germplasm has been used extensively in breeding programmes to improve yield and quality of cultivated chickpea (Croser *et al.*, 2003).

Agronomic considerations

Reduction of time to flowering and maturity has made a major contribution towards increasing and stabilizing chickpea productivity in the tropics. Super-early chickpea genotypes like ICCV 96029 have reduced crop duration from 160 to <130 days in the subtropics (Kumar and Rao, 2001). Short-duration cultivars are making a large impact in extending chickpea cultivation to rice fallows. A large biomass accumulation associated with the increase of N fixation occurs from winter sowing with cultivars tolerant to cold and biotic stresses due to the reduction of terminal heat stress and drought stress effect (Erskine and Malhotra, 1997).

Focus of research has recently shifted to more basic and strategic aspects. Much effort has been put on developing genetic populations for studying inheritance, determining linkage groups and ultimately developing a genome map of chickpea, which will also facilitate pyramiding of genes for single and multiple traits and enhance the pace of progress for higher yields and stable performance of chickpea in the future.

Conclusion

Considerable information has been collected on chickpea–*Rhizobium* symbiosis over the last two decades on host specificity, taxonomy, ecology, inoculant development and management practices, which may accelerate the ongoing efforts for harnessing N_2 fixation in chickpea for sustainable production. Understanding the key molecular factors and steps in chickpea–*Rhizobium* specificity and interaction is of crucial importance for the development of *Rhizobium* strains and chickpea cultivars with high N_2-fixation potential. Better N_2 fixation can be achieved by selecting rhizobial strains of superior N_2-fixing capacity and competitive ability against the native rhizobia under natural conditions. Mutations are important and the genes responsible for efficient N fixation and wider adaptability need to be identified and cloned using molecular markers to be incorporated to create efficient *Rhizobium* strains. As the efficiency of symbiosis also depends on the chickpea cultivars, the symbiotic characteristics need to be considered in breeding for better symbiosis using mutants and exotic germplasm. Profusely nodulating cultivars with selective affinity for a certain rhizobial strain and/or intrinsic capacity of fixing N in the presence of high NO_3 levels should be an objective in breeding programmes.

Adequate information is already available on the effects of environmental conditions and management practices on N fixation. The available evidence indicates the advantages of rhizobial inoculations for increasing yields and reducing fertilizer consumption in chickpea-based rotations with promising results from granular inoculants for soil applications under variable conditions. Complementary effects of dual or multiple inoculations of *Rhizobium* with associative N-fixing bacteria, VA Micor and phosphate-solubilizing bacteria may enhance success and, consequently, acceptability of inoculants among farmers.

References

Aouni, M.E. (1998) La symbiose *Rhizobium* légumineuses: contraintes de l'environment sur la nodulation, la croissance et la compétitivité. *Thèse de Doctorat d'Etat ès Sciences en Biologie*, Faculté des Sciences de Tunis, Tunis.

Aouni, M.E., Mhamdi, R., Jebra, M. and Amarger, N. (2001) Characterization of rhizobia nodulating chickpea in Tunisia. *Agronomie* 21, 577–581.

Aslam, M., Mahmood, I.A., Peoples, M.B., Schwenke, G.D. and Herridge, D.F. (2003)

Contribution of chickpea nitrogen fixation to increased wheat production and soil organic fertility in rainfed cropping. *Biology and Fertility of Soils* 38, 59–64.

Ayaz, S., McKenzie, B.A., Hill, G.D. and McNeil, D.L. (2004) Nitrogen distribution in four grain legumes. *Journal of Agricultural Science* 142, 309–317.

Beck, D.P. (1992) Yield and nitrogen fixation of chickpea cultivars in response to inoculation with selected rhizobial strains. *Agronomy Journal* 84, 510–516.

Bottomley, P. (1991) Ecology of *Rhizobium* and *Bradyrhizobium*. In: Stacey, G., Burris, R.H. and Evans, H.J. (eds) *Biological Nitrogen Fixation*. Chapman & Hall, New York, pp. 292–347.

Broughton, W.J., Jabbouri, S. and Perret, X. (2000) Keys to symbiotic harmony. *Journal of Bacteriology* 182, 5641–5652.

Cadahia, E., Leyvar, A. and Ruiz-Argueso, T. (1986) Indigenous plasmids and cultural characteristics of rhizobia nodulating chickpeas (*Cicer arietinum* L.). *Archives in Microbiology* 146, 239–244.

Catroux, G., Hartmann, A. and Revellin, C. (2001) Trends in rhizobial inoculant production and use. *Plant and Soil* 230, 21–30.

Chakrabarti, S.K., Mishra, A.K. and Chakrabatty, P.K. (1986) DNA homology studies of rhizobia from *Cicer arietinum* L. *Canadian Journal of Microbiology* 32, 524–527.

Chalal, P.P.K. and Chalal, V.P.S. (1991) Effect of rhizobium and root knot nematodes on nitrogen fixation and nitrate utilization in chickpeas (*Cicer arietinum* L.). *Journal of Ravishankar University* 4–5(B), 43–45.

Chandra, R. and Pareek, R.P. (1991) Comparative performance and competitiveness of chickpea (*Cicer arietinum* L.) *Rhizobium* strains in acidic and neutral soils. *Indian Journal of Pulses Research* 4(2), 189–194.

Copeland, L., Lee, H.S. and Cowlishaw, N. (1995) Carbon metabolism in chickpea nodules. *Soil Biology and Biochemistry* 27, 381–385.

Croser, J.S., Ahmad, F., Clarke, H.J. and Siddique, K.H.M. (2003) Utilization of wild *Cicer* in chickpea improvement: progress, constraints, and prospects. *Australian Journal of Agricultural Research* 54, 429–444.

Dadarwal, K.R. (1980) Host bacterium factors involved in legume symbioses. *Indian Journal of Microbiology* 20, 152–172.

Dangaria, C.J., Parameshwarappa, R., Salimath, P.M. and Annigeri, B.S. (1994) Genetic divergence for nodulating characters in chickpea. *Legume Research* 17, 32–36.

Daniel, J. (2004) Infection and invasion of roots by symbiotic nitrogen-fixing rhizobia during nodulation of temperate legumes. *Microbiology and Molecular Biology Reviews* 68(2), 280–300.

Djordjevic, M.A., Gabriel, D.W. and Rolfe, B.G. (1987) *Rhizobium*: the refined parasite of legumes. *Annual Review of Phytopathology* 25, 145–168.

Dorosinskii, L.M., Afanas'eva, L.M. and Kutin, A.A. (1979) Nitrogen fixing activity of pea nodule bacteria during different phases of host plant development. *Mikrobiologiia* 48(3), 495–501.

Elsheikh, E.A.E. and Wood, M. (1990) Effect of salinity on growth, nodulation and nitrogen yield of chickpea (*Cicer arietinum* L.). *Journal of Experimental Botany* 41(231), 1263–1269.

Erskine, W. and Malhotra, R.S. (1997) Integration of winter-sown chickpea and lentil into cropping systems in the Middle East. *Grain Legumes* 16, 20–21.

Felton, W., Marcellos, H. and Martin, R.J. (1995) A comparison of three fallow management strategies for the long term productivity of wheat in northern New South Wales. *Australian Journal of Experimental Agriculture* 35, 915–921.

Gaur, Y.D. and Sen, A.N. (1979) Cross inoculation group specificity in *Cicer–Rhizobium* symbiosis. *New Phytology* 83, 745–754.

Gupta, B.R., Prasad, S.N. and Pathak, A.N. (1988) Studies on the *Rhizobium* inoculation on chickpea in Uttar Pradesh. *Farm Science Journal* 3(1), 24–31.

Hafeez, Y.F., Idris, M., Malik, K.A., Asad, S. and Malik, I.L. (1987) Screening of chickpea (*Cicer arietinum* L.) and lentil (*Lens culinaris* Medic) for their nitrogen fixation potential. *Modern Trends in Plant Science Research Pakistan*, 140–144.

Hafeez, F.Y., Ahmad, T., Hameed, S., Danso, S.K.A. and Malik, K.A. (1998) Comparison of direct and indirect methods of measuring nitrogen fixation in field grown chickpea genotypes. *Pakistan Journal of Botany* 30(2), 199–207.

Hameed, S., Yasmin, S., Malik, K.A., Zafar, Y. and Hafeez, F.Y. (2004) *Rhizobium, Bradyrhizobium* and *Agrobacterium* strains isolated from cultivated legumes. *Biology and Fertility of Soils* 39(3), 179–185.

Hameed, S., Mubeen, F., Malik, K.A. and Hafeez, F.Y. (2005) Nodule co-occupancy of *Agrobacterium* and *Bradyrhizobium* with potential benefit to legume host. In: Wang, Y.P., Lin, M., Tian, Z.X., Elmerich, C. and Newton, W.E. (eds) *Biological Nitrogen Fixation, Sustainable Agriculture and Environment: Proceedings of 14th International Nitrogen Fixation Congress*. Springer, Dordrecht, The Netherlands, pp. 295–296.

Hirsch, A.M., Lum, M.R. and Downie, J.A. (2001) What makes the rhizobia–legume symbiosis so special? *Plant Physiology* 127, 1484–1492.

Horn, C.P., Dalal, R.C., Birch, C.J. and Doughton, J.A. (1996) Sowing time and tillage practice affect chickpea yield and nitrogen fixation. 2. Nitrogen accumulation, nitrogen fixation and soil nitrogen balance. *Australian Journal of Experimental Agriculture* 36(6), 701–706.

Jarvis, B.D.W., Pankhurst, C.E. and Patel, J.J. (1982) *Rhizobium loti*, a new species of legume-root nodule bacteria. *International Journal of Systematic Bacteriology* 32, 378–380.

Jarvis, B.D.W., van Berkum, P., Chen, W.X., Nour, S.M., Fernandez, M.P. and Cleyet-Marel, J.C. and Gillis, M. (1997) Transfer of *Rhizobium loti, Rhizobium huakuii, Rhizobium ciceri, Rhizobium mediterraneum*, and *Rhizobium tianshanense* to *Mesorhizobium* gen. nov. *International Journal of Systematic Bacteriology* 47, 895–898.

Jordan, D.C. (1984) Rhizobiaceae. In: Kreig, N.R. (ed.) *Bergey's Manual of Systematic Bacteriology*, Vol. 1. Williams & Wilkins, Baltimore, Maryland, pp. 234–256.

Kantar, F., Elkoca, E., Ogutcu, H. and Algur, O.F. (2003) Chickpea yields in relation to *Rhizobium* inoculation from wild chickpea at high altitudes. *Journal of Agronomy and Crop Science* 189, 1–7.

Karim, H.A., Aslam, N., Hameed, S., Malik, K.A. and Hafeez, F.Y. (2003) Molecular characterization of typical and atypical nodule isolates from *Cicer arietinum*. In: *Proceedings of 7th International Conference on Trends in Biochemistry and Molecular Biology*. 2–5 April 2003. Lahore, Pakistan, p. 47.

Kumar, J. and Rao, B.V. (2001) Registration of ICCV 96029, a super early and double podded chickpea germplasm. *Crop Science* 41, 605–606.

Kyei-Boahen, S., Slinkard, A.E. and Walley, F. (2002a) Evaluation of rhizobial inoculation methods for chickpea. *Agronomy Journal* 94, 851–859.

Kyei-Boahen, S., Slinkard, A.E. and Walley, F. (2002b) Isotopic fractionation during N_2 fixation by chickpea. *Soil Biology and Biochemistry* 34, 417–420.

Laranjo, M., Branco, C., Soares, R., Alho, L. and Carvalho, M.D.E. (2002) Comparison of chickpea rhizobial isolates from diverse Portuguese natural populations based on symbiotic effectiveness and DNA fingerprint. *Journal of Applied Microbiology* 92, 1–8.

Laranjo, M., Machado, J., Young, J.P. and Oliveira, S. (2004) High diversity of chickpea *Mesorhizobium* species isolated in a Portuguese agriculture region. *FEMS Microbiology Ecology* 48, 101–107.

Levin, B.R. and Bergstrom, C.T. (2000) Bacteria are different: observation, interpretations, speculation, and opinions about mechanism of adaptive evolution in prokaryotes. *Proceedings of the National Academy of Sciences of the United States of America* 97, 6981–6985.

Lochner, H.H., Strijdom, B.W. and Steyn, P.L. (1991) Limitations of colony morphology and antibiotic resistance in the identification of a *Bradyrhizobium* sp. (*Lotus*) strain in soil. *Biology and Fertility of Soils* 11, 128–134.

Long, S.R. (2001) Genes and signals in the *Rhizobium*–legume symbiosis. *Plant Physiology* 125, 69–72.

Maatallah, J., Berraho, E.B., Sanjuan, J. and Lluch, C. (2002) Phenotypic characterization of rhizobia isolated from chickpea (*Cicer arietinum*) growing in Moroccan soils. *Agronomie* 22, 321–329.

Malhotra, R.S., Erskine, W. and Koropka, J. (1999) Genetic resources of chickpea. *Grain Legumes* 25, 16–17.

Martinez-Romero, E., Caballero-Mellado, J., Gandara, B., Rogel, M.A., Lopez Merino, A., Wang, E.T., Fuentes-Ramirez, E., Toledo, I., Martinez, L., Hernandez-Lucas, I. and Martinez-Romero, J. (2000) Ecological, phylogenetic and taxonomic remarks on diazothrophs and related genera. In: Pedrosa, F.O. (ed.) *Nitrogen Fixation: From Molecules to Crop Productivity*. Kluwer Academic, Dordrecht, The Netherlands, pp. 155–160.

Mefti, H. (2003) Réponse à l'inoculation de trois souches de rhizobium exotiques évaluées à travers la fixation symbiotique de l'azote et le rendement du pois chiche (*Cicer arietinum* L.). *Les Colloques* 100, pp. 81–87.

Naseem, S., Aslam, A., Malik, K.A. and Hafeez, F.Y. (2005) Understanding the genetic instability in *Cicer arietinum* root nodule bacteria. In: Wang, Y.P., Lin, M., Tian, Z.X., Elmerich, C. and Newton, W.E. (eds) *Biological Nitrogen Fixation, Sustainable Agriculture and Environment: Proceedings of 14th International Nitrogen Fixation Congress*. Springer, Dordrecht, The Netherlands, p. 319.

Nour, S.M., Fernandez, M.P., Normand, P. and Cleyet-Marel, J.C. (1994) *Rhizobium ciceri* sp. nov., consisting of strains that nodulate chickpeas (*Cicer arietinum* L.). *International Journal of Systematic Bacteriology* 44, 511–522.

Nour, S.M., Cleyet-Marel, J.C., Normand, P. and Fernandez, M.P. (1995) Genomic heterogeneity of strains nodulating chickpeas (*Cicer arietinum* L.) and description of *Rhizobium mediterraneum* sp. nov. *International Journal of Systematic Bacteriology* 45(4), 640–648.

Ounane, S.M., Irekti, H. and Bacha, F. (2003) Effet du déficit hydrique sur la fixation de l'azote et la biomasse chez le pois chiche (*Cicer arietinum*) inoculé avec différentes souches de *Mezorhizobium ciceri*. *Les Colloques* 100, 69–79.

Pal, S.C. (1996) Effect of heavy metals on legume–*Rhizobium* symbiosis. In: *Proceedings of the International Symposium on Biological Nitrogen Fixation Associated with Rice Production*. Bangladesh Agricultural Research Council, Dhaka, pp. 21–29.

Pareek, R.P. and Chandra, R. (2003) Chickpea: microbiology and nitrogen fixation. In: Masood, A., ShivaKumar, B.G. and Singh, N.B. (eds) *Chickpea Research in India*. Indian Institute of Pulses Research, Kanpur, India, pp. 167–193.

Peix, A., Rivas-Boyero, A.A., Mateos, P.F., Rodrigues-Barrueco, C., Martinez-Molina, E. and Veleaquez, E. (2001) Growth promotion of chickpea and barley by a phosphate solubilizing strain of *Mesorhizobium mediterranean* under growth chamber conditions. *Soil Biology and Biochemistry* 33, 103–110.

Pena-Cabriales, J.J. and Castellanos, J.Z. (1993) Effect of water stress on N_2 fixation and grain yield of *Phaseolus vulgaris* L. *Plant and Soil* 152, 151–155.

Peoples, M.B., Ladha, J.K. and Herridge, D.F. (1995) Enhancing legume N_2 fixation through plant and soil management. *Plant and Soil* 174, 83–101.

Perret, X., Staehelin, C. and Broughton W.J. (2000) Molecular basis of symbiotic promiscuity. *Microbiology and Molecular Biology Reviews* 64(1), 180–201.

Provorov, N.A. and Vorobyov, N.I. (2000) Population genetics of rhizobia: construction and analysis of an 'infection and release' model. *Journal of Theoretical Biology* 205, 105–199.

Rajasekhar, E., Thippannawar, C.M. and Sreenivasulu, N. (1991) Antirhizobial activity of seedborne fungal culture filtrate in chickpea. *Current Research University of Agricultural Sciences, Bangalore* 20(10), 222–223.

Rengel, Z. (2002) Breeding for better symbiosis. *Plant and Soil* 245, 147–162.

Rivas, R., Gutierreze, C., Abril, A., Mateos, P.F., Martinez-Molina, E., Ventosa, A. and Velazquez, E. (2005) *Paenibacillus rhizosphaerae* sp. Nov., isolated from the rhizosphere of *Cicer arietinum*. *International Journal of Systematic Evolutionary Microbiology* 55, 1305–1309.

Roberts, F.M. and Schmidt, E.L. (1983) Response of three indigenous serogroups of *Rhizobium japonicum* to the rhizosphere of pre-emergent seedlings of soybeans. *Soil Biology and Biochemistry* 17, 579–580.

Roughley, R.J., Gemell, L.G., Thompson, J.A. and Brockwell, J. (1993) The number of *Bradyrhizobium* sp. (*Lupinus*) applied to seed and its effect on rhizosphere colonization, nodulation and yield of lupin. *Soil Biology and Biochemistry* 25, 1453–1458.

Rupela, O.P. (1992) Natural occurrence and salient characters of non-nodulating chickpea plants. *Crop Science* 32, 349–352.

Rupela, O.P. (1994) Screening for intracultivaral variability of nodulation in chickpea and pigeonpea. In: *Linking Biological Nitrogen Fixation Research in Asia: Report of Meeting of the Asia Working Group on Biological Nitrogen Fixation in Legumes.* 6–8 December 1993. ICRISAT Asia Centre, Hyderabad, India, pp. 75–83.

Sattar, M.A., Quadar, M.A. and Danso, S.K.A. (1993) Nodulation nitrogen fixation and yield of chickpea as influenced by host cultivar and *Bradyrhizobium* strain differences. *Soil Biology and Biochemistry* 27(4–5), 725–727.

Sharma, R.D., Pareek, R.P. and Chandra, R. (1995) Residual effect of phosphate and Rhizobium inoculation in chickpea on succeeding maize and fodder sorghum. *Journal of the Indian Society of Soil Science* 43(4), 600–603.

Soussi, M., Ocaña, A. and Lluch, C. (1998) Effects of salt stress on growth, photosynthesis and nitrogen fixation in chickpea (*Cicer arietinum* L.). *Journal of Experimental Botany* 49, 1329–1337.

Strijdom, B.W. and Deschodt, C.C. (1976) Carriers of Rhizobia and the effects of prior treatment on the survival of Rhizobia. In: Nutman, P.S. (ed.) *Symbiotic Nitrogen Fixation in Plants*. Cambridge University Press, Cambridge, pp. 151–168.

Sullivan, J.T. and Ronson, C.W. (1998) Evolution of rhizobia by acquisition of a 500-kb symbiosis island that integrates into a *phe-tRNA* gene. *Proceedings of the National Academy of Sciences of the United States of America* 95, 5145–5149.

Sullivan, J.T., Patrick, H.N., Lowther, W.L., Scott, D.B. and Ronson, C.W. (1995) Nodulating strains of *Rhizobium loti* arise through chromosomal symbiotic gene transfer in the environment. *Proceedings of the National Academy of Sciences of the United States of America* 92, 8985–8989.

Sylvester-Bradley, R., Thornton, P. and Jones, P. (1988) Colony dimorphism in *Bradyrhizobium* strains. *Applied and Environmental Microbiology* 54, 1033–1038.

Tejera, N., Soussi, M. and Lluch, C. (2006) Physiological and nutritional indicators of tolerance to salinity in chickpea plants growing under symbiotic conditions. *Environmental and Experimental Botany* 58(1–3), 17–24.

Thavarajah, D., Rosalind, A., Ball, R.A. and Schoenau, J.J. (2005) Nitrogen fixation, amino acid, and ureide associations in chickpea. *Crop Science* 45, 2497–2502.

Thies, J.E., Holmes, E.M. and Vachot, A. (2001) Application of molecular techniques to studies in *Rhizobium* ecology: a review. *Australian Journal of Experimental Agriculture* 41, 299–319.

Truchet, G., Debelle, F., Vasse, J., Terzaghi, B., Garnerone, A.M., Rosenberg, C., Batut, J., Maillet, F. and Denarie, J. (1985) Identification of a *Rhizobium meliloti* pSym2011 region controlling the host specificity of root hair curling and nodulation. *Journal of Bacteriology* 164, 1200–1210.

Vincent, J.M. (1974) Root nodules symbiosis with *Rhizobium*. In: Quispel, A. (ed.) *The Biology of Nitrogen Fixation*. North-Holland, Amsterdam, pp. 381–429.

Walsh, K.B. (1995) Physiology of legume nodule and its response to stress. *Soil Biology and Biochemistry* 27, 637–655.

Zhang, X.X., Kosier, B. and Priefer, U.B. (2001) Symbiotic plasmid rearrangement in *Rhizobium leguminosarum* bv. *viciae* VF39SM. *Journal of Bacteriology* 183, 2141–2144.

Zurayk, R., Adlan, M., Baalbaki, R. and Saxena, M.C. (1998) Interactive effects of salinity and biological nitrogen fixation on chickpea (*Cicer arietinum* L.) growth. *Journal of Agronomy and Crop Science* 180(4), 249–258.

9 Chickpea in Cropping Systems

A.F. Berrada,[1] B.G. Shivakumar[2] and N.T. Yaduraju[3]

[1]*Colorado State University, Arkansas Valley Research Center, Rocky Ford, Colorado, USA;* [2]*Division of Agronomy, Indian Agricultural Research Institute, Pusa, New Delhi 110012, India;* [3]*National Agricultural Technology Programme, KAB II, Pusa, New Delhi 110012, India*

Introduction

Chickpea (*Cicer arietinum* L.) is one of the most important pulse crops cultivated in more than 11 million hectares spread across the globe with 22 countries cultivating it in an area of more than 10,000 ha (FAOSTAT, 2005). It is a major pulse crop in South Asia, the Middle East, East Africa, western Mediterranean, Australia and Mexico. Chickpea acreage has increased steadily in Canada and parts of the USA in recent years. Chickpea and other pulse crops such as lentil (*Lens culinaris* Medik.), dry pea (*Pisum sativum* L.) and dry bean (*Phaseolus vulgaris* L.) are a major source of protein in human diets, particularly in low-income countries. As legumes, pulse crops play an important role in sustainable cropping systems due to their ability to fix atmospheric nitrogen (N_2). Increasing demand for 'natural' foods has led to renewed interest in sustainable farming, which relies more on biological systems and less on external inputs, such as pesticides and synthetic fertilizers, to optimize crop production. When grown in rotation with cereals or other crops, legumes help break the cycle of diseases, insects and weeds. In this chapter, we will examine the role of chickpea in cropping systems around the world. The bulk of the information included here is based on published results, most of which came out of research programmes in India, Australia, North America and countries served by the International Center for Agricultural Research in Dry Areas (ICARDA), Aleppo, Syria, and the International Crops Research Institute for the Semi-Arid Tropics (ICRISAT), Hyderabad, India.

Cropping Systems Involving Chickpea

Chickpea is cultivated during different times of the year in different countries depending upon the climatic conditions. It is sown in spring (from late March to mid-April in Turkey and the USA, and from February to April in the Mediterranean) when the ground has warmed or when the rains recede (from mid-September to November, rarely later in India and Pakistan; from September to January or April in Ethiopia) depending on the region (Smithson *et al.*, 1985). Efforts are also being made to grow chickpea during winter in many Mediterranean and North African countries, as well as in Australia (Tutwiler, 1995; Siddique *et al.*, 1999; Yau Sui-Kwong, 2005). Chickpea is cultivated in all sorts of cropping systems, e.g. as a sole crop, mixed or intercropped with barley (*Hordeum vulgare* L.), grasspea (*Lathyrus sativus* L.), linseed (*Linum usitatissimum* L.), mustard (*Brassica juncea* L.), peas, maize (*Zea mays* L.), coffee (*Coffea Arabica* or *C. robusta*), safflower (*Carthamus tinctorius* L.), potato (*Solanum tuberosum* L.), sweet potato (*Ipomoea batatas* L.), sorghum (*Sorghum bicolor* L.) or wheat (*Triticum aestivum* L.). In rotation, it often follows wheat, barley, rice (*Oriza sativa* L.) or tef (*Eragrostis tef*) (van der Maesen, 1972). In India, chickpea is also grown as a catch crop in sugarcane (*Saccharum officinarum* L.) fields and often as a second crop after rice.

Continuous cropping of chickpea is rare due to diseases such as ascochyta blight caused by *Ascochyta rabiei* and soil erosion hazards. Due to the aggressive nature of ascochyta blight, careful consideration must be given to crop rotation and field selection. It is recommended that chickpea should not be planted in the same field more than once in 4 years to allow for the breakdown of chickpea residue on which the disease survives. Planting chickpea on chickpea stubble may result in total crop failure and can also increase the disease risk to neighbouring fields (http://www.agr.gov.sk.ca/docs/crops/integrated_pest_management/disease/ascochytaonChickpeas.asp, verified 13 June 2006).

Sequential cropping

The growing demand for food grains, declining arable land and expanding irrigation facilities have resulted in the cultivation of crops in sequence in many developing countries. Chickpea has been successfully cultivated after the preceding crops both in irrigated conditions and with conserved moisture under rainfed conditions. There are many sequential cropping systems with cereals, oilseeds, etc. as base crops and chickpea as succeeding crop (Ahlawat *et al.*, 2003). Successful sequential cropping systems have also been reported from many countries. Wheat–chickpea in Ethiopia and Spain, and tef–chickpea and fodder oat–chickpea in Ethiopia are emerging sequential cropping systems, whereas growing chickpea in rice fallow in Nepal, Bangladesh and eastern India is also gaining popularity (López-Bellido *et al.*, 1998, 2001; Zewdu, 2002; Harris *et al.*, 2005). The introduction of chickpea in sequential cropping systems has led to increased cropping intensity in Bangladesh (Hassan *et al.*, 2003). Malik *et al.* (2002) reported higher monetary benefit under both conventional and zero tillage with rice–chickpea sequence among several popular cropping sequences tried in Pakistan. However, Zahid *et al.* (2004) reported that, in Pakistan, chickpea cannot be grown after rice under irrigation due to late maturity of the crop, which affects rice plantation in the next season. A more suitable crop rotation is rice–wheat. The major constraints of chickpea cultivation in rice–chickpea were: (i) high weed infestation; (ii) high insect (pod borer – *Helicoverpa armigera*) attack; (iii) wet conditions and poor drainage of the soil due to its clayey nature; (iv) excessive vegetative growth followed by less pod bearing; and (v) more income from rice and wheat, which are higher-yielding and safer crops.

Berrada (2004) reported a unique cropping system in south-western Colorado, whereby kabuli (white seed coat) chickpea is grown in 3–4 consecutive years after dry bean and after 7 years of lucerne (*Medicago sativa* L.). Lucerne serves as a soil booster in this dryland environment. Most chickpea in the USA and Canada, however, is grown after spring wheat, winter wheat or spring barley. The crop is predominantly of the kabuli type (420–550 mg/seed), as opposed to desi (coloured seed coat) type (60–300 mg/seed), which is grown mostly in South Asia and East Africa. In a review of pulse-crop adaptation in the Northern Great Plains of the USA and Canada, Miller *et al.* (2002) concluded that chickpea production practices were the least understood. Release of ascochyta blight–resistant cultivars in recent years has led to increased chickpea production in North America.

Advances in no-till (zero-tillage) technology have allowed farmers in the Great Plains and elsewhere to intensify cropping systems, by replacing fallow in cereal-based systems with alternative crops such as maize, sorghum, millet (pearl millet – *Pennisetum glaucum* L., proso millet – *Panicum miliaceum* L. and others) and pulse crops (Farahani *et al.*, 1998; Lyon *et al.*, 2004). The extent of cropping intensification will vary with soil and climatic conditions, resource availability, government programmes, cost-to-benefit ratio and other factors. Zentner *et al.* (2002) reported that rotating grain cereals with oilseed and pulse crops contributed to a higher and more stable net farm income in most soil-climatic regions of the Canadian prairies. However,

> in the very dry Brown soil zone and drier regions of the Dark Brown soil zone where the production risk with stubble cropping is high, the elimination of summer fallow from the cropping system may not be economically feasible under present and near-future economic conditions. The use of conservation tillage practices in the management of mixed cropping systems is highly profitable in the more moist Black and Gray soil zones (compared with conventional mechanical tillage methods) because of significant yield advantages and substantial resource savings that can be obtained by substituting herbicides for the large amount of tillage that is normally used.
>
> (Zentner *et al.*, 2002)

In India, there has been a large increase in mixed and sequential cropping involving chickpea, partly due to the availability of short-duration cultivars. Ahlawat *et al.* (2003) listed the following major crop sequences in India:

1. *Arid ecosystem*
   (a) Rainfed upland: pearl millet–chickpea
   (b) Irrigated upland: sorghum–chickpea
2. *Semiarid ecosystem*
   (a) Irrigated upland: maize–chickpea
   (b) Irrigated lowland: rice–chickpea

3. *Subhumid ecosystem*
   (a) Rainfed upland: cotton–chickpea
   (b) Irrigated upland: maize–chickpea
4. *Humid ecosystem*
   (a) Lowland: maize–chickpea

Survey results show that groundnut–chickpea rotation was adopted by 36.7% of the farmers interviewed in Gujarat, India, followed by fallow–chickpea (20.8%), paddy–chickpea (15%), sorghum–chickpea (12.9%) and maize–chickpea (10%) (Shiyani *et al.*, 2001). In the northern hilly zone of Madhya Pradesh where water resources are limited, irrigation is usually reserved for rice or wheat (Bajpai *et al.*, 1991). Soils with high water content are generally sown with wheat following rice. Pulse crops are grown on comparatively marginal lands. Under limited water supply, chickpea is more profitable than wheat, pea or lentil following rice (Bajpai *et al.*, 1991).

Wheat is the most common crop after cotton (*Gossypium hirsutum* L.) in Ethiopia, although, legumes such as faba bean (*Vicia faba* L.) and chickpea have become popular of late (Hulugalle *et al.*, 2001). Lint yield and fibre quality were decreased by rotating cotton with pulse crops.

> Management constraints such as lack of effective herbicides, insect damage, harvesting damage, and availability of suitable marketing options were greater with legumes than with wheat.
>
> (Hulugalle *et al.*, 2001)

In the irrigated cotton–chickpea cropping system in India, chickpea planting is often delayed, leading to low productivity (Ahlawat *et al.*, 2003). In the Campiña region of southern Spain, wheat is typically rotated with sunflower (*Helianthus annuus* L.), and to a lesser extent with faba bean or chickpea (López-Bellido *et al.*, 2000). In Kenya, there is evidence that management of short-rains fallow with legumes could reduce weed infestation and enhance N availability to the following crop (Cheruiyot *et al.*, 2003).

Intercropping

In subsistence farming areas of Asia and Africa with low precipitation, intercropping provides a hedge against crop failure. Chickpea is a traditional *rabi* (winter) season crop in India; hence, intercropping systems involve mostly *rabi* crops such as wheat, *rabi* maize, mustard, safflower, linseed, barley, coriander (*Coriandrum sativum* L.) and *rabi* sorghum. Chickpea is also grown as a catch crop in sugarcane fields (Ahlawat *et al.*, 2003).

The selection of crops for intercropping with chickpea often depends on the current needs of the farmers and not on a preset system. Because planting time and resource availability in developing countries fluctuate greatly, the component crops are selected based on the ease of finding seeds, farmers' own needs and anticipated success of the intercropping system. Furthermore, in new areas, the intercropping systems depend on the local traditional crops during that season.

Sharma *et al.* (1993) reported high returns when chickpea, pea or French bean were intercropped with spring-planted sugarcane in Madhya Pradesh, India. In Karnataka, the yield of chickpea intercropped with safflower was similar to that of sole-crop chickpea (Hiremath *et al.*, 1991). Chickpea/safflower row ratios of 2:1, 3:1 and 3:2 gave the highest total yields. In Madhya Pradesh, Thakur *et al.* (2000) demonstrated that chickpea + safflower intercropping in 3:1 and 6:2 row ratios was superior to pure stands of either crop components and to chickpea + mustard and chickpea + linseed (Table 9.1). Chickpea + linseed intercropping is one of the important systems in central India.

Chickpea yield was adversely affected by intercropping with Indian mustard, barley and linseed in New Delhi (Ahlawat *et al.*, 2005). Chickpea yield increased as the proportion of chickpea in the mixture increased from 2:1 to 4:1. Sole Indian mustard productivity, as measured in chickpea-equivalent yield (CEY), was highest, followed by chickpea + Indian mustard (2:1). Chickpea + linseed and sole chickpea recorded similar CEY. The recommended intercropping systems based on relative crowding coefficient were chickpea + barley in all row proportions and chickpea + linseed in 3:1 and 4:1 row proportions. Chickpea + mustard in 4:1 row proportion gave the highest net return over a 2-year period in Faizabad, India (Singh and Yadav, 1992). Chickpea as sole crop or mixed (1:1) with barley, mustard or wheat was the least profitable. Mandal *et al.* (1986) found that wheat + chickpea intercropping was most efficient with one irrigation application in terms of land-equivalent ratio, relative crowding coefficient and economic return.

Apart from the Indian subcontinent, intercropping systems with chickpea have also been reported from many other countries. Chickpea and sunflower intercropping was successful in Italy. The success of intercropping systems at low levels of interspecific competition was explained in terms of a more

Table 9.1. Productivity and economic benefit of chickpea-based intercropping systems at Chhindwara, India, averaged over two seasons. (Adapted from Thakur *et al.*, 2000.)

Treatment	Seed yield (kg/ha)		Chickpea-equivalent yield (kg/ha)	Land-equivalent ratio	Benefit/cost ratio
	Main crop	Intercrop			
Chickpea	1656	–	1656	1.00	2.79
Mustard	–	750	919	1.00	1.85
Safflower	–	2206	1983	1.00	4.26
Linseed	–	1235	1846	1.00	3.63
Chickpea + mustard (3:1)	1383	252	1691	1.17	2.99
Chickpea + mustard (6:2)	1226	294	1588	1.13	2.80
Chickpea + safflower (3:1)	1380	834	2132	1.21	3.83
Chickpea + safflower (6:2)	1261	948	2118	1.19	3.78
Chickpea + linseed (3:1)	1232	242	1600	0.94	2.84
Chickpea + linseed (6:2)	1168	322	1658	0.97	2.93
CD (P = 0.05)	251	68	70	0.04	0.16

balanced and efficient use of soil moisture due to different water requirements of the two species at different times (Campiglia and Caporali, 1992). In Egypt, intercropping of chickpea with sugarcane at 1:1 row ratio recorded the highest total income despite reduced yield in chickpea as compared to sole crop (El Gergawi and Abou Salama, 1994). Hussein and El Deeb (1999) also demonstrated successful intercropping of chickpea in sugarbeet (*Beta vulgaris* L.).

The selection of compatible genotypes and appropriate planting patterns helps minimize competition for resources and boosts the productivity of the system. In a 1998–2000 experiment in Kanpur, India, 'BG 256' chickpea intercropped with 'Neelam' linseed in a 6:2 ratio gave the highest return and income-equivalent ratio (Mishra *et al.*, 2001). Ahlawat *et al.* (2003) stated that 'unlike spreading types of chickpea, which are vulnerable to incidence of pests and diseases, the newly identified tall, upright and high-yielding cultivars do not interfere with the growth of inter-crops'. Ali and Mishra (2002) showed a significant effect of chickpea cultivar on the productivity of chickpea + Indian mustard intercropping. Furthermore, north–south planting resulted in a significant increase in the number of pods or plants of chickpea over east–west planting, but not in seed yield of either chickpea or mustard. The success of intercropping systems involving chickpea relies not only on the appropriate selection of cultivars and component crops but also on effective management practices such as balanced nutrition, adequate irrigation scheduling and intercultural operations fulfilling the requirement of component crops in the system.

Nitrogen Contribution of Chickpea

A major impetus for including legumes such as chickpea in cropping systems stems from their ability to fix N_2, which ultimately reduces the need for commercial fertilizers to meet the crops' N requirements. Estimates of the contribution of legumes to soil N balance vary greatly due to numerous factors including crop yield, soil and climatic conditions, and also due to variations in the methods used to budget N. Khan *et al.* (2003) reported an N balance (N fixed – grain N) of 94 kg/ha for chickpea when belowground N was included in the budget, compared to 1 kg/ha when only aboveground N was included – an increase of 93 kg/ha.

Beck (1992) reported average N_2 fixation by eight kabuli chickpea cultivars of 68 kg N/ha in the first year and 27 kg N/ha in the second (drier) year, at the ICARDA station of Tel Hadya, Syria. Inoculation with the best rhizobial strain increased the portion of N derived from fixation from 52% to 72%. In another study in Syria, Pilbeam *et al.* (1998) reported chickpea N_2 fixation of 16–48 kg N/ha, depending on the season, but concluded that 'a negative soil N budget was likely because the amount of N removed in the grain was usually greater than the amount of atmospheric N_2 fixed'. Campiglia *et al.* (1998) observed that seed inoculation with a suitable rhizobial strain gave significant increases in nodule numbers only during the wetter season due to a significant genotype by growing season interaction. The application of N fertilizer had a small effect on seed yield but decreased nodule numbers about 21% during the drier season.

In Queensland, Australia, autumn-sown chickpea usually fixed more N than winter-sown chickpea (Horn *et al.*, 1996). Initial soil NO_3–N concentration and other factors like crop growth also affect the amount of N_2 fixed by chickpea.

In a review of Australian data, Evans *et al.* (2001) reported a wide variation in the net effects of grain legumes on N balance. The difference between N fixed and N harvested in the grain (N balance) ranged from –67 to 102 kg/ha (mean of 6 kg/ha) with chickpea, –46 to 181 kg/ha (mean of 40 kg/ha) with pea and –29 to 247 kg/ha (mean of 80 kg/ha) with lupin (*Lupinus angustifolius* L.). N balance was related to the amount (N fixation) and proportion (% of N fixation) of crop N derived from N_2 fixation, but not to legume grain yield (GY). When N fixation exceeded 49 (chickpea), 39 (pea) and 30 (lupin) kg N/ha, N balance was often positive, averaging 0.60 kg/kg of N fixed. Increases in shoot dry matter (SDM) and % of N fixation usually increased the legume's effect on N balance. The additive effects of SDM, % of N fixation and GY explained 87% of the variation in N balance.

> Using crop-specific models based on these parameters the average effects of grain legumes on soil N balance across Australia were estimated to be 88 (lupin), 44 (pea), and 18 (chickpea) kg N/ha.
>
> (Evans *et al.*, 2001)

The benefit to wheat from the prior legume (chickpea, faba bean, lupin) was at least equivalent to that obtained from applying 50 kg N/ha as ammonium nitrate (Marcellos, 1984).

Marcellos *et al.* (1998) reported average N_2 fixation by chickpea of 73 kg N/ha (range 4–116 kg/ha) in wheat-based cropping systems of northern New South Wales (NSW) in Australia. The wide range in N_2 fixation was associated with variable % of N fixation (mean 57%; range 4–79%). Chickpea spared nitrate, relative to wheat (mean 18 kg N/ha; range 1–35 kg N/ha), and mineralized additional nitrate during the summer fallow (mean 18 kg N/ha; range 5–40 kg N/ha). Nitrate–N in the top 1.2 m of soil at wheat-planting averaged 89 kg/ha after chickpea (range 63–113 kg/ha) and 56 kg/ha after wheat (range 33–94 kg/ha). Marcellos *et al.* (1998) concluded that 'chickpea did not fix sufficient N_2 or mineralize sufficient N from residues either to maintain soil fertility or to sustain a productive chickpea–wheat rotation without inputs of additional fertilizer N.' Another study in northern NSW found smaller benefits of chickpea in terms of yield, protein content and fertilizer N requirement on the following wheat crop, compared to that of lucerne and clover (*Trifolium subterraneum* L.) (Holford and Crocker, 1997). The smaller positive effects of chickpea were probably due to 'soil N sparing or rapid mineralization from crop residues rather than any net contribution of N fixation to soil N accretion'. Reported values of legume contribution to N balance may be underestimated, according to Khan *et al.* (2003).

Soil nitrate–N increased by 35 kg/ha after 6 months of fallow following chickpea (85 kg N/ha) compared with continuous wheat (50 kg N/ha), on a vertisol in Queensland, Australia (Dalal *et al.*, 1998). This resulted in additional N in the wheat grain (20 kg/ha) and aboveground plant biomass (25 kg/ha). In Thailand, the available soil N pool was increased by about 32 kg/ha following chickpea than following wheat (Patwary *et al.*, 1989). In Manitoba, Canada, including field pea in the rotation reduced N fertilizer need for the following wheat crop by ~30 kg N/ha (http://www.gov.mb.ca/agriculture/research/ardi/projects/98-019.html, verified 20 January 2006). Newer pulse crops such as dry bean and chickpea contributed less than 15 kg N/ha. Likewise, Miller and Holmes (2005) reported a smaller N margin for chickpea than for pea in Montana, USA. Nitrogen margin for chickpea and pea averaged, respectively, 47 and 84 kg N/ha greater than for wheat, which was a reflection of greater pea than chickpea grain yields.

The N dynamics in intercropping systems are often more complex than in sequential cropping systems. As crops having dissimilar nutrient requirements are grown together, the biological N fixation of chickpea alone may not be sufficient to meet the N requirements of the component crops. This requires judicious N management of the cropping system so as not to adversely affect (e.g. if high rates of N fertilizers are applied) N fixation by chickpea. Often, under the misconception that chickpea will meet the N requirement of the system, no N fertilizer is applied, reducing the productivity of the system. Research is lacking in this area.

Rotational Effects of Chickpea on Grain Yield, Grain Protein Concentration and Water Availability

Nitrogen fixation by legumes often results in higher grain yield and/or grain protein concentration of the subsequent cereal crop compared to continuous cropping. Hossain *et al.* (1996) reported an increase in wheat yield of 810 kg/ha in 1989 and 1360 kg/ha in 1990 following chickpea than following wheat. N uptake by wheat increased by 27 kg/ha in 1989 and 32 kg/ha in 1990 following chickpea, which was less than the increase obtained after lucerne and medic leys or grass and legume leys. Averaged over the two seasons, soil mineral N to a depth of 1.2 m increased by 93, 91, 68 and 37 kg/ha following grass or legume, lucerne and medic leys, and chickpea, respectively, than following wheat. Wheat grain protein concentration was substantially higher following all pasture leys (11.7–15.8%) than following wheat (8.0–9.4%) or chickpea (9.4–10.1%). Dalal *et al.* (1998) reported a wheat yield increase of 17–61% (mean of 40% or 825 kg/ha) following chickpea than following wheat. Wheat grain protein concentration increased from 9.4% in wheat–wheat to 10.7% in chickpea–wheat (Table 9.2).

Water-use efficiency (WUE) by wheat increased on average from 9.2 kg grain/ha/mm in wheat–wheat to 11.7 kg grain/ha/mm in chickpea–wheat (Dalal *et al.*, 1998). Available soil moisture at wheat-seeding was similar in both rotations in all years except 1996. However, wheat in chickpea–wheat used about 20 mm more water than in wheat–wheat. The beneficial effects of chickpea on wheat yield and grain protein were attributed to additional nitrate-N following chickpea, which also enhanced WUE. Dalal *et al.* (1998) suggested that the beneficial effects of chickpea in rotation with wheat in Queensland, Australia, could be enhanced by early sowing (May to mid-June) of chickpea, zero tillage

Table 9.2. Grain yield and protein concentration of wheat in chickpea–wheat and wheat–wheat at Warra, Queensland, from 1988 to 1996; no crop was planted in 1991 due to drought. (Adapted from Dalal *et al.*, 1998.)

	Grain yield (kg/ha)			Grain protein (%)		
Year	Chickpea–wheat	Wheat–wheat	LSD (P = 0.05)	Chickpea–wheat	Wheat–wheat	LSD (P = 0.05)
1988	4620	3076	311	9.4	8.3	NS
1989	2875	2073	467	10.1	8.0	1.2
1990	3586	2234	243	9.4	8.3	1.0
1992	4230	3476	543	12.4	10.8	0.7
1993	2198	1883	NS	11.8	9.6	1.1
1994	1604	1023	152	10.1	8.7	1.1
1995	1761	1198	220	12.1	11.8	NS
1996	3023	2265	756	10.1	10.2	NS
Mean	2942	2117	125	10.7	9.4	0.5

LSD: Least significant difference at P = 0.05; NS: Non-significant difference at P = 0.05.

and supplementary N application to the following wheat crop. Marcellos *et al.* (1998) found no consistent differences between chickpea and wheat in their effects on soil water content, either before harvest or after summer fallow. Armstrong *et al.* (2003) reported lower plant-available water content of the soil at wheat sowing after chickpea than after wheat. Angadi *et al.* (1999; reported by Miller *et al.*, 2002) found that chickpea extracted soil water from a depth of 1.2 m compared to 0.9 m for lentil and dry pea. All three pulse crops used less water than spring wheat but the difference between chickpea and wheat was small. WUE of chickpea (6.2 kg/ha/mm) was lower than that of dry pea (8.5 kg/ha/mm) and spring wheat (7.5 kg/ha/mm) in the Northern Great Plains of the USA and Canada (Miller *et al.*, 2002). Higher wheat yields were generally obtained after pea than after chickpea due to lower water use.

Wheat following chickpea produced 25% higher grain yield and four times higher gross margin than wheat following wheat over a 2-year period, when no additional N was applied to wheat (http://www.apsru.gov.au/apsru/Projects/WFS/pdffiles/Broadcaster/balbr0699.pdf, verified 20 January 2006). When N fertilizer was applied to wheat, the advantage of chickpea–wheat over continuous wheat shrunk to 9% and 7% higher wheat yield and gross margin, respectively. Barley (malting grade) following chickpea had 40% higher grain yield and six times higher gross margin than barley (feed grade) following wheat or barley. No N was added to any of the crops. In Saskatchewan, Canada,

> adjusting fertilizer N rates to account for estimated total N contribution from the previous pulse crop (chickpea, lentil, and pea) effectively neutralized the benefits on wheat yield and protein compared with the effects following mustard.
>
> (Miller *et al.*, 2003a)

In NSW, Australia, wheat yield, grain N and ear number increased by as much as 97%, 39% and 58%, respectively, when wheat followed legumes (chickpea, faba bean or lupin) compared to when it followed wheat. There were no significant differences among the legumes (Marcellos, 1984).

The rotational benefits of pulse crops may vary with soil type, precipitation amount and other factors. Miller *et al.* (2003b) reported that the benefits of chickpea, lentil or pea to the following wheat crop were less on a clay soil than on a silt loam in Saskatchewan, Canada. On the loam soil, pea yield was similar to that of spring wheat and 39% and 34% greater than that of lentil and desi chickpea, respectively. On the clay soil, pea and chickpea yields were equal but 26% less than wheat yield and 29% greater than lentil yield. The apparent N margin for pea averaged 40 and 32 kg/ha greater than for lentil and chickpea, respectively, indicating superior N_2 fixation. All three pulse-crop stubbles had greater soil NO_3–N than wheat in spring. They averaged 28 and 12 kg N/ha greater at the clay and loam soil sites, respectively. Postharvest soil water content to 1.2 m depth was 25–49 mm greater under the clay soil, and 12–31 mm greater under the loam soil for all pulse crops compared with wheat. The difference occurred primarily below 61 cm but disappeared by spring (Miller *et al.*, 2003b). In another study, durum wheat (*T. turgidum* L.) grain yield increased by 7% and grain protein concentration by 11% when grown after pulse crops (chickpea, lentil and dry pea) than after spring wheat (Gan *et al.*, 2003). Pulse and oilseed crops (mustard – *B. juncea* L. or canola – *B. napus* L.) grown for the previous 2 years increased durum grain yield by 15% and protein concentration by 18% compared with continuous wheat. It was reported that fall residual soil NO_3–N and available soil water accounted for 3–28% of the increased durum yield in 2 of 5 site-years, whereas those two factors accounted for 12–24% of the increased grain crude protein concentration (GCPC) in 3 of 5 site-years. Durum grain yield was negatively related to GCPC. The relationship was stronger when durum was preceded by oilseeds compared with pulses. Gan *et al.* (2003) concluded that 'broadleaf crops in no-till cropping systems provide significant rotational benefits to durum wheat in the semiarid northern Great Plains'.

Under average rainfall, barley, durum wheat, spring wheat and winter wheat yields following flax (linseed), pea and chickpea ranged from 84% to 101% of the fallow control and were greater than the following wheat in seven of nine cases (Miller and Holmes, 2005). Cereal yields following chickpea were 8% and 11% greater than following wheat in 2 of 3 years. Grain protein concentrations of the cereal crops were lowest following wheat and chickpea but were above the critical value of 14.7% for N sufficiency for spring wheat, except at one of five locations. The reduced cereal grain protein concentration following chickpea may be due to the low chickpea yields, which may have limited N fixation and subsequent N cycling.

> Under severe drought, cereal test crop yields following broadleaf crops ranged from 21 to 41% of the fallow control and were equal or less than those following wheat.
>
> (Miller and Holmes, 2005)

In the Campiña region of southern Spain, wheat in rotation with faba bean resulted in the highest grain yields during a 4-year period (López-Bellido *et al.*, 2000). Wheat–sunflower, wheat–chickpea and wheat–fallow produced similar grain yields, whereas continuous wheat resulted in the lowest yields. Similar results were also reported by Giambalvo *et al.* (2004) from Italy, where continuous wheat recorded the lowest grain yield but no differences were found in wheat grown after three legume crops including chickpea. In the Campiña region study, differences among rotations were minimal during wet years (3 of the 4 study years) since water was not limiting. When a prior drier period was included in the analysis, wheat yield was impacted as follows: wheat–faba bean > wheat–fallow > wheat–chickpea = wheat–sunflower > continuous wheat. Faba bean had a positive effect on wheat yield throughout the study period (1987–1998), whereas chickpea had little effect (López-Bellido *et al.*, 2000). Wheat yield increased significantly with the application of up to 100 kg N/ha in the wet years but did not respond to N in the dry years. Nitrogen fertilizer had a more marked effect on wheat yield for the rotations with no legumes.

In West Asia and North Africa, both of which have a Mediterranean climate, the average WUE and potential transpiration efficiency are usually lower for chickpea than for cereal crops. Despite this, the rotation benefits and higher economic returns provide the potential for chickpea to replace fallow or to break continuous cereal cropping in the region's farming system (Zhang *et al.*, 2000). In peninsular India, inclusion of chickpea in cropping systems increased system productivity and WUE. The net income, cost-to-benefit ratio and production efficiency of crop sequences were in the order: maize–chickpea > maize–safflower > maize–wheat (Khot and Yargattikar, 1998).

Insect and Disease Management Considerations

Chickpea is susceptible to many insect pests and diseases, the intensity and frequency of which varies with cropping systems (Table 9.3). Pest occurrence is usually less when chickpea is grown in rotation or in association (mixed cropping) with other crops compared to sole cropping, which may be due to changes in micro climate, disruption of pest's cycle, etc. The reduction in pest infestation or damage will vary with the cropping system and type of insect or disease.

Sequential cropping systems

Breaking the regular sequence of crops brings about disruption in the pest's cycle by changed host plants (Ahlawat *et al.*, 2003). The advantages of cropping sequences are felt both in component crops and in chickpea. Fusarium wilt damage to safflower was less severe where chickpea was grown as component crop, compared to continuous cultivation of sorghum and safflower in rotation (Ahlawat *et al.*, 2003). None the less, appropriate crop rotations can help prevent disease proliferation, particularly by soil or residue-borne pathogens (Krupinsky *et al.*, 2002). Rotation to non-host crops for a sufficient period of time will allow for the decomposition of infested crop residues and the reduction in pathogen viability. For example, rotating chickpea with non-leguminous crops can easily control chickpea cyst nematode caused by *Heterodera ciceri*

Table 9.3. Examples of problem pests in chickpea cropping systems.

Cropping system	Pests/diseases affected	Reference
Sequential cropping		
Rice–chickpea	Nematodes in soil: *Tylenchorhynchus mashhoodi, Helicotylenchus indicus, Hoplolaimus indicus (Basirolaimus indicus), Meloidogyne incognita, Rotylenchulus reniformis, Pratylenchus zeae*	Haidar *et al.* (2001)
Wheat–chickpea	Crown rot in wheat caused by *Fusarium pseudograminearum*	Moore *et al.* (2003)
Chickpea–okra–chilli	Nematodes in soil: *M. incognita, H. indicus, H. indicus, Tylenchus filiformis, T. brassicae and R. reniformis*	Wani (2005)
Intercropping		
Chickpea + mustard	Chickpea blight: *Ascochyta rabiei*	Gaur and Singh (1994)
Chickpea + wheat/barley/ mustard	*Helicoverpa armigera* in chickpea	Prasad and Chand (1989)
Chickpea + coriander	*H. armigera* in chickpea	Sekhar *et al.* (1995)
Chickpea + wheat/mustard/ safflower	*H. armigera* in chickpea	Das (1998)
Chickpea + wheat/mustard	*H. armigera* in chickpea	Altaf Hossain *et al.* (1999)

(Ahlawat *et al.*, 2003). Likewise, the incidence of root-knot nematode caused by *Meloidogyne artiellia* can be reduced by alternating chickpea with non-host crops (http://nematode.unl.edu/pest20.htm, verified 23 January 2006). Host crops include barley, oat (*Avena sativa*), sorghum, wheat (*T. durum* and *T. vulgare*), cabbage (*B. oleravea*), turnip (*B. rapa*), lucerne, broad bean (*V. faba*), clovers (*T. incarnatum* and *T. repens*) and vetch (*V. sativa*). Root-knot nematode can cause as much as 80% yield loss to chickpea and wheat in rotation (http://nematode.unl.edu/pest20.htm, verified 23 January 2006). Nene *et al.* (1991) reported a high incidence of wet root knot and collar rot disease of chickpea following rice.

Krupinsky *et al.* (2002) stated that 'the inclusion of a pulse crop in rotations, especially with no tillage, enhances the population and activity of beneficial soil organisms and minimizes the impact of cereal root diseases'. The incidence of crown rot in wheat was less after chickpea (range 1–36%, mean 12%) than after wheat (range 5–52%, mean 30%) in the northern grains region of NSW (Felton *et al.*, 1998). Modelling showed significant interactions between prior crop and total water (pre-plant soil water plus in-crop rainfall) on the incidence of crown rot. 'When wheat followed chickpea, incidence of the disease declined sharply with increasing water. When wheat followed wheat, there was a marginal decline in disease incidence with increasing water'. Kirkegaard *et al.* (2004) reported that 'brassicas (canola and mustard) may provide an excellent alternative rotation crop to chickpea for high value durum wheat (in northern NSW) due to an apparent capacity to more effectively reduce the severity of crown rot infection in subsequent crops'.

Ascochyta blight is one of the most serious diseases of chickpea. It persists in crop residues, seeds and volunteer plants, and spreads rapidly under cool moist conditions. Control measures include resistant cultivars, removal and/or burning of infested residues, and rotation to non-host crops. Chickpea should not be planted in the same field more often than every 3 of 4 years (Wiese *et al.*, 1995). Also, 'in order to reduce the risk of sclerotinia (white mould) infection, chickpea should not be rotated in a short cycle with sunflower, dry bean, lentil, pea, or canola, especially under irrigation' (Margheim *et al.*, 2004).

Intercropping or mixed cropping systems

Intercropping or mixed cropping systems also bring about reduction in the incidence of insect pests and diseases both of chickpea and the component crops. This is mainly due to the dilution effect by which the insect pests and pathogens are forced to search for the host plant unlike in sole cropping. Intercropping offers an entirely new approach in pest management by adversely affecting insect pests both physically and biologically (Prasad *et al.*, 1987). It limits the spreading and proliferation of pests compared to sole cropping. Mehto *et al.* (1988) reported that intercropping generally delayed the appearance of the major pests of chickpea and reduced their incidence. Some minor pests were not at all observed in intercrops. Chickpea should preferably be intercropped to manage pod borer, which would result in higher yields and returns compared to sole cropping (Altaf Hossain, 2003).

Soil Erosion Hazard

Chickpea and other pulse crops provide less crop residue than, for example, grain cereals, which increase the potential for soil erosion. Chickpea provided about one-half the amount of crop residue on the soil surface as spring wheat grown under the same conditions in southern Saskatchewan (unpublished data cited by Miller *et al.*, 2002).

> The lower residue production (of pulse crops) combined with rapid residue decomposition can make for disastrous situations.
>
> (Miller *et al.*, 2002)

Azizi *et al.* (2003) working in north-east Khorramabad, Iran, reported less water infiltration and higher soil erosion due to runoff with chickpea than with Barrel medic (*M. truncatula*). Consequently, it is important to grow chickpea in association with crops that produce adequate amounts of residues and to maximize crop residue retention by using conservation-tillage practices such as no-till, particularly on highly erodible lands. Erosion does not appear to be a major concern when chickpea is cultivated on deep vertisols in India.

Research Needs

Ahlawat *et al.* (2003) pointed to the need for greater genotypic compatibility in intercropping systems and improved management practices. Moreover, 'the development of suitable cultivars and associated techniques could greatly expand the area in chickpea in rainfed lowland of India' (Ahlawat *et al.*, 2003). For example, chickpea in rice–chickpea rotation is prone to soil compaction, collar wet root rot, pod borer and water stress. Alleviation of these constraints will enhance chickpea yield and make it more attractive to farmers.

Several authors reported small or even negative contribution of chickpea to soil N balance, even though chickpea generally had a positive effect on the succeeding crop. Work by Khan *et al.* (2003) indicates that earlier studies may not have included belowground biomass N in their calculations, thus grossly underestimating legume N contribution. Improved and standardized methods of N budgeting and reporting will increase our knowledge of the role of legumes in cropping systems. Moreover, nutrient management in mixed cropping systems, where the component crops may have different N requirements, needs to be worked out. More often, farmers in developing countries do not apply N fertilizers, thinking that chickpea will supply the N needs of the intercrops through biological fixation, which is usually not the case, although applying too much N to the system will inhibit N fixation by chickpea. Published results generally show a smaller N contribution of chickpea compared to other pulse crops such as pea or lupin, but it is not always clear whether the difference is due to greater N_2 fixation, biomass production or both. This warrants clarification. None the less, the need to develop improved chickpea cultivars and management practices for the purpose of maximizing symbiotic N fixation is as important as ever. Furthermore, with growing mechanization of crop production, planting geometry becomes a major impediment in intercropping systems. Development of machine-friendly cultural practices, particularly for sowing and harvesting, needs to be given serious thought.

Summary and Conclusion

Chickpea is a leading pulse crop in several countries around the globe. It is cultivated at different times of the year depending on the climatic conditions and soil types. Thus, a multitude of chickpea cropping systems have been reported. In the Indian subcontinent comprising India, Pakistan, Bangladesh and Nepal, where most of the chickpea is grown (over 80% of chickpea acreage worldwide), sequential cropping and intercropping systems with a large variety of component crops are common. In other parts of the world such as North America, chickpea cropping systems are still evolving. The advantages and disadvantages of chickpea-based cropping systems are often location-specific. Hence, there cannot be generalized cropping systems that are suitable in all situations, given the great diversity of the physical and socio-economic environments in which chickpea is cultivated.

When grown as an intercrop or in rotation with other crops, chickpea enhances soil N and helps break the cycle of diseases and insect pests, often leading to enhanced productivity of the system. However, chickpea produces little crop residue, and thus should be cultivated using conservation tillage, especially on highly erodible soils. No-till practices have allowed farmers in the Great Plains to intensify and diversify cropping systems due to water savings. Probably the most common crop rotation involving chickpea is wheat–chickpea, although several other crop rotations have been reported including rice–chickpea, maize–chickpea, sorghum–chickpea and cotton–chickpea. Constraints include delayed chickpea planting after cotton and soil compaction after rice.

Intercropping is common in South Asia and provides a hedge against crop failure. Chickpea is intercropped with a large number of crops, including wheat, maize, safflower, linseed and mustard. Productivity of the system generally increases when the proportion of chickpea relative to that of the companion crop increases. Taller and more upright chickpea cultivars compete better with intercrops and are usually less prone to diseases compared to spreading-type cultivars. Low yields and limited marketing options were listed as impediments to the expansion of chickpea in developing countries.

Estimates of chickpea contribution to soil N balance vary greatly depending on the methods used to measure and budget N, soil and climatic conditions, cultivar, rhizobial strain, etc. There is evidence that some estimates may be too low if they did not include belowground biomass N. The average N contribution of chickpea across Australia was estimated at 18 kg N/ha, less than that of lupin (88 kg N/ha) or pea (44 kg N/ha). Several studies concluded that even though chickpea enhanced N availability to the succeeding crop, it was not enough to sustain high productivity of the system without additional input of N. None the less, cereal grain yield and/or protein concentration was often higher following chickpea than following another cereal crop. Because of its deeper root system, chickpea extracts more soil moisture than, for example, pea, and may be less suitable as a rotation crop in dry years or environments.

Chickpea is susceptible to many diseases and insect pests, the incidence of which diminishes when chickpea is grown in rotation or association (intercropping and mixed cropping) with other crops. This is mostly due to the disruption in the pest's cycle. For example, rotating chickpea with non-host crops for an adequate period of time will allow for the decomposition of infested crop residues and a decline in pathogen population. Similarly, intercropping will generally delay the appearance and may reduce the incidence of major chickpea pests.

Research needs include better estimates of chickpea contribution to N balance of cropping systems and adapted cultivars and management practices, particularly in mixed or intercropping systems.

References

Ahlawat, I.P.S., Masood, A. and Shivakumar, B.G. (2003) Cropping systems research in chickpea. In: Masood, A., Shiv Kumar and Singh, N.B. (ed.) *Chickpea Research in India*. Indian Institute of Pulse Research (IIPR), Kanpur, India, pp. 119–131.

Ahlawat, I.P.S., Gangaiah, B. and Ompal Singh (2005) Production potential of chickpea (*Cicer arietinum*)-based intercropping systems under irrigated conditions. *Indian Journal of Agronomy* 50(1), 27–30.

Ali, M. and Mishra, J.P. (2002) Genotypic compatibility with an interplay of row orientation in gram (*Cicer arietinum*) + Indian mustard (*Brassica juncea*) under irrigated conditions of Indo-Gangetic alluviums. *Indian Journal of Agricultural Sciences* 72(2), 97–100.

Altaf Hossain, M. (2003) Management of chickpea pod borer, *Helicoverpa armigera* (Hubner) through intercroppings and insecticide spraying. *Thai Journal of Agricultural Science* 36(1), 51–56.

Altaf Hossain, M., Rahman, M.M. and Islam, M.N. (1999) Effect of intercropping and sowing dates on the incidence of pod borer, *Helicoverpa armigera*, yield and yield loss of chickpea. *Annals of Bangladesh Agriculture* 9(1), 57–64.

Angadi, S., McConkey, B., Ulrich, D., Cutforth, H., Miller, P., Entz, M., Brandt, S. and Volkmar, K. (1999) Developing viable cropping options for the semiarid prairies. Project Report Agricultural Agri-Food Canada, Swift Current, Saskatchewan, Canada.

Armstrong, R.D., Millar, G., Haplin, N.V. and Standley, J. (2003) Using zero tillage, fertilizers and legume rotations to maintain productivity and soil fertility in opportunity cropping systems on a shallow vertosol. *Australian Journal of Experimental Agriculture* 43, 141–153.

Azizi, K., Ghalavand, A., Sharifabad, H.H. and Sanavy, S.A.M. (2003) The effects of cropping systems (annual medics [*Medicago* spp.], fallow and chickpea) on run-off, erosion and soil moisture. *Iranian Journal of Range and Desert Research* 10(4), 429–444, 493.

Bajpai, R.P., Singh, V.K., Purohit, K.K. and Sisodia, R.I. (1991) Performance of pulses and wheat in rice based cropping systems under limited irrigation. *Indian Journal of Pulses Research* 4(2), 177–179.

Beck, D.P. (1992) Yield and nitrogen fixation of chickpea cultivars in response to inoculation with selected rhizobial strains. *Agronomy Journal* 84, 510–516.

Berrada, A. (2004) Developing sustainable dryland cropping systems in SW Colorado and SE Utah using conservation tillage and crop diversification: 2002 and 2003 results. *Agricultural Experimenl Station Technical Bulletin* TB04-02, Colorado State University, Fort Collins, Colorado.

Campiglia, E. and Caporali, F. (1992) Effect of spatial arrangement of plants, sowing methods and nitrogen fertilizer applications on sunflower (*Helianthus annuus* L.)–chickpea (*Cicer arietinum* L.) intercrops. Note II. Complementarity of water use. *Rivista di Agronomia* 26(4), 508–516.

Campiglia, E., Saccardo, F., Mancinelli, R. and di Bonito, R. (1998) Yield and nodulation of chickpea genotypes in response to inoculation with selected rhizobial strains. *Italus Hortus* 5(3), 19–24.

Cheruiyot, E.K., Mumera, L.M., Nakhone, L.N. and Mwonga, S.M. (2003) Effect of legume-managed fallow on weeds and soil nitrogen in following maize (*Zea mays* L.) and wheat (*Triticum aestivum* L.) crops in the Rift Valley highlands of Kenya. *Australian Journal of Experimental Agriculture* 43, 597–604.

Dalal, R.C., Strong, W.M., Weston, E.J., Cooper, J.E., Wildermuth, G.B., Lehane, K.J., King, A.J. and Holmes, C.J. (1998) Sustaining productivity of a vertisol at Warra, Queensland, with fertilizer, no-tillage, or legumes: 5. Wheat yields, nitrogen benefits and water-use efficiency of chickpea-wheat rotation. *Australian Journal of Experimental Agriculture* 38, 489–501.

Das, S.B. (1998) Impact of intercropping on *Helicoverpa armigera* (Hub.): incidence and crop yield of chickpea in West

Nimar Valley of Madhya Pradesh. *Insect Environment* 4(3), 84–85.

El Gergawi, A.S.S. and Abou Salama, A.M. (1994) Intercropping of autumn planted sugarcane with chickpea and lentil in Middle Egypt. *Assiut Journal of Agricultural Sciences* 25(5), 47–56.

Evans, J., McNeill, A.M., Unkovich, M.J., Fettell, N.A. and Heenan, D.P. (2001) Net nitrogen balances for cool-season grain legume crops and contributions to wheat nitrogen uptake: a review. *Australian Journal of Experimental Agriculture* 41, 347–359.

FAOSTAT (2005) Food and Agricultural Organization Statistical Databases, Rome.

Farahani, H.J., Peterson, G.A., Westfall, D.G., Sherrod, L.A. and Ahuja, L.R. (1998) Soil water storage in dryland cropping systems: the significance of cropping intensification. *Soil Science Society of America Journal* 62, 984–991.

Felton, W.L., Marcellos, H., Alston, C., Martin, R.J., Backhouse, D., Burgess, L.W. and Herridge, D.F. (1998) Chickpea in wheat-based cropping systems of northern New South Wales II. Influence on biomass, grain yield, and crown rot in the following wheat crop. *Australian Journal of Agricultural Research* 49, 401–407.

Gan, Y.T., Miller, P.R., McConkey, B.G., Zentner, R.P., Stevenson, F.C. and McDonald, C.L. (2003) Influence of diverse cropping sequences on durum wheat yield and protein in the semiarid northern great plains. *Agronomy Journal* 95, 245–252.

Gaur, R.B. and Singh, R.D. (1994) Cropping system in relation to chickpea blight (*Ascochyta rabiei*) I: effect of inter cropping on ascochyta blight spread. *Indian Journal of Mycology and Plant Pathology* 24(1), 33–37.

Giambalvo, D., Stringi, L., Durante,G., Amato, G. and Frenda, A.S. (2004) Nitrogen efficiency component analysis in wheat under rainfed Mediterranean conditions: effects of crop rotation and nitrogen fertilization. *Options Mediterraneennes Serie A Seminaires Mediterraneens* 60, 169–173.

Haidar, M.G., Askary, T.H. and Nath, R.P. (2001) Nematode population as influenced by paddy based cropping sequences. *Indian Journal of Nematology* 31(1), 68–71.

Harris, D., Khanal, N., Joshi, K.D., Musa, A.M., Johansen, C., Rao, J.V.D.-K.K., Kankal, M. and Rashid, A. (2005) Impacts of replacing rice fallows with productive crops in Bangladesh, Nepal and eastern India. *Aspects of Applied Biology* 75, 61–63.

Hassan, A.A., Karim, N.N., Hamid, A. and Salam, M.A. (2003) Soil water management and conservation practices towards a new cropping pattern in drought prone areas of Bangladesh. *Pakistan Journal of Agronomy* 2(2), 77–84.

Hiremath, S.M., Hosmani, M.M. and Chittapur, B.M. (1991) Performance of safflower (*Carthamus tinctorius*) + chickpea (*Cicer arietinum*) intercropping systems. *Indian Journal of Agronomy* 36(3), 407–409.

Holford, I.C.R. and Crocker, G.J. (1997) A comparison of chickpeas and pasture legumes for sustaining yields and nitrogen status of subsequent wheat. *Australian Journal of Agricultural Research* 48(3), 305–316.

Horn, C.P., Dalal, R.C., Birch, C.J. and Doughton, J.A. (1996) Nitrogen fixation in chickpea as affected by planting time and tillage practice. In: *Proceedings of the 8th Australian Agronomy Conference*. 30 January–2 February 1996. Toowoomba, Queensland, Australia, pp. 317–320.

Hossain, S.A., Strong, W.M., Waring, S.A., Dalal, R.C. and Weston, E.J. (1996) Comparison of legume-based cropping systems at Warra, Queensland. II. Mineral nitrogen accumulation and availability to the subsequent wheat crop. *Australian Journal of Agricultural Research* 37(3), 245–261.

Hulugalle, N.R., Entwistle, P.C., Scott, F. and Kahl, J. (2001) Rotation crops for irrigated cotton in a medium-fine, self-mulching, grey vertosol. *Australian Journal of Soil Research* 39, 317–328.

Hussein, A.H.A. and El Deeb, M.A. (1999) Evaluation of intercropping faba bean, chickpea and lentil with sugar beet in middle Egypt. *Arab Universities Journal of Agricultural Sciences* 7(2), 475–482.

Khan, D.F., Peoples, M.B., Schwenke, G.D., Felton, W.L., Chen, D. and Herridge, D.F. (2003) Effects of below-ground nitrogen on N balances on field-grown fababean,

chickpea, and barley. *Australian Journal of Agricultural Research* 54, 333–340.
Khot, A.B. and Yargattikar, A.T. (1998) Performance of maize based crop sequences under various irrigation levels in vertisols. *Mysore Journal of Agricultural Sciences* 32(4), 313–316.
Kirkegaard, J.A., Simpfendorfer, S., Holland, J., Bambach, R., Moore, K.J. and Rebetzke, G.J. (2004) Effect of previous crops on crown rot and yield of durum and bread wheat in northern NSW. *Australian Journal of Agricultural Research* 55, 321–334.
Krupinsky, J.M., Bailey, K.L., McMullen, M.P., Gossen, B.D. and Turkington, T.K. (2002) Managing plant disease risk in diversified cropping systems. *Agronomy Journal* 94, 198–209.
López-Bellido, R.J. and López-Bellido, L. (2001) Effects of crop rotation and nitrogen fertilization on soil nitrate and wheat yield under rainfed Mediterranean conditions. *Agronomie* 21(6/7), 509–516.
López-Bellido, L., Fuentes, M., Castillo, J.E. and López-Garrido, F.J. (1998) Effects of tillage, crop rotation and nitrogen fertilization on wheat-grain quality grown under rainfed Mediterranean conditions. *Field Crops Research* 57(3), 265–276.
López-Bellido, L., López-Bellido, R.J., Castillo, J.E. and López-Bellido, F.J. (2000) Effects of tillage, crop rotation and nitrogen fertilization on wheat under rainfed Mediterranean conditions. *Agronomy Journal* 92, 1054–1063.
Lyon, D.J., Baltensperger, D.D., Blumenthal, J.M., Burgener, P.A. and Harveson, R.M. (2004) Eliminating summer fallow reduces winter wheat yields, but not necessarily system profitability. *Crop Science* 44, 855–860.
Malik, M.A., Saleem, M.F., Sana, M. and Aziz, A. (2002) Agro-economic expression of different relay crops after rice harvest under conventional and zero tillage. *International Journal of Agriculture and Biology* 4(2), 277–278.
Mandal, B.K., Dasgupta, S. and Ray, P.K. (1986) Yield of wheat, mustard and chickpea grown as sole and intercrops with four moisture regimes. *Indian Journal of Agricultural Sciences* 56(8), 577–583.
Marcellos, H. (1984) Influence of prior crops of chickpea, fababeans, and lupins on wheat. *The Journal of the Australian Institute of Agricultural Science* 50, 111–113.
Marcellos, H., Felton, W.L. and Herridge, D.F. (1998) Chickpea in wheat-based cropping systems of Northern New South Wales: I. N_2 fixation and influence on soil nitrate and water. *Australian Journal of Agricultural Research* 49(3), 391–400.
Margheim, J.F., Baltensperger, D.D., Wilson, R.G., Lyon, D.J., Hein, G.L., Harveson, R.M., Burgener, P., Krall, J.M., Cecil, J.T., Rickersten, J.R., Merrigan, A.P., Watson, M.H. and Hansen, B.J. (2004) Chickpea production in the High Plains. South Dakota State University Extension Fact Sheet 922, University of Wyoming Bull. B1153, University of Nebraska Cooperative Extension. EC04-183.
Mehto, D.N., Singh, K.M. and Singh, R.N. (1988) Influence of intercropping on succession and population build up of insect pests in chickpea, *Cicer arietinum* Linn. *Indian Journal of Entomology* 50(3), 257–275.
Miller, P.R. and Holmes, J.A. (2005) Cropping sequence effects of four broadleaf crops on four cereal crops in the Northern Great Plains. *Agronomy Journal* 97, 189–200.
Miller, P.R., McConkey, B.G., Clayton, G.W., Brandt, S.A., Staricka, J.A., Johnson, A.M., Lafond, J.P., Schatz, B.G., Baltensperger, D.D. and Neill, E.K. (2002) Pulse crop adaptation in the Northern Great Plains. *Agronomy Journal* 94, 261–272.
Miller, P.R., Gan, Y., McConkey, B.G. and McDonald, C.L. (2003a) Pulse crops for the Northern Great Plains: II. Cropping sequence effects on cereal, oilseed, and pulse crops. *Agronomy Journal* 95, 980–986.
Miller, P.R., Gan, Y., McConkey, B.G. and McDonald, C.L. (2003b) Pulse crops for the Northern Great Plains: I. Grain productivity and residual effects on soil water and nitrogen. *Agronomy Journal* 95, 972–979.
Mishra, J.P., Ali, M. and Arya, R.L. (2001) Genotypic compatibility in relation to row ratio in the intercropping of linseed (*Linum*

usitatissimum) and gram (*Cicer arietinum*) under rainfed conditions. *Indian Journal of Agricultural Sciences* 71(6), 359–362.

Moore, K., Backhouse, D., Simpfendorfer, S. and Verrell, A. (2003) Crop rotation reduces crown rot in wheat. *Update of research in progress at the Tamworth Agricultural Institute 2002*. p. 43.

Nene, Y.L., Reddy, M.V., Haware, M.P., Ghanekar, A.M. and Amin, K.S. (1991) Field diagnosis of chickpea diseases and the control. Information Bulletin No. 28, ICRISAT, Hyderabad, India.

Patwary, S.U., Haque, Q. and Badruddin, M. (1989) Role of legume on nitrogen balance and A-value of soil under different sequential cropping systems. *Thai Journal of Agricultural Science* 22(3), 213–221.

Pilbeam, C.J., Wood, M., Harris, H.C. and Tuladhar, J. (1998) Productivity and nitrogen use of three different wheat-based rotations in North West Syria. *Australian Journal of Agricultural Research* 49, 451–458.

Prasad, D. and Chand, P. (1989) Effect of intercropping on the incidence of Heliothis armigera (Hub.) and grain yields of chickpea. *Journal of Research Birsa Agricultural University* 1(1), 15–18.

Prasad, D., Singh K.M., Katiyar, R.N. and Singh, R.N. (1987) Impact of intercropping on the plant growth, pest incidence and crop yield of pea, *Pisum sativum* Linn. *Indian Journal of Entomology* 49(2), 153–172.

Sekhar, P.R., Rao, N.V., Venkataiah, M. and Rajasri, M. (1995) Influence of intercrops of pigeonpea and chickpea on pod borer incidence. *Indian Journal of Pulses Research* 8(1), 41–44.

Sharma, R.K., Bangar, K.S., Sharma, S.R., Gwal, H.B. and Verma, H.D. (1993) Studies on intercropping of pulses in spring planted sugarcane. *Indian Journal of Pulses Research* 6(2), 161–164.

Shiyani, R.L., Joshi, P.K and Bantilan, M.C.S. (2001) Impacts of chickpea research in Gujarat. In: *Impact Series No. 9*. ICRISAT, Hyderabad, India.

Siddique, K.H.M., Loss, S.P., Regan, K.L. and Jettner, R.L. (1999) Adaptation and seed yield of cool season grain legumes in Mediterranean environments of south-western Australia. *Australian Journal of Agricultural Research* 50(3), 375–387.

Singh, D.K. and Yadav, D.S. (1992) Production potential and economics of chickpea (*Cicer arietinum*)-based intercropping systems under rainfed condition. *Indian Journal of Agronomy* 37(3), 424–429.

Smithson, J.B., Thompson, J.A. and Summerfield, R.J. (1985) Chickpea (*Cicer arietinum* L.). In: Summerfield, R.J. and Roberts, E.H. (eds) *Grain Legume Crops*. Collins, London.

Thakur, N.S., Pannase, S.K. and Sharma, R.S. (2000) Production potential of gram (*Cicer arietinum*)-based intercropping systems under rainfed condition. *Indian Journal of Agronomy* 45(3), 534–539.

Tutwiler, R.N. (1995) The great chickpea challenge: introducing winter sowing in the Mediterranean region. *ICARDA-Social-Science-Papers* 4, p. 30.

van der Maesen, L.J.G. (1972) A monograph of the genus with special reference to the chickpea (*Cicer arietinum* L.): its ecology and cultivation. Communication of Agricultural University, Wageningen, Dordrecht, The Netherlands.

Wani, A.H. (2005) Effect of cropping sequences and ploughing on plant parasitic nematodes and plant growth in field. *Indian Journal of Nematology* 35(1), 63–67.

Wiese, M.V., Kaiser, W.J., Smith, L.J. and Muehlbauer, F.J. (1995) Ascochyta blight of chickpea. CIS 886. University of Idaho, College of Agriculture, Cooperative Extension System and Agricultural Experimental Station, Moscow.

Yau Sui-Kwong (2005) Optimal sowing time and seeding rate for winter-sown, rain-fed chickpea in a cool, semi-arid Mediterranean area. *Australian Journal of Agricultural Research* 56(11), 1227–1233.

Zahid, M.A., Khan, H.R., Bakhsh, A. and Iqbal, S.M. (2004) Chickpea cultivation in rice-growing area of Punjab province of Pakistan: potential and constraints. *International Chickpea and Pigeonpea Newsletter* 11, 32–34.

Zentner, R.P., Wall, D.D., Nagy, C.N., Smith, E.G., Young, D.L., Miller, P.R., Campbell, C.A., McConkey, B.G., Brandt, S.A.,

Lafond, G.P., Johnston, A.M. and Derksen, D.A. (2002) Economics of crop diversification and soil tillage opportunities in the Canadian Prairies. *Agronomy Journal* 94, 216–230.

Zewdu, T. (2002) Double cropping of forage oats and chickpea at Wereta vertisol conditions, northwestern Ethiopia. *AgriTopia* 17(1), 10–11.

Zhang, H., Pala, M., Oweis, T. and Harris, H. (2000) Water use and water-use efficiency of chickpea and lentil in a Mediterranean environment. *Australian Journal of Agricultural Research* 51(2), 295–304.

10 Nutrient Management in Chickpea

I.P.S. Ahlawat,[1] B. Gangaiah[1] and M. Ashraf Zahid[2]

[1]*Division of Agronomy, Indian Agricultural Research Institute, New Delhi 110012, India;* [2]*Pulses Program, National Agricultural Research Centre, PO NIH, Park Road, Islamabad 45500, Pakistan*

Introduction

Chickpea (*Cicer arietinum* L.), the third most important pulse crop in the world after beans and field peas, is cultivated mostly in semiarid regions of South Asia with some area concentrated in the Mediterranean. In recent times, its cultivation has picked up enormously in Australia and Saskatchewan, Canada, to cater to the growing export market demands mainly from India and Pakistan. Although it is suitable for production in low-fertility soils, its productivity is often limited by mineral nutrient deficiencies in many chickpea-growing regions of the world. It has been estimated that 3.95 billion hectares of the world's ice-free land are prone to mineral nutrient stresses, of which 14% is subjected to potential micronutrient stresses (Gettier *et al.*, 1985). Dulal (1976), however, opined that 20% of the total arable lands are prone to mineral stresses. Of the two chickpea varieties, the desi type accounts for 85% and the kabuli type for 15% of the area. These two types differ in their yield potential and hence in nutrient requirements. In general, each tonne of chickpea grain removes 121.9 kg (67.3 N + 6.6 P + 48 K) of primary nutrients (Prasad *et al.*, 2002), 34.7 kg (18.7 Ca + 7.3 Mg + 8.7 S) of secondary nutrients and ~1000 g of four micronutrients (38 Zn + 868 Fe + 70 Mn + 11.3 Cu; Aulakh *et al.*, 1985). Although, data on the uptake of two other micronutrients (B and Mo) are not available, based on mean concentration, it is estimated to be ~35 g (B) and 1.5 g (Mo). The magnitude of yield losses due to nutrient deficiency also varies among the nutrients (Ali *et al.*, 2002). N, P, Fe, B and S may cause yield losses up to 10%, 29–45%, 22–90%, 100% and 16–30%, respectively, in chickpea depending on soil fertility, climate and plant factors. Yield is also limited by nutrient toxicities, which are more common with micronutrients. However, studies on this aspect are limited.

Soils

Chickpea, which is a hardy crop, has the ability to come up well even in marginal soils. It is grown on a wide range of soils varying in texture and soil fertility. The soil texture varies from sandy to clay. Sandy soils include dunes in Thal, Pakistan and western Rajasthan in India, while clay soils include deep vertisols spread over peninsular India, West Asia and Ethiopian highlands. In general, the soils of chickpea-growing areas have low organic carbon content, which is indicative of low soil fertility. Chickpea can be grown successfully in soils with a pH ranging from 6 to 9. Low pH (<4.6), besides limiting some micronutrient availability, causes some toxicity problems and poor nodulation. In acidic soils, increased incidence of fusarium wilt has also been reported (Kay, 1979). Liming is helpful for enhancing soil pH and alleviating problems associated with acidic soils. On the other hand, in soils with pH > 10, seedlings fail to establish due to salinity, alkalinity and sodicity (common problem in parts of India and the Nile Valley), and thus the productivity of chickpea is adversely affected.

Soil salinity is one of the major limiting factors of chickpea production in semiarid regions, and chloride salinity has a more depressive effect on yield than sulphate salinity. High exchangeable sodium percentage (ESP) and resultant deficiency of Fe, Cu and Zn in sodic soils causes poor crop establishment and performance (Table 10.1). Therefore, selection of salt-tolerant genotypes for cultivation in soils with electrical conductivity (ECe) > 4 dS/m is recommended. The ability to exclude Na and Cl^- from shoots has been used as a criterion for selecting salt-tolerant genotypes. The release of salt-tolerant cultivars like Karnal Chana (CSG 8962) in India and others elsewhere is a boon to chickpea cultivation in salt-affected soils. *Rhizobium* inoculation, mineral N fertilization and gypsum application have also been found effective in alleviating salinity problems. The nutrient-related limitations of chickpea productivity and their management are discussed in the following sections.

Macronutrients

Nitrogen

Chickpea, which is a legume, has the inherent ability to fix atmospheric nitrogen (N_2) in symbiotic association with *Mesorhizobium ciceri*. The rich N content of

Table 10.1. Effect of sodicity on performance of chickpea. (From Kumar, 1985.)

Exchangeable sodium percentage	Emergence at 20th day	Leaf area (cm^2/plant)	Seed yield (g/plant)
15	77.4	761.3	125.0
20	71.5	382.0	70.6
24	67.4	222.2	41.6
28	57.6	132.6	10.9
CD (*P* = 0.05)	2.0	151.0	8.7

chickpea grain coupled with soil N meets the seedling stage demand, while biologically fixed N takes care of the needs at later stages of crop under favourable conditions. Thus, the crop rarely shows N deficiency symptoms, but it does show the symptoms of deficiency with failure of biological nitrogen fixation (BNF) and in N-deficient soils. The typical N deficiency symptoms appear initially in older leaves as chlorosis. Other symptoms include pink pigmentation of stems and on the upper surfaces of older leaves. The deficiency symptoms of N in chickpea are depicted in Fig. 10.1. To alleviate N deficiency arising between the expiry of seed N and effective nodule formation, 20 kg N/ha is recommended as a starter dose. This starter dose is more essential in sandy loam soils, which have low soil N content. Temperatures as high as 30°C in the top 10 cm of the soil profile may hinder early development of nodules (Rawsthorne *et al.*, 1985) and therefore would require a starter dose of N. Dry seedbeds are not conducive for BNF in the Saskatchewan province of Canada. In such soils, desi types require 30–45 kg N/ha, whereas kabuli types are usually nonresponsive (Walley *et al.*, 2005). This kind of behaviour has been ascribed to differences in phenotype, genotype and maturity duration (Muehlbauer and Singh, 1987). However, significant responses of kabuli types up to 35 kg N/ha have been observed at Faisalabad in Pakistan (Saeed *et al.*, 2004).

A higher starter dose of N (>20 kg/ha) at seedling phase may limit glutamine availability for glutamine synthetase and glutamate synthase pathways (Swahney *et al.*, 1985). However, chickpea grown in rice fallows, sown late (in December) and grown after exhaustive crops like maize and sorghum may respond up to 40 kg N/ha. The *Rhizobial* activity is reduced at low temperatures in rice fallows and also in late-sown situations, where short periods are available for growth and development and greater depletion of soil N by exhaustive

Fig. 10.1. Nitrogen-deficient chickpea plant.

Table 10.2. Effect of chickpea intercropping and NP fertilization in autumn-planted sugarcane. (From Singh *et al.*, 2001.)

Cropping system (fertilizer kg/ha)	Cane yield (t/ha)	Chickpea yield (t/ha)
Sugarcane sole (225 N)	108.54	–
Sugarcane + 1 row chickpea (no additional fertilizer)	103.78	1.26
Sugarcane + 1 row chickpea (15 N + 20 P_2O_5 additional fertilizer)	107.11	1.22
CD (P = 0.05)	NS	NS

preceding crops are reasons for the response of chickpea to higher doses of N. Similarly, a reduction in fertilizer N requirement by ~50% after heavily fertilized crops like potato may be possible, probably due to higher residual soil N. Chickpea intercropped with autumn-planted sugarcane may not require additional fertilization (N and P), and fertilizers applied to sugarcane are adequate (Table 10.2) to meet the needs of both the crops.

In soils with poor population of symbiotic bacteria, *Rhizobial* inoculation of seed or soil has been recommended. The critical period for N application to prevent yield losses due to failure of inoculation was found to be within 6 weeks of seeding at Montana, Canada, where spectral reflectance was used to assess inoculation failure and resulting plant N deficiency, and the same was used as criteria of N fertilization (McConnell *et al.*, 2002). Further, under dryland conditions of Stavropol province, chickpea does not nodulate; hence, presowing application of 60 kg ammophos/ha has been recommended for higher seed yields (Sysoev, 1977). Singh *et al.* (2005) estimated the fertilizer N requirements of chickpea quantitatively for alluvial soil. The equations of N requirement of chickpea with and without manure were: fertilizer N (kg/ha) = 34.71 (target yield; in t/ha)–0.19 (soil test N; in kg/ha) and 20.91(target yield; in t/ha)–0.11 (soil test N; in kg/ha)–0.19 (manure N applied; in kg/ha), respectively.

In Mediterranean climates, chickpea experiences terminal drought that reduces leaf photosynthesis and N mobilization to reproductive parts. Further, during the postflowering period, nodules become ineffective due to their degeneration. Under these situations, postflowering N fertilization is recommended. Foliar spray of urea from the beginning of flowering up to 50% flowering stage has been found to enhance chickpea yield and seed N yield (Table 10.3). Similar positive responses to postflowering N fertilization as foliar spray of 2% urea at the time of pod formation have also been reported (Bharud, 2001). As the nodule activity decreases from peak flowering stage, the crop also experiences N stress. Hence, split application of N may be promising to basal N fertilization in some cases. However, postflowering N fertilization to soil may also adversely affect the nitrogenase activity. An increase in yield of ~70% with 50 kg N/ha over no fertilizer N was reported under Sudan conditions (El-Hadi and El-Sheikh, 1999). In cooler climates, N fertilization may improve the cold susceptibility.

Table 10.3. Effect of timing of foliar application of N[a] on chickpea. (From Palta *et al.*, 2005.)

Time of foliar spray	Seed yield (g/plant)	Seeds/pod	N content at maturity (g/plant)	Seed N yield (g/plant)
Ist flower	5.6	1.4	445.2	91.9
50% flowering	5.1	1.3	439.4	92.4
50% pod set	4.2	1.0	379.6	69.8
End of podding	4.1	1.0	385.5	71.3
Control	4.0	1.0	335.2	61.5
CD ($P = 0.05$)	0.8	0.2	55.1	28.2

[a]114 g N^{15}/plant were applied.

In P-deficient soils, chickpea responds poorly to N fertilization without P fertilization. However, 30–60 kg P_2O_5/ha along with 15 kg N/ha may give significant response. Combined use of P fertilizer with starter dose of N and *Rhizobium* inoculation may prove more beneficial. Application of farmyard manure (FYM) along with 20 kg N/ha was found to be effective in enhancing yield of chickpea over FYM application alone in New Delhi.

Of the total N (103 kg/ha) present in chickpea crop at physiological maturity, 60%, 35% and 5% are obtained from BNF, soil and fertilizer, respectively in conditions of no N fertilization (Kurdali, 1996). This clearly shows lesser importance of N fertilizer in chickpea nutrition compared with other legume. The agronomic N efficiency of the starter dose (15–20 kg N/ha) may vary from 8.5 to 19.5 kg grain/kg N applied (Prasad and Subbiah, 1982), while the physiological N efficiency may vary from 25 to 27 kg seed/kg N absorbed (Singh *et al.*, 1999). The varietal differences in N use efficiency have also been reported, and may range from 3.54 to 11.65 kg seed/kg N (Bhattacharya and Ali, 2002).

Phosphorus

Phosphorus (P) is the most critical nutrient limiting chickpea production, and is deficient in the majority of chickpea-cultivated tracts including India. Legume fertilization is often P-based, since it is highly essential for intensive N fixation. Thus, P, by way of its role in energy transformation and enhancing root growth, is essential for nodulation and effective N fixation. The deficiency symptoms include dark green foliage and red-purple pigmentation on stem and upper surface of leaflets of the lower leaves. The leaflets later become yellow-green or buff-green. The typical P deficiency symptoms of chickpea are shown in Fig. 10.2. The requirement of P in kabuli types is usually higher (40 kg P_2O_5/ha) than in desi types (20 kg P_2O_5/ha). Further, winter-sown chickpea responds better than spring-sown chickpea with a higher yield of 50–100%.

The response to P fertilization depends on soil moisture status, as soil moisture stress may decrease the availability of applied P, resulting in poor biomass production and reduced P uptake. Thus, in alfisols (with low water-holding capacity and kaolinite clay), chickpea is less responsive to P fertilization than

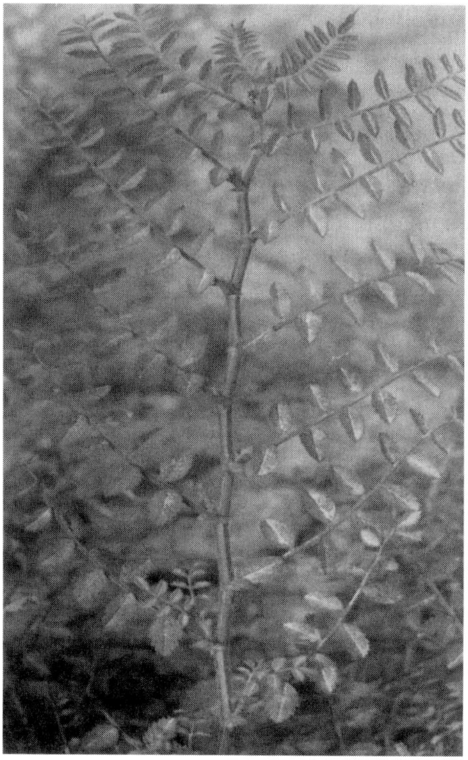

Fig.10.2. Phosphorus deficiency in chickpea.

in vertisols (with high water-holding capacity and montmorillonite clay). In drylands, water is often a limiting factor of chickpea productivity, and the effectiveness of applied fertilizer would depend on its availability. Efforts have been made to overcome this problem through aqua-fertilization of P, which has a distinct advantage over its dry placement in chickpea (Table 10.4).

The need for P fertilization is usually more in acidic soils than in others due to higher P fixation. P needs of chickpea may also vary with soil P status. In soils with <15, 15–22.5 and >22.5 kg available P/ha, kabuli types required 60 and 30 kg P_2O_5/ha and no P, respectively (AICRP, 2003). In terrarosa soils of north Syria with <2.5 ppm available P, 50 kg P_2O_5/ha as triple superphosphate (TSP) was found effective (Saxena, 1984), while in Canadian prairies with P content (μg/g) of 6.6 to 31.4 in 7 sites (3 in 1996 and 4 in 1997), 20 kg P_2O_5/ha was found to be optimum (Walley *et al.*, 2005). Fertilizer P requirements of chickpea were quantitatively estimated for alluvial soil with and without manure by Singh *et al.* (2005) as: fertilizer P (kg/ha) = 16.64 (target yield; in t/ha)–0.81 (soil test P; in kg/ha) and 15.32(target yield; in t/ha)–1.22 (soil test P; in kg/ha)–0.23 (manure P applied; in kg/ha), respectively.

With data from 2100 trials on P in chickpea in India, Tandon (1987) reported an agronomic efficiency of 18 kg grain/ kg P. The apparent fertilizer P recoveries usually vary from 10% to 15% in chickpea depending upon soil type and rate of P fertilization. Further, percentage of P utilized from fertilizer source decreases with increasing rates of P fertilization. The percentage of P uptake and utiliza-

Table 10.4. Effect of aqua-fertilization of phosphorus on yield and P uptake of chickpea under dryland conditions. (From Pramanik and Singh, 2004.)

Mode of P fertilization	Grain yield (t/ha) Pooled over two seasons	P uptake (kg/ha) Mean of 2 years
Dry placement of P in soil	2.195	7.40
Aqua-fertilization of P	2.307	8.09
CD (P = 0.05)	0.054	0.50

tion is usually higher with its placement at 5 cm below than that applied at soil surface. Similarly, chickpeas derived relatively more P from basal dose than from crop side-dressing of P at 10–12cm depth in furrows 25 days after sowing (Singh, 1995) stage. At Londrina (Brazil), pit (pit is like a pot in which experiments conducted) fertilization of P was found to enhance chickpea yield by 31% over row fertilization (Voss *et al.*, 1998). Similarly, Din *et al.* (1999) noted that band placement may reduce the P requirement of chickpea over broadcast application, affecting an economy of 71% P fertilizer. In heavy soils, deep placement of P (15 cm) proves better than mixing with soil. Foliar spray of 2% diammonium phosphate (DAP) may also be promising in enhancing the yield of chickpea.

The critical P concentration for maximum grain yield in chickpea ranged from 0.18% to 0.27% in young whole shoots (<30 cm tall), while in recently matured leaves at flower initiation it ranged from 0.34% to 0.39% (Din *et al.*, 1999).

In general, all the sources of P are equally effective in chickpea, except rock phosphate. Rock phosphate is a poor source of P in neutral to alkaline soils, but a fairly good source for acidic soils. Rock phosphate and organic manure along with arbuscular mycorrhizal fungi have been found as effective and ecologically sound alternatives to intensive use of P fertilizers for chickpea grown on acidic soils (Alloush *et al.*, 2000). Phosphate-solubilizing microorganisms are found effective in solubilization of soil P through excretion of hydroxyl acids. Some of these acids may form chelates with cations such as Ca^{2+} and Fe^{2+}, resulting in effective solubilization of phosphates. The benefits of these microorganisms were more with rock phosphate. Enhanced crop performance with P alone and in combination with *Bacillus polymyxa* (Saad and Sharma, 2003), *Aspergillus awamori* (Sekhar and Aery, 2001), *Glomus calospora*, *G. mossae* and *Acaulospora* sp. (Meshram *et al.*, 1999) has also been reported in chickpea. Among the sources of P, ammonium polyphosphates, DAP and single superphosphate (SSP) are equally effective in chickpea. Chickpea root exudes citrate and other organic acids, which may enhance the solubility of Ca–P. Therefore, chickpea is efficient in utilizing P from alkaline than acid soils (Ae *et al.*, 1991).

Potassium

Potassium (K) requirement of chickpea varies from 14.3 to 57.7 kg/t of grain depending on soil type, but its use is limited due to sufficient levels of soil K in most regions to support the crop. On the basis of 3.65 million soil samples tested in 1998–1999 in India, 12.9% and 36.7% soils were categorized as low and medium for K status. The deficiency is, however, widespread in eastern and north-eastern states and fairly common in the states of Jammu and Kashmir, Kerala and Uttar Pradesh. The highest loss up to 50% due to omission of K from fertilization has been reported in grey flood plain soils at Bogra, Bangladesh (Islam *et al.*, 1995). The typical K deficiency symptoms of chickpea include chlorosis of margins and tips of older leaves, reddish pigmentation on leaflets and necrosis, first on the tips of leaflets and later covering the whole leaflet turning it to light brown in colour.

A regular K fertilization in alfisols and maintenance of K dose in inceptisols is essential for chickpea production in India due to low non-exchangeable K and low exchangeable K, respectively. In alluvial soils of Uttar Pradesh, the critical level of soil K for chickpea has been estimated as 137 kg/ha (Subbarao, 1993), and the sufficiency level of K in the leaf blade of chickpea varied from 1.3% to 1.6% at 30 days after sowing (DAS) and 1.4–1.7% at 70 DAS. Aini and Tang (1998) estimated critical concentration of K in chickpea as 1.4–1.5% in youngest fully expanded leaves, 2.7–2.8% in the first and second leaf petioles and 2.1–2.2% in whole shoots. In alluvial soils, responses to 15–30 kg K_2O/ha have been observed (Tiwari, 1988). Fertilizer K requirements of chickpea were quantitatively estimated for alluvial soil with and without manure by Singh *et al.* (2005) as: fertilizer K (kg/ha) = 55.18 (target yield; in t/ha)–0.24 (soil test K; in kg/ha) and 25.28 (target yield; in t/ha)–0.095 (soil test K; in kg/ha)–0.32 (manure K applied; in kg/ha), respectively. The agronomic efficiency of applied K (15 kg K_2O/ha) varied with soil K status from 18 kg grain/kg K_2O in medium K soil to 9.8 kg grain/kg K_2O in high K soils (Shukla and Prasad, 1979). As stover accounts for 50–90% of total K uptake by chickpea, its recycling could compensate for 70–90% of K removed from soil (Pieri and Oliver, 1986). However, it would not be possible in areas where straw is used as fodder for livestock.

Calcium and magnesium

The base elements, calcium (Ca) and magnesium (Mg), are generally deficient in acidic soils. Soils with <1.5 meq exchangeable Ca/100 g soil or <25% of cation exchange capacity (CEC), and <1.0 meq exchangeable Mg/100 g soil or <4–15% of CEC are usually considered as Ca- and Mg-deficient.

The deficiency symptoms of Ca include arrested development of rachis and browning of the leaflets of young expanding leaves. Leaflets also show necrosis. While older plant parts remain dark green, the death of growing points enhances development of auxillary buds. In case of Mg deficiency, young and middle-aged leaves show light green chlorosis initially, which becomes more severe on the distal half of the leaflets of fully expanded leaves. Under severe deficiency, leaflet tips and margins show necrosis.

The Mg requirement of legumes is generally higher than that of cereals and oilseeds. In neutral soil with 1.6 cmol (p+) Mg/kg soil, Mg fertilization increased chickpea yield by 22.5%. Kimberlite (inorganic waste of diamond mines) rich in CaO (5–9.8%) and MgO (19.5–22%) applied at 2 t/ha enhanced chickpea yield by increasing soil pH and meeting Ca and Mg needs of chickpea. Liming also enhanced chickpea yield by 159% in acidic soils with pH 5.5 (Tandon, 1991). Ca nitrate application has been found effective in the control of collar rot disease (*Sclerotium rolfsü* var. Sojae) in chickpea (Foster, 1999).

Sulphur

The importance of sulphur (S) in chickpea, which requires ~9 kg for producing 1 t of grains, arises due to the fact that legumes are rich in S-containing amino acids (methionine, cystine and cysteine), and because it is a constituent of nitrogenase enzyme, which aids in BNF. Thus, S has become the fourth most important nutrient of crop production. The need for S fertilization has also been identified because of the increased use of S-free fertilizers and higher productivity of crops associated with greater uptake of S. The loss due to omission of S from fertilization was up to 15% in grey flood plain soils at Bogra, Bangladesh (Islam *et al.*, 1995).

Coarse-textured soils that are poor in organic carbon are also poor in S. S deficiency is also seen in chickpea grown in calcareous soils with pH >8. S deficiency symptoms of chickpea become visible if its concentration is <0.75% at 45–55 DAS. S-deficient plants appear erect with premature drying and withering of young leaves. Eventually the entire foliage turns chlorotic, nodulation is severely restricted and finally the setting of seeds is also limited.

Although all sources of S are equally effective, in acidic soils and those in which leaching of S as SO_4^{-2} occurs, gypsum proves superior to other sources of S. Phosphogypsum, a by-product of gypsum (with 15% S, 21% Ca and 0.2–1.2% P_2O_5), also proves to be an effective source of S in acidic soils. The response to S fertilization is usually observed up to 20–30 kg/ha. However, in soils with coarse texture and low organic matter content, responses up to 60 kg S/ha have been observed. The interaction of P and S is usually synergistic at its low rate of application (40 kg/ha), but antagonistic at high rates (>80 kg S/ha) (Table 10.5). Average responses of 5.3 kg grain/kg S were reported in S-deficient soils in India, with value cost ratio of 12. Of the total S uptake by chickpea, 8.5–8.7% is only retained in the grain, which indicates the richness of straw in S content. Thus, their recycling may substantially reduce the need for S fertilization.

Micronutrients

Intensive cropping coupled with high crop yields has gradually led to micronutrient deficiencies. Although these nutrients are micro in terms of uptake, their contributions are as important as those of macronutrients. The necessity of all the six micronutrients in chickpea and their extent of yield limitations have been well established in pot/solution culture and in some cases under field situations. The micronutrients limiting chickpea productivity in the order of importance

Table 10.5. Performance of chickpea as influenced by interaction effect of P and S. (From Shinde and Saraf, 1992.)

	Grain yield (t/ha)			
	S (kg/ha)			
P_2O_5 (kg/ha)	0	40	80	Mean
0	1.24	1.61	1.82	1.56
PSB	1.52	1.91	1.94	1.79
30	1.68	2.29	2.28	2.08
30 + PSB	2.11	2.37	2.38	2.29
60	2.03	2.51	2.30	2.28
60 + PSB	2.24	2.55	2.34	2.38
Mean	1.80	2.21	2.17	–
SEM±	$P = 0.017$	$S = 0.038$	$P \times S = 0.043$	–

PSB = phosphate-solubilizing bacteria.

are: Zn>Fe>B. Quite often, micronutrient deficiencies are misdiagnosed as pest-related damage. This misdiagnosis often proves costly by way of reducing efficiency of fertilizers and other inputs. Hence, there is a need to develop appropriate diagnosis of micronutrient deficiencies for alleviating the associated problems in crop production.

Zinc

Zinc (Zn) deficiency is common among chickpea-growing regions of the world, but is widespread in the Mediterranean countries like Turkey and Asia. Zn fertilization enhances water use efficiency (Khan *et al.*, 2003), amino acid (lysine, histidine and arginine) and crude protein contents of chickpea (Nayyar *et al.*, 1990). The pH or degree of acidity/alkalinity of the root medium greatly influences Ca availability and thus its role in plant nutrition. To overcome this limitation of Ca availability for plants, liming (acidic soils) and gypsum application (saline/sodic soils) is practiced. Alkalinity forces Zn out of solution, often leading to severe deficiencies of this metal despite the use of chelating agents to help keep them in solution. In sodic soils, combined application of Ca and Zn fertilizers were found beneficial to plants.

Zn deficiency symptoms are first noticed in the upper one-third of the leaflets, which become light yellow in colour and turn pinkish brown later, followed by bronzing and necrosis. Drying of leaf margins of older leaves and twisting of emerging leaves and terminal bud (interveinal necrosis) are usual symptoms of Zn deficiency. The leaflet size is reduced and plants have few branches with stunted growth. Leaf lamina is abnormally thick and stiff. In Zn-stressed sodic soils, no grain formation is observed. The deficiency symptoms of Zn in chickpea are shown in Fig. 10.3.

The critical concentrations of Zn in soils vary from 0.48 ppm in non-calcareous soils to 1.75–2.5 ppm in loams. Critical Zn concentration in plants

Fig. 10.3. Zinc-deficient chickpea plant.

is 15–20 ppm at 45 DAS in top leaf. The critical limit of Zn concentration in both soil and plant decreases with advancement of crop growth stage.

Significant yield responses up to 25 kg $ZnSO_4$/ha have been observed, but 10 kg $ZnSO_4$/ha seem to be the optimum in most of the chickpea-growing areas. The benefits of Zn fertilization include enhanced root growth, nodulation and N content of nodules. Misra *et al.* (2002) observed an increase of 55% in root nodulation and 26% in N content of nodules with 20 mg Zn/kg soil. Zn fertilization in Zn-stressed sodic soils was found to restore failed grain formation by way of narrowing Na/K and Na/Ca + Mg ratios (Misra, 2001). Zn fertilization increases not only Zn uptake, but also Fe and P uptake (Dravid and Goswami, 1987). With data from 14 trials, Rattan *et al.* (1997) reported an average yield response of 200 kg/ha (with a range of 40–390 kg/ha) due to Zn fertilization in chickpea. Soil fertilization of Zn is common as it is available to the crop at an early stage before the plants express the usual deficiency symptoms, but foliar spray of 0.5% $ZnSO_4$ (4–8 weeks after sowing) along with 0.25% lime is also effective in correcting deficiency quickly. A differential behaviour of chickpea cultivars to Zn deficiency has also been observed (Table 10.6). This needs to be explored in alleviating Zn-induced stress in this crop (nutrient uptake stages of plants differ significantly; for some it is mainly in the seedling phase and hence later supply is not useful).

Of the total Zn recovered by chickpea crop, ~41% is found in seeds. However, the recovery of applied Zn fertilizer is only 0.85–0.90%. The interaction effect of P × Zn is usually antagonistic. At higher rates of P fertilization, the availability of Zn in soil is reduced due to formation of relatively insoluble $ZnPO_4$. However, this theory has been questioned based on the fact that $ZnPO_4$ is as good a source of Zn as other sources. Similarly, antagonistic effects of Zn × S have also been reported.

Table 10.6. Genotypic behaviour of chickpeas to Zn fertilization on yield and water use of chickpea. (From Khan *et al*., 2004.)

	Shoot biomass (g/plant)			Water use efficiency (g/l)		
Genotype	–Zn	+Zn	Mean	–Zn	+Zn	Mean
Tyson	1.06	3.50	2.28	2.80	3.60	3.20
ICC 4958	2.33	3.92	3.10	3.47	3.72	3.60
T 1587	1.75	3.77	2.80	3.02	3.71	3.40
NIFA 88	1.26	3.07	2.20	2.47	3.13	2.80
Mean	1.60	3.60	–	2.90	3.50	–
CD (P = 0.05)	Genotypes	0.17			0.21	
	Zn	0.12			0.15	
	Interaction	0.24			NS	

Iron

Iron (Fe) deficiency is a complex physiological disorder of plants grown in calcareous soils with high pH that have low available Fe and inadequate uptake or because of impaired transport and metabolism of Fe induced by other ions. Fe deficiency is commonly noticed in South Asia, especially in calcareous vertisols, and West Asia and North Africa (WANA) region. Although mild Fe deficiency may not cause economic yield loss, severe deficiency may affect the crop yields considerably in susceptible cultivars. The yield reduction due to Fe deficiency was as high as 44% in Syria and Lebanon (Saxena and Sheldrake, 1980). Significant yield reductions have also been reported in India (Sakal *et al*., 1987). At the International Centre for Agricultural Research in Dry Areas (ICARDA) in Syria, the deficiency was more pronounced in winter-sown than spring-sown chickpeas. Similarly late-maturing varieties are more susceptible to Fe chlorosis than early cultivars. Waterlogging has been found to accentuate Fe chlorosis.

A typical Fe deficiency symptom is chlorosis of the youngest leaves, which become bleached and thin with acute deficiency, and often wither and fall off as deficiency advances. The crop deficiency symptoms are shown in Fig. 10.4. An Fe concentration of 80–85 ppm at 50 DAS in the top leaf is critical in chickpea. Soil application of Fe is usually uneconomical due to reversion to unavailable forms. Hence, foliar spray of 250 l of 1% ferrous salt has been found promising. However, responses to both soil and foliar fertilization of Fe in chickpea have been reported (Takkar *et al*., 1989) (Table 10.7). Foliar fertilization of Fe as 0.5 kg $FeSO_4$/ha has been found to enhance yield of chickpea by 450 kg/ha over control (Takkar and Nayyar, 1986), indicating an agronomic efficiency of 900 kg grain/kg Fe.

There have been efforts to tackle the Fe deficiency through a breeding approach since Fe susceptibility is governed by a single recessive gene. Hence, a negative selection breeding strategy for discarding susceptible lines has been adopted. This approach has successfully resolved Fe deficiency in chickpea (Saxena *et al*., 1990). Several tolerant chickpea genotypes (T 1, BD 93, Chaffa and G 20) have been identified for successful cultivation of chickpea in Fe-deficient soils in India. Irrigation-induced chlorosis was reported to be governed by two homozygous

Fig. 10.4. Genotypic variation in iron deficiency in chickpea.

recessive genes y_1 and y_2 (Gumber *et al.*, 1997). This negative selection approach has been effective in solving Fe deficiency in Mediterranean climates. However, it remains unaddressed in South Asia.

Boron

Boron (B) toxicity is a major concern in arid and semiarid regions of the world, particularly Australia (Hobson *et al.*, 2001). However, B deficiency has been reported in Indian states of Orissa, Bihar, Uttar Pradesh and Gujarat. Soils derived from granite, igneous, acidic and metamorphic sedimentation are often poor in B. Similarly, strongly weathered (acrisols and ferrasols), coarse-textured (arenosols) and shallow (lithosols) soils are also low in B status. The critical concentration of B in soils is 0.5 ppm, while the critical concentration in chickpea plants is 20–30 ppm in the top leaves at 50 DAS.

B-deficient leaves show severe chlorosis and bleaching, which is followed by tissue necrosis, and curling and drying of leaflets. There is a marked reduction in the number of flowers, which also lack pigmentation. The flower drop and failure in setting of pods in kabuli varieties prevalent in Chitwan,

Table 10.7. Grain yield of chickpea as influenced by mode and rate of iron fertilization. (From Takkar *et al.*, 1989.)

Mode of application	Rate	Grain yield (t/ha)
Soil	10 kg Fe/ha	1.4
	20 kg Fe/ha	1.7
Foliar	1% $FeSO_4$ solution	1.4
	1% $FeSO_4$ solution	1.4
Control	–	1.2
CD (P = 0.05)	–	0.3

Makwanpur and Nawalpuri districts of Nepal have been ascribed to B deficiency. B application at 0.5 kg/ha has been found effective in reducing flower drop and enhancing grain yield. An increase in grain yield up to 750 kg/ha due to 2 kg B/ha fertilization with an agronomic efficiency of 375 kg grain/kg B has been recorded (Sakal *et al.*, 1998). The optimum dose of B depends on soil B status. In soils with <0.35 ppm B, 3 kg B/ha was the optimum, whereas in soils with >0.35 ppm B, 2 kg B/ha was sufficient to meet the crop needs (Sakal *et al.*, 1990). In saline soils, higher amounts of B may hamper germination and growth of seedlings due to enhanced uptake of Na to the levels of toxicity with a corresponding decrease in K uptake. A decrease in B concentration in chickpea due to chloride salinity could also be noticed because of antagonistic effect of B with chlorine.

Cultivar differences to B deficiency have been observed, and in low B soils, cultivation of tolerant cultivars (DHG 82-12 and C 235) seems promising to susceptible (DHG 82-10) ones.

The boron toxicity symptoms in chickpea include yellowing or drying of the tips and margins of the leaves, with the older leaves being more severely affected than younger leaves (Wayne *et al.*, 2003). At high concentrations, even plant death is common. The toxicity of boron is dependent on soil moisture status. In years of sufficient moisture for plant growth, toxicity symptoms may not be evident as roots have access to ample water supply without extending roots into subsoils. However, in dry conditions, symptoms of toxicity may appear, as roots take up stored water from subsoils containing toxic levels of boron (Hobson *et al.*, 2001).

Molybdenum

The importance of molybdenum (Mo) in legumes in general and chickpea in particular arises from the fact that it is a constituent of nitrate reductase and nitrogenase enzymes, which play an important role in N fixation. The Mo availability is low in high clay soils (vertisols of Madhya Pradesh and Gujarat in India) and laterites, but high in saline and alkaline soils. Thus, both deficiency and toxicity of Mo are encountered. However, toxicity symptoms appear at an earlier stage (32 DAS) than deficiency symptoms (45 DAS). In Mo-deficient chickpea, the flowers produced are less in number, smaller in size and many of them fail to open or mature, consequently leading to lower seed yield. The deficiency, threshold of deficiency and threshold of toxicity values for chickpea leaves are 0.38, 1.2 and 15 μg/g, respectively. Substantial yield losses (20%) due to omission of Mo from fertilization have been observed in grey flood plain soils at Bogra, Bangladesh (Islam *et al.*, 1995).

The beneficial effects of soil application of 1 kg Mo/ha in Bangladesh (Islam *et al.*, 1995), seed treatment with 3.5 g sodium molybdate in India (Deo and Kothari, 2002) and foliar fertilization of 15–30 ppm Mo in Egypt (Sawries, 2001; Shetaia and Soheir, 2001) have been reported. Yield enhancements were higher with seed treatment (16.76%) than with soil application (13.13%). The response to Mo was greater when applied along with P and *Rhizobium* inoculation (Mudholkar and Ahlawat, 1979).

Manganese

Soils with 1 ppm DTPA-extractable manganese (Mn) may support crop growth successfully. In coarse-textured soils where rice is grown continuously, Mn gets leached, and crops after rice usually experience Mn deficiency. In India, Mn deficiency is widespread in soils of Karnataka.

Chickpea shows Mn deficiency symptoms at 6 weeks after sowing as chlorosis of leaflets in apical margins that curl inwards. Leaf size is reduced with young leaves exhibiting purplish discoloration. The terminal buds show withering and necrosis.

In general, soil application of 20 kg Mn/ha or foliar spray of 1–2.5 kg Mn/ha may mitigate the deficiency of Mn. Soil application needs larger quantities of fertilizer, which makes it uneconomical. It is also reversed to unavailable forms when applied to soil. Hence, foliar spray of 0.5% $MnSO_4$ is recommended. Genotypic differences for Mn deficiency have also been observed (Abdul Rashid *et al.*, 1990: Nayyar *et al.*, 1990). In Mn-deficient genotypes, adventitious roots may develop in place of tap roots, and this could be exploited for overcoming the deficiency problems of Mn.

Copper

Copper (Cu) deficiency has been reported in newly cleared soils of South-west Australia with neutral to alkaline pH. Isolated cases of Cu deficiency in India have also been reported in pulse-growing regions of Karnataka and Gujarat. The critical soil Cu concentration is 0.2 ppm.

In Cu-deficient chickpea, young leaves at 5 weeks after sowing show paling and twisting. Plants are more flaccid and appear as wilted even with adequate water supply. Male flower sterility, delayed flowering and senescence are other most important symptoms of Cu deficiency. In youngest tissue of spring-sown chickpea, a critical concentration of 2.6 ppm has been reported (Brennan and Bolland, 2003), while in youngest mature leaves it was 4.3 ppm, which is adequate for growth (Weir and Cresswell, 1982). Toxicity of Cu in soil may inhibit nodulation, lower protein content and yields of chickpea. In coarse-textured and problematic soils, application of 5–10 kg $CuSO_4$/ha or foliar spray of 0.5–1.0 kg $CuSO_4$/ha on soil has been found promising for improving chickpea productivity (Ahlawat, 1990). Application of relatively high levels of N and P fertilizers may induce Cu deficiency in plants grown in low Cu soils.

Nutrient Management in Cropping Systems

Chickpea, mainly a winter crop, is also grown in autumn and spring seasons in different parts of the world in various sequential and intercropping systems. The residual effect of previous crops and differential requirement of nutrients of component crops demand efficient management for higher system productivity. The management requirements of cropping systems are discussed below.

In the Mediterranean, a 2-year rotation of wheat–fallow was less economical in both yield and soil organic matter than a wheat–chickpea rotation (Table 10.8). The relative wheat grain yields, compared with yield in the wheat–fallow rotation, were 70% and 45% following chickpea and wheat, respectively. The wheat yields in fallow–wheat rotation were obtained only once in 2 years (ICARDA, 1997). Green manuring of chickpea is usually beneficial in the succeeding rice crop than N fertilization alone. Stover recycling from crops in chickpea–maize cropping system was found to economize 50% of the recommended dose of NPK fertilizers (Ahlawat *et al.*, 2005). Similarly, P fertilization of chickpea in a rice–chickpea system may affect an economy of 20–30 kg P_2O_5/ha in succeeding rice over its fertilization to rice. In sandy soils of Jharkhand in India, K fertilization to both rice and chickpea enhances system productivity over fertilization of both the crops. The recommended dose of ferilizers in rice–chickpea and urdbean–chickpea sequences is usually essential for higher system productivity (Sharma *et al.*, 1992). However, in a soybean–chickpea sequence (Singh *et al.*, 1993), 50% fertilizer economy (10–25–0 Kg/ha N–P_2O_5–K_2O) may be affected without any reduction in productivity, when soybean receives 100% recommended fertilizers (20–80–20 kg/ha N–P_2O_5–K_2O). N fertilizer economy up to 60–70 kg/ha in succeeding maize (Ahlawat *et al.*, 1981) and 40 kg/ha in pearl millet (De and Gautam, 1987) after chickpea has been reported possibly due to sparing effect coupled with BNF. The soil N balances (crop N–grain N) of chickpea were positive (80–135 kg/ha), whereas those of wheat were negative (from −20 to −66 kg/ha) (Turfin *et al.*, 2002), which shows the cause for beneficial effect of chickpea on succeeding cereals compared with a rotation of cereal–cereal. Legume roots add substantially to N recycling. It was found that chickpea roots account for 29% of total plant N (Turfin *et al.*, 2002).

Intercropping of chickpea with cereals like wheat, barley and *Brassicas*, as well as linseed and safflower under rainfed condition, is common in South Asia, especially in India. The component crops may require higher N fertilizers (40–60 kg N/ha). However, higher initial fertilizer N up to flowering could be detrimental for BNF. Hence, split application of N fertilizer (20 kg/ha as basal and rest at flowering stage of crops) has been recommended to take the maximum benefit from BNF. In chickpea–wheat intercropping with various row proportions, the fertilizer dose for wheat (40–50–20 kg N–P_2O_5–K_2O) was found to be sufficient for both the crops. Similarly, chickpea grown as intercrop in autumn with sugarcane usually does not require additional fertilizers over the sole crop of sugarcane.

References

Ae, N., Arihara, J. and Okada, K. (1991) Phosphorus response of chickpea and evaluation of phosphorus availability in Indian Alfisols and Vertisols. In: Johansen, C., Lee, K.K. and Sahrawat, K.L. (eds) *Phosphorus Nutrition in Grain Legumes in the Semi-arid Tropics*. ICRISAT, Hyderabad, India, pp. 33–41.

Ahlawat, I.P.S. (1990) Diagnosis and alleviation of mineral nutrient constraints in chickpea. In: *Chickpea in the Nineties: Proceedings of the 2nd International Workshop on Chickpea Improvement*. ICRISAT, Hyderabad, India, pp. 93–99.

Ahlawat, I.P.S., Singh, A. and Saraf, C.S. (1981) Effects of winter legumes on the nitrogen economy and productivity of succeeding cereals. *Experimental Agriculture* 17, 57–62.

Ahlawat, I.P.S., Gangaiah, B. and Ompal Singh (2005) Effect of fertilizer and stover management on productivity and soil fertility in chickpea (*Cicer arietinum*) – maize (*Zea mays*) cropping system. *The Indian Journal of Agricultural Sciences* 75, 400–403.

AICRP (All India Co-ordinated Research Project) (2003) Annual Report, All India Co-ordinated Research Project on improvement of pulses. Indian Institute of Pulses Research, Kanpur, India.

Aini, N. and Tang, C. (1998) Diagnosis of potassium deficiency in faba bean and chickpea by plant analysis. *Australian Journal of Experimental Agriculture* 38, 503–509.

Ali, M.Y., Krishnamurthy, L., Saxena, N.P., Rupela, O.P., Jagadish Kumar, and Johansen, C. (2002) Scope for genetic manipulation of mineral acquisitions in chickpea. *Plant and Soil* 245, 123–134.

Alloush, G.A., Zeto, S.K. and Clark, R.B. (2000) Phosphorus source, organic matter and arbuscular mycorrhiza effects on growth and mineral acquisition of chickpea grown in acidic soil. *Journal of Plant Nutrition* 23, 1351–1369.

Aulakh, M.S., Sidhu, B.S., Arora, B.R. and Singh, B. (1985) Content and uptake of nutrients by pulses and oilseed crops. *Indian Journal of Ecology* 12, 238–242.

Bharud, R.W. (2001) Effect of foliar application of NAA and urea on growth and yield of chickpea. *National Symposium on Pulses for Sustainable Agriculture and Nutritional Security*. Indian Society of Pulses Research and Development, New Delhi, pp. 35–36.

Bhattacharya, A. and Ali, M. (2002) Physiological traits and seed yield in chickpea under high nitrogen input. *Legume Research* 25, 9–14.

Brennan, R.F. and Bolland, M.D.A. (2003) Comparing copper requirements of faba bean, chickpea and lentil with spring wheat. *Journal of Plant Nutrition* 26, 883–889.

De, R. and Gautam, R.C. (1987) Management practices to increase and stabilize pearl millet production in India. In: Witcombe, J.R. (ed.) *Proceedings of the International Pearlmillet Workshop*. ICRISAT, Hyderabad, India, pp. 247–253.

Deo, C. and Kothari, M.L., (2002) Effect of modes and levels of molybdenum application on grain yield, protein content and nodulation of chickpea grown on loamy sand soil. *Communications in Soil Science and Plant Analysis* 33, 2905–2915.

Din, J., Rashid, A. and Ahmed, J. (1999) Diagnosis and correction of phosphorus deficiency in chickpea grown in rainfed calcareous soils of Pakistan. *Journal of Indian Society of Soil Science* 47, 504–509.

Dravid, M.S. and Goswami, N.N. (1987) Effect of salinity, fertility and variety on growth and nutrient uptake by chickpea (*Cicer arietinum* L.). *Journal of Nuclear Agriculture and Biology* 16, 69–72.

Dulal, R. (1976) Inventory of the major soils of the world with special reference to mineral stresses. In: Wright, M.J. (ed.) *Plant Adaptation to Mineral Stresses in Problem Soils*. Cornell University Agricultural Experiment Station, Ithaca, New York, p. 3.

El-Hadi, E.A. and El-Sheikh, E.A.E. (1999) Effect of Rhizobial inoculation and nitrogen fertilization on yield and protein content of six chickpea (*Cicer arietinum* L.) cultivars in marginal soils under irrigation. *Nutrient Cycling in Agroecosystems* 54, 57–63.

Foster, N.R. (1999) Calcium nitrate to control collar rot during chickpea germination. *Seed Technology* 21, 41–43.

Gettier, S.W., Martens, D.C. and Brumback, T.B., Jr. (1985) Timing of foliar manganese application for correction of manganese deficiency in soybean. *Agronomy Journal* 77, 627–630.

Gumber, R.K., Singh, S., Gill, J.S. and Rathore, P.K. (1997) Genetics of irrigation induced iron chlorosis in chickpea. *International Pigeonpea and Chickpea Newsletter* 4, 10–11.

Hawthrone, W., Hannay, J. and Heinjus D. (2003) *Growing Chickpeas*. Fact sheet (FS No. 20/99). www.pir.sa.gov.au/factsheets

Hobson, K.B., Seymour, L.R., Armstrong, R.D., Connor, D.J., Brand, J.D. and Materne, M. (2001) Improving the adaptation, profitability and reliability of pulses growing in hostile alkaline sub soils. http://www.regional.org.au/asa/2001/p/3/hobson.htm.

ICARDA (International Centre for Agricultural Research in the Dry Areas) (1997) Annual Report. ICARDA, Aleppo, Syria, pp. 39–40.

Islam, M.A., Saha, U.K., Islam, M.S. and Saha, R.R. (1995) Fertilizing chickpea for yield with special emphasis on boron and molybdenum in grey flood plain soils of Bangladesh. *Annals of Bangladesh Agriculture* 5, 43–49.

Kay, D.E. (1979) Chickpea (*Cicer arierinum*). In: *Crop Production Digest No. 3, Food Legumes*. Tropical Products Institute, London, pp. 48–71.

Khan, H.R., McDonald, G.K. and Rengel, Z. (2003) Zinc fertilization improves water use efficiency, grain yield and seed Zn content in chickpea. *Plant and Soil* 249, 349–400.

Khan, H.R., McDonald, G.K. and Rengel, Z. (2004) Zinc fertilization and water stress affects plant water relations, stomatal conductance and osmotic adjustments in chickpea (*Cicer arietinum* L.). *Plant and Soil* 267, 271–284.

Kumar, D. (1985) Emergence, establishment and seed yield of chickpea as affected by sodicity. *Annals of Arid Zone* 24, 334–340.

Kurdali, F. (1996) Nitrogen and phosphorus assimilation, mobilization and partitioning in rainfed chickpea (*Cicer arietinum* L.). *Field Crops Research* 47, 81–92.

McConnell, J.T., Miller, P.R., Lawrence, R.L., Engel, R. and Nielsen, G.A. (2002) Managing inoculation failure of field pea and chickpea based on spectral responses. *Canadian Journal of Plant Science* 82, 273–282.

Meshram, A.T., Jadhav, A.C., Konde, B.K. and Wani, P.V. (1999) Effect of VAM fungi and P sources on nutrient uptake and yield of chickpea. *Journal of Maharashtra Agricultural Universities* 24, 340–341.

Misra, S.K. (2001) Salt tolerance of different cultivars of chickpea as influenced by zinc fertilization. *Journal of the Indian Society of soil Science* 49, 135–140.

Misra, S.K., Upadhyay, R.M. and Tiwari, V.N. (2002) Effect of salt and zinc on nodulation leghaemoglobin and nitrogen content of rabi legumes. *Indian Journal of Pulses Research* 15, 145–148.

Mudholkar, N.J. and Ahlawat, I.P.S. (1979) Response of bengal gram to nitrogen, phosphorus and molybdenum. *Indian Journal of Agronomy* 24, 61–65.

Muehlbauer, F.J. and Singh, K.B. (1987) Genetics of chickpea. In: Saxena, M.C. and Singh, K.B. (eds) *The Chickpea*. CAB International, Wallingford, UK, pp. 99–126.

Nayyar, V.K., Takkar, P.N., Bansal, R.L., Singh, S.P., Kaura, N.P. and Sadana, U.S. (1990) *Micronutrients in Soils and Crops of Punjab*. Research Bulletin, Department of Soils, Punjab Agricultural University, Ludhiana, India.

Palta, J.A., Nandwal, A.S., Kumari, S. and Turner, N.C. (2005) Foliar nitrogen applications increase the seed yield and protein content in chickpea (*Cicer arietinum* L.) subject to terminal drought. *Australian Journal of Agricultural Research* 56, 105–112.

Pieri, C. and Oliver, R. (1986) In: Nutrient balances and the need for potassium. *Proceedings of 13th Congress International Potash Institute*, Bern, Switzerland. pp. 73–92.

Pramanik, K. and Singh, R.K. (2004) Effect of levels and mode of phosphorus and biofertilizers on chickpea (*Cicer arietinum*) under dryland conditions. *Indian Journal of Agronomy* 48, 294–296.

Prasad, R. and Subbiah, B.V. (1982) Nitrogen, the key plant nutrient in Indian Agriculture. *Fertilizer News* 27, 27–42.

Prasad, R., Sharma, S.N., Singh, S. and Shivay, Y.S. (2002) Relative productivity, profitability and NPK removal from soil by some rice based cropping systems. *Journal of Sustainable Agriculture* 19, 31–37.

Rashid, A., Jalal-Ud-Din and Bashir, M. (1990) Manganese deficiency in chickpea grown on a calcareous soil of Pakistan. *International Chickpea Newsletter* 22, 21–22.

Rattan, R.K., Datta, S.P., Saharan, N. and Katyal, J.C. (1997) Zinc in Indian agriculture: a look forward. *Fertilizer News* 42, 75–89.

Rawsthorne, S., Hadley, P., Roberts, E.H. and Summerfield, R.J. (1985) Effects of supplemental nitrate and thermal regime on the nitrogen nutrition of chickpea (*Cicer arietinum* L.). I. Growth and development. *Plant and Soil* 83, 265–277.

Saeed, M., Akram, H.M., Iqbal, M.S., Allah, Y. and Abbas, A. (2004) Impact of fertilizer on the seed yields of chickpea genotypes.

International Journal of Agriculture and Biology 6, 108–109.
Sakal, R., Sinha, R.B. and Singh, A.P. (1987) Response of chickpea to iron application in a calcareous soil. *International Chickpea Newsletter* 16, 14–15.
Sakal, R., Sinha, R.B. and Singh, A.P. (1990) Response of chickpea to boron application in calcareous soil. *International Chickpea and Pigeonpea Newsletter* 22, 28–29.
Sakal, R., Sinha, R.B., Singh, A.P. and Bhogal, N.S. (1998) Responses of some *rabi* pulses to B, Zn and S application in farmers fields. *Fertilizer News* 43, 39–43.
Sawries, E.S. (2001) Effect of phosphorus fertilization and micronutrients on yield and yield components of chickpea (*Cicer arietinum* L.). *Annals of Agricultural Science* (Cairo) 46, 155–164.
Saxena, M.C. and Sheldrake, A.R. (1980) Iron chlorosis in chickpea (*Cicer arietinum* L.) grown on high pH calcareous vertisols. *Field Crops Research* 3, 211–214.
Saxena, N.P. (1984) In: Goldswothy, P.R. and Fisher, N.M. (eds) The chickpeas. *The Physiology of Tropical Field Crops*. Wiley, Chichester, UK, pp. 419–452.
Saxena, M.C., Mahrotra, R.S. and Singh, A.K. (1990) Iron deficiencies in chickpea in the Mediterranean region and its control through resistant genotypes and nutrient application. *Plant and Soil* 123, 251–254.
Sekhar, D.M.R. and Aery, N.C. (2001) Phosphate rock with farm yard manure as P fertilizer in neutral and weakly alkaline soils. *Current Science* 80, 1113–1115.
Sharma, H.M., Roysharma, R.P. and Mishra, S.S. (1992) Fertilizer management in legumes and cereal based cropping system. *Indian Journal of Pulses Research* 5, 40–44.
Shetaia, A.M. and Soheir, A.M. (2001) Yield and yield components responses of chickpea (*Cicer arietinum* L.) to phosphorus fertilization and micronutrients. *Arab Universities Journal of Agricultural Sciences* 9, 235–243.
Shinde V.S. and Saraf, C.S. (1992) Sulphur and phosphorus interactions in the presence of phosphate solubilizing bacteria in chickpea. *International Chickpea Newsletter* 26, 27–29.
Shukla, U.C. and Prasad, K.G. 1979. Sulphur–zinc interaction in groundnut. *Journal of the Indian Society of Soil Science* 27, 60–64.
Singh, S. (1995) Effect of P application methods on yield of chickpea (*Cicer arietinum* L.) and P utilization. *Journal of Nuclear Agriculture and Biology* 24, 210–214.
Singh, K., Billore, S.D. and Bargale, M. (1993) Economy in fertilizer and energy use through double cropping soybean (*Glycine max*)-chickpea (*Cicer arietinum*). *Indian Journal of Agronomy* 38, 630–632.
Singh, J., Bajaj, J.C. and Pathak, H. (2005) Quantitative estimation of fertilizer requirement for maize and chickpea in the alluvial soils of the Indo-Gangetic plains. *Journal of the Indian Society of Soil Science* 53, 101–106.
Singh, K., Gumber, R.K. and Singh, S. (2001) Exploring chickpea as intercrop in sugarcane. *International Chickpea and Pigeonpea Newsletter* 8, 21–22.
Singh, U., Patil, S.K., Das, R.O., Padila, J.L., Singh, V.P. and Pal, A.R. (1999) Nitrogen dynamics and crop growth on an alfisol and a vertisol under rainfed lowland rice-based cropping system. *Field Crops Research* 61, 237–252.
Subbarao, A. (1993) Analysis for soils for available major nutrients. In: Tandoan, H.L.S. (ed.) *Methods of Analysis of Soils, Plants, Waters and Fertilizers*. Fertilizer Development and Consultation Organization, New Delhi, pp. 13–35.
Swahney, J.S., Amarjeet and Singh, R. (1985) Effect of applied nitrate on enzymes of ammonia assimilation in nodules of *Cicer arietinum* L. *Plant and Soil* 86, 241–248.
Sysoev, Yu A. (1977) Effect of fertilizers on yield of chickpea. *Nauchnye-Trudy-Stavropolskii-Selskokhozyaistvennyi-Institut* 40, 19–23.
Takkar, P.N. and Nayyar, V.K. (1986) Integrated approach to combat micronutrient deficiency. In: *Proceedings of FAI Seminar on Growth and Modernization of Fertilizer Industry, PS II/2*. Fertilizer Association of India, New Delhi, pp. 1–16.
Takkar, P.N., Chhibba, I.M. and Mehta, S.K. (1989) *Twenty Years of Co-ordinated Research on Micronutrients in Soils and*

Plants. Bulletin I. Indian Institute of Soil Science, Bhopal, India.

Tandon, H.L.S. (1987) *Phosphorus Research and Agricultural Production in India*. Fertilizer Development and Consultation Organization, New Delhi.

Tandon, H.L.S. (1991) *Secondary andMmicro-nutrients in Agriculture: A Guide Book-cum-Directory*. Fertilizer Development and Control Organization, New Delhi.

Tiwari, K.N. (1988) Potassium in Soils and Crop Response to potassium Application in U.P. Annual Report, PRII Sponsored Project, Kanpur. Department of Soils and Agricultural Chemistry, Chandra Sekhar Azad University of Agriculture and Technology, Kanpur, India.

Turfin, J.E., Herridge, F. and Robertson, M.J. (2002) Nitrogen fixation and soil nitrate interactions in field-grown chickpea (*Cicer arietinum*) and faba bean (*Vicia faba*). *Australian Journal of Agricultural Research* 53, 599–608.

Voss, M., Parra, M.S. and Campos, A.D. (1998) Yield of commonbean and chickpea as a function of soluble phosphate applied in pits or in the seeding row. *Revista-Brasileira-de-Ciencia-do-Solo* 22, 163–168.

Walley, F.L., Kyei-Boahen, S., Hnatowich, G. and Stevenson, C. (2005) Nitrogen and phosphorus fertility management for desi and kabuli chickpea. *Canadian Journal of Plant Sciences* 85, 73–79.

Weir, R.G. and Cresswell, G.C. (1982) *Plant Nutrient Disorders*, Vol. 4: *Pastures and Field Crops*. Inkata Press, Sydney, Australia.

11 Weed Management in Chickpea

J.P. Yenish

Extension Weed Scientist, Washington State University, Pullman, Washington, USA

Introduction

Chickpea competes very poorly with weeds. Yield reductions of 23–87% due to competition from weeds have been shown to occur in chickpea crops (Bhan and Kukula, 1987). Mohammadi *et al.* (2005) estimated 10% reduction in chickpea seed yield for every additional 26 g/m^2 of weed dry weight (Fig. 11.1). Additional losses due to weeds are often seen with reduced harvest efficiency and reduced crop quality (McKay *et al.*, 2002). Yield loss will vary due to the intensity of infestations and varying species of weeds between locations (Bhan and Kukula, 1987). However, regardless of the environment in which it is grown chickpea will likely be the most poorly competitive crop in the rotation system.

Generally, weeds that emerge before, or at the time of, crop emergence have a greater competitive advantage than those emerging after the crop (O'Donovan *et al.*, 1985; Dieleman *et al.*, 1995; Knezevic *et al.*, 1995; Bosnic and Swanton, 1997). Slow emergence, short plant height and late canopy cover of cool season legumes in general, and chickpea specifically, allow weeds to compete effectively against these crops. A study by Moes and Domitruk (1995) showed average emergence for chickpea, lentils, dry peas and wheat of 23, 21, 18 and 17 days after planting, respectively, in a no-tillage system. Moreover, the same study showed average canopy closure of 69, 62, 56 and 53 days after planting for the same respective species. In a seeding study conducted for more than 3 years in Lebanon, emergence for November seeding dates ranged from 44 to 72 days after planting, whereas date of emergence for January and February seeding dates ranged from 30 to 48 and 23 to 32 days after planting, respectively (Yau, 2005). Slower emergence of the crop provides opportunity for weeds to germinate and establish without suppression by the crop. The earlier the weeds emerge relative to the crop, the greater the crop yield loss if they are left uncontrolled (Kropff *et al.*, 1992). Furthermore, delayed canopy closure

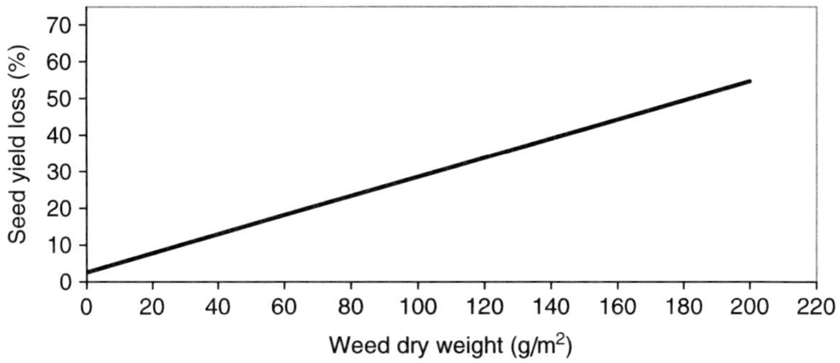

Fig. 11.1. Relationship between weed dry weight and percentage chickpea yield loss for average of Tabriz and Kermanshah, Iran, in 2002 and 2003, respectively. Linear regression model, $y = 2.67 + 0.2599x$, $R^2 = 0.65$. (After Mohammadi *et al.*, 2005.)

exposes the soil surface to sunlight for a longer period allowing additional weed germination later into the season. Later emerging weeds will compete with the crop as it produces seed and can directly reduce seed size and quality. Seed quality and yield may also be reduced indirectly by weeds interfering with the harvest of the crop. Heavy wet immature weed vegetation may interfere with mechanical or hand harvest of chickpea and weeds containing latex, such as prickly lettuce (*Lactuca serriola*), which may cause further problems as the latex disrupts normal machinery operation. Delayed emergence and canopy closure can be overcome to a certain extent by increasing the crop seeding rate. However, crop seed size is important in chickpea, and increasing seeding rate to increase crop competitiveness with weeds may be detrimental to harvested crop quality (Gan *et al.*, 2002; Yau, 2005). The importance of seed size varies between chickpea markets, and the impact of seeding rate on seed size varies with cropping systems.

The critical weed-free period is defined as the period of crop growth during which the crop must be kept weed-free to prevent yield loss due to weed interference (Weaver and Tan, 1987; Van Acker *et al.*, 1993). Mohammadi *et al.* (2005) indicated that in two locations in Iran, emergence of chickpea occurred at 5 and 8 days after planting. Weed-removal studies by these researchers indicated that the critical weed-free period was from 48 to 49 days after emergence (Fig. 11.2). A study in Tunisia estimated the critical weed-free period for chickpea at 10 weeks after emergence, for a location with low to medium severity of weed infestation, and 4 weeks for a separate location where the infestation was described as severe (Knott and Halila, 1986). In the Iranian study, the critical weed-free period lasted until the crop was in the early to full flowering stage of growth (Mohammadi *et al.*, 2005). In reality, the critical weed-free period is an estimate and will vary with environment (Table 11.1).

In Australia, the percentage of chickpea loss due to weed stand density was shown to follow a rectangular hyperbolic model (Fig. 11.3) (Whish *et al.*, 2002). There were slight differences in the yield loss response curve due to

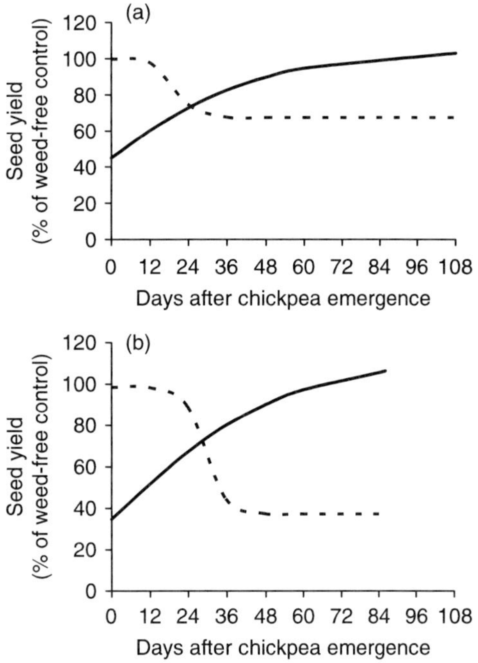

Fig. 11.2. Chickpea seed yield response to increasing length of weed-free period (—) and duration of weed-infested period (---) in days after chickpea emergence at (a) Tabriz and (b) Kermanshah, Iran, in 2002 and 2003, respectively. (After Mohammadi *et al.*, 2005.)

chickpea row spacing and variability between experiment locations. Moreover, the yield loss response curve differed when chickpea was grown in competition with a broadleaf (*Rapistrum rugosum*) or a grass (*Avena sterilis* ssp. *ludoviciana*) species. However, yield loss due to weed stand density did follow the rectangular hyperbolic model described by Cousens (1985a,b).

Table 11.1. Critical period of weed interference in chickpea obtained by various researchers.

Authors	Start of critical period Days after crop emergence	End of critical period
Mohammadi *et al.* (2005)	17	49
Al-Thahabi *et al.* (1994)	35	49
Ali (1993)	0	56
Bhan and Kukula (1987)	30	60
Ahlawat *et al.* (1981)	28	42
Saxena *et al.* (1976)	30	60
Mean	25	53

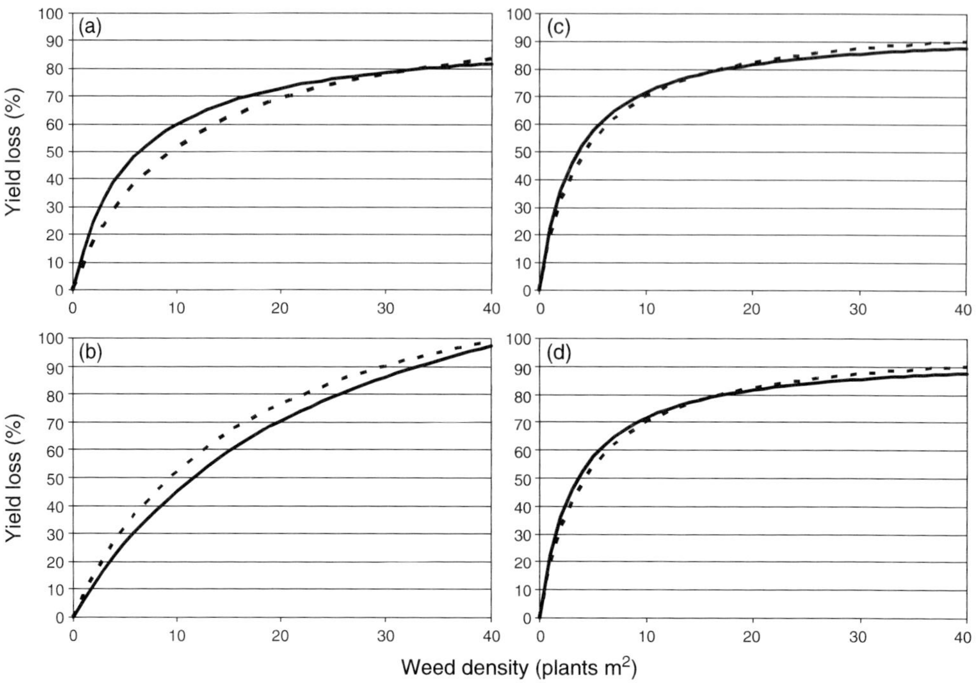

Fig. 11.3. Hyperbolic curves of chickpea grain yield loss fitted to the weed density and row spacing data. Letters denote treatments at two locations in Australia: (a) wide (---) and narrow (—) rowed chickpea under competition with *Rapistrum rugosum* at Tamworth, New South Wales; (b) wide and narrow rows under competition with *Avena sterilis* at Tamworth; (c) wide and narrow rows under competition with *R. rugosum* at Warialda, New South Wales; and (d) wide and narrow rows under competition with *A. sterilis* at Warialda. (From Whish *et al.*, 2002.)

Problem Weeds

The species of weeds that commonly infest chickpea fields vary with changing production environments. Soil characteristics, moisture amount, precipitation pattern, crop rotation, temperature, latitude, altitude, fertility, weed control technology and other factors interact to determine weed flora and intensity.

As with most weeds in a particular crop, weeds affecting chickpea have a similar ecology and biology. Generally, cool-season broadleaf weeds are the most difficult to control in chickpea. Problematic annual broadleaf species in chickpea include *Brassicacae, Asteraceae, Chenopodiaceae, Fabaceae* and *Polygonaceae,* among other families (Bhan and Kukula, 1987). Although many annual grass species can be selectively and effectively controlled with herbicides, there are some grass (*Poaceae*) species commonly reported in chickpea that are resistant, e.g. *Phalaris* and *Avena* among others. Generally, most grass weeds that are present in chickpea are classified as cool-season species. These types of grass species are often categorized as C3 plants based on their photosynthetic system. Additionally, perennial weeds can be a problem in chickpea production systems.

Perennial species of weeds that have been reported to occur include those of the *Polygonaceae*, *Convolvulaceae*, *Asteraceae*, *Poaceae* and other families.

Typically, weeds that cause problems in chickpea are problematic within a crop rotation and not to the chickpea crop alone. An exception to this is parasitic plants, which are partially or completely obligate to the chickpea crop. *Orobanche* and *Cuscuta* spp. have been reported to be parasitic weeds of chickpea (Cubero *et al.*, 1986). These weeds directly draw nutrients from growing chickpea plants, reducing yield and seed quality. Typically, seeds of these parasitic species can lie dormant for long periods of time until induced to germinate by host plant exudates (ter Borg, 1986). However, in the case of *Cuscuta* spp., the long soil seed life is largely due to its hard seed coat, which requires scarification by biotic or abiotic means prior to seed germination, rather than in response to a host plant stimulus (Benvenuti *et al.*, 2005). Control of parasitic plants in chickpea is a particular challenge since the plants not only have similar growth habit, but also coexist at some point in their life cycle.

Methods of Weed Control

There are five recognized weed control techniques (Anderson, 1983): preventive, cultural, mechanical, chemical and biological. Mixing two or more of these weed management principles is the basis for integrated weed management. Each of the five principles is equally good at managing weeds in chickpea with the exception of biological control. Few biological methods of weed control have proven to be effective in annual cropping systems, and individual weed species that are problematic in chickpea are largely a function of cropping systems or rotation rather than of the individual crop of chickpea.

Preventive weed control

The most cost-effective form of weed control is the preventive technique. A quote attributed to the 13th-century English judge Henry de Bracton, 'an ounce of prevention is worth a pound of cure', is very true in the realm of weed management. The definition of preventive weed management is to eliminate the introduction or spread of specified weed species in an area not currently infested with these plant species (Radosevich *et al.*, 1997). The area may vary from individual fields to entire nations. Although much of preventive weed control could be regulatory in nature, there is still much that an individual farmer can do to prevent the introduction of weeds. In fact, individual farmer practices are more likely to be effective in preventing weeds than any regulatory practices.

Individual farmers concentrating on preventing weed introduction often only need to plant weed-free seed; use manure generated from weed-free hay, feed or bedding; clean field equipment, particularly harvest equipment; and prevent weeds that are encroaching on their property from spreading seed or other vegetative propagules onto their land (Young *et al.*, 2000). Generally, weed-free seed is easy to obtain, due to the relatively large seed size. Cleaning may be done after seed purchase if clean seed is in short supply. Few weed seeds are as large as developed chickpea varieties, particularly kabuli (coloured seed coat) types. However, farmers need to adequately destroy weed seeds during screenings through composting, hammer mill or other methods. Livestock manure can also be a source of weed seed entering a farming operation. Feed and hay should be inspected prior to feeding or mixing. Even if feed, hay and bedding are found to be weed-free, manure should be composted to further eliminate weed seed through the temperatures generated. Temperature as high as 83°C may be required to adequately reduce weed seed viability, although some species may lose viability at lower temperatures (Wiese *et al.*, 1998). Acquired animals should also be quarantined to allow the digestive system to be completely free of foreign feed. Additionally, hair, wool, hides and hooves should also be inspected to make sure no weed seed is trapped either directly or encased in mud or other foreign matter.

Equipment should be cleaned prior to removal from, and entry into, fields to prevent weed dispersal (Anderson, 1983). Methods for cleaning equipment may be as simple as hand removal of seeds, rhizomes, etc., or as high-tech as radiation treatment. Special attention needs to be paid to harvest equipment since weed seed matures at the same time as the crop, and modern harvest equipment is an effective means to disperse weed seed within and between fields. In cases where residue is removed from crop fields for the purpose of offsite separation of grain and stover, the farmer can dispose that residue at the processing site rather than transport it back to his own fields for disposal. Mixing of stover from multiple fields and then returning a portion to each field will result in the spread of weeds.

Irrigation water is also a common source of weed seed influx. Kelley and Burns (1975) in a detailed study of the Pacific North-west's Columbia River calculated that as much as 15,000 seed/ha could be disseminated onto fields by unscreened irrigation waters. Incoming irrigation water should be screened or other apparatus used to remove weed seed from the water prior to applying it to a field.

Controlling encroaching populations of weeds is also important in preventing new weed infestations. Most weed seed falls <1 m from the mother plant (Radosevich *et al.*, 1997). Thus, patches of weeds tend to creep across a given area if allowed to disperse naturally. At the same time, long-range dispersal of weed seeds is important in the colonization of new weed patches. Therefore, the nearest patches of weeds that are encroaching on a particular area are the most threatening to the crops in the adjacent area, while weed seeds travelling long distances are a potential future problem after they colonize into weed patches.

Preventive weed control methods could also be applied to weed species that are already present in fields within the farming operation. This is particularly important in regions where chemical weed control is a large component of weed management programmes. Preventing external influxes of weed seed is important to keep herbicide-resistant weed biotypes out of the farming operation. Moreover, preventive weed management could isolate weed infestation to one field or a portion of one field within a grower operation. In areas with high herbicide use, a weed seed in screenings means a seed contaminant probably survived a herbicide application to the field in which it was produced. Therefore, there is a greater potential for that seed to be resistant to the herbicide than the wild type. Although the mother plant may have emerged after the herbicide application due to poor application or equipment problems, there is no increased probability that the plant is herbicide-resistant. However, the potential exists to bring herbicide resistance into the farming operation by using screening methods or contaminated seed, or to move resistant weeds within and between fields.

Cultural weed control

Cultural methods of weed management utilize practices common to good crop management (Smith and Martin, 1995). The goal is to manage a crop or cropping system to maximize competition with weeds. In other words, make the crop as healthy and competitive as possible to lessen or better tolerate weed competition.

Tools for cultural weed control include competitive varieties, timeliness of planting, cover crops, competitive rotational crops or rotational crops of different life cycles. Planting crops of differing life cycles prevents population increases of any single species of weed (Radosevich *et al.*, 1997). Additionally, practices such as optimal fertilizer placement and timing, optimal irrigation timing and other input timing or placement benefit the crop at the expense of the weeds (Di Tomaso, 1995). Cultural practices are generally applied within, between and throughout the crop rotation. Unfortunately, chickpea is usually the weakest competitor in the rotations in which they are grown. Typically, cultural weed control in rotational crops keeps weed populations low, and benefits are realized in the yield and quality of chickpea. Chickpea provides little or no opportunity for cultural weed control within the rotation. Ultimately, the value of chickpea as a cultural weed control tool within a crop rotation is directly related to differences in the ecology and biology between chickpea and other crops in the rotation.

The primary benefit to chickpea weed control due to rotation is in the management of parasitic plants. However, with the seeds of *Orobanche* and *Cuscuta* spp. that are able to lie dormant for 14 (Lopez-Granados and Garcia-Torres, 1999) and 60 years (British Columbia Ministry of Agriculture, Food, and Fisheries, 2002), respectively, the need for such a long period between chickpea and other susceptible crops is impractical.

There are several cultural practices commonly involved with chickpea production that make them more prone to losses from weed competition. Fall planting of chickpea in the Mediterranean and other regions tends to result in greater losses due to weeds, because of slower crop growth and development, than with spring planting (Yau, 2005). Moreover, weed competition may be equal or greater from winter annual and cold weather–tolerant weeds than from spring annual weeds in a spring-seeded crop. While winter chickpea production may provide a yield advantage and an opportunity to relay crops within a year, weed competition has all but prevented fall sowing of chickpea.

Seed size is an important quality component of chickpea. Decreasing chickpea plant density is a cultural practice that helps ensure the largest seed possible under growing conditions (Gan *et al.*, 2002; Yau, 2005). Decreasing the density of chickpea plants may also reduce crop disease due to a more open canopy, which reduces humidity in and around chickpea foliage (Krupinsky *et al.*, 2002). However, decreasing plant density of an already poor competitor like chickpea makes losses due to weeds even worse and requires greater intensity of weed control efforts.

Rotational crops in chickpea production systems vary greatly within and between production areas. In some areas of the Indian subcontinent, chickpea may be grown in rotation with millet, sorghum, maize, cotton, guar, sesame or rice (Saxena, 1987). In other areas, chickpea may be grown in rotation with wheat, barley, forages or as a crop in the understorey of an orchard or vineyard. While the most common practice of growing chickpea is as a sole crop, there are instances in which they are grown as part of a mixed crop. In other situations, the crop may be grown in relay with other crops, which allows for two harvested crops per year. Regardless of the crop rotation, chickpea is an attractive crop due to the disease-free, fertility and monetary benefits they bring to the rotation. They are not grown to provide a weed control benefit. Moreover, they could become a liability in regard to weed control due to increased weed seed production because of poor competition or limitations of the use of soil-persistent herbicides in rotational crops.

Cultural control of weeds is an important part of cropping systems. Managing weeds by cultural methods means reduced expenditures on pesticides, fuel and labour. While cultural management of weeds is an important part of crop rotations that include chickpea, the crop is more often the beneficiary of positive aspects of cultural weed control than a benefactor.

Mechanical weed control

Mechanical methods are the primary approach to weed control in chickpea. These practices range from tractor-powered tillage to weed removal by hand (i.e. hoeing or pulling). Pulling weeds by hand or removing by human-powered equipment such as a wheel hoe is common in less industrialized nations (Knott and Halila, 1986; Bhan and Kukula, 1987). However, even in these nations the labour cost is becoming prohibitive (Solh and Pala, 1990). In a more industrialized production, mechanical control of weeds is limited by the narrow row spacing commonly used. Row spacing is too narrow to allow between-row cultivation without damaging the crop. Mechanical weed control is limited to aggressive and multiple tillage operations prior to planting with ploughs, cultivators or disks and post-plant to early post-emergence use of a harrow, cultipacker or rotary hoe.

In order to be effective, mechanical weed control must be repeated throughout the critical weed-free period. This often means that the initial weeding may occur before chickpea emerges. Light tillage with a cultipacker, harrow or rotary hoe can be effective to some degree in removing small weed seedlings without excessive damage to chickpea seedlings. Generally, more aggressive tillage will damage crops and may result in reduced yields compared to unweeded fields. Thus, mechanical control of weeds in chickpea after crop emergence is largely limited to hand weeding only.

The frequency of weed removal by mechanical methods varies with production systems and environment. Experimentally, up to five mechanical weed removals have been necessary to ensure weed-free conditions in chickpea or other legumes (Knott and Halila, 1986). However, competition from weeds may occur as weeds are allowed to develop into a stage that permits proper identification and grasping for removal. Weeds are in competition with the crop during this time and often there is physical damage to the crop as weeds are removed by hand or machine.

Chemical weed control

Chemical weed control is synonymous with using herbicides to control weeds. Herbicides are regulated by various government agencies and laws. Products available for use vary greatly between regulatory entities. Further complicating matters are the limited number of herbicide manufacturers and limited markets for potential chickpea herbicides. Moreover, herbicides that are effective for controlling the weed spectrum in one chickpea production system in a particular geographic area may be completely worthless against weeds in another production system or limited in their use due to soil persistence. Thus, discussing specific herbicides across the board is pointless as recommendations for one country may be ineffective or illegal in another country, or even in different regions of the same nation. Therefore, discussion in this section will be general in nature and specific only to make a point. Readers should follow all local laws and regulations, seeking specific recommendations from local agronomists, extension personnel, or other qualified individuals or entities.

Non-selective herbicides may be used alone or as an aid to tillage in controlling weeds prior to planting or emergence of chickpea (McKay *et al.*, 2002). Herbicides such as glyphosate or paraquat do not have activity once they come into contact with the soil (Weed Science Society of America, 2002). These herbicides are commonly used in no-tillage or reduced tillage production systems to ensure control of weeds with no or minimal tillage prior to chickpea planting (Baker *et al.*, 1996). More typically, control of weeds prior to planting is done using tillage operations, which are applied specifically for weed control or which provide effective weed control while preparing a suitable seedbed (Knott and Halila, 1986).

Often residual herbicides are applied prior to planting chickpea (Bhan and Kukula, 1987). These residual herbicides will provide control against weeds from a few days to several weeks. The exact duration of weed control will vary with herbicide characteristics, rate, environmental conditions or other factors. Ideally, residual herbicides will provide effective weed control from the time of application until the chickpea crop develops beyond the critical weed-free period or the full bloom stage as noted earlier. Herbicides that have provided effective broadleaf weed control with no or acceptable crop injury include several dinitroanalines (trifluralin, pendemethalin, etc.), triazines (metribuzin, simazine), acetanalides (metolachlor) (Bhan and Kukula, 1987; Solh and Pala, 1990), aryl triazinone (sulfentrazone), isoxazole (isoxaflutole) and imidazolinones (imazethapyr) (Lyon and Wilson, 2005). These herbicides may be mechanically incorporated with sweep cultivators, disks, harrow or other tillage tools prior to planting, or applied following planting, but prior to crop emergence. In some instances, residual herbicides may be tank-mixed with glyphosate or other non-selective herbicides. It is often necessary to combine or tank-mix one or more herbicides in order to broaden the spectrum of the weeds controlled (Bruff and Shaw, 1992; Hydrick and Shaw, 1994; Zhang *et al.*, 1995). Post-emergence herbicides available for chickpea are very limited relative to the number of products available for more widely grown legumes such as soybean (Zollinger, 2006). Grass weeds can be effectively controlled by aryloxyphenoxy propionate or cyclohexanedione herbicides with excellent crop safety (McKay *et al.*, 2002). However, resistant grass species have developed through heavy use or selection pressure of this class of chemistry in legumes or crops grown in rotation with legumes (Delye, 2005). Herbicides for post-emergence broadleaf control are extremely limited for chickpea and include only a few herbicides with legal uses changing with regulating agencies between countries. Crop safety is often limiting with post-emergence broadleaf herbicides in chickpea.

Controlling weeds or desiccating a crop to aid in harvest can be done by mechanical or chemical options (McKay *et al.*, 2002). Although mowing or swathing the crop is possible, it is not always the best method of control since chickpea often does not cure well in the swath (Hnatowich, 2000). Similarly, a crop with uneven maturity will have great variability in seed size and quality when swathed or mowed prior to harvest. Chemical desiccation can be achieved using products such as glyphosate or paraquat (McKay *et al.*, 2002). As weed control options are limited (particularly for broadleaf weeds), chemical desiccation is often necessary to aid in the mechanical aspects of harvest or the timeliness to ensure successful dry-down of weeds and crops prior to harvest.

Integrated Weed Management

CAB International defines integrated pest management as:

> A pest management system that, in the context of the management associated environment and the population dynamics of the pest species, utilizes all suitable techniques and methods in as compatible a manner as possible to maintain the pest populations at levels below those causing economically unacceptable damage or loss. Consideration is given to social acceptability, ecological stability, environmental safety, and human resource development.
>
> (Maredia, 2003)

More simply, integrated weed management uses all weed control strategies to effectively control weeds in a safe, cost-effective and environmentally sound manner. Currently, all or nearly all varieties of chickpea are grown under some form of integrated pest management. Integrated weed management in chickpea is a necessity due to its poor competitiveness, lack of effective herbicides, the need to rotate crops due to disease or other pest management and other reasons. Generally, chickpea is successfully grown due to integrated pest management throughout the rotation. Chickpea, by itself, has limited benefit in a truly integrated cropping system.

Summary

Chickpea is a very poor competitor against weeds. Most of the compendiums published earlier on chickpea production note lack of effective herbicides and the need to expand available herbicides. Most herbicides labelled for use in chickpea were initially developed for the soybean market and were safe for chickpea. However, the introduction and phenomenal success of glyphosate-resistant soybeans have largely reduced the development of herbicides for use in soybeans. Moreover, although glyphosate-resistant or the development of other herbicide-resistant technology was mentioned by some as a method to provide outstanding weed control in minor crops, development of herbicide-resistant minor crops is non-existent (Devine, 2005). Given that the reduction in new herbicide molecules is being discovered and labelled for use in any crop, it is unlikely that new herbicides for use in chickpea will be developed without the development of herbicide-resistant varieties.

Even if herbicide-resistant technology develops for use in chickpea, growers must learn to effectively control weeds using integrated weed management. Herbicide-resistant chickpea might serve to limit variety development of the crop and reduce germplasm with resistance to certain diseases or other pests (Devine, 2005). Heavy use of a single herbicide might be increased through the use of herbicide-resistant crop systems and the development of herbicide-resistant weeds or weed species shift (Ball, 1992) would likely occur. Thus, the development of additional herbicides for chickpea through transgenic means or otherwise may not produce a sustainable system of chickpea production by itself. Integrating existing practices is the likely answer to sustained chickpea production within a cropping system.

References

Ahlawat, I.P.S., Singh, A. and Safar, C.S. (1981) It pays to control weeds in pulses. *Indian Farming* 31, 11–13.

Ali, M. (1993) Studies on crop-weed competition in chickpea (*Cicer arietinum* L.)/mustard (*Brassica juncea* (L.) Czern and Coss) intercropping. In: *Proceeding 1993 of an Indian Society of Weed Science International Symposium.* Integrated Weed Management for Sustainable Agriculture, Hisar, India, pp. 18–20.

Al-Thahabi, S.A., Yasin, J.Z., Abu-Irmaileh, B.E., Haddad, N.I. and Saxena, M.C. (1994) Effect of weed removal on productivity of chickpea (*Cicer arietimum* L.) and lentil (*Lens culinaris* Med.) in a Mediterranean environment. *Journal of Agronomy and Crop Science* 5, 333–341.

Anderson, W.P. (1983) *Weed Science: Principles*, 2nd edn. West Publishing Company, St. Paul, Minnesota.

Baker, C.J., Saxton, K.E. and Ritchie, W.R. (1996) *No-Tillage Seeding: Science and Practice*. CAB International, Wallingford, UK.

Ball, D.A. (1992) Weed seedbank response to tillage, herbicides, and crop rotation sequence. *Weed Science* 40, 654–659.

Benvenuti, S., Dinelli, G., Bonetti, A. and Catizone, P. (2005) Germination ecology, emergence, and host detection in *Cuscuta campestris*. *Weed Research* 45, 270–278.

Bhan, V.M. and Kukula, S. (1987) Weeds and their control in chickpeas. In: Saxena, M.C. and Singh, K.B. (eds) *The Chickpea*. CAB International, Wallingford, UK, pp. 319–328.

Bosnic, A.C. and Swanton, C.J. (1997) Influence of barnyardgrass (*Echinochloa crus-galli*) time of emergence and density on corn (*Zea mays*). *Weed Science* 45, 276–282.

British Columbia Ministry of Agriculture, Food, and Fisheries (2002) *Guide to Weeds in British Columbia*. British Columbia Ministry of Agriculture, Food, and Fisheries, British Columbia.

Bruff, S.A. and Shaw, D.R. (1992) Tank-mix combinations for weed control in stale seedbed soybean (*Glycine max*). *Weed Technology* 6, 45–51.

Cousens, R.D. (1985a) A simple model relating yield loss to weed density. *Annals of Applied Biology* 107, 239–252.

Cousens, R.D. (1985b) An empirical model relating crop yield to weed and crop density and a statistical comparison with other models. *Journal of Agricultural Science* 105, 513–521.

Cubero, J.I., Pieterse, A., Saghir, A.R. and ter Borg, S. (1986) Parasitic weeds on cool season food legumes. In: Summerfield, R.J. (ed.) *World Crops: Cool Season Food Legumes*. Kluwer Academic, Dordrecht, The Netherlands, pp. 549–561.

Delye, C. (2005) Weed resistance to acetyl coenzyme A carboxylase inhibitors: an update. *Weed Science* 53, 728–746.

Devine, M.D. (2005) Why are there not more herbicide-tolerant crops? *Pest Management Science* 61, 312–317.

Di Tomaso, J.M. (1995) Approaches for improving crop competitiveness through the manipulation of fertilization strategies. *Weed Science* 43, 491–497.

Dieleman, A., Hamill, A.S., Weise, S.F. and Swanton, C.J. (1995) Empirical models of pigweed (*Amaranthus* spp.) interference in soybean (*Glycine max*). *Weed Science* 43, 612–618.

Gan, Y.T., Miller, P.R., McConkey, B.G., Zentner, R.P., Liu, P.H. and McDonald, C.L. (2002) Optimum plant population density for chickpea and dry pea in a semiarid environment. *Canadian Journal of Plant Science* 82, 531–537.

Hnatowich, G. (2000) *Pulse Production Manual*. Saskatchewan Pulse Growers, Saskatoon, Saskatchewan, Canada.

Hydrick, D.E. and Shaw, D.R. (1994) Effects of tank-mix combinations of non-selective foliar and selective soil-applied herbicides on three weed species. *Weed Technology* 8, 129–133.

Kelley, A.D. and Burns, V.F. (1975) Dissemination of weed seeds by irrigation water. *Weed Science* 23, 486–493.

Knezevic, S.Z., Weise, S.F. and Swanton, C.J. (1995) Comparison of empirical models depicting density of *Amaranthus retroflexus* L. and relative leaf area as predictors of yield loss in maize (*Zea mays* L.). *Weed Research* 35, 207–214.

Knott, C.M. and Halila, H.M. (1986) Weeds in food legumes: problems, effects, control. In: Summerfield, R.J. (ed.) *World Crops: Cool Season Food Legumes*. Kluwer Academic, Dordrecht, The Netherlands, pp. 535–548.

Kropff, M.J., Weaver, S.E. and Smits, M.A. (1992) Use of ecophysiological models for crop-weed interference: relations amongst weed density, relative time of weed emergence, relative leaf area, and yield loss. *Weed Science* 40, 296–301.

Krupinsky, J.M., Bailey, K.L., McMullen, M.P., Gossen, B.D. and Turkington, T.K. (2002) Managing plant disease risk in diversified cropping systems. *Agronomy Journal* 94, 198–209.

Lopez-Granados, F. and Garcia-Torres, L. (1999) Longevity of crenate broomrape (*Orobanche crenata*) seed under soil and laboratory conditions. *Weed Science* 47, 161–166.

Lyon, D.J. and Wilson, R.G. (2005) Chemical weed control in dryland and irrigated chickpea. *Weed Technology* 19, 959–965.

McKay, K., Miller, P., Jenks, B., Riesselman, J., Neill, K., Buschena, D. and Bussan, A.J. (2002) *Growing Chickpea in the Northern Great Plains*. North Dakota State University Extension Bulletin A-1236, Fargo, North Dakota.

Maredia, K.M. (2003) Introduction and overview. In: Maredia, K.M., Kakouo, D. and Mota-Sanchez, D. (eds) *Integrated Pest Management in the Global Arena*. CAB International, Wallingford, UK, pp. 1–8.

Moes, J. and Domitruk, D. (1995) Relative competitiveness of no-till sown crops. In: Lafond, G.P., Plas, H.M. and Smith, E.G. (eds) *Bringing Conservation Technology to the Farm*. Prairie Agricultural Research Initiative – Agriculture and Agri-Food Canada. Available at: http://paridss.usask.ca/factbook/cfarms/moes.html

Mohammadi, G., Javanshir, A., Khooie, F.R., Mohammadi, S.A. and Zehtab Salmasi, S. (2005) Critical period of weed interference in chickpea. *Weed Research* 45, 57–63.

O'Donovan, J.T., de St. Remy, E.A., O'Sullivan, P.A., Dew, D.A. and Sharma, A.K. (1985) Influence of the relative time of emergence of wild oat (*Avena fatua*) on yield loss of barley (*Hordeum vulgare*) and wheat (*Triticum aestivum*). *Weed Science* 33, 498–503.

Radosevich, S., Holt, J. and Ghersa, C. (1997) *Weed Ecology: Implications for Management*, 2nd edn. Wiley, New York.

Saxena, M.C. (1987) Agronomy of chickpea. In: Saxena, M.C. and Singh, K.B. (eds) *The Chickpea*. CAB International, Wallingford, UK, pp. 207–232.

Saxena, M.C., Subramaniyan, K.K. and Yadav, D.S. (1976) Chemical and mechanical control of weeds in gram. *Pantnagar Journal of Research* 1, 112–116.

Smith, A.E. and Martin, L.D. (1995) Weed management systems for pastures and hay crops. In: Smith, A.E. (ed.) *Handbook of Weed Management Systems*. Marcel Dekker, New York, pp. 477–517.

Solh, M.B. and Pala, M. (1990) Weed control in chickpea. *Options Mediterraneennes – Serie Seminaries* 9, 93–99.

ter Borg, S.J. (1986) Effects of environmental factors on Orobanche-host relationships: a review and some recent results. In: *Proceedings of a Workshop on Biology and Control of Orobanche*. LH/-VPO, Wageningen, The Netherlands, pp. 57–69.

Van Acker, R.C., Weise, S.F. and Swanton, C.J. (1993) The critical period of weed control in soybeans (*Glycine max* (L.) Merr.). *Weed Science* 41, 194–200.

Weaver, S.E. and Tan, C.S. (1987) Critical period of weed interference in transplanted tomatoes and its relation to water stress and shading. *Canadian Journal of Plant Science* 67, 575–583.

Weed Science Society of America (2002) *Herbicide Handbook*, 8th edn. WSSA, Lawrence, Kansas.

Whish, J.P.M., Sindell, B.M., Jessop, R.S. and Felton, W.L. (2002) The effect of row spacing and weed density on yield loss of chickpea. *Australian Journal of Agricultural Research* 53, 1335–1340.

Wiese, A.F., Sweeten, J.M., Bean, B.W., Salisbury, C.D. and Chenault, E.W. (1998) High temperature composting of cattle feedlot manure kills weed seed. *Applied Engineering in Agriculture* 14, 377–380.

Yau, S.K. (2005) Optimal sowing time and seeding rate for winter-sown, rain-fed chickpea in a cool, semi-arid Mediterranean area. *Australian Journal of Agricultural Research* 56, 1227–1233.

Young, F.L., Matthews, J., Al-Menoufi, A., Sauerborn, J., Pieterse, A.H. and Kharrat, M. (2000) Integrated weed management for food legumes and lupines. In: Knight, R. (ed.) *Linking Research and Marketing Opportunities for Pulses in the 21st Century*. Kluwer Academic, Dordrecht, The Netherlands, pp. 481–490.

Zhang, J., Hamill, A.S. and Weaver, S.E. (1995) Antagonism and synergism between herbicides: trends from previous studies. *Weed Technology* 9, 86–90.

Zollinger, R.K. (2006) *2006 North Dakota Weed Control Guide*. NDSU Extension Service Bulletin W-253, North Dakota State University, Fargo, North Dakota. Available at: www.ag.ndsu.nodak.edu/weeds/w253/w253-1c.htm

12 Irrigation Management in Chickpea

H.S. Sekhon and G. Singh

Department of Plant Breeding, Genetics and Biotechnology, Punjab Agricultural University, Ludhiana 141004, India

Introduction

Water is a primary component of biological processes. It plays a vital role in agriculture. Plant tissues contain 80–90% water by weight, although its content varies with plant parts. Thus, water is a critical input in enhancing crop production as it directly or indirectly influences all the physiological processes of plants.

Chickpea (*Cicer arietinum* L.) is the third most important pulse crop in the world. Approximately 73% of the global chickpea area is in South and South-east Asia where it is grown largely as rainfed crop in the post-rainy season (FAO, 2003). However, the crop yields are low and unstable because the crop usually suffers from receding soil moisture at the time of sowing and terminal drought. The drought (rainfall deficit) is often aggravated under the conditions of the semi-arid tropics by erratic and unpredictable rainfall, occurrence of high temperatures, prolonged solar radiation and poor soil characteristics. Plant responses to drought stress are affected by time of occurrence, duration and intensity. Besides these factors, climate, edaphic and agronomic factors also make the water stress a complex phenomenon. The degree to which soil water deficit influences the crop growth and productivity depends not only on the degree of aridity of the atmosphere, but also on a number of plant characteristics that influence water uptake by the plant from the soil and the rate of transpiration. However, cultivars within crop species may differ in their response to the magnitude, timing, sequencing and combinations of these environmental factors (Miller *et al.*, 2002). On the other hand, the drought in the wet tropics may create a flood in the arid zone. Excess soil moisture or waterlogging due to heavy and continuous rains and faulty irrigation coupled with poor drainage conditions can cause several changes in the plant and lead to its death (Uppal *et al.*, 2002). Anoxia can be one consequence of waterlogging and submergence of plants. Anoxia in plant tissues reduces the rate of energy production by 65–97% compared with aeration (Visser *et al.*, 2003).

Plant is sandwiched between soil and atmosphere. Water is loosely held in soil, with –0.1 to –20 bars water potential, but its retentivity in atmosphere is very low, with –100 to –2000 bars water potential. So soil acts as a source of water and atmosphere acts as a sink, while plants act as a connecting channel, with –5 to –50 bars water potential; this is referred to as soil–plant–atmosphere continuum (SPAC) (Ritchie, 1981). The three main phases of plant–water relationship are water absorption, water conduction and water loss processes. Thus, to understand the importance of water in crop productivity, the thorough understanding of these phases is very essential.

On-farm irrigation to chickpea is applied in different ways, but in each case water management or concept of efficient water use is very important. Therefore, there is a great need to integrate water management for achieving higher productivity. In this chapter, the approaches towards water requirement, effect of drought, waterlogging, irrigation scheduling, factors affecting water use, effect of poor-quality water, rainwater conservation and harvesting or recycling are discussed briefly.

Water Requirement

Water requirement of a crop is the quantity of water, regardless of source, needed for normal growth and yield of a crop during a period of time under field condition at a particular place. It may be supplied by precipitation, irrigation or both. Crop water requirement is driven by evaporative demand of the atmosphere, expressed as pan-evaporation, prevailing during the growing season. Water is needed mainly to meet the demands of evaporation (E), transpiration (T) and the metabolic need of the plant, which are together termed as consumptive use. Consumptive water use of the crop is less than its water requirement. Water requirement includes losses during the application of irrigation water in the field (percolation, seepage and runoff) and water required for special operations such as land preparation, transplantation and leaching. Crop water requirement depends on various factors like genotype, growth stage, crop duration, plant density, growing season, soil factors (i.e. texture, structure, depth and topography) and climatic factors (rainfall, temperature, relative humidity and wind velocity), as well as crop management practices like tillage and weeding.

Earlier findings revealed that water requirement of chickpea is location- and season-specific and also varies with the method and scheduling of irrigation. (Prihar and Sandhu, 1968; Singh and Pannu, 1998). The water use vis-à-vis irrigation requirement based on irrigation water/cumulative pan-evaporation (IW/CPE) ratio and critical growth stages. Studies conducted at the International Center for Agricultural Research in the Dry Areas (ICARDA), Aleppo, Syria, showed that chickpea required 1 m^3 of water to produce 0.5 kg grains (ICARDA, 2004).

Effect of Drought

Drought means a rainfall deficit or a prolonged period of scanty rainfall. It effects the functioning of plant and crop yield. As the soil dries due to lack of water, there are changes in the physical condition of the soil such as increases in strength and the formation of air gaps between root and soil. Soil drying inhibits normal development of the nodal root system. Hence, there is reduction in the amount of contact, which can lead to restricted water and nutrient uptake.

Reduction of photosynthesis is considered to be major yield-limiting factor as it influences leaf area expansion, leaf senescence, partitioning of assimilates and biomass production. So the yield component affected most by drought is dry matter accumulation largely as a consequence of poor canopy expansion. Wery *et al.* (1993) have shown the effect of drought stress on the processes involved in the grain yield (Table 12.1).

Mechanism of drought stress resistance

Drought is an abiotic stress that plants usually experience in semiarid tropics. Drought stress may vary in intensity and time during the crop cycle. In North India, terminal moisture stress is more common while in South India both early and late stresses are experienced. In order to characterize drought stress, the most precise measurement is water potential at soil or plant level. During stress, 'potential climatic deficit' is equal to 'potential evapotranspiration' minus rainfall. A positive value indicates that the crop suffers from drought stress unless it has some mechanism of resistance. For example, legume can use more water stored in the soil by reducing its leaf area index (i.e. smaller leaflets and lower shoot growth rate) as reported by Wery *et al.* (1993).

In grain legumes, the time at which drought occurs in the plant life cycle usually affects grain yield and its components (total biomass, vegetative biomass, number of grains, grain weight and water content). The life cycle of grain legumes can be divided into five phases: (i) from seedlings emergence to beginning of flowering; (ii) from beginning of flowering to beginning of grain formation; (iii) from beginning of grain formation to the end of grain formation;

Table 12.1. Effect of drought stress on the processes involved in grain yield and nitrogen balance of grain legumes. (From Wery *et al.*, 1993.)

Stages showing increasing drought stress	Physiological activity affected	Crop parameter affected	Yield component affected
Full turgor potential	Cell elongation Nodule activity	Organ size Nitrogen fixation	Total biomass Nitrogen
Beginning of stomata closure	Photosynthesis	Carbon balance	–
Increase in canopy temperature	Translocation	Grain filling Death of plant organs (leaves, flowers and pods)	Grain – number yield – weight – vigour
Wilting	–	–	–
Death of the plant	–	–	–

(iv) from end of grain formation to physiological maturity; and (v) from physiological maturity to harvest. In chickpea, data revealed that phase II reduces the number of grains per pod as well as the 100-grain weight. Drought in phase (v) adversely affects grain quality. Table 12.1 shows that drought stress does not affect all plant functions similarly; cell elongation and nitrogen (N) fixation are more susceptible than photosynthesis and transpiration.

Most drought resistance mechanisms, except tolerance to dehydration, can be understood at the plant level, unlike resistance to thermal stresses which is located mainly at the cell level (Blum, 1988). Actually, adaptive mechanisms are more important because in the field drought is established more slowly than freezing or heating. The classification based on plant water potential is mainly used to analyse different mechanisms of drought.

Turner (1986) gave three mechanisms of drought resistance:

1. *Drought-escaping* – Plant that completes its life cycle before serious water deficit develops. Drought escape involves plants with early flowering and pod initiation. Siddique *et al.* (1999) reported that these plants attain early vigour, faster canopy development, less number of days to flowering, podding, seed filling and maturity. They avoid high risks of frost damage and low yield due to poor fertilization associated with cool temperatures. Drought avoidance can also be achieved by reducing the intercepted non-photosynthetic radiation. This increase in leaf reflectance can be obtained with glandular hairs present in chickpea (Khanna-Chopra and Sinha, 1987).

2. *Drought-tolerant with high water potential (dehydration postponement)* – Plant that avoids drought by avoiding tissue dehydration. In this mechanism, plants combat drought by stomatal control, by increasing water use through deep root system or by accumulation of solutes to maintain turgor as water storage develops, i.e. osmotic adjustment. Open stomata result in transpirational cooling of leaf. For more extraction of soil moisture the requirements are: (i) deeper roots; (ii) high root density; and (iii) adequate longitudinal conductance in main roots (Fischer *et al.*, 1982).

3. *Drought-tolerant with low water potential (dehydration tolerance)* – Plant that endures rainfall deficits at low tissue water potential. In this case, plant cells have the ability to continue metabolism at low leaf water status. This mechanism involves stability of membrane, which can be assessed through electrolyte leakage from desiccated tissue, but is not often correlated with dehydration tolerance. Accumulation of proline in cell in response to water deficit is another mechanism protecting protein structures as cell dehydrators and as an organic N source. Dehydration tolerance is related to the degree of osmotic adjustment (Hsiao *et al.*, 1984).

Osmotic adjustment and root growth dynamics are very important to combat drought in plants.

Osmoregulation or osmotic adjustment

Osmoregulation is a key mechanism for increasing crop yields subjected to drought conditions (Zhang *et al.*, 1999). Osmotic adjustment helps in maintaining stomatal conductance and photosynthesis at low leaf water potential.

The mechanism also helps in delaying leaf senescence, reducing flower abortion and improving root growth and water extraction from soil (Morgan, 1984; Ludlow and Muchow, 1990). The higher benefits of osmotic adjustment occur when plants are subjected to severe water deficits and much focus is on plant survival. Indeed, the oft-cited benefit of turgor maintenance in cells is likely to result in crop behaviour that is exactly opposite to what is beneficial to the crop. When soil water availability becomes limiting to physiological processes, crops are more likely to benefit from a conservative approach in the use of remaining soil water, rather than rapidly using water by keeping leaves turgid, stomata open and tissue growing. So for beneficial yield responses to osmotic adjustment, the main emphasis should be on stimulation of root development.

High root biomass maintains high leaf turgor and leaf water potential besides reducing canopy temperature and providing accessibility to deeper soil, thereby enhancing the vegetative biomass. Root size, morphology, depth, length, density, hydraulic conductance and function are basic requirements to meet the transpirational demands of the shoot (Passioura, 1982). Serraj and Sinclair (2002) emphasized that osmotic adjustment has more significance for root growth than any definite relation with grain yield. Recent studies conducted at the Indian Institute of Pulses Research (IIPR), Kanpur, India, showed genotypic variation for osmotic adjustment in chickpea (Basu and Singh, 2003). Genotype K 850 showed maximum variation in osmotic potential, indicating the extent of flexibility in it (Table 12.2). Singh (2002) reported that high osmotic adjustment chickpea cultivars (C 214, G 130 and H 208) absorbed 20–30 mm more water from subsoil layers than the cultivar (P 324) of low osmotic adjustment class. Thus, osmotic adjustment is not only related to the grain yield, but also has an internal physiological consistency for various intervening steps related to growth process and yield.

The inheritance of osmotic adjustment in chickpea suggests that this characteristic seems to be under the control of few genes (Serraj *et al.*, 2004). In general, very little attention has been given to genetic improvement of adaptation to intermittent drought.

Table 12.2. Variation in osmotic potentials among different chickpea genotypes. (From Basu and Singh, 2003.)

	Osmotic potential (MPa) Soil moisture at 60 cm depth		
Genotype	13%	10%	7%
K 850	–0.98	–1.40	–1.85
C 214	–1.30	–1.40	–1.70
ICCV 10	–1.20	–1.38	–1.57
BG 364	–1.20	–1.28	–1.40
BG 256	–1.10	1.35	–1.50
ICC 4958	–1.35	–1.38	–1.52
IPC 94-132	–1.22	–1.35	–1.55
Annegiri	–1.20	–1.37	–1.60

Root growth dynamics

The gains from soil moisture are likely to occur from maximum extraction of the limited amount of water to make it available for transpiration (Summerfield *et al.*, 1981). This can be achieved through the mechanism of adaptation associated with the root system. Hussain *et al.* (1990) found that proportionately increased partitioning of assimilates to roots and reduced leaf area index was the most important mechanism when faba bean was subjected to drought. Thus, more information is needed to select genotypes of appropriate root/shoot ratios and their realization to grain yield under drought environments. Genotypic differences in root length and spread of root systems are reported by Sheldrake and Saxena (1979) and Minchin *et al.* (1980). Rooting depth and density are among the main drought avoidance traits identified to confer grain yield under terminal drought environments (Passioura, 1982; Ludlow and Muchow 1990; Subbarao *et al.*, 1995; Turner *et al.*, 2001). Minchin *et al.* (1980) reported that short-duration genotypes of chickpea have faster initial growth rate, but are not able to sustain it during grain filling. The drought stress at podding aggravates the loss of active roots, which coincides with reproductive phase (Summerfield *et al.*, 1984).

In the rainfed environments, the depth of rooting is often cited as an important criterion to supply soil moisture from deeper soil zones. Deep rooting was positively correlated with crop growth grain yield, cooler canopy temperature and soil water extraction in bean (Sponchiado *et al.*, 1989). Gregory (1988) concluded that depth of rooting is a genetically controlled trait. Heritability of genotypic differences in root traits seems to be very low in chickpea. The small leaf size in chickpea has a simple inheritance and is governed by a pair of recessive genes (Saxena *et al.*, 1993). Thus, deep-rooted genotypes must be included particularly in breeding programmes that extract substantial amount of water available in the subsoil at maturity. Recent efforts in molecular mapping of genes and marker-assisted selection for root traits in chickpea will facilitate the identification of alternate sources to widen the genetic base for crop drought avoidance programmes.

Efforts made at the International Crops Research Institute for the Semi-Arid Tropics (ICRISAT), Hyderabad, India, to find sources for deep and large root system led to the identification of the chickpea variety ICC 4958 (Saxena *et al.*, 1993) and later to the development of drought-tolerant varieties by incorporating the deep and large root system of chickpea into a well-adapted genetic background (Saxena, 2002). In the mini core germplasm collection studies on chickpea for drought avoidance, genotypic variation was observed for root and shoot growth. The absence of a significant difference in root or shoot growth between extremely late-planted Annigeri and ICC 4958 emphasizes the need for further comprehensive investigation of the whole germplasm collection to choose the best parental lines to identify quantitative trait loci (QTLs) for the root traits across different growth conditions. A comparison of the two genotypes ICCV 2 and JG 62 exhibited a contrasting temporal interaction for root mass. The root growth of ICCV 2 at early stages was good and, therefore, some of the existing RILs of JG 2 × ICCV 2, although not conclusively bred, can also be expected to possess a better root system (Krishnamurthy *et al.*, 2003).

Water stress and microbial activity

Water stress may affect survival of bacteria and *Rhizobium*–legume symbiosis (Marshal 1964; Penacarbiales and Alexander, 1979). The soil moisture that is adequate for seed germination is also adequate for bacterial movement and nodule(s) formation. However, this condition changes with time and may not be optimum for subsequent nodulations and their potential activities. It has also been noticed that following successful infection, due to reduced water supply, can retard nodule development, accelerate senescence of nodules and lead to lower N fixation rate (Gallacher and Sprent, 1978). Yield limitation in chickpea could result from suboptimal carbon or N fixation, which is dependent upon ready supply of photosynthates. In general, a large number of nodules are confined to 10–15 cm depth of soil, and in chickpea are attached to the primary root under rainfed conditions. However, when the soil zone (15–20 cm) is exposed to severe water stress, most of the nodules disintegrate. The rate and duration of nodules is markedly influenced by water and temperature. On the other hand, waterlogging affects N fixation due to anoxia (Sprent *et al*., 1983).

The major pathway of water into nodules is through vascular connections with the root. The surface of the nodule is the main area for gaseous exchange and is therefore more adapted to water loss than uptake. Nodules always need an efficient water supply to export the product of fixation. Water-stressed plants transpire at a lower rate than unstressed plants. Water stress is quickly reflected as changes in hormonal content (Hsiao, 1973). The nodules are an active site of synthesis of auxins and cytokinins. Therefore, it is likely that nodules, besides the supply of organic N, are a source of cytokinins that makes the plant more tolerant to moisture stress (Phillips and Torrey, 1973).

Effect of Waterlogging

Waterlogging is the saturation of the soil root zone with water. It may be due to flooding, seepage, over irrigation or poor drainage. Grain legumes are an integral part of agriculture. When a soil is temporarily flooded, a number of physical and chemical changes may profoundly influence the soil properties. The soil pore space, normally filled with air, is filled with water and the aeration status of the soil decreases below critical limits. Aerobic systems are O_2-sufficient and contain 0.20–0.28 mol/m^3 O_2. Hypoxic systems contain reduced O_2 (0.05–0.11 mol/m^3). Anaerobic or anoxic systems are O_2-depleted with <0.05 mol/m^3 O_2. Low O_2 concentration in the soil leads to reduced growth of plants due to reduced availability of nutrients and poor photosynthesis. Subsequent changes include decreased permeability of roots, reduced mineral uptake, alteration in growth hormone balance, leaf epinasty, chlorosis and abscission (Pezeshki, 1994).

As the soil is O_2-depleted, the soil microflora changes. Denitrification and increased reduction of Mn, Fe and S occur together with an increase in N gases and methane production by anaerobic bacteria. Products from the attenuated respiration of organic substrates, such as lactic acid, ethanol, acetic acid and bituric acid, can accumulate. In addition, with fermentation of organic matter, amino acids are converted to NH_2, organic acids (including pyruvic acid), CO_2, methane, amines and H_2S. Ethylene, an extremely physiologically active organic substance, increases in concentration through production of both soil microorganisms and plant roots. Under waterlogging conditions, root growth becomes more horizontal or even upward rather than downward. When conditions become anoxic, root extension ceases (Jackson and Drew, 1984; Clive and Mary-Jane, 1994; Noble and Rogers, 1994).

The effect of waterlogging on root growth is more severe than on shoot growth. Under aerobic conditions, the metabolic energy required for root growth is generated primarily in mitochondria, where electrons are transferred to O_2 from sugars by enzyme catalysis. Under waterlogged conditions, the depletion of O_2 deprives the plant of its primary energy source and root growth declines. Many compounds accumulated under anaerobic conditions are potentially toxic. On the other hand, waterlogging can inhibit leaf and stem expansion and cause rapid wilting and necrosis of plant leaves. Carbon fixation per unit leaf area can be depressed when plants are waterlogged (Trought and Drew, 1980) and is associated with inhibition of ribulose biphosphate carboxylase and oxygenase production as well as stomatal closure (Bradford, 1982). The adverse effect of stagnation of water after irrigation in rice soil on chickpea productivity is shown in Table 12.6.

Effect on nutrition uptake

Legumes are highly sensitive to waterlogging. Generally N accumulation is slowed by cataloguing due to reduced nodulation and decreased specific activity of nitrogenase (Minchin and Pate, 1975).The formation of nodules is retarded by both CO_2 and ethylene (Goodlass and Smith, 1979). In prolonged waterlogging, nodules and outer root tissues slough off causing a temporary N deficiency. Waterlogging also results in disappearance of leghaemoglobin in nodules and cessation of N fixation activities, reaching almost zero during the 3 days of waterlogging.

Irrigation Scheduling

Although chickpea can be grown under limited moisture conditions, the crop requires adequate supply of moisture for proper growth and development. The irrigation application may vary with soil, climate and genotype. The scheduling of irrigation at the right time and in the right quantity are the most important factors for realizing high grain yields of chickpea. Proper scheduling of irrigation is very essential for efficient use of irrigation water and production inputs, saving of irrigation water and energy, higher crop yield and lower production cost. In addition, under adequate availability of water the main aim is to increase the water use efficiency (WUE). However, in limited moisture conditions, the emphasis is on rationalizing the limited water by applying it at moisture-sensitive stage(s) of crop growth (Ahlawat and Rana, 2005).

Table 12.3. Optimum irrigation schedule and irrigation requirement of chickpea. (From Ahlawat and Rana, 2002.)

Optimum IW/CPE ratio	0.4–0.8
Depletion of available soil moisture	50–75%
Irrigation depth (cm)	6–8
Number of irrigations	1–4
Requirement of irrigation (cm)	8–24
Critical stages	Vegetative and pod formation

For obtaining high yield potentials and quality produce, crops vary in requirement of soil moisture level that has to be maintained during crop growth and development. Near field capacity, most of the plants can use soil moisture and nutrients efficiently. Due to continuous loss of water through evapotranspiration, there is a need for irrigation and the quantity of water required at each irrigation is determined by the quantity of usable water stored in the crop root zone and the rate at which this supply is depleted.

There are several approaches for scheduling irrigation, but a relatively more practical meteorological approach is irrigation water/evaporation (IW/E), irrigation water/cumulative pan-evaporation (IW/CPE) or irrigation depth/cumulative pan-evaporation (ID/CPE), i.e. ratio between a fixed amount of irrigation water (IW) and cumulative open pan-evaporation minus rainfall since previous irrigation as the basis for scheduling of irrigation. Smaller ratio means irrigation at larger intervals and vice versa. The internal plant water status directly influences physiological processes, which ultimately affect quantity of product. Therefore, any plant characteristic, which relates directly or indirectly to the plant water status, growth and yield, may serve as a criterion for timing of irrigation field crops.

Ahlawat and Rana (2002) reported optimum IW/CPE ratio for chickpea, i.e. 0.4–0.8, irrigation depth 6–8 cm and requirement of irrigation water 8–24 cm (Table 12.3). However, irrigation application at critical stage(s) is more important.

Irrigation experiments conducted on chickpea in a vertisol at ICRISAT showed that irrigation applied at vegetative and pod-filling stages gave a yield of 2280 kg/ha, while without irrigation the grain yield was 1250 kg/ha (Table 12.4). Irrigation applied at pod-filling stage was not significantly better than yield without irrigation as growth of the crop was poor. Irrigation applied at vegetative and flowering stages was inferior to vegetative and pod-filling stages because the former delays pod initiation, due to which the pod development period reduces (Johansen *et al.*, 2002).

Factors Affecting Water Use and Crop Productivity

Soil factors

The application of irrigation depends on soil characteristics. Soil governs only depth and frequency factor in irrigation (Dastane *et al.*, 1970). In light-textured

Table 12.4. Response to irrigation applied at different growth stages of chickpea on vertisol. (From Johansen *et al.*, 2002.)

Growth stage	Total dry matter (kg/ha)	Grain yield (kg/ha)	Harvest index (%)
No irrigation	2170	1250	57
Vegetative (31 DAS)	3240	1770	55
Flowering (52 DAS)	2940	1660	56
Pod fill (73 DAS)	2270	1300	57
Vegetative + flowering	3580	1720	48
Vegetative + pod fill	3970	2280	57
Vegetative + flowering + pod fill	4070	1740	43
SE	124	106	2.3

soils, light but frequent irrigations are required. Light irrigation on loamy sand soil not only improves WUE, but also restricts the leaching of applied nutrients. The irrigation depth varies widely at different stages of crop depth. Chickpea responded significantly to irrigation on light-textured soils. At Hisar, Kanpur and New Delhi crops required one irrigation at pod formation stage where soil was sandy loam, while at Ludhiana, Parbhani and Sriganganagar two irrigations at vegetative and pod formation stages were required because of light-textured soils. However, at Navsari and Powarkheda clay loam soils required three irrigations at branching, flowering and pod development stages, which gave statistically higher yields than without irrigation (Ahlawat and Rana, 2005). Other studies carried out at Indore, Jabalpur, Rahuri, Dharwad, Kharagpur and Srigupta indicated higher yield with 3–4 irrigations applied at ID/CPE of 0.6–0.8 (Prihar and Sandhu, 1968; Sharma *et al.*, 1996). On heavy black clay soils at Indore, Dharwad and Rahuri, light irrigations of 6 cm were found better than heavy (8 cm) irrigation.

Plant factors

Poor plant stand can result from abiotic stress. There are indications that lack of adequate moisture in the seedbed is an important constraint. Keatinge and Cooper (1983) reported that rains after seed germination not only contribute to adequate soil moisture reserves, but are also important in establishing uniform and vigorous plant stands. Drought stress during vegetative stage results in poor plant growth. Wilting of leaf, drooping and rolling of leaves are the visual indicators of water need of plants. Water deficit adversely affected cell elongation, cell division, NO_3 reductase activity and photosynthesis. Water stress causes primarily stomatal closure, decreases assimilation and growth thereafter. Dehydration also causes mechanical rupture of protoplasm, protein denaturation and gene mutations (Prabha *et al.*, 1985). Rupela and Kumar Rao (1987) reported that rhizobia are unable to survive under drought. Moreover, limitation of water interferes with infection by rhizobia and nodule initiation, as well as formation. There may be genotypic variability for symbiosis under drought conditions.

Drought also causes root senescence. A damaged root system influences water uptake. Movement of water in the root, through cortex, pericycle and endodermis offers resistance to water absorption. Transpiration is the most important process of soil water absorption. Its extent is determined by the gradient between the atmosphere and plant through the leaves. The stomata present on leaf regulates both water loss and exchange of CO_2. In monocots, stomatal frequency of the upper and lower surface of leaf is almost same. However, in dicots, the lower surface of leaf has higher frequency than the upper surface. Change in stomatal frequency is associated with leaf expansion. Under water stress conditions, the size of the leaf reduces but the number of stomata per capita increases. Water deficits cause increase in leaf or canopy temperature. There is also a decrease in photosynthesis due to reduction in source size (leaf size). The poor leaf area to harvest solar radiation, coupled with poor translocation of assimilates between plant parts, leads to poor growth and lesser grain yield under water-stressed conditions.

Genotypes of chickpea may differ greatly under drought conditions. Deshmukh *et al.* (2004) showed that Phule G 96006 had minimum reduction in grain yield due to moisture stress along with least reduction for leaf area at different stages. Further, this genotype had the highest drought tolerance efficiency (DTE), least susceptible index (disease severity index [DSI]) and minimum reduction in grain yield due to stress (Table 12.5). It also maintained the highest harvest index under moisture stress and irrigated conditions, indicating that the genotype Phule G 96006 may be rated as tolerant genotype for moisture stress conditions.

Climatic factors

The atmospheric evaporative demand of a place mainly estimates the amount of water loss when the crop canopy covers the soil adequately and moisture is freely available. The important environmental factors affecting evapotranspiration are temperature, wind velocity, light intensity, atmospheric vapour pressure and soil water supply to roots. An increase in temperature increases water

Table 12.5. Grain yield and drought characteristics influenced by different chickpea genotypes. (From Deshmukh *et al.*, 2004.)

Genotype	Grain yield (kg/ha)		Percentage reduction in yield	Drought tolerance efficiency (%)	Disease severity index (%)	Membrane injury index		Harvest index	
	I_o	I_i				I_o	I_i	I_o	I_i
BGD 127	878	1444	31.1	60.8	1.05	0.18	0.13	41.0	42.7
RSG 888	961	1500	35.9	64.0	0.97	0.21	0.22	36.7	42.3
Phule G 96006	1261	1767	28.6	71.3	0.78	0.19	0.16	44.0	51.7
Phule G 5	1122	1767	39.3	60.6	1.05	0.22	0.25	37.1	49.4
Phule G 9414-7	739	1344	45.0	54.9	1.22	0.32	0.36	32.5	45.0

I_o: moisture stress condition; I_i: irrigation condition.

loss from crop canopy because it enhances steepness of vapour pressure gradient from crop canopy to outside air. The higher temperatures that are often associated with greater vapour pressure deficit will lead to a greater evapotranspiration, rapid depletion of soil water and early development of plant water deficit. A crop suffers from two types of heat shock: lethal temperatures even for a short period (ranging from a few minutes to a few hours, usually around midday); and moderate heat, a long period of temperature higher than the optimum. The first type affects flowers and pods, while the second has a cumulative effect on the plant carbon balance. Heat stress has a negative effect on protein structure and activity and plasmic and chloroplastic membranes, and decreases photosynthesis (Blum, 1988). In chickpea, heat stress takes place mainly in the last part of the grain-filling phase. So water availability at this stage is very important. Reproductive organs, especially flowers, are very susceptible to heat. In chickpea, N fixation is also more susceptible to heat (35°C) than the processes related to grain production (Rawasthorne *et al.*, 1985).

Agronomic management

Seed priming

Better germination and crop establishment seldom occur in the rainfed conditions and marginal environments of the semiarid tropics. Unpredictable and erratic rainfall and light soils contribute to a situation in which good crop establishment is often the exception, rather than the rule. Seed priming is very important for better germination and establishment of rainfed chickpea crops. Standard methods of seed priming developed by specialist companies are energy- and technology-intensive, costly and unavailable to marginal farmers. Harris (1996) proposed a low-cost, low-risk and simple technique, i.e. 'on-farm seed priming'. In this technique, seed is soaked in tap water for a given period prior to sowing. On-farm seed priming studies conducted in chickpea under rainfed conditions at Rajasthan, Gujarat and Madhya Pradesh in India and in Bangladesh showed faster emergence, better crop stands and lower incidence of re-sowing, vigorous plants, better drought tolerance, earlier flowering and 50% more grain yield than the non-primed seed. The optimum seed soaking time of chickpea was 8 h (Harris et al., 1999). Musa *et al.* (2001) reported an increase in the grain yield of chickpea with seed priming mainly due to better crop stands and more nodulation. Kaur *et al.* (2002) showed an increase in shoot length and biomass with water-primed seed of chickpea. The grain yield in primed and non-primed treatments was 5.05 and 3.61 g/plant, respectively. The activities of amylases, invertases, sucrose synthatase and sucrose phosphate were higher in the primed seedlings than in the non-primed ones. The production of enzymes is responsible for better growth and yield in primed seeds.

Depth of sowing

Dropping seed at a proper depth is an important factor, which influences the emergence and establishment of crop especially under rainfed conditions.

Various factors influencing seeding depth include soil texture, date of sowing, climatic conditions and seed size. Studies conducted in Australia for comparing the effect of depth of sowing on chickpea showed that deep sowing at 10 cm, particularly in early sowing when soil moisture from summer rainfall is stored in the soil, is proved highly beneficial for achieving good plant stand and higher grain yield. Soil moisture was more at 10 cm depth than at shallower depths (Siddique and Loss, 1996). They further suggested that deeper sowing also improves the survival of *Rhizobium* inoculation as soil temperature decreases and moisture availability increases with increasing depth. Experiments conducted at the Punjab Agricultural University, Ludhiana, India, also showed beneficial effects of deep sowing (10–12 cm) on the growth and yield of chickpea (PAU, 1981). Furthermore, deep sowing not only improves root and shoot growth, but also reduces wilt attack under rainfed conditions.

Time of sowing

In arid and semiarid regions, time of sowing is an important cultural practice in WUE. The main reason for choosing the optimum sowing time is to ensure good germination by placing the seed in the optimum moisture zone. After seed germination, the moisture should be optimum for root growth and root penetration, to envelope maximum soil volume for nutrient uptake. The critical moisture requirement level of seed germination differs between crops. Compared with pea, lentil and faba bean, chickpea has a relatively higher moisture requirement for seed germination (Hadas and Stibbe, 1973). Nevertheless, the critical soil moisture required for seed germination and seedling emergence is well below field capacity (Saxena, 1984). Saxena (1987) reported that the adverse effects of suboptimal seedbed moisture content on plant stand have been reported for chickpea. Delaying the planting time of spring chickpea to late March at ICARDA, when seedbed moisture was suboptimal, resulted in a very poor establishment, contributing to the failure of crop (Saxena, 1980). Similar observations were recorded under Indian conditions when planting is done late after the monsoon. In North Indian conditions, 10–25 October is the best time for sowing chickpea under rainfed conditions (Sekhon *et al.*, 1994).

Under late sowing conditions, particularly after cotton, chickpea is highly benefited with irrigation (Dumbre and Deshmukh, 1983). Studies conducted at Hisar, India, showed that two irrigations at pre-flowering and pod development stages gave 16.4% and 7.4% more grain yield over no irrigation and one irrigation at pre-flowering stage, respectively. Highest consumptive use was recorded with two irrigations, with maximum (7.75 kg/ha mm) WUE in the control (7.01 kg/ha mm) (Singh *et al.*, 2004).

Planting method

Chickpea is usually sown on light-textured soils by flat sowing method. When chickpea is sown after rice or on heavy soils, the major problem faced by farmers is that the crop damages badly when irrigated due to failure of rains during the crop season. Soon after irrigation, the crop turns pale yellow and plants start dying thereafter. In on-farm trials, the application of irrigation to chickpea resulted in drastic reductions in the grain yield of crop sown on flat bed (Table 12.6). In the

Table 12.6. Effect of planting methods and irrigation on the grain yield of chickpea sown after rice in on-farm trials. (From Sekhon *et al*., 2004.)

	Grain yield (kg/ha)			
Planting method	Kothe Rehlan (2000/01)	Sidhwanbet (2000/03)	Kothe Rehlan (2001/02)	Mean
Flat bed, no irrigation	1872	1500	2175	1849
Flat bed, irrigation at pod initiation	365	640	488	497
Raised bed, no irrigation	1695	1268	1890	1618
Raised bed, irrigation at pod initiation	2022	1674	2292	1996

case of 67.5 cm wide raised beds, sowing of two rows without irrigation showed 12.5% decrease in yield than the flat bed plots without irrigation. However, crop sown on raised beds with one irrigation at pod initiation gave 7.9% higher yield over that sown on flat bed (Sekhon *et al*., 2004). Patel *et al*. (1987) and Chandrakar *et al*. (1991) also reported that on heavy soils, excessive moisture under field conditions reduced the growth, nodulation, root growth and yield of chickpea drastically. At Hisar, chickpea in paired row planting (30/60 cm) in ridge-furrow system recorded 30% saving of irrigation water and higher yield over flat sowing (Agarwal *et al*., 1997).

Synchronizing fertilizer use with irrigation

Synchronous use of water and fertilizer application may improve each other's efficiency. Agarwal and Khanna (1983) showed that the WUE in chickpea could be increased significantly by adopting improved water management practices and balanced use of fertilizers. The crop response to applied nutrients is determined by the soil moisture. For example, in Punjab, chickpea responded significantly up to 40 kg P_2O_5/ha under irrigated conditions, but only up to 20 kg P_2O_5/ha under rainfed conditions (Pasricha *et al*., 1987). Thus, the judicious use of costly inputs like water and fertilizers is very essential to attain higher WUE in chickpea.

Method of irrigation

Irrigation can be applied to crop by several methods (check basin, furrow, subsurface and sprinkler, etc.) The choice of a particular method depends on surface topography, soil type, crops to be grown, irrigation source and depth of water application. Gowande *et al*. (1997) reported 66.6% water economy in chickpea at Akola, Maharashtra, India, with four-row, broad-bed-furrows over border method. However, furrow irrigation showed 30–33% water economy in paired row, 30 + 60 cm in ridge-furrow system (Singh, 2002). Sprinkler irrigation revealed 56% saving of water more than surface irrigation (Sivanappan, 1995).

Use of antitranspirants

Antitranspirants are the chemicals that are employed to reduce transpiration. The work on antitranspirants has been reviewed by Gale and Hagan

(1966), Cheema and Uppal (1979) and Uppal *et al.* (2002). According to them antitranspirants can be classified into three categories: (i) film-forming compounds; (ii) metabolic inhibitors; and (iii) reflectants.

FILM-FORMING COMPOUNDS These compounds act mechanically as the waterproof films or barriers on leaf surface and reduce the escape of water vapour. These films formed from plastic or wax emulsions exhibit a certain degree of permeability, which is more with CO_2 and O_2 than with water vapour. These antitranspirants under dry and hot conditions have been reported to reduce transpiration without influencing crop growth. Various kinds of films are cetyl alcohol Folicote, wax emulsion, Mobileaf, Vaporgard and Wilt pruf (di-1-p-menthen).

METABOLIC INHIBITORS These metabolically active compounds are of two types: the first type, abscisic acid (ABA), changes the osmotic potential of the guard cells, thereby affecting the turgor of these cells; the second type, phenyl mercuric acetate (PMA) and alkenyl succinic acid (ASA), modifies the permeability of plasma membrane in such a way that it results in stomatal closure. ASA is a plant hormone that closes stomata and is also non-toxic. ABA closes stomata by efflux of potassium to the surrounding cells and is found effective in increasing WUE in wheat and barley (Jones and Mansfield, 1970). PMA contains mercury, which combines with the protein, and changes the permeability behaviour of guard cell plasma membrane. This results in efflux of water and closure of stomata due to loss of turgidity. Nagarajah and Ratnasooriya (1977) reported that though PMA is a very effective antitranspirant, but is not used in food legumes as it is toxic.

REFLECTANTS These reduce the energy load on the leaf by increasing the albedo, thus reducing the leaf temperature. Seginer (1969) reported 15–30% reduction in net energy by applying reflective coating. Kaolinite and kaolin are the important reflectants. The effect of reflectants is more on crops like soybean, which is a low-light-saturating plant. Lemeur and Rosenberg (1976) explained that Kaolin was most effective in increasing the reflection coefficient of soybean in the wave band of photosynthetically active radiation (PAR; 400–700 nm), but the effect on near-infrared (NIR; 750–1550 m) was small.

Application of antitranspirants is not popular in chickpea as it is a low-value and risky crop. Single application of these chemical compounds is not effective in chickpea. Since several applications are needed, the treatment of antitranspirants becomes uneconomical. The major problem lies in maintaining control over microclimate. It is also not possible to compare the results of experiments until the growth of the plant is not taken into account by way of an analytical approach. Therefore, results obtained cannot be generalized over a large area and under different agroclimatic conditions.

INTERCROPPING In drought-prone conditions, intercropping offers advantages in increasing the crop productivity, decreasing variability and risk in production. The degree of advantage depends on minimizing competition for maximization of complementarity of light, water and nutrients (Ofori and Stern, 1987). In the case of water use, crop combinations in which individual components use water predominantly at different times or extract it from different layers of the soil profile are most advantageous (Johansen *et al.*, 2002). Singh and Chaudhary (1998) showed that chickpea grown on sandy loam soil extracted water from 200 cm soil depth. The major proportion of water was absorbed from 30 to 150 cm soil depth (>80%), while 0–30 and 150–180 cm soil layers contributed 7–10% of total water use.

Use of Poor-quality Water

Soil salinity is a problem encountered in arid and semiarid lands where irrigation is practised. Salt accumulation in soils is generally caused by lack of appropriate drainage combined with seepage from water delivery systems, as well as by inappropriate irrigation management. In saline soils, high Na^{+}/Ca^{2+} ratios adversely affect the soil structure. So saline water should not be used for irrigation as it leads to excessive accumulation of salts and other toxic elements up to levels that may affect the growth and yield of a crop. Chickpea is sensitive to salt stress. Even under marginal conditions of salinity, crop growth is stunted without showing any visible symptoms. So the choice of land and irrigation water is greatly restricted. With increasing levels of salinity, a crop may exhibit symptoms of chlorosis or necrosis. Legumes like chickpea are very sensitive to soil sodicity and the yield decreases significantly even when the soil exchangeable sodium percentage (ESP) is <15.

The degree of soil alkalinity depends on quality of irrigation water, frequency of irrigation, soil type and its permeability, salt tolerance characteristics of the plant and climatic conditions. Generally, light-textured soils are less salinized than medium- and heavy-textured soils. Salt tolerance near the root zone is governed by the combined effect of the quality of irrigation water, irrigation management, climate and water transmission properties of the soil during the cropping period. Chickpea can withstand a wide range of soil pH (5.5–8.6), but it is very sensitive to saline and alkaline soils.

Rainwater Conservation

During monsoon season, due to heavy rains most of the rainwater is lost in the form of runoff. Singh (2002) reported that in Central India up to 60% of rainwater is lost as runoff in vertiosols during the monsoon season. Where the crop was established, runoff was reduced to 91%. In Hyderabad, vertisols utilize only 38% of the rainfall for evapotranspiration during the cropping season. Approximately 29% of the precipitation was lost as runoff, 24% evaporated from the bare fallow and 9% was lost by percolation. Therefore, there is a dire need for efficient rainwater management, which is possible through water conservation and recycling of rainwater.

In fields kept fallow during monsoon, conservation of soil moisture ensures a successful rainfed chickpea crop. The following measures can be adopted to conserve maximum amount of rainwater in the soil:

1. Level the field properly before the monsoon season.
2. Divide each field into small plots and make strong bunds to prevent runoff.
3. Do not allow weeds to grow as they deplete the field soil moisture and nutrients.
4. Deep plough the land for better water absorption; at the end of the monsoon season, ploughing should be invariably followed by planking.
5. Plough the land only once before sowing. However, if the soil appears to be deficient in moisture, run a roller before sowing. This will facilitate in bringing moisture near the soil surface for good germination.

Rainwater Harvesting and Recycling

There are two basic components of rainwater harvesting and recycling: one is catchment or source area over which runoff is maximized and the other is storage facility or use area. Water storage techniques for holding the water collected from a catchment area are of two types: (i) the soil profile or monolith and (ii) tanks or ponds. The stored water acts as a reserve for the crops in the rainfall-deficit periods. Conservation of rainwater and its effective utilization are the important aspects of rainwater management.

A watershed-based water-harvesting and recycling system has been developed at ICRISAT. Similarly, a model watershed programme was undertaken by the Government of India in the 1980s. After the inception of the model watershed programme, there has been considerable increase in the number of wells in most watersheds due to improvement of groundwater level. The increased water availability in wells has enabled farmers to provide life-saving irrigation. The B:C ratio of all crops was quite high in watershed compared with non-watershed villages. The most important fact to keep in mind is that water stored in ponds should be utilized at critical stage of crop growth by the most efficacious method (Singh, 2002).

Conclusion

Although chickpea thrives well under limited water conditions, poor grain yields coupled with a great variability are recorded in genotypes under rainfed conditions. Since drought stress is a complex phenomenon, attempts to measure the degree of tolerance with a single parameter have limited value. Several physiological, morphological and phonological traits may play a significant role in crop adaptation to drought. Thus, drought stress is a great challenge for both agronomic and genetic management strategies. To improve productivity of chickpea, there is a dire need to develop ideotypes that have a more prolific root system which can extract more available soil moisture and make efficient use of it in crop establishment, growth and yield. To achieve the goal of higher and stable productivity, a multidisciplinary team of scientists (plant breeder, agronomist and crop physiologist) should work in coordination. The development of ideotypes and management technology should be location- and season-specific with major emphasis on the improvement of WUE. Before making recommendations, the improved genotypes and technology should be thoroughly tested in on-farm trials. As chickpea is mainly grown under rainfed conditions, the future thrust should be on the following aspects:

1. Genetic advancement and identification of genotypes having large and deep root systems can help in drought-prone environments.
2. Research for incorporating relevant drought resistance traits through biotechnological approaches must be intensified. There seems to be much scope for improving traits using quantitative trait loci (QTLs) and molecular breeding techniques, transgenic with enhanced drought tolerance.
3. Agronomic manipulations in time of sowing, method of sowing, crop geometry and plant density, weed control, nutrient management and mulches should be exploited to improve WUE.
4. Watershed technology for recycling rainwater can contribute greatly to boost the productivity of chickpea.
5. Land levelling and proper drainage of excess rainwater are very important factors to avoid soil erosion for proper resource conservation use.
6. Rapid seed multiplication of drought-tolerant varieties and their proper distribution are very essential.
7. Organization of farmers' training camps can provide the knowledge of new rainfed technology.
8. Where irrigation is available the application of water in right quantity and right time with right method should be applied. Faulty irrigation management may lead to rise in water table and salinization.

References

Agarwal, M.C. and Khanna, S.S. (1983) Efficient soil and water management in Haryana. In: *Annual Report of Haryana Agricultural University*. Hisar, India, p. 118.

Agarwal, M.C., Dhaliwal, A.S., Jaiswal, C.S., Parbhakar, A. and Ahuja, M.S. (1997) *Status of Research on Agricultural Water Management of Northern Region*. AICRP on Water Management, Directorate of Water Management Research, ICAR, Patna, India, p. 138.

Ahlawat, I.P.S. and Rana, D.S. (2002) Agronomic practices and crop productivity. In: Singh, G., Kolar, J.S. and Sekhon, H.S. (eds) *Recent Advances in Agronomy*. Indian Society of Agronomy, New Delhi, India.

Ahlawat, I.P.S. and Rana, D.S. (2005) Concept of efficient water use in pulses. In: Singh, G., Sekhon, H.S. and Kolar, J. (eds) *Pulses*. Agrotech Publishing Academy, Udaipur, India, pp. 313–340.

Basu, P.S. and Singh, D.N. (2003) Physiology and abiotic stresses in chickpea. In: Ali, M., Shiv Kumar and Singh, N.B. (eds) *Chickpea Research in India*. IIPR, Kanpur, India, pp. 137–166.

Blum, A. (1988) *Plant Breeding for Stress Environments*. CRC Press, Boca Raton, Florida.

Bradford, K.J. (1982) Regulation of shoot responses to root stress by ethylene, abscisic acid and cytokinin. In: Waveing, P.F. (ed.) *Plant Growth Substances*. Academic Press, New York, pp. 599–608.

Chandrakar, B.T., Chandravanshi, B.R. and Taunk, S.K. (1991) Effect of irrigation methods on yield potential of chickpea (*Cicer arietinum* L.) in a rice (Oryza sativa) -based cropping system. *Indian Journal of Agronomy* 36 (Suppl.), 1–4.

Cheema, S.S. and Uppal, H.S. (1979) Antitranspirants. In: Malik, C.P. (ed.) *Annual Review of Plant Sciences*, Vol. 1. Kalyani publishers, New Delhi, India, pp. 115–140.

Clive, L.N. and Mary-Jane, E.R. (1994) Response of temperate forage legumes to waterlogging and salinity. In: Passarakli, M. (ed.) *Hand Book of Plant and Crop Stress*. Marcel Dekker, New York, pp. 473–496.

Dastane, N.G., Singh, M., Hukkeri, S.B. and Vamadevan, V.K. (1970) Review of work done on Water Management Research. Navbharat Prakashan, Poona, India, pp. 106.

Deshmukh, D.V., Mhase, L.B. and Jamadagni, B.M. (2004) Evaluation of chickpea genotypes for drought tolerance. *Indian Journal of Pulses Research* 17, 47–49.

Dumbre, A.D. and Deshmukh, R.B. (1983) Yield of gram varieties as affected by sowing dates, fertilizer application and irrigation. *Journal of Maharashtra Agricultural Universities* 8, 300.

FAO (2003) *Production Year Book, 2002*. Food and Agriculture Organization of the United Nations, Rome, Italy.

Fischer, K.S., Johnson, E.C. and Edmeadas, G.O. (1982) Breeding and selection for drought resistance in tropical maize. In: *Drought Resistance in Crops with Emphasis on Rice*. IRRI, Manila, Philippines, pp. 377–399.

Gallacher, A.E. and Sprent, J.I. (1978) The effect of different water regimes on growth and development of green house growth of *V. faba*. *Journal of Experimental Botany* 28, 412–423.

Gale, J. and Hagan, R.M. (1966) Plant antitranspirants. *Annual Review of Plant Physiology* 17, 269–282.

Goodlass, G. and Smith, K.A. (1979) Effects of ethylene on root extension and nodulation of pears (*Pisum sativum* L.) and white clover (*Trifoluim repens*). *Plant and Soil* 51, 387–395.

Gowande, R.L., Jambhulkar, R.D., Bhinge, A.B., Khan, I. and Ramteke, A.S. (1997) Comparison of broad furrow, check basin and border method irrigation. *Punjab Rao Krishi Vidhyapeeth Research Journal* 21, 91–93.

Gregory, P.J. (1988) Root growth of chickpea, faba bean, lentil and pea. Effects of water and salt stresses. In: Summerfield, R.J. (ed.) *World Crops: Cool Season Food Legumes*. Kluwer Academic, Dordrecht, The Netherlands.

Hadas, A. and Stibbe, E. (1973) An analysis of soil water movement towards seedlings prior to emergence. In: Hadas, A., Swartzendruber, D., Rijtema, P.E., Fuchs, M. and Yaron, B. (eds) *Ecological Studies and Physics of Soil Water and Salt*. Chapman & Hall, London.

Harris, D. (1996) The effects of manure, genotype, seed priming and date of sowing on the emergence and early growth of *Sorghum bicolor* L. Moench in semi-arid Botswana. *Soil and Tillage Research* 40, 73–78.

Harris, D., Joshi, A., Khan, P.A., Gothkar, P. and Sodhi, P.S. (1999) On-farm seed priming in semi-arid agriculture: development and evaluation in maize, rice and chickpea in India using participatory methods. *Experimental Agriculture* 35, 15–29.

Hsiao, C.T. (1973) Plant responses to water stress. *Annual Review of Plant Physiology* 24, 519–570.

Hsiao, T.C., O'Toole, J.C., Yambao, E.B. and Turner, N.C. (1984) Influence of osmotic adjustment on leaf rolling and tissue death in rice (*Oryza sativa* L.). *Plant Physiology* 95, 338–341.

Hussain, M.M., Reid, J.B., Othman, H. and Gallaghar, J.N. (1990) Growth and water use of faba bean (*Vicia faba*) to be a subhumid climate. I. Root and shoot adaptations to drought stress. *Field Crops Research* 23, 1–17.

ICARDA. (2004) *National Resource Management. Annual Report*. ICARDA, Aleppo, Syria, pp. 42–44.

Jackson, M.B. and Drew, M.C. (1984) Effects of flooding on growth and metabolism of herbaceous plants. In: Kozlowski, T.T. (ed.) *Flooding and Plant Growth*. Academic Press, New York, pp. 47–128.

Johansen, C., Singh, D.N., Krishnamurthy, L., Saxena, N.P., Chauhan, Y.S. and Kumar Rao, J.V.D.K. (2002) Options for alleviating moisture stress in pulses crops. In: Asthana, A.N. and Ali, M. (eds) *Recent Advances in Pulses Research*. Indian Society of Pulses Research and Development, IIPR, Kanpur, India, pp. 425–442.

Jones, R.J. and Mansfield, T.A. (1970) Suppression of stomatal opening in leaves treated with abscisic acid. *Journal of Experimental Botany* 21, 714–719.

Kaur, S., Gupta, N.K. and Kaur, N. (2002) Effect of osmo and hydropriming of chickpea seeds on crop performance in field. *International Chickpea and Pigeonpea Newsletter* 9, 15–16.

Keatinge, J.D.G. and Cooper, P.J.M. (1983) Kabuli chickpea as a winter sown crop in Syria: moisture relations and crop productivity. *Journal of Agricultural Science* 100, 667–680.

Khanna-Chopra, R. and Sinha, S.K. (1987) Chickpea: physiological aspects of growth and yield. In: Saxena, M.C. and Singh, K.B. (eds) *The Chickpea*. CAB International, Wallingford, UK, pp. 163–190.

Krishnamurthy, L., Kashiwagi, J., Upadhyaya, H.D. and Serraj, R. (2003) Genetic diversity of drought avoidance root traits in the mini-core germplasm collection of chickpea. *International Chickpea and Pigeonpea Newsletter* 10, 21–28.

Lemeur, R. and Rosenberg, N.J. (1976) Reflectant induced modification of soybean canopy radiation balance. III. A comparison of the effectiveness of elite and Kaolinite reflectants. *Agronomy Journal* 68, 30–35.

Ludlow, M.M. and Muchow, R.C. (1990) A critical evaluation of traits for improving crop yields in water-limited environments. *Advances in Agronomy* 43, 107–153.

Marshal, M.C. (1964) Survival of root-nodule bacteria in dry soils exposed to high temperatures. *Australian Journal of Agricultural Research* 15, 273–281.

Miller, P.R., McConkey, G., Clayton, G.W., Brandt, S.A., Staricka, A.J., Johnston, A.M., Layfond, G.P., Schatz, B.G., Baltensperger, L.D. and Neill, K.E. (2002) Pulse crop adaptation in the northern great plains. *Agronomy Journal* 94, 261–272.

Minchin, F.R. and Pate, J.S. (1975) Effect of water aeration and salt regime on nitrogen fixation in a nodulated legume-definition of an optimum root environment. *Journal of Experimental Botany* 26, 60–69.

Minchin, F.R., Summerfield, R.J., Hadley, P. and Roberts, H.E. (1980) Growth, longevity and nodulation of roots in relation to seed yield in chickpea (*Cicer arietinum* L.). *Experimental Agriculture* 3, 241–261.

Morgan, J.M. (1984) Osmoregulation and water stress in higher plants. *Annual Review of Plant Physiology* 35, 299–319.

Musa, A.M., Harris, D., Johansen, C. and Kumar, J. (2001) Short duration chickpea to replace fallow after Aman rice: the role of on-farm seed priming in the high Barind tract of Bangladesh. *Experimental Agriculture* 37, 509–521.

Nagarajah, S. and Ratnasooriya, G.B. (1977) Studies with antitranspirants on tea (*Camellia sinesis* L.). *Plant and Soil* 48, 185–197.

Noble, C.L. and Rogers, M.J.E. (1994) Response of temperate forage legumes to waterlogging and salinity. In: Pessarakli, M. (ed.) *Hand Book of Plant and Crop Stress*. Marcel Dekker, New York, pp. 473–498.

Ofori, C. and Stern, W.R. (1987) Cereal-legume intercropping systems. *Advances in Agronomy* 41, 41–90.

Pasricha, N.S., Aulakh, M.S., Bahl, G.S. and Baddesha, H.S. (1987) Nutritional requirements of oilseed and pulse crops in Punjab (1976–86). *Research Bulletin 15*, Punjab Agricultural University, Ludhiana, Punjab, India.

Passioura, J.B. (1982) The role of root system characteristics in the drought resistance of crop plants. In: *Drought Resistance in Crops with Emphasis on Rice*. Manila, Philippines.

Patel, R.G., Joshi, R.S. and Raman, S. (1987) Effect of water stagnation and nitrogen on growth and yield of chickpea. *Indian Journal of Agronomy* 32, 12–14.

PAU (1981) *Present Status and Future Strategies of Pulse Research in Punjab*. Pulses Section, Punjab Agricultural University, Ludhiana, Punjab, India, p. 16.

Penacarbiales, J.J. and Alexander, M. (1979) Survival of *Rhizobium* in soils under growing drying. *Soil Science American Proceedings* 43, 962–966.

Pezeshki, S.R. (1994) Plant responses to flooding. In: Wilkinson, R.E. (ed.) *Plant–Environment Interactions*. Marcel Dekker, New York, pp. 289–323.

Phillips, D.A. and Torrey, J.G. (1973) Research in grain legume improvement. *Span* 16, 9–11.

Prabha, C., Arora, Y.K. and Wagle, D.S. (1985) Phospholipids of wheat chloroplasts and its membranes under water stress. *Plant Science* 38, 13.

Prihar, S. and Sandhu, B.S. (1968) *Irrigation of Field Crops: Principles and Practices*. ICAR, New Delhi, India, p. 142.

Rawasthorne, S., Headley, P., Roberts, E.N. and Summerfield, R.J. (1985) Effects of supplemental nitrate and thermal regime on the nitrogen nutrition of chickpea (*Cicer arietinum* L.). I. Growth and development. *Plant and Soil* 83, 265–267.

Ritchie, J.T. (1981) Water dynamics in the soil–plant atmosphere. *Plant and Soil* 58, 81–96.

Rupela, O.P. and Kumar Rao, J.V.D.K. (1987) Effect of drought and salinity on symbiotic nitrogen fixation in legumes with emphasis on chickpea. In: *Adaptation of Chickpea and Pigeonpea to Abiotic Stress: Proceedings Consultants Workshop*. 19–21 December 1984. ICRISAT, Hyderabad, India, pp. 123–131.

Saxena, M.C. (1980) Recent advances in chickpea agronomy. In: *Proceedings of International Workshop on Chickpea Improvement*. February/March 1979, ICRISAT, Hyderabad, India.

Saxena, M.C. (1984) Agronomic studies on winter chickpeas. In: Saxena, M.C. and Singh, K.B. (eds) *Proceedings of the Workshop on Ascochyta Blight and Winter Sowing of Chickpeas*. 4–7 May 1981. ICARDA, Aleppo, Syria, pp. 127–137.

Saxena, M.C. (1987) Agronomy of chickpea. In: Saxena, M.C. and Singh, K.B. (eds) *The Chickpea*. CAB International, Wallingford, UK, pp. 207–232.

Saxena, N.P. (2002) Management of drought in chickpea: a holistic approach. In: Saxena, N.P. (ed.) *Management of Agricultural Drought: Agronomic and Genetic Options*. Oxford & IBH Publishing, New Delhi, India, pp. 103–122.

Saxena, N.P., Krishnamurthy, L. and Johansen, C. (1993) Registration of a drought resistant chickpea germplasm. *Crop Science* 33, 1424.

Seginer, I. (1969) The effect of albedo on the evapotranspiration rate. *Agricultural Meteorology* 6, 5–31.

Sekhon, H.S., Singh, G., Gill, A.S., Gumber, R.K., Kumar, K., Singh, D. and Randhawa, A.S. (1994) Influence of environment on chickpea yield. *International Chickpea and Pigeonpea Newsletter* 1, 15–16.

Sekhon, H.S., Singh, G., Chandi, J.S., Sardana, V., Singh, I. and Ram, H. (2004) Effects of planting methods and irrigation on the productivity of chickpea sown after rice. *International Chickpea and Pigeonpea Newsletter* 11, 22–25.

Serraj, R. and Sinclair, T.R. (2002) Osmolyte accumulation: can it really help increase crop yield under drought conditions. *Plant Cell Environment* 25, 333–341.

Serraj, R., Buhariwalla, H.K., Sharma, K.K., Gaur, P.M. and Crouch, J.M. (2004) Crop improvement of drought resistance in pulses: a holistic approach. *Indian Journal of Pulses Research* 17, 1–13.

Serraj, R., Krishnamurthy, K., Kumar, J., Chandra, S. and Crounch, J.H. (2004) Variation in root traits of chickpea (*Cicer arietinum* L.) grown under terminal drought. *Indian Journal of Pulses Research* 88, 115–127.

Sharma, S.K., Tomer, S.S., Srivastava, S.P. and Agarwal, S.B. (1996) Effect of irrigation and phosphorus on the performance of chickpea plants on different dates. *Advances in Agricultural Research in India* 6, 51–58.

Sheldrake, A.R. and Saxena, N.P. (1979) The growth and development of chickpea seedlings under progressive moisture stress. In: Mussell, H.W. and Staples, R.C. (eds) *Stress Physiology in Crop Plants*. Wiley-Interscience, New York.

Siddique, K.H.M. and Loss, S.P. (1996) Deep seeding is beneficial to chickpea in Western Australia. *International Chickpea and Pigeonpea Newsletter* 3, 31–32.

Siddique, K.H.M., Loss, S.P., Regan, K.L. and Jettener, R. (1999) Adaptation of cool season grain legumes in Mediterranean-type environments of south-western Australia. *Australian Journal of Agricultural Research* 50, 375–387.

Singh, D.P. (2002) Drought management in field crops. In: *Recent Advances in Agronomy*. Indian Society of Agronomy, New Delhi, India, pp. 253–277.

Singh, D.P. and Chaudhary, B.D. (1998) Drought tolerance in field crops: from understanding to realization. In: Behl, R.K., Singh, D.P. and Lodhi, G.P. (eds) *Crop Improvement for Stress Tolerance*. CCS Haryana Agricultural University, Hisar and Max Mullar Bhavan, New Delhi, India, pp. 33–36.

Singh, D.P. and Pannu, R.K. (1998) Irrigation management in field crops. In: Yadav, R.L., Singh, P., Prasad, R. and Ahlawat, J.P.S. (eds) *Fifty Years of Agronomic Research in India*. Indian Society of Agronomy, New Delhi, India, pp. 87–140.

Singh, S., Saini, S.S. and Singh, B.P. (2004) Effect of irrigation, sulphur and seed inoculation on growth, yield and sulphur uptake of chickpea (*Cicer arietinum*) under late-sown conditions. *Indian Journal of Agronomy* 49, 57–59.

Sivanappan, R.K. (1995) Need for holistic approach: water management. *The Hindu Survey of India Agriculture*, 156–159.

Sponchiado, B.N., White, J.W., Castillo, J.A. and Jones. P.G. (1989) Root growth of common bean cultivars in relation to drought tolerance in environments with contrasting soil types. *Experimental Agriculture* 25, 249–250.

Sprent, J.I., Minchin, F.R. and Thomas, R.J. (1983) Environmental effects on physiology of nodulation and nitrogen fixation. In: Jones, D.G. and Davies, D.R. (eds) *Temerate Legumes: Physiology, Genetics and Nodulation*. Pitman Books, London, pp. 269–318.

Subbarao, G.V., Johnasen, C., Slinkard, K.E., Rao, R.C.N., Saxena, N.P. and Chauhan, Y.S. (1995) Strategies for improving drought resistance in grain legumes. *Critical Review in Plant Sciences* 14, 469–523.

Summerfield, R.J., Minchin, F.R., Roberts, E.H. and Hadley, P. (1981) Adaptation to contrasting aerial environments in chickpea (*Cicer arietinum* L.). *Tropical Agriculture* 2, 97–113.

Summerfield, R.J., Hadley, P., Roberts, E.H., Minchin, F.R. and Rawsthorne, S. (1984) Sensitivity of chickpea (*Cicer arietinum*) to hot temperatures during the reproductive period. *Experimental Agriculture* 20, 77–93.

Trought, M.C.T and Drew, M.C. (1980) The development of waterlogging damage in wheat seedlings (*Triticum aestivum* L.). I. Shoot and root growth in relation to changes in the concentration of dissolved gases and solutes in the soil solution. *Plant and Soil* 54, 77–94.

Turner, N.C. (1986) Crop water deficits: a decade of progress. *Advances in Agronomy* 39, 1–51.

Turner, N.C., Wright, G.C. and Siddique, K.H.M. (2001) Adaptation of grain legumes (pulses) to water limited environments. *Advances in Agronomy* 14, 193–231.

Uppal, H.S., Mathauda, S.S. and Mahey, R.K. (2002) Water management and crop productivity. In: Singh, G., Sekhon, H.S. and Kolar, J.S. (eds) *Recent Advances in Agronomy*. Indian Society of Agronomy, IARI, New Delhi, India, pp. 278–313.

Visser, E.J.W., Voesenek, L.A.C.J., Vertapetian, B.B. and Jacson, A.R. (2003) Flooding and plant growth. *Annals of Botany* 91, 107–109.

Wery, J., Turc, O. and Lecoeur, J. (1993) Mechanisms of resistance to cold, heat and drought in cool-season food legumes, with special reference to chickpea and pea. In: *Breeding for Stress Tolerance in Cool-Season Food Legumes*. Wiley, Chichester, UK, pp. 271–291.

Zhang, J., Nguyen, H.T. and Blum, A. (1999) Genetic analysis of osmotic adjustment in crop plants. *Journal of Experimental Botany* 50, 291–302.

13 Integrated Crop Production and Management Technology of Chickpea

S. Pande,[1] Y. Gan,[2] M. Pathak[1] and S.S. Yadav[3]

[1]*International Crops Research Institute for the Semi-Arid Tropics (ICRISAT), Hyderabad 502324, India;* [2]*Agriculture and Agri-Food Canada, Airport Road East, Box 1030, Swift Current, Saskatchewan, S9H 3X2, Canada;* [3]*Pulse Research Laboratory, Pulse Research Laboratory, Division of Genetics, Indian Agricultural Research Institute, New Delhi 110012, India*

Introduction

Chickpea (*Cicer arietinum* L.) is a high-value crop and plays a vital role in crop diversification and economic viability of farming. This annual legume is grown in more than 45 countries across five continents. Many countries such as India, Turkey, Pakistan, Bangladesh, Nepal, Iran, Mexico, Myanmar, Ethiopia, Australia, Spain, Canada, Syria and Morocco contribute more than 90% to the global production. Chickpea is cultivated over an area of 10.4 million hectares with a production of 8.57 million tonnes and an average productivity of 826 kg/ha globally. Chickpea is the premier food legume crop in India, covering ≥8.0 million hectares with a production of 7.77 million tonnes. It is a hardy crop and can be grown in marginal lands where the high-input crops fail to give economic returns. Even on marginal lands, higher economic returns from chickpea are possible by the use of integrated crop management (ICM) practices. Canada, Mexico and Australia have obtained an average yield of more than 1.2 t/ha for many years, whereas the global average yield never exceeded 0.8 t/ha. Similar gain in chickpea yield is possible in other countries with improved crop management. The demand for chickpea in 2010 is estimated to be at 11.1 million tonnes. This is a major challenge to the chickpea scientific community, policymakers and extension agencies. A combination of productivity enhancement and area expansion may help achieve this target. The vast rice fallow in South Asia offers an opportunity for expansion of areas for chickpea production (Joshi *et al.*, 2001).

Crop Production Systems and Chickpea

Development of varieties specific to intercropping or sequential cropping systems, incorporation of desired traits and improved adaptability to diverse agroclimatic conditions has made chickpea an attractive proposition for different cropping systems. The numerous alternatives available for intercropping or sequential cropping make chickpea suitable for both rainfed and irrigation farming systems. There is a distinct possibility of introducing chickpea in improved cropping systems to augment the productivity in selected regions.

Continuous monocropping of cereals causes deterioration of soil fertility and quality. There is greater opportunity for crop diversification with chickpea to break the cereal monocropping, and to increase productivity at a systems level. Crop diversification provides continuous income and a variety of food items for family consumption while ensuring optimum utilization of fertilizer, labour and water. However, the progress of diversification depends mainly on its inputs and market constraints, and financial profitability to the farmer. The adoption of crop diversification schemes using chickpea is dictated by a combination of several factors including land capability, rainfall patterns, water quality, crop suitability and available technology options and economic factors.

To make intensive cropping systems an everlasting preposition, diversification of cereals, oilseeds and other cash crops – like cotton and sugarcane – with chickpea has increased tremendously because of its soil amelioration properties and minimum reliance on the external source of nitrogen and augmentation of total productivity. Development of short-duration varieties insulated against major biotic stresses with appropriate production technology has resulted in the identification of several remunerative and more productive cropping systems. They have already demonstrated their yield potential and encouraged the introduction of chickpea in new niches. Cotton–chickpea and rice–chickpea sequential cropping are examples of diversification through chickpea. An estimated 2.5 million hectares of additional area can be utilized for chickpea production through crop system manipulation, crop diversification and multicropping systems. This can lift the national average of cropping intensity significantly. In addition, there is scope for introduction of chickpea in new niches such as wasteland, reclaimed soils and rice fallow by efficient watershed management and as a replacement of less remunerative crops.

Chickpea production in different cropping systems in Asia: the Indian subcontinent

The prevalent cropping systems in the Indian subcontinent are the outcome of the technological innovations, household needs, reflection of government policies, and availability of production inputs, market forces and socio-economic compulsion. Chickpea is grown in a variety of cropping system such as monocropping, double cropping, relay cropping and mixed cropping. In central and peninsular India, where rainfall is not adequate to grow two crops in a year, monocropping is practised. In these areas, chickpea is rotated with sorghum in 2-year cycles. Monocropping is also practised in flood and *Tal* affected areas (diara land) of eastern Uttar Pradesh, Bihar, West Bengal and Orissa, where chickpea is planted after receding of flood waters. Double cropping is practised in areas having assured rainfall (eastern Uttar Pradesh, Bihar, West Bengal) or irrigated facilities (Punjab, Haryana, western Uttar Pradesh). In these areas, chickpea is grown after kharif maize, sorghum, pearl millet, short-duration cotton and early rice. After the harvest of rice, the user of low-energy tillage methods confines the cultivation to a shallow depth giving rise to a compact layer, which may affect root growth of the crop, particularly in clayey soils. However, the major production constraints of the system are limited choice of improved varieties, non-availability of cold-tolerant varieties, more incidence of pod borer and foliar diseases such as ascochyta blight (*Ascochyta rabiei*), and botrytis grey mould (BGM) (*Botrytis cinerea*), besides poor soil tilth and nodulation.

In the Indian subcontinent, including Myanmar (Burma), chickpea is entirely autumn- or winter-sown and the desi (coloured seed coat) types predominate. Several major producing zones may be distinguished, each with its own characteristic cropping patterns and constraints to production:

1. The alluvial soils of the Indo-Gangetic Plain (IGP);
2. The rice fallows of north-eastern India, Nepal, Bangladesh and Myanmar (Burma);
3. The sandy soils of southern Punjab and Haryana, and northern Rajasthan in India, and the Thal region of Pakistan;
4. The black soils of central India;
5. The black soils of peninsular India.

Chickpea is sown in October and November and matures from February to April. Crop duration ranges from ~110 days in peninsular India to 170 days in north-western areas, and is curtailed by increasing temperatures and water stress. Soils are neutral to alkaline, and coarse sands to heavy clays.

In the north-eastern rice fallows, *khesari* or grass pea (*Lathryus sativus*), lentils (*Lens culinaris*) and chickpea are broadcast into standing crops of rice ('utera' or 'paira' cultivation). Chickpea also offers an alternative to black gram (*Vigna mungo*) for a similar situation in the rice fallows of the Krishna and Godavari flood plains of eastern Andhra Pradesh, India. In sandy soils, seeds are sown very deep (up to 25 cm) so as to place them into a moist seedbed.

Chickpea follows various rainy season crops. On the alluvial soils of the IGP, they are sown after pearl millet (*Pennisetum glaucum*), sorghum (*Sorghum bicolor*), maize (*Zea mays*), sugarcane (*Saccharum officinarum*) or guar (*Cyamopsis tetragonoloba*), and eastwards, after sesame (*Sesamum indicum*) or early rice (*Oryza sativa*). They succeed rice in the rice fallows. In the sandy soils of Rajasthan, India, and surrounding areas they commonly follow pearl millet. On the black soils of central and peninsular India, chickpea is usually sown after a period of fallow, as these heavy soils are difficult to work on during the rainy season. Fallowing the rainy season is also common on sandy and alluvial soils, cultivation and 'planking' (i.e. dragging a flat, heavy implement across the soil surface) are practised to conserve moisture, and heavy rolling is used to raise moisture to the surface.

Chickpea is frequently grown as a sole crop but is also intercropped in alternate rows with sugarcane or between wide rows of rape or mustard (*Brassica campestris* var. sarson prair), with linseed (*Linum usitatissimum*), in the IGP; with 'taramira' (*Eruca sativa*) in the sandy soil areas, or with safflower

(*Carthamus tinctorius*) in peninsular India. The crop is also mixed with wheat (*Triticum aestivum*), barley (*Hordeum vulgare*) or linseed in the northern and central areas of the subcontinent and with sorghum in peninsular India.

The *ghed* area of Gujarat, India, deserves special mention. In coastal alluvium, inundated during the rainy season, the entire area supports continuous post-rainy-season chickpea crops, which give seed yields as large as 3000 kg/ha. The *bhal* area, also in Gujarat, with medium black soils inundated for 4 months of the year, supports mainly cotton in the post-rainy season but chickpea is a very productive alternative crop.

Constraints to production also differ between areas. In north-western India and Pakistan, the main limitation is ascochyta blight; further eastwards BGM becomes important. Another major cause of yield instability, and hence the failure of the chickpea to compete with wheat (*Triticum aestivum*) in irrigated areas in North India, is a tendency to produce excessive vegetative growth, which causes crops to lodge, aggravates disease problems and reduces seed yields. In the north-east, the early onset of rains during the reproductive period can inhibit fruiting, delay maturity and interfere with harvesting. Dry seedbeds are an important constraint in Rajasthan and other areas with sandy soils. In peninsular India, the limited availability of water throughout crop growth is a major limiting factor and seed yields have been consistently doubled by timely irrigation. Soil salinity and alkalinity are important limitations in parts of the IGP, sandy soils of Rajasthan and the *bhal* and *ghed* tracts of Gujarat.

The last four decades have witnessed a phenomenal growth of cereal crop productivity in the developing world, particularly rice and wheat in Asia, triggered by the Green Revolution. Increased irrigation allowed rapid intensification, and cereal crops became the primary source of food supply for Asia's escalating population. Rice–wheat cropping system in South Asia emerged as the most important source of food supply (Kataki *et al.*, 2001).

In South Asia, a substantial proportion of land is under a single crop, usually rainy season rice, with the land remaining fallow during the following post-rainy season. This largely occurs for rainfed rice, where irrigation facilities for rice or a post-rice crop are not available. Nevertheless, residual soil moisture can support growth of a legume crop after rice. There are greater opportunities for popularization of chickpea in the cereal-based cropping systems (CBCS) by farmer-participatory development of ICM (Integrated pest management (IPM), integrated disease management (IDM) and integrated nutrient management (INM)) and farmer-preferred cultivars of chickpea that are adapted to local growing conditions. Possible niches for the inclusion of legumes along with chickpea in rice-based cropping systems in Asia are summarized in Table 13.1.

Major breakthroughs in chickpea technologies in Asia

The major abiotic constraints in Asia include drought or heat stress, and cold stress. Pod borer (*Helicoverpa armigera*), ascochyta blight, fusarium wilt (*Fusarium oxysporum* f. sp. *ciceris*), root rots (*Rhizoctonia solani* and *Macrophomina phaseolina*), stunt (bean leaf roll virus), and BGM are the major biotic constraints. Economic losses due to abiotic constraints are generally larger than those from biotic constraints (Ryan, 1995). Chickpea is traditionally grown

Table 13.1. Possible niches for inclusion of legumes along with chickpea in the cereal-based cropping systems in South-east Asia, West Asia and North Africa (WANA) and West Africa.

Country	Cropping systems
South-east Asia	
Bangladesh	Aus rice[a]/jute–fallow–winter legume[b]; aman rice–legumes; aman rice[a]–wheat; mung bean–aman rice–legume (upland)
China	
Central China	Groundnut–rice; rice–groundnut–soybean; groundnut–wheat or rape seed; soybean–rice in second year
Southern China	Groundnut–rice–soybean–rapeseed; faba bean, pea or sweet potato; rice–groundnut–wheat or sweet potato; Soybean–rice seedlings–groundnut
India	
North India	Rice–wheat
Eastern and peninsular India	Rice–chickpea/grass pea/lentil/pea; rice–rice
South-eastern coastal regions	Rice–mung bean/black gram; rice–rice
Indonesia	Rice–rice–groundnut; rice–groundnut–groundnut; Rice–soybean–groundnut or groundnut + maize
Myanmar	Rice–legumes (chickpea, lentil, groundnut); mung bean–rice–legume; sesame–rice–legume
Nepal	Rice–legume (chickpea, lentil, grass pea)–fallow; rice–legume–early rice; rice–wheat–mung bean; rice–fallow–groundnut (spring)
Pakistan	
Sindh	Rice–wheat; rice–chickpea
Baluchistan	Rice–chickpea/grass pea; rice–flax–coriander/pea
Punjab	Rice–wheat; rice–chickpea/lentil/pea/barseem clover
Philippines	Rice–mung bean/groundnut/soybean; rice tobacco/vegetables
Sri Lanka	Rice–other field crops; rice–cowpea/mung bean; rice–vegetables (onion, chillies, etc.); rice–cowpea/mung bean/other field crops
Thailand	Sesame/mung bean/groundnut/soybean/jute–rice; rice–mung bean/black gram/groundnut
Vietnam	Rice–groundnut; rice–mung bean
West Asia and North Africa	
Turkey	Rainfed wheat system
West Africa	
Nigeria	Millet/sorghum–winter cowpea
Ghana	Maize–cowpea

[a]'Aus' rice is sown early in the rainy season; 'aman' rice is sown after the beginning of the main rainy season.
[b]Winter legumes include chickpea, lentil, grass pea and field pea.

in areas where temperatures are cool. This restricted cultivation of the medium- and long-duration varieties to high-latitude areas. These varieties matured far too late when planted in the tropics and succumbed to heat, drought and disease pressure. The development of short-duration varieties that escape the above constraints has enabled expansion of chickpea cultivation to lower latitudes where the crop is not traditionally grown. In these areas, the new varieties have proved profitable, leading to expansion of cultivated areas, e.g. Andhra Pradesh in South India. Success stories of short-duration tropical chickpea are also available from a number of other semiarid states of India. Improved varieties have found niches in the high Barind region of Bangladesh, greening the drylands of Barind, which were earlier left fallow. Following hundreds of field demonstrations, adoption increased from 200 ha in 1984 to 10,000 ha by 1998 (Musa *et al.*, 1998). In Myanmar the local chickpea cultivars and land races suffered heavy losses in production due to fusarium wilt, drought and heat stress. The introduction and release of fusarium-resistant lines and early maturing cultivars have changed the situation (ICRISAT, 2000). In northern latitudes (e.g. North India and Pakistan), varieties with resistance to ascochyta blight have been developed and this has led to increased productivity of chickpea there.

Chickpea production in cropping systems in West Asia and North Africa, and Europe

Chickpea is exclusively spring-sown (except in the Nile Valley) throughout an extensive and ecologically diverse region, extending westwards from Afghanistan through the Middle and Near East, into North Africa and southern Europe. Crops are kabuli (white seed coat) types except in a small area in Iran where desi types are grown. Several major chickpea-producing zones can be distinguished:

1. Mountainous areas of the Middle East (northern Afghanistan, Iran and Turkey);
2. Lowlands of the Middle East (Iraq, Syria, Jordan, Lebanon and Israel);
3. Nile Valley;
4. North Africa;
5. Southern Europe.

Combined production of pulses in these five regions is less compared with that in the Indian subcontinent, and chickpea is superseded in importance by lentils in the Near East, by faba beans (*Vicia faba*) in the Nile Valley and North Africa, and by these and other species in southern Europe.

Chickpea is sown between February and May, timed to coincide with the rising temperatures in spring and the cessation of winter rains. Crop durations vary from 3 to 4 months and growth is terminated by increasing temperatures and water stress in summer. In the Nile Valley, in contrast, chickpea is sown from September to November and matures in spring. The soils are clayey to sandy, and neutral to alkaline.

Chickpea is sole-cropped throughout the region, although it is frequently seen as undercrops in olive groves in northern Syria. It is grown generally where annual rainfall exceeds 400 mm, in two-course (annual) or longer rotations with winter cereals (bread, durum wheat or barley), summer crops (melons) and forages. At the drier end of the range, three-course rotations include a fallow year, which is replaced by a summer crop or forage as rainfall increases. In wetter zones, cereals and chickpea are grown in alternative years.

Crops are rainfed, except in parts of the Nile Valley. Phosphatic fertilizers are frequently applied and, in the Near East, the use of fungicides and insecticides to control ascochyta blight and helicoverpa pod borer, respectively, is uncommon.

Tall genotypes, with fruits borne well above the ground, is being evaluated, as they will facilitate mechanical harvesting. Improved *Ascochyta* and frost resistance, weed control and cold tolerance (in highland areas) will be necessary if the practice of winter sowing is to be widely adopted as an alternative to seedling in the spring.

The West Asia and North Africa (WANA) region extends from Morocco in the west to Pakistan and Afghanistan in the east, and from Turkey in the north to Ethiopia and Yemen in the south. WANA is the centre of origin of wheat, barley, lentils and chickpea, and also of sheep and goats. It is also the birthplace of Judaism, Christianity and Islam. WANA held the non-European parts of the Greek, Roman, Ottoman and Persian empires. The region comprises the Arab countries and several of today's largest non-Arab Islamic countries. This 'social ecology' of WANA must be respected and remembered in any discussion of 'eco-regions', particularly when it comes to matters of management and policy in agriculture. WANA farming systems thus differ from areas in Australia, California, Chile and South Africa with similar crops and climates. 'Farming does not take place in a cultural vacuum' (Fadda, 1992). WANA is characterized by high population growth, low and erratic rainfall, limited areas of arable land, some of the world's biggest and harshest deserts and limited water resources for irrigation. Climates vary from Mediterranean to monsoonal and from temperate to tropical (Tutwiler and Bailey, 1991). Food and feed production prospects are contrasted in 18 of the 25 countries considered by the International Centre for Agricultural Research in Dry Areas (ICARDA), Aleppo, Syria, as part of its WANA region: Afghanistan, Algeria, Egypt, Ethiopia, Iran, Iraq, Jordan, Lebanon, Libya, Morocco, Oman, Pakistan, Saudi Arabia, Sudan, Syria, Tunisia, Turkey and Yemen. Other sources (TAC/CGIAR, 1992) consider Ethiopia and Sudan to be part of sub-Saharan Africa, and Pakistan as part of Asia, so these countries are considered separately here. Chickpea is one of the most important proteinous crops in the WANA region and the Indo-Pak subcontinent. Although the yield potential is comparably high in the developed cold-tolerant varieties, the area under cultivation has not increased substantially. Ascochyta blight has been identified as one of the major limiting factors of cold-tolerant chickpea production in the WANA region. Kabuli-type chickpea is the preferred variety in the WANA region; desi types are more common in the Indo-Pak subcontinent.

Major breakthroughs in chickpea technologies in West Asia and North Africa

Drought is one of the major constraints to chickpea productivity mainly because chickpea is grown in rainfed areas under receding residual soil moisture. In

Turkey, the area under chickpea expanded into the wheat fallow rotations due to the availability of improved varieties and management practices, and supportive government policies (Acikgoz *et al.*, 1994). In other countries in the WANA region chickpea is a spring-grown crop. In WANA where rainfall is in winter, due to susceptibility to low temperatures and ascochyta blight, chickpea is sowed in spring and the crop grows under receding residual soil moisture, which results in low yields. Advancing sowing from spring to winter ensures that most of the vegetative and reproductive growth would occur under moisture-assured conditions and would thus result in high yields. Blending of genetic improvement (incorporation of cold and ascochyta blight resistance) and appropriate agronomic practices allowed the crop to be grown in late autumn or early winter, thus escaping terminal stress. This has substantially increased the yields (Saxena *et al.*, 1996).

Chickpea production in cropping systems in eastern and western Africa

Chickpea is cultivated at altitudes between 1400 and 2300 m in a zone extending from Ethiopia in the north, through East Africa, to Zambia and Malawi in the south. It is the chief food legume in the northern and central highlands of Ethiopia, but has been only recently introduced to the southern regions where it is of minor importance. Sowing generally coincides with the end of the rainy season, from August to September in Ethiopia (harvesting in January to February), and southwards from February to April (harvesting in June to August). Where rainfall is bi-modal, chickpea can be sown at the end of both the rainy periods (e.g. in April and October in Kenya). The crop is usually sown in clayey soils, which are neutral to alkaline. They are almost always grown as mono crops but are also found mixed with safflower and sorghum in Ethiopia, and with maize in Kenya. In Ethiopia, chickpea occurs in rotation with wheat, barley and tef (*Eragrostis tef*). Crops are seldom irrigated or fertilized and pesticide use is minimal. The main problem appears to be the wilt–root rot complex, which causes crop losses up to 80%.

Major breakthroughs in chickpea technologies in eastern and western Africa

The major constraints include drought, helicoverpa pod borer, fusarium wilt, root rot, low yield potential and stunt. Chickpea is usually grown under residual soil moisture. Development of varieties with short to extra-short duration has also enabled chickpea crop to avoid terminal drought and give high yields. Ethiopia and Sudan have released some of these varieties. Ethiopia has also released varieties with resistance to fusarium wilt.

Chickpea production in the Americas and Australia

The Americas and Australia constitute an extensive zone of considerable diversity in climate and cropping practices. Chickpea is a minor crop except in Mexico, which is by far the largest producer in that region. The crop is mainly kabuli type, but desi type is grown in Australia and comprises 80% of Mexico's chickpea production for animal feed. In South and Central America, chickpea is sown in autumn or winter and in rotation with maize, soybean (*Glycine max*), sesame, wheat or pasture. In the USA, Canada and Australia chickpea has been introduced comparatively recently. The main concentrations are in the states of California, Washington and Idaho in the USA, western Canada, and south-eastern Queensland in Australia. Inputs are large and operations are completely mechanized from sowing to harvesting, the latter being successful even though cultivars are of conventional plant habit. Current production is very small, but chickpea offers promise as an alternative to spring cereals in areas where rainfall is marginal.

Major breakthroughs in chickpea technologies in the Americas and Australia

In the early 1990s, Washington State University, USA, released the early maturing, ascochyta blight-resistant desi variety 'Myles', along with a few kabuli types, which has expanded dramatically in western Canada during 1998–2001 as a result of which chickpea production reached 400,000 ha. Desi chickpea breeding and selection programme in Western Australia started in 1992, and by 1997 two high-yielding and high-quality varieties 'Heera' and 'Sona' were released. These varieties dominated the industry until ascochyta blight emerged in 1999. The Department of Agriculture, Western Australia (DAWA), in collaboration with the Centre for Legumes in Mediterranean Agriculture (CLIMA) addressed the new threat, and moderately ascochyta-resistant, cold-tolerant and high-yielding desi lines WACPE 2075 and WACPE 2095 were released in 2004. Simultaneously hundreds of promising breeding lines with a high degree of resistance to *Ascochyta* have been developed and are now being evaluated for seed yield and quality. Substantial effort is also ongoing within the CLIMA research alliance on germplasm enhancement of chickpea for disease resistance, chilling tolerance, adaptation and drought resistance using advanced biotechnology (double haploidy, interspecific hybridization, molecular markers, genetic resources, etc.). CLIMA and DAWA collaboration, with investment from the Council of Grain Grower Organizations Limited (COGGO), has resulted in a planned release of high-quality, ascochyta-resistant, large-seeded kabuli chickpea in Australia in early 2005.

Production Constraints

In the post-Green Revolution period, per capita availability of chickpea declined due to the widening demand–supply gap caused by mismatch in population and production growths. The major constraints that limit production include non-availability of quality seeds of improved varieties, poor crop management, and biotic and abiotic stresses prevalent in the chickpea-growing areas. There is a need for a major breakthrough in the new technology front to increase yield levels through morpho-physiological changes in plant type and successful hybrid technology, development of multiple disease-resistant varieties and tolerance to abiotic stresses and insect pests. The potential of existing technologies in raising the production has not been realized in the farmers' field because of non-availability of quality seeds, poor management practices and low priority given to the crop by farmers.

To achieve sustained production of chickpea, there is a need to understand and address the following productivity constraints: (i) lack of high-yielding varieties adapted to diverse growing conditions; (ii) cultivation of chickpea under resource constraints, particularly under rainfed, marginal and low input environments; (iii) biotic and abiotic stress affecting chickpea; (iv) poor response to high inputs; (v) emerging deficiencies of micronutrients; (vi) non-adoption of improved production technology; and (vii) administrative machinery for replacement of low-yielding varieties with high-yielding, pest- and disease-resistant varieties.

The total yield losses to chickpea by abiotic stresses may exceed those caused by biotic stresses. The important abiotic stresses affecting chickpea productivity are drought, heat, cold and salinity:

Drought – Terminal drought is the most important abiotic stress for chickpea as it is grown under rainfed conditions on residual soil moisture in the post-rainy season. The crop often experiences increasing drought stress during the reproductive growth stage, resulting in early senescence, and reduction in pod and seed development. Two strategies that are employed in chickpea for drought management are escape and tolerance. Developing early maturing cultivars to escape terminal drought is the most effective strategy as it enables the crop to complete its life cycle before the onset of severe drought.

Salinity – Chickpea is very sensitive to salinity and the extent of yield losses depends on the level of soil salinity. Yield losses occur due to reduction in germination, plant growth (biomass) and seed size. In chickpea-growing areas, saline soils are common in West and Central Asia and Australia. Some salinity-tolerant lines have also been identified in India (Dua and Sharma, 1995) and Pakistan (Asharf and Waheed, 1992). A salt-tolerant variety Karnal Chana 1 (CSG 8963), which can be grown in saline soils with electrical conductivity up to 6 dS/m, has been released in India (Dua *et al.*, 2001).

Cold – Chickpea has been traditionally grown in spring in the WANA region. As the crop is grown on residual moisture under rainfed conditions, it is often exposed to high temperature and moisture stress during the pod-filling stage. Advancement of sowing to winter can help escape these stresses and prolong crop duration. Resistance to ascochyta blight and cold tolerance both at seedling and flowering stages are needed in the cultivars for winter sowing in WANA. Chickpea is particularly sensitive to chilling temperatures during reproductive growth as these adversely affect pollen germination, fertilization and pod setting. Considerable yield losses in chickpea can occur due to cold stress in North India, Canada and some parts of Australia.

Diseases and insect pests are two important biotic stresses affecting chickpea productivity.

Diseases – The following are the most important diseases in chickpea production worldwide: Ascochyta blight caused by *A. rabiei* (Pass) Labr.; BGM caused by *B. cinerea* Pers.; fusarium wilt caused by *F. oxysporum* f. sp. *ciceri*; dry root rot caused by *M. phaseolina*; collar rot caused by *Sclerotium rolfsii*; and phytophthora root rot caused by *Phytophthora medicaginis*. The ICM is responsible for disease management. For detailed management strategies and practices, see Chapter 24 (Singh *et al.*, this volume).

Insect pests – Pod borer caused by *H. armigera* Hubner and leaf miner caused by *Liriomyza ciceriana* Rondani are the important insect pests of chickpea in major production areas, whereas seed beetle or bruchid caused by *Callosobruchus* spp. is the most problematic storage pest of chickpea. Farm managers must include insect management in their ICM strategies and practices. For more details, see Chapter 25 (Sharma *et al.*, this volume).

Integrated Crop Production and Management of Chickpea

In the last three decades, attempts to overcome biotic constraints of chickpea production have mainly focused on the use of chemical pesticides and/or host-plant resistance. These single-factor management strategies to combat biotic and/or abiotic constraints were studied in isolation from each other. As a result, the yield losses caused by pest or disease epidemics, along with poor agronomy, remained alarming and significant. There is a greater opportunity to combine best technologies that combat insect pests and diseases with improved agronomical practices and emerge with an ICM package. The ICM package of chickpea provides greater scope and need for its validation, scale-up and scale-out with the involvement of farmers.

The ICM is a holistic approach that coordinates available crop and pest management technologies in an economically and ecologically sound manner. One major component of ICM is the development and use of cultivars with high-yielding, better disease and insect tolerance, and improved adaptability. Other major components include seed treatment with fungicides and insecticides, improved agronomical practices such as use of *Rhizobium*, seed priming (soaking seeds for 8 h in water before sowing), optimum plant population density, optimum seeding time, use of phosphorous (P) fertilizer in low-P soils, use of proper row spacing and applying need-based spray of pesticides to manage biotic constraints.

It is a holistic guiding principle that encompasses all the activities from selection of crop to harvest and storage. Broadly speaking, however, ICM strategies are based on three main pillars: (i) prevention, (ii) monitoring and (iii) intervention.

Most of the ICM activities emphasize heavily on the preventive measures especially before the crop is sown, i.e. at the field preparation or seed treatment stages, since the initial inoculum of the propagule is largely responsible for the subsequent high pest build-up. Agronomic practices such as deep summer ploughing, soil solarization, cleansing of crop refuse, elimination of weed hosts, crop rotation, water management, intercropping, trap crop, border cropping, nutrition management, as well as genetic sources of resistance along with habitat management are the major tools of the prevention.

Regular monitoring is very crucial for pest management as it is one of the most important decision-making tools. An efficient monitoring programme can pay huge dividends in lowering pest control costs. An agro-ecosystem analysis (AESA) based on crop health at different stages of growth, population dynamics of pests and natural enemies, soil conditions, climatic factors and farmers' past experiences is considered for decision making. Field scouting, use of sticky traps, pheromone traps and soil sample analysis for soil-borne plant pathogens are usually employed as monitoring tools. Diagnosis techniques, economic threshold level (ETL) and pest forecasting models are now available to assist in the proper timing of ICM interventions.

Various ICM interventions are devised to reduce the effects of economically damaging pest populations to acceptable levels. Mechanical, biological, cultural and chemical control measures are applied individually or in combination. Some of the principal ICM interventions include cultural and physical measures, monitoring of pests, bio-control agents (bio-pesticides, natural enemies), host-plant resistance (HPR), botanicals and target-specific, less hazardous chemical pesticides.

Since none of the available management options is effective in controlling the major biotic constraints, integration of different available options is desirable to produce an economically viable chickpea crop. An ICM technology for management of chickpea diseases and insect pests was developed at the International Crops Research Institute for the Semi-Arid Tropics (ICRISAT), Hyderabad, India, and the National Centre for Integrated Pest Management (NCIPM), New Delhi, for effective management of ascochyta blight, BGM and pod borer in the IGP of India and Nepal. The IPM technology consisted of the combined use of a moderate level of HPR (such as Avarodhi), wider row spacing, seed dressing with Bavistin + Thiram (1:1) at 3 g/kg seed, and need- or weather-based foliar application of fungicide (Bavistin) and insecticide (monocrotophos). Selected farmers were educated on disease and insect pest identification and adoption of ICM by conducting farmers' orientation camps several times during a crop season. The recommended technology resulted in an increase of grain yield by 96–124% and of net income by 100–400%. The technology was successfully used in rehabilitating chickpea in the rice fallows of the IGP of Nepal and India (Pande *et al.*, 2005).

Farmers' Participatory Chickpea Production Innovations: Success Stories on Impact and Expansion

Farmers' participatory varietal selection

Participatory varietal selection (PVS) is one of the approaches adopted to make popular new varieties. PVS research provides feedback on farmers' preferred traits that can be included in participatory plant breeding (PPB). This increases the efficiency of formal breeding programmes in developing and popularizing varieties appropriate for resource-poor farmers. The current multilocational trial system can be modified without increasing farmers' participation. Alternatively, the problems can be addressed by introducing a major component of participatory varietal testing. Redesigning the trials by decentralization or increased farmer participation is the most efficient solution. Decentralization or greater participation can provide solutions mainly by allowing more replications, particularly those that increase the number of test sites. Adding more researcher-managed test sites in a decentralized testing programme is expensive. Adding farmers in a participatory testing programme is cheaper because there are many farmers who are willing to collaborate with minimal costs involved (Witcombe, 2001).

Realizing the enormous potential of available varieties, and production and protection technologies for breaking stagnation in the production and productivity of chickpea crop, the Indian Council of Agricultural Research (ICAR), New Delhi, launched a National Pulses Development Programme (NPDP) on transfer of pulse production technology by organizing frontline demonstrations in 1989/90. From August 1990, the NPDP has been brought under the ambit of Technology Mission on Oilseeds and Pulses (TMOP) and is being operated in 304 districts of the country. A major thrust in the frontline demonstration programme has been the demonstration of productivity potentials and benefits of latest improved varieties, production and protection technologies, and productive and profitable cropping systems involving chickpea as a component in mixed, relay, sequential, sole and intercropping systems directly by the scientists of All India Coordinated Research Project (AICRP) on chickpea under real farm situations in India.

Rapid strides made over the last two decades in the genetic improvement of chickpea have extended the frontiers of geographic adaptation of the crop well beyond previous delineations, resulting in novel cropping systems both in the traditional and non-traditional regions of cultivation. None the less, agronomic adaptation continues to pose problems due to persisting sensitivity of the crop to temperature and, to a lesser degree, to photoperiod. Unlike many tropical and subtropical legumes, chickpea's developmental responses to temperature and photoperiod can be quite dramatic.

Pande *et al.* (2005) reported the revival of chickpea production in Nepal through ICM and PVS. Chickpea farmers in Nepal have adopted the practice of seed multiplication and sale (cultivar Avarodhi) for the next cropping season. The farmer-to-farmer sale of chickpea seed is quite popular and remunerative in many villages. In a village-level seed system, participating farmers were trained especially to minimize the damage during seed storage, for example, the farmers' seed production association in the villages of Pithuwa, and Lalbandi, Nepal, initiated production of good-quality seed for selling to farming communities. About 5 t of seeds of the improved varieties were harvested in 2003/04 and sold as seed in the subsequent crop season. Similarly, Bardibas village in the Mohathari district of Nepal was identified as a chickpea seed village in 2002/03. In the subsequent season, seed sales spread to different villages within a radius of 10–45 km.

Farmers' participatory ICM success stories on impact and expansion: Bangladesh

The IDM is an integral component of ICM, which is a holistic approach and takes into account all the interventions right from field preparation for the crop to the storage of the grain. A good knowledge of disease epidemiology and control options available to farmers in a given area is needed before an integrated package can be implemented. Shahjahan *et al.* (2003) reported several options to combat BGM disease and helicoverpa pod borer in Bangladesh. The integrated package consisted of a BGM-tolerant and erect variety ICCL 87322; seed treatment (2 g/kg seed) with a mixture of commercial fungicides (Thiram + Bavistin in a 1:1 ratio); sowing chickpea crop in wider row spacing; application of *Rhizobium* inoculum (210 g/ha); diammonium phosphate (DAP) (100 kg/ha), and need- and weather-based foliar spray with pesticides (fungicide and insecticides) to control BGM and helicoverpa pod borer. Most of the participating farmers recorded a yield increase of 80% (up to 2.5 t/ha) by adopting the integrated package over local practice.

Chickpea integrated package adoption in Barind tracts of Bangladesh was very slow in the beginning, but the progression increased with the release of eight improved chickpea varieties in the Barichola series (Barichola 1–8), several of which (especially Barichola 1, 3, 4 and 5) were found suitable under rainfed conditions in the Barind (>2.5 t/ha yield). The recommended rainfed cropping system for the Barind comprising short-duration T-aman rice + green manure (*Sesbania*, grown during April–June) + rabi (winter) crop (chickpea) prompted more adoption of chickpea in the post-rice lands. In addition, transfer of seed-priming technology ensured 25% higher seedling emergence and 47% higher yields of chickpea in the Barind. Most of the farmers who adopted the integrated package recorded a yield increase of 80% over local practice in chickpea production. Today the area seeded with chickpea in the Barind has increased many folds. The present estimate is nearly 10,000–12,000 ha of chickpea in the Barind. Farmers who were earlier unable to produce anything during rabi season are now harvesting a crop worth as much as, or more than, rice, essentially with no additional monetary inputs.

Farmers' participatory ICM success stories on impact and expansion: India

In ICM, farmers are made frontline researchers. ICM in farmers' field schools (FFSs) is aimed to improve farmers' understanding of pest problems, and to develop strategies for overcoming them through experimentation in their own fields. A good knowledge of insect life cycle and disease epidemiology, and control options available to farmers in a given area is necessary before an integrated package, including good agronomy and plant protection, can be implemented. This ICM of chickpea involved improved cultivars with HPR, fungicide seed treatment, foliar application of recommended fungicides and insecticides for foliar diseases, as well as pod borer control.

The NCIPM validated the ICM module in chickpea at different locations representing diverse agroclimatic zones under different cropping systems such as rice–wheat cropping system in Uttar Pradesh, traditional chickpea-growing area in Rajasthan and cotton-based cropping system in Maharashtra during 2001–2003 so as to reduce the losses due to pest infestation in fields under farmers' participatory mode. The ICM module mainly comprised of seed treatment with *Trichoderma* + vitavax, installation of pheromone traps for monitoring of *Helicoverpa*, fixing of wooden T-shaped bird perches, spray of Ha NPV followed by crude neem seed extract (5%) and spray of endosulfan as a last resort. In Maharashtra, sowing of coriander after the tenth row was included as one of the components of ICM. In Rajasthan, seed treatment with chlorpyriphos was included in the module for termite management. The area under IPM varied from 2 ha in Uttar Pradesh to 25 ha in Rajasthan. The ICM module has been further accepted by several farmers and expanded to several villages successfully.

Farmers' participatory ICM success stories on impact and expansion: Nepal

Farmers' participatory on-farm holistic ICM was expanded in RWCS in the Terai region of eastern IGP (Nepal), as a collaborative activity between ICRISAT and Nepal Agricultural Research Council. About 3000 participating and 8000 non-participating farmers from 13 villages in 11 districts were involved in conducting the on-farm research from 1998/99 to 2002/03. The trial consisted of ICM and non-ICM treatments. The ICM package consisted of improved cultivar Avarodhi, seed treatment with a mixture of thiram and carbendazim (1:1) 2 g/kg seed and *Rhizobium* culture 500 g/ha, wider row spacing, soil application of DAP 100 kg/ha and need-based foliar application of carbendazim or chlorothalonil and monocrotophos or endosulfan for control of BGM and pod borer. The non-ICM package consisted of a local cultivar with none of the inputs of ICM.

During 1998/99 trials, there were 110 farmers who adopted ICM technology. In the following year there was a fivefold increase in the number of farmers. The good news kept spreading and 1100 farmers adopted IPM technology by 2000/01. The best result was that ICM technology was firmly adopted by 7000 farmers in 2001/02 and by 11,000 in 2002/03. Incidence of wilt was reduced to 2% in ICM plots across all locations. In these plots, damage due to BGM and pod borer was significantly ($P = 0.05$) lower than for non-ICM plots and the grain yield was 80–120% higher than non-ICM plots across locations throughout.

Pande *et al.* (2005) reported successful demonstration and adoption of ICM technology in Nepal. The ICRISAT–NARC–Natural Resources Institute (NRI) collaborative programme was initiated in 1998 with 110 farmers and increased to 20,000 farmers in 2004/05 crop season on a total area of ~6500 ha. The impact of ICM technology on the livelihood of resource-poor farmers in Nepal was estimated through structured surveys. Data were collected from 200 farmers in mid-western Terai and central Terai. The ICM technology worked as a catalyst to improve the livelihood of poor farmers. The ultimate impacts included an increase of: (i) family income by 80–100%; (ii) protein consumption of the farmer families by 40%; (iii) brick and morter houses by 22%; (iv) labour use in agriculture and non-agricultural jobs by 20%; (v) household expenditure by 45% and payback of debts; and (vi) livestock ownership by 30%.

Chickpea production also economically benefited the farmers by additional wealth generation as follows: (i) seed transaction benefits by selling the excess seed of improved cultivar Avarodhi; average household seed transaction is about 127 kg of seed, which farmers sold to other farmers and also to non-governmental organizations (NGOs) at $0.45/kg; (ii) sale of surplus products; (iii) consumption of chickpea to combat malnutrition; (iv) reduced fertilizer inputs in subsequent rice crop compared with rice–wheat rotation/rice–chickpea rotation reduced the requirements of urea from 71 to 24 kg/ha, DAP from 85 to 61 kg/ha and compost from 7.6 to 5.2 t/ha; and (v) increase in yield of rice in the next season. These economic benefits ranged up to $5000/village. In Lalbandi village, there was a chickpea revolution among the farmers and some earned profits up to $1000 in one season, a large amount in the rural areas of Nepal.

Future Thrust: Research and Development

Research needs and opportunities

Susceptibility of cultivars to biotic stresses adversely affects yield stability in chickpea. In recent years, cultivars resistant to one or more stresses have been bred, which increased the stability of chickpea production. However, single-gene-based resistance proved to be ephemeral in nature due to the emergence of increasingly more virulent races or pathotypes. Therefore, insulation of varieties against major biotic stresses through pyramiding of genes should be taken up with the help of both conventional and molecular tools to meet the challenge posed by highly virulent pathogens like fusarium wilt, ascochyta blight and BGM. Stable resistant sources for many diseases and insect pests besides precise information on important aspects such as identification and characterization of races or pathotypes, rate of emergence of new races or pathotypes, and genetic control are immediately sought for directed improvement in resistance breeding.

The ICRISAT, IARI, National Chemical Laboratory (NCL) and IIPR in India have comprehensive research programmes for the development of transgenic plants possessing resistance to *H. armigera* (Hubn.). At the IIPR, transformed callus and plantlets possessing *npt*II gene have been obtained through *Agrobacterium*-mediated transformation (Mishra and Singh, 2002). However, most of the transformation in chickpea is limited to the transfer of marker genes. Recently, the NCL has reported transmission of a recombinant gene to the progeny of transformed chickpea. However, rooting of transgenic shoots combined with poor shoot growth following rooting has been identified as the major barrier, and needs to be resolved. The ICRISAT has developed transgenics for resistance to *H. armigera* (Hubn.) using Bt genes, which is in the advanced stage (T4) of testing under contained facilities.

An integrated molecular map of chickpea genome has recently been established with a total of 354 markers including 18 STMSs, 96 DAFs, 70 AFLPs, 37 ISSRs, 17 RAPDs, 8 isozymes, 3cDNAs, 2 SCARs and 3 loci, which confer resistance against different races of fusarium wilt (Winter *et al.*, 2000).

Using recombinant inbred lines, genes conferring resistance to race 1 and 4 of fusarium wilt have been identified as closely linked to each other and to two random amplified polymorphic DNA markers, CS 27 and UBC 170. Reports that the genes for resistance to races 1, 4 and 5 of fusarium wilt are located on a linkage group and the gene for resistance to race zero is independent from them indicate clustering of several resistance genes (Winter *et al.*, 2000).

Recently, two quantitative trait loci (QTLs) conferring resistance to ascochyta blight are identified, and molecular markers closely associated with these loci are also located (Santra *et al.*, 2000). Efforts are on to map QTLs associated with tolerance to abiotic stresses and earliness. Selective genotyping for cold tolerance using amplified fragment length polymorphisms is being investigated to identify putative genomic segments that contribute to cold tolerance. The first ever intraspecific genome map has been developed through collaboration between the ICRISAT and Washington State University, USA.

The above-mentioned research areas will contribute towards the effective ICM of chickpea.

Seed systems: bottleneck

It is known that investment in fertilizers, water, plant protection chemicals and other inputs does not yield economical returns without the use of quality seed. Quality seed is the basic, critical and cheapest input for ICM, and for enhancing agricultural productivity and increasing higher net monetary returns per unit area. Returns on investment depend on varietal purity and physical quality of the seed that is used in the production of crops. For centuries, farmers have used their own seed by selecting and saving part of their harvest. These practices are still followed by most of them through organized seed production. Globally, chickpea is mostly cultivated on marginal lands with low fertility and moisture stress as farmers consider that chickpea does not have high potential as a cash crop. Farmers seldom use good-quality seeds and adopt new production technologies. New improved varieties and technologies can potentially increase the production with proper management conditions.

An efficient seed production system that can provide quality seeds at economically viable costs is the backbone of any crop production technology. Seed cost plays an important role in the adoption of ICM technology. The technology itself and other crop management practices are critical in determining the production costs.

Improved seed is one of the major inputs to increase production and improve the grain quality for export. Even though the production area of chickpea has doubled during the last two decades, the quality of seeds available to farmers is far below the actual needs. Suitable varieties for varied agroecological domains are still not available to them.

Quality seed production is a prerequisite for securing production, expanding production areas and increasing productivity of improved cultivars through the ICM. The entire 'seed chain' – nucleus seed → breeder seed → foundation seed → certified seed production and distribution – needs attention. In India, public institutions produce the nucleus seed, breeder seed and foundation seed, whereas various seed-producing agencies produce the certified seed.

A new concept for seed production through the Seed Village Programme (SVP) was implemented in Punjab (India) and Bangladesh. In both countries, this programme has achieved great success. The basic concept of SVP is to produce the seed locally using the following procedures:

- SVP should be organized in a compact area comprising a few villages.
- Selected village(s) should produce enough seed to fulfil the need of a particular area, i.e. block or district.
- Selected farmers should be provided training in seed production.
- Selected area should be irrigated for both summer and kharif (rainy).
- For production of certified seed, foundation seed should be provided. In case of shortage of certified seed, production may be organized from certified stage 1 to certified seed.

Pande *et al.* (2005) reported successful revival of chickpea production in Nepal through the adoption of ICM practices along with seed production through SVP. Chickpea farmers in Nepal have adopted the practice of seed storage for the next season, and some of the farmers preserve the seed to sell to other farmers. The chickpea seed market business is thriving in several villages. Participating farmers store about 15 kg of seed and contact farmers store 25 kg of seed on average for use in the next crop season. The farmer-to-farmer sale of chickpea seed is quite popular and remunerative in many villages. Few enthusiastic young farmers and farmer groups showed interest in large-scale seed production of the improved cultivar Avarodhi due to its low disease susceptibility and high yield. In a village-level seed system, participating farmers were trained especially to minimize the damage caused during seed handling and storage. The farmers' seed production association at village level and at individual level initiated production of good-quality seed for selling to farm communities in Nepal, Bangladesh and parts of India. In the subsequent season, seed sales spread to different villages within a radius of 10–45 km of the seed village.

Socio-economic opportunities

Improved technologies for chickpea production have certainly helped in area expansion in selected pockets but could not raise the overall average yields significantly. The movement to marginal area has increased the risk of production. Chickpea yields in India are more variable than wheat yields (Kelly and Parthasarathy Rao, 1994). Although development of improved varieties is the key to a technical breakthrough in yields, they should be targeted to specific niches in cropping patterns, especially lines with early maturity to escape drought and heat stress, and resistance to major insect pests and diseases. An important priority is to characterize the existing and potential cropping patterns in terms of both agroclimatic and socio-economic factors (Byerlee and White, 2000).

A combination of yield and area expansion would be a possible option in meeting the projected demand in 2010. Demand for chickpea in 2010 is estimated on the basis of growth rates in population and income, and expenditure elasticity of demand (keeping all other factors constant). The chickpea demand in 2010 is estimated at 11.1 million tonnes up from 8.0 million tonnes in 1996/97, an increase of 38%. Approximately 85% of the additional demand is coming from India.

Chickpea area could be expanded in the north-eastern plains of India in rice fallows, lower Myanmar and Bangladesh in the high Barind tract where large areas remain fallow in winter if cultivars with tolerance to high temperature and acidic soils were available. Adoption of available technology should be increased through improved seed multiplication system, improved crop management practices, resistance to biotic and abiotic stresses, and more importantly wide demonstration to farmers and policymakers on the advantage of incorporating food legumes in the farming systems.

Finally, high-yielding varieties should also target the traditional chickpea-growing areas to allow substitution of cereals by food legumes. For example, in India an option would be to enable chickpea to shift to the traditional growing areas in favoured environments to sustain the cereal-based systems, which are presently facing severe soil and environmental constraints.

Government policies and support

Food legumes including chickpea play a key role in soil improvement and income generation. Government interventions play a major role to sustain R&D activities, which are vital for the industrial production of food legumes. Few Asian governments are currently promoting policies to prioritize and promote food legume crops by investing in R&D for improved varieties and agronomic management for food legumes, and providing credit for food legume farmers. Crop diversification as a strategy was adopted by few countries like the Philippines to promote and accelerate agricultural growth, maximize use of land and optimize farm productivity and availability. Countries such as India, Nepal, Bangladesh, Pakistan and the Philippines have included food legumes in diversifying CBCS (rice and wheat). The Ministry of Agriculture and Rural Development (MARD) of Vietnam is implementing a project on 'Developing seed of groundnut and soybean in Vietnam during 2000–2005' with a layout of $1.5 million. This opens up a new orientation to food legume research and development programme in Vietnam. The expansion of areas under food legume production has been identified in the strategy of agricultural development for restructuring cropping systems in several Asian countries. Diversification with food legumes should be considered an important national policy so as to increase productivity in low-productivity areas, because food legumes need less water than cereals.

The government should implement the following policies to achieve the goal of sustainable integrated chickpea crop production and management:

- Establishment of improved seed and dissemination systems;
- Increase in support price and provision of inputs on time;
- Encouragement of private sector in seed production business;
- Availability and quality control of pesticides through legislation;
- Marketing systems should be made favourable to the farmers by taking more steps;

- Technological advantage in cereal cultivation has been further reinforced by strong policy support as the Government of India provided remunerative and assured prices for wheat and rice. Therefore, farmers cultivating cereals not only enjoyed productivity advantages but also benefited from stable and assured economic returns, which raised the crop's relative profitability. Consequently, there was a large shift in land and other resources towards cultivation of these two cereal crops. A similar policy support is urgently required in the case of legumes, especially chickpea, because a cereal–legume rotational system has been proved to be the key in continuously increasing productivity at a systems level.

Several policy measures have been recently initiated by the Government of India to boost the production of legumes including chickpea in the country. In the past, the investment on legumes research had been very low as compared to cereals primarily due to national policy for attaining food security. During the Seventh Five Year Plan, research investment on legumes was Rs 101.2 million, which was raised to Rs 301.9 million in the Eighth Plan. The financial outlay for the Ninth Five Year Plan was more than Rs 850 million. Similarly, the financial outlay for the Tenth Five Year Plan has been raised to Rs 1200 million. Recently, the Indian Ministry of Agriculture has launched an Integrated Scheme on Oilseeds, Pulses, Oil Palm and Maize (ISOPOM) envisaging not only to promote developmental efforts but also to finance research activities, which are vital for improving legumes production in India.

Similarly, the World Bank has also granted ~$250 million (Rs 1000 cr) to the Government of India under the National Agricultural Innovative Project (NAIP) to promote diversification of research activities in India, which will play a vital role in improving the area and production of legumes in the country.

Institutional development

Socio-economic and institutional constraints in chickpea include inadequate extension services and motivation for production technology dissemination due to insufficient funds and mobility, poor technical knowledge among farmers, competition from other profitable or less risky crops, non-availability of seeds of improved varieties and low priority accorded by the government. In spite of recent technological advances, a major breakthrough in chickpea productivity is yet to be witnessed. The government needs to provide adequate policy and institutional support for the production of chickpea in semiarid areas to complement the efforts of scientists in raising productivity levels in this crop. There is an urgent need to shift the yield frontier of chickpea upwards in a wide range of environments. This calls for target-oriented and location-specific R&D efforts, which currently have a low profile. On the basis of the preceding scenario analysis, the future strategies and policies to enhance production and productivity of chickpea in relation to institutional development are:

- Establishment of farmer seed production and distribution systems for proper dissemination;

- Involving the private sector in seed production and distribution;
- Training farmers and extension workers in production technologies;
- Motivating extension services in technology dissemination by providing them sufficient training, funds and mobility;
- Implementing the latest recommendations by researchers and the agricultural department for crop production;
- Availability and quality control of pesticides through legislation;
- Efficient market channel development.

Conclusion

Chickpea is an important component of cropping systems and provides an opportunity to produce diversified food products for humans. Inclusion of this annual legume in cropping systems will provide an opportunity to intensify and diversify cereal-based agricultural systems, making cropping systems more sustainable. Rotation of chickpea with cereal crops makes cereal production more economic and sustainable. Development of second-generation green revolution technologies is crucial for chickpea production. Large variations in yields are currently experienced, which is compounded by intense biotic and abiotic stresses. Policymakers and farmers in the developing world must be aware that monocropping, with conventional cereals, is unsustainable and often presents great risks associated with growing conditions, market and economic investment. Scientists, farmers and policymakers will need to find niches within the existing systems for chickpea production. Public and private sectors should work together to implement a crop diversification strategy for a specific area to enhance chickpea productivity to meet the increasing demand. This could be achieved by integration of efforts from multidisciplinary teams of researchers. The development of new varieties with improved traits is a key component in the promotion of ICM in chickpea. Varieties with resistance to major insect pests and diseases, and responsive to improved management practices, could bring a highly significant change in chickpea production. Development of low-cost, effective and eco-friendly integrated crop production (ICP) and management technologies to enhance chickpea production is desirable.

The global average yield of chickpea (0.75 t/ha) is low compared to that of other legumes. Concerted efforts are needed to enhance the production potential through participatory technology development. There is a need to further refine the ICM strategies for different agroecosystems. The models for prediction of abiotic and biotic stresses responsible for yield reduction have to be developed and validated. The IPM options already established for ascochyta blight, BGM and pod borer need to be further refined and popularized among farmers. A vast opportunity exists for expansion of chickpea area in the rice fallows (about 14 million hectares) of South Asia. However, techniques for sowing in rice fallow still need to be refined to get good plant establishment in such conditions.

The non-availability of sources of resistance for many stresses (drought, ascochyta blight, BGM and pod borer) has been a major constraint in the development of resistant varieties. Efforts need to be strengthened to identify new sources of resistance in the cultivated and wild *Cicer* germplasm, pyramid resistance genes from diverse sources and develop varieties with multiple resistances. The wild *Cicer* spp. should be exploited for enhancing the genetic base of the cultigen and for introgression of genes for resistance or tolerance to stresses. Rapid genetic enhancement can be achieved by using a combination of conventional breeding and biotechnological approaches. Some of the advances made in this field are embryo rescue for interspecific hybridization, double haploid technology for achieving homozygosity, marker-assisted selection and transgenic traits that cannot be improved through conventional methods. Challenges are many, but chickpea scientists should convert them into sustained opportunities for research effort to enhance the science and production of chickpea.

References

Acikgoz, N., Karaca, M., Er, C. and Meyveci, K. (1994) Chickpea and lentil production in Turkey. In: Muehlbauer, F.J. and Kaiser, W.J. (eds) *Expanding the Production and Use of Cool Season Food Legumes*. Kluwer Academic, Dordrecht, The Netherlands, pp. 388–398.

Asharf, M. and Waheed, A. (1992) Screening chickpea (*Cicer arietinum* L.) for salt tolerance. *Tropenlandwirt* 93, 45–55.

Byerlee, D. and White, R. (2000) Agriculture systems intensification and diversification through food legumes: technological and policy options. In: Knight, R. (ed.) *Linking Research and Marketing Opportunities for Pulses in the 21st Century*. Kluwer Academic, Dordrecht, The Netherlands, pp. 31–46.

Dua, R.P. and Sharma, P.C. (1995) Salinity tolerance of kabuli and desi chickpea genotypes. *International Chickpea and Pigeonpea Newsletter* 2, 19–22.

Dua, R.P., Chaturvedi, S.K. and Shiv, S. (2001) *Reference Varieties of Chickpea for IPR Regime*. Indian Institute of Pulses Research, Kanpur, India, p. 34.

Fadda, N.R. (1992) Foreword. In: Ryan, J. and Matar, A. (eds) *Fertilizer Use Efficiency Under Rain-Fed Agriculture in West Asia and North Africa (ICARDA-067)*. International Centre for Agricultural Research in the Dry Areas, Aleppo, Syria.

ICRISAT (2000) *Bringing Hope to Marginal Environments: Chickpea Improvement at ICRISAT*. ICRISAT, Hyderabad, India, p. 50.

Joshi, P.K., Parthasarathy Rao, P., Gowda, C.L.L., Jones, R.B., Silim, S.N., Saxena, K.B. and Jagdish Kumar (2001) *The World Chickpea and Pigeonpea Economics: Facts, Trends, and Outlook*. ICRISAT, Hyderabad, India, p. 68.

Kataki, P.K., Hobbs, P. and Adhikary, B. (2001) The rice–wheat cropping system of South Asia: trends, constraints and productivity – a prologue. *Journal of Crop Production* 3(2), 1–26.

Kelly, T.G. and Parthasarathy Rao, P. (1994) Chickpea competitiveness in India. *Economic and Political Weekly* 29(26), 89–100.

Mishra, S.K. and Singh, N.P. (2002) Transgenic: a reality for mitigating yield losses due to biotic stresses in pulse crops. In: Ali, M., Chaturvedi, S.K. and Gurha, S.N. (eds) *Proceedings of the National Symposium on Pulses for Sustainable Agriculture and Nutritional Security*. Indian Society of Pulses Research and Development, Kanpur, India, pp. 82–86.

Musa, A.M., Shahjahan, M., Kumar, J. and Johansen, C. (1998) *Farming Systems in the Barind Tract of Bangladesh*. Presented at the Farming Systems Symposium at the International Congress of Agronomy, Environment and Food Security for the 21st Century, 23–27 November 1998. Indian Society of Agronomy, New Delhi, India.

Pande, S., Stevenson, P., Rao, J.N., Neupane, R.K., Chaudhary, R.N., Grzywacz, D., Baurai, V.A. and Kishore, G.K. (2005)

Reviving chickpea in Nepal through integrated crop management, with emphasis of botrytis gray mold. *Plant Disease* 89(12), 1252–1262.

Ryan, J.G. (1995) A global perspective on pigeonpea and chickpea sustainable production systems: present status and future potential. In: Sinha, S.K. and Paroda, R.S. (eds) *Production of Pulse Crops in Asia and the Pacific Region*. Regional Office for Asia and the Pacific, Food and Agriculture Organization of the United Nations, RAPA/FAO Publication No. 1995/8, Bangkok, Thailand, pp. 225–248.

Santra, D.K., Tekeoglu, M., Ratnaparkhe, M.L., Kaiser, W.J. and Muehlbauer, F.J. (2000) Identification and mapping of QTLs conferring resistance to ascochyta blight in chickpea. *Crop Science* 40, 1606–1612.

Saxena, N.P., Saxena, M.C., Johansen, C., Virmani, S.M. and Harris, H. (eds) (1996) *Adaptation of Chickpea in the West Asia and North Africa Region*. ICRISAT, Hyderabad, India; ICARDA, Aleppo, Syria, p. 270.

Shahjahan, M., Razzaque, M.A., Rahman, M.M., Bakr, M.A., Hamid, A. and Pande, S. (2003) *Twenty-Five Years of Bangladesh NARS-ICRISAT Partnership: Research and Development in Chickpea, Pigeonpea and Groundnut*. BARC, Dhaka, Bangladesh; ICRISAT, Hyderabad, India.

TAC/CGIAR (1992) *Review of CGIAR Priorities and Strategies*, Part 1. TAC Secretariat, Food and Agriculture Organization, Rome, Italy (mimeo).

Tutwiler, R.N. and Bailey, E. (1991) Agricultural growth and sustainability: conditions for their compatibility in the semi-arid tropics (Middle East). Paper prepared for DSE/IFPRI conference on 'Agricultural sustainability, growth and poverty alleviation: issues and policies', 23–27 September 1991. Feldafing, Germany.

Winter, P., Benko-Iseppon, A.M., Huttel, B., Ratnaparkhe, M., Tullu, A., Sonnante, G., Pfaff, T., Tekeoglu, M., Santra, D., Sant, V.J., Rajesh, P.N., Kahl, G. and Muehlbauer, F.J. (2000) A linkage map of the chickpea (*Cicer arietinum* L.) genome based on recombinant inbred lines from a *C. arietinum* × *C. reticulatum* cross: localization of resistance genes for Fusarium wilt races 4 and 5. *Theoretical and Applied Genetics* 10, 1155–1163.

Witcombe, J.R. (2001) Participatory variety selection in high potential production systems. In: *Proceedings of a Workshop: An Exchange of Experiences from South and Southeast Asia: International Symposium on Participatory Plant Breeding and Participatory Plant Genetic Resource Enhancement*. 1–5 May 2000. Pokhara, Nepal, pp. 19–27.

14 Commercial Cultivation and Profitability

A.A. Reddy,[1] V.C. Mathur,[2] M. Yadav[3] and S.S. Yadav[4]

[1]*Indian Institute of Pulses Research, Kanpur 208024, Uttar Pradesh, India;* [2]*Division of Agricultural Economics, Indian Agricultural Research Institute, New Delhi 110012, India;* [3]*College of Business Administration, Tarleton State University, (Texas A&M University System), Stephenville, TX 76402, USA;* [4]*Pulse Research Laboratory, Division of Genetics, Indian Agricultural Research Institute, New Delhi 110012, India*

Introduction

Chickpea (*Cicer arietinum* (L.)), commonly known as bengal gram or garbanzo, belongs to the family *Leguminosae* and is a major pulse crop that contributes ~20% of the world pulse production after dry beans and dry peas. There are two types of chickpea: the desi type (mostly brown seeded), traditionally grown in warmer climates of Asia and Africa, mostly in India, Pakistan, Bangladesh, Myanmar and Ethiopia; and the kabuli type (white-seeded), a large-seeded variety more suited to the temperate climates of Turkey, Mexico, the USA, Afghanistan and Iran (West Asia, North Africa and America). Some countries like Canada and Australia produce both desi and kabuli types. Kabuli type constitutes ~15% of global chickpea production and desi type constitutes the remaining 85%.

Production

Globally, chickpea is grown in approximately 40 countries across all continents. In 2005, the world chickpea production was 9.2 million tonnes on area of 11.2 million hectares. Most of the chickpea is produced in developing countries; in 2005, these countries accounted for ~97% of the world area and 96% of the world production. Asia accounts for 91% of the global area and 90% of the global production. Africa and North and Central America contribute ~4% of the global production, each from 4% and 2% of global area, while Oceania accounts for 2% of the area and production. South America contributes a mere 0.08% of global production and 0.06% of the global area (FAOSTAT database, 2006).

During the last 30 years, global production of chickpea has increased by ~64% from 5.2 million tonnes in 1975 to 9.2 million tonnes in 2005. Although the area under chickpea cultivation has increased by 18% during the same period, i.e. from 9.5 million hectares in 1975 to 11.2 million hectares in 2005, much of this came from productivity increases. World chickpea yields have increased ~38% during 1975–2005, rising from 591 (1975) to 818 kg/ha (2005). Although this crop is cultivated in all the continents around the world, India, Pakistan, Myanmar, Australia, Canada, Iran, Turkey, Syrian Arab Republic, Mexico and Ethiopia are the major producing countries, together accounting for ~94% of the average global production during 2003–2005. Each of these countries, excluding Syrian Arab Republic, has a share of 1% or more in world chickpea production; the share of Syria in world production is slightly less than 1%, but more than 0.75%. Table 14.1 shows the temporal changes as well as the growth in average production and productivity of chickpea during the last decade in these major chickpea producing countries.

India is the largest producer of chickpea in the world, accounting for 64% of the global production during 2003–2005. Pakistan, Turkey and Iran are the other major chickpea-producing countries, accounting for 8.65%, 7.35% and 3.73%, respectively. The other Asian countries that produce significant amounts of chickpea are Myanmar (2.73%) and Syria (<1%).

Although these countries account for ~87% of the world production, a relatively low growth in production was observed during 1996–2005. The compound annual rates of growth in production are low for all these Asian countries except for Myanmar, where chickpea production increased at an annual rate of 15.28% (Table 14.1). In India, production has remained stagnant as indicated by an annual rate of growth of −0.57%. The stagnation or low growth in production and productivity can be attributed to increased production and productivity of cereals like rice and wheat, as well as other irrigated crops, to ensure food security to the neglect of pulses, which continue to be grown in fragile environments. Among these countries, Myanmar is the only exception with a high annual compound rate of growth in productivity of 8.6%. In recent years, China has emerged as an important chickpea producer with production and productivity growth at 15% and 8%, respectively, during the last decade.

Exports

Sources of world exports of chickpea are widely dispersed across the continents. Asia accounted for a little more than 40% of the average global exports during 2002–2004, of which the share of South Asia was ~2%. North and Central America contributed ~31% of the global exports, of which developed countries of North America shared 14%. Oceania accounted for ~17% of the global exports during 2002–2004, followed by Africa (7%) and Europe (4%). The major exporting countries are Australia, Canada, Iran, Mexico, Tanzania and Turkey, which together accounted for 83% of the global exports. Other important exporting countries are Ethiopia, Morocco, Pakistan, Russia, Syrian Arab Republic, United Arab Emirates (UAE) and the USA, all of which had an

Table 14.1. Temporal changes and growth in production and productivity of chickpea in major producing countries. (Computed by the authors using FAOSTAT database, 2006.)

	Production				Productivity			
Country	Total estimate 1993–1995 ('000 t)	Total estimate 2003–2005 ('000 t)	CAGR 1996–2005 (percentage/ annum)	Coefficient of variation 1996–2005 (%)	Average yield 1993–1995 (kg/ha)	Average yield 2003–2005 (kg/ha)	CAGR 1996–2005 (percentage/ annum)	Coefficient of variation 1996–2005 (%)
Australia	183	169	–4.53	24.98	990	1154	3.06	29.70
Bangladesh	64	10	–20.83	88.42	733	729	–0.13	5.63
Canada	1	72	26.38	106.11	1385	1306	–0.79	14.38
Ethiopia	86	128	0.33	16.31	598	806	–1.40	9.20
India	5278	5300	–0.59	16.71	774	784	0.56	6.38
Iran, Islamic Republic	316	310	1.81	18.61	485	411	1.06	10.25
Mexico	163	240	2.36	24.90	1522	1600	0.21	5.80
Myanmar	78	229	15.28	49.25	625	1104	8.60	27.64
Pakistan	439	718	–0.34	24.91	421	706	0.98	20.34
Syrian Arab Republic	48	63	2.02	32.10	634	671	–0.20	18.30
Turkey	707	610	–1.49	10.75	928	953	0.10	5.76
World	7677	8302	–0.13	5.63	729	781	0.74	4.23

CAGR: Compound annual growth rate.

average share of >1% during 2002–2004. India, though the largest producer of chickpea in the world, only accounted for 0.78% of the global exports.

The shares of some important chickpea-producing countries are shown in Table 14.2. The cumulative share of the various countries shown in Table 14.2 is 93% of world exports. Three South Asian countries, India, Pakistan and Bangladesh, account for 50% of world imports. This indicates that even though South Asia produces ~73% of world production, it is still deficit in chickpea production (Fig. 14.1). Sources of world exports widely spread across continents, i.e. Turkey, Australia, Mexico, Iran, Canada and Tanzania, account for 80% of world chickpea exports. Developing countries are net importers, while developed countries are net exporters. Chickpea production in Asia, Africa and Europe is deficient, whereas it is in surplus in America and Oceania. Yield levels are high in America, Oceania and Europe compared with developing countries (Fig. 14.1). Many new countries entered into chickpea production considering the continuing opportunities in exports and diversified uses like animal feed and the major ingredient of snack and other food items. Production loss due to biotic (helicoverpa pod borer [$500 million], fusarium wilt [$250 million], ascochyta blight [$250 million], botrytis gray mold [$100 million]) and abiotic (drought $1 billion) stresses is very high, as it is mostly grown in semiarid tropics by resource-poor farmers. As a result, world chickpea production is very unstable with a coefficient of variation of 10%. The major stresses for chickpea yield reduction cause huge loss ($2 billion).

Table 14.2. Share of important producing and exporting countries of chickpea in global exports during 2002–2004. (Computed by the authors from FAOSTAT database, 2006.)

Country	Average exports (t)	Share in world exports (%)	Exports as proportion of domestic production (%)	Unit value ($/t)
India	5,790	0.78	0.11	571
Australia	129,206	17.44	85.38	316
Canada	89,473	12.08	97.5	375
Ethiopia	17,123	2.31	11.76	303
Iran	104,017	14.04	Neg	529
Mexico	122,436	16.53	51.37	627
Morocco	9,524	1.29	20.96	381
Pakistan	9,460	1.23	1.72	312
Russia	11,331	1.53	84.98	223
Syrian Arab Republic	12,719	1.72	17.26	435
Tanzania	24,370	3.29	77.76	290
Turkey	142,448	19.23	22.85	468
UAE	12,375	1.67	–	333
USA	16,776	2.64	–	534
World	740,877	100.00	9.25	451

Neg: negligible; –: proportion could not be computed as data on domestic production in UAE and the USA are not available.

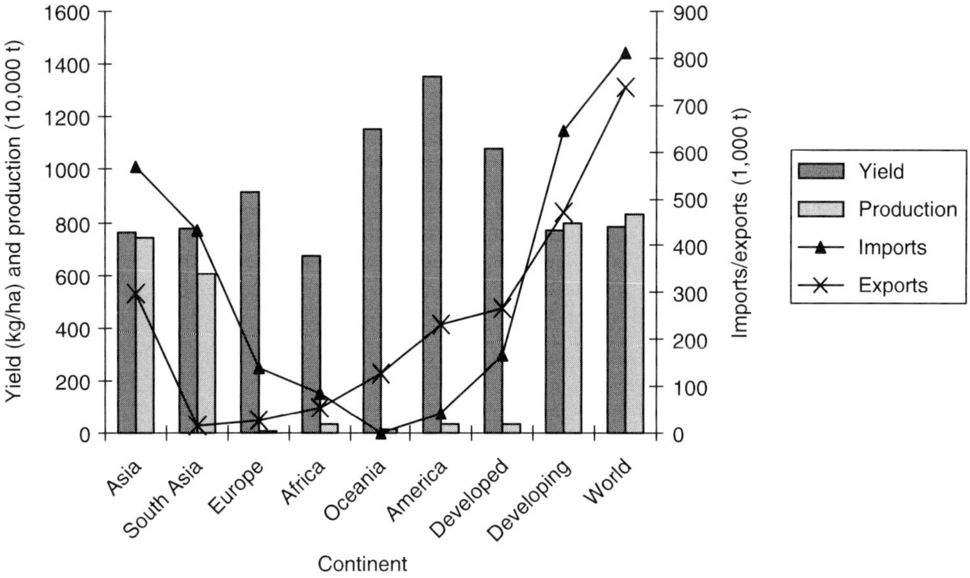

Fig. 14.1. World production, yield, import and export scenario (2005).

Imports

During 2002–2004, 19 countries around the world accounted for 87% of global chickpea imports (Table 14.3). Bangladesh, India, Pakistan and Sri Lanka in South Asia accounted for 51% of the global imports. Although India is the largest producer of chickpea in the world, it also accounts for nearly a quarter of the world imports. Algeria, Jordan, Tunisia, Saudi Arabia, UAE, Spain, Italy, the UK and the USA are some of the other major chickpea-importing countries.

Export Prices

Unit values of exports per tonne vary considerably between countries. The average unit value of exports for the world as a whole was $451/t during 2002–2004 (Table 14.2). For other important exporting nations, unit values of exports ranged between $223 and $627. For several European and South-east Asian countries, the realized export prices per tonne were considerably higher than those of the major exporting countries. High export prices may be attributed to the differences in quality of the chickpea, particularly in terms of the large-sized seeds as in the case of the kabuli type.

The data presented above show that there is significant diversity in the geographical distribution of production, exports and imports of chickpea. In recent years, the share of the top three exporters (Turkey, Australia and Mexico) has declined substantially and many new countries have started producing and exporting chickpea in international markets. The recent trend of reduction in

Table 14.3. Share of important importing countries of chickpea in global imports during 2002–2004. (Computed by the authors from FAOSTAT database, 2006.)

Country	Average imports (tonnes) total estimate 2002–2004	Share in world imports (%) total estimate 2002–2004	Imports as proportion of domestic production (%) total estimate 2002–2004
Algeria	44,625	5.34	273
Bangladesh	81,803	9.79	767
Colombia	10,354	1.24	–
Egypt	5,958	0.71	49
France	10,430	1.25	–
Greece	4,829	0.57	–
India	203,103	24.31	4
Italy	23,841	2.85	391
Jordan	22,721	2.72	1187
Lebanon	8,568	1.03	368
Pakistan	124,880	14.95	23
Portugal	11,637	1.4	1030
Saudi Arabia	21,342	2.55	–
Spain	56,714	6.79	95
Sri Lanka	19,861	2.38	–
Tunisia	19,302	2.31	162
UAE	28,156	3.37	–
UK	18,587	2.23	–
USA	11,925	1.43	–
World	835,317	100.00	–

–: could not be computed as data on domestic production are not available.

international prices of chickpea is significant as it reflects a shift in production from major consuming countries to the more competitive low-cost producers such as Iran, Canada, Tanzania and Ethiopia as in other agricultural commodities. The world chickpea markets are highly segmented in line with its final use. One is for large-sized kabuli chickpea for human consumption, for which demand is limited, but there is high premium for such extra large-seeded kabuli type; the other is the desi type market for human consumption as dhal and flour, for which competition is mainly through low price and low cost of production for reasonable quality of chickpea. Another is for feed purpose, for which the price determines the competitiveness of the producer. There is limited scope of expansion of demand for large kabuli types as the market is small and demand is inelastic. Demand for desi type is somewhat elastic, as at lower prices it can substitute vegetables and cereals in many developing countries where per capita income is low. Demand for desi types for animal consumption is highly elastic, as at lower prices it can substitute dry peas, soybean meal and other animal feeds in both developing and developed countries.

Competitiveness and Profitability

Given the highly diversified production and consumption scenarios across countries and the increase in world chickpea trade (global trade in chickpea [exports plus imports] increased by 129% and 43% in quantum and value terms, respectively, between 1995 and 2004), there is a need to examine the competitiveness and profitability of chickpea cultivation in major producing and consuming countries. The structure of production and marketing costs and quality are the major factors that determine a country's competitiveness in the international market. The acreage under chickpea is determined not only by international competitiveness, but also by the profitability of competing crops within the country. This section deals with the production costs and profitability of chickpea in some important countries. For comparison, five developed countries, Australia, Canada, the USA, Spain and France, and four developing countries, India, Turkey, Algeria and Jordan, have been selected for detailed study. These countries represent a mix of chickpea producers, exporters and importers. As India is the largest producer and consumer of chickpea in the world, it has been selected as a case to examine the availability of improved production technology and constraints in its adoption, potential and actual yields and yield gaps and cost of production.

Developed Countries

The five developed countries considered for the purposes of comparison are Australia from Oceania, Canada and the USA from North America, and Spain and France from Europe.

Australia

Australia produces ~180,000 t of chickpea annually. It produces both desi and kabuli types. Approximately 85% of the total production is exported. Australian chickpea production is characterized by high variability in yield and acreage due to frequent outbreaks of ascochyta blight. Recently, there has been significant reduction in the production of chickpea. During 1996–2005, although yields increased at an annual rate of 3.06%, growth in production was negative at –4.53% per annum. Variability was high in both productivity and production at ~30% and 25%, respectively. Chickpea harvest often clashes with wheat harvest and traditionally wheat has been given priority in Australia. Average yield of chickpea during 2003–2005 was 1154 kg/ha, while the yield of wheat was 2119 kg/ha (FAOSTAT database, 2006), which indicates that chickpea price should at least be double that of wheat in order to be competitive.

Table 14.4 depicts the details of cost–benefit analysis of chickpea in medium- and low-risk ascochyta blight regions in Australia. Like in the USA and Canada, farmers use all the recommended practices suggested by agricultural scientists. As average farm size is more than 500 ha, farms are completely

Table 14.4. Cost of cultivation ($/ha) of Australian desi chickpea in 2004. (From Wayne *et al.*, 2003. Available at: www.pir.sa.gov.au/factsheets)

	Medium-risk ascochyta blight			Low-risk ascochyta blight		
Item	Number of operations	Cost/unit	Cost/ha	Number of operations	Cost/unit	Cost/ha
Glyphosate fallow spray	3 × 1.2 l/ ha	5.50/l	19.8	3 × 1.2 l/ha	5.50/l	19.8
Planting	1	4.50/ha	4.5	1	4.50/ha	4.5
Ground rig (own equipment)	8	2/ha	16	6	2/ha	12
Harvesting (own equipment)	1	10/ha	10	1	10/ha	10
Transport (100 km contract)	1 × 1.5 t/ha	10.00/t	15	1 × 1.3 t/ha	10.00/t	13
Labour cost	1	10/ha	10	1	10/ ha	10
Planting seed (treated)	45 kg/ha	1.20/kg	54	45 kg/ha	1.20/kg	54
Inoculation	1	3/100 kg seed	1.35	1	3/100 kg seed	1.35
Simazine (PSPE)	1.5 l/ha	6.00/l	9	1.5 l/ha	6.00/l	9
Dithane DF fungicide	2 × 1 kg/ha	7.50/kg	15	–	–	–
NPV insecticide + Aminofeed	1 × 375 ml/ha	70.00/l	27	1 × 0.375 l/ha	70.00/l	27
Larvin 375®	–	–	–	1 × 0.75 l/ha	25.00/l	18.75
Desiccation (glyphosate + ally)	1.2 l + 5 g/ha	9.60/ha	9.6	–	–	–
Agronomist	1	4.00/ha	4	1	4.00/ ha	4
Total variable cost	–	–	196	–	–	183
Yield (1.5 t/ha)	–	–	–	–	1.3 t/ha	–
Price ($300/t)	–	–	–	–	–	–
Gross income	–	–	450	–	–	390
Net return over variable cost	–	–	254	–	–	207
Net return over total cost assuming fixed cost ($100/ha)	–	–	154	–	–	107

mechanized. Economies of scale operate at farm level, resulting in low operating cost per hectare of $183 and $196/ha, respectively, in medium- and low-risk areas. Farmers use herbicides, pesticides, fungicides and inoculants as per recommended practice. The yield reported is 1500 kg/ha in medium-risk zones and 1300 kg/ha in low-risk zones. At a selling price of $0.3/kg, the net profit over variable cost is approximately $254 and $207/ha. If we assume a total fixed cost of $100/ha, the net profit over total cost is $154 and $107/ha in medium- and low-risk areas, respectively. Profits are lower in Australia when compared with the USA, because desi types are largely cultivated here, which fetch a lower price of $0.3/kg as against $0.49/kg in the USA.

In the international market, Australian varieties are regarded as inferior to both Canadian and US varieties. There is need to increase area under kabuli type to realize higher prices in the international market. It is likely that kabuli chickpea material bred at the International Center for Research in Dry Land Agriculture (ICARDA), Aleppo, Syria, may have considerable potential for Australia, especially considering the disease-resistant traits of some of the germplasm (Siddique *et al.*, 2004). Ascochyta blight is a serious problem and is the major cause of reduction in area under chickpea in recent years. Australia is expected to enjoy a net benefit of $0.89 million annually for the next 20 years, if disease-resistant varieties are propagated (Siddique *et al.*, 2004). Given the favourable climatic conditions in South Australia and the current chickpea production deficits in Asia and Africa, there is considerable potential for chickpea exports from Australia.

Canada

Chickpea acreage in Canada constitutes a small share of the total cropped area in Canada and appears to be declining in recent years. Area under chickpea declined from 283,000 ha in 2000 to 65,000 ha in 2005. Approximately 90% of chickpea produced in Canada is of kabuli type. Cultivation includes the large kabuli type with a seed size of 8–9 mm and the small kabuli type with a more uniform seed size of ~7 mm. Yields of the desi and small-seeded kabuli types are ~20% higher than that of the large-seeded kabuli types. The maturity period is 95–105 days for desi type and 100–110 days for kabuli type. Cultivation of kabuli type is more dispersed and therefore shows less variability than desi type. The Canadian chickpea harvest generally occurs during the period mid-August to early October (Agriculture and Agri-Food Canada, 2005).

Chickpea acreage in Canada ranges from 50 to 500 ha. All farm operations including land preparation, sowing, harvesting and threshing are done mechanically. The farmers cultivate modern improved varieties and apply herbicides, insecticides and fertilizers as recommended. Due to large-scale adoption of improved production, plant protection practices and mechanization, farm produce is of uniform quality and fetches high prices in international markets. Uniform quality also facilitates easy grading, postharvest operations and branding.

However, in recent years, area under chickpea decreased by 55% due to: (i) lower potential returns from kabuli type; (ii) increased costs and risks of ascochyta blight disease; and (iii) higher risk associated with this crop as compared with some of the alternative crops such as wheat. Even though the gap between yield of wheat (2599 kg/ha) and chickpea (1509 kg/ha) is narrow in Canada, producers feel that chickpea production is highly risky in relation to both productivity and price. In order to compete in international markets, desi types are replaced by kabuli types in recent years. Iran, Mexico and Syria are the major competitors of Canada for kabuli-type chickpea exports. Although the yields of kabuli types are nearly 20% lower than that of desi types, the international price premiums are sufficient to compensate lower yields. Chickpea farm gate prices range between $0.22–0.40/kg for desi type and $0.48–0.88/kg for 9 mm kabuli type. The international demand scenario of main product segments in which Canada is interested is given below (Agriculture and Agri-Food Canada, 2004; Saskatchewan Pulse Growers, 2005).

Prices

Canadian prices are largely determined in the international markets because Canada exports most of its production. Although prices of the large-seeded kabuli are higher than prices of the desi, they are also more volatile. Prices of the large-seeded kabuli increase as the seed size increases from 7 to 10 mm. The producer receives a weighted average price for kabuli chickpea based on the percentage of various seed sizes. The price of the small-seeded kabuli type is generally higher than that of the desi type, but lower than the weighted average price of the large-seeded kabuli type. Prices are negotiated directly between the dealers and customers based on supply and demand factors for each type of chickpea. The prices negotiated could be for immediate delivery or for delivery at some future date.

Large-calibre chickpea (8–9 mm)

Canada directly competes with Mexico for export of large-calibre kabuli chickpea to India, Europe, North Africa and the Middle East. Mexico is best suited for large-calibre chickpea production but, since kabuli chickpea prices have been in a slump, other crops have tended to replace it, resulting in significant contraction in area under kabuli type in recent years. The USA also produces large-calibre kabuli types, but its production has shrunk in response to low prices.

Small-calibre chickpea (7 mm)

Canada exports small-calibre chickpea mostly to Pakistan. Hence, increased demand for Canadian small-calibre chickpea depends upon production prospects in Pakistan, demand and supply position of large-calibre kabuli chickpea in Canada and competition from Australia. Competitive harvest cycles in importing countries and competing exporting countries are important for exploring enhanced export opportunities. India and Pakistan typically harvest their winter crop in March–April; Mexico also tends to harvest in March–April. Turkey and Syria harvest chickpea in May–June. In the USA, the European Union (EU) and Ukraine, harvest time is July. Canada harvests chickpea from mid-August to early October. In Australia, chickpea is harvested during December and January. Canada and Australia can easily adjust their acreage according to the production trends in Pakistan and India. Sowing time in Canada is March–April and in Australia is July–August, by which time production figures of Pakistan and India are available.

Ascochyta blight is a severe problem in Canada also. Estimates of 2005 reported that only 70% of seeded area under this crop was harvested (Saskatchewan Pulse Growers, 2005). In addition, chickpea production must compete with wheat. Even though the yield gap between wheat and chickpea is low, wheat producers get government support in terms of loans and crop insurance as substantial incentives that are not currently available to many of the chickpea farmers.

The Canadian chickpea-processing industry is efficient in terms of both cost and quality. Large-scale processors make the Canadian chickpea competitive in global markets with high-end consumers. Agro-industries purchase, transport, grade, pack and deliver chickpea to domestic ports or directly to importing port destinations. Buyers generally purchase chickpea from farmers on a free-on-board (FOB) basis at the farm gate. Farmers sometimes deliver produce to packing units and are compensated accordingly. This imparts considerable flexibility to producers and variability in prices received by them. As the cost structure of Canada is similar to that of the USA, this is discussed in detail in the following section.

USA

The USA is also a typical example of the production system in a developed country characterized by large landholdings, high levels of mechanization, high adoption rate of modern technology and high yield per hectare (Brester *et al.*, 2003). The average farm size is 180 ha, but operational farm holding ranges from 200 to 800 ha depending upon size of leased land. Using a full-costing method (including costs of inputs, machinery use, labour, management and land), it is estimated that the total economic cost of production of dryland chickpea is \$554/ha (Table 14.5). Capital depreciation (27% of total cost), land rent (20%), wages (10%), seed (10%), fuel and electricity (8%), fertilizer (6%) and chemicals (3%) constitute a major chunk of the total cost of production (South Dakota State University Extension, 2005).

Tables 14.5 and 14.6 show the costs and returns of chickpea produced under dryland (rainfed) and irrigated cropping systems, respectively. The average seed yields realized are 1475 and 2497 kg/ha in dryland and irrigated land, respectively. With these yields, net return per hectare is \$109/ha in dryland and \$282/ha in irrigated land. The net profit, as proportion of costs, is ~20% both

Table 14.5. Cost and return for dryland chickpea at different yield levels. (From South Dakota State University Extension, 2005.)

Item	Dryland chickpea grain yield (kg/ha)				
Grain price (\$/kg)	2043	1702	1475	1021	681
Large (65%) \$0.52/kg	696	580	499	349	232
Medium (25%) \$0.37/kg	189	158	136	95	63
Small (10%) \$0.19/kg	38	32	28	20	13
Total revenue (\$/ha)	923	769	663	462	308
(–) Total cost (\$/ha)	554	554	554	554	554
(=) Net return (\$/ha)	369	215	109	–92	–246

Actual average yield realized is 1475 kg/ha.

Table 14.6. Cost and return for irrigated chickpea at different seed yields. (From South Dakota State University Extension, 2005.)

Item	Irrigated chickpea grain yield (kg/ha)				
Grain price ($/kg)	2724	2497	2043	1703	1362
Large (85%) $0.52/kg	1214	1104	910	759	608
Medium (10%) $0.37/kg	101	92	76	63	50
Small (5%) $0.19/kg	25	24	20	15	13
Total revenue ($/ha)	1334	1220	1001	834	667
(–) Total cost ($/ha)	938	938	938	938	938
(=) Net return ($/ha)	397	282	63	–104	–271

Actual average yield realized is 2497 kg/ha.

for dryland and irrigated land. These average figures indicate that profitability is reasonably high for both irrigated and dryland chickpea farmers.

The revenues presented in these tables for different assumed grain yields are based on a three-tiered pricing schedule. For example, in dryland system, on an average, 65% of production is categorized as large seeds, 25% as medium seeds and 10% as small seeds. The contract price is $0.52/kg for large-seeded, $0.37/kg for medium-seeded and $0.19/kg for small-seeded types. Data in Tables 14.5 and 14.6 indicate that at these contract prices, net returns can be adversely affected by even a small reduction in grain yield, especially in dryland chickpea. The farmers incur losses when yields fall below 1220 kg/ha in drylands and 1914 kg/ha in irrigated lands. In addition to grain yield reductions, there are also potential losses associated with reduced seed size that are not quantified in these tables. The economic viability of a specialty crop such as chickpea will ultimately depend on several factors including market development, contract and seed pricing, supply conditions in large consumers like India, pesticide availability and production capability. Price (2002) suggests that competitiveness of chickpea in the USA is also determined by several other factors that are summarized below:

- US pulses are high-quality commodities, commanding price premiums. Many price-sensitive segments of foreign markets are unwilling to pay significant premiums for such quality, especially when lower-cost pulses from other countries are plentiful. For example, India imports many of its pulses from Myanmar, Canada and Australia, where both prices and quality are lower than in the USA.
- US exporters bag and containerize shipments to preserve quality. Although this results in less product damage, the process is more costly than bulk shipping.
- US transportation costs are relatively high. Long distances lead to high trucking costs, particularly in the Northern Plains. Rail freight rates to ports are also high.
- The high value of the US dollar relative to other currencies makes US exports more expensive than those from other countries.

Europe

Europe is the net importer of chickpea. The average imports of chickpea by Europe during 2002–2004 were 140,000t while the average exports during the same period were merely 28,000t. Most of the EU chickpea imports come from Mexico, Turkey, Canada and the USA. Spain, Italy, France, Portugal and the UK are the main importing countries. The EU generally imports large-seeded kabuli types.

Spain

Spain is the only major chickpea producer in the EU, with an average production of ~59,000t during 2002–2004. The country mostly produces the kabuli type and is therefore an important kabuli chickpea importing country. During 2002–2004, it imported ~57,000t of chickpea, implying that nearly half of its annual consumption is met through imports. Most of its imports comprise large-seeded kabuli types (11–12mm) from Mexico and Canada (Saskatchewan Pulse Growers, 2005). The last 10 years marked a drastic reduction in chickpea area in the country with acreage declining from 104,000ha in 1995 to 63,000ha in 2005. The occurrence of ascochyta blight coupled with imports of cheap-quality chickpea from Mexico adversely affected chickpea cultivation in Spain. The average yield was 708kg/ha during 2002–2004 and just 281kg/ha in 2005 (FAOSTAT database, 2006). By merely using rational systems of cultivation, some farmers have been able to double their yields, even with their traditional landraces (Gubero *et al.*, 1990). Experimental results suggest that shifting of chickpea acreage from traditional spring to winter season can give four times more yield. The lack of adequate transfer of technology from research institutions to farmers is the primary cause of low yield.

Prospects of chickpea cultivation in Spain

There is a good potential for expansion of area under chickpea in Spain for feed purposes as most of the food demand is met from imports of large-seeded kabuli chickpea from Mexico. The Spanish animal feed industry is mostly based on soybean meal, and chickpea as an alternative animal feed has to compete with the currently used animal feeds like soybean meal. Further, good-quality large-seeded chickpea preferred for human consumption fetches three times higher prices than that used for animal feed. As such, farmers are reluctant to cultivate chickpea for animal feed. Although some areas not suitable for cultivation of high-quality chickpea for human consumption need to be diverted to the cultivation of chickpea for animal feed, an alternative for increasing area under chickpea is by increasing yield to compensate for low price. There is a need for right policy options to increase chickpea consumption as feed and increase cropped area under high-yielding varieties with proper extension and breeding programmes.

France

Current demand for chickpea in France is mainly for human consumption and the country imports more than 10,000t/year for the purpose. With a yield of 2000kg/ha, 5000ha is enough to saturate the French market for kabuli chickpea (Pluvinage, 1990). Thus, the current demand is low and on the basis of the above yield and a price of $0.57/kg, as in the USA, the expected gross revenue from chickpea is approximately $1140/ha. Assuming the normal cultivation cost of $938/ha as in the USA under dry-land or irrigated conditions, farmers can receive ~21% net profit over costs.

The situation is less clear for chickpea grown for animal consumption, which should however benefit from a bigger market for crops containing protein as has happened for the fodder pea. At a price of $0.23/kg, which is identical to that of the fodder pea, a yield of 5500kg/ha must be attained with chickpea to obtain the same margin of profit. This appears to be possible on the basis of the results obtained at the research stations. Chickpea can therefore become a highly competitive crop for the rotation system in the dry cereal areas of the French Mediterranean, as it is more suitable and less costly than sunflower (Pluvinage, 1990).

It is obvious that the future of this crop is linked to its yield, as the market is relatively favourable. However, it is necessary, from the economic point of view, to work in accordance with several related considerations of opportunity cost as compared with the other crops. In the countries north of the Mediterranean Sea (Spain and France), the development of chickpea production techniques aimed at human consumption will rapidly encounter some constraints due to a limited market. On the other hand, if the cultivars with higher yield potential are developed, they could help in alleviating the shortage of animal feed as has occurred with fodder peas in the temperate zone. The market for this type of chickpea could be substantial, and it could permit restoration of some area in the region to a traditional Mediterranean crop.

The experiences of both Spain and France indicate that in European countries, demand of chickpea for human consumption is very limited and it is met substantially through import of kabuli chickpea from Turkey, Mexico, Canada and Australia. Scope for expansion of area under feed-quality chickpea exists, provided it is produced at low cost compared to soybean meal, dry peas and other fodder crops, as the demand for animal feed is increasing but is highly price-elastic due to availability of substitutes.

Developing Countries

Among developing countries, India from South Asia, Turkey and Jordan from the Mediterranean and Algeria from Africa have been included for detailed study.

India

In India, area under chickpea is ~6.72 million hectares and production is 5.30 million tonnes (total estimate 2003–2005 average), which constitute 63% and

64% of world chickpea area and production, respectively, with an average yield of 784 kg/ha. Chickpea is a major pulse crop of the country, accounting for ~30% of the total pulse area and 40% of the total pulse production. In terms of area of cultivation, it ranks second after dry beans, while in terms of production it ranks first among pulses. However, in the last 10 years, at the national level, a certain amount of stagnation has crept into the production, productivity and area under chickpea, as indicated by the triennium averages for these variables during 1993–2005. Since chickpea is cultivated in several parts of the country, it would be interesting to examine the region-wise trends in area, production and productivity.

In India, chickpea is grown in the north-east zone (NEZ) comprising Bihar, Orissa and West Bengal; the north-west zone (NWZ) comprising Gujarat, Punjab, Haryana, Himachal Pradesh, Rajastan and Uttar Pradesh; the central zone (CZ) comprising Madhya Pradesh and Maharashtra; and the south zone (SZ) comprising Andhra Pradesh, Karnataka and Tamil Nadu. Region-wise trends indicate that a significant reduction in area occurred in the NWZ and NEZ (Table 14.7) between 1971 and 2004. Significant increases in area are reported in the CZ and SZ (Fig. 14.2), which resulted in significant additional production from the CZ and SZ (Fig. 14.3). Substantial increases in yield also occurred in the CZ and SZ, though the NWZ still had higher average yield at 815 kg/ha. These trends clearly indicate that the CZ and SZ of the country play an important role in the expansion of area and production of chickpea. Increase in area in CZ and SZ is due to: (i) lack of alternate crops in winter season as wheat cultivation is not possible in South India due to high temperature; (ii) expansion of area in CZ as the region has drylands without irrigation that are not suitable for wheat cultivation; and more importantly (iii) introduction of short-duration varieties into large areas of rice fallows.

Yield gap analysis

Since area under chickpea cultivation virtually declined during 1974–2004, gains in production were obviously the result of improvements in productivity.

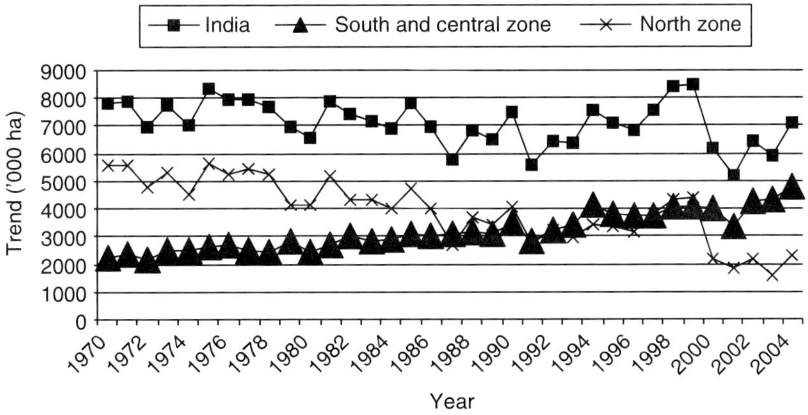

Fig. 14.2. Trends in area of chickpea ('000 ha) in India during 1970–2004.

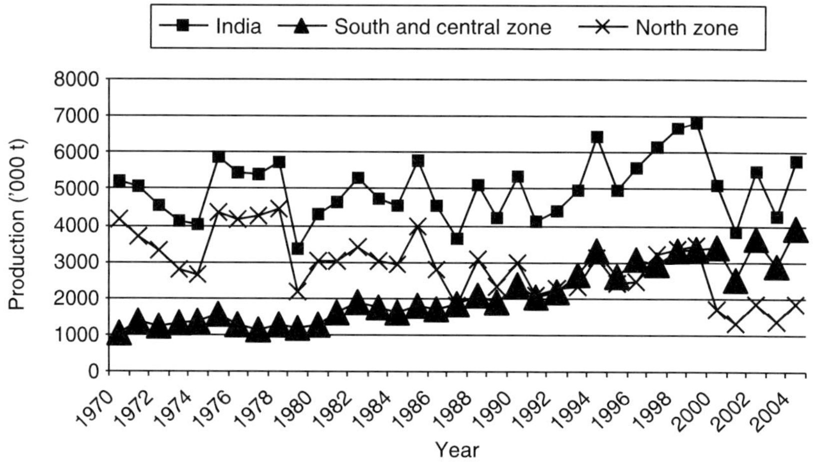

Fig. 14.3. Trends in production of chickpea ('000 t) in India during 1970–2004.

At the national level, a 31% increase in yield was observed in 2004 over 1974 (Table 14.7). However, intrazonal increases in yield varied over the period. Mean yields of All India Coordinated Trials (research station) indicate that the CZ reported highest potential yield (2070 kg/ha) for desi type, while the NWZ reported the highest potential yield (1462 kg/ha) for kabuli type (Table 14.8). Yield gap analysis (YGA) shows that significant yield differences exist between desi and kabuli cultivars on research stations. These differences range from 330 kg/ha in the NEZ to 783 kg/ha in the CZ. Yield gap I, which is the gap between research station and on-farm trial yields, is highest in the SZ (30%) for desi variety and lowest (17%) in the NWZ. Yield gap II, which is the gap between on-farm trials and zonal average yields, is very large in all zones, ranging from 64% in the NEZ to 148% in the CZ. Wider yield gap II is an indication that

Table 14.7. Chickpea production trends during 1971–2004. (From IIPR, 2006.)

Item	Year/change	NEZ	NWZ	CZ	SZ	India
Area ('000 ha)	1971–1974	383	4678	2137	226	7420
	2001–2004	314	1818	3236	786	6158
	Percentage change	−18	−61	51	248	−17
Yield (kg/ha)	1971–1974	570	663	467	421	598
	2001–2004	749	815	702	773	782
	Percentage change	31	23	50	84	31
Production ('000 t)	1971–1974	223	2883	1234	83	4433
	2001–2004	243	1460	2531	596	4838
	Production percentage change	9	−49	105	614	9

NEZ: north-east zone (Bihar, Orissa and West Bengal); NWZ: north-west zone (Gujarat, Punjab, Haryana, Himachal Pradesh, Rajastan and Uttar Pradesh); CZ: central zone (Madhya Pradesh and Maharashtra); SZ: south zone (Andhra Pradesh, Karnataka and Tamil Nadu).

Table 14.8. Zone-wise yield gaps in chickpea production in India during 2000–2005. (From Annual Report of All India Coordinated Research Projects [AICRP] on chickpea, 2005.)

	Yield (kg/ha)						
Zone	Research station – desi	Research station – kabuli	Difference desi and kabuli	On-farm trial	Zonal average	Yield gap I (%)	Yield gap II (%)
NEZ	1491.2	1161.6	329.6	1230.0	749.0	21.2	64.2
NWZ	1825.6	1461.8	363.8	1558.0	815.0	17.2	91.2
CZ	2069.6	1286.2	783.4	1745.0	702.0	18.6	148.6
SZ	1676.8	1316.8	360.0	1290.0	773.0	30.0	66.9

there is a large gap between on-farm demonstration yield and zonal average yield, which can be bridged by wider adoption of technology by farmers. The existing technology has the potential of doubling production at national level without increasing area under chickpea if farmers adopt recommended package of practices.

India: micro-level evidences

Field surveys were conducted during 2002–2005 in Hamirpur and Fatehpur districts of Uttar Pradesh to examine the economics of chickpea cultivation at farm level. The survey covered a total of 120 farmers including 60 adopters who participated in on-farm trials of technology developed at the Indian Institute of Pulses Research, Kanpur, India, and 60 who were non-adopters of the technologies developed. Average area under chickpea of sample farmers was only 0.4 ha accounting for ~40% of area under cultivation in winter season.

Cost structure

The cost structure of chickpea production of sample farmers (pooled) is presented in Table 14.9 for 2004. The total cost is $289/ha, which is relatively lower than many developed and even developing countries. Variable costs contribute to ~68% of the total costs. Most of the operations are performed manually, except land preparation. Seed rate is 60 kg/ha, which is lower than recommended rate of 70–80 kg/ha. Low seed rate is the main cause for low plant population, which ultimately determines yield. Fertilizer is either applied in suboptimal doses or not applied at all. Generally, farmers do not spray any insecticides and fungicides, but under severe attack of pod borer, farmers apply suboptimal doses of pesticides. Many farmers practise manual weeding. Most of the operations are carried out by family labour with the help of neighbouring farmers on exchange basis. Only harvesting and threshing is done on contract basis, with payment in kind (1:11 of the harvested grain has to be given for harvesting and threshing). A few farmers provide only a life-saving irrigation at flowering or pod-formation stage. Most of the farmers use their own seeds or procure them from neighbouring farmers. There is little availability of certified seeds at private seed shops or government seed agencies. The average yield

Table 14.9. Cost of cultivation of chickpea per hectare on farmers' fields in 2004. (From field survey.)

Item	Physical	$/unit	Total $
Land preparation (tractor hours)	7.7	4.89	39.11 (13.5)
Seed (kg)	59.8	0.56	33.33 (11.5)
Sowing labour (man days)	7.5	1.11	8.89 (3.1)
Weeding/uprooting (man days)	24	1.11	26.67 (9.2)
Insecticide (l)	0.5	6.22	16.89 (5.8)
Irrigation (hours)	9.8	1.56	15.56 (5.4)
Fertilizer (kg)	20.5	0.22	4.44 (1.5)
Harvesting (man days)	19.7	–	22.22 (7.7)
Threshing (man days)	16	–	19.11 (6.6)
Interest on variable cost	–	–	11.17 (3.9)
Total variable cost (A)	–	–	197.39 (68.1)
Rental value of land	–	–	81.46 (28.1)
Depreciation on implements	–	–	5.64 (1.9)
Interest on fixed capital	–	–	5.22 (1.8)
Total fixed cost (B)	–	–	92.32 (31.9)
Total cost (A + B)	–	–	289.71 (100)
Grain yield (kg/ha)	938	0.4	375
Value of by-product	–	–	11.1
Gross return (Rs/ha)	–	–	386.1
Net profit (Rs/ha)	–	–	96.3
Benefit/cost ratio	–	–	1.33
Total man days	70	–	–

Figures within parentheses in the last column indicate percentage of total cost.

obtained in the study area is 938 kg/ha. At a selling price of $0.4/kg, the gross revenue is about $386/ha and net profit over total cost is $96/ha.

Yield gap analysis of sample farmers

On-farm demonstrations are mainly conducted on improved cultural practices, integrated nutrient management (INM), integrated pest management (IPM) and difference in variety. It is recommended to use INM (9 Kg N/ha +23 Kg P_2O_5/ha +10 Kg S/ha + *Rhizobium* culture) and IPM (seed treatment with *Trichoderma* + Vitavax + spray of NPV 250 l/ha and one spray of Endosulphan 0.07%), as well as one life-saving irrigation at flowering or pod-formation stage. On-farm trials were conducted on more than 200 farmers' fields. The detailed methodology for decomposition of yield gap II has been given in the Appendix. The average yield realized in on-farm trials is 1510 kg/ha, while in farmers' fields it is 938 kg/ha. The yield gap between on-farm and local yield is ~61%, with 29% being attributable to the differences in cultural practices (Table 14.10).

Economics of chickpea cultivation

On the input side, the use of improved seeds accounted for a substantial 12.5% difference in yields. Examination on costs and returns from chickpea for adopters

Table 14.10. Sources of yield gap in chickpea. (From field survey.)

Source of difference	Yield gap (%)
Total difference in output (yield gap II)	60.8
Sources contributing to yield gap	
Differences in cultural practices	28.6
Input use gap	
Labour	5.7
Seed	12.5
Nutrients	3.7
Pesticides	1.2
Irrigation	9.1
All inputs combined	32.2

and non-adopters of certified seeds indicated that for adopters average yield is 1270 kg/ha (Table 14.11), while for non-adopters, it was ~920 kg/ha. The average cost per hectare in the case of adopters was estimated at $304/ha, which is higher than that of non-adopters at $281/ha. However, net profit is also higher in the case of adopters ($204/ha) than that of non-adopters ($87/ha). There is no significant difference between farm sizes with respect to profits, but cost per hectare is high for large farm category. These results indicate that net profit can increase by more than 130% with the adoption of improved certified seeds.

Profitability of chickpea compared with other competing crops

Wheat, barley, lentils and mustard are the competing crops of chickpea in India. Even though kabuli chickpea yield is 20% less than that of desi chickpea (Table 14.12), the market price for kabuli type is attractive ($0.78/kg). The highest net profit per hectare was estimated for kabuli chickpea followed by wheat and barley. The benefit/cost ratio was also highest for kabuli chickpea at 1.56, followed by barley and lentils. Profitability of desi chickpea was low in comparison with wheat and barley on account of low yields. Chickpea yield is about one-third of wheat and barley. Even though the price received for desi chickpea

Table 14.11. Economics of chickpea cultivation of adopters and non-adopters of quality seed. (From field survey.)

	Adopters				Non-adopters			
Farm size	Cost ($/ha)	Yield (kg/ha)	Gross revenue ($/ha)	Profit ($/ha)	Cost ($/ha)	Yield (kg/ha)	Gross revenue ($/ha)	Profit ($/ha)
Small	316	1300	520	204	282	950	380	98
Medium	296	1290	516	220	274	845	338	64
Large	308	1200	480	172	300	940	376	76
All groups	304	1270	508	204	281	920	368	87

Table 14.12. Relative profitability of chickpea and competing crops in India. (From field survey.)

Item	Wheat	Chickpea (desi)	Chickpea (kabuli)	Barley	Lentil	Mustard
Gross return ($/ha)	490	385	585	352	318	326
Variable cost ($/ha)	342	289	375	238	254	253
Net profit over variable cost ($/ha)	148	96	210	114	64	73
Benefit/cost ratio ($/ha)	1.43	1.33	1.56	1.48	1.25	1.29
Yield/ha	2724	938	750	2705	816	814
Price ($/kg)	0.18	0.41	0.78	0.13	0.39	0.40

is more than twice that of wheat and barley, it is not sufficient to compensate for low yields. As a result, chickpea is substituted by wheat wherever irrigation facilities are available, and by barley in some unirrigated tracts. These figures indicate that there is good scope for expansion of area under kabuli chickpea in India, provided adequate quality seed is available.

Profitability of rice–chickpea system

With the data of the field survey, it was also estimated that replacement of wheat with kabuli chickpea in the rice–wheat system of North India can generate additional net returns of 83% (Table 14.13) for the producers, mainly on account of lower production costs per hectare, besides reducing water and fertilizer requirement by ~35% and 40%, respectively. Replacing wheat by desi chickpea also significantly reduced water and fertilizer requirement and cost per hectare, but resulted in lower net returns.

Import–export scenario of India

Although India is the world's largest producer of chickpea, it also imports chickpea to meet domestic requirements. Iran, Turkey and Mexico are the major sources of kabuli chickpea, while Canada, Australia and Myanmar are the main sources of desi chickpea (Price *et al.*, 2003). Approximately 40% of the total chickpea imports to India are of the kabuli type. Mexican kabuli chickpea

Table 14.13. Rice–wheat vs rice–chickpea cropping system in Uttar Pradesh, India. (From field survey.)

System	Gross revenue ($/ha)	Cost ($/ha)	Grain production (kg/ha)	Water (number of irrigations)	Plant nutrients (kg/ha)	Net return ($/ha)
1. Rice–wheat	1020	703.2	5400	8	250	316.8
2. Rice–chickpea (desi)	956	647.8	3740	5	140	308.2
3. Rice–chickpea (kabuli)	1251	670.8	3500	5	150	580.2
Gain: 2 over 1 (%)	−6.3	−7.9	−30.7	−35	−44	−2.7
Gain: 3 over 1 (%)	22.6	−4.6	−35.2	−35	−40	83.1

Paddy yield = 2700 kg/ha, price = $0.20/kg; wheat yield = 2700 kg/ha, price = $0.18/kg; chickpea (desi) yield = 1040 kg/ha, price = $0.4/kg; chickpea (kabuli) yield = 800 kg/ha, price = $0.89/kg.

Table 14.14. Indian imports of chickpea and exports of dhal in 2001/02. (From Government of India, 2005.)

Imports (chickpea)			Exports (dhal)		
Country	Quantity (t)	Unit value ($/kg)	Country	Quantity (t)	Unit value ($/kg)
Canada	103,617	0.32	USA	1,189	0.56
Iran	70,305	0.38	UK	596	0.59
Australia	75,736	0.34	Malaysia	371	0.61
Myanmar	33,112	0.34	Kuwait	381	0.57
Turkey	20,258	0.41	UAE	364	0.54
Mexico	9,061	0.62	Sri Lanka	463	0.42
Others	55,097	0.33	Others	908	0.59
Total	367,186	0.35	Total	4,276	0.57

is of the extra-large type and commands a premium of more than 50% over Turkish chickpea (Table 14.14). Canadian, Australian and Myanmar varieties are cheaper. Importers prefer Myanmar chickpea compared with Canadian and Australian chickpea due to similarities in taste and cooking quality to locally produced chickpea, as well as lower transport and other logistics costs, as Myanmar is closer to India. However, in case of supply shortages during off-season, imports are also made from Australia and Canada.

India also exports chickpea dhal mostly to the USA, UK, Kuwait, UAE and other countries where Indians reside (Table 14.14). These exports fetch a premium in these countries, as they cater to the demand of ethnic Indians.

Turkey

Turkey is a net food-exporting country. Annual production of chickpea in Turkey is 600,000 t, of which 23% is exported mainly to Europe, Arab markets and India. As it is one of the three crops promoted by the country for limiting fallow lands, chickpea has received considerable attention and large additional areas have come under its cultivation during the 1980s and 1990s. However, due to reduced attention by government and competition from other exporting countries, area under chickpea has reduced significantly in recent years. Production has declined by ~20% from 755,000 to 603,000 t during the last 10 years mainly due to reduction in area by ~18% and in yield by ~3%. Mean yield during 2000–2005 was 935 kg/ha.

Per capita consumption of chickpea in Turkey is high at 6.65 kg/year (FAOSTAT database, 2006). Domestic preference is for kabuli chickpea, which is used in a number of ways in Turkish kitchens in items such as chickpea bread, main dishes, humus and pastry as well as for garnishing meat and rice plates (Acikgoz, 1990). It is also consumed as a snack in various forms including the sweetened ones. However, domestic demand is not sufficient to boost production, as the population is small. The only way to do this is by increasing exports. Mostly medium-sized kabuli chickpea (8–9 mm) is produced in Turkey. There is a need for increasing yield to make chickpea price competitive in the international market.

The annual variability in national production of chickpea (coefficient of variation = 10.75%), though slightly higher than average variability in world production during the last decade, is still lower than the variability seen in many of the other important producing countries. Similarly, the variability in productivity is also relatively low compared with other producers. This is indicative of higher stability in chickpea production in Turkey. When compared to production, exports are highly unstable with a coefficient of variation >40% during 1995–2004. Due to unstable export demand, limited domestic demand and recent withdrawal of government support system, chickpea acreage has declined with areas shifting to wheat, for which domestic as well as international demands are more stable.

The main yield-limiting factors for chickpea are the lack of the registered varieties for different ecological regions and the cultivars resistant to diseases, especially ascochyta blight. Most of the chickpea-growing regions are prone to ascochyta blight. Although large-scale region-wise yield losses are not frequent, localized epidemics occur quite often (Ismail, 2003). In order to avoid the damage from blight disease, growers prefer late spring sowing, which sometimes lasts until the second fortnight of May. Late sowing causes yield decreases due to shorter vegetative growth. Although there are some recommendations for controlling the disease such as fungicidal seed dressing, foliar applications and some cultural practices, these are not generally practised by farmers due to high cost and labour shortage. A 20% reduction in yield due to blight will reduce yield from 935 to 748 kg/ ha, leading to losses amounting to $130/ha at a market price of $0.7/kg. Therefore, development of blight-resistant varieties is very important for stable returns to farmers. Another cause of concern is tough competition in exports of kabuli chickpea from Mexico, Iran, Canada and the USA. This increasing competition poses the biggest risk for future expansion of area under chickpea. Small farm size, low level of mechanization and inadequate market infrastructure, low domestic demand, frequent attack of ascochyta blight and unavailability of improved varieties for different agroecological zones are some of the limiting factors to increase area and enhance production of chickpea in this country. Another important factor is that the Turkish chickpea exports are positioned as high-end quality products, but postharvest processes are not comparable with those in competing exporting countries like the USA, Canada and Australia, thus leading to a slackening of demand in recent years. Further, the significant yield gap between wheat and chickpea in Turkey, with average yield of 2258 kg/ha for wheat and 968 kg/ha for chickpea, makes wheat much more profitable in domestic markets even though chickpea price is more than twice that of wheat.

With the dismantling of the agricultural incentive system in Turkey recently, fertilizer and pesticide subsidies have been curtailed and remaining price supports have been gradually converted to floor prices. In order to make kabuli chickpea of Turkey competitive in international markets, the country needs to strengthen agricultural extension and research services. It has to develop trained manpower to advice on chickpea cultivation and a system to provide market information, besides developing efficient transportation and postharvest processes, as Turkey kabuli chickpea is mostly imported by developed countries or targeted to high-end consumers in developing countries. Recent government initiatives to promote private seed sector, organic production, market-oriented and integrated research activities as well as improved technology transfer systems can enhance chickpea competitiveness (Ismail, 2003). The Exporters Union of Turkey established an Exporters Union Seed and Research Company called Industrial Technology Advisors (ITAs) in 1998 to improve the production and quality of food legumes. ITAs has launched an innovative and integrated technology transfer project, which has resulted in increase in area and production in a very short period of time (Ismail, 2003).

Algeria

Algeria is one of the largest countries in Africa, but only 3.5% of its total area is used for agricultural production. Cereals are the predominant crops grown by Algerian farmers, covering ~40% of total agricultural land. Scarcity of underground water resources, low and erratic rainfall, drought recurrence, high temperatures and salinity are the key constraints to agricultural production.

In Algeria, production of chickpea declined from 20,000 to 15,000 t in the last decade. During the same period, yield increased from 457 to 735 kg/ha, but area decreased from 45,000 to 21,000 ha despite a remunerative price ($0.7/kg in 2004). The main reason for decrease in area under chickpea is that it has to compete with wheat, whose yield of 3714 kg/ha is about five times that of chickpea (750 kg/ha). Fodder crops, which are very remunerative because of high prices of meat, also impart strong competition to chickpea. As a result, Algeria imports ~273% of its domestic production annually, mostly consisting kabuli chickpea (FAOSTAT database, 2006).

Production cost of chickpea

Production costs are derived from published secondary sources. Direct production costs of the crop during the early 1990s were estimated at $333/ha with mechanical harvesting and at $416/ha with manual harvesting (Pluvinage, 1990; ICARDA, 2003). The loss of seed with mechanical harvesting was ~100 kg/ha, which makes the two methods equivalent in terms of costs, since the selling price of kabuli chickpea was $0.7/kg. It is thus necessary to harvest at least 480 kg/ha (mechanical harvesting) or 590 kg/ha (manual harvesting) to cover the direct costs, without taking into account the general running costs. At the current average yield of 735 kg/ha, gross revenue is $514/ha with manual harvesting and $444/ha after adjusting for 100 kg loss with mechanized harvesting. The profit over total cost is $98/ha (23%) and $111 (33%), respectively. Although profit is low when compared with Jordan and some developed countries like the USA and Australia, it is in line with profit realized in India.

The major cause of low profitability and reduction in area under chickpea is that the adoption of modern high-yielding varieties is <50% for chickpea in both Asia and Africa and most of the developing countries, but adoption rate is

~90% for wheat. Susceptibility of high-yielding varieties to various biotic and abiotic stresses further restricts the area expansion approach. As a result, Asia and Africa remain chickpea-deficient continents (ICARDA, 2003).

Jordan

Jordan, though a developing country, depends mostly on oil exports, with agriculture contributing merely 3% to the gross domestic product (GDP). Contribution of agriculture is low on account of limited cultivable land. Jordan is the net importer of food, including its staple wheat. Jordan imports more than ten times its domestic production of chickpea. Chickpea rotates with cereals, mostly wheat, in rainfed areas where annual rainfall is usually 300–400 mm. It has also diverted a major chunk of land for cultivation of wheat, barley and fruit crops for export purpose. As a result, the average area under chickpea reduced from 3000 ha during 1987–1995 to 1000 ha during 2000–2005. Production of chickpea was stagnant at ~2000 t/year during 1987–2005 (FAOSTAT database, 2006). In the same period, chickpea yield increased significantly from 860 to 1627 kg/ha. Even though chickpea yields are competitive in comparison with wheat average yields of 1358 kg/ha, wheat and other crops have replaced chickpea in recent years. In the case of wheat and barley, most of the operations are mechanized with low cost, while most of the operations of chickpea cultivation are done manually, as small acreage and short plant type of chickpea do not permit mechanization. Farmers are also shifting to remunerative fruit crops like almonds, grapes and apricots for export market and some are cultivating vegetables like tomato and cucumber to increase profits per unit area.

Problems in adoption of improved technology

To ensure good reserves of soil moisture control weeds, and to avoid the occasional incidence of ascochyta blight disease, farmers usually delay chickpea sowing to spring, usually until March. However, spring sowing of chickpea leads to reduction in yields due to short growing season and terminal drought. Farmers still use the traditional cultural practices for different reasons, mainly due to lack of capital, even though they are relatively aware of its usefulness. Most of the time they spend money on seed and fertilizer only. Weak agricultural extension network further limits efficient transfer of technology.

Until now, research does not offer optimum means of mechanical harvest, especially for small farms. The average size of the agricultural holding is ~3.9 ha; the acreage under chickpea is still less, which does not permit mechanized production and concerned long-term research efforts. Furthermore, the cost resulting from production inputs such as drilling and levelling, weed and pest control and fertilization is relatively high and farmers are not willing to spend money for a risky crop like chickpea. Jordan is a Mediterranean country characterized by highly variable and irregular rainfall, which increases crop and economic loss. In the past, the government was backing the prices in favour of the farmers by offering them a buying price for their produce higher than that of the market. The recent liberalization policies resulted in the withdrawal of government support, which increased cost of production and domestic prices, and ultimately increased imports.

Cost factors

The FAO data indicate that the national average yield over the recent period was 1627 kg/ha. Most of the farmers use recommended seed rate and fertilizer doses, but only suboptimal doses of herbicides and pesticides. Mostly all operations except threshing are performed by manual labour. Chickpea cultivation is done with relatively high input conditions to maximize returns. Table 14.15

Table 14.15. Cost of cultivation and returns of chickpea per hectare in Jordan. (From Nidal and Nanish, 2002.)

		Farmers' practice		Recommended practice	
Item	Unit	Quantity	Value ($)	Quantity	Value ($)
Seed	(kg)	120	88.9	120	88.9
Fertilizer	(kg)	100	28.2	100	28.2
Herbicide	(l)	–	–	1	35.3
Pesticide	(l)	–	–	1	28.2
Bags	(No.)	10	9.9	10	9.9
Total	–	–	127.0	–	190.5
Machine labour					
Land preparation including harrowing to cover the seeds	(h)	0.3	3.5	1.7	26.7
Seed drilling and fertilization	(h)	–	–	0.6	14.2
Spraying (herbicide and pesticide)	(h)	–	–	0.8	4.5
Threshing	(h)	4	45.1	4	45.1
Sieving	(t)	1600	7.8	2100	10.6
Transport	(t)	1600	4.7	2100	4.7
Total	–	–	61.1	–	105.8
Labour work					
Hand sowing (spreading)	(h)	17	12.0	–	–
Fertilizer spreading	(h)	17	12.0	–	–
Weeding	(h)	100	105.8	–	–
Hand-harvest	(h)	120	108.7	120	108.7
Straw racking, packing, loading and unloading, others	(h)	14	10.6	14	10.6
Total	–	–	249.1	–	119.3
Interest (6.0%)	–	–	25.5	–	25.0
Land rent	–	–	282.1	–	282.1
Grand total	–	–	744.8	–	722.7
Average seed yield (kg/ha)	–	1600	–	2100	–
Average straw yield (kg/ha)	–	2500	–	3409	–
Total income ($/ha)	–	–	1222	–	1617
Net return ($/ha)	–	–	477.2(64)	–	895 (123)

Figures within parentheses in the last row indicate percentage of total cost.

presents farmers' practice of cultivation of chickpea along with recommended practice. The cost of production under farmers' practice is ~$745/ha (Nidal and Nanish, 2002). Cost of production is high compared with other developing countries due to high input cost.

Although chickpea productivity and profitability is high, area under chickpea declined mainly due to the withdrawal of government support, scarcity of land and high level of import dependence. While cultivation of crops like chickpea faces many risks from increased imports at lower prices, decline in domestic prices and production loss due to disease outbreak and drought and demand uncertainty, the competing wheat crop has a triple advantage in a small country like Jordan because of its stable domestic demand, relatively higher insulation from international price fluctuations through government support and its use as a staple food. As a result, farmers go for wheat cultivation even though chickpea is comparatively more profitable. Farmers are also shifting to the cultivation of fruits and vegetable crops due to high and stable export prices and rising export and domestic demand. Since scope for expanding area under chickpea is limited in Jordan, the alternative for increased production is the adoption of improved varieties and other practices to increase yield.

Conclusions

Among developed countries, cost of production is high in France and the USA and low in Australia. The USA produces mostly large, high-quality kabuli type of chickpea that fetches higher prices in the international market. In all these developed countries, farm size is large, which facilitates mechanization of all farm operations. Most of the farmers use fungicides, fertilizers and pesticides in optimal doses as advised by extension specialists. However, cultivation of chickpea does not get adequate support from government as in the case of wheat. All these countries produce chickpea mostly for export purposes. The USA and, to some extent, Canada specialize in producing and exporting kabuli types, while Australia produces and exports both desi and kabuli types. The competitive advantage for developed countries comes from large-scale production, processing and marketing. In European countries, the production base is very limited except in Spain. In Europe, demand for human consumption is limited and most of the demand is met through imports of extra-large kabuli chickpea from Mexico, Canada and the USA. However, there is large scope for expansion of area under chickpea for use as animal feed, if yields can be improved to match those of dry peas and soybean meals by breeding high-yielding varieties and developing appropriate management practices.

Among the developing countries considered, only Turkey is a frequent and large exporter of chickpea. While Jordan imports about ten times its domestic production, Algeria imports more than 200%. India imports ~5% of domestic production. Turkey is facing competition from other exporters like Canada, Mexico, the USA, Australia and Iran to share the same pie. As Turkey's exports are positioned as high-quality supplies, quality produce from the USA, Canada and Mexico, with superior postharvest infrastructure, gets preference over

Turkish chickpea. In this scenario, Turkey has to increase its productivity as well as postharvest processing efficiency to increase or stabilize market share. Jordan is a food-deficient country and imports about ten times its domestic production of chickpea. In spite of high chickpea yields, Jordan faces problems in production due to land limitation and competition for the existing land from competing crops on account of high export potential of horticultural commodities and price support for wheat, and has to import to meet domestic consumption requirements. In the case of Algeria, chickpea yield levels are too low to compete for area with alternate crops like wheat, and its domestic chickpea production is not able to compete with imported chickpea. Chickpea also competes with fodder crops as demand for fodder has increased in recent years due to increased demand for meat, thus making fodder crops more remunerative. There is a need in Algeria to at least double chickpea yields through concerted research and extension efforts. There is also a possibility of cultivating chickpea as fodder crop if yield can be increased to the level of dry peas in view of increasing demand for fodder.

Being the largest producer and consumer of chickpea, India occupies special status in world chickpea markets. In recent years, large tracts of rice fallows are replaced by rice–chickpea cropping system in both Central and South India due to introduction of short-duration varieties. There is considerable scope for expansion of area under chickpea by replacing wheat with chickpea in water-deficient and resource-poor regions. This will help to reduce imports of chickpea and may even reverse India's current position of a net importer. Scope exists for increasing area under kabuli chickpea production, as there is a ready domestic and international market for it. There should be greater emphasis on development of improved seeds that give higher and stable yields, as farmers are willing to pay a premium price for better seeds.

The imbalance in technology adoption in developed and developing countries needs to be addressed to bridge the gap in productivity. There is a need to increase seed replacement rate in most of the developing countries. With the increase in labour cost, there is an urgent need for mechanization of peak season operations like harvesting and threshing in developing countries, which will increase scope of area expansion and also reduce cost of production.

References

Acikgoz, N. (1990) Chickpea production in Turkey. In: *Options Méditerranéennes*. CIHEAM, Paris, pp. 167–170.

Agriculture and Agri-Food Canada (2004) Chickpea situation and outlook. *Market Analysis Division* (bi-weekly bulletin) 17(15), 14 September.

Agriculture and Agri-Food Canada (2005) European Union: pulse crops situation and outlook. *Market Analysis Division* (bi-weekly bulletin) 18(12), 17 June. Available at: http://www.agr.gc.ca/maddam/e/bulletine/v18e/v18n12_e.htm

Annual Report (2005) *All India Coordinated Research Project on Chickpea*. Indian Institute of Pulses Research, Kanpur, India.

Brester, G.W., Buschena, D. and Gray, K. (2003) Economic issues related to chickpea production in the northern plains. Agricultural Marketing Policy Center Linfield Hall, Montana State University Bozeman, Briefing No. 33.

FAOSTAT Database (2006) Food and Agriculture Organization, Rome. Available at: www.fao.org

Government of India (2005) Ministry of Commerce, DGCI&S.

Gubero, J.I., Moreno, M.T. and Gil, T.J. (1990) Chickpea breeding in Spain. In: *Options Méditerranéennes*. CIHEAM, Paris, pp. 157–161.

ICARDA (2003) Algeria and ICARDA, Twenty-five years of collaboration in scientific agricultural research between the Democratic and Popular Algerian Republic (DPAR) and the International Center for Agricultural Research in the Dry Areas (ICARDA): Ties that Bind. No. 20. Institut Technique des Grandes Cultures (ITGC), Algérie. Available at: http://www.icarda.org/Publications/Donors/Algeria/Algeria.pdf

IIPR (2006) *Data Base on Production of Pulse Crops*. Indian Institute of Pulses Research, Kanpur, India.

Ismail, K. (2003) Participatory transfer of integrated technology: a promising approach to increase food legume production in Turkey. Seed Info No. 23, ICARDA, Aleppo, Syria, July 2003. Available at: http://www.icarda.org/News/Seed%20Info/SeedInfo_25/Research.htm

Nidal, M.A. and Nanish, H. (2002) Lentil and chickpea production in Jordan Food Legume Improvement Project at National Center for Agricultural Research and Technology Transfer (NCARTT), Jordan. Available at: http://www.solutions-site.org/artman/publish/article_8.shtml

Pluvinage, J. (1990) Chickpea in Mediterranean production systems: two contrasting examples of possible developments in Algeria and France. In: *Options Méditerranéennes*. CIHEAM, Paris, pp. 133–136.

Price, G.K. (2002) Will the Farm Act get pulses racing? *Agricultural Outlook AGO-296* November, 18–21.

Price, G.K., Landes, R. and Govindan, A. (2003) *India's Pulse Sector Results from the Field Research*. Electronic Outlook Report from the Economic Research Service. United State Department of Agriculture, WRS-03-01. Available at: www.ers.usda.gov

Saskatchewan Pulse Growers (2005) *Overview of Chickpea Markets*. Available at: http://www.saskpulse.com/media/pdfs/market-overview-chickpea.pdf

Siddique, K.H.M., Regan, K.L. and Baker, M.J. (2004) New ascochyta blight resistant, high quality kabuli chickpea varieties for Australia. In: *Proceedings of the 4th International Crop Science Congress*. Brisbane, Australia. Available at: http://www.cropscience.org.au/icsc2004/symposia/6/1/467_siddique.htm

South Dakota State University Extension (2005) *Chickpea Production in the High Plains*, Fact Sheet No. 922.

Wayne, H., John, H. and Heinjus, D. (2003) *Growing Chickpeas*. Primary Industries and Resources of South Australia, Fact Sheet No. 20/99. Available at: www.pir.sa.gov.au/factsheets

Appendix 1
Comparative cost and benefit of chickpea in different countries.

	Yield (kg/ha)	Gross return ($/ha)	Cost ($/ha)	Net return ($/ha)	Net return over cost (%)
Turkey	935	–	–	–	–
Jordan	1600	1222	745	477	64.0
Algeria	735	514	333	181	54.4
Canada	1509	–	–	–	–
USA	1475	723	462	261	56.5
Australia	969	450	296	154	52.0
Spain	620	–	–	–	–
France	2000	1140	782	358	45.8
India	938	386	289	97	33.6

Appendix 2
Comparative yields of wheat and chickpea (kg/ha). (From FAOSTAT database, 2006.)

	Wheat	Chickpea	Ratio
Algeria	3714	750	5.0
Australia	2119	969	2.2
Canada	2599	1509	1.7
India	2738	833	3.3
Jordan	1358	1454	0.9
Mexico	5000	1600	3.1
Spain	1689	281	6.0
France	6982	–	–
Turkey	2258	968	2.3
Syria	2452	663	3.7
UAE	3429	–	–
USA	2823	–	–
Iran	2339	411	5.7
Ethiopia	1375	794	1.7
Italy	3539	1179	3.0

Appendix 3
World imports and exports (000 tonnes) of chickpea (2002–2004). (From FAOSTAT database, 2006.)

	Imports 2002–2004				Exports 2002–2004		
Country	Mean	Share	CV 1995–2004 (%)	Country	Mean	Share	CV 1995–2004 (%)
India	203.1	25.1	89.6	Turkey	142.4	19.3	40.2
Pakistan	124.9	15.4	75.9	Australia	129.2	17.5	55.2
Bangladesh	81.8	10.1	79.2	Mexico	122.4	16.6	28.8
Spain	56.7	7.0	16.7	Iran	104	14.1	74.8
Algeria	44.6	5.5	32.3	Canada	89.5	12.1	100.4
UAE	28.2	3.5	68.3	Tanzania	24.4	3.3	117.9
Italy	23.8	2.9	18	Ethiopia	17.1	2.3	295.3
Jordan	22.7	2.8	33.3	USA	16.8	2.3	54
Saudi Arabia	21.3	2.6	21.4	Syria	12.7	1.7	102.6
Sri Lanka	19.9	2.5	39.7	UAE	12.4	1.7	69.5
Others	208.3	22.6	11.6	Others	70.0	9	42.2
World	835.3	100	35.9	World	740.8	100	28.3

15 Genetics and Cytogenetics

D. McNeil,[1] F. Ahmad,[2] S. Abbo[3] and P.N. Bahl[4]

[1]*Victorian Department of Primary Industries, PMB 260, Horsham, VIC 3400, Australia;* [2]*Brandon University, Brandon, MB R7A 6A9, Canada;* [3]*The Levi Eshkol School of Agriculture, The Hebrew University of Jerusalem, Rehovot 76100, Israel;* [4]*A-9, Nirman Vihar, New Delhi 110092, India*

Introduction

In the last three decades extensive research has been done worldwide for the improvement of chickpea and its production. As a result of these efforts, genetic enhancement has been widely reported for agronomic characters, particularly grain yield and its components. This has been achieved mainly by using a combination of Mendelian and quantitative genetics aligned with selection either towards an ideotype or based on performance in the target growing, processing or marketing environment. Many characteristics such as yield are based entirely on quantitative selection, whereas others (e.g. *Ascochyta* resistance) appear to be controlled by both quantitative and qualitative factors. To date exploitation has primarily been of the existing genetic variability present for these agronomic traits both in desi (coloured seed coat) and kabuli (white seed coat) groups within the cultivated species of chickpea. Recently exploitation has shifted to include genes from species closer to chickpea (of which there are a large number of species in the genus; Sharma *et al.*, 2005b) and insertion via genetic modification (GM) of foreign genes into them (Sarmah *et al.*, 2004). The particular interest in gaining novel genes from outside the cultivated species has arisen from the identification of a substantial narrowing of the genetic base for chickpea during its domestication (Abbo *et al.*, 2003). Both the approaches have been utilized in conjunction with non-GM molecular approaches such as marker-assisted selection (Sharma *et al.*, 2004), hybridity testing (Winter *et al.*, 1999), molecular mapping (Flandez-Galvez *et al.*, 2003) background selection (Pittock *et al.*, 2004) and gene identification. These molecular approaches will be considered in more detail in Chapter 21 (Rajesh *et al.*, this volume). However, in the current scenario, chickpea quality has become an increasingly important concern in marketing and profitability. Therefore, quality characteristics like seed mass, seed volume, hydration capacity (before and after soaking), swelling capacity (before and after soaking), storability, cooking quality and digestibility now deserve, and are receiving, greater attention for their genetic improvement because of their role and importance in value addition. There is also an increasing interest in improving nutritional characteristics as quality determinants in chickpea with considerable interest in selecting lines enhanced for sterols, isoflavones, flavonoids and other nutraceuticals.

Qualitative Traits

The genetic basis of qualitative traits in chickpea

Qualitative traits are characterized by distinct phenotype groups that can be separated and inheritance patterns that can be determined using Mendelian genetics. This allows a phenotypic identification of a specific set of alleles for specific gene loci. Comprehensive lists of qualitative traits and their gene nomenclature are given by Muehlbauer and Singh (1987) and Pundir *et al.* (1985). They were further updated as a descriptor list by the International Board for Plant Genetic Resources (IBPGR) in 1993. Currently, an international initiative is being undertaken involving the collections at the International Center for Agricultural Research in the Dry Areas (ICARDA), Aleppo, Syria; International Crops Research Institute for the Semi-Arid Tropics (ICRISAT), Hyderabad, India; United States Department of Agriculture (USDA); and the Australian Temperate Field Crops Collection (ATFCC) in conjunction with the International Rice Research Institute (IRRI), Philippines, to incorporate the evaluation data for both quantitative and qualitative traits of chickpea into a single web-searchable International Crop Information System (ICIS) database (Balachandra, 2005). These traits include leaf shape and arrangement, plant habit, stem and foliage characteristics, flower colour, seed and cotyledon characteristics, podding, nodulation and disease resistance.

Recently, the use of molecular technologies such as molecular mapping of the quantitative trait loci (QTLs) and development of linked and/or perfect DNA markers has blurred the distinction between qualitative and quantitative traits. These tools have allowed dissection of at least part of a quantitative trait effect into qualitative effects resulting from allele changes at specific gene loci. The molecular techniques also allow qualitative inheritance of DNA markers to be followed instead of using phenotyping for the characteristic actually being followed. These systems have been extensively used for studying disease resistances that are modified by a suite of genes with differential effects. An example of a particular resistance gene with a qualitative effect as part of the overall quantitative gene management of *Ascochyta* is provided by Cho *et al.* (2005). They postulated that an allele of the flavanone 3-hydroxylase gene gives a form of resistance to *Ascochyta*. Equally complex genetic linkage maps have been developed (Cho *et al.*, 2002; Flandez-Galvez *et al.*, 2003) that include qualitative information on DNA sequences representing known chickpea morphological or phenotype genes, biochemical genes (e.g. isozymes), analogues to genes in other species (e.g. lentils and pea) and DNA sequences without known function. The result of these developments has been an explosion of qualitative genes across all crop species. However, further discussion will be restricted to qualitative genes of interest to breeding programmes, which are still followed, at least in part, using non-DNA-based phenotypic expression.

With the exception indicated earlier for mixed qualitative and quantitative disease resistances there have been few qualitative genes that have conferred substantial benefits upon chickpea. Rubio *et al.* (2004) identified situations in which erect or bushy, single or double and early or late qualitative genes benefited chickpea productivity across a range of environments. Other qualitative genes (frequently as 2-gene systems) that may prove beneficial for chickpea have been identified. They include resistance to iron chlorosis (Gumber *et al.*, 1997), seed coat thickness (Gil and Cubero, 1993), vigour (Sabaghpour *et al.*, 2003) and speed of germination (Dahiya *et al.*, 1994). In addition to these potentially beneficial phenotypes, a broad range of visual, biochemical and physiological phenotypes have been identified. Broadly these are detrimental to the chickpea and will not be discussed further except to indicate that many such phenotypes have been created by induced mutations. Examples include albinism, fasciation, leaf necrosis, non-nodulation and giant pod.

Numerous potentially beneficial phenotypes, which may be qualitatively inherited, have been identified in species closely related to chickpea (Croser *et al.*, 2003). However, to date few have been confirmed as being qualitatively inherited as phenotypes once incorporated into chickpea backgrounds. Examples of genes of interest include virus, insect, disease and nematode resistances. One gene of interest has been identified as a dehydrin in *Cicer pinnatifidum* (Bhattarai and Fettig, 2005), which may confer drought resistance.

There has been considerable interest in conferring qualitative genes of benefit by GM methods in chickpea. The most advanced at this stage are Bt genes for *Helicoverpa* resistance (Sanyal *et al.*, 2005). However, to date commercial release has not yet taken place.

Quantitative Traits

The genetic basis of quantitative traits in chickpea

Quantitative traits are characterized by a phenotypic continuum, in a way that prevents classification into easily defined categories. Environmental effects (including intrinsic organismal factors) are also capable of creating a wide phenotypic range, thereby interfering with classification even in the case of Mendelian traits. In such cases, biometrical tools are very helpful in quantifying the genetic component of the phenotypic variation vs the environmental component, regardless of the actual number of genes involved.

As it is of low economic importance for most industrialized nations, chickpea genetics is lagging behind crops like wheat, soybean or maize (Abbo *et al.*, 2003), and its quantitative genetics is no exception. Indeed, only few studies were devoted strictly to analysing the quantitative traits in chickpea. However, indirect evidence for quantitative genetic control of certain traits can be drawn from a number of studies that deal with important agronomic traits.

Polygenic traits

Among the first to use biometrical genetic tools in chickpea were Moreno and Cubero (1978) and Martinez *et al.* (1979). These authors studied the genetic basis of a number of morphological traits like leaflet length and width, pod length and seed weight, as well as seed yield and yield components, in the two botanical cultivar groups, *microsperma* (desi) and *macrosperma* (kabuli). Dominance (of lower values) was observed for leaflet length, width and shape index, whereas dominance of higher values was recorded for seed per pod. Overdominance (of higher values) was recorded for pod number, seed number and seed yield per plant (Moreno and Cubero, 1978; Martinez *et al.*, 1979).

The most comprehensive documentation of quantitative traits in chickpea was given in the ICRISAT chickpea germplasm catalogue (Pundir *et al.*, 1988). After screening thousands of entries, Pundir *et al.* (1988) obtained bell-shaped frequency distribution curves (or nearly so) for the following traits: days to flowering, duration of flowering, plant height, canopy width, seed weight, days to physiological maturity, number of primary and secondary branches, number of pods per plant, seed yield and seed protein concentration. The existence of such frequency distribution curves based on a worldwide germplasm collection is a good indication for polygenic control. Moreover, other more focused studies of agronomic traits corroborate this interpretation.

Seed yield is a highly complex trait integrating genetic and environmental effects occurring at all stages of crop development from germination to harvest maturity, making it a polygenic trait by definition. Singh *et al.* (1990) studied the association between seed yield and 14 other agronomic traits among 3200 field-grown chickpea cultivars. On the basis of such a large number of independent cultivars these correlations are good approximations of the genetic correlations between the studied traits. In this work, seed yield had a close association with biological yield, canopy width, harvest index, seed weight, plant height, days to flowering and maturity, flowering duration and protein content. Aiming at devising strategies for improving chickpea yield Singh *et al.* (1992) analysed 28 diallel trials carried out over 8 years, and estimated the genetic variance for several agronomic traits. Although days to flowering, plant height and seed size were predominantly affected by additive gene action, both additive and non-additive effects were important for seed yield (Singh *et al.*, 1992). This observation is partly in line with the earlier work of Jaiswal and Singh (1989), who reported preponderance of non-additive gene effects on yield and yield components among *C. arietinum* × *C. reticulatum* hybrid progeny. Consequently, Jaiswal and Singh (1989) anticipated poor yield gain under selection in such crosses. However, impressive yield improvement following introgression in such interspecific crosses was reported by Singh and Ocampo (1997) pointing to the importance of additive effects on yield in chickpea.

Many chickpea lines are vulnerable to low temperature stress during reproduction (flower development, pollen germination and zygote development) probably due to the ancient adaptation of the cultigen into a summer crop (Abbo *et al.*, 2003). The ability to set pods in low temperatures involves multiple developmental events and is therefore a complex (quantitative) trait. Unfortunately, despite progress in breeding for improved pod set at low temperatures by Clarke *et al.* (2004), no genetic information has been made available on this complex phenotype. The genetic basis of freezing tolerance at the seedling stage was investigated, and additive, non-additive as well as their respective interaction gene effects were identified (Malhotra and Singh, 1991).

Across most of its growing areas chickpea is a dryland crop relying on residual soil moisture. Therefore, its deep and prolific root system may contribute to yield improvement through drought avoidance (Kashiwagi *et al.*, 2006). Recent investigations of root traits using recombinant inbred lines point to a polygenic control of root length density and root mass with broad-sense heritability values of 0.23 and 0.27, respectively (Serraj *et al.*, 2004; Kashiwagi *et al.*, 2006). Additive inheritance of root length and speed of radicle emergence – two important adaptive traits for rainfed farming – were reported by Waldia *et al.* (1993).

The ability to maintain dry matter accumulation in the seeds is a major adaptive trait under terminal drought conditions (Palta *et al.*, 2004). Leport *et al.* (1999) studied the yield physiology of a number of Australian chickpea cultivars in water-limited conditions with and without irrigation after flowering. These authors reported that desi cultivars were superior to kabuli cultivars under terminal drought conditions, and were able to redirect assimilates into seeds rather than into vegetative organs. Interestingly, cultivars derived from a desi × kabuli cross occupied an intermediate position between the tested desi and kabuli cultivars (therein). This observation is in line with multifactorial inheritance of the measured phenotype (dry matter distribution after flowering).

Oligogenic traits

Ascochyta blight response

Mendelian inheritance of ascochyta blight response was reported in chickpea (e.g. Vir *et al.*, 1975); however, many research groups have recently opted to apply quantitative genetic tools for better understanding of this phenotype. One such early report is by van Rheenen and Haware (1994), who stressed the quantitative nature of the disease response. Lichtenzveig *et al.* (2002) used biometric tools to better understand the relationship between ascochyta blight response and time to flowering genes, and reported negative genetic correlation between the two traits ($r = -0.19$ to -0.44) pointing to the problem of combining good resistance with early flowering in chickpea. Tekeoglu *et al.* (2000) and Santra *et al.* (2000) identified several QTLs for ascochyta blight response in several chickpea crosses. Flandez-Galvez (2003) and Collard *et al.* (2003a,b) identified QTLs for ascochyta blight resistance in interspecific and intraspecific chickpea crosses, thereby corroborating previous reports on the oligogenic nature of this trait (e.g. Santra *et al.*, 2000; Tekeoglu *et al.*, 2000). Pathotype specific resistance QTLs were defined on the chickpea map by Udupa and Baum (2003) using simple sequence repeat (SSR) markers, and further work was done by Millan *et al.* (2003), who identified randomly amplified polymorphic DNA (RAPD) markers linked with ascochyta blight resistance QTLs.

Fusarium wilt resistance

Fusarium wilt resistance in chickpea was also studied using both Mendelian (e.g. Upadhyaya *et al.*, 1983; Kumar, 1998; Tulllu *et al.*, 1999) and quantitative approaches (e.g. Winter *et al.*, 2000; Jana *et al.*, 2003; and Cobos *et al.*, 2005). The linkage groups that were established using molecular markers suggest that a cluster of *Fusarium* resistance genes exists in linkage group 2 (Winter *et al.*, 2000 nomenclature). However, the gene for race zero response was identified on linkage group 5 (same nomenclature) by Cobos *et al.* (2005). Recent work has indicated that although the overall resistance to *Fusarium* is an oligogenic trait, the resistance to each individual race is controlled by a single gene (Sharma *et al.*, 2005a).

Seed quality traits

Seed coat thickness is an important parameter for processing, with desi seeds having thicker coats compared with those of kabuli type (Kumar and Singh, 1989). A number of genes appear to control this trait, with thick seed coat partly dominant to thin seed coat (therein).

As Calcium (Ca) is an essential mineral, Ca concentration is an important quality factor. With the frequency distribution of Ca concentration in hybrid progeny, this phenotype appears to be governed by a number of genes, and is partly independent from seed weight loci (Abbo *et al.*, 2000). Carotenoid concentration is another quality parameter that was dealt with using QTL methodology. Narrow-sense heritability values of 0.5–0.9 were reported with negative correlation with seed weight (Abbo *et al.*, 2005). A number of QTLs for pro-vitamin A carotenoids were identified on the chickpea genetic map, with one linked to a QTL for seed weight as expected from the correlation analysis (therein).

Cytology of *Cicer* Species

Taxonomy of *Cicer*

The genus *Cicer*, which includes the cultivated chickpea, has been taxonomically placed in the monogeneric tribe *Cicereae Alef.* of the family *Fabaceae* (Kupicha, 1981). As it currently stands, the genus *Cicer* consists of 43 species, divided into four sections – *Monocicer, Chamaecicer, Polycicer* and *Acanthocicer* – based on their morphological characteristics, life cycle and geographical distribution (van der Maesen, 1987; further reviewed by Croser *et al.*, 2003; Ahmad *et al.*, 2005). Eight of these annual and some of the perennial *Cicer* spp. are of particular interest as genetic resources to plant breeders (Ahmad *et al.*, 2005). Of the nine annual *Cicer* spp., eight are classified within the *Monocicer* section and one, *C. chorassanicum*, within the *Chamaecicer* section. Of the remaining species 33 are known to be perennial, while *C. laetum* Rass. and Sharip have an unspecified life cycle (van der Maesen, 1987).

Species relationships and gene pools in *Cicer*

Species relationships in the genus *Cicer* have recently been reviewed by Ahmad *et al.* (2005). The general consensus is that *C. reticulatum* and *C. echinospermum* are the wild species closely related to the domesticated *C. arietinum*. Additionally, *C. bigugum*, *C. pinnatifidum* and *C. judaicum* show a closer relationship among themselves and appear to form a group that is closest to the first group containing the cultivated species. The rest of the annual species – *C. chorassanicum*, *C. yamashitae* and *C. cuneatum* – share an even more distant relationship with the cultivated species. However, this relationship holds true only generally, as the five annual species that are not a part of the group with the cultivated species appear to change positions for relatedness to each other and to *C. arietinum*. Recently the perennial *Cicer* spp. have been subjected to phylogenetic analysis (reviewed by Ahmad *et al.*, 2005) and, with some exceptions, they have generally shown a distant relationship with the cultivated species. The relationship between the perennial species *C. anatolicum* and domesticated chickpea still remains a controversial issue (Kazan and Muehlbauer, 1991; Iruela *et al.*, 2002).

On the basis of Harlan and de Wet's (1971) definition, and in consideration of the results obtained from crossability, biochemical or molecular diversity, and karyotypic studies, a recently revised model of the wild annual *Cicer* gene pools has been proposed (Croser *et al.*, 2003). However, if one were to follow Harlan and de Wet's (1971) definition alone, the primary gene pool of *Cicer* would consist of *C. arietinum* and only one wild species, the wild annual progenitor *C. reticulatum*. The other closely related species, *C. echinospermum*, does cross quite readily with the cultivated species; however, it shows varying levels of sterility in the F_1 and F_2 generations. The secondary gene pool thus consists of *C. echinospermum* only. *C. bijugum*, *C. pinnatifidum* and *C. judaicum*, which have been reported to give hybrids readily when crossed with the cultivated species (Verma *et al.*, 1990; Singh *et al.*, 1994; Singh *et al.*, 1999a,b), have been placed in the secondary gene pool by Croser *et al.* (2003). This, however, is an issue that is not fully resolved, due to lack of any hybridity evidence for at least two of the three interspecific hybrid combinations. A general scepticism exists in the scientific community regarding the authenticity of these partially fertile hybrids (R.P.S. Pundir, ICRISAT, Hyderabad, India, 1994, personal communication; J.I. Cubero and B. Ocampo, Spain, 2004, personal communication; F.J. Muehlbauer, USDA, Pullman, 2004, personal communication). Ahmad *et al.* (2005) have recently proposed that the above three species should be placed in the tertiary gene pool of chickpea, along with the remaining annual species *C. chorassanicum*, *C. yamashitae* and *C. cuneatum*. The lack of availability of perennial *Cicer* spp. has greatly restricted the assessment of their crossability with the cultivated species. Given the current situation with perennial *Cicer* spp., and until proven otherwise, these perennial species should be appropriately placed in the tertiary gene pool along with the six other annual wild species.

Cytogenetics of *Cicer* species

Chickpea is a crop that is not amenable to cytogenetic studies. Hence, most studies are limited to chromosome counts and feulgen-stained studies of karyotypes. No cytogenetic stocks, other than tetraploids, are available in chickpea (Bahl, 1987; Gupta and Sharma, 1991) and except for one (Vlacilova *et al.*, 2002), linkage groups have not yet been associated with respective chromosomes.

Karyotypic studies in Cicer

Karyotype analysis is considered an important method for genome analysis and it is anticipated that such identification should further define interspecific relationships, help identify interspecific hybrids, lay a foundation for identifying wild species chromosome(s) in *C. arietinum* × wild species hybrid derivatives, and provide a basis for any relationship between karyotypic similarities and interspecific crossability (Ahmad, 2000). A considerable number of cytological studies have been conducted on *C. arietinum* (reviewed by Ahmad *et al.*, 2005) but not as much on the related wild species. Approximately 43% of the chickpea genome has been characterized as heterochromatic (Tayyar *et al.*, 1994; Venora *et al.*, 1995a; Galasso *et al.*, 1996). Additionally, evolution in the genus *Cicer* has not involved a concurrent increase in heterochromatin content, but has rather resulted in rearrangement of heterochromatin among the different chromosomes of the genome (Tayyar *et al.*, 1994; Galasso *et al.*, 1996).

The nine annual and nine of the remaining 34 species are confirmed diploids with $2n = 2x = 16$ chromosomes (Ladizinsky and Adler, 1976a,b; Ahmad, 1989; Pundir *et al.*, 1993; Ohri, 1999; Ahmad, 2000; Ahmad and Chen, 2000; Ahmad *et al.*, 2005). The cultivated chickpea karyotype consists of a long chromosome pair that is submetacentric and distinctly satellited, six pairs of medium-sized metacentric to submetacentric chromosomes and a pair of very short metacentric chromosomes (Ocampo *et al.*, 1992; Ahmad and Hymowitz, 1993; Ahmad, 2000). The identification of some of the chromosomes in *Cicer* spp. presents some difficulty due to their small size, which affects both the clear discrimination of the arm ratios and the total chromosome length. This could, hopefully, be alleviated by identifying physical landmarks of molecular markers on specific chromosomes and chromosome arms (Abbo *et al.*, 1994; Gortner *et al.*, 1998; Staginnus *et al.*, 1999).

The karyotypes of *C. arietinum*, *C. reticulatum*, *C. echinospermum*, *C. songaricum* and *C. anatolicum* show much similarity to each other and form the first group, while the second group contains all the remaining species of the genus. The karyotype of the first group of species is characterized as being asymmetrical, a ratio of the longest to shortest chromosomes close to 2.5 or higher, large coefficients of interchromosomal size variation, and one of the long chromosomes being satellited (Ohri and Pal, 1991; Ocampo *et al.*, 1992;

Ohri, 1999; Ahmad, 2000), although some lines of *C. reticulatum* may possibly have two pairs of satellited chromosomes. Species of the second group (consisting of the remainder of the annual species) have a rather symmetrical karyotype with little interchromosomal size variation, as well as a small longest to shortest chromosome ratio, and satellites located on the smaller chromosome of the complement. A circumstantial relationship between interspecific karyotypic similarity and crossability has recently been speculated (Ahmad, 2000; Croser *et al.*, 2003; Ahmad *et al.*, 2005), but the hypothesis remains to be thoroughly tested.

Pachytene analysis of chromosomes provides detailed information about chromosome morphology. For example, detailed pachytene chromosome analysis has shown the third chromosome to be clearly satellited in *C. arietinum* (Ahmad and Hymowitz, 1993), whereas the less refined somatic karyotype shows the first chromosome pair to be satellited (Ocampo *et al.*, 1992; Ahmad, 2000). Such refined pachytene analysis has made individual chromosome identification more conclusive and added a further dimension to chickpea cytology (Ahmad and Hymowitz, 1993). Preliminary studies have shown details of pachytene karyotypes in *C. reticulatum*, *C. bijugum* and *C. chorassanicum* (Sharma and Gupta, 1984, 1986).

Meiotic associations in Cicer

A few incomplete and rather scattered studies have been conducted to determine the morphology and behaviour of individual chromosomes in the cultivated chickpea and related *Cicer* spp. (Ladizinsky and Adler, 1976a,b; Sharma and Gupta, 1986; Ahmad and Hymowitz, 1993; Pundir *et al.*, 1993; Ahmad and Chen, 2000). Meiosis in *Cicer* spp. is characterized by a rather diffused prophase, and precocious separation of one or two longer homologous pairs is occasionally observed in some pollen mother cells (PMCs) (Ahmad and Chen, 2000). The shorter chromosomes frequently form ring bivalents, whereas the longer chromosomes form rod bivalents (Sharma and Gupta, 1986; Ahmad and Chen, 2000). In the genus *Cicer*, chiasma frequency per unit genome length is primarily a function of the diploid genome length ($r^2 = 0.86$) (Ahmad and Chen, 2000).

Chromosome pairing in interspecific hybrids in Cicer

Owing to early operating and strong post-fertilization crossability barriers (Bassiri *et al.*, 1987; Ahmad *et al.*, 1988; Stamigna *et al.*, 2000; Ahmad and Slinkard, 2003, 2004; Ahmad, 2005), only a few authentic interspecific hybrids are known in the genus *Cicer*. Hence, there is limited information available on chromosome pairing in interspecific *Cicer* hybrids. Moreover, gene introgression in the cultivated chickpea, by utilizing interspecific hybridization, is limited mainly to the progenitor species *C. reticulatum*.

Ladizinsky and Adler (1976a,b) studied meiotic chromosome associations in six interspecific hybrids of *Cicer*. The interspecific hybrid *C. arietinum* × *C. reticulatum* is easy to make, develops normally, has regular meiosis with eight bivalents and is fully fertile (Ladizinsky and Adler, 1976a,b). In contrast, the hybrid *C. arietinum* × *C. echinospermum* is characterized by six bivalents plus a quadrivalent (Ladizinsky and Adler, 1976a,b), which along with other cryptic structural hybridity renders the F_1 or F_2 plants highly sterile. A reciprocal translocation also differentiates the genomes of *C. reticulatum* and *C. echinospermum*, which results in complete sterility (Ladizinsky and Adler, 1976a,b). Chromosome association data indicate a close chromosome homology between *C. bijugum*, *C. judaicum* and *C. pinnatifidum* (Ladizinsky and Adler, 1976a,b; Ahmad, 2005). Univalent formation was lowest in *C. pinnatifidum* × *C. bijugum* and highest in *C. judaicum* × *C. pinnatifidum*. The only hybrid between *C. judaicum* and *C. chorassanicum* has been reported by Ahmad *et al.* (1987), and is characterized by a high number of univalents and very low pollen fertility. Authentic interspecific hybrids of *C. arietinum* × *C. pinnatifidum* (Badami *et al.*, 1997), and putative hybrids, *C. chorassanicum* × *C. pinnatifidum* and *C. chorassanicum* × *C. yamashitae* (Ahmad, 2005), have been produced but no chromosome pairing data are available due to the albino nature of the hybrids.

Chromosome banding and molecular cytogenetic studies in Cicer

There is a lack of understanding of the cytogenetic map in chickpea and related *Cicer* spp. Individual chromosomes of chickpea are reported to be identifiable by C-banding and fluorochrome staining (Galasso and Pignone, 1992; Tayyar *et al.*, 1994; Galasso *et al.*, 1996). The related annual wild species have also been subjected to a solitary C-banding analysis (Tayyar *et al.*, 1994) and the general conclusion drawn is that the heterochromatic C-bands are located proximally around the centromere with only occasional bands in intercalary and distal positions. Pachytene chromosome analysis of the cultivated species (Ahmad and Hymowitz, 1993) has further corroborated such distribution of heterochromatin. Although individual pairs of chromosomes within a species were identified (Tayyar *et al.*, 1994), its applicability in an interspecific hybrid situation remains to be explored.

Given the poor state of chickpea cytology, the progress in molecular cytogenetics in the genus is quite impressive. Individual chickpea chromosomes have been successfully sorted by flow cytometry (Vlacilova *et al.*, 2002) and utilized for mapping specific DNA sequences and genes to individual chromosomes. Thus, specific genes (coding for various rRNA loci), major random repetitive DNA sequences, sequence-tagged microsatellite site (STMS) markers, En/Spm-like transposon sequences, simple sequence repeats and *Arabidopsis*-type telomeric sequences have been successfully hybridized to, and localized on, the chickpea chromosomes by fluorescent *in situ* hybridization (FISH) (Abbo *et al.*, 1994; Galasso *et al.*, 1996; Gortner *et al.*, 1998; Staginnus *et al.*, 1999, 2001; Vlacilova *et al.*, 2002; Valarik *et al.*, 2004). Using polymerase chain reaction (PCR) and FISH, Vlacilova *et al.* (2002) have successfully associated two STMS markers (belonging to linkage group 8 of Winter *et al.*, 2000) to the shortest chromosome of the chickpea genome. Recently, FISH analysis on super-stretched flow sorted chickpea chromosomes has revealed spatial resolution of neighbouring loci that has not been obtained by any other method (Valarik *et al.*, 2004). It is expected that further technical advances will lead to the development in genome mapping of chickpea and associate all genetic linkage groups to specific well-defined chromosomes.

Role of Genetics in Value Addition in Chickpea

The consumption of chickpea in different countries varies depending on dietary pattern. It is mostly consumed in the form of whole seed, dhal (decorticated split cotyledons) or dhal flour (*besan*). In the Indian subcontinent about 75% of chickpea is utilized either as dhal or dhal flour and the remaining as whole seed, whereas in most of the other countries including Australia, Ethiopia, Mexico, Sudan, Tanzania, Turkey, the USA and UK chickpea is consumed as a whole seed (Jambunathan and Singh, 1989). Processing of chickpea in the Indian subcontinent includes dehusking and milling, which essentially involve removal of the seed coat from the grain and recovery of cotyledons in the form of dhal. Removal of the seed coat reduces roughage, improves storability and palatability for consumption in various forms. It also improves cooking quality and digestibility. The average yield of dhal in chickpea varies from 67% to 74% (Singh and Iyer, 1998; Kuldeep and Singh, 2003). The rest of the grain material becomes by-products comprising husk, powder, and large and small broken, shrivelled and underprocessed grains (Kurien and Parpia, 1968). These by-products can also be an important source for value addition by replacing other sources of fibre or as potential sources of nutraceuticals.

Market price of chickpea depends on consumer acceptability traits like seed size, seed colour, cooking quality and yield of dhal after milling, rather than on nutritional considerations (Pushpamma, 1975). Genetic enhancement of the consumer acceptance traits will depend on the extent of variability present in the germplasm of desi and kabuli types. From the value addition point of view, the single most important characteristic that fetches better market price is larger seed size, particularly in the case of kabulis. Fortunately, there is a large range of variation available in the chickpea germplasm for seed size, which in terms of 100-seed weight varies from 2 g to more than 60 g (Williams and Singh, 1986) with appreciable variation both within desi and kabuli germplasm. Also, 100-seed weight is a highly heritable trait and a number of researchers have reported additive gene action controlling this characteristic (Bahl and Salimath, 1996). These attributes associated with 100-seed weight are conducive for developing large-seeded varieties by a suitably designed breeding programme. A number of large-seeded genotypes both in desi and kabuli groups with 100-seed weight up to 45 g have been developed in the breeding programme, undertaken for improvement of quality traits, at the Division of Genetics, IARI, New Delhi (S.S. Yadav, New Delhi, 2005, personal communication). The large-seeded kabuli types obtained as a result of this programme get at least 20–25% premium in the market over other commercially released kabuli types.

Kabuli–desi introgression programmes are underway at several research stations where desi germplam is introgressed into kabuli types and vice versa. Improved kabulis and desis obtained from such programmes can be used to identify genotypes for different dietary purposes. Kabuli types, reported to be nutritionally superior to desi types in terms of cooking time, biological value and sensory properties (Singh *et al.*, 1991), particularly large-seeded ones developed through introgressive breeding, can be used to identify genotypes for areas or regions where chickpea is consumed as a whole seed. On the other hand, medium-sized desis, developed from this breeding approach, showing minimum loss as broken grains, can be used in dhal milling industry.

In addition to seed size, there are several other consumer quality traits like seed mass, seed volume before and after soaking, as well as swelling and hydration capacity, which show considerable phenotypic and genotypic variation (Kumar *et al.*, 1998; Yadav *et al.*, 2003). Research at the University of Melbourne and the Victorian Department of Primary Industries in Australia has been characterizing the inheritance of seed colour, shape and size, as well as a range of nutraceutical characteristics such as cotyledon and seed coat flavonoids (D. McNeil, Victoria, 2006, personal communication) and carotenoids (Abbo *et al.*, 2005), and seed Ca concentration (Abbo *et al.*, 2000). Chickpea breeders can exploit this variability to achieve value addition through genetic enhancement of these traits. Mehla *et al.* (2001) working on 55 kabuli germplasm lines reported varietal variation in cooking time and suggested the use of genotypes with short cooking time in the breeding programmes aimed for improving the cooking quality of kabulis, thus adding value to the product. Nutritional benefits have been observed from consumption of chickpea (Zulet *et al.*, 1998), and variation for the compounds thought to produce these benefits are currently being sought.

References

Abbo, S., Miller, T.E., Reader, S.M., Dunford, R.P. and King, I.P. (1994) Detection of ribosomal DNA sites in lentil and chickpea by fluorescent *in situ* hybridization. *Genome* 37, 713–716.

Abbo, S., Grusak, M.A., Tzuk, T. and Reifen, R. (2000) Genetic control of seed weight and calcium concentration in chickpea seed. *Plant Breeding* 119(5), 427–431.

Abbo, S., Berger, J. and Turner, N.C. (2003) Viewpoint: evolution of cultivated chickpea: four bottlenecks limit diversity and constrain adaptation. *Functional Plant Biology* 30(10), 1081–1087.

Abbo, S., Molina, C., Jungmann, R., Grusak, M.A., Berkovitch, Z., Reifen, R., Kahl, G., Winter, P. and Reifen, R. (2005) Quantitative trait loci governing carotenoid concentration and weight in seeds of chickpea (*Cicer arietinum* L.). *Theoretical and Applied Genetics* 111(2), 185–195.

Ahmad, F. (1989) The chromosomal architecture of *Cicer anatolicum* Alef., a wild perennial relative of chickpea. *Cytologia* 54, 753–757.

Ahmad, F. (2000) A comparative study of chromosome morphology among the nine annual species of *Cicer* L. *Cytobios* 101, 37–53.

Ahmad, F. (2005) An analysis of interspecific crossability among the six tertiary gene pool annual species of *Cicer* L. In: Kharkwal, M.C. (ed.) *4th International Food Legume Research Conference-Abstracts*. IARI, New Delhi, India, p. 45.

Ahmad, F. and Chen, Q. (2000) Meiosis in *Cicer* L. species: the relationship between chiasma frequency and genome length. *Cytologia* 65, 161–166.

Ahmad, F. and Hymowitz, T. (1993) The fine structure of chickpea (*Cicer arietinum* L.) chromosomes as revealed by pachytene analysis. *Theoretical and Applied Genetics* 86, 637–641.

Ahmad, F. and Slinkard, A.E. (2003) Limitations to bridge-species facilitated alien gene transfers in chickpea: pre-fertilization events. *Journal of Genetics and Breeding* 57, 69–74.

Ahmad, F. and Slinkard, A.E. (2004) The extent of embryo and endosperm growth following interspecific hybridization between *Cicer arietinum* L. and related annual wild species. *Genetic Resources and Crop Evolution* 51, 765–772.

Ahmad, F., Slinkard, A.E. and Scoles, G.J. (1987) The cytogenetic relationship between *Cicer judaicum* Boiss. and *Cicer chorassanicum* (Bge.) M. Pop. *Genome* 29, 883–886.

Ahmad, F., Slinkard, A.E. and Scoles, G.J. (1988) Investigations into the barrier(s) to interspecific hybridization between *Cicer arietinum* L. and eight other annual *Cicer* species. *Plant Breeding* 100, 193–198.

Ahmad, F., Gaur, P.M. and Croser, J.S. (2005) Chickpea (*Cicer arietinum* L.). In: Singh, R.J. and Jauhar, P.P. (eds) *Genetic Resources, Chromosome Engineering, and Crop Improvement*, Volume 1. *Grain Legumes*. CRC Press, Boca Raton, pp. 187–217.

Badami, P.S., Mallikarjuma, N. and Moss, J.P. (1997) Interspecific hybridization between *Cicer arietinum* and *Cicer pinnatifidum*. *Plant Breeding* 116, 393–395.

Bahl, P.N. (1987) Cytology of chickpea. In: Saxena, M.C. and Singh, K.B. (eds) *The Chickpea*. CAB International, Wallingford, UK, pp. 83–97.

Bahl, P.N. and Salimath, P.M. (1996) *Genetics, Cytogenetics and Breeding of Crop Plants*, Volume 1. *Pulses and Oilseeds*. Science Publishers, Lebanon, New Hampshire.

Balachandra, R. (2005) Available at: http://www.icis.cgiar.org:8080/workshops/report/2005/reports/3_Wed_06_Ranjan_Presentation_23-02-2004.pdf

Bhattarai, T. and Fettig, S. (2005) Isolation and characterization of a dehydrin gene from Cicer pinnatifidum, a drought-resistant wild relative of chickpea. *Physiologia Plantarum* 123(4), 452–458.

Bassiri, A., Ahmad, F. and Slinkard, A.E. (1987) Pollen grain germination and pollen tube growth following *in vivo* and *in vitro* self- and interspecific pollinations in annual *Cicer* species. *Euphytica* 36, 667–675.

Cho, S.H., Kumar, J., Shultz, J.L., Anupama, K., Tefera, F. and Muehlbauer, F.J. (2002) Mapping genes for double podding and other morphological traits in chickpea. *Euphytica* 28(2), 285–292.

Cho, S.H., Chen, W.D. and Muehlbauer, F.J. (2005) Constitutive expression of the flavanone 3-hydroxylase gene related to pathotype-specific ascochyta blight resistance in *Cicer arietinum* L. *Physiological and Molecular Plant Pathology* 67(2), 100–107.

Clarke, H.J., Khan, T.N. and Siddique, K.H.M. (2004) Pollen selection for chilling tolerance at hybridisation leads to improved chickpea cultivars. *Euphytica* 139, 65–74.

Cobos, M.J., Fernandez, M., Rubio, J., Kharrat, M., Moreno, M.T., Gil, J. and Millan, T. (2005) A linkage map of chickpea (*Cicer arietinum* L.) based on populations from Kabuli × Desi crosses: location of genes for resistance to fusarium wilt race 0. *Theoretical and Applied Genetics* 110, 1347–1353.

Collard, B.C.Y., Pang, E.C.K., Ades, P.K. and Taylor P.W.J. (2003a) Preliminary investigation of QTLs associated with seedling resistance to ascochyta blight from *Cicer echinospermum*, a wild relative of chickpea. *Theoretical and Applied Genetics* 107(4), 719–729.

Collard, B.C.Y., Pang E.C.K. and Taylor P.W.J. (2003b) Selection of wild *Cicer* accessions for the generation of mapping populations segregating for resistance to ascochyta blight. *Euphytica* 130(1), 1–9.

Croser, J.S., Ahmad, F., Clarke, H.J. and Siddique, K.H.M. (2003). Utilisation of wild *Cicer* in chickpea improvement: progress, constraints, and prospects. *Australian Journal of Agricultural Research* 54, 429–444.

Dahiya, O.S., Tomer, R.P.S., Dahiya, B.S. and Kumar, A. (1994) Genetics for speed of germination under laboratory conditions in chickpea (*Cicer arietinum* L.). *Seed Science and Technology* 22(3), 629–632.

Flandez-Galvez, H., Ford, R., Pang, E.C.K. and Taylor, P.W.J. (2003) An intraspecific linkage map of the chickpea (*Cicer arietinum* L.) genome based on sequence tagged microsatellite site and resistance gene analog markers. *Theoretical and Applied Genetics* 106(8), 1447–1456.

Galasso, I. and Pignone, D. (1992) Characterization of chickpea chromosomes by banding techniques. *Genetic Resources and Crop Evolution* 39, 115–119.

Galasso, I., Frediani, M., Maggiani, M., Cremonini, R. and Pignone, D. (1996)

Chromatin characterization by banding techniques, *in situ* hybridization, and nuclear DNA content in *Cicer* L. (*Leguminosae*). *Genome* 39, 258–265.

Gil, J. and Cubero, J.L. (1993) Inheritance of seed coat thickness in chickpea (*Cicer arietinum* L.) and its evolutionary implications. *Plant Breeding* 111(3), 257–260.

Gortner, G., Nenno, M., Weising, K., Zink, D., Nagl, W. and Kahl, G. (1998) Chromosomal localization and distribution of simple sequence repeats and the Arabidopsis-type telomere sequence in the genome of *Cicer arietinum* L. *Chromosome Research* 6, 97–104.

Gumber, R.K., Sarvjeet Singh Gill, J.S. and Rathore, P.K. (1997) Genetics of irrigation-induced iron chlorosis in chickpea. *International Chickpea and Pigeonpea Newsletter* 4, 10–11.

Gupta, P.K. and Sharma, P.C. (1991) Cytogenetics and related aspects in some pulse crops. In: Tsuchiya, T. and Gupta, P.K. (eds) *Chromosome Engineering in Plants: Genetics, Breeding, Evolution, Part B*. Elsevier Science, Amsterdam, The Netherlands, pp. 1–31.

Harlan, J.R. and de Wet, J.M.J. (1971) Toward a rational classification of cultivated plants. *Taxon* 20, 509–517.

IBPGR (1993) *Descriptors for Chickpea* (*Cicer arietinum* L.). International Board for Plant Genetic Resources, Rome, Italy.

Iruela, M., Rubio, J., Cubero, J.I., Gil, J. and Millan, T. (2002) Phylogenetic analysis in the genus *Cicer* and cultivated species using RAPD and ISSR markers. *Theoretical and Applied Genetics* 104, 643–651.

Jaiswal, H.K. and Singh, B.D. (1989) Analysis of gene effects for yield and certain yield traits in crosses between *Cicer arietinum* L. and *C. reticulatum* Ladiz. *Indian Journal of Genetics* 49, 9–17.

Jambunathan, R. and Singh, U. (1989) Present status and prospects for utilization of chickpea. In: van Rheenen, H.A. and Saxena, M.C. (eds) *Proceedings of Second International Workshop on Chickpea Improvement.* ICRISAT, Hyderabad, India, pp. 41–46.

Jana, T., Sharma, T.R., Prasad, R.D. and Arora, D.K. (2003) Molecular characterization of *Macrophomina phaseolina and Fusarium* species by a single primer RAPD technique. *Microbiological Research* 158, 249–257.

Kashiwagi, J., Krishnamurthy, L., Crouch, J.H. and Serraj, R. (2006) Variability of root length density and its contributions to seed yield in chickpea (*Cicer arietinum* L.) under terminal stress. *Field Crops Research* 95, 171–181.

Kazan, K. and Muehlbauer, F.J. (1991) Allozyme variation and phylogeny in annual species of *Cicer* (*Leguminosae*). *Plant Systematics and Evolution* 175, 11–21.

Kuldeep, A. and Singh, G. (2003) Physico-chemical and milling quality of some improved varieties of chickpea (*Cicer arietinum*). *Journal of Food Science and Technology* 40(4), 439–442.

Kumar, J. and Singh, U. (1989) Seed coat thickness in chickpea: variation and inheritance in a desi × kabuli cross. *Indian Journal of Genetics* 49, 245–249.

Kumar, S. (1998) Inheritance of resistance to *Fusarium* wilt (race 2) in chickpea. *Plant Breeding* 117(2), 139–142.

Kumar, R., Verma, B.S., Ahlawat, T.R. and Yadav, O.P. (1998) Grain quality in desi chickpea (*Cicer arietinum* L.). *Legume Research* 21(1), 33–36.

Kupicha, F.K. (1981) *Cicereae* Alefeld. In: Polhill, R.M. and Raven, P.H. (eds) *Advances in Legume Systematics,* Part 1. Royal Botanic Gardens, Kew, UK, p. 382.

Kurien, P.P. and Parpia, H.A.B. (1968) Pulse milling in India – I. *Journal of Food Science and Technology* 5, 203–207.

Ladizinsky, G. and Adler, A. (1976a) The origin of chickpea *Cicer arietinum* L. *Euphytica* 25, 211–217.

Ladizinsky, G. and Adler, A. (1976b) Genetic relationships among the annual species of *Cicer* L. *Theoretical and Applied Genetics* 48, 197–203.

Leport, L., Turner, N.C., French, R.J., Barr, M.D., Duda, R., Davies, S.L., Tennant, D. and Siddique, K.H.M. (1999) Physiological response of chickpea genotypes to terminal drought in a Mediterranean-type environment. *European Journal of Agronomy* 11, 279–291.

Lichtenzveig, J., Shtienberg, D., Zhang, H.B., and Bonfil, D.J. (2002) Biometric analyses of the inheritance of resistance to *Didymella rabiei* in chickpea. *Phytopathology* 92(4), 417–423.

Malhotra, R.S. and Singh, K.B. (1991) Gene action for cold tolerance in chickpea. *Theoretical and Applied Genetics* 82, 598–601.

Martinez, A., Coiduras, A., Moreno, M.T. and Cubero, J.J. (1979) Quantitative inheritance and barriers to crossability in *Cicer arietinum* L. *Theoretical and Applied Genetics* 55, 5–16.

Mehla, L.S., Waldia, R.S. and Dahiya, S.S. (2001) Variation and relationship among cooking quality attributes across environments in kabuli chickpea (*Cicer arietinum* L.). *Journal of Food Science and Technology* 38(3), 283–286.

Millan, T., Rubio, J., Iruela, M., Daly, K., Cubero, J.I. and Gil, J. (2003) Markers associated with ascochyta blight resistance in chickpea and their potential in marker-assisted selection. *Field Crops Research* 84(3), 373–384.

Moreno, M.T. and Cubero, J.I. (1978) Variation in *Cicer arietinum* L. *Euphytica* 27, 465–485.

Muehlbauer, F.J. and Singh, K.B. (1987) Genetics of chickpea. In: Saxena, M.C. and Singh, K.B. (eds) *The Chickpea*. CAB International, Wallingford, UK, pp. 99–126.

Ocampo, B., Venora, G., Errico, A., Singh, K.B. and Saccardo, F. (1992) Karyotype anaylsis in the genus *Cicer. Journal of Genetics and Breeding* 46, 229–240.

Ohri, D. (1999) Cytology of *Cicer soongaricum* Steph. ex DC, a wild relative of chickpea. *Genetic Resources and Crop Evolution* 46, 111–113.

Ohri, D. and Pal, M. (1991) The origin of chickpea (*Cicer arietinum* L.): karyotype and nuclear DNA amount. *Heredity* 66, 367–372.

Palta, J.A., Turner, N.C. and French, R.J. (2004) The yield performance of lupin genotypes under terminal drought in a Mediterranean-type environment. *Australian Journal of Agricultural Research* 55, 449–459.

Pittock, C., Materne, M. and Taylor, P.W.J. (2004) Molecular breeding: defect elimination in chickpea. (Poster). *5th European Conference on Grain Legumes, Dijon, June, 2004 – Conference Handbook* (abstracts of posters and workshops).

Pundir, R.P.S, Rao, N.K. and van den Maesen, L.J.G. (1985) Distribution of qualitative traits in the world germplasm of chickpea (*Cicer arietinum* L.). *Euphytica* 34(3), 697–703.

Pundir, R.P.S., Reddy, K.N. and Mengesha, M.H. (1988) *ICRISAT. Chickpea Germplasm Catalog: Evaluation and Analysis*. ICRISAT, Hyderabad, India.

Pundir, R.P.S., Mengesha, M.H. and Reddy, G.V. (1993) Morphology and cytology of *Cicer canariense*, a wild relative of chickpea. *Euphytica* 69, 73–75.

Pushpamma, P. (1975) Evaluation of nutritional value, cooking quality and consumer preferences of grain legumes. In: *Proceedings of the International Workshop on Grain Legumes*. ICRISAT, Hyderabad, India, pp. 213–220.

Rubio, J., Flores, F., Moreno, M.T., Cubero, J.I. and Gil, J. (2004) Effects of the erect/bushy habit, single/double pod and late/early flowering genes on yield and seed size and their stability in chickpea. *Field Crops Research* 90(2/3), 255–262.

Sabaghpour, S.H., Kumar, J. and Rao, T.N. (2003) Inheritance of growth vigour and its association with other characters in chickpea. *Plant Breeding* 122(6), 542–544.

Santra, D.K., Tekeoglu, M., Ratnaparkhe, M., Kaiser, W.J. and Muehlbauer, F.J. (2000) Identification and mapping of QTLs conferring resistance to ascochyta blight in chickpea. *Crop Science* 40, 1606–1612.

Sanyal, I., Singh, A.K., Kaushik, M. and Amla, D.V. (2005) Agrobacterium-mediated transformation of chickpea (*Cicer arietinum* L.) with *Bacillus thuringiensis* cry1Ac gene for resistance against pod borer insect *Helicoverpa armigera. Plant Science* 168(4), 1135–1146.

Sarmah, B.K., Moore, A., Tate, W., Molvig, L., Morton, R.L., Rees, D.P., Chiaiese, P., Chrispeels, M.J., Tabe, L.M. and Higgins, T.J.V. (2004) Transgenic chickpea seeds expressing high levels of a bean α-amylase inhibitor. *Molecular Breeding* 14(1), 73–72.

Serraj, R., Krishnamurthy, L., Kashiwagi, J., Kumar, J., Chandra, S. and Crouch, J.H. (2004)

Variation in root traits of chickpea (*Cicer arietinum* L.) grown under terminal drought. *Field Crops Research* 88, 115–127.

Sharma, P.C. and Gupta, P.K. (1984) Pachytene chromosome morphology in *Cicer bijugum* Rech. *Perspectives in Cytology and Genetics* 4, 485–489.

Sharma, P.C. and Gupta, P.K. (1986) Cytogenetics of legume genera *Cicer* L. and *Lens* L. In: Gupta, P.K. and Bahl, J.R. (eds) *Genetics and Crop Improvement*. Rastogi & Co, Meerut, India, pp. 321–340.

Sharma, K.D., Winter, P. and Muehlbauer, F.J. (2004) Molecular Mapping of *Fusarium Oxysporum* F. Sp. Ciceris Race 3 Resistance Gene in Chickpea. *Journal of Theoretical and Applied Genetics* 108, 1243–1248.

Sharma, K.D., Chen, W.D. and Muehlbauer, F.J. (2005a) Genetics of chickpea resistance to five races of Fusarium wilt and a concise set of race differentials for *Fusarium oxysporum* f. sp. *ciceris*. *Plant Disease* 89, 385–390.

Sharma, H.C., Pampapathy, G., Lanka, S.K. and Ridsdill-Smith, T.J. (2005b) Exploitation of wild *Cicer reticulatum* germplasm for resistance to *Helicoverpa armigera*. *Journal of Economic Entomology* 98, 2246–2253.

Singh, U. and Iyer, L. (1998) Dehulling chickpea (*Cicer arietinum* L.): a comparative study on laboratory mills, pre-treatment and genotypes. *Journal of Food Science and Technology* 35, 499–503.

Singh, K.B. and Ocampo, B. (1997) Exploitation of wild Cicer species for yield improvement in chickpea. *Theoretical and Applied Genetics* 95, 418–423.

Singh, O., Gowda, C.L.L., Sethi, S.C., Dasgupta, T. and Smithson, J.B. (1992) Genetic analysis of agronomic characters in chickpea. *Theoretical and Applied Genetics* 83(8), 956–962.

Singh, K.B., Malhotra, R.S., Halila, M.H., Knights, E.J. and Verma, M.M. (1994) Current status and future strategy in breeding chickpea for resistance to biotic and abiotic stresses. *Euphytica* 73, 137–149.

Singh, A., Singh, N.P. and Asthana, A.N. (1999a) Genetic potential of wide crosses in chickpea. *Legume Research* 22, 19–25.

Singh, N.P., Singh, A. and Asthana, A.N. (1999b) Studies on inter-specific crossability barriers in chickpea. *Indian Journal of Pulses Research* 12, 13–19.

Singh, U., Subrahmanyam, N. and Kumar, J. (1991) Cooking quality and nutritional attributes of some newly developed cultivars of chickpea. *Journal of Food Science and Agriculture* 55, 37–46.

Staginnus, C., Winter, P., Desel, C., Schmidt, T. and Kahl, G. (1999) Molecular structure and chromosomal localization of major repetitive DNA families in the chickpea (*Cicer arietinum* L.) genome. *Plant Molecular Biology* 39, 1037–1050.

Staginnus, C., Huttel, B., Desel, C., Schmidt, T. and Kahl, G. (2001) A PCR-based assay to detect En/Spm-like transposon sequences in plants. *Chromosome Research* 9, 591–605.

Stamigna, C., Crinò, P. and Saccardo, F. (2000) Wild relatives of chickpea: multiple disease resistance and problems to introgression in the cultigen. *Journal of Genetics and Breeding* 54, 213–219.

Tayyar, R.I., Lukaszewski, A.J. and Waines, J.G. (1994) Chromosome banding patterns in the annual species of *Cicer*. *Genome* 37, 656–663.

Tekeoglu, M., Santra, D.K., Kaiser, W.J. and Muehlbauer, F.J. (2000) Ascochyta blight resistance inheritance in three chickpea recombinant inbred line populations. *Crop Science* 40, 1251–1256.

Tullu, A., Kaiser, W.J., Kraft, J.M. and Muehlbauer, F.J. (1999) A second gene for resistance to race 4 of Fusarium wilt in chickpea and linkage with a RAPD marker. *Euphytica* 109(1), 43–50.

Udupa, S.M. and Baum, M. (2003) Genetic dissection of pathotype-specific resistance to ascochyta blight disease in chickpea (*Cicer arietinum* L.) using microsatellite markers. *Theoretical and Applied Genetics* 106(7), 1196–1202.

Upadhyaya, H.D., Smithson, J.B., Haware, M.P. and Kumar, J. (1983) Resistance to wilt in chickpea. II. Further evidence for two genes for resistance to race 1. *Euphytica* 32, 749–755.

Valarik, M., Barto, J., Kovajova, P., Kubalakova, M., de Jong, J.H. and Dole, J. (2004) High-resolution FISH on super-stretched flow-sorted plant chromosomes. *The Plant Journal* 37, 940–950.

van der Maesen, L.J.G. (1987) Origin, history and taxonomy of chickpea In: Saxena, M.C. and Singh, K.B. (eds) *The Chickpea*. CAB International, Wallingford, UK, pp. 11–34.

Venora, G., Galasso, I. and Pignone, D. (1995a) Quantitative heterochromatin determination by means of image analysis. *Journal of Genetics and Breeding* 49, 187–190.

Verma, M.M., Sandhu, J.S., Rrar, H.S. and Brar, J.S. (1990) Crossability studies in different species of *Cicer* (L.). *Crop Improvement* 17, 179–181.

Vir, S., Grewal, J.S. and Gupta, V.P. (1975) Inheritance of resistance to Ascochyta blight in chickpea. *Euphytica* 24(1), 209–211.

Vlacilova, K., Ohri, D., Vrana, J., Cihalikova, J., Kubalakova, M., Kahl, G. and Dolezel, J. (2002) Development of flow cytogenetics and physical genome mapping in chickpea (*Cicer arietinum* L.). *Chromosome Research* 10, 695–706.

Williams, P.C. and Singh, U. (1986) Nutritional quality and the evaluation of quality in breeding programmes. In: Saxena, M.C. and Singh, K.B. (eds) *The Chickpea*. CAB International, Wallingford, UK, pp. 329–356.

Waldia, R.S., Chhabra, A.K., Tomar, R.P.S. and Solanki, I.S. (1993) Inheritance of speed of radicle emergence and root length in chickpea (*Cicer arietinum* L.). *Journal of Genetics and Breeding* 47, 1–4.

Winter, P., Pfaff, T., Udupa, S.M., Hüttel, B., Sharma, P.C., Sahi, S., Arreguin-Espinoza, R., Weigand, F., Muehlbauer, F.J. and Kahl, G. (1999) Characterization and mapping of sequence-tagged microsatellite sites in the chickpea (*Cicer arietinum* L.) genome. *Molecular and General Genetics* 262, 90–101.

Winter, P., Benko-Iseppon, A.M., Hüttel, B., Ratnaparkhe, M., Tullu, A., Sonnante, G., Pfaff, T., Tekeoglu, M., Santra, D., Sant, V.J., Rajesh, P.N., Kahl, G. and Muehlbauer, F.J. (2000) A linkage map of the chickpea (*Cicer arietinum* L.) genome based on recombinant inbred lines from a *C. arietinum* × *C. reticulatum* cross: localization of resistance genes for *fusarium* wilt races 4 and 5. *Theoretical and Applied Genetics* 101, 1155–1163.

Yadav, H.S., Singh, O.P. and Agrawal, S.C. (2003) Assessment of heritable variation and selection of genotypes for consumers' quality traits in chickpea. *Indian Journal of pulses Research* 16(1), 14–16.

Zulet, M.A., Macarulla, M.T., Portillo, M.P., Barrio, A., del Rocandio, A., Garcin, H., Higueret, P., Noel-Suberville, C. and Martinez, J. (1998) Healthy properties of a diet containing chickpea on rat altered lipid profiles and lecithin: cholesterol acyltransferase activity (LCAT). In: *Proceedings of 3rd European Conference on Grain Legumes*. November 1998. Valladolid, Spain, pp. 326–327.

16 Utilization of Wild Relatives

S. Abbo,[1] R.J. Redden[2] and S.S. Yadav[3]

[1]*The Levi Eshkol School of Agriculture, The Hebrew University of Jerusalem, Rehovot 76100, Israel;* [2]*Australian Temperate Field Crops Collection, Department of Primary Industries, Victorian Institute for Dryland Agriculture, Horsham, Victoria 3401, Australia;* [3]*Pulse Research Laboratory, Division of Genetics, Indian Agricultural Research Institute, New Delhi 110012, India*

Introduction

The crop plants are the derivatives of wild taxa (Harlan, 1992) that underwent further evolution under domestication (Ladizinsky, 1998). The process of domestication in the incipient farmers' fields involved a certain degree of reduction in the allelic repertoire (or genetic erosion) usually referred to as the domestication 'founder effect' (Ladizinsky, 1985) or 'evolutionary bottlenecks' (e.g. Abbo *et al.*, 2003). This means that from the very early days of agriculture, the domesticated stocks encompassed only a limited fraction of the genetic variation present in the wild progenitors that gave rise to crop plants (Ladizinsky, 1985, 1998). A further reduction in the genetic variation of many crop plants occurred during the last century with the introduction of modern plant-breeding methodologies and the replacement of traditional (and genetically diverse) local landraces with modern (and genetically uniform) high-yielding crop varieties (Tanksley and McCouch, 1997).

As long as plant breeders are able to find the required allelic variation for yield increase and for combating biotic and abiotic stresses within adapted genetic backgrounds, they will mostly employ such germplasm in their crossing schemes. This is because the majority of hybrid progeny from domesticated × wild crosses lack the basic agronomic features required from adapted cultivars. Under certain circumstances, conventional breeding is likely to further reduce the genetic variation of the respective crop. However, if adequate allelic variants are not identified within modern varieties, breeders will seek solution by first screening and crossing with exotic landraces, and then investigate wild accessions from the crop's wild progenitor or other closely related taxa.

In this chapter, we review the potential of wild *Cicer* taxa for chickpea improvement using conventional breeding methodology and propose a new framework for progress based on comparative genetics and physiology of cross-incompatible species.

Ecogeography of Wild *Cicer* Species, with Emphasis on the Annual Taxa

The taxonomy of the genus *Cicer* is given by van der Maesen *et al.* (Chapter 2, this volume), and summaries of the cytology and morphological comparisons of the domestic with the primary (*Cicer reticulatum*) and secondary (*C. echinospernum*) (Fig. 16.1) wild gene pools (GPs) are given by Redden and Berger (Chapter 1, this volume). The descriptions and the botanical key to all *Cicer* spp. are given by van der Maesen *et al.* (Chapter 2, this volume), along with a molecular diversity analysis indicating that the annual tertiary GP relatives are closer to the domestic GP than the perennial wild relatives, with *C. canariense* and *C. cuneatum* more distantly related to all other *Cicer* spp.

In the past, cultivated chickpea (*C. arietinum* L.) was considered to have originated from the southern Caucasus and northern Persia (Iran) regions (van der Maesen, 1972). However, with the discovery of the wild progenitor *C. reticulatum* by Ladizinsky (1975), present-day south-eastern Turkey is considered as the most likely origin of cultivated chickpea (Ladizinsky, 1995). This is consistent with the very limited distribution of the *C. reticulatum* wild progenitor species and of the closely related *C. echinospermum* in south-eastern Turkey (Ladizinsky, 1975; Berger *et al.*, 2003).

Recent discoveries have raised the number of *Cicer* spp. to 44, with 34 perennials and 9 annuals (van der Maesen, 1987). These are distributed in

Fig. 16.1. *Cicer echinospermum* as a prostrate weed in a traditional chickpea field, in its typical basaltic vertisol habitat near Hilvan, eastern Turkey. The pods of the wild species are similar in size to the pods of the erect Turkish landrace. (Photo by S. Abbo.)

isolated pockets from the Canary Isles, North Africa, Ethiopia, Egypt, Greece, the Fertile Crescent from Israel through Syria, Turkey, northern Iraq, western Iran, North India and from the Caucasus throughout the mountain ranges of Central Asia (van der Maesen, 1984; Santos-Guerra and Lewis, 1986).

The *Cicer* spp. range from close to sea level (*C. judaicum*) to more than 5000 m altitude in the Himalayas (*C. microphyllum*). The domestic chickpea, *C. arietinum*, only occurs in cultivation, and does not independently colonize (van der Maesen, 1984). *C. reticulatum* in GP1 grows on soils derived from limestone bedrock dominated by oak shrubs (Ladizinsky, 1975). *C. echinospermum* in GP2 occurs on basaltic vertisols that are most suitable for cultivation. Therefore, many of the accessions were collected from plants that appear as weeds in chickpea fields (Fig. 16.1), or in fallow fields, and along adjacent rocky habitats. The tertiary GP (GP3) annuals, *C. bijugum*, *C. pinnatifidum*, *C. judaicum* and *C. cuneatum*, occur mainly in primary plant formations, but may appear as weeds under special circumstances in traditional cultivation regimes (e.g. Ben David *et al.*, 2006), while other tertiary annuals, *C. chorassinicum* and *C. yamashitae*, mainly occur on rocky slopes and in scrub vegetation (Table 16.2; van der Maesen, 1984).

The most widely distributed species from the eastern Mediterranean to Armenia is *C. pinnatifidum* (Fig. 16.2); a smaller range from Syria to south-eastern Turkey and Iraq occurs for *C. bijugum*, and *C. judaicum* is confined to the eastern Mediterranean (Berger *et al.*, 2003). Shan *et al.* (2005) have described geographic regions of maximum genetic variation as Israel, Palestinian Authority for *C. judaicum* and different areas of south-east Turkey for *C. pinnatifidum*,

Fig. 16.2. *Cicer pinnatifidum* in its typical stony habitat near Besni, eastern Turkey. (Photo by S. Abbo.)

Table 16.1. Geographical distribution and ecogeography of *Cicer* spp. (From van der Maesen, 1987.)

Species	Geographical distribution	Ecogeography
Cicer acanthophyllum Boriss.	Afghanistan, Pakistan and Pamirs	Rubble slopes, dry lakes
C. anatolicum Alef	Turkey, north-west and west Iran, north Iraq and Armenia	Sandy or rocky slopes, rubble, calcareous metamorphic rocks, in scrub or pine forest, with junipers, near water or in dry places
C. arietinum L.	Mediterranean region stretching to Burma, Ethiopia, Mexico, Chile and cooler parts of the tropics	A cool-season crop of the semiarid tropics, spring crop of the warm temperate zones; sandy to clay-loam soils
C. atlanticum Coss. Ex Maire	Morocco and Atlas mountains	Alpine vegetation on rubbles slopes
C. bijugum K.H. Rech	South-east Turkey, north Syria and north Iraq	Weed in orchards, abandoned fields and grazing areas
C. canariense Santos Guerra & Lewis	Canary Islands: La Palma and Tenerife	Dry meso-canarian zone with canarian pine, close to streams
C. chorassanicum (Bge) M. Pop.	North and central Afghanistan, north and north-east Iran	Rocky and rubble slopes, calcareous or granitic, dry valleys
C. cuneatum Hochst. ex Rich	Ethiopia, south-east Egypt and north-east Sudan: Jebel Elba, Saudi Arabia	Cultivated fields, open vegetation, sometimes a weed in sorghum
C. echinospermum P.H. Davis	Turkey, east Anatolia and north Iraq: Jebel Sinjar	Rocky slopes, vineyards, fallow and cultivated field, grass vegetation
C. fedtschenkoi Lincz	South Kyrgyzstan, Tadzhikistan, north and north-east Afghanistan	Dry stone slopes and granite valleys, south-exposed, near lakes and streambeds
C. flexuosum Lipsky	South Kyrgyzstan and Tadzhikistan: Tian-Shan	Rocks, rubble slopes, near riverbeds, tree and scrub vegetations
C. floribundum Fenzl.	Turkey and south Anatolia	Mountain slopes, forest of oriental beech (*Fagus orientalis*) and *Quercus coccifera*
C. graecum Orph.	Greece and Peloponnesus	Hedges, maize fields, pine forests, calcareous soils
C. heterophyllum Contandr. *et al.*	Turkey and Toros Daglari	Pine and oak forests
C. incanum Korotk.	South Tadzhikistan and Pamirs	Calcareous rocks
C. incisum (Willd.) K. Maly	Greece–Crete, Turkey, Iran, Lebanon, Armenia and Georgia	Subalpine vegetation, rubble slopes, calcareous and igneous soils
C. isauricum P.H. Davis	Turkey	Grey volcanic rocks, rubble slopes, *Abies cilicia* and *Pinus nigra* forest

Continued

Table 16.1. *Continued*

Species	Geographical distribution	Ecogeography
C. judaicum Boiss.	Lebanon, Israel, Jordan and Palestinion Authority	Rocky places and fallow fields, sometimes a weed in crops
C. kermanense Bornm.	South-east Iran	Mountains and rocky places
C. korshinskyi Lincz.	Tadzhikistan and north-west Pamirs	Dry rock slopes
C. macracanthum M. Pop.	Afghanistan, India, Pakistan, Tadzhikistan and Uzbekistan	Dry streambeds, valley, rubble slopes
C. microphyllum Benth.	East Afghanistan, China: West Tibet, India, Pakistan and Pamirs	Rubble slopes, in dry riverbeds, in pasture, open vegetation or near subalpine forest of birch (*Betula*)
C. montbretii Jaub. & Sp.	Albania, Bulgaria, Turkey and west Anatolia	Rocky and loam hill slopes, *P. nigra* and *Quercus* forest, near water
C. multijugum van der Maesen	Afghanistan	Rubble slopes and limestone hillsides
C. nuristanicum Kitamura	Afghanistan, India-Kasmir and Pakistan-Chitral	Abies fores, pastures, shady humid places, limestone rocks
C. oxyodon Boiss. & Hoh.	Iran, Afghanistan and north Iraq	Rubble and earth slopes, cultivated field, *Rhus–Quercus* forests
C. paucijugum (M. Pop.) Nevski.	Tadzhikistan	Stone slopes, *Juniperus salina* vegetation
C. pinnatifidum Jaub. & Sp.	Cyprus, north Iraq, Syria, Turkey-Anatolia and Armenia	Rocky and rubble places, vineyards, *P. brutia* and scrub vegetation
C. pungens Boiss.	Afghanistan and Tadzhikistan	Stony and rubble slopes, volcanic ashes, limestone, alpine meadows
C. rechingeri Podlech	Afghanistan	Dry slopes, granitic rubble, glacial moraine
C. reticulatum Ladiz.	East Turkey	Limestone hills, oak scrub vegetation, weedy habitats
C. songaricum Steph. ex DC.	Kazakhstan, Kyrgyzstan and Tadzhikistan	Rubble slopes, near streams, but also dry places
C. spiroceras Juab. & Sp.	Iran	Rubble slopes, oak forests
C. stapfianum K.H. Rech.	Iran	Mountains
C. subaphyllum Boiss.	Iran	Rubble slopes
C. tragacanthoides Jaub. & Sp.	Iran–Turkmenistan	Mountain, alpine region, dry rocky rubble slopes
C. yamashitae Kitamura	Afghanistan	Rocks, rubble slopes

C. bijugum and *C. reticulatum*. They also identified surrounding regions that are prospective for the collection of new accessions. More restricted distributions for *C. cuneatum* occur in Ethiopia, Eritrea, Egypt and Saudi Arabia, *C. yamashitae* in Afghanistan, *C. chorassanicum* from Iraq to Afghanistan, while most restricted distributions of *C. echinospermum* and *C. reticulatum* appear in south-east Turkey (Berger *et al.*, 2003). The annual wild species are better characterized than the perennials for traits of interest to breeders and are more prospective as sources for introgression of novel genes (Berger *et al.*, 2003; Ladizinsky and Adler, 1976).

The perennial *Cicer* spp. tend to occur on rocky slopes, sometimes in forests and/or in open vegetation near streams (Table 16.1). *Cicer* plants are poor competitors, and are therefore better protected from grazing amongst the rubble, where they suffer less from more aggressive annuals like wild cereals. Certain species like *C. montbrettii* and *C. floribundum* that are adapted to forests may prefer shade and have both a tap root and surface roots (van der Maesen, 1984). Berger *et al.* (2003) found that *C. anatolicum* was widely distributed above 36°N latitude from Turkey to Iran. Although few wild perennials are also widely distributed (*C. incisum*, *C. macranthum*, *C. microphyllum* and *C. montbretii*), most of them have a narrow regional distribution (Table 16.1).

Berger *et al.* (2003) have emphasized the very low numbers of unique accessions (124) of wild annual chickpea species and perennial *C. anatolicum* in collections worldwide, and the potential for greater utilization of wild diversity. Wild habitats are under severe threat of genetic erosion due to the combined pressures of increased human activity and animal grazing, causing ecological degradation and climate change (IPCC, 2001; Larghetti *et al.*, 2003). Our ability to utilize the wild *Cicer* GPs for chickpea improvement is entirely dependent upon the application of specific conservation measures (Berger *et al.*, 2003).

Evolutionary Constraints on Chickpea Production

Chickpea (*C. arietinum* L.) is considered one of the 'founder crops' of the 'Neolithic Revolution' in the Near East that took place *c.*10,000 years ago (Lev-Yadun *et al.*, 2000; Zohary and Hopf, 2000). This crop package included einkorn wheat (*Triticum monococcum* L.), emmer wheat (*T. turgidum* L.), barley (*Hordeum vulgare* L.), lentil (*Lens culinaris* Medic), pea (*Pisum sativum* L.), chickpea (*C. arietinum* L.), bitter vetch (*Vicia ervilia* (L.) Willd.) and flax (*Linum usitatissimum* L.), the first fibre crop (therein). The distribution of the wild progenitors of the above-mentioned West Asian crops is relatively wide, extending throughout the eastern Mediterranean basin into West Asia, and in some cases as far as Central Asia (Zohary and Hopf, 2000). In contrast, the wild progenitor of cultivated chickpea, *C. reticulatum*, is a rare species, currently reported from only 18 locations (37.3–39.3°N, 38.2–43.6°E) in south-eastern Turkey (Berger *et al.*, 2003). By definition, a narrowly distributed species such as *C. reticulatum* can harbour only limited adaptive genetic variation compared with those spanning a much larger ecogeographic range such as the wild progenitors of wheat, barley, pea or lentil (Abbo *et al.*, 2003). The narrow ecogeographic distribution of the wild progenitor *C. reticulatum* is in fact a bottleneck in the evolutionary history of the chickpea crop.

Genetic evidence suggests that domestication in many crops involved a limited number of founding genotypes (e.g. Ladizinsky, 1985; Abbo *et al.*, 2001). This selection of a small number of individuals out of diverse original populations (founder effect), which later became widespread as domesticated forms, is an almost universal phenomenon known as the 'domestication bottleneck', and represents further narrowing of the genetic base of chickpea. A number of groups provided evidence supporting a founder effect in chickpea, from isozyme diversity studies (Labdi *et al.*, 1996; Tayyar and Waines, 1996), sequence-tagged microsatellite sites (STMS) (Choumane *et al.*, 2000), randomly amplified polymorphic DNA (RAPD) and inter-simple sequence repeat (ISSR) markers (Iruela *et al.*, 2002). In all cases, the wild progenitor *C. reticulatum* was more diverse than the cultigen *C. arietinum*.

The life cycle of most annual eastern Mediterranean flora is autumn germination, late winter or early spring flowering and early summer maturation (Zohary, 1966, 1972; Feinbrun-Dothan, 1978, 1986). Likewise, most crops of West Asian origin and their wild progenitors are characterized by a similar phenology. Chickpea is the exception to this rule, undergoing a third genetic bottleneck when it was transformed into a summer crop in West Asian and Mediterranean environments, probably to avoid the devastating effect of ascochyta blight caused by the fungus *Didymella rabiei* (Pass.) Lab. (Kumar and Abbo, 2001). This change in sowing practices must have required the selection of vernalization-insensitive genotypes, since the wild progenitor is vernalization responsive (Abbo *et al.*, 2002). In contrast, no vernalization response can be detected in cultivated chickpea (Summerfield *et al.*, 1989), suggesting that the selection of genotypes that suit summer cropping has further eroded genetic diversity in chickpea – a third evolutionary bottleneck.

Chickpea is an exception among other West Asian crops, both in terms of restricted distribution of its wild progenitor and the conversion of its growing cycle. This succession of evolutionary bottlenecks has had a profound effect on domesticated chickpea. Studies using genetic markers have demonstrated low levels of polymorphism in chickpea relative to other crops. For example, morphological and restriction fragment length polymorphism (RFLP) markers were polymorphic enough to establish complete linkage maps using cultivated crosses in pea (Ellis *et al.*, 1992) and barley (Kleinhofs *et al.*, 1993). Due to a lower level of DNA polymorphism, this approach was inadequate in chickpea. Only the application of highly polymorphic simple sequence repeat (SSR) markers produced a detailed interspecific genetic map of chickpea (Winter *et al.*, 2000), and recently an intraspecific map (Flandez-Galvez *et al.*, 2003). There is a relatively small number of pest or disease resistance genes in chickpea (Singh, 1997) compared with the cereal crops (GrainGenes, 2001). For example, ascochyta blight resistance is particularly rare in chickpea (Singh *et al.*, 1994). Similarly, there is low variability of phenology alleles in domesticated chickpea, as evidenced by the lack of a vernalization response (Summerfield *et al.*, 1989), and the high proportion of strongly day length-sensitive types among kabuli cultivars (Or *et al.*, 1999; Kumar and Abbo, 2001).

Unlike pea, barley and wheat, which originated from their West Asian core area and became staple crops across the temperate zones of Europe and Asia, the ancient selection of chickpea genotypes to suit summer cropping has enabled the dissemination of the crop to subtropical regions of India and East Africa, where it is mostly a post-rainy-season crop. Consequently, domesticated chickpea is poorly adapted to the high rainfall, temperate regions of Asia and Europe. Chickpea was introduced into wheat-based cropping systems of the USA and Canada as a summer crop. Chickpea never became a major autumn-planted rotation crop in any winter-wheat growing area (Ladizinsky, 1995). Multilocation agronomic evaluation showed that in Australian Mediterranean-like areas, chickpea is less productive than other legumes of Near East origin such as faba bean, pea, common vetch (*Vicia sativa* L.) and narbon bean (*V. narbonensis* L.) (Siddique *et al.*, 1999).

The relatively poor performance of autumn-sown chickpea stems mainly from late phenology, which exposes the crop to terminal drought during the pod-filling stage (Turner *et al.*, 2001). However, eliminating this limiting factor by using early flowering types reveals other inherent yield limitations of domesticated chickpea, such as sensitivity to low temperatures during the reproductive phase, a complex phenomenon expressed as early flower bud abortion, poor pollen germination and consequently high rates of inflated pod abortion (probably zygote failure). Even early flowering stocks may fail to escape terminal drought due to unfruitful early flowering (Savithri *et al.*, 1980; Srinivasan *et al.*, 1998, 1999; Croser *et al.*, 2003; Clarke *et al.*, 2004). In chickpea-growing areas characterized by short cold winters followed by rapid desiccation, this is a major yield-limiting factor (Morgan *et al.*, 1991; Turner *et al.*, 2001; Croser *et al.*, 2003). A detailed review of chickpea evolutionary history and its agronomic implications for productivity in semiarid systems was recently provided by Kumar and Abbo (2001). Taking into account the above factors and considering chickpea global distribution and production statistics (FAO, 2005) supports the notion that its yield potential reflects its evolutionary history.

The Potential of Wild *Cicer* Taxa to Overcome Production Limitations

The low diversity in domesticated chickpea does not provide agronomists, physiologists and breeders with sufficient contrasting genotypes for study, which results in limited understanding of chickpea biology (e.g. Singh, 1991; Singh *et al.*, 1997; Davies *et al.*, 2000). The work of Summerfield *et al.* (1989), which demonstrated lack of vernalization response in domesticated chickpea, is an excellent illustration of this point. Interestingly, using *C. arietinum* × *C. reticulatum* progeny, Abbo *et al.* (2002) were able to demonstrate the existence of vernalization-responsive alleles, suggesting that domesticated chickpea is probably monomorphic at the respective loci. Similar limitations apply to genetic analyses for other agronomic and economic traits such as drought and disease response, temperature response, day length response or nutritional value traits.

Further improvement of domesticated chickpea for higher yield potential either directly or mediated by resistance to biotic and abiotic stresses depends on breeders' abilities to identify promoting allelic variation. Genetic analyses based on domesticated chickpea segregating populations or recombinant inbred lines (RILs) can only expose existing polymorphism; thus, monomorphic loci will remain obscure, as in the case of the vernalization response. Yet another illustration of this point is the relative rarity of domesticated genotypes resistant to biotic and abiotic stresses summarized in Singh *et al.* (1994). Another subtle but very important aspect of the low genetic diversity in domesticated chickpea is related to the breeding methodology. Since the narrow genetic base reduces the range of adaptive alleles displayed by the crop, it follows that the potential solutions sought for by breeders are selected from a limited repertoire. In other words, without genetic information on the existence of potentially promoting alleles or adaptive strategies, breeders are unlikely to attempt to expose the relevant loci in their crossing programmes.

As long as plant breeders are able to find the required allelic variation for whatever purpose within adapted genetic backgrounds they will mostly employ such germplasm in their crossing schemes. This is because the majority of hybrid progeny from domesticated × wild crosses lack the basic agronomic features required from adapted cultivars. In the absence of adequate allelic variants among modern varieties, breeders will seek solutions by first screening and crossing with exotic landraces, and if necessary with wild accessions from the crop's wild progenitor or other closely related taxa.

As discussed earlier, wild *Cicer* taxa that are cross-compatible with the cultigen (GP1, *C. reticulatum,* and GP2, *C. echinospermum*) have narrow ecogeographic amplitude, and may therefore offer only a limited array of adaptive allelic variation. The use of more distantly related GP3 wild *Cicer* spp. that grow in other world regions and that have the potential of extending the crop's allelic repertoire has been advocated for some time (e.g. Badami *et al.*, 1997; Mallikarjuna, 1999). However, it should be borne in mind that crosses between domesticated chickpea and any wild *Cicer* spp. other than the GP1 or GP2 species may involve difficulties in a number of processes bearing on the prospects for introgression. Usually, pollen germination and subsequent pollen tube growth and fertilization are not a problem in interspecific crosses among annual *Cicer* spp. (Ahmad *et al.*, 1988). However, outside the GP1, following interspecific *Cicer* crosses, abnormal embryo development and later collapse occur between 10 and 14 days after pollination (Badami *et al.*, 1997; Ahmad and Slinkard, 2004). In the *C. arietinum* × *C. pinnatifidum* cross, cultured embryos developed into albino F_1 plants (Badami *et al.*, 1997). Furthermore, upon improvement of the hormonal status of such *Cicer*-rescued hybrids, they survived to reach the reproductive stage, albeit with complete microspore sterility (Badami *et al.*, 1997; Mallikarjuna, 1999). Therefore, it appears that in the genus *Cicer* an array of at least three isolation mechanisms exists: embryo abortion, albinism and microspore sterility (Badami *et al.*, 1997; Mallikarjuna, 1999). Another crossing barrier was observed in *Cicer*: abnormal flower development with protruding stigma in F_1 of *C. judaicum* × *C. pinnatifidum* (Ladizinsky and Adler, 1976) and among F_2 of *C. judaicum* × *C. bijugum* crosses (S. Abbo, Israel, unpublished data). This abnormal flower morphology completely prevents pod set in the F_1 and F_2 generations, and requires an extensive backcrossing scheme to obtain a minimal number of recombinant progeny.

Hence, it becomes clear that due to the differential crossing compatibility, reduced chromosome pairing, gamete fertility or differential seed set in the segregating progeny, only a fraction of all possible gamete and zygote combinations are employed in crosses outside the chickpea GP1. This means that not all the alleles from the donor wild taxa are at the breeders' disposal because some alleles will not take part in chromosomal crossing-over or by mere chance will be lost through aborted gametes by any of the above-mentioned genetic isolation mechanisms. Moreover, under such circumstances it is impossible for the breeder to calculate the likelihood of transferring a specific allele of interest in cross-combinations outside GP1.

Examples for Utilization of Wild *Cicer* in Chickpea Breeding and Research

The successful utilization of wild *Cicer* for practical breeding purposes like change in plant architecture and improvement in biotic and abiotic stress resistance to ultimately increase the yield potential has been very limited throughout the world. Desirable GPs showing resistance to various biotic and abiotic stresses have been reported in wild species of *Cicer* and their hybrid derivatives by various workers (Yadav *et al.*, 2002; Juan *et al.*, 2003; Knights *et al.*, 2003; Malhotra *et al.*, 2003; Nguyen *et al.*, 2005). These include resistance to drought, cold, heat, cyst nematodes, phytophthora root rot, ascochyta blight and fusarium wilt resistance among other useful characteristics. However, successful commercialization of wild genes is not yet a routine in this crop. Chickpea breeders are not able to utilize most of the *Cicer* taxa due to complex crossing barriers operating between the cultigen and all wild taxa except from *C. reticulatum* and *C. echinospermum*. Even in crosses involving the latter two taxa, linkage drag is a problem. Although successful transfer of single gene traits was achieved, the actual release of new cultivars and their use by farmers was rarely reported.

Successful Utilization of Wild *Cicer reticulatum*

Impressive yield increase (up to 39%) compared with the domesticated parents was reported by Singh and Ocampo (1997). These authors used crosses involving the above two wild taxa and an ascochyta-resistant cultivar ILC482, and employed a pedigree selection procedure to advance their material.

Yadav *et al.* (2002) developed a protocol for using the wild progenitor for successful utilization of wild *C. reticulatum* for chickpea improvement. Crosses were made using adapted cultivars as female parent and *C. reticulatum* as male parent and one backcross with a well-adapted cultivar. In segregating F_2 to F_6 populations, superior single plants for desirable quantitative traits were selected under multiple sick-plot procedure. Preliminary yield test of promising lines in F_7 generation and advanced yield test of selected lines in F_8 generation were carried out. Multilocation yield and pathological testing of superior genotypes were carried out across growing environments in India. After 4 years of comprehensive yield testing, variety BG-1103 was released for commercial cultivation in North India in 2005.

Likewise, Malhotra *et al.* (2003) also reported successful utilization of *C. reticulatum* for the development of commercial genotypes. E.J. Knights, (Australia, 2005, personal communication) also developed commercial material using *C. echinospermum* germplasm to improve nematode, *Phytophora*, wet root rot and resistance to abiotic stresses.

Selection in Segregating Generations

Considering linkage drag (and crossability to a certain extent) as a big barrier in wide *Cicer* crosses, the advancement and selection in segregating generations, i.e. F_2 to F_6, is very important to select a promising segregant. For this purpose, it is best to plant the F_2 seeds in a spaced nursery, maintaining 20×45 cm distance under multiple sick-plot environments. F_2 populations should be between 2000 and 5000. Simultaneously, superior single plant selection may be carried out for multitraits. The selected plants should possess resistance against soil-borne diseases like fusarium wilt, dry root rot and wet root rot, capacity to produce high biomass, early vigour, medium early maturity, larger seed size, higher number of pods per plant and other traits according to local and breeder requirements. After harvest, selection for superior seed traits may be exercised in the laboratory. Thus, superior yielding single plants showing promising traits can be bulked for F_3 plantings to accommodate maximum desirable GPs. It is desirable to follow the same procedures in F_3, F_4 and F_5 generations to save time, labour and resources. Populations in these generations should also be maintained between 2000 and 4000. After the F_5 generations, bulking is not recommended. Thus in F_6, superior single plant progenies should be marked and harvested separately. The superior yielding and resistant lines possessing desirable traits may be promoted for preliminary yield test.

The preliminary yield test of superior lines showing 20% yield superiority over best-adapted cultivars should be conducted under normal conditions as per local recommendations with at least two replications. In preliminary yield trials, all crop protection measures should be utilized during cropping season to protect the crops from insects and pests. After harvest, significantly superior yielding lines with good-quality seeds shall be promoted for further multilocation yield testing in advance trials. Figure 16.3 summarizes the selection procedure of Yadav *et al.* (2002).

A New Framework for Progress with Cross-incompatible Taxa

In recent studies Abbo *et al.* (2003) and Ben-David and Abbo (2005) have advocated a new strategy for utilizing wild *Cicer* spp. for chickpea improvement. Such an approach was extensively used by cereal researchers, and hence will be equally applicable to legume research (Evans and Wardlaw, 1976; Evans,

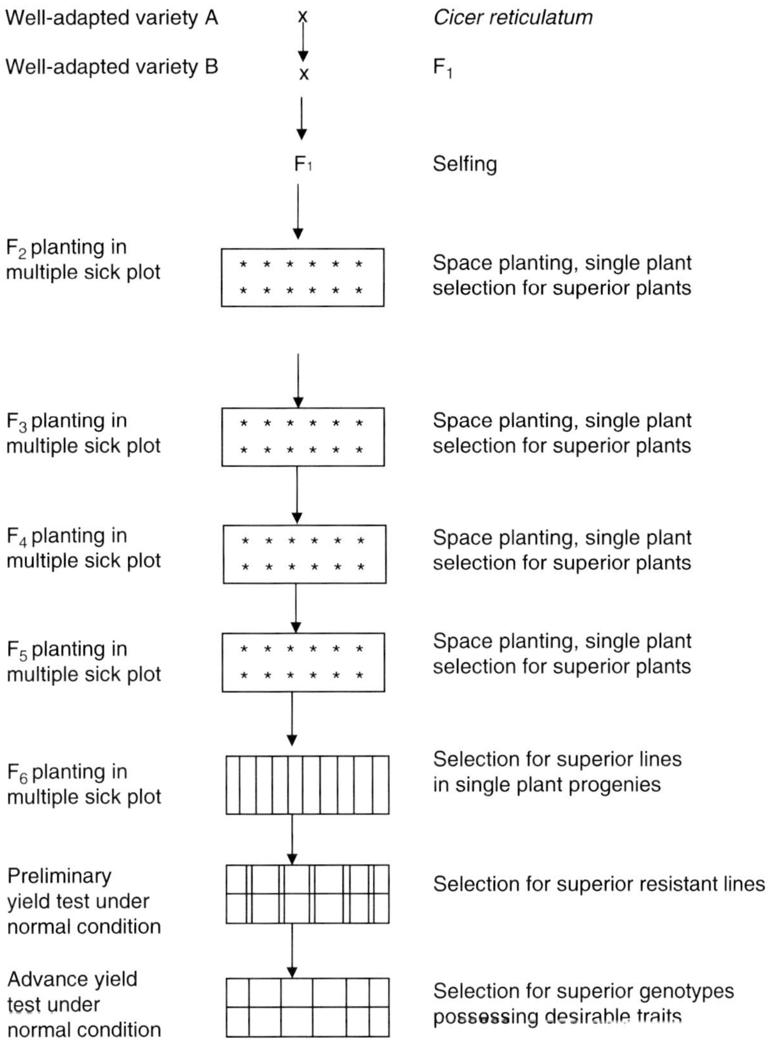

Fig. 16.3. Hybridization and advancement of segregating populations in wild crosses.

1993). This approach involves comparative genetic and physiological studies of an array of related species regardless of their crossability relations. Such an experimental system is useful in identifying the existence of adaptive loci that may be obscured (due to low polymorphism) in domesticated chickpea. At the same time, genetic mapping should be conducted in intraspecific populations derived from crosses between (wild × wild) contrasting genotypes originating from different ecogeographic niches. Such crosses are usually free from differential gamete and/or zygote viability problems, and are therefore expected to yield accurate genetic information on the linkage relationship between key physiological traits, genes of interest and genetic markers (see review by Abbo *et al.*, 2003). Assembling genetic mapping data from such populations (e.g.

progeny of *C. judaicum* × *C. judaicum*, or *C. bijugum* × *C. bijugum*) together with known genetic mapping data based on the *C. arietinum* × *C. reticulatum* RIL population of Winter *et al.* (2000) and domesticated × domesticated maps (e.g. Flandez-Galves *et al.*, 2003) will enable inference of the existence and location of any locus of interest relevant for chickpea improvement. Resorting to Vavilov's 'law of homologous series in variation' (Vavilov, 1922) will clarify this point. Given the close phylogenetic relationship between the different *Cicer* spp., it is most likely that any adaptive locus found in a given species will have its homologous counterpart with its variant allelomorphs in the other congeneric *Cicer* taxa.

Collection of information on the function and linkage relationship of an allele in a wild *Cicer* sp. outside GP1 is only the first, but very important, step forward. It is anticipated that in the near future chickpea breeders will be able to take advantage of recent developments in genetic mapping, genomics and sequence data accumulation of a number of legume species as well as comparative intergeneric genetic mapping among food and feed legumes (e.g. Penmetsa and Cook, 2000; Choi *et al.*, 2004). The fast technological innovations in sequencing technology probably mean that complete sequence data for the model legume *Medicago truncatula* will be commonly available very soon. Therefore, new options of utilizing sequence homology in legume-coding regions and a variety of genomic tools (e.g. genetic maps, bacterial artificial chromosome [BAC] libraries and expressed sequence tags [ESTs]) from *Cicer* taxa outside GP1 will increase dramatically.

Another exciting area of rapid progress involves the technology of homologous recombination (HR). HR is the process by which highly similar or identical DNA sequences recombine. Therefore, HR is seen as one of the fundamental processes of life (see recent review by Schuermann *et al.*, 2005). In certain organisms like yeast, geneticists are regularly using HR to induce targeted sequence changes at loci of their choice. In most plants, however, this technology is still in its infancy because all the molecular and genetic components of the processes are not fully understood (e.g. Terada *et al.*, 2002; Schuermann *et al.*, 2005). However, it is anticipated that once the molecular basis of HR is deciphered, one or more of the approaches to increase the rate of somatic (and meiotic) recombination could be optimized. HR, in conjunction with modern transformation vectors based on precise sequence information of loci of interest, would thus enable routine successful allelic changes between currently cross-incompatible *Cicer* spp. Until the above vision becomes a reality, chickpea breeders' capacity for manoeuvre will remain restricted within GP1 or GP2.

References

Abbo, S., Lev-Yadun, S. and Ladizinsky, G. (2001) Tracing the wild genetic stocks of crop plants. *Genome* 44, 309–310.

Abbo, S., Lev-Yadun, S. and Galwey, N. (2002) Vernalization response of wild chickpea. *New Phytologist* 154, 695–701.

Abbo, S., Berger, J. and Turner, N.C. (2003) Evolution of cultivated chickpea: four genetic bottlenecks limit diversity and constrain crop adaptation. *Functional Plant Biology* 30, 1081–1087.

Ahmad, F. and Slinkard, A.E. (2004) The extent of embryo and endosperm growth following interspecific hybridization between *Cicer arietinum* L. and related annual wild species. *Genetic Resources and Crop Evolution* 51, 765–772.

Ahmad, F., Slinkard, A.E. and Scoles, G.J. (1988) Investigations into the barrier(s) to interspecific hybridization between *Cicer arietinum* L. and eight other wild annual *Cicer* species. *Plant Breeding* 100, 193–198.

Badami, P.S., Mallikarjuna, N. and Moss, J.P. (1997) Interspecific hybridization between *Cicer arietinum* and *C. pinnatifidum. Plant Breeding* 116, 393–395.

Ben-David, R. and Abbo, S. (2005) Phenological variation among Israeli populations of *Cicer judaicum* Boiss. *Australian Journal of Agricultural Research* 56, 1219–1225.

Ben-David, R., Can, C., Lev-Yadun, S. and Abbo, S. (2006) Ecogeography and demography of *Cicer judaicum*, a wild annual relative of domesticated chickpea. *Crop Science* 46, 1360–1370.

Berger, J.D., Abbo, S. and Turner, N.C. (2003) Ecogeography of annual wild *Cicer* species: the poor state of the world collection. *Crop Science* 43, 1076–1090.

Choi, H.-K., Mun, J.-H., Kim, D.-J., Zhu, H., Beak, J.-M., Mudge, J., Roe, B., Ellis, N., Doyle, J., Kiss, G.B., Young, N.D. and Cook, D.R. (2004) Estimating genome conservation between crop and model legume species. *Proceedings of the National Academy of Sciences Of the United States of America* 101, 15289–15294.

Choumane, W., Winter, P., Weigand, F. and Kahl, G. (2000) Conservation and variability of sequence tagged microsatellite sites (STMSs) from chickpea (*Cicer aerietinum* L.) within the genus *Cicer. Theoretical and Applied Genetics* 101, 269–278.

Clarke, H.J., Khan, T.N. and Siddique, K.H.M. (2004) Pollen selection for chilling tolerance at hybridisation leads to improved chickpea cultivars. *Euphytica* 139, 65–74.

Croser, J.S., Clarke, H.J., Siddique, K.H.M. and Khan, T.N. (2003) Low-temperature stress: implications for chickpea (*Cicer arietinum* L.) improvement. *Critical Reviews in Plant Science* 22, 185–219.

Davies, S.L., Turner, N.C., Palta, J.A., Siddique, K.H.M. and Plummer, J.A. (2000) Remobilisation of carbon and nitrogen supports seed filling in chickpea subjected to water deficit. *Australian Journal of Agricultural Research* 51, 855–866.

Ellis, T.H.N., Turner, L., Hellens, R.P., Lee, D., Harker, C.L., Enard, C., Domoney, C. and Davies, D.R. (1992) Linkage maps in pea. *Genetics* 130, 649–663.

Evans, T.L. (1993) *Crop Evolution, Adaptation and Yield.* Cambridge University Press, Cambridge.

Evans, L.T. and Wardlaw, I.F. (1976) Aspects of comparative physiology of grain yield in cereals. *Advances in Agronomy* 28, 301–359.

FAO (2005) Available at: http://apps.fao.org

Feinbrun-Dothan, N. (1978) *Flora Palaestina.* The Israel Academy of Sciences and Humanities, Jerusalem.

Feinbrun-Dothan, N. (1986) *Flora Palaestina.* The Israel Academy of Sciences and Humanities, Jerusalem.

Flandez-Galvez, H., Ford, R., Pang, E.C.K. and Taylor, P.W.J. (2003) An intraspecific linkage map of the chickpea (*Cicer arietinum* L.) genome based on sequence tagged microsatellite site and resistance gene analog markers. *Theoretical and Applied Genetics* 106, 1447–1456.

GrainGenes (2001) Grains database. Available at: http://wheat.pw.usda.gov/ggpages/maps.html

Harlan, J.R. (1992) *Crops and Man*, 2nd edn. American Society of Agronomy, Madison, Wisconsin.

IPCC (2001) Climate change 2001: the scientific basis. In: Houghton, J.T., Ding, Y., Griggs, D.J., Noguer, M., van der Linden, P.J. and Xiaosu, D. (eds) *Contribution of Working Group I to the Third Assessment Report of the Intergovernmental Panel on Climate Change (IPCC).* Cambridge University Press, Cambridge, p. 944.

Iruela, M., Rubio, J., Cubero, J.I., Gil, J. and Millan, T. (2002) Phylogenetic analysis in the genus *Cicer* and cultivated chickpea using RAPD and ISSR markers. *Theoretical and Applied Genetics* 104, 643–651.

Juan, G., Rubio, J., Cobos, M.J., Domingues, F., Millan, T., Moreno, M.T. and Cubero, J. (2003) Mapping populations derived from crosses between *Cicer arietinum* and *C. reticulatum* using morphological and molecular markers. In: *Chickpea Research for the Millennium: Proceedings of the International Chickpea Conference*. 20–22 January 2003. Chattisgarh, India, pp. 49–52.

Kleinhofs, A., Kilian, A. *et al.* (1993) A molecular, isozyme and morphological map of the barley (*Hordeum vulgare* L.) genome. *Theoretical and Applied Genetics* 86, 705–712.

Knights, E.J., Southwell, R.J. and Schwinghamer, M.W. (2003) Evaluation of wild *Cicer* species for resistance to Phytophthora root rot. In: *Chickpea Research for the Millennium: Proceedings of the International Chickpea Conference*. 20–22 January 2003. Chattisgarh, India, pp. 54–57.

Kumar, J. and Abbo, S. (2001) Genetics of flowering time in chickpea and its bearing on productivity in semi-arid environments. *Advances in Agronomy* 72, 107–138.

Labdi, M., Robertson, L.D., Singh, K.B. and Charrier, A. (1996) Genetic diversity and phylogenetic relationships among the annual *Cicer* species as revealed by isozyme polymorphism. *Euphytica* 88, 181–188.

Ladizinsky, G. (1975) A new *Cicer* from Turkey. *Notes Royal Botanical Garden Edinburgh* 34, 201–202.

Ladizinsky, G. (1985) Founder effect in crop plant evolution. *Economic Botany* 39, 191–199.

Ladizinsky, G. (1995) Chickpea. In: Smartt, J. and Simmonds, N.W. (eds) *Evolution of Crop Plants*. Longman Scientific & Technical, Harlow, UK, pp. 258–261.

Ladizinsky, G. (1998) *Plant Evolution Under Domestication*. Kluwer Academic, Dordrecht, The Netherlands.

Ladizinsky, G. and Adler, A. (1976) Genetic relations among the annual species of *Cicer* L. *Theoretical and Applied Genetics* 48, 197–203.

Larghetti, G., Perrino, P., Cifarelli, S. and Hammer, K. (2003) Collecting landraces and wild relatives in Tremeti Islands, Italy. *Plant Genetic Resources Newsletter* 133, 1–7.

Lev-Yadun, S., Gopher, A. and Abbo, S. (2000) The cradle of agriculture. *Science* 288, 1602–1603.

Malhotra, R.S. *et al.* (2003) Interspecific hybridization in chickpea. In: *Chickpea Research for the Millennium: Proceedings of the International Chickpea Conference*. 20–22 January 2003. Chattisgarh, India, pp. 39–49.

Mallikarjuna, N. (1999) Ovule and embryo culture to obtain hybrids from interspecific incompatible pollinations in chickpea. *Euphytica* 110, 1–6.

Morgan, J.M., Rodriguez-Maribona, B. and Knights, E.J. (1991) Adaptation to water-deficit in chickpea breeding lines by osmoregulation: relationship to grain-yields in the field. *Field Crops Research* 27, 61–70.

Nguyen, T.T., Taylor, P.W.J., Redden, R.J. and Ford, R. (2004) Genetic diversity in *Cicer* using AFLP analysis. *Plant Breeding* 123, 173–179.

Nguyen, T.T., Taylor, P.W.J., Redden, R.J. and Ford, R. (2005). Resistance to *Ascochyta rabiei* (Pass.) Lab. in a wild *Cicer* germplasm collection. *Australian Journal Experimental Agriculture* 45(10), 1291–1296.

Or, E., Hovav, R. and Abbo, S. (1999) A major gene for flowering time in chickpea. *Crop Science* 39, 315–322.

Penmetsa, R.V. and Cook, D.R. (2000) Production and characterization of diverse developmental mutants of *Medicago truncatula*. *Plant Physiology* 123, 1387–1397.

Santos-Guerra, A. and Lewis, G.P. (1986) A new species of *Cicer* from the Canary Islands. *Kew Bulletin* 41, 459–62.

Savithri, K.S., Ganapathy, P.S. and Sinha, S.K. (1980) Sensitivity to low tolerance in pollen germination and fruit set in *Cicer arietinum* L. *Journal of Experimental Botany* 31, 475–481.

Schuermann, D., Molinier, J., Fritsch, O. and Hohn, B. (2005) The dual nature of homologous recombination in plants. *Trends in Genetics* 21, 172–181.

Shan, F., Clarke, F.C., Plummer, J.S. and Yan, G. (2005) Geographical patterns of genetic variation in world collections of wild annual *Cicer* characterised by amplified fragment length polymorphism. *Theoretical and Applied Genetics* 110, 381–391.

Siddique, K.H.M., Loss, S.P., Regan, K.L. and Jettner, R.L. (1999) Adaptation and seed yield of cool season grain legumes in Mediterranean environments of south-western Australia. *Australian Journal of Agricultural Research* 50, 375–387.

Singh, K.B. (1997) Chickpea (*Cicer arietinum* L.). *Field Crops Research* 53, 161–170.

Singh, K.B. and Ocampo, B. (1997) Exploitation of wild *Cicer* species for yield improvement in chickpea. *Theoretical and Applied Genetics* 95, 418–423.

Singh, K.B., Malhotra, R.S., Halila, M.H., Knights, E.J. and Verma, M.M. (1994) Current status and future strategy in breeding chickpea for resistance to biotic and abiotic stresses. *Euphytica* 73, 137–149.

Singh, K.B., Malhotra, R.S., Saxena, M.C. and Bejiga, G. (1997) Superiority of winter sowing over traditional spring sowing of chickpea in the Mediterranean region. *Agronomy Journal* 89, 112–118.

Singh, P. (1991) Influence of water deficits on phenology growth and dry matter allocation in chickpea (*Cicer arietinum*). *Field Crops Research* 28, 1–16.

Srinivasan, A., Johansen, C. and Saxena, N.P. (1998) Cold tolerance during early reproductive growth of chickpea (*Cicer arietinum* L.): characterization of stress and genetic variation in pod set. *Field Crops Research* 57, 181–193.

Srinivasan, A., Saxena, N.P. and Johansen, C. (1999) Cold tolerance during early reproductive growth of chickpea (*Cicer arietinum* L.): genetic variation in gamete development and function. *Field Crops Research* 60, 209–222.

Summerfield, R.J., Ellis, R.H. and Roberts, E.H. (1989) Vernalization in chickpea (*Cicer arietinum* L.): fact or artifact? *Annals of Botany* 64, 599–604.

Tanksley, S.D. and McCouch, S.R. (1997) Seed banks and molecular maps: unlocking genetic potential from the wild. *Science* 277, 1063–1066.

Tayyar, R.I. and Waines, J.G. (1996) Genetic relationships among annual species of *Cicer* (Fabaceae) using isozyme variation. *Theoretical and Applied Genetics* 92, 245–254.

Terada, R., Urawe, H., Inagaki, Y., Tsugane, K. and Iida, S. (2002) Efficient gene targeting by homologous recombination in rice. *Nature Biotechnology* 20, 1030–1034.

Turner, N.C., Wright, G.C. and Siddique, K.H.M. (2001) Adaptation of grain legumes (pulses) to water limited environments. *Advances in Agronomy* 71, 193–231.

van der Maesen, L.J.G. (1972) *Cicer* L. A Monograph of the Genus, with special reference to the chickpea (*Cicer arietinum* L.), its Ecology and Cultivation. Wageningen 7–10, 1–342.

van der Maesen, L.J.G. (1984) Taxonomy, distribution and evolution of the chickpea and its wild relatives. In: Witcombe, J.R. and Erskine, W. (eds) *Genetic Resources and Their Exploitation: Chickpeas, Faba Beans and Lentils*. Martinus Nijhoff/Dr W. Junk Publishers, The Hague, The Netherlands, pp. 95–104.

van der Maesen, L.J.G. (1987) Origin, history and taxonomy of chickpea. In: Saxena, M.C. and Singh, K.B. (eds) *The Chickpea*. CAB International, Wallingford, UK, pp. 11–34.

Vavilov, N.I. (1922) The law of homologous series in variation. *Journal of Genetics* 12, 47–89.

Winter, P., Benko-Iseppon, A.-M. *et al.* (2000) A linkage map of the chickpea (*Cicer arietinum* L.) genome based on recombinant inbred lines from a *C. arietinum* × *C. reticulatum* cross: localization of resistance genes for *fusarium* wilt races 4 and 5. *Theoretical and Applied Genetics* 101, 1155–1163.

Yadav, S.S., Turner, N.C. and Kumar, J. (2002) Commercialization and utilization of wild genes for higher productivity in chickpea. In: *Plant Breeding for the 11 Millennium:*

Proceedings of the 12th Australian Plant Breeding Conference. 15–20 September 2002. Perth, Western Australia, pp. 155–160.

Zohary, D. and Hopf, M. (2000) *Domestication of Plants in the Old World*. Clarendon Press, Oxford.

Zohary, M. (1966) *Flora Palaestina*. The Israel Academy of Sciences and Humanities, Jerusalem.

Zohary, M. (1972) *Flora Palaestina*. The Israel Academy of Sciences and Humanities, Jerusalem.

17 Biodiversity Management in Chickpea

R.J. Redden,[1] B.J. Furman,[2] H.D. Upadhyaya,[3] R.P.S. Pundir,[3] C.L.L. Gowda,[3] C.J. Coyne[4] and D. Enneking[1]

[1]*Australian Temperate Field Crops Collection, Department of Primary Industries, Victorian Institute for Dryland Agriculture, Horsham, Victoria 3401, Australia;* [2]*International Centre for Agricultural Research in Dry Areas (ICARDA), PO Box 5466, Aleppo, Syria;* [3]*Genetics Resources Divisions, International Crops Research Institute for the Semi-Arid Tropics (ICRISAT) Asia Center, Hyderabad 502 324, India;* [4]*Plant Introduction Unit, 303 Johnson Hall, Washington State University, Pullman, WA 99164, USA*

Morphological Diversity

Variation in cultivated chickpea is mostly noted in the separation of the kabuli large-seeded and the desi small-seeded gene pools, each with associated patterns of traits (Moreno and Cubero, 1978). The desi types have seed size generally in the range 10–25 g/100 seed, and are associated with small leaves or leaflets, and small pods with 1–3 seeds that are angular-shaped or beaked. Seeds vary in colour from cream to orange, dull green, various shades of brown to black, and flowers usually vary from pink to red, blue veined and purple. In the desi type, pod surface may be smooth, wrinkled or tuberculate, and anthocyanin occurrence on leaves and stems is associated with coloured flowers (van der Maesen, 1973).

Chickpea has an air-sac-inflated pod surrounding the seed, and pods and leaves have glandular pubescence.

There are small-seeded kabuli types of 25–35 g/100 seed, but the large-seeded kabuli range from 40 to 60 g/100 seed, with owl's head-shaped and generally cream coloured seeds and white flowers (Moreno and Cubero, 1978). Other seed colours of pink, red or black may occur at low frequencies. Leaves or leaflets and pods also tend to be large. Morphological diversity is narrower amongst the kabulis than amongst the desis from which they were derived in relatively recent millennia since there are few cytoplasmic differences (Moreno and Cubero, 1978).

Diversity analyses with randomly amplified polymorphic DNA (RAPD) and intersimple sequence repeat (ISSR) markers showed separate clustering of kabuli from desi types (Iruela *et al.*, 2002). This study showed that cultivated chickpea also had less polymorphism for molecular markers than wild species, consistent with other diversity analyses using protein and isozyme analyses.

In a collection of more than 16,000 mainly desi landraces at the International Crops Research Institute for the Semi-Arid Tropics (ICRISAT), Hyderabad, India, a core collection was identified. The most variable traits were: days to flowering and to maturity, plant height and width, number of pods per plant and yield (Upadhyaya *et al.*, 2001). Unimodal and bimodal distributions of traits for 150 accessions including desi and kabuli displayed: rachis length 3.5–7 cm; leaflets per leaf 11–16; leaflet length 10–17 mm; leaflet width 6–14 mm bimodal with peaks at 8 and 11 mm; pod length 17–30 mm with bimodal peaks at 20 and 26 mm; pod width 8–14 mm with bimodal peaks at 10 and 13 mm; pod thickness 8–14 mm with bimodal peaks at 9 and 13 mm; pods per plant from 10–150 with a mean of 75; seeds per pod 1–2 skewed towards 1; seed size 10–70 g/100 seed with bimodal peaks at 20 and 40 g; and number of branches 1–4 (Moreno and Cubero, 1978). They found high correlations between leaf rachis, leaflets, pods and seed size traits; these are reflected in the separation of the desi (*microsperma*) and kabuli (*macrosperma*) gene pools, which are also geographically separated.

Desi types are grown mainly in the Indian subcontinent, Asia, East and South Africa, while kabuli types are found in the Mediterranean and are imported to Chile (Malhotra *et al.*, 2000). The gene bank at the International Centre for Agricultural Research in the Dry Areas (ICARDA), Aleppo, Syria, has mainly kabuli germplasm, which has been assessed for phenological and growth diversity in winter and spring sowings, and diversity similarly assessed by the ICRISAT gene bank (Malhotra *et al.*, 2000). Earlier at ICRISAT, the range of phenological expression was wider than at ICARDA, and hence the ranges and maximum values for height and width were also greater. The geographic distribution of trait expressions largely follows the desi vs kabuli distributions, with intermediate seed types found in Ethiopia and Iran (Malhotra *et al.*, 2000). Fusarium wilt resistance is found in the Indian subcontinent and Ethiopia, dry root rot resistance in India and Iran, ascochyta blight resistance in India, Iran, Mexico and Turkey, grey mould resistance in Iran, helicoverpa resistance in India, and high seed protein in Pakistan and Sudan. Additionally a range of morphological mutants are also described by Malhotra *et al.* (2000).

Germplasm Collections

The global chickpea germplasm collection is housed at ICRISAT, which has a total of 17,258 accessions in its gene bank, including 17,123 accessions of cultivated chickpea (*Cicer arietinum*) and 135 accessions comprising 18 wild *Cicer* spp., from 43 different countries. Of the entire cultivated chickpea accessions, 75.2% are of desi type, 21% kabuli type and 3.8% intermediate type (Upadhyaya, 2003). ICARDA has the Central and West Asia, and North Africa (CWANA) regional mandate for chickpea improvement and houses a total of 12,448 accessions in its gene bank, including 12,180 accessions of cultivated chickpea and 268 accessions, comprising 10 wild *Cicer* spp., from 60 different countries. The ICARDA collection focuses on kabuli types and, as such, has the largest global collection of this variety. The frequency of the ICRISAT and ICARDA collections by country of origin of cultivated chickpea is given in Table 17.1; holdings by species are given in Table 17.2. A recent focus of ICARDA is the acquisition of material from Central Asia and the Caucasus (CAC), an area historically underrepresented.

In addition to the Consultative Group on International Agricultural Research (CGIAR) centres, other major chickpea collections are conserved at the Seed and Plant Improvement Institute (SPII), Iran (4925 accessions); the United States Department of Agriculture's (USDA) Western Regional Plant Introduction Station (WRPIS) (4662); the Centre for Legumes in Mediterranean Agriculture (CLIMA), Australia (4351); the National Bureau of Plant Genetic Resources (NBPGR), India (3830); and the N.I. Vavilov Institute of Plant Industry (VIR) in St Petersburg, Russia (2113).

Table 17.1. Frequency by country of origin in ICRISAT and ICARDA holdings.

Country origin	ICRISAT	ICARDA
Afghanistan	702	1,030
Algeria	16	60
Armenia	–	56
Australia	3	4
Azerbaijan	–	119
Bangladesh	170	1
Bulgaria	9	217
Chile	139	349
China	24	34
Colombia	1	9
Cuba	–	4
Cyprus	44	40
Czech Republic	8	7
Ecuador	–	2
Egypt	53	64
Ethiopia	933	110
France	2	21
Georgia	–	11
Germany	11	2
Greece	25	21
Guatemala	–	2
Hungary	4	3
India	7,081	439
Iran	4,856	1,782
Iraq	18	37
Israel	51	–
Italy	45	87
Jordan	25	156
Kazakhstan	–	14
Kenya	1	6
Kyrgyzstan	–	9
Latvia	–	3

Continued

Table 17.1. *Continued*

Country origin	ICRISAT	ICARDA
Lebanon	20	28
Libyan Arab Jamahiriya	–	5
Malawi	81	3
Mexico	397	160
Moldova, Republic of	–	21
Morocco	249	231
Nepal	80	8
Nigeria	3	1
Pakistan	484	270
Palestine	–	64
Peru	3	4
Portugal	84	179
Romania	–	4
Russian Federation	133	167
Slovakia	–	7
Spain	122	293
Sri Lanka	3	2
Sudan	12	14
Syrian Arab Republic	226	533
Tajikistan	–	170
Tanzania	97	–
Tunisia	33	271
Turkey	478	965
Turkmenistan	–	14
Ukraine	–	81
Uganda	1	–
UK	–	8
USA	82	124
Uzbekistan	–	304
Venezuela	–	1
Yugoslavia	6	6
Unknown	308	224
Breeding lines		3,319
Total	17,123	12,180

At ICARDA, each sample in its gene bank is documented according to internationally agreed upon descriptor lists jointly published by the International Plant Genetic Resources Institute (IPGRI), ICRISAT and ICARDA, and made up of various categories of passport, characterization and evaluation data. Passport data are recorded in a database at the time of introduction of each accession, and characterization and evaluation information is added when available. The documentation system also assists the germplasm curators in controlling seed stock, seed viability and distribution of germplasm samples to users.

There are two main types of collections stored at the genetic resources unit (GRU): an active collection, which is available for distribution and includes

Table 17.2. *Cicer* spp. and accession numbers held at ICRISAT and ICARDA.

Species	ICRISAT holdings	ICARDA holdings
Cicer anatolicum	3	2
C. arietinum	17,123	12,180
C. bijugum	8	45
C. chorassanicum	3	3
C. cuneatum	1	5
C. echinospermum	4	15
C. flexuosum	–	1
C. floribundum	1	–
C. judaicum	22	77
C. macracanthum	5	–
C. microphyllum	52	–
C. montbretii	2	–
C. multijugum	1	–
C. nuristanicum	2	–
C. pinnatifidum	11	55
C. pungens	9	–
C. rechingeri	1	–
C. reticulatum	6	60
C. yamashitae	3	5
C. anariense	1	–
Total	17,258	12,448

all accessions; and a base collection, kept under conditions where the seeds remain viable for up to 100 years, containing ~86% of the total holdings. The active collection is stored in plastic containers in an atmosphere of 0°C and 15–20% relative humidity (RH); the base collection is stored in hermetically (vacuum) sealed aluminum foil packets and kept at –20°C. In addition, a duplicate collection, comprising 96% of the holdings, is housed at ICRISAT and other donor institutes for safety purposes.

Core and Mini Core Collections and Their Use for Diversity Studies and New Breeding Goals

With the awareness of crop germplasm exploration and collection during the 1970s and later, numerous gene banks were established. At many such places, the numbers of collections are quite large and researchers have not been able to make good use of the germplasm resources. Such vast collections will not be of much use unless the individual accessions are characterized for relevant traits, and summarized and maintained in a user-friendly manner. Some of the traits of agronomic importance are influenced to a great extent by genotype × environments interactions, and require replicated multilocational evaluations.

This is a very costly and resource-demanding task owing to the large size of the germplasm collection. To overcome this, Frankel (1984) proposed sampling of the collection to a manageable sample or 'core collection'. A core collection contains a subset of accessions from the entire collection, which captures most of the available diversity of species (Brown, 1989). Following the concept, a chickpea core collection was developed at ICRISAT, which had 1956 accessions (about 10% of the entire collection), but represented almost the full diversity of ~17,000 accessions of the species (Upadhyaya *et al.*, 2001). Similarly, a chickpea core of 505 accessions was developed from 3350 accessions by the scientists at the USDA–Pullman, Washington (Hannan *et al.*, 1994).

When the size of the entire collection is large, even a core collection becomes unwieldy for evaluation by breeders. To overcome this, ICRISAT scientists developed a two-stage seminal strategy to develop a mini core collection, which consists of 10% accessions of the core collection (only 1% of the entire collection) (Upadhyaya and Ortiz, 2001) and represents the diversity of the entire core collection. The first stage involves developing a representative core subset (about 10%) from the entire collection using all the available information on origin, geographical distribution, and characterization and evaluation data of accessions. The second stage involves evaluation of the core subset for various morphological, agronomic and quality traits, as well as selecting a further subset of about 10% accessions from the core subset. At both stages standard clustering procedure should be used to separate groups of similar accessions. At ICRISAT, the scientists have already developed mini core collections of chickpea consisting of 211 accessions (Upadhyaya and Ortiz, 2001).

Diversity assessment in the core subset: the chickpea core consisting of 1956 accessions had 1465 desi, 433 kabuli and 58 intermediate seed types. This core collection was evaluated for seven morphological and 15 agronomic traits to estimate phenotypic diversity. The three groups differed significantly for flower and plant colour, dots on seed testa, seed testa texture, plant width, days to maturity, pods per plant, 100-seed weight and plot yield. The kabuli and intermediate types were not significantly different for growth habit and seed colour; however, they differed significantly from desi types for both the traits. Desi, kabuli and intermediate types were significantly different for plant width, days to maturity, pods per plant, 100-seed weight and plot yield. Kabuli plants had broader width, matured late, lowest number of pods, highest seed weight and lowest plot yield. The average phenotypic diversity index was highest in the intermediate types (0.2653) and lowest in the kabuli types (0.1490) (Upadhyaya *et al.*, 2002).

Identification of sources for highly useful traits: due to reduced size, the core collection can be evaluated extensively to identify the useful parents for crop improvement. By evaluating core collection of chickpea, we identified new sources of important traits, namely, early maturity (28 accessions) and large-seeded kabuli (16 accessions) types. The core collection was tested for agronomic traits through the yield trials conducted during two seasons (1999–2001) at ICRISAT along with two control cultivars – Annigeri and L 550. The controls were the released cultivars and were well adapted to the region's conditions. The scientists did not expect many germplasm accessions to be found significantly superior-yielding compared to the controls. However, they found that seven chickpea accessions (ICCs 6122, 8324, 12197, 13124, 14230, 16862 and 16934) were superior in respect to the four important yield traits (days to flowering, pod number, grain yield and 100 seed weight) compared to control Annigeri. Similarly, nine kabuli accessions (ICCs 3410, 5644, 6160, 6246, 7200, 8042, 10755, 10783 and 15763) were superior in respect to the four yield traits mentioned earlier compared to the control L 550 (Table 17.3). The evaluation of chickpea core at Pullman, Washington, revealed high levels of resistance to ascochyta blight and pre-emergence damping-off diseases (Hannan *et al.*, 1994).

To gain benefits, the mini core collection of chickpea was evaluated to identify useful traits for use in crop improvement. From the chickpea mini core, 18 accessions having traits related to drought tolerance (Kashiwagi *et al.*, 2005) and 29 accessions having moderate to high tolerance to soil salinity (Serraj *et al.*, 2004) were identified. Similarly, Pande *et al.* (2005) screened the mini core collection for resistance to various diseases and identified 67 accessions

Table 17.3. Chickpea accessions found superior compared to respective desi and kabuli type controls in core collection of chickpea, mean of 1999/2000 and 2000/01 seasons, ICRISAT, Hyderabad, India.

Accession[a]	Country of origin	Biological status	Days to 50% flowering	Pods per plant	Seed yield (kg/h)	100-Seed weight (g)
Desi						
ICC 6122	India	Landrace	40	64	1832	34.6
ICC 8324	India	Landrace	46	72	2251	21.7
ICC 12197	India	Breeding line	44	72	2178	28.6
ICC 13124	India	Landrace	48	53	2188	33.8
ICC 14230	India	Landrace	49	55	2166	33.6
ICC 16862	India	Landrace	45	74	2288	25.2
ICC 16934	India	Landrace	48	72	2212	23.1
Kabuli						
ICC 3410	Iran	Landrace	54	65	2138	21.8
ICC 5644	India	Landrace	61	60	2138	23.3
ICC 6160	Syria	Landrace	59	66	1996	40.5
ICC 6246	Tunisia	Landrace	63	68	1898	21.8
ICC 7200	Egypt	Breeding line	62	59	2171	21.5
ICC 8042	Iran	Landrace	59	55	2075	30.8
ICC 10755	Turkey	Landrace	61	58	2014	31.4
ICC 10783	Turkey	Landrace	61	60	2167	35.6
ICC 15763	Morocco	Landrace	59	69	1886	26.0
Control						
Annigeri (desi)	India	Cultivar	50	70	2057	21.3
L550 (kabuli)	India	Cultivar	63	64	1858	19.9

[a]Accessions were significantly superior for yield in all classes except the early group.

resistant or highly resistant to fusarium wilt, 3, 55, and 6 accessions moderately resistant to ascochyta blight, botrytis grey mould and dry root rot, respectively. ICC 11284 was the only accession moderately resistant to both the foliar diseases (ascochyta blight and botrytis grey mould). Four accessions were found with combined resistance to dry root rot and fusarium wilt, and 15 accessions to fusarium wilt and botrytis grey mould. The evaluation of chickpea mini core at the Indian Institute of Pulses Research (IIPR), Kanpur, India during 2002/04 revealed 12 very promising accessions. Of these, six accessions, namely ICCs 14194, 14196, 14197, 14199, 12034 and EC 381882, were involved in hybridization to develop large-seeded kabuli types.

Researchers at the ICRISAT with those at the ICARDA have developed a global composite collection of 3000 accessions (Table 17.4) that will be molecular profiled using 50 polymorphic simple sequence repeat (SSR) markers (Upadhyaya *et al.*, 2006). The data generated will be used to define the genetic structure of the global composite collection and to select a reference sample of 300 accessions representing the maximum diversity for the isolation of allelic variants of candidate gene associated with beneficial traits. It is then expected that molecular biologists and plant breeders will have opportunities to use diverse lines in functional and comparative genomics, in mapping and

Table 17.4. Characteristics of germplasm included in global composite collection of chickpea.

Germplasm/ traits	Number of accessions	Germplasm/ Traits	Number of accessions	Germplasm/ traits	Number of accessions
ICRISAT collection		ICRISAT collection		ICRISAT collection	
Core collection	1956	Tolerance to leaf miner	5	Large seed size	18
Cultivars/ breeding lines	39	Resistance to nematode	8	Double pods	8
Resistance to ascochyta blight	13	Tolerance to cold	12	High seed protein	10
Black root rot	8	Heat	4	Morphological diversity	35
Botrytis grey mould	8	Drought	10	Wild *Cicer* spp.	3
Collar rot	9	Salinity	4	ICARDA collection	
Dry root rot	6	I Input responsive	4	Characterization and evaluation data	212
Fusarium wilt	50	High BNF	8	Random selection	497
Stunt	8	Early maturity	25	Wild *Cicer* spp.	17
Tolerance to pod borer	16	Multiseeded pods	7	Total	3000

cloning gene(s), and in applied plant breeding to diversify the genetic base of the breeding populations, which will lead to the development of broad-based elite breeding lines/cultivars with superior yields and enhanced adaptation to diverse environments

Conservation and Documentation

Storage

Conservation of *Cicer* is almost wholly as seed, from annuals and also the wild perennials.

The principal conservers of diversity are as outlined earlier, the ICRISAT and ICARDA gene banks, and national gene banks in various countries from the Middle East centre of origin to Russia and the USA. However national breeding programmes in many countries have also associated working collections kept for short to medium term utility, in which storage conditions may not be as stringent for maintenance of viability as in major gene banks.

Viability will decline in all stored seed, as a function of initial viability, seed moisture and storage temperature (Roberts and Ellis, 1984). Seed harvest, drying, processing, and pre-storage temperature and humidity conditions may all affect the initial viability, which is usually close to 100% for freshly harvested seed from well-grown plants. Each seed lot has a viability constant, affected by both genotype and pre-storage conditions, which affects seed longevity. Seed survival curves conform to a negative cumulative normal distribution, which are transformed to straight line slopes with probit percentage viability plotted against time (Roberts and Ellis, 1984). They report that this slope is unaffected by either genotype within species or pre-storage conditions, but is affected by logarithmic constants for seed moisture, temperature and species. This enables seed longevity to be predicted for given storage conditions and initial viability, within a wide range of operational storage conditions from ambient to various levels of reduced temperature and seed moisture within biological limits (seeds will die if completely desiccated). With regular storage at 5 ± 1% seed moisture and within a temperature of −20°C and −30°C, the occurrence of mutations during storage appears to be a very minor concern (Roberts and Ellis, 1984).

The recommended storage facilities for gene banks (FAO/IPGRI, 1994), would contain a base collection of more than 100 seeds (FAO/IPGRI recommend more than 1000 seeds) in −18°C to −20°C, and an active collection of up to 1000 seed at 2–5°C, respectively for long-term conservation, and for immediate usage and distribution. Drying of seed before storage at 10–25°C and 10–15% RH is recommended. The base collection should aim to maintain the genetic integrity of the seed samples of accessions as originally received and enable long-term conservation of genetic diversity. It is also used to replenish seed stocks in the active collection in every fourth regeneration cycle (FAO/IPGRI, 1994). A duplicate collection kept in long-term storage is also recommended for safety purposes, also referred to as a back-up collection.

Further investigation of chickpea seed longevity is in progress at the Australian Temperate and Field Crops Collection (ATFCC) with storage trials of two genotypes at five different temperatures (40°C, 22°C, 15°C, 2°C and −18°C) and three seed moisture levels (6–7%, 10–11%, 12–13%) begun with seed harvest in December 2002, postharvest storage at 15°C, and initiation of treatments in February 2003 (Redden, Horsham, 2006, personal communication). An additional eight genotypes are being tested at 22°C. This study will examine whether the seed moisture and temperature storage constants for *Cicer* are invariant for genotype and conform to a logarithmic formula. Only the 40°C treatments have run the course from full viability to termination, and of themselves do not provide enough data to test the predictions, though initial indications are for genotypic variation and for greater longevity than predicted.

Such longevity studies are important to enable better planning of intervals between regeneration cycles and to reduce the staff as well as physical resources required to otherwise monitor seed viability with germination tests. Seed in the active collection may possibly retain high viability for at least 20 years, given high initial viability, seed desiccation and immediate deposit in active storage. Optimization of storage conditions is a cheap and efficient means of extending seed longevity, thereby lengthening the regeneration intervals and reducing the annual costs for maintenance of high levels of seed viability in the active collections.

Seeds should be stored in moisture-proof containers; heat sealing of plastic lined foil envelopes is one of the common gene banks procedures. Opening and resealing of these packets for seed removal and distribution is best carried out in a drying room, such as 15°C and 15% RH as described earlier.

A decision guide for seed regeneration, and precautions to maintain genetic integrity are outlined by Sackville Hamilton and Chorlton (1997).

Documentation

There are three main activities, described in the IPGRI handbook for germplasm collections (Reed *et al.*, 2004):

1. Germplasm and site characterization (passport) for an accession collected as a traditional landrace *in situ*, for placement in an *ex situ* gene bank elsewhere;
2. Gene bank inventory with accession identifiers, taxonomic descriptors, country and site passport data, dates of accession, collector and donor details, synonyms for accessions acquired from other gene banks;
3. Accession characterization data in the case of chickpea standard trait descriptors are provided in IBPGR/ICRISAT and ICARDA, 1993.

Full records are very important for checking through synonyms for duplication of accessions, precise collection site data with referencing by latitude, longitude and altitude if possible, and for quality assurance on identity. This can be an issue in gene banks with multiple sources of accessions, and large numbers of accessions and species.

Equally important are characterization and evaluation data. Better targeted and more efficient utilization of germplasm by plant breeders and researchers can be achieved if the trait characteristics of accessions are known. This can include the agronomic, disease reaction, yield and quality data of accessions in a particular study, and over the different studies for each accession. These data can now be retrieved with International Crop Information Systems (ICIS) platforms for digital search and retrieval across relational databases for all relevant studies. An example is the ICIS evaluation database for chickpea, which combines databases of the ICARDA, ICRISAT, USDA and ATFCC enabling a more comprehensive search of genetic resources over large collections for multiple trait expressions, is available at: http://www.iris.irri.org/ranjanweb/SiteMain.jsp.

A number of gene banks provide well-documented and up-to-date taxonomic guides for the *Cicer* genus, e.g. the germplasm resources information network (GRIN) database of the USDA, available at: http://www.ars-grin.gov/cgi-bin/npgs/html/genform.pl

Taxonomic classifications may differ between gene banks; some may tend to split species into subspecies in comparison with GRIN, which describes alternative nomenclature.

Germplasm Enhancement

Pre-breeding describes the transfer of desirable traits from exotic germplasm into elite breeding lines, whereas germplasm enhancement describes this activity as a process in the context of utility and sustainability of crops. In practice, germplasm enhancement is a long-term undertaking to utilize traits from unadapted and even wild germplasm sources to obtain traits that are not available in elite or domesticated lines. Most work in this area is carried out by public plant breeders and scientists because a private investment in the generation of enhanced germplasm is not profitable in the short term (Ortiz, 2002).

For chickpea, most germplasm enhancement has been carried out by the ICRISAT and ICARDA. Large-scale characterization and evaluation of chickpea germplasm collections (Singh and Malhotra, 1984a,b) has resulted in the publication of informative and detailed catalogues (Pundir *et al.*, 1988; Singh *et al.*, 1991; Robertson *et al.*, 1995). Much early work was focused on improving disease resistance, cold tolerance and the introgression of desi germplasm into the less diverse kabuli gene pool (Singh and Reddy, 1991; Singh, 1993). Germplasm developed from these activities is distributed annually through international nurseries to national breeding programmes. This has facilitated the use of enhanced germplasm in many countries. In fact, this material helped to revive the Australian kabuli industry after the devastating ascochyta blight epidemic of 2001, since resistant germplasm could be readily identified in the previous year's international nursery growing under epidemic conditions, near Horsham in 2001 (T. Bretag, Horsham, 2005, personal communication). International nurseries have become the predominant source of new germplasm in the Australian breeding programme for kabuli chickpea (K. Hobson, personal communication).

Since publication of the last monograph on chickpea (van der Maesen, 1973; Saxena and Singh, 1987), great advances have been made in the identification of chickpea germplasm with useful traits. Recent work has led to refinement of standardized methods for *Ascochyta* screening by development of differential tester sets (Chen *et al.*, 2004) and identification of tolerance to *Meloidogyne javanica* (Ansari *et al.*, 2004).

Molecular DNA marker technology is gaining quickly in resolution, much of it owing to syntenic genetic relationships with other crops for which genomic libraries are being developed. Genomic sequence resources facilitate the generation and testing of increasing numbers of DNA markers to map traits of interest with more precision (Winter *et al.*, 2003). The increasing availability of molecular markers promises to speed up the identification of useful traits during germplasm enhancement. For example, Sharma *et al.* (2004) identified DNA markers for *Fusarium oxysporum* f. sp. *ciceris* race 3 resistance. The establishment of sufficiently high throughput and cost-effective molecular breeding programmes still awaits realization. In anticipation of such future activity, several recent and current germplasm enhancement projects include a molecular marker component (Santra *et al.*, 2000).

Efforts are also being made to improve the utilization of the wild gene pool. Ever since useful disease and abiotic stress resistances were identified in the secondary and tertiary gene pool of chickpea, there has been a recognized need to transgress infertility barriers to facilitate the transfer of such traits to cultivated chickpea. Wild species are increasingly recognized as worthwhile sources of biotic and abiotic stress tolerance with recent releases of desi cultivars in India demonstrating that productive genotypes can be developed from wide crosses (Yadav *et al.*, 2006).

Following initial genetic characterization of wild species (Nguyen *et al.*, 2004) their exploitation for reproductive cold tolerance is a priority in Australia (H. Clarke, Perth, 2005, personal communication). While the availability of unique wild germplasm in *ex situ* collection is very limited (Berger *et al.*, 2003), existing resources are receiving attention as potential donors of useful traits such as resistance to *H. armigera* (Sharma *et al.*, 2005).

At present, a project in Horsham is aiming to identify tolerance to elevated levels of subsoil Boron and to develop enhanced germplasm for salinity tolerance utilizing accessions identified by Maliro *et al.* (2004), additional sources identified using eco-geographic principles and continued systematic screening of germplasm.

References

Ansari, M.A., Patel, B.A., Mhase, N.L., Patel, D.J., Douaik, A. and Sharma, S.B. (2004) Tolerance of chickpea (*Cicer arietinum* L.) lines to root-knot nematode, *Meloidogyne javanica* (Treub) Chitwood. *Genetic Resources and Crop Evolution* 51, 449–453.

Brown, A.H.D. (1989) Core collections: a practical approach to genetic resources management. *Genome* 31, 818–824.

Berger, J., Abbo, S. and Turner, N.C. (2003) Ecogeography of annual wild *Cicer* species: the poor state of the world collection. *Crop Science* 43, 1076–1090.

Chen, W., Coyne, C.J., Peever, T.L. and Muehlbauer, F.J. (2004) Characterization of chickpea differentials for pathogenicity assay of Ascochyta blight and identification of chickpea accessions resistant to *Didymella rabiei*. *Plant Pathology* 53(6), 759–769.

FAO/IPGRI (1994) Genebank Standards. Food and Agriculture Organization of the United Nations, Rome, International Plant Genetic Resources Institute, Rome.

Frankel, O.H. (1984) Genetic perspectives of germplasm conservation. In: Arber, W., Llimensee, K., Peacock, W.J. and Starliger, P. (eds) *Genetic Manipulations: Impact of Man and Society*. Cambridge University Press, Cambridge, pp. 161–170.

Hannan, R.M., Kaiser, W.J. and Muehlbauer, F.J. (1994) Development and utilization of the USDA chickpea germplasm core collection. *1994 Agronomy Abstracts*. American Society of Agronomy, Madison, Wisconsin, p. 217.

IBPGR/ICRISAT and ICARDA (1993) Descriptors for chickpea (*Cicer arietinum* L.). International Board for Plant Genetic Resources, Rome, Italy.

Iruela, M., Rubio, J., Cubero, J.I., Gil, J. and Millan, T. (2002) Phylogenetic analysis in the genus *Cicer* and cultivated chickpea using RAPD and ISSR markers. *Theoretical and Applied Genetics* 104, 643–651.

Kashiwagi, J., Krishnamurthy, L., Upadhyaya, H.D., Krishna, H., Chandra, S., Vandez, V. and Serraj, R. (2005) Genetic variability of drought-avoidance root traits in the mini-core germplasm collection of chickpea (*Cicer arietinum* L.). *Euphytica* 146, 213–222.

Malhotra, R.S., Pundir, R.P.S. and Kaiser, W.J. (2000). *Cicer* species – conserved resources, priorities for collection and future prospects. In: Knight, R. (ed.) *Linking Research Marketing Opportunities for Pulses in the 21st Century*. Kluwer Academic, Dordrecht, The Netherlands, pp. 603–611.

Maliro, M.F.A., McNeil, D., Kollmorgen, J., Pittock, C. and Redden, R.J. (2004) Screening chickpea (*Cicer arietinum* L.) and wild relatives germplasm from diverse country sources for salt tolerance. In: *Proceedings of the 4th International Crop Science Congress*. 26 September–1 October 2004. Brisbane, Australia, p. 274.

Moreno, M. and Cubero, J.I. (1978) Variation in *Cicer arietinum* L. *Euphytica* 27, 465–485.

Nguyen, T.T., Taylor, P.W.J., Redden, R.J. and Ford, R. (2004) Genetic diversity estimates in *Cicer* using AFLP analysis. *Plant Breeding* 123, 173–179.

Ortiz, R. (2002) Germplasm enhancement to sustain genetic gains in crop improvement. In: Engels, J.M.M., Ramanatha Rao, V. and Jackson, M.T. (eds) *Managing Plant Genetic Diversity*. IPGRI, Rome, pp. 275–290.

Pande, S., Kishore, G.K., Upadhyaya, H.D. and Rao, J.N. (2005) Identification of sources of multiple fungal diseases resistance using mini-core collection in chickpea. *Plant Disease* 90, p. 1024.

Pundir, R.P.S., Reddy, K.N. and Mengesha, M.H. (1988) *ICRISAT Chickpea Germplasm Catalog: Evaluation and Analysis*. ICRISAT, Hyderabad, India.

Reed, B.M., Engelmann, F., Dulloo, M.E. and Engels, J.M.M. (2004) *Technical Guidelines for the Management of Field and In Vitro Germplasm Collections*. IPGRI Technical Handbook No 7.

Roberts, E.H. and Ellis, R.H. (1984) The implications of the deterioration of orthodox seeds during storage for genetic resources conservation. In: Holden, J.H.W. and Williams, J.T. (eds) *Crop Genetic Resources: Conservation and Evaluation*. Gheorge Allen & Unwin, London, pp. 18–37.

Robertson, L.D., Singh, K.B. and Ocampo, B. (1995) *A Catalog of Annual Wild Cicer Species*. ICARDA, Aleppo, Syria.

Sackville Hamilton, N.R. and Chorlton, K.K. (1997) Regeneration of accessions in seed collections: a decision guide. In: Engels, J. (ed.) *Handbook for Genebanks*, No 5. International Plant Genetic Resources Institute, Rome, Italy, pp. 1–72.

Santra, D.K., Takeoglu, M., Ratnaparkhe, M., Kaiser, W.J. and Muehlbauer, F.J. (2000) Identification and mapping of QTLs conferring resistance to Ascochyte blight in chickpea. *Crop Science* 40, 1606–1612.

Saxena, M.C. and Singh, K.B (1987) *The Chickpea*. CAB International, Wallingford, UK.

Serraj, R., Krishnamurthy, L. and Upadhyaya, H.D. (2004) Screening of chickpea mini-core germplasm for tolerance to soil salinity. *International Chickpea and Pigeonpea Newsletter* 11, 29–32.

Sharma, H.C., Pampapathy, G., Lanka, S.K. and Ridsdill-Smith, T.J. (2005) Antibiosis mechanism of resistance to pod borer, *Heliocoverpa armigera* in wild relatives of chickpea. *Euphytica* 142, 107–117.

Singh, K.B. (1993) Problems and prospects for stress resistance breeding in chickpea. In: Singh, K.B. and Saxena, M.C. (eds) *Breeding for Stress Tolerance in Cool-Season Food Legumes*. Wiley, Chichester, UK, pp. 17–35.

Singh, K.B. and Malhotra, R.S. (1984a) Collection and evaluation of chickpea genetic resources. In: Witcombe, J.R. and Erskine, W. (eds) *Genetic Resources and Their Exploitation – Chickpeas, Faba Beans and Lentils*. Martinus Nijhoff/Dr W. Junk Publishers (for ICARDA and IBPGR), The Hague, Boston, Lancaster, pp. 105–122.

Singh, K.B. and Malhotra, R.S. (1984b) Exploitation of chickpea genetic resources. In: Witcombe, J.R. and Erskine, W. (eds) *Genetic Resources and Their Exploitation – Chickpeas, Faba Beans and Lentils*. Martinus Nijhoff/Dr W. Junk Publishers (for ICARDA and IBPGR), The Hague, Boston, Lancaster, pp. 123–130.

Singh, K.B. and Reddy, M.V. (1991) Advance in disease-resistance breeding in chickpea. *Advances in Agronomy* 45, 191–222.

Singh, K.B., Holly, L. and Beijiga, G.A. (1991) A catalog of kabuli chickpea germplasm. ICARDA, Aleppo, Syria.

Sharma, K.D., Winter, P., Kahl, G. and Muehlbauer, F.J. (2004) Molecular mapping of *Fusarium oxysporum* f. sp. *Ciceri* race 3 resistance gene in chickpea. *Theoretical and Applied Genetics* 108, 1243–1248.

Upadhyaya, H.D. (2003) Geographical patterns of variation for morphological and agronomic characteristics in the chickpea germplasm collection. *Euphytica* 132, 343–352.

Upadhyaya, H.D. and Ortiz, R. (2001) A mini core subset for capturing diversity and promoting utilization of chickpea genetic resources. *Theoretical and Applied Genetics* 102, 1292–1298.

Upadhyaya, H.D., Bramel, P.J. and Singh, S. (2001) Development of a chickpea core subset using geographic distribution and quantitative traits. *Crop Science* 41, 206–210.

Upadhyaya, H.D., Furman, B.J., Dwivedi, S.L., Udupa, S.M., Gowda, C.L.L., Baum, M., Crouch, J.H., Buhariwalla, H.K. and Sube Singh (2006) Development of a composite collection for mining germplasm possessing allelic variation for beneficial traits in chickpea. *Plant Genetic Resources: Characterization and Utilization* 4(1), 13–19.

Upadhyaya, H.D., Ortiz, R., Bramel, P.J. and Sube Singh. (2002). Phenotypic diversity for morphological and agronomic characteristics in chickpea core collection. *Euphytica* 123, 333–342.

van der Maesen, L.J.G. (1973) *Cicer* L.: a monograph of the genus, with special reference to the chickpea (*Cicer arietinum* L.), its ecology and cultivation. Communication Agricultural University, Wageningen, The Netherlands, 72(10), 342 pp.

Yadav, S.S., Kumar, J., Turner, N.C., Berger, J., Redden, B., McNeil, D., Materne, M., Knights, T. and Bahl, P.N. (2006) Recent developments in breeding for wide adaptation and drought tolerance in grain legumes. *Plant Genetic Resources* 4(3) 181–187.

Winter, P., Staginnus, C., Sharma, P.C. and Kahl, G. (2003) Organisation and genetic mapping of the chickpea genome. In: Jaiwal, P.K. and Singh, R.P. (eds) *Improvement Strategies of Leguminosae Biotechnology – Focus on Biotechnology*, Vol. 10A. Springer, Berlin, pp. 303–352.

18 Conventional Breeding Methods

P.M. SALIMATH,[1] C. TOKER,[2] J.S. SANDHU,[3] J. KUMAR,[4] B. SUMA,[1] S.S. YADAV[4] AND P.N. BAHL[5]

[1]Department of Genetics and Plant Breeding, University of Agricultural Sciences, Dharwad, Karnataka, India; [2]Department of Field Crops, Faculty of Agriculture, Akdeniz University, TR-07059 Antalya, Turkey; [3]Department of Plant Breeding, Genetics and Biotechnology, Punjab Agricultural University, Ludhiana 141004, India; [4]Pulse Research Laboratory, Division of Genetics, Indian Agricultural Research Institute, New Delhi 110012, India; [5]A-9, Nirman Vihar, New Delhi 110092, India

Introduction

Chickpea is a predominantly self-pollinated crop. Cross-pollination to the extent of 0–1% has been reported (Smithson *et al.*, 1985; Singh, 1987). The extent of cross-pollination varies with environments. Gowda (1981) studied the extent of natural crossing at International Crops Research Institute for the Semi-Arid Tropics (ICRISAT), Hyderabad, India, in four F_2 populations of crosses of simple × compound leaf where simple leaf character is recessive to the normal compound leaf. Cross-pollination was estimated to be 1.92% and confirmed the low level of outcrossing in chickpea. This estimate is very close to 1.58% reported by Niknejad and Khosh-Khui (1972). Natural cross-pollination was verified by Malhotra and Singh (1986) under the Mediterranean environment at Tel Hadya in northern Syria, when a high population of bees was artificially released for pollinating faba bean in nearby fields.

Breeding Objectives

Higher and more stable yields are the major goals of chickpea-breeding programme. Like any other legume, the productivity of chickpea is generally low with the average yields ranging from 570 to 766 kg/ha in Asia, 600–660 kg/ha in Africa, 1600 kg/ha in North and Central America and average ~750 kg/ha in Europe (Food and Agricultural Organization, 1994). On an average, chickpea yields range from 400 to 600 kg/ha, but can exceed 2000 kg/ha.

These figures underline the need for genetic enhancement of productivity and achieving greater stability. Following are the major objectives:

1. Breeding for higher yield: chickpea is indeterminate in its growth habit, photo-sensitive as well as thermo-sensitive (Bahl *et al.*, 1979) and is characterized by poor partitioning of the photosynthates resulting in very low harvest index values (Jain, 1975). The reasons for low productivity have been summarized very well by Bahl (1981) and Singh (1987). High yield potential can be achieved by improving biomass and its favourable partitioning (Singh *et al.*, 1990). Another way of enhancing yield potentiality is by breeding cultivars responsive to fertilizer and irrigation. This is a long-term objective since there is no variability for this trait in the existing germplasm and therefore some novel techniques including the modern biotechnological tools need to be used for realizing this objective.
2. Breeding for extended adaptation of chickpea in space and time: the possibility and ways of achieving this objective have been very well outlined by Bahl *et al.* (1990) and Singh (1990). They have discussed the ways of breeding cultivars suited for winter sowing in the Mediterranean and for double cropping in the irrigated areas of the Indian subcontinent.
3. Breeding for resistance to biotic stress: the serious damage caused by disease and pest results in yield instability. Ascochyta blight, wilt and root rot are the major diseases; pod borer and leaf minor are the major pests; and cyst, root knot and root lesion nematodes are the major soil microorganisms that cause considerable yield losses. Obviously, breeding for resistance to these stresses will help stabilize chickpea production.
4. Breeding for resistance to abiotic stress: considerable yield losses occur due to abiotic stresses like drought, salinity, cold and frost. Resistance or tolerance to these stresses is more complex. However, resistant germplasm against these stresses has been identified. Therefore, breeding for resistance or tolerance to abiotic stresses is also an important objective.
5. Identification of stable form of male sterility: although exploitation of heterosis is difficult in chickpea due to many problems including difficulty in producing large quantity of hybrid seeds coupled with very high seed rate, male sterility will be useful to employ population improvement schemes. In the absence of stable male sterility, development of stable and useful male sterility becomes an important objective in chickpea. This may however be considered as a long-term objective.

Conventional Approaches

Since chickpea is a predominantly self-pollinated crop, development of true breeding and homozygous genotypes with desirable traits are the main objectives in the breeding programmes. Three steps of breeding programme are as follows:

1. Creation of genetic variation;
2. Selection within that variation;
3. Evaluation of selected lines.

Thus, variation forms the basis of a breeding programme. The success of a breeding programme depends on the nature and magnitude of variability. Creation of genetic variation becomes crucial for a sound and successful breeding programme, through:

1. Introduction of cultivars and segregating material from other sources within or outside the country;
2. Hybridization;
3. Mutation.

Plant introduction

Plant introduction as a method involves obtaining and evaluating genotypes from within and outside the country and identifying desirable genotype adapted to the local environment with high productivity or any other desired specific trait. The success of plant introduction therefore depends on the nature of the introduced material. Generally, the material introduced from locations with similar soil characteristics and climatic conditions is expected to be successful.

Introduction is generally facilitated by the following ways:

1. Exchange of material with fellow plant breeders;
2. Exploration of areas showing rich variation of the species;
3. Obtaining generic resources from international institutes/organizations like ICRISAT, International Center for Agricultural Research in the Dry Areas (ICARDA), Aleppo, Syria, Food and Agricultural Organization (FAO), Rome, Italy and the US Department of Agriculture (USDA) plant introduction stations.

The introduced material may be homozygous and stable like pure lines, or it may be heterogeneous mixed with landraces or heterozygous segregating populations. If the introduced material is a pure line, it will be immediately accepted as it shows good adaptability. But with mixtures (landraces) or segregating material, it will be necessary to identify either a productive line or a line with a specific desirable trait that will be useful in a recombination breeding programme.

Although plant introduction is employed at the beginning of a breeding programme, it is nevertheless a continuous process. It is the cheapest and fastest way of developing cultivars. Therefore, it becomes an important breeding procedure in a country where there is financial constraint or lack of trained staff, and also in those countries where the area under the crop is relatively small. There are several examples of successful introduction in chickpea. A bruchid-resistant line (G109-1) is a selection from an exotic variety introduced to India from Turkey (Saxena and Raina, 1970).

Pure line selection

As the number of local landraces and exotic germplasm increased, purification and evaluation of work was initiated more systematically. These landraces were variable and purification was made through selection. These selections were evaluated in different states of India and cultivars were developed (Table 18.1).

Table 18.1. Chickpea varieties developed by pure line selection. (From Singh, 1987; Dua *et al.*, 2001.)

State	Cultivars
Punjab	Pb7, G24, S26
Uttar Pradesh	T1, T2, T3, T87, K4, K468, KWR108, KGD1168, Pragati, DPC92-3
Madhya Pradesh	JGG1, JG5, JG62, JG221, JG315, Ujjain21, Ujjain24, Gwalior2
Maharashtra	Chaffa, D8, BDN9-3
Rajasthan	RS10, GNG146
Tamil Nadu	Co1, Co2
Andhra Pradesh	Warangal, Jyoti
Bihar	BR17, BR77, BR78, RAU15
Haryana	CSG8962
Karnataka	Annegeri
Gujarat	Dahod yellow

Hybridization

The objective of hybridization is to combine desirable traits from two or more parents into a single cultivar. This follows after the existing natural variation is exhausted.

Selection of parents

The success or failure of hybridization programme is largely determined by proper choice of the parents. When the aim is to replace the existing variety with a superior one, the existing variety with adaptation to the local environment is a logical choice as one parent. The second parent must be so chosen that it complements the first parent. If creation of variation for the desired traits or broadening of the genetic base is the objective, then diverse parents are selected. Published evaluation reports of the germplasm from national gene banks, e.g. National Bureau of Plant Genetic Resources (NBPGR) in India, international institutes and other national research reports as well as periodicals serve as valuable sources to select parents for a crossing programme based on specific objectives.

To ensure rational and scientific basis in selection of the parents, biometrical approaches are employed to analyse diversity, to be involved in hybridization and the combining ability of genotypes to be involved in hybridization.

Crossing techniques

Crossing in chickpea is a tedious job and the success of the artificial hybridization ranges from 10% to 50% depending upon the weather, particularly temperature and humidity, besides the genotypes involved. Since chickpea flowers are small and delicate, emasculation and pollination generally cause damage to the floral parts, thus reducing the success rate. Artificial hybridization techniques have been reviewed by Argikar (1970), van der Maesen (1972), Smart (1976) and Auckland and van der Maesen (1980). Pollen grains germinate in

~30 min after pollination. The germinated pollen tube takes 4 h to reach the base of ovary (Malti and Shivanna, 1983). However, it takes 24 h for fertilization after pollination. Functional specialization of stigmatic cells has been reported by Turano Margaret *et al.* (1983).

For emasculation, the flower bud that is due to open the following day is selected. The selected flower bud is held at the base between the thumb and forefinger. The front sepal is stripped off with the help of forceps and the keel is pushed downwards. The exposed anthers are then gently removed. After that a coloured cotton thread or a tag is tied around the internodes below the emasculated bud. Pollination is generally done in the morning on the following day. Half-opened flowers are the best source of matured pollen. Such flower buds are collected from the intended male parent in a petri dish or a glass tube. At the time of pollination, the standard and wings of the selected buds are removed and the upper portion of the keel is lightly pressed between the thumb and the forefinger to shed enough pollen on the stigma of the emasculated flower. In order to increase the success rate of artificial hybridization the following points are taken care of:

1. Selection of large flower buds;
2. Selection of lateral buds rather than the terminal ones since the success is reported to be better on the lateral buds (Sindhu *et al.*, 1981);
3. Avoiding mechanical injury to the floral parts at the time of emasculation and pollination;
4. Attempting hybridization after the formation of the first pod (Bahl and Gowda, 1975).

Timing of pollination and fertilization is also important in deciding the success rate. Under the conditions of low temperature, emasculation in the afternoon followed by pollination in the following morning gives better results. However, when the temperatures are high, morning emasculation followed by immediate pollination or pollination in the same afternoon is reported to be successful (Khosh-Khui and Niknejad, 1972; Singh and Auckland, 1975; Bejiga and Tessema, 1981; Pundir and Reddy, 1984).

Recombination breeding in chickpea was taken up in the 1930s with the objective of transferring the desirable attributes of donors into the local Indian cultivars. The first effort was made to transfer the attribute of disease resistance and the first ascochyta blight-resistant variety C12/34 was developed from the cross Pb7 × F8 in 1946. F8 was used as a resistant donor for this. In recombination breeding, selections are made as per pedigree/bulk methods or their modifications. Hybridization involves single crosses, three-way crosses or multiple crosses depending upon the objective of programme.

SINGLE CROSSES These are extensively used in chickpea to develop new cultivars. Through single crosses, efforts were made to select the superior recombinants than parents used in the cross. The major thrust remained to transfer the resistance against biotic and abiotic stresses. However, cultivars were also developed from single crosses for wide adaptation, suitable for rainfed/irrigated/late-sown conditions. The new varieties developed in India from single crosses between 1960 and 2004 included: 26 for wilt resistance, 9 for large seed, 8 for blight resistance, 8 for wide adaptation over rainfed and sometimes irrigated conditions, 6 for high yield mainly of kabuli types and 5 for late sowing, usually in combination with other key traits (Singh, 1987; Dua *et al.*, 2001).

THREE-WAY CROSSES These involve three different parents and each parent possesses a specific desirable trait, with the aim of recombining the traits together in a new cultivar. In the first step, an F_1 is developed between the parents of choice and then third parent is crossed with the F_1 to develop a three-way cross. These crosses provide additional opportunity for gene interaction. Three-way crosses can be managed by pedigree or bulk-breeding methods. The progenies of three-way crosses are more variable with wide genetic base than single crosses. Thus, it takes more time to isolate uniform progenies. In chickpea, a large number of three-way crosses were attempted and some of them led to the development of new cultivars (Table 18.2).

MULTIPLE CROSSES In these, more than three parents are involved, which are crossed in various ways. These crosses are used to broaden the genetic base and to enhance the opportunities for recombination. The segregating generations can be managed by following pedigree or bulk-breeding methods. The cultivars developed from multiple crosses are expected to have wider adaptation for a range of environments. Details of cultivars bred by following multiple crosses are given in Table 18.3.

Handling of segregating populations

It is always desirable to attempt well-planned crosses based on diversity and combining ability of parents. Since heterosis of F_1 hybrids has been shown to be correlated well with the performance of the segregating populations later, poor

Table 18.2. Varieties developed from three-way crosses in chickpeas. (From Singh, 1987; Dua *et al.*, 2001.)

Name of variety	Pedigree	Year of release	Important traits
BG244	([850-3/27 × P922] × P9847 kabuli type)	1985	Blight and stunt-resistant, cold-tolerant
BGD72	([BG256 × E100Y] × BG256)	1999	High yielding
C214	([G24 × IP58] × G24)	1971	Drought-resistant and cold-tolerant
H208	([26 × G24] × C235)	1977	Drought-resistant
HK94-134	([H82-2 × E100Y] × Bhim)	2001	Large-seeded kabuli
ICCC37	([P481 × JG62] × P1630)	2001	Wilt and dry root rot-resistant
JG11	([PhuleG 5 × Narsinghpur bold] × ICCC37)	1989	Wilt-resistant and moderately root rot-resistant
JG130	([PhuleG 5 × Narsinghpur bold] × JG74)	2000	Wilt-resistant, dry root rot and pod borer-tolerant
JKG-1	([ICCV2 × Surutato] × ICC7344)	2002	Large-seeded kabuli

Table 18.3. Varieties developed from multiple crosses in chickpea. (From Singh, 1987; Dua *et al*., 2001.)

Name of variety	Pedigree	Year of release	Important traits
BG256	([JG62 × 850-3/27] × [L550 × H208])	1984	Wilt, blight-tolerant and large-seeded
ICCV2	([K850 × GW5/6 × P48] × [L550 × Guamchil 2])	1993	Wilt and botrytis grey mould (BGM)-resistant, early maturing
Phule G 95311	([ICCC32 × ICCL 80004] × [ICCC49 × FLIP 82K] × ICCV3)	2002	Large-seeded
SG2	([E100Y × P436] × [L550 × F378])	1987	Wilt-tolerant, double-podded

performing F_1 crosses can be rejected which enables handling of few but promising hybrids so that sufficiently large populations of F_2 from selected F_1 populations can be handled for achieving better results. Further, the performance of F_2 populations can also be taken as a guide for their further advancement. Usually, F_1 hybrids showing high heterosis and less in-breeding depression are the ideal ones to pursue with since additive gene effects seem to govern the expression of traits in such populations, which can be fixed by simple selection.

Usefulness of early generation selection is often debated. It was reported to be advantageous by Dahiya *et al*. (1984b), and Salimath and Bahl (1985b), while Rahman and Bahl (1985) reported it to be useful for seeds per pod and seed weight, but not for pod number and seed yield. Yield, which is a complex trait, is generally less heritable. Therefore, selection for yield per se particularly in the early segregating generations may not give better results. However, indirect selection for component traits related to productivity has been found to be useful. Bisen *et al*. (1985) and Salimath and Bahl (1985b) observed that selection for seed size and seeds per pod may be better. Kumar and Bahl (1992) compared yield performance of F_6 lines derived by exercising selection pressure for four successive generations (F_2–F_5) and concluded that indirect selection for yield by pod number and seed weight was more efficient than direct selection for yield.

Pedigree method of selection is the most widely used method of handling segregating population in self-pollinated crops. However, there are various reports for and against the pedigree method of selection. Lal *et al*. (1973) support pedigree method of breeding for high yield in chickpea, whereas other researchers found this method less suitable and advocated bulk method (Byth *et al*., 1980) or the single-seed descent method (Singh and Auckland, 1975). While comparing different methods of handling segregating populations in chickpea, Rahman and Bahl (1985) concluded that mass selection is superior to random bulk and single-seed descent methods. Singh and Waldia (1994) found that the bulk method of breeding is advantageous for the development of high-yielding and short-duration varieties. Singh (1987) has summarized the usefulness of various breeding methods in respect of specific traits, e.g.

pedigree method for resistance to biotic stresses; bulk-pedigree method for traits such as drought tolerance and winter hardiness, modified bulk method for abiotic stresses, seed size, earliness and plant type, etc.; backcross method for interspecific hybridization and limited backcross for desi × kabuli introgression and resistance breeding.

Mutation breeding

The use of traditional breeding methods for a longer period might have narrowed genetic variability. Further improvement in a crop species requires increasing novel genetic variability. From this point of view, mutations may be fruitful in plant-breeding programmes (Micke, 1988a,b).

History

The first mutation idea, traced back to 1900, was given by de Vries (1901) in *Die Mutationstheorie*. The oldest description of spontaneous mutants appeared around 300 BC in China and was described especially in cereal crops in an ancient book, *Lulan* (van Harten, 1998), while induced mutations were discovered in fruit fly (*Drosophila melanogaster* L.) by Muller (1927). After the discovery of x-ray-induced mutations, Stadler (1928) confirmed the power of x-rays to induce mutation and showed its application to barley (*Hordeum vulgare* L.). After pioneering researches, induced mutations have been successfully exploited by many scientists all over the world. But mutation works were gradually reduced towards the 1960s due to many of its negative effects (Allard, 1960). In 1964, when the Food and Agricultural Organization/International Atomic Energy Agency (FAO/IAEA) Division of Nuclear Techniques in Food and Agriculture was established, plant breeders were encouraged to use mutation techniques for beneficial purposes.

Mutagens

Mutation can be briefly defined as a sudden heritable change in the characteristic of a living organism. Agents that induce mutations are called mutagens. Mutagens are mainly of two types: (i) physical radiation and (ii) chemical mutagens (Table 18.4). Mutagens are not only beneficial to create genetic variability in a crop species, but also useful for effective control of pests during postharvest storage. Toker *et al.* (2005) introduced a simple and reliable method to detect gamma-irradiated *Cicer* seeds including *C. bijugum* P.H. Rech. and *C. reticulatum* Ladiz. Recently, mutagenesis has received great attention for use as a promising new technique known as 'targeting induced local lesions in genomes' (TILLING).

CONSIDERATIONS FOR INDUCED MUTATIONS Following considerations should be taken into account before starting mutation-breeding programmes:

1. Mutations are mostly recessive and they can be selected in the second generation, M_2. Unlike recessive, dominant mutations occur in very small frequencies and they can be selected in M_1. Selection of polygenic traits should be started in individual plant progenies of M_3.

Table 18.4. Mutagens induced mutations.

Physical mutagens (radiation)	Chemical mutagens
Ionizing radiation	Alkylating agents, i.e. sulphur and nitrogen mustards, epoxides, ethylene imine (EI), ethylmethane sulphonate (EMS), methylmethane sulphonate (MMS), diazoalkanes, *N*'-methyl-*N*-nitro-*N*-nitroso-guanidine (MNNG)
Particulate radiation, i.e. alfa rays, beta rays, fast and thermal neutrons	Acridine dyes, i.e. acriflavine, proflavine, acridine orange, acridine yellow, ethidium bromide
Non-particulate radiation, i.e. x-rays and gamma rays	Base analogues, i.e. 5-bromouracil, 5-chlorouracil
Non-ionizing radiation, i.e. ultraviolet (UV) radiation	Others, i.e. nitrous acid, hydroxyl amine, sodium azide

2. Mutations are beneficial with very low frequencies (~0.1%), while treatments reduce germination, growth rate, vigour and pollen and ovule fertility in the living organisms.
3. Mutations are randomly induced and they might occur in any gene(s). However, some gene(s) can be more induced than others.
4. Mutations can be recurrent. The same gene(s) in a crop plant species may be induced again.
5. Mutations have generally pleiotropic effects due to closely linked gene(s).

Selection of variety, mutagen and dose

Mutation-breeding programme should be clearly planned and well defined, and large enough to select desirable mutations as well. The variety selected for mutagenesis should be one of the best varieties released recently and used for at least two varieties. It will be useful in improving specific characteristics of well-adapted and high-yielding varieties, which had one or two deficient traits. Some output from well-defined mutation-breeding programmes are given in Tables 18.5 and 18.6. Haq and Singh (1994) and Cagirgan and Toker (2004) selected cold-tolerant and ascochyta blight-resistant chickpea mutants. Kharkwal (1998) used three doses of gamma rays, fast neutrons and two different chemical mutagens on four chickpea genotypes with treatments included. It was concluded that chemical mutagens were more efficient than physical in inducing viable and more number of mutations. Among the chemicals, *N*-nitroso-*N*-methyl urea (NMU) was the most effective, while among the physical mutagens gamma rays were the most effective (Kharkwal, 1999). Gamma rays are the most often used mutagen to change gene(s) in released chickpea mutants for commercial production. Effective dose ranged from 100 to 600 Gy, and 200 Gy was the most frequently used dose to obtain beneficial mutations (Table 18.4). Micke and Donini (1993) recommended 10–16 krad x-rays for chickpea. Similarly, the 50% growth reduction (LD_{50}) of primary shoots and useful dose range for chickpea were given by 18–26 and 12–18 gamma rays,

Table 18.5. Mutant chickpeas released for commercial production.

Mutant variety	Parent variety	Mutagen(s)	Main characters induced	Released	
				Country	Year
Hyprosola (M-699)	Faridpur-1	Gamma rays, 200 Gy	Earliness	Bangladesh	1981
Binasola-3 (L-84)	G-97 (Binasolo-2)	Gamma rays, 200 Gy	Earliness	Bangladesh	2005
Plovdiv-8[a]	–	Gamma rays	Yield and ascochyta blight resistance	Bulgaria	1979
Line 3	NCL #055	Gamma rays, 50 Gy + EMS 0.025%	Yield	Egypt	1992
Kiran (RSG-2)	RSG10	Fast neutrons, 4.5×10^{12}	Erectoid type	India	1984
Ajay (Pusa 408)	G-130	Gamma rays, 600 Gy	Yield	India	1985
Atul (Pusa 413)	G-130	Gamma rays, 600 Gy	Yield	India	1985
Girnar (Pusa 417)	BG-203	Gamma rays, 600 Gy	Yield	India	1985
RS11[b]	RSG10	–	Suitable for irrigated conditions	India	–
C-72	6153	Gamma rays, 150 Gy	Ascochyta blight resistance	Pakistan	1983
CM-88	C-727	Gamma rays, 100 Gy	Ascochyta blight and fusarium wilt	Pakistan	1994
CM-98 (CM 31-1/85)	6153	Gamma rays, 200 Gy	Yield	Pakistan	1998
NIFA-88 (CM-1918)	6153	Gamma rays, 100 Gy	Ascochyta blight resistance	Pakistan	1990
NIFA-95	Line 6151	Gamma rays, 200 Gy	Ascochyta blight resistance	Pakistan	1995
TAEK-SAGEL	–	Gamma rays, 100/150 Gy	Yield	Turkey	2005

[a]Singh (1987); Sharma and Varshney (1999).
[b]Singh (2005).

Table 18.6. Nodulation mutants in chickpea.

Mutant variety	Gene(s)	Parent variety	Mutagen(s)	Main characters	Reference
Induced					
PM 233	*rn*$_1$[a]	P 502 (ICC 640)	Gamma rays	Nod-, mr	Davis *et al.* (1985, 1986)
PM 665	*rn*$_2$	P 502 (ICC 640)	Gamma rays	Ref32Nod-, mr	Davis (1988, 1991)
PM 679	*rn*$_3$	P 502 (ICC 640)	Gamma rays	Nod-, mr, linked to simple leaves	
PM 405	*rn*$_4$	P 502 (ICC 640)	Gamma rays	Nod+ fix-, mr, small white nodule	
PM 796	*rn*$_5$	P 502 (ICC 640)	Gamma rays	Nod+ fix-, mr, green nodule	
PM 233B	*rn*$_1$	ICC 640	Gamma rays	Nod-, blockage of bacteria infection	Mathew and Davis (1990)
PM 638	*rn*$_7$	ICC 640	Gamma rays	Nod-, mr	Paruvangada and Davis (1999)
Spontaneous					
ICC 435M	*rn*$_6$	ICC 435	Nod-, mr, non-allelic to PM 233		Rupela and Sundarshana (1986); Rupela (1992); Singh *et al.* (1992a,b)
ICC 4918M		ICC 4918	Nod-		
ICC 4993M		ICC 4993	Nod-		Singh and Rupela (1991)
ICC 5003M		ICC 5003	Nod-		
Rabat NN	*rn*$_8$	Rabat	Nod-, mr		Singh and Rupela (1998)
Annigeri NN		Annigeri	Nod-,		Singh and Rupela (1998)
P319-1NN		Annigeri	Nod-, mr, allelic to Annigeri NN		
2100257, 2400106		Aydin-92	Nod++		Canci *et al.* (2004b)

[a]rn: root nodule; Nod-: no nodules; mr: monogenic recessive; Nod+: nodulation; Nod++: hypernodulation; fix-: ineffective nodulation; NN: non-nodulating.

respectively (Conger *et al.*, 1977). LD_{50} and a dose close to LD_{50} are considered optimum by many researchers. This will be producing maximum frequency of mutations with minimum hazard. An optimum dose can be determined with a preliminary treatment. Overdoses of mutagens are lethal to many plants, while underdose will produce less mutation frequency. Some factors that influence the effects of mutagens are biological (nuclear volume, chromosome volume and DNA content of variety), environmental (oxygen, water status and temperature) and chemical (Sigurbjönsson, 1983).

Parts of chickpea to be treated

Although air-dried seeds are the most frequently used part of chickpea, pollen grains may be used directly, and vegetative organs for *in vitro* mutagenesis may be used as well. However, pollen grains are used infrequently because emasculation and pollination are very difficult, time-consuming and pollen viability is short. Irradiation of pollen grains can be beneficial to overcome prefertilization and post-fertilization problems especially in intraspecific hybridizations. Treatments of interspecific and intraspecific hybrids can produce translocations. Mutations do not induce chimeras when pollens are irradiated.

Advantages of mutation breeding

Mutation breeding not only creates variability in a crop species, but also shortens time taken for the development of cultivars through induced mutation compared with that by hybridizations. The average time elapsed from the beginning of mutation treatment to the release of the mutant cultivars was approximately 9 years, while this time was 18 years for cultivar arising from crossing programmes (Brock, 1977). Moreover, mutations are induced in both qualitative and quantitative characters in a short time altering new alleles of known and previously unknown genes, and modify linkage (Konzak *et al.*, 1977). Further, desirable variability could be brought about as variability in family *Leguminosae* (*Fabaceae*) through induced mutations (Toker and Cagirgan, 2004), whereas it is limited through hybridization. It is expected to be induced in accordance with Vavilov's law of parallel variation. It can create all kind of variation in a family. Gaur and Gour (2002) reported mutant with nine flowers per node, while chickpea in general possesses one or two flowers per node. Furthermore, mutations both induced and spontaneous are one of the most important sources of evolution and will be helpful in studying genetics of novel characters created (Table 18.7). Special problems of chickpea have been solved through mutation breeding. For example, M16119 is both cold-tolerant and cold-resistant to ascochyta blight (Haq and Singh, 1994). Some chickpea mutants (CM 88 and CM 98) could result in gains of $9.6 million in Pakistan (Ahloowalia *et al.*, 2004).

Disadvantages of mutation breeding

The frequency of desirable mutations is very low, i.e. ~0.1%. Success of mutation breeding depends on the methodology, effective screening techniques, population grown in M_1 and successive generations. Larger the population in M_1, more is the success in selection of desirable mutants. Breeders have

Table 18.7. Induced and spontaneous mutant chickpeas in literature.

Mutants	Reference
Bushy mutant	Singh and Dahiya (1974)
A spontaneous polycarpellary mutant	Pundir and van der Maesen (1981)
BGM-401	Kharkwal (1983)
Lobed vexillium mutant	Rao and Pundir (1983)
Spontaneous polycarpellary mutant (HMS 6-1, E 100Y, H 79-100 [M])	Dahiya *et al.* (1984)
1D 37, 1D 331, 1D 450, 1D 451, 1D 687, 1D 435, 1D 478-1, 1D 700, 1D 44, 2D 86, 2D 681, 2D 510-1, 2D 510-2, 2D 693, 2D 707	Bhatnagar *et al.* (1985)
Spontaneous (L 550-PM)	Gowda and Singh (1986)
Annigeri-TP	Rupela and Mengesha (1988)
CG 276-2 MGG	Gomez (1988)
Twin flower peduncles and polycarpy, ICC 17101 and Glabrous (hairless) induced mutant	Pundir *et al.* (1988); Pundir and Reddy (1989, 1997)
Spontaneous mutant of Amethyst (fasciation)	Knights (1989, 1993)
Ref68 giant pod mutant	Kumar *et al.* (1989)
WM 5-3, WM 297-11, WM 297-13, WM 297-24, WM 297-19	Kshirsagar and Patil (1989)
Induced gigas variants	Davis *et al.* (1990); Fagerberg *et al.* (1990)
Plant type	Sandhu *et al.* (1990)
Gigas	Davis *et al.* (1990); Fagerberg *et al.* (1990)
CM 1, RC 32, CM 72, C 141, CM 663, CM 687, E 1289, CM 1913	Hassan and Khan (1991)
Ref139 determinate growth habit	van Rheenen *et al.* (1994); van Rheenen (1996)
Different induced 18 desi and 5 kabuli mutants	Gumber *et al.* (1995)
FLIP 84-92C-1, FLIP 90-73C-1, FLIP 90-73C-2, ILC 5901-1, ILC 5901-2	Omar and Singh (1995)
H 82-2 (M)	Lather *et al.* (1997)
Nine flower-producing spontaneous mutant	Gaur and Gour (2002, 2003)
CM 1762/99, CM 1965/99, CM 3268/99, CM 3339/99, CM 3358/99, CM 3513/99	Atta *et al.* (2003)
BGM 403, BGM 405, BGM 408, BGM 413, BGM 415, BGM 416, BGM 417, BGM 418, BGM 419, BGM 421, BMG 424, BMG 425, BMG 426, BMG 427, BMG 428, BMG 429, BMG 430, BGM 547, CM 68, CM 359, CM 438, desi mutants, FLIP 94-501, FLIP 94-502, G-293, G-302, kabuli mutants, mutant no. 16119	Maluszyski (2003)
Seven different induced mutants (dwarf, compact, prostrate, gigas, white flower, non-flowering and sterile)	Khan *et al.* (2004)
Cold-tolerant and ascochyta blight-resistant, a new simple leaf type, hypernodulating mutations and root mutants	Cagirgan and Toker (2004); Canci *et al.* (2004a); Toker and Cagirgan (2004)
CM 2000	www.niab.org.pk

to screen large population for desirable mutations. The screening procedures in large populations will take considerable time, labour and other resources. Some mutations involve pleiotropic effects due to linked gene(s), other mutations, chromosomal aberrations and deletions. These mutants often have to be backcrossed to parents or adapted varieties.

Mutant chickpeas

The number of mutant varieties officially released and recorded in FAO/IAEA Mutant Varieties Database was 2252 before the end of 2000 (Maluszynski *et al.*, 2000; Ahloowalia *et al.*, 2004; Kharkwal *et al.*, 2005). From these mutant varieties, 265 mutant grain legume cultivars have been released in 32 countries (Bhatia *et al.*, 2001a). Gamma rays were the most frequently used technique to alter genes accounting for 64% of the radiation-induced mutant varieties (Maluszynski *et al.*, 2000; Ahloowalia *et al.*, 2004). Sanilac, a variety of common bean (*Phaseolus vulgaris* L.), was the first released legume mutant in Michigan in 1956 (van Harten, 1998). In the case of chickpea, mutation induction works were probably initiated in Indian subcontinent (see in Sharma and Varshney, 1999). The works have been encouraged by FAO/IAEA projects in order to create useful variation (Khan and Shakoor, 1977; Shaikh, 1977; Sharma and Kharkwal, 1982, 1983; Haq *et al.*, 1983, 1986; Shaikh *et al.*, 1983a,b). These works created many unique mutations for use in plant-breeding programmes. Approximately 15 mutants in chickpea were released for commercial production as seen in Table 18.5 (Singh, 1987; Sharma and Varshney, 1999; Bhatia *et al.*, 2001a,b,c; Maluszynski, 2003; Singh, 2005). A Turkish chickpea cultivar irradiated with gamma rays will be released very soon (D. Bilhan, Turkey, 2006, personal communication). Mutation frequency of some spontaneous non-nodulating mutants at the rate of 120 and 490/million plants was observed (Rupela, 1992). Spontaneous and induced mutant chickpeas are presented in Table 18.7 (Bhatia *et al.*, 2001a,b,c). More than 150 chickpea mutants have been mentioned in literature till date.

In conclusion, genetic variation in available germplasm collections has been widely used to combat biotic and abiotic stresses. Apart from few stress-resistant characteristics (Singh, 1997), desirable sources of resistance have not been found in the cultivated collections with the exception of wilds (Singh *et al.*, 1998; Croser *et al.*, 2003; Toker, 2005). Therefore, mutagenesis is the only efficient tool to create a new desirable genetic variability in chickpea.

Population improvement

The principle procedures followed in the improvement of self-pollinated crops are the modified conventional breeding methods, such as pedigree, bulk and backcross breeding. Such procedures, though significant and productive in their own right, impose restriction on the chance of better recombination because of larger linkage blocks associated with rapid progress towards homozygosity and low genetic variability (Clegg *et al.*, 1972). Further, negative correlations among yield components and high genotype × environment interactions prevent full exploitation of genetic variability for character like yield. If these propositions are to be accepted, we must re-evaluate our breeding procedures.

Since chickpea is predominantly a self-pollinated crop, the presence of linkage blocks and inverse relations among the correlated characters are more common. Under such circumstances, the biparental mating in an appropriate segregating population is effective in breaking larger linkage blocks and provides more chances of recombination than the selfing series (Gill, 1987). It is a useful system of mating for rapid generation of variability and may appropriately be applied where lack of desired variation is the immediate bottleneck in the breeding programmes.

The close linkages leading to unfavourable associations need to be broken in order to make selection more effective and useful. In self-pollinated crops like chickpea, the conventional methods like pedigree or bulk, which handle the segregating populations, do not provide any opportunity for the continued reshuffling of the genes. Hence, any unfavourable associations observed in an early segregating generations like in F_2 are liable to persist through the filial generations. However, breeders can alter such associations by resorting to approaches like biparental mating in the segregating populations. Such approaches have been tried and positive results have been reported in some crops like wheat and safflower (Nanda *et al.*, 1990; Parameshwarappa *et al.*, 1997).

For implementing successful population improvement programme, the basic prerequisite is the case with which large number of crosses can be attempted in a segregating population. As chickpea is a self-pollinated crop and is difficult to cross, the implementation of population improvement programme appears to be difficult. Therefore, for successful implementation of such programmes, various mechanisms that facilitate large-scale hybridization should be examined.

Artificial crossing without emasculation is advantageous and also enables the breeder to increase the number of crosses per day and obtain more hybrid seeds with the same efforts. Artificial crossing without emasculation can be attempted because the stigma becomes receptive before anthers dehisce (Retig, 1971; Dahiya, 1974; Singh, 1987).

Arora and Singh (1990) reported only ~58.2% of success whereas Singh (2001) reported 86% of success. However, Pundir and Reddy (1984) indicated that these techniques will be advantageous if there are morphological differences between the parents, which will enable roguing of selfed plants. This approach will be more useful in population improvement programmes/biparental population (BIP) matings, where large number of crosses need to be attempted in segregating populations.

Modifications in the floral structure that encourage large-scale successful hybridization are also important in this context. The occurrence of an 'open flower' mutant (Dahiya *et al.*, 1984a) reduces the time needed for manual cross-pollination. Pundir and Reddy (1998) reported an 'open flower' natural mutant in a good agronomic background.

Male sterility is another mechanism, which is useful in attempting large-scale artificial hybridization. This has been successfully used in different crops like soybean, cowpea and pigeon pea. Male sterility has also been reported a number of times in chickpea (Singh and Shyam, 1958; Chaudhary *et al.*, 1970;

Sethi, 1979). However, usable male sterility has not been reported. Research is required to develop and identify stable male sterility in chickpea. In the absence of such mechanism, efforts have been made to find out the possibility of chemical hybridizing agents (CHAs). Mathur and Lal (1999) found fluoro oxanyl to be quite useful in inducing male sterility in chickpea. They observed 70–80% sterility in both kabuli and desi genotypes although the latter were more sensitive. Chakraborty and Devakumar (2006) working on the CHA in chickpea screened a number of CHAs at different concentrations to find out the effective chemical. Pryidone derivatives were found to be most effective.

Self-incompatibility has been reported by Dahiya *et al.* (1990). Lather and Dahiya (1992) reported frequent occurrence of self-incompatibility in the F_2 generation of biparental mating of a cross (H82-5 × E100YM) × Bheema. They postulated that seeds failed to develop in these plants as a result of prezygotic phenomenon of self-incompatibility where genes for self-incompatibility might have either restricted pollen germination or prevented the growth of the pollen tube within the incompatible pistil. This mechanism needs some more alteration before it is put to use in chickpea.

Another feature required to effect cross-pollination, and thus achieve enhanced utilization of heterosis in chickpea, is the transfer of pollen grains to stigma of the desired genotype. It is possible that an increased bee population around chickpea plants can do this job; however, this needs to be investigated. Assuming that the male sterility and bees as pollen vectors work, the research gains in chickpea could possibly be expedited following recurrent selection or the production of hybrid seeds.

Kampali *et al.* (2002a,b) reported that the BIP had better mean performance than the F_3 self for all the characters under study. Sufficiently high genetic variation was maintained in the BIP population for most of the characters except for secondary branches. BIP also exhibited improved estimates of heritability and genetic advances.

The same author observed a higher frequency of transgressive segregants in the segregating populations of ICCV-10 × BG-256 advanced by following selection of biparental mating compared with random bulk and selective bulk methods. On the contrary, increased variability was not realized in the cross involving genotypes Bheema and ICCV-10 (Nijagun, 2005).

There are no other reports on the effectiveness of biparental approach in chickpea. However, the aforementioned reports indicate that biparental approach if applied in a specific cross combination will give better results.

Summary

Considering the conventional approaches and breeding methods employed by different chickpea breeders at various national and international organizations, it is evident that the population improvement for higher productivity, multiple resistance and desirable plant type are still a myth. Although various options for population improvement are available, a breakthrough in plants like wheat and rice is still way ahead. Therefore, to achieve the real breakthrough for desirable traits it is suggested that diverse sources through multiple hybridization, to incorporate multiple resistance, are required.

References

Ahloowalia, B.S., Maluszynski, M. and Nichterlein, K. (2004) Global impact of mutation-derived varieties. *Euphytica* 135, 187–204.

Allard, R.W. (1960) *Principles of Plant Breeding*. Wiley, New York.

Argikar, G.P. (1970) Gram. In: Kachroo, P. (ed.) *Pulse Crops of India*. ICAR, New Delhi, pp. 54–135.

Arora, P.P. and Singh, I.S. (1990) Crossing without emasculation in chickpea. *International Chickpea Newsletter* 23, 11.

Atta, B.M., Haq, M.A., Shah, T.M., Sadiq, M., Hassan, M. and Syed, H. (2003) Induced flower color mutations in chickpea. *International Chickpea and Pigeonpea Newsletter* 10, 6–7.

Auckland, A.K. and van der Maesen, L.J.G. (1980) Chickpea. In: Fehr, W.R. and Hadely, H.H. (eds) *Hybridization of Crop Plants*. American Society of Agronomy, Madison, Wisconsin, pp. 249–259.

Bahl, P.N. (1981) Yield potential of improved cultivar and plant type concept in chickpea. In: *Proceedings of the FAI Group Discussion on Increasing Pulse and Oilseed Production in India*. New Delhi, pp. 32–40.

Bahl, P.N. and Gowda, C.L.L. (1975) Pod setting in crosses of Bengal gram. *Indian Journal of Genetics* 35, 13–16.

Bahl, P.N., Singh, S.P., Hayat, R., Raju, D.B. and Jain, H.K. (1979) Breeding for improved plant architecture and high protein yields. *FAO/IAEA International Symposium on Seed Improvement in Cereals and Grain Legumes*. Vienna, Austria, pp. 297–307.

Bahl, P.N., Salimath, P.M., Malik, B.A., Dahiya, B.S., Deshmukh, R.B. and Rangaswamy, P. (1990) New approaches in chickpea breeding and prospects in the nineties. In: *Chickpea in the Nineties: Proceedings of the 2nd International Workshop on Chickpea Improvement*. 4–8 December 1989. ICRISAT, Hyderabad, India, pp. 225–232.

Bejiga, G. and Tessema, T. (1981) Identification on the optimum time of emasculation and pollination to increase percentage of hybrid seed set in chickpea. *Journal of Agricultural Sciences* 3, 129–134.

Bhatia, C.R., Maluszynski, M., Nichterlein, K. and van Zanten, L. (2001a) Grain Legume cultivars derived from induced mutations, and mutation affecting nodulation. *Mutation Breeding* 13, 1–44.

Bhatia, C.R., Nichterlein, K. and Maluszynski, M. (2001b) Mutation affecting nodulation in grain legumes and their potential in sustainable cropping systems. In: Maluszynski, M. and Kasha, K.J. (eds) *Mutations. In Vitro and Molecular Techniques for Enviornmentally Sustainable Crop Improvement*. Kluwer Academic, Dordrecht, The Netherlands, pp. 201–226.

Bhatia, C.R., Nichterlein, K. and Maluszynski, M. (2001c) Mutation affecting nodulation in grain legumes and their potential in sustainable cropping systems. *Euphytica* 120, 415–432.

Bhatnagar, C.P., Handa, D.K. and Misra, A. (1985) Prospects of utilizing mutation breeding in inducing nematode resistance in chickpea. *International Chickpea Newsletter* 12, 5–7.

Bisen, M.S., Singh, S.P. and Rao, S.K. (1985) Effectiveness of selection methods in chickpea *Cicer arietinum* L. under different environments. *Theoretical and Applied Genetics* 70(6), 661–666.

Brock, R.D. (1977) Prospects and perspectives in mutation breeding. In: Muhammed, A., Aksel, R. and Von Borstel, R.C. (eds) *Genetic Diversity in Plants*. Plenum Press, New York, pp. 117–132.

Byth, D.E., Green, J.M. and Hawtin, G.C. (1980) ICRISAT/ICARDA chickpea breeding strategies. In: *Proceedings of the International Workshop on Chickpea Improvement*. 28 February–2 March1979. ICRISAT, Hyderabad, India, pp. 11–27.

Cagirgan, M.I. and Toker, C. (2004) Selection of chickpea mutants for cold tolerance and ascochyta blight resistance. *International Chickpea and Pigeonpea Newsletter* 11, 14–16.

Canci, H., Cagirgan, M.I. and Toker, C. (2004a) Genotypic variations for root and shoot growth at seedling stage in chickpea. *International Chickpea and Pigeonpea Newsletter* 11, 11–12.

Canci, H., Toker, C. and Cagirgan, M.I. (2004b) Preliminary screening for nodulation in chickpea mutants. *International Chickpea and Pigeonpea Newsletter* 11, 13–14.

Chakraborty, K. and Devakumar, C. (2006) Chemical hybridizing agents for chickpea (*Cicer arietinum* L.): leads from QSAR analysis of ethyl ozanilates and pyridones. *Journal of Agricultural and Food Chemistry* 54(5), 1868–1873.

Chaudhary, M.S., Chandra, S. and Tomar, Y.S. (1970) Genetical studies on a variant form in Bengal gram (*Cicer arietinum*). *Indian Journal of Heredity* 2, 133–138.

Clegg, M.T., Allard, R.W. and Kahlar, A.L. (1972) Is the gene unit of selection? Evidence from two experimental plant populations. *Proceedings of the National Academy of Sciences of the United States of America* 19, 2474–2478.

Conger, B.V., Konzak, C.F. and Nilan, R.A. (1977) Radiation sensitivity and modifying factors. *Manual on Mutation Breeding*, 2nd edn. Technical reports series No. 119. International Atomic Energy Agency, Vienna, Austria, pp. 40–50.

Croser, J.S., Ahmad, F., Clarke, H.J. and Siddique, K.H.M. (2003) Utilization of wild *Cicer* in chickpea improvement – progress, constraints, and prospects. *Australian Journal of Agricultural Research* 54, 429–444.

Dahiya, B.S. (1974) Crossing without emasculation in chickpea. *Indian Journal of Genetics* 34, 206–207.

Dahiya, B.S., Lather, U.S., Solanki, I.S. and Kumar, R. (1984) Useful spontaneous mutants in chickpea. *International Chickpea Newsletter* 11, 4–7.

Dahiya, B.S., Lather, V.S., Solanki, I.S. and Kumar, R. (1984a) Useful spontaneous mutants in chickpea. *International Chickpea Newsletter* 11, 4–7.

Dahiya, B.S., Waldia, R.S., Kaushik, L.S. and Solanki, I.S. (1984b) Early generation yield testing versus visual selection in chickpea (*Cicer arietinum* L.). *Theoretical and Applied Genetics* 68(6), 525–529.

Dahiya, B.S., Lather, V.S. and Kumar, R. (1990) A brachytic mutant of chickpea. *International Chickpea Newsletter* 22, 42.

Davis, T.M. (1988) Two genes that confer ineffective nodulation in chickpea (*Cicer arietinum* L.). *Journal of Heredity* 79, 476–478.

Davis, T.M. (1991) Linkage relationships of genes for leaf morphology, flower color and root nodulation in chickpea. *Euphytica* 54, 117–124.

Davis, T.M., Foster, K.W. and Philips, D.A. (1985) Nonnodulation mutants in chickpea. *Crop Science* 25, 345–348.

Davis, T.M., Foster, K.W. and Philips, D.A. (1986) Inheritance and expression of three genes controlling root nodule formation in chickpea. *Crop Science* 26, 719–723.

Davis, T.M., Matthews, L.J. and Fagerberg, W.R. (1990) Comparison of tetraploid and single gene-induced gigas variants in chickpea (*Cicer arietinum* L.). I. Origin and genetic-characterization. *American Journal of Botany* 77, 295–299.

de Vries, H. (1901) *Die Mutationstheorie*. Veit & Co., Leipzig, Germany.

Dua, R.P., Chaturvedi, S.K. and Shiv, S. (2001) Reference varieties of chickpea for IPR regime. Indian Institute of Pulses Research, Kanpur duration genotypes in chickpea (*Cicer arietinum* L.). *Legume Research* 21(2), 121–124.

Fagerberg, W.R., Davis, T.M. and Matthews, L.J. (1990) Comparison of tetraploid and single gene-induced gigas variants in chickpea (*Cicer arietinum* L.). II. Morphological and anatomical leaf characters. *American Journal of Botany* 77, 300–304.

Food and Agriuclture Organization of the United Nations (1994) *Production Yearbook*. FAO, Rome, Italy.

Gaur, P.M. and Gour, V.K. (2002) A gene producing one to nine flowering node in chickpea. *Euphytica* 128, 231–235.

Gaur, P.M. and Gour, V.K. (2003) Broad-few-leaflets and outwardly curved wings: two mutants of chickpea. *Plant Breeding* 122, 192–194.

Gill, K.S. (1987) Recent concept in breeding self pollinated crops. In: Gill, K.S., Khehra, A.S. and Verma, M.M. (eds) *Proceedings of the 1st Symposium on Crop Improvement*, Vol. 1. Crop Improvement Society of India, Ludhiana, India, pp. 83–129.

Gomez, G.R.M. (1988) Spontaneous mutation in line CG 276-2, a kabuli-type chickpea. *International Chickpea Newsletter* 19, 3–4.

Gowda, C.L.L. (1981) National crossing in chickpea. *International Chickpea Newsletter* 5, 6.

Gowda, C.L.L. and Singh, O. (1986) Inheritance of an upright mutant characteristic in chickpea. *International Chickpea Newsletter* 15, 8–9.

Gumber, R.K., Singh, S. and Singh, K. (1995) Frequency and spectrum of mutations induced by gamma rays in desi and kabuli chickpea. *International Chickpea and Pigeonpea Newsletter* 2, 8–9.

Haq, M.A. and Singh, K.B. (1994) Induction of cold tolerance in kabuli chickpea (*Cicer arietinum* L.) through induced mutations. *Mutation Breeding Newsletter* 41, 6–7.

Haq, M.A., Sadiq, M. and Hassan, M. (1983) Use of induced mutations for the induction of resistance against *Ascochyta* blight in chickpea (*Cicer arietinum* L.). In: *Induced Mutations for Improvement of Grain Legumes Production III, IAEA-TECDOC-299*. International Atomic Energy Agency, Vienna, Austria, pp. 55–64.

Haq, M.A., Sadiq, M., Hassan, M., Sadiq, M.S. and Altaf, N. (1986) Development of blight resistant varieties of chickpea. In: Siddiqui, K.A. and Faruqui, A.F. (eds) *New Genetical Approaches to Crop Improvement*. Atomic Energy Agricultural Research Center, Tando Jam, Pakistan, pp. 629–635.

Hassan, S. and Khan, I. (1991) Improvement in chickpea production through induced mutations. *International Chickpea Newsletter* 25, 12–13.

Jain, H.K. (1975) Breeding for yield and other attributes in grain legumes. *Indian Journal of Genetics* 35, 169–187.

Kampali, N., Salimath, P.M. and Kajjidoni, S.T. (2002a) Genetic variability created through biparental mating in chickpea (*Cicer arietinum* L.). *Indian Journal of Genetics* 62(2), 128–130.

Kampali, N., Salimath, P.M. and Kajjidoni, S.T. (2002b) Transgressive segregants in populations derived from different mating and selection schemes in chickpea. *Crop Research* 23(1), 95–98.

Khan, M.A. and Shakoor, A. (1977) Grain legumes in Pakistan. In: *Induced Mutations for the Improvement of Grain Legumes in South East Asia (1975), IAEA-203*. International Atomic Energy Agency, Vienna, Austria, pp. 21–28.

Khan, S., Wani, M.R., Bhat, M. and Parveen, K. (2004) Induction of morphological mutants in chickpea. *International Chickpea and Pigeonpea Newsletter* 11, 6–7.

Kharkwal, M.C. (1983) Mutations breeding in chickpea improvement. *International Chickpea Newsletter* 9, 4–5.

Kharkwal, M.C. (1998) Induced mutations in chickpea (*Cicer arietinum* L.) I. Comparative mutagenic effectiveness and efficiency of physical and chemical mutagens. *Indian Journal of Genetics* 58, 159–167.

Kharkwal, M.C. (1999) Induced mutations in chickpea (*Cicer arietinum* L.) III. Frequency and spectrum of viable mutations. *Indian Journal of Genetics* 59, 451–464.

Kharkwal, M.C., Haq, M.A., Gopalakrishna, T. and Pawar, S.E. (2005) Mutation breeding for grain legume improvement. In: Kharkwal, M.C. (ed.) *4th International Food Legumes Research Conference, Food Legumes for Nutritional Security and Sustainable Agriculture*. New Delhi, India, p. 9.

Khosh-Khui, M. and Niknejad, M. (1972) Plant height and width inheritance and their correlation with some of the yield components in chickpea. *Journal of Agricultural Sciences* 78, 37–38.

Knights, E.J. (1989) Fasciation in chickpea. *International Chickpea Newsletter* 20, 2.

Knights, E.J. (1993) Fasciation in chickpea: genetics and evolution. *Euphytica* 69, 163–166.

Konzak, C.F., Nilan, R.A. and Kleinhofs, A. (1977) Artificial mutagenesis as a aid in overcoming genetic vulnerability of crop plants. In: Muhammed, A., Aksel, R. and von Borstel, R.C. (eds) *Genetic Diversity in Plants*. Plenum Press, New York, pp. 163–177.

Kshirsagar, A.R. and Patil, A.J. (1989) Agronomic potential of chemically induced mutation in chickpea. *International Chickpea Newsletter* 21, 26.

Kumar, J. and Bahl, P.N. (1992) Direct and indirect selection for yield in chickpea. *Euphytica* 60, 197–199.

Kumar, J., Bahl, P.N. and Raju, D.B. (1989) Giant pod mutant and its inheritance in chickpea. *International Chickpea Newsletter* 21, 6–7.

Lal, S., Singh, S.N., Lal, S.B. and Tripathi, B.D.H. (1973) Improvement in gram through pedigree method of breeding. In: *Proceedings of the 10th Workshop on Rabi Pulses* AICRP (ICAR), Bose Institute. 10–14 September. ICAR, New Delhi, India, pp. 70–80.

Lather, V.S. and Dahiya, B.S. (1992) Self-incompatibility in chickpea. *International Chickpea Newsletter* 26, 6.

Lather, V.S., Waldia, R.S. and Mehla, I.S. (1997) Early vigor spontaneous mutant in chickpea. *International Chickpea and Pigeonpea Newsletter* 4, 11–12.

Malhotra, R.S. and Singh, K.B. (1986) Natural cross-pollination in chickpea. *International Chickpea Newsletter* 14, 4–5.

Malti and Shivanna, K.R. (1983) Pollen pistil interaction in chickpea. *International Chickpea Newsletter* 9, 10–11.

Maluszynski, M. (2003) Index Issue No. 21–44. *Mutation Breeding Newsletter* 46, 1–80.

Maluszynski, M., Nichterlein, K., van Zanten, L. and Ahloowalia, B.S. (2000) Officially released mutant varieties – the FAO/IAEA Database. *Mutation Breeding Review* 12, 1–84.

Mathew, L.J. and Davis, T.M. (1990) Anatomical comparison of wild type and nonnodulating mutant chickpea (*Cicer arietinum* L.). *Canadian Journal of Botany* 68, 1201–1207.

Mathur, D.S. and Lal, S.K. (1999) Chemical induction of male sterility in chickpea. *Indian Journal of Genetics* 59(3), 379–380.

Micke, A. (1988a) Genetic improvement of food legumes in developing countries by mutation induction. In: Summerfield, R.J. (ed.) *World Crops: Cool Season Food Legumes*. Kluwer Academic, Dordrecht, The Netherlands, pp. 1031–1047.

Micke, A. (1988b) Improvement of grain legumes production using induced mutation. An overview. In: *Proceedings of a Workshop on the Improvement of Grain Legume Production Using Induced Mutations*, FAO/IAEA Division, Pullman, Washington, DC. International Atomic Energy Agency, Vienna, Austria, pp. 1–51.

Micke, A. and Donini, B. (1993) Induced mutations. In: Hayward, M.D., Bosemark, N.O. and Romagosa, I. (eds) *Plant Breeding: Principles and Prospects*. Chapman & Hall, London, pp. 53–62.

Muller, H.J. (1927) Artificial transmutation of the gene. *Science* LXVI, 84–87.

Nanda, G.S., Singh, G. and Gill, K.S. (1990) Efficiency of intermating in F_2 generation of an intervarietal cross in bread wheat. *Indian Journal of Genetics and Plant Breeding* 50(4), 364–368.

Nijagun, H.G. (2005) Studies on genetic enhancement of productivity and wilt resistance in chickpea. PhD thesis, University of Agricultural Sciences, Dharwad, Karnataka, India.

Niknejad, M. and Khosh-Khui, M. (1972) *Indian Journal of Agricultural Sciences* 42(4), 273–274.

Omar, M. and Singh, K.B. (1995) Development of early mutants with resistance to ascochyta blight or leaf miner. *International Chickpea and Pigeonpea Newsletter* 2, 10–11.

Parameshwarappa, K.G., Kulkarni, M.S., Gulganji, G.G., Kubsad, V.S. and Mallapur, C.P. (1997) An assessment of genetic variability created through biparental mating in safflower (Carthamus *tinctorius*). *5th International Safflower Conference*. 2–7 June 1997. Bari, Italy, pp. 238–239.

Paruvangada, V.G. and Davis, T.M. (1999) A dominant, host plant mutation conferring ineffective nodulation in the chickpea–*Rhizobium* symbiosis. *Journal of Heredity* 90, 297–299.

Pundir, R.P.S. and Reddy, G.U. (1998) Two new traits – open flower and small leaf in chickpea (*Cicer arietinum* L.). *Euphytica* 102, 357–361.

Pundir, R.P.S. and Reddy, G.V. (1989) Induction, genetics and possible use of glabrousness in chickpea. *Euphytica* 42, 141–144.

Pundir, R.P.S. and Reddy, G.V. (1997) Occurrence of new characteristics in chickpea – purplish stem with white flowers. *International Chickpea and Pigeonpea Newsletter* 4, 8–9.

Pundir, R.P.S. and Reddy, K.N. (1984) Comparison of cross-pollination methods in chickpea. *International Chickpea Newsletter* 11, 9–11.

Pundir, R.P.S. and van der Maesen, L.J.G. (1981) A spontaneous polycarpellary mutant in chickpea (*Cicer arietinum* L.). *International Chickpea Newsletter* 4, 7–8.

Pundir, R.P.S., Mengesha, M.H. and Reddy, K.N. (1988) Occurrence and genetics of a natural mutant of chickpea having twin flower peduncles and polycarpy. *Journal of Heredity* 79, 479–481.

Rahman, M.A. and Bahl, P.N. (1985) Comparison of single seed descent, mass selection and random bulk methods in chickpea. *Indian Journal of Genetics* 45(2), 186–193.

Rao, N.K. and Pundir, R.P.S. (1983) Inheritance and linkage relationships of new lobed vexillium mutant chickpea. *Journal of Heredity* 74, 300–300.

Retig, G. (1971) Hybridization method in chickpea (*Cicer arietinum* L.). II. Crossing without emasculation. *Israel Journal of Agricultural Research* 21(3), 113.

Rupela, O.P. (1992) Natural occurrence and salient characters of non-nodulating chickpea plants. *Crop Science* 32, 349–352.

Rupela, O.P. and Mengesha, M.H. (1988) An induced twin-podded mutant from a wilt-resistant Annigeri chickpea cultivar. *International Chickpea Newsletter* 18, 3–4.

Rupela, O.P. and Sundarshana, S.R. (1986) Identification of a nonnodulating spontaneous mutant in chickpea. *International Chickpea Newsletter* 15, 13–14.

Salimath, P.M. and Bahl, P.N. (1985b). Early generation selection in chickpea. III Predicted and realized gains. *Experimental Genetics* 1(2), 59–62.

Sandhu, J.S., Brar, H.S. and Verma, M.M. (1990) Inheritance of plant type mutant and its utilization in chickpea. *Euphytica* 48, 111–112.

Saxena, H.P. and Raina, A.K. (1970) A bruchid resistant strain of Bengal gram. *Current Science* 39, 189–190.

Sethi, S.C. (1979) Male sterility. *International Chickpea Newsletter* 1, 4.

Shaikh, M.A.Q. (1977) Grain legumes in Bangladesh. In: *Induced Mutations for the Improvement of Grain Legumes in South East Asia (1975), IAEA-203*. International Atomic Energy Agency, Vienna, Austria, pp. 61–70.

Shaikh, M.A.Q., Bhuiya, A.D., Saha, C.S. and Rahman, L. (1983a) Performance and agronomic potential of radiation-induced and collected genotypes of chickpea (*Cicer arietinum* L.). In: *Induced Mutations for Improvement of Grain Legumes Production III, IAEA-TECDOC-299*. International Atomic Energy Agency, Vienna, Austria, pp. 47–54.

Shaikh, M.A.Q., Khanum, S., Begum, S., Ahmed, Z.U., Majid, M.A. and Zaman, K.M.S. (1983b) Effects of chemical mutagens on four species of grain legumes. In: *Induced Mutations for Improvement of Grain Legumes Production III, IAEA-TECDOC-299*. International Atomic Energy Agency, Vienna, Austria, pp. 77–85.

Sharma, B. and Kharkwal, M.C. (1982) Induced mutations in grain legumes. In: *Induced Mutations for the Improvement of Grain Legumes Production II, IAEA-TECDOC-260*. International Atomic Energy Agency, Vienna, Austria, pp. 59–64.

Sharma, B. and Kharkwal, M.C. (1983) Mutation studies and mutation breeding in grain legumes. In: *Induced Mutations for Improvement of Grain Legumes Production III, IAEA-TECDOC-299*. International Atomic Energy Agency, Vienna, Austria, pp. 65–75.

Sharma, P.C. and Varshney, R.K. (1999) Induced mutations in chickpea improvement. In: Siddiqui, B.A. and Khan, S. (eds) *Breeding in Crop Plants, Mutations and In Vitro Mutation Breeding*. Kalyani Publishers, Ludhiana, India, pp. 57–67.

Sigurbjönsson, B. (1983) Induced mutations. In: Wood, D.R. (ed.) *Crop Breeding*. American Society of Agronomy and Crop Science Society of America, Madison, Wisconsin, pp. 153–176.

Sindhu, J.S., Lal, S.B. and Singh, R.P. (1981) Studies on the factors determining crossing success in chickpea (*Cicer arietinum*). *Pulse Crops Newsletter* 1, 21–22.

Singh, B.D. (2005) Mutations in crop improvement. In: Singh, B.D. (ed.) *Plant Breeding, Principles and Methods*. Kalyani Publishers, Ludhiana, India, pp. 698–731.

Singh, D. and Shyam, R. (1958) Sterility in gram (*Cicer arietinum*). *Current Science* 27, 491–492.

Singh, H. and Dahiya, B.S. (1974) Note on inheritance of bushy mutant in chickpea. *Current Science* 43, 731–732.

Singh, K.B. (1987) Chickpea breeding. In: Saxena, M.C. and Singh, K.B. (eds) *The Chickpea*. CAB International, Wallingford, UK, pp. 127–162.

Singh, K.B. (1997) Chickpea (*Cicer arietinum* L.). *Field Crops Research* 53, 161–170.

Singh, K.B. (1990) Winter chickpea: problems and potential in the Mediterranean region. In: Saxena, M.C., Cubro, J.I. and Wery, J. (eds) *Present Status and Future Prospects of Chickpea Crop Production and Improvement in the Mediterranean Countries. Options Mediterraneannes*, Series A: Seminares Mediterraneous, No. 9. CIHEAM, Zaragoza, Spain, pp. 25–34.

Singh, K.B. and Auckland, A.K. (1975) Chickpea breeding at ICRISAT. In: *Proceedings of the International Workshop on Grain legumes*. ICRISAT, Hyderabad, India, pp. 3–17.

Singh, K.B., Ocampo, B. and Robertson, L.D. (1998) Diversity for abiotic and biotic stress resistance in the wild annual *Cicer* species. *Genetic Resources and Crop Evolution* 45, 9–17.

Singh, N. (2001) Crossing technique without emasculation in chickpea. *Indian Journal of Genetics* 61(4), 366.

Singh, O. and Rupela, O.P. (1991) Genetic relationship among the newly identified non-nodulating chickpea lines. *International Chickpea Newsletter* 25, 11–12.

Singh, I. and Waldia, R.S. (1994) Comparison of selection methods in chickpea under late sown condition, crop research, Hissar. 8(3), 508–511.

Singh, O. and Rupela, O.P. (1998) A new gene that controls root nodulation in chickpea. *Crop Science* 38, 360–362.

Singh, O., Rupela, O.P. and van Rheenen, H.A. (1992a) Genetics of nonnodulation in chickpea. *International Chickpea Newsletter* 15, 5–6.

Singh, O., van Rheenen, H.A. and Rupela, O.P. (1992b) Inheritance of a new nonnodulation gene in chickpea. *Crop Science* 32, 41–43.

Smartt, J. (1976) Comparative evolution of pulse crops. *Euphytica* 25, 139–143.

Smithson, J.B., Thompson, J.A. and Summerfield, R.J. (1985) Chickpea (*Cicer arietinum*) In: R.J. Summerfield and E.H. Roberts (eds) *Grain Legume Crops*. Collins, London, UK. pp. 312–390.

Stadler, L.J. (1928) Mutations in barley induced by x-rays and radium. *Science* 68, 186–187.

Toker, C. (2005) Preliminary screening and selection for cold tolerance in annual wild *Cicer* species. *Genetic Resources and Crop Evolution* 52, 1–5.

Toker, C. and Cagirgan, M.I. (2004) Spectrum and frequency of induced mutations in Chickpea. *International Chickpea and Pigeonpea Newsletter* 11, 8–10.

Toker, C., Uzun, B., Canci, H. and Ceylan, F.O. (2005) Effects of gamma irradiation on the shoot length of *Cicer* seeds. *Radiation Physics and Chemistry* 73, 265–367.

Turano, M.J., Baird, L.M. and Webster, B.D. (1983) Characteristics of the stigma of chickpea. *Crop Science* 23, 1033–1036.

van Harten, A.M. (1998) *Mutation Breeding: Theory and Practical Applications*. Cambridge University Press, Cambridge.

van der Maesen, L.J.G. (1972) *Cicer* L. A Monograph of the Genes, with Special Reference to the Chickpea (*Cicer arietinum*), Its Ecology and Cultivation. Mededlingen Landbouwhogeschool (Communication Agricultural University), Wageningen, The Netherlands.

van Rheenen, H.A. (1996) Determinate growth in chickpea. *International Chickpea and Pigeonpea Newsletter* 3, 18.

van Rheenen, H.A., Pundir, R.P.S. and Miranda, J.H. (1994) Induction and inheritance of determinate growth habit in chickpea (*Cicer arietinum* L.). *Euphytica* 78, 137–141.

19 Breeding Achievements

P.M. Gaur,[1] C.L.L. Gowda,[1] E.J. Knights,[2]
T. Warkentin,[3] N. Açıkgöz,[4] S.S. Yadav[5] and
J. Kumar[5]

[1]*Genetics Resources Divisions, International Crops Research Institute for the Semi-Arid Tropics (ICRISAT), Hyderabad 502 324, India;* [2]*NSW Department of Primary Industries, Tamworth Agricultural Institute, Calala, New South Wales 2340, Australia;* [3]*Crop Development Centre, University of Saskatchewan, Saskatoon, Saskatchewan S7N 5A8, Canada;* [4]*Aegean Agricultural Research Institute (AARI), PO Box 9, Menemen 35661, Izmir Turkey;* [5]*Pulse Research Laboratory, Division of Genetics, Indian Agricultural Research Institute, New Delhi 110012, India*

Introduction

The first systematic efforts on chickpea breeding probably started about 100 years ago in India after the establishment of Imperial (now Indian) Agricultural Research Institute at Pusa, Bihar (now at New Delhi) in 1905. More than 350 improved varieties have since been released globally, and about half of these have been released in India, which accounts for about two-thirds of the global chickpea production and has the largest national chickpea-breeding programme in the world. The two international institutes established by the Consultative Group on International Agricultural Research (CGIAR), the International Crops Research Institute for the Semi-Arid Tropics (ICRISAT) (established in 1972), Hyderabad, India, and the International Centre for Agricultural Research in the Dry Areas (ICARDA) (established in 1977), Aleppo, Syria, have provided a boost to chickpea-breeding programmes of the national agricultural research systems (NARS) globally through supply of germplasm and improved breeding materials. The breeding materials supplied by these two institutes have contributed to the release of more than 150 varieties in more than 40 countries during the last three decades.

Breeding efforts have contributed substantially to improving yield potential [global yield increased from 591 kg/ha in 1975 to 818 kg/ha in 2005 (FAOSTAT data, 2006)], regional adaptation through resistance or tolerance to abiotic and biotic stresses, plant type and grain characteristics. This chapter attempts to highlight important chickpea-breeding achievements, particularly of the last two decades.

An Overview of Breeding Approaches

In the early phase of chickpea breeding, most varieties were developed through selection from landraces that were either collected in the country or introduced from other countries. At present, hybridization is invariably being used to increase genetic variability in the breeding material, and thus most of the recent varieties are cross-bred.

The pedigree method, earlier used at many institutes, is not in much practice at present in its original form, because it is cumbersome and can handle only a limited number of crosses. The bulk method, with modifications, is now the most common selection method used after hybridization in chickpea. It is well established that individual plant selection for yield in early segregating generations is not effective in chickpea. Thus, selection for simple traits (e.g. seed traits, maturity, resistance to diseases) is done in early segregating generations (F_2 and F_3), whereas single plant selection for yield usually starts from F_4 or later generations. A recombinant-derived family method that uses early generation selection for yield in F_2-derived F_4 or later generation families to eliminate inferior crosses and inferior F_2-derived families has also been suggested (Slinkard *et al.*, 2000) and is used at the University of Saskatchewan, Canada.

Many breeding programmes prefer to take more than one generation per year to reduce the time in development of a variety. Off-season nurseries or greenhouse facilities are used to take one or two generations in the off-season. As long days induce early flowering in chickpea, extended photoperiod (15–16 h) through artificial lighting can be provided until flower initiation for rapid generation turnover (Sethi *et al.*, 1981). The single seed descent (SSD) method is commonly used whereby one or more early generations are advanced without selection in the greenhouse. Depending on greenhouse and other resources, SSD can often accommodate large populations.

Most chickpea-breeding programmes have been confined to intraspecific hybridization that includes desi × desi, kabuli × kabuli or desi × kabuli crosses. Crosses between desi and kabuli parents have been used extensively to exploit genes present in one group. For example, desi parents have contributed important genes for fusarium wilt and ascochyta resistance and drought tolerance in kabuli-type breeding programmes; conversely, kabuli parents have been a source of improved seed quality, especially large seed size, in desi-type breeding programmes. Desi × kabuli crosses have consistently produced high-yielding progenies and have been the source of many new cultivars (Yadav *et al.*, 2004a).

Efforts have been made to use interspecific crosses for enhancing genetic variability and introgressing useful genes into the cultigen from wild *Cicer* spp. Only two annual wild species, *Cicer reticulatum* and *C. echinospermum*, have so far been exploited in breeding programmes, as the crossing of the cultigen with other species has remained a challenge even with embryo rescue techniques. There is a need to continue efforts on exploiting wild *Cicer* spp. of the tertiary gene pool, as these contain many useful genes, particularly resistance to some biotic and abiotic stresses (Ahmad *et al.*, 2005). One variety, PUSA 1103, developed by the Indian Agricultural Research Institute (IARI), New Delhi,

India, from an interspecific cross of *C. arietinum* with *C. reticulatum*, has been released for commercial cultivation in North India.

Mutation breeding has also been used in chickpea improvement for creating variability. Some mutants have been directly released as varieties, whereas many others have been used as parents in crossing programmes. At least 12 varieties have been developed through mutation breeding. These include seven (Pusa 408, Pusa 413, Pusa 417, Pusa 547, RS 11, RSG 2 and WCG 2) developed by IARI and State Agricultural Universities in India; four (CM 72, CM 88, CM, 98 and CM 2000) by the Nuclear Institute of Agriculture and Biology, Faisalabad, Pakistan; and one (Hyprosola) by the Bangladesh Institute of Nuclear Agriculture, Mymensingh, Bangladesh.

Recent advances in the transformation and plant regeneration protocols for chickpea now mean that transgenic technology can be used to improve those traits for which adequate variability is not available in the primary gene pool. These include resistance to pod borer and other biotic and abiotic stresses, as well as sulphur-containing amino acids.

Marker-assisted selection (MAS) is being considered for improving the precision and efficiency of conventional plant-breeding methods. MAS would be useful for improving those traits that are difficult or inconvenient to select directly (e.g. root traits for drought avoidance, antinutritional factors, quality traits), for pyramiding resistance genes from different sources when the resistance is polygenically controlled (e.g. resistance to ascochyta blight), for bringing together genes conferring different resistance mechanisms (e.g. antixenosis, antibiosis and tolerance for pod borer), and for combining resistance to two or more stresses (e.g. resistance to fusarium wilt and to pod borer). MAS will also be used for tracking introgression of resistance genes from transgenics to cultivars or elite breeding lines.

Breeding Achievements

Early maturity

Chickpea is grown in a wide range of environments and cropping systems, and its maturity ranges from 80 to 180 days depending on genotype, soil moisture, temperature, latitude and altitude. However, in at least two-thirds of the global chickpea area the growing season available for chickpea is short (90–120 days) due to risk of drought or frost at the end of the season (pod-filling stage of the crop). More than 70% of the chickpea area is in South and South-east Asia. Here the crop is grown in the post-rainy season on receding soil moisture, and often experiences end-of-season drought. Early maturity is important for these areas and also for the autumn-sown rainfed crop in Mediterranean-type environments (e.g. Australia) to escape terminal drought. Early maturity, associated with early flowering and double-podded traits, has been found to be important in Canada (temperate environments) to escape frost at the end of the season (Anbessa *et al.*, 2006a,b).

Development of early-maturing varieties is one of the major objectives in chickpea-breeding programmes of ICRISAT and the national programmes of several countries including India, Myanmar, Bangladesh, Ethiopia, Australia and Canada. The first breakthrough from ICRISAT's breeding programme was the development of an extra-short-duration kabuli variety ICCV 2, which flowers in about 30 days and matures in about 85 days at Hyderabad, India (altitude: 545 m above the mean sea level, latitude: 17°27'N). It was instrumental in expanding the kabuli chickpea area from cool, long-season environments (e.g. North India) to warmer, short-season environments (e.g. South India and Myanmar). Collaborative efforts between ICRISAT and the Indian national programmes have led to the development of many other short-duration varieties. Some of the more popular ones are ICCC 37 and JG 11 (desi) and KAK 2 and JGK 1 (kabuli). Several early-maturing kabuli varieties (PUSA 1053, PUSA 1088, PUSA 1105 and PUSA 1108) have also been developed by IARI and released for commercial cultivation in Central and North India.

ICCV 96029 and ICCV 96030 are two super-early and cold-tolerant desi lines developed by ICRISAT that mature in 75–80 days in South India (Kumar and Rao, 1996). These lines were evaluated for their suitability as short-duration vegetable crop (immature green seed used as vegetable) in North India. ICCV 96029 set pods during cold months and produced 3.6 t/ha green pods in 70 days after sowing, while the control variety PBG 1 took more than 120 days to initiate podding (Sandhu *et al.*, 2002). The super-early chickpea lines are extensively being used in crossing programmes by NARS in India and Canada for development of early-maturing varieties. Recently, IARI has also developed several extra-early genotypes of both desi (e.g. BG 1044) and kabuli types (e.g. BG 1054, DG 5005, 5006) which provide additional sources of earliness for chickpea-breeding programmes.

Adaptation to late-planting conditions

The chickpea area under late-sown conditions is increasing in South Asia, particularly in India. The Indian government has set a target of bringing an additional area of more than 2 million hectares under chickpea cultivation, which will mainly be available in conditions subjected to late planting. About 14 million hectares in South Asia remains fallow in winter season after harvest of rainy season rice due to lack of irrigation (Subbarao *et al.*, 2001). These rice fallows offer enormous potential for expansion of chickpea area in South Asia.

The sowing in rice fallows is often delayed due to late harvest of rice crop. The studies conducted on promotion of chickpea in rice fallows of Bangladesh indicate that short-duration chickpea cultivars have better chances of success in rice fallows as the crop often experiences end-of-season drought stress (Musa *et al.*, 2001). Identification of suitable early-maturing varieties has made it possible to bring more than 10,000 ha of rice fallows in Barind, north-western Bangladesh, under chickpea cultivation. Similarly, the extra-short-duration variety ICCV 2 is popular in Myanmar under rice–chickpea cropping system (ICRISAT, 2003).

Efforts are being made to expand chickpea cultivation in rice fallows of India, which has nearly 82% of the rice fallows available in South Asia. Large-scale trials were conducted in five states of India (Chhattisgarh, Jharkhand, Orissa, West Bengal and eastern Madhya Pradesh), and in Nepal and Bangladesh during 2002–2005. Short-duration varieties were found suitable and the farmers preferred the *kabuli* varieties ICCV 2, KAK 2 and JGK 1 (Gaur and Gowda, 2005).

A few varieties (e.g. Pusa 372, H 82-2, PBG 1 and KPG 59) were released earlier for late-planting conditions of North India. Recent experiments conducted on cultivation of chickpea after rice harvest in North India (Uttar Pradesh and Haryana) showed chickpea yield potential of about 2.5 t/ha after a rice crop (Yadav *et. al.*, 2005). PUSA 1103 developed through interspecific hybridization [(BG 256 × *C. reticulatum*) × BG 362] has been released for rice-based cropping system under late plantings in North India. It is tolerant to drought and resistant to soil-borne diseases and stunt virus (Yadav *et al.*, 2002a,b).

Tolerance to abiotic stresses

Drought

Terminal (end-of-season) drought is the most important constraint to chickpea production, accounting for 40–50% yield reduction globally (Ryan, 1977). Development of early-maturing varieties has been the most effective strategy for escape from terminal drought. However, it is necessary to match the crop maturity with the crop season length to obtain high yield.

The conventional plant-breeding approach has been generally used to develop drought-tolerant varieties. It is based on selection for yield and its components under a given drought environment. The large environmental variation necessitates evaluation of material at several locations and/or over years. Thus, the need for trait-based selection has been emphasized. Efforts have been made to identify plant traits for drought tolerance and incorporate these traits in well-adapted varieties. The drought tolerance traits of genotype ICC 4958 were thoroughly studied and it was concluded that high root mass was responsible for drought tolerance in this genotype. ICC 4958 had 30% more root volume than the popular variety Annigeri (Saxena *et al.*, 1993). Recent studies at ICRISAT on root growth parameters of 12 contrasting chickpea genotypes showed a clear positive relationship between root biomass and seed yield under moisture stress conditions (Kashiwagi *et al.*, 2006). Promising drought-tolerant lines with fusarium wilt resistance and high yield have been developed using trait-based selections in the progenies of a three-way cross involving ICC 4958 (high root mass), Annigeri (local adaptation) and ICC 12237 (fusarium wilt resistance) (Saxena, 2003). Several lines that combine the traits of large root mass and fewer leaflets with increased wilt tolerance were obtained (Saxena, 2003).

Breeding for high root mass is very difficult due to the laborious methods involved in digging and measuring root length and density. Molecular markers closely linked with major quantitative trait loci (QTL) controlling root traits can facilitate MAS for root traits. ICRISAT scientists have identified molecular markers for a major QTL that accounts for 33% of the variation for root weight and root length (Chandra *et al.*, 2004). Recently, the chickpea mini-core collection was evaluated for root traits, and parents showing larger variation than that found between Annigeri and ICC 4958 were identified for development of new mapping populations (Kashiwagi *et al.*, 2005). These include ICC 8261 for a large root system and ICC 283 and ICC 1882 for smaller roots. Recombinant inbred line (RIL) populations have been developed from two crosses (ICCV 283 × ICC 8261, ICCV 4958 × ICC 1882) and are being studied to identify additional QTLs for root traits (Gaur *et al.*, 2007). IARI scientists are also working on development of RILs from PUSA 362 × SBD 377 for mapping QTLs for drought tolerance (Aqeel *et al.*, 2006).

Another approach being used for enhancing drought tolerance in chickpea is the development of transgenics, using promoters regulating either a single gene or a cascade of genes involved in stress response. ICRISAT scientists have developed transgenics having a dehydration responsive element construct (*DREB1A* gene attached to a drought-responsive promoter rd 29A) known to regulate a number of genes involved in the response to drought and other stresses, such as salinity and cold. Transgenic plants are undergoing molecular characterization. The *P5CSF-129A* gene that increases proline accumulation in the plant and improves its tolerance to osmotic stress has also been used in the genetic transformation of chickpea. Some selected transgenic plants have shown up to fivefold increase in proline content (ICRISAT, 2004).

Low temperature

Freezing (mean daily temperature <–1.5°C) and chilling temperatures (mean daily temperature between –1.5°C and 15°C) are important constraints to chickpea production in some regions (reviewed by Croser *et al.*, 2003). Winter-sown chickpea in the West Asia and North Africa (WANA) region often experiences freezing temperatures during the seedling and early vegetative stages, and chilling temperatures at the early reproductive stage. Freezing temperature reduces growth vigour and vegetative biomass, whereas chilling temperature at flowering causes flower and pod abortion. Therefore, cultivars for winter sowing in WANA need to have cold tolerance both at seedling and flowering stages. Screening of germplasm at ICARDA has identified several cold-tolerant lines from the cultivated (Singh *et al.*, 1995) and wild species (Robertson *et al.*, 1995). Collaborative research efforts of ICARDA and NARS in WANA to advance the sowing date of kabuli chickpea to winter, by developing cultivars resistant to ascochyta blight with tolerance to cold, have led to potential yield increases of 50–100%. National programmes in the region have released more than 40 chickpea cultivars adapted to winter sowing, and the technology is being increasingly adopted by farmers.

Chilling temperatures during early reproductive growth cause yield losses in chickpea in many parts of the world, including the Indian subcontinent and Australia. Clarke and Siddique (2004) demonstrated that low temperature (mean daily temperature below 15°C) affects both the development and function of reproductive structure of chickpea flowers leading to flower abortion. Several breeding lines (e.g. ICCV 88502, ICCV 88503, ICCV 88506, ICCV 88510 and ICCV 88516) that set pods at lower temperature have been developed at ICRISAT (ICRISAT, 1994; Srinivasan *et al.*, 1998). Similarly, IARI scientists have developed several genotypes (BGD 112 green, BG 1100, BG 1101, PUSA 1103, BGD 1005, PUSA 1108, DG 5025, DG 5027, DG 5028, DG 5036 and DG 5042) that are able to produce high biomass and continue to produce flowers and set pods at mean temperatures between 8°C and 10°C (S.S. Yadav, New Delhi, 2006, personal communication).

A pollen selection method was developed in Australia and applied to select for chilling tolerance in crosses involving ICCV 88516 (the most chill-tolerant line identified in Australia) and chill-sensitive cultivars Amethyst and Tyson, leading to development and release of chill-tolerant cultivars Sonali and Rupali (Clarke *et al.*, 2004). Restriction fragment length polymorphism (RFLP) markers for chilling tolerance were identified and subsequently converted to sequence characterized amplified region (SCAR) markers. These were used successfully to select chill-tolerant progeny from a cross between Amethyst and ICCV 88516 but were ineffective in other crosses (Millan *et al.*, 2006).

Salinity

Chickpea, like many other legumes, is sensitive to soil salinity. As chickpea is usually grown under residual soil moisture, the salinity stress is often combined with terminal drought. Soil salinity is a major abiotic constraint to chickpea productivity in many parts of India, Pakistan, Iran, West Asia, North Africa and Australia.

Limited efforts to identify salinity tolerance within chickpea indicated low genotypic variation (Saxena, 1984). However, some genotypes with low to moderate tolerance were subsequently identified (Singh and Singh, 1984; Asharf and Waheed, 1992; Dua and Sharma, 1995; Kathiria *et al.*, 1997; Maliro *et al.*, 2004). A desi chickpea variety Karnal Chana 1 (CSG 8963), which can be grown in saline soils with ECe up to 6 dS/m, was released in India. In Australia, the ICRISAT breeding line ICCV 96836 (released as *Genesis 836*) was among the most salt-tolerant lines identified in pot trials (Maliro *et al.*, 2004).

A recent screening of 263 diverse chickpea genotypes (including the mini-core collection, breeding lines, wild relatives and some earlier reported salt-tolerant lines) at ICRISAT showed a sixfold range of variation for seed yield under salinity, with several genotypes yielding 20% more than the salt-tolerant cultivar Karnal Chana 1 (Vadez *et al.*, 2006). No significant relation was found between biomass at the late vegetative stage and final seed yield under salinity. Performance of seed yield under salinity was explained in part by the yield potential under control conditions, and a salinity tolerance component. The parents of ICCV 2 × JG 62 RIL mapping population showed good contrast for salinity tolerance and this population is being used to identify markers for salinity tolerance QTLs (Vadez *et al.*, 2006).

Resistance to diseases

Fusarium wilt

Fusarium wilt, caused by *Fusarium oxysporum* f. sp. *ciceri*, is the most important root disease of chickpea, particularly in the semiarid tropics (SAT) where the chickpea-growing season is dry and warm. Seven races of the fungus have so far been identified – races 1 to 4 from India (Haware and Nene, 1982), races 0 and 5 from Spain (Jimenez-Diaz *et al.*, 1989) and race 6 from the USA (Phillips, 1988).

Effective field, greenhouse and laboratory procedures for resistance screening have been developed (Nene *et al.*, 1981). As development of a wilt-sick plot for field screening is fairly easy, such plots are available at many research stations around the world. Several sources of absolute resistance have been identified. A collection of more than 13,500 germplasm accessions from 40 countries was evaluated for race 1 of *F. oxysporum* at ICRISAT and 160 resistant accessions (150 desi and 10 kabuli) were identified (Haware *et al.*, 1992). Although resistance to fusarium wilt in kabuli germplasm is limited, some accessions with high resistance have been identified (Gaur *et al.*, 2006).

Simple and effective field-screening techniques and the availability of many highly resistant sources have led to excellent progress in breeding cultivars resistant to fusarium wilt, particularly for races 1 to 4. Several cultivars with durable and stable resistance to fusarium wilt have been released in India and in several other countries. Resistance to fusarium wilt is mandatory for any variety released in India.

Germplasm lines and cultivars are available that possess resistance to multiple races. For example, WR 315 is resistant to all races, except race 3; and JG 74 is resistant to all races, except races 2 and 5 (Haware, 1998). The resistance to fusarium wilt seems to be very stable. Excellent progress has been made in understanding the genetic control of fusarium wilt resistance and mapping genes resistant to fusarium wilt (reviewed by Millan *et al.*, 2006). Molecular markers closely linked with some of the genes conferring resistance to various fusarium wilt races have been identified and can be used for pyramiding resistance genes for these races and, in countries where fusarium wilt is not present (e.g. Australia), pre-emptive resistance breeding has been taken up.

Ascochyta blight

Ascochyta blight, caused by *Ascochyta rabiei* (Pass.) Labr., is a highly devastating foliar disease of chickpea in West and Central Asia, North Africa, North America and Australia. In the Indian subcontinent, the disease occurs in northwest India and Pakistan. It occurs mainly in areas where cool, cloudy and humid weather prevails during the crop season.

Considerable effort has been expended in identifying resistant germplasm and breeding for resistance. More than 13,000 germplasm accessions were evaluated at ICARDA and 11 kabuli (ILC 72, ILC 196, ILC 201, ILC 202, ILC 2506, ILC 2956, ILC 3274, ILC 3279, ILC 3346, ILC 3956, ILC 4421) and 6 desi (ICC 3634, ICC 4200, ICC 4248, ICC 4368, ICC 5124, ICC 6981) accessions were identified as resistant (Reddy and Singh, 1984). Singh *et al.* (1984) tested 112 lines in 11 countries and found four lines (ILC 72, ILC 191, ILC 3279 and ILC 3856) resistant in eight countries. Singh and Reddy (1993) reported that three desi accessions (ICC 4475, ICC 6328 and ICC 12004) and two kabuli accessions (ILC 200 and ILC 6482) showed resistance to six races in repeated greenhouse and field screening during 1979–1991. Some important resistant lines directly released as varieties in different countries include ILC 72, ILC 195, ILC 482 and ILC 3279.

A lack of highly resistant germplasm and a highly variable pathogen have precluded the development of varieties having high level and durable resistance (Pande *et al.*, 2005). However, ICARDA scientists did develop more than 3000 lines with moderate ascochyta blight resistance and these were widely distributed to NARS (Malhotra *et al.*, 2003). Good progress has also been made in combining large seed size with improved ascochyta blight resistance in kabuli chickpea. ICRISAT has concentrated on improvement of ascochyta blight resistance in desi chickpea. Its strategy has been to enhance resistance by accumulating resistance genes from different sources. Some progenies from multiple crosses (e.g. ICCV 04512, ICCV 04514, ICCV 04516) have shown high levels of resistance (score of 3–4 on a rating scale of 1–9, where 1 = resistant and 9 = susceptible) to multiple isolates (Gaur and Gowda, 2005).

Good progress has been made in breeding cultivars with improved ascochyta blight resistance by NARS in many countries. Promising varieties released in different countries include PUSA 261, PBG 1, GNG 469 and Gaurav in India; Dasht, NIFA 88, CM 72, CM 88, CM 98 and CM 2000 in Pakistan; Dwelley, Sanford, Myles, Evans and Sierra in USA; CDC Frontier, CDC Anna, CDC Cabri, CDC Desiray, CDC Nika and Amit in Canada; and Sonali, Rupali, Genesis 508, Genesis 090, Genesis 836, Yorker, Flipper, Nafice and Almaz in Australia.

Millan *et al.* (2006) recently reviewed progress on identification of markers for ascochyta blight resistance QTLs. Molecular markers are now available for at least two major QTLs involved in ascochyta blight resistance and it seems feasible to use MAS for pyramiding ascochyta blight resistance genes. However, it may not be possible to develop varieties with absolute resistance to ascochyta blight. Therefore, integrated diseases management will remain the key for successful cultivation of chickpea in ascochyta blight prone areas (Gan *et al.*, 2006).

Botrytis grey mould

Progress in breeding for resistance to botrytis grey mould (*Botrytis cinerea*. Pars.) has been hampered by a lack of highly effective host resistance in chickpea (Pande *et al.*, 2006). Of more than 12,000 germplasm accessions and breeding lines screened at ICRISAT, none was found to be highly resistant (Pande *et al.*, 2002). However, moderate resistance has been observed in genotypes in various field and controlled environment tests in India, and ICC 14344 was released as cultivar Avarodhi (IIPR, 1997). Such host-plant resistance can provide useful protection against the disease. Under high disease pressure in Australia, two moderately resistant desi cultivars yielded 70–75% more than a susceptible cultivar (Knights and Siddique, 2002). Resistance to *B. cinerea* has been attributed to a single dominant gene (Rewal and Grewal, 1989) or complementary dominant genes (Rahul *et al.*, 1995). Although a higher level of resistance has been identified in some wild *Cicer* spp., including closely related *C. echinospermum*, this resistance has yet to be incorporated in the cultivated species (Pande *et al.*, 2002).

Selection for plant types that maximize canopy aeration can be an important adjunct to host resistance. A high correlation between disease level and lodging was demonstrated in genotype trials in Australia (Knights and Siddique, 2002): in general, the disease was lowest in those genotypes having good lodging resistance. In Bangladesh an upright branching habit and open canopy were also suggested as reasons for a lower disease incidence in the breeding line ICCL 87322 (Bakr *et al.*, 2002).

Viral diseases

Viral diseases have been reported to cause sporadic but significant yield loss occasionally in some pockets. Major symptoms include discolouring (yellow, orange or brown) of foliage, browning of phloem and stunting of growth. Many viruses have been identified that can cause stunt disease. Chickpea chlorotic dwarf monogeminivirus (CCDV) and chickpea luteovirus (CpLV) are prevalent in India and Pakistan; faba bean necrotic yellows nanovirus (FBNYV) in Syria, Turkey and Lebanon; and bean leaf roll luteovirus (BLRV) in Algeria (Horn *et al.*, 1995, 1996; Yahia *et al.*, 1997). These are transmitted by a number of aphid vectors, including *Aphis craccivora* Koch and *Myzus persicae* Subzer (Kaiser and Danesh, 1971).

Efforts have been made to identify sources of resistance and develop virus-resistant cultivars. ICRISAT scientists screened more than 10,000 germplasm accessions and breeding lines for resistance to virus at Hisar in North India, a hot spot for CCDV. Two lines, GG 669 and ICCC 10, showed field resistance. However, these were susceptible to luteoviruses when screened at Junagargh, Gujarat, India (Haware, 1998). Chalam *et al.* (1986) used mechanical inoculation for screening 143 chickpea germplasm lines against cucumber mosaic virus (CMV) and bean yellow mosaic virus (BYMV). Two accessions (ICC 1781, 8203) were found resistant (no visible symptoms) to CMV and nine accessions (ICC 607, 1468, 2162, 2342, 3440, 3598, 4045, 6999, 11550) resistant to BYMV. A screening of more than 500 germplasm lines for resistance to CMV was carried out under late plantings at IARI during 2004/05 and several resistant lines (BG 1100, BG 1101, PUSA 1103 and BGD 112) were identified (Kumar *et al.*, 2005a). These lines can be used as donors in virus resistance breeding. A few available cultivars have showed some level of resistance to virus, e.g. Pusa 244, Girnar, Udai, SAKI 9516 in India (Dua *et al.*, 2001).

Other diseases

Compared to the major diseases minimal effort has been taken to breed for resistance to the large range of other root and foliar diseases that affect chickpea. Progress for resistance to root diseases has been made often in tandem with selection for resistance to fusarium wilt. The presence of other root disease pathogens in dedicated fusarium-screening nurseries has enabled progress in breeding for resistance to diseases such as dry root rot (*Rhizoctonia bataticola*), wet root rot (*R. solani*), black root rot (*F. solani*) and collar rot (*Sclerotium rolfsii*). Using pathogen-specific resistance screening, varying levels of field resistance to many of these root diseases were identified. Although no immunity to dry root rot has been identified within the cultigen, some germplasm accessions and breeding lines were found to have useful resistance (Pande *et al.*, 2004). Resistance to *R. bataticola* is determined by a single dominant gene (Rao and Haware, 1987).

A more focused effort has been made to breed for resistance to phytophthora root rot in Australia where the disease is a major production constraint. Effective field resistance was identified in an accession from Tajikistan (Brinsmead *et al.*, 1985). This resistance was subsequently incorporated in a series of cultivars released in north-eastern Australia where yield increases of 56–112%, relative to susceptible cultivars, were recorded (Singh *et al.*, 1994). Significantly better resistance was subsequently identified in accessions of the wild relative *C. echinospermum* (Knights *et al.*, 2003) and this resistance is being transferred to domesticated backgrounds.

There are no reported cases of breeding specifically for resistance to minor foliar diseases. However, useful resistance to both sclerotinia stem rot (*Sclerotinia sclerotiorum*) (Gurha *et al.*, 1998) and alternaria blight (*Alternaria alternata*) (Yadav and Udit, 1993; Gurha *et al.*, 2002) has been found.

Resistance to insect pests

Pod borer

Pod borer (*Helicoverpa armigera* Hübner) is the most important insect pest of chickpea globally. The insect is highly polyphagous and sources with high levels of resistance are not available in chickpea germplasm. Therefore, it has not been possible to breed varieties with adequate host resistance.

More than 14,800 germplasm accessions were screened for tolerance to pod borer in unsprayed areas of ICRISAT during 1984–1990. ICC 506 showed a mean of 8.6% pod damage as against 29.9% in the popular cultivar Annigeri (Lateef and Pimbert, 1990). Many other germplasm accessions, breeding lines and cultivars having some resistance have been identified (see Sharma *et al.*, 2003 for a review). Because of predominance of fixable (additive) genetic variance and high heritability, pedigree selection was expected to enhance pod borer resistance in early-maturing desi types. On the other hand, both additive and dominance variance are important for pod borer resistance in medium- and long-duration desi and kabuli chickpea. Hence, it was suggested that selection for pod borer resistance should be delayed until F_5 generation (Gowda *et al.*, 2005)

More than 160 accessions of annual wild *Cicer* spp. have been screened at ICRISAT for *Helicoverpa* resistance. Larval growth was slow on 21 of these accessions and this phenomenon of antibiosis was unique to the wild species (Sharma *et al.*, 2005). Efforts are being made to combine the non-preference (antixenosis) mechanism of resistance identified in the cultigen (e.g. ICC 506 EB) and antibiosis mechanism of resistance identified in *C. reticulatum*. The preliminary screening of some of perennial wild *Cicer* spp. revealed that *C. microphyllum* and *C. canariense* had pod borer rating as low as 1.0, whereas *C. judaicum*, reported earlier as a source of resistance, had a damage rating of 4.0, and cultivated chickpea genotypes had leaf and pod damage rating of 8.5 and 9.0, respectively (Sharma *et al.*, 2006). Thus, these two species offer the best source of *Helicoverpa* resistance in chickpea. However, these are not accessible currently due to crossability barriers with the cultigen.

About 3000 germplasm lines were evaluated against *Helicoverpa* damage at IARI during 1999–2000. All lines were damaged by *Helicoverpa*; however, the intensity of damage varied from genotype to genotype. Spreading types were more susceptible than erect types, extra-large-seeded lines were more susceptible than small-seeded lines, kabuli types were more susceptible than desi types and damage under irrigated conditions was more than under rainfed conditions. These results suggest that *Helicoverpa* incidence is highly dependent on plant type and growing environment (Yadav *et al.*, 2006).

Development of transgenics has the most potential for enhancing resistance to *Helicoverpa*. Transgenics using *Bacillus thuringiensis* (Bt) *Cry1Ab* and *Cry1Ac* have been developed at ICRISAT and at many other institutions. Insect bioassays and evaluation of transgenics in contained field trials are in progress. Chickpea transgenics developed using the truncated native *Cry1Ac* gene at National Botanical Research Institute, Lucknow, India, showed effective resistance to *Helicoverpa*. Transgenics expressing Cry1Ac protein above 10 ng/mg soluble protein caused more than 80% insect mortality (Sanyal *et al.*, 2005).

Other insect pests

Leaf miner (*Liriomyza cicerina* Rondani) is an important insect pest of chickpea in WANA and southern Europe. Efforts have been made to identify sources of resistance in the cultivated and wild species. Thirty-one tolerant lines were identified at ICARDA from screening of 6800 kabuli chickpea germplasm lines under natural field infestation. Of these, only four (ILC 726, 1776, 3350, 5901) were found to be promising resistance sources. Most of the leaf miner–tolerant lines had smaller leaflets and seed. The most tolerant line ILC 5901 had multipinnate leaves (Malhotra *et al.*, 1996). High resistance (score 2 on 1–9 scale, where 1 = plants free from damage and 9 = very highly susceptible) has been identified in some accessions of wild species. These accessions belong to the species C. *echinospermum* (ILWC 35), *C. bijugum* (ILWC 227), *C. judaicum* (ILWC 206, 207, 210, 211, 223), *C. pinnatifidum* (ILWC 9, 97, 226, 252), *C. cuneatum* (ILWC 40, 185, 187, 232) and *C. chorassanicum* (ILWC 147) (Robertson *et al.*, 1995). Limited progress has been made in breeding for resistance to leaf miner.

Seed beetle or bruchid (*Callosobruchus* spp.) is the most important storage pest of chickpea. No source of resistance has been identified in the cultivated species. Some of the accessions of wild species *C. reticulatum* (ILWC 104), *C. echinospermum* (ILWC 39, 179, 181), *C. bijugum* (ILWC 65, 67, 68, 70, 73, 74, 75, 83, 177), *C. judaicum* (ILWC 46, 54, 173, 174 176 189) and *C. cuneatum* (ILWC 187) showed no damage (score 1 on 1–9 scale, where 1 = seeds free from damage and 9 = very highly susceptible) by this insect (Roberson *et al.*, 1995). However, there are no reports of successful transfer of bruchid resistance from wild to the cultivated species. Chickpea transgenics carrying the bean α-amylase inhibitor 1 (αAI1) gene showed very high protection against two species of *Callosobruchus*, *C. maculatus* and *C. chinensis* (Higgins *et al.*, 2004). Scientists at IARI observed that green-, black- and rough-seeded genotypes were more resistant in comparison to white- and brown-seeded genotypes (S.S. Yadav, IARI, New Delhi, 2006, personal communication).

Resistance to nematodes

There has not been a systematic screening of chickpea germplasm for resistance to the two most important root-knot nematode species, *Meloidogyne incognita* and *M. javanica*. However, depending on the subset of genotypes and methodologies used, numerous greenhouse and field studies in India have indicated significant genotypic variation for resistance and tolerance. For example, in a field test Ali and Ahmad (2000) found no immunity amongst 600 lines screened against *M. javanica*, but seven entries sustained only minimal populations; for *M. incognita*, Hassan and Devi (2004) reported 34 of 72 genotypes to be highly resistant, whereas Mhase *et al.* (1999) found all 108 genotypes tested susceptible. No resistance has been reported for a third species, *M. artiella* (Sharma *et al.*, 1994). There are limited efforts on studying inheritance of resistance to nematodes. Resistance to reniformis nematode has been reported to be controlled by a single dominant gene (Kumar *et al.*, 2005b).

Absence of effective host resistance to cyst nematode within the cultigen (Di Vito *et al.*, 1988) led to a search amongst the wild *Cicer* spp.: high resistance was found in three of them including *C. reticulatum*. This species was subsequently used as a resistance source in crosses with a kabuli parent and resistant segregants were recovered (Malhotra *et al.*, 2002). A backcrossing programme has enabled recovery of resistant lines with acceptable plant and seed type (R.S. Malhotra, ICARDA, 2006, personal communication). In Australia *C. reticulatum* and *C. echinospermum* have also been used as sources of superior resistance to two root-lesion nematode species, *Pratylenchus thornei* and *P. neglectus*. Greenhouse tests showed both wild species to be significantly more resistant than a range of chickpea genotypes. In a targeted breeding programme, superior resistance was subsequently recovered in fixed interspecific lines and BC_1F_2 progeny (J.P. Thompson, Queensland 2006, personal communication).

Resistance to multiple stresses

The chickpea crop is generally attacked by more than one stress in most chickpea-growing regions and, thus, cultivars with multiple stress resistance are needed to enhance and stabilize chickpea production. While numerous cultivars possessing good resistance to individual stress are available, limited numbers have been found resistant to multiple stresses. Breeding for multiple stress resistance is a complex task, particularly when donors with multiple stress resistance are not available. In such cases, resistance genes from different sources need to be combined using multiple crosses.

It is important to identify donors for multiple stress resistance, so that the number of crosses required to develop multiple stress tolerance cultivars can be minimized. Pundir *et al.* (1988) compiled data available up to 1983 on evaluation of ICRISAT's chickpea germplasm. Several accessions were identified to have resistance to two diseases, and a few accessions had resistance to more than two diseases. For example, ICC 12237 was resistant to fusarium wilt, dry root rot and black root rot, and ICC 11321 was resistant to fusarium wilt, ascochyta blight and botrytis grey mould. More than 200 accessions resistant to fusarium wilt and root rot were tested at 24 locations in 11 countries during a 5-year period to identify broad-based and stable resistance to fusarium wilt and root rots (Nene *et al.*, 1989). Three accessions originating from India (ICC10803, ICC 11550 and ICC 11551) and three from Iran (ICC 2862, ICC 9023, ICC 9032) were resistant at about two-thirds of the locations indicating their broad-based resistance to fusarium wilt and root rots (Nene *et al.*, 1989).

Excellent sources of multiple resistance have been identified in wild species. For example, accessions ILWC 70 and ILWC 73 of *C. bijugum* have resistance to ascochyta blight (score 3), fusarium wilt (score 1), bruchid (score 1) and cyst nematode (score 2) and tolerance to low temperature stress (score 2–3) (Roberson *et al.*, 1995). Thus, it would be more efficient to use wild species in development of multiple stress resistance cultivars.

Multiple crosses have been used to develop cultivars with resistance to multiple stresses. Several multiple-resistant cultivars have been developed at IARI (Kumar *et al.*, 2003; Yadav *et al.*, 2004a). For example, BG 1063 and BG 1075 have resistance to fusarium wilt, root rots and stunt; PUSA 1088 (kabuli), BG 1089 (kabuli), BG 1093 and BG 72 have resistance to fusarium wilt and root rots and tolerance to moisture stress; and BGD 112 and PUSA 362 have resistance to fusarium wilt and root rots and tolerance to moisture stress and extremes of temperatures (chilling and heat). ICRISAT scientists have recently developed four desi chickpea breeding lines (e.g. ICCV 05529–05532), which combine resistance to ascochyta blight (score 3–4), botrytis grey mould (score 4) and fusarium wilt (<10% plant mortality) (P.M. Gaur, ICRISAT, 2006, personal communication).

Only two annual wild species (*C. reticulatum* and *C. echinospermum*), which can be easily crossed with the cultigen, have been used in breeding programmes. One variety (PUSA 1103) has been developed by IARI using *C. reticulatum* as one of the parents in crosses. This variety is tolerant to drought stress and resistant to root diseases and stunt virus (Yadav *et al.*, 2002a,b).

Development of multiple disease–resistant varieties through conventional breeding approaches is cumbersome and time-consuming due to difficulties in selecting plants with desired combination of genes in segregating generations. MAS can provide an effective and efficient breeding tool for detecting, tracking and combining stress-resistant genes in segregating generations and, thus, significantly reduces the time and the size of the populations required to develop a multiple stress–resistant cultivar.

Wide Adaptation

Most national breeding programmes aim to develop cultivars for specific adaptation rather than wide adaptation as developing specific varieties for each target region, instead of widely adapted varieties, can exploit positive genotype × location (GL) interactions to increase crop yields (Annicchiarico *et al.*, 2005).

Many national programmes initially evaluate promising breeding lines at multilocations and later identify and release varieties for a particular region or subregion based on specific adaptations. Wide adaptation of a variety is challenged by large genotype × environment (GE) interactions. It has been suggested that only GL interaction, rather than all kinds of GE interaction, can be exploited by selecting for specific adaptation or by growing specifically adapted genotypes (Annicchiarico, 2002).

GE interactions of crossover type between the testing locations may occur when the locations are climatically very diverse. Selection of the genotypes with wide adaptation or low GE interaction (good average yield across locations and years) often discards lines with good performance in unfavourable sites and poor response to favourable conditions because of their low average grain yield, though these would be the ideal lines for farmers in unfavourable locations. Therefore, breeding for difficult environments must be based on the exploitation of specific adaptation, and this in turn can only be done by selecting directly in the target environments.

The nurseries supplied globally by the international institutes provide an opportunity to assess performance of genotypes at diverse locations across countries. Some widely adapted genotypes have been identified through these nurseries, as evidenced by their release in two or more countries. Varieties that have been released in three or more countries include FLIP 84-92C, FLIP 84-79C (Algeria, Morocco, Tunisia), ICCV 2 (India, Sudan, Myanmar) and ICCV 88202 (Australia, Myanmar, India).

A comprehensive study on GE interactions on chickpea genotypes in India was recently conducted (Berger *et al.*, 2006). Forty-six genotypes (41 of Indian, 3 of Australian and 2 of Mediterranean-basin origin) were evaluated for 3 years at seven locations covering the major chickpea-growing areas. The consistent association between germplasm origin and specific adaptation to North and South India suggests that the varieties developed by state-based breeding programmes generally have specific adaptation to local environments. The most consistently high-yielding varieties were ICCV 10, BG 391 and BG 1006.

IARI scientists are making concerted efforts to develop varieties with wide adaptation. The breeding strategy being used involves development of multiple crosses using diverse sources of desirable donors (including from wild species), advancing segregating populations under stress environment and selecting for multiple traits simultaneously (Yadav and Kumar, 2005). Several high-yielding, widely adapted varieties have been developed using this approach. A large-seeded kabuli variety PUSA 1108 was developed from a multiple cross [(BG 315 × ILC 72) × (ICCV 13 × Flip 85-11)] × (ICCV 32 × Surutoto 77) and released in 2006 for North India.

ICRISAT has extensively used multiple crosses in its breeding programme over the last two decades. Several varieties were developed from crosses involving more than four parents in the pedigree [e.g. ICCV 2, ICCV 3, ICCV 95311 (released as KAK 2), and ICCV 95418 (released as Virat)]. However, the widely adapted varieties have been found to originate from both single (e.g. ICCV 10, ICCV 88202) and multiple (ICCV 2) crosses.

Plant Type

Donald (1968) defined crop ideotype as an idealized plant type with a specific combination of characteristics that helps to maximize community performance of plants for higher grain yield per unit area. The basic assumption is that a high-yielding plant in isolation would perform equally well in a population. The relationship between the individual plant and plants in population remains reasonably valid in many crop species, particularly cereals or determinate crops, where the yield of the main shoot is the major contributing factor. However, this does not necessarily happen in indeterminate species such as chickpea (Sinha and Swaminathan, 1984). A sharp decline in yield per plant occurs in chickpea with increasing plant population.

Traditionally, chickpea has been grown mostly as a rainfed crop. The breeding programmes have generally focused on development of varieties suitable to rainfed conditions. The available varieties do not respond favourably to irrigated conditions. Although one or two irrigations considerably enhance the yield, particularly in moisture stress conditions, a decline in yield occurs at higher irrigation levels (Sinha *et al.*, 1985). Excessive vegetative growth is often seen in cooler long-season environments (e.g. North India) when an unstressed crop is irrigated or receives rains at flowering stage.

The concept of ideotype would be different for each geographic region and growing condition, as the plant traits required for specific adaptation to these environments would be different. Lower population density would be desirable for rainfed conditions to enable plants to reach maturity before the soil moisture is depleted, with a provision that the plant has plasticity for growth if moisture availability is enhanced. Plant type that can resist excessive vegetative growth is needed for favourable environments, particularly irrigated conditions.

Some attempts have been made to define chickpea ideotypes for different growing conditions. Saxena and Johansen (1990) suggested that an ideotype for drought stress environments should have early maturity, deep root system and smaller leaf size. In the Mediterranean climate, the crop experiences a cool and wet winter followed by rapid warming in spring leading to terminal drought. Sedgley *et al.* (1990) contended that the ideotype for Mediterranean environments should include early flowering and tolerance to cold during flowering. In irrigated and high-fertility conditions, a compact plant type with erect growth habit and short internodes may be useful as it could resist excessive vegetative growth and accommodate more plants per unit area (Dahiya and Lather, 1990).

An effort has been made to breed for erectness combined with tall growth by hybridization between conventional spreading and tall types (Bahl, 1980). Many tall-type progenies have been obtained but they maintain a poor harvest index because their internodes are long and hence substantial amounts of assimilates are spent on structural parts of the plant. Reducing internodal length and increasing the number of podded nodes may be useful approaches to improving harvest index in the plant.

A spontaneous mutant with short internodes and compact growth habit (E100YM) has been used to develop promising progeny which can be grown at high plant intensity (Lather, 2000). An induced mutant with short internodes and compact growth habit (JGM 1) has also been obtained. The genes for compact growth habit in E100YM and JGM 1 are not allelic (P.M. Gaur, ICRISAT, 2006, personal communication), so JGM 1 provides an alternative source of short internodes and compact growth habit.

Some of the wild species (e.g. *C. reticulatum* and *C. echinospermum*) produce profuse branching and podding. These traits have been introgressed in the cultivated species to improve yield potential. Singh and Ocampo (1997) obtained high-yielding kabuli lines from crosses of cultigen with *C. reticulatum* and *C. echinospermum*. Singh *et al.* (2005) also recovered interspecific derivatives from crosses with *C. reticulatum* that were higher-yielding than the parents. A high-yielding desi chickpea variety (PUSA 1103) developed using *C. reticulatum* as one of the parents (Yadav *et al.*, 2002b) has recently been released for commercial cultivation in North India.

Seed Traits

The market price of chickpea is primarily decided by the appearance (size, shape and colour) and size of the grain. Kabuli chickpea is generally used as whole grain and most desis are used for making splits (dhal) and flour (besan), so the preferred seed traits for these two types of chickpea vary considerably.

Desi type

Most markets prefer small to medium desi seeds (100 seed weight 16–22 g) and pay only modest premiums for the large grades. There is preference for yellow to light brown seed coat colour, and small niche markets exist for green- and black-seeded types. More than 70% of desi chickpea is used for making dhal and a portion is further processed into flour (besan). High milling efficiency (dhal recovery) is therefore an important trait.

Breeding programmes have given greater emphasis to market-preferred seed traits in recent years. Most desi varieties now have 100 seed weight approaching 22 g. However, some large-seeded varieties (>25 g/100 seed) have also been developed to meet a specific demand for this category, e.g. Pusa 256, Pusa 362, Pusa 72, Pusa 1103, K 850, Vishwas and Vishal in India; CDC Cabri and CDC Nika in Canada; and Gully, Lasseter and Kyabra in Australia. Previously many varieties with dark brown seed coat were released, but there is now increasing emphasis on a lighter colour (yellow to light brown).

In the Indian subcontinent a small proportion of desi chickpea is used for parching (roasting) to produce the product *Puthana*. Two varieties (JG 5 and JGG 1) with pea-shaped seed (intermediate between desi and kabuli types) and attractive pink seed coat colour (called *gulabi* in Hindi) were released in India specifically for use in parching. Other desi-type varieties are also widely used for parching and there is a premium market for large-seeded types with 9–10 mm seed size. Recently, desi chickpea–breeding genotypes (DG 5025 to DG 5051) have been developed by IARI which have seed size between 40 and 52 g/100 seed (S.S. Yadav, IARI, New Delhi, 2006, personal communication).

Kabuli type

Seed size is the most important quality trait for kabuli chickpeas. In general, larger seeds get a high price premium. There is generally a preference for white or beige seed coat colour and ram's head seed shape. As the bulk of kabuli chickpea is cooked as whole grain, cooking time and seed volume expansion (on soaking) are considered important quality traits.

There is increasing demand for large kabuli chickpea in South Asia. Most kabuli-breeding programmes in India have shifted emphasis from medium (~25 g/100 seed) to large (>30 g/100 seed) seed size, as no kabuli variety at present is recommended for release that has seed size less than 25 g/100 seed. Many recently released varieties have seed size >30 g/100 seed or 8–9 mm diameter (e.g. PUSA 1053, PUSA 1088, PUSA 1105, PUSA 1108, BG 2024, JGK 1, JGK 2, JGK 3, KAK 2, LBeG 7, Vihar and Virat). None of the released varieties in India has seed size greater than 40 g/100 seed. However, several breeding lines with seed size between 40 and 55 g/100 seed are being tested in All India Coordinated Research Project (AICRP) trials. Some of the genotypes (e.g. DG 5000 to DG 5024, DG 5050 and 5051) developed at IARI in this seed size category combine early maturity, high yield potential and resistance to multiple stresses (S.S. Yadav, IARI, New Delhi, 2006, personal communication).

There is also increasing demand for extra-large kabuli (>55 g/100 seed) chickpea, as this variety commands very high premiums. Some varieties released in this category are Kimberley Large (60 g/100 seed) in Australia and Dylan (56 g/100 seed) in the USA. The demand for extra-large kabulis in the Indian subcontinent is met through import, mainly from Mexico and Turkey. The extra-large kabuli germplasm introduced in India from other countries was found poorly adapted (Yadav *et al.*, 2004 b,c). The Indian National Programme, in partnership with ICRISAT, has recently launched a project to fast-track development of fusarium wilt–resistant, extra-large-seeded kabuli varieties adapted to local environments.

Nutritional Quality

Chickpea is an important supplementary protein source in cereal and starchy diets for millions of people in the developing countries of Asia, Africa and the Middle East. In addition to having high protein content (20–22%), chickpea is rich in fibre and minerals (phosphorus, calcium, magnesium, iron and zinc) and its lipid fraction is high in unsaturated fatty acids. It contains higher amounts of carotenoids (such as β-carotene, cryptoxanthin, lutein and zeaxanthin) than the genetically engineered 'golden rice' (Abbo *et al.*, 2005). While it is a cheap source of protein and energy in the developing world, it is gaining popularity as health food in the developed countries.

Chickpea does not contain any specific major antinutritional factors, such as ODAP in grasspea, vicin in faba bean and trypsin inhibitors in soybean. It has lower content of phytic acid (9.6mg/g) as compared to mungbean (12.0), pigeonpea (12.7), urdbean (13.7) and soybean (36.4) (Chitra *et al.*, 1995). The only negative factor ascribed to chickpea consumption is more flatulence due to a higher concentration of raffinose family oligosaccharides (RFOs) than other dry edible legumes.

There has been negligible input into the improvement of nutritional quality of chickpea. Although some studies have indicated genetic variability for different quality traits, this variation has yet to be exploited in breeding programmes.

Protein content of existing cultivars is generally in the range of 18–22% but much larger variability (12.4–32.5%) exists in the cultivated and wild species, and this could be exploited to breed higher protein (~25%) varieties. The nutritional value of off-grade chickpeas was found to be acceptable in ruminant and pig diets (Mustafa *et al.*, 2000). The sulphur-containing amino acids methionine and cystine are the limiting amino acids. Transgenic technology is being used to enhance the level of sulphur-containing amino acids because the required variation is absent from the primary gene pool. Transgenics developed by introducing a seed-specific chimeric gene encoding sunflower seed albumin (SSA) produced 24–94% higher methionine, but 10–15% lower cysteine than comparable non-transgenic chickpea (Higgins *et al.*, 2004).

There is a need to assess genetic variability for various nutritional and antinutritional traits in the germplasm of cultivated and wild species. Studies are also needed on GE interactions and genetics of these traits. Identification of contrasting parents for the content of each nutritional and antinutritional trait would be required for development of genetic populations needed for mapping of genes controlling these traits. Availability of such basic information on genetic variability and genetics of nutritional and antinutritional traits would help in development of breeding strategies (both conventional and biotechnological approaches) for improvement of these traits in chickpea.

Conclusions

Progress in chickpea improvement has come exclusively from conventional breeding approaches. There has been rapid progress in developing biotechnological tools and approaches in the recent past. The technology for development of chickpea transgenics is well established and the genome map of chickpea is rapidly expanding. Molecular markers have been identified for some traits and for many other traits the research is in progress. It is expected that the combined application of biotechnological and conventional approaches will lead to rapid progress in the coming years.

Historically chickpea-breeding programmes have given very little emphasis to improving industrial and nutritional quality of chickpea. The use of chickpea in ready-to-eat foods, products requiring less cooking time, fast foods, snacks, functional foods, nutraceuticals and dietary supplements is rapidly increasing. In the future, breeding programmes are likely to take into account the industry's emerging quality requirements and also give due emphasis to the enhancement of nutritional quality.

References

Abbo, S., Molina, C., Jungmann, R., Grusak, M.A., Berkovitch, Z., Reifen, R., Kahl, G., Winter, P. and Reifen, R. (2005) Quantitative trait loci governing carotenoid concentration and weight in seeds of chickpea (*Cicer arietinum* L.). *Theoretical and Applied Genetics* 111, 185–195.

Ahmad, F., Gaur, P.M. and Croser, J. (2005) Chickpea (*Cicer arietinum* L.). In: Singh, R.J. and Jauhar, P.P. (eds) *Grain Legumes, Vol. 1: Genetic Resources, Chromosome Engineering, and Crop Improvement.* CRC Press, Boca Raton, Florida, pp. 187–217.

Ali, S.S. and Ahmad, R. (2000) Screening of chickpea germplasm against nematode. *International Chickpea and Pigeonpea Newsletter* 7, 8–9.

Anbessa, Y., Warkentin, T.D., Vandenberg, A. and Ball, R. (2006a) Inheritance of time to flowering in chickpea in a short season temperate environment. *Journal of Heredity* 97, 55–61.

Anbessa, Y., Warkentin, T.D., Vandenberg, A. and Bandara, M. (2006b) Heritability and predicted gain from selection in components of crop duration in divergent chickpea cross populations. *Euphytica* 152, 1–8.

Annicchiarico, P. (2002) *Genotype × Environment Interactions: Challenges and Opportunities for Plant Breeding and Cultivar Recommendations.* FAO, Rome, Italy.

Annicchiarico, P., Bellah, F. and Chiari, T. (2005) Defining subregions and estimating benefits for a specific-adaptation strategy by breeding programs: a case study. *Crop Science* 45, 1741–1749.

Asharf, M. and Waheed, A. (1992) Screening chickpea (*Cicer arietinum* L.) for salt tolerance. *Tropenlandwirt* 93, 45–55.

Bahl, P.N. (1980) Kabuli–desi introgression and genesis of new plant type in chickpea. In: *Proceedings of International Workshop on Chickpea Improvement.* ICRISAT, Hyderabad, India, pp. 75–80.

Bakr, A.M., Ali, S.H., Ali, M.A. and Rahman, M.A. (2002) Chickpea status and production constraints in Bangladesh. In: Bakr, M.A., Siddique, K.H.M. and Johansen, C. (eds) *Integrated Management of Botrytis Grey Mould of Chickpea in Bangladesh and Australia: Summary Proceedings of a Project Inception Workshop.* 1–2 June 2002, Bangladesh Agriculture Research Institute, Joydebpur, Gazipur, Bangladesh; Centre for Legumes in Mediterranean Agriculture, Western Australia, Australia, pp. 19–32.

Berger, J.D., Ali, M., Basu, P.S., Chaudhary, B.D., Chaturvedi, S.K., Deshmukh, P.S., Dharmaraj, P.S., Dwivedi, S.K., Gangadhar, G.C., Gaur, P.M., Kumar, J., Pannu, R.K., Siddique, K.H.M., Singh, D.N., Singh, D.P., Singh, S.J., Turner, N.C., Yadava, H.S. and Yadav, S.S. (2006) Genotype by environment studies demonstrate the critical role of phenology in adaptation of chickpea (*Cicer arietinum* L.) to high and low yielding environments of India. *Field Crops Research* 98, 230–244.

Brinsmead, R.B., Rettke, M.L., Irwin, J.A.G., Ryley, M.J. and Langdon, P.W. (1985) Resistance in chickpea to *Phytophthora megaperma* f. sp. *Medicaginis. Plant Disease* 69, 504–506.

Chalam, T.V., Reddy, M.V., Subbayya, J., Nene, Y.L. and Beniwal, S.P.S. (1986) Screening of chickpea for resistance to cucumber mosaic and bean yellow mosaic viruses. *International Chickpea Newsletter* 14, 5–26.

Chandra, S., Buhariwalla, H.K., Kashiwagi, J., Harikrishna, S., Sridevi, K.R., Krisnamurthy, L., Serraj, R. and Crouch, J.H. (2004) Identifying QTL-linked markers in marker-deficient crops. Poster in International Crop Science Congress, 26 September–1 October 2004, Brisbane, Australia. Available at: http://www.cropscience.org.au/icsc2004/poster/3/4/1/795_chandras.htm

Chitra, U., Vimala, V., Singh, U. and Geervani, P. (1995) Variability in phytic acid content and protein digestibility of grain legumes. *Plant Foods for Human Nutrition* 47, 163–172.

Clarke, H. and Siddique, K.H.M. (2004) Response of chickpea genotypes to low temperature stress during reproductive development. *Field Crops Research* 90, 323–334.

Clarke, H., Khan, T.N. and Siddique, K.H.M. (2004) Pollen selection for chilling tolerance at hybridization leads to improved chickpea cultivars. *Euphytica* 139, 65–74.

Croser, J.S., Clarke, H.J., Siddique, K.H.M. and Khan, T.N. (2003) Low temperature stress: implications for chickpea (*Cicer arietinum* L.) improvement. *Critical Reviews in Plant Sciences* 22, 185–219.

Dahiya, B.S. and Lather, V.S. (1990) A breakthrough in chickpea yields. Paper 1. *International Chickpea Newsletter* 22, 6–8.

Di Vito, M., Greco, N., Singh, K.B. and Saxena, M.C. (1988) Response of chickpea germplasm lines to *Heterodera ciceri* attack. *Nematologia Mediterranea* 16, 17–18.

Donald, C.M. (1968) The breeding of crop ideotypes. *Euphytica* 17, 385–403.

Dua, R.P. and Sharma, P.C. (1995) Salinity tolerance of kabuli and desi chickpea genotypes. *International Chickpea and Pigeonpea Newsletter* 2, 19–22.

Dua, R.P., Chaturvedi, S.K. and Sewak, S. (2001) *Reference varieties of chickpea for IPR regime*. Indian Institute of Pulses Research, Kanpur, India.

FAOSTAT data (2006) Available at: http://faostat.fao.org/faostat/ (last updated 24 January 2006).

Gan, Y.T., Siddique, K.H.M., MacLeod, W.J. and Jayakumar, P. (2006) Management options for minimizing the damage by ascochyta blight (*Ascochyta rabiei*) in chickpea (*Cicer arietinum* L.). *Field Crops Research* 97, 121–134.

Gaur, P.M. and Gowda, C.L.L. (2005) Trends in world chickpea production, research and development. In: Knights, E.J. and Merrill, R. (eds) *Proceedings of Focus 2005: Chickpea in the Farming Systems*. 21–23 September 2005, Goondiwindi, Qld, Australia. Pulse Australia, Toowoomba, Australia, pp. 8–15.

Gaur, P.M., Pande, S., Upadhyaya, H.D. and Rao, B.V. (2006) Extra-large kabuli chickpea with high resistance to fusarium wilt. *International Chickpea and Pigeonpea Newsletter* 13, 5–7.

Gaur, P.M., Krishnamurthy, L. and Kashiwagi, J. (2007) Current status and challenges of improving root traits in chickpea (*Cicer arietinum* L.). *Plant Production Science* (in press).

Gowda, C.L.L., Ramesh, S., Chandra, S. and Upadhyaya, H.D. (2005) Genetic basis of pod borer (*Helicoverpa armigera*) resistance and grain yield in desi and kabuli chickpea (*Cicer arietinum* L.). *Euphytica* 145, 199–214.

Gurha, S.N., Singh, V. and Naimuddin, R.A. (1998) Management of stem rot disease in chickpea and rajmash. *Indian Journal of Pulses Research* 11, 127–129.

Gurha, S.N., Ghosh, A., Singh, R.A. and Naimuddin, V.D. (2002) Seed-borne nature and possible resistance against Alternaria blight in chickpea. *Annals of Plant Protection Science* 10, 399–400.

Hassan, M.G. and Devi, S. (2004) Evaluation of mungbean, urdbean, pea, lentil and chickpea germplasm for resistance to root-knot nematode (*Meloidogyne incognita*). *Indian Journal of Nematology* 34, 228–229.

Haware, M.P. (1998) Diseases of Chickpea. In: Allen, D.J. and Lenne, J.M. (eds) *The Plant Pathology of Food and Pasture Legumes*. CAB International, Wallingford, UK, pp. 473–516.

Haware, M.P. and Nene, Y.L. (1982) Races of *Fusarium oxysporum* f. sp. *ciceri*. *Plant Disease* 66, 809–810.

Haware, M.P., Nene, Y.L., Pundir, R.P.S. and Narayana Rao, J. (1992) Screening of world chickpea germplasm for resistance to fusarium wilt. *Field Crops Research* 30, 147–154.

Higgins, T.J., Chiaiese, P., Sarmah, B., Molvig, L. and Tabe, L. (2004) Genetic improvement of chickpeas: seed composition and protection. In: *Proceedings 5th European Conference on Grain Legumes and 2nd International Conference on Legume Genomics and Genetics*. 7–11 June 2004, Dijon, France, pp. 21–22.

Horn, N.M., Makkouk, K.M., Kumari, S.G., van den Heuvel, J.F.J.M. and Reddy, D.V.R. (1995) Survey of chickpea (*Cicer arietinum* L.) for chickpea stunt disease and associated viruses in Syria, Turkey and Lebanon. *Phytopathologie Mediterranea* 34, 192–198.

Horn, N.M., Reddy, S.V., van den Heuvel, J.F.J.M. and Reddy, D.V.R. (1996) Survey of Chickpea (*Cicer arietinum* L.) for chickpea stunt disease and associated viruses in India and Pakistan. *Plant Disease* 80, 286–290.

ICRISAT (1994) Cold-tolerant chickpea varieties ICCV 88503, ICCV 88506, and ICCV 88510. *Plant Material Description no. 53*. ICRISAT, Hyderabad, India.

ICRISAT (2003) Yellow gold in Myanmar. In: *Building a Strong ICRISAT*. Annual Report 2003. ICRISAT, Hyderabad, India, pp. 12–13.

ICRISAT (2004) *Archival Report: Global Theme Biotechnology 2004*. ICRISAT, Hyderabad, India.

IIPR (Indian Institute of Pulses Research) (1997) *Package of practices for pulse crops in India*. Indian Institute of Pulses Research, Kanpur, India.

Jimenez-Diaz, R.M., Basallote-Ureba, M.J. and Rapoport, H. (1989) Colonization and pathogenesis in chickpea infected by races of *Fusarium oxysporum* f. sp. *ciceri*. In: Tjamos, E.C. and Beckman, C. (eds) *Vascular Wilt Disease of Plants*. NATO ASI Series, Vol. H28. Springler-Verlag, Berlin, Germany, pp. 113–121.

Kaiser, W.J. and Danesh, D. (1971) Etiology of virus-induced wilt of *Cicer arietinum*. *Phytopathology* 61, 453–457.

Kashiwagi, J., Krishnamurthy, L., Upadhyaya, H.D., Krishna, H., Chandra, S., Vincent, V. and Rachid, S. (2005) Genetic variability of drought-avoidance root traits in the mini-core germplasm collection of chickpea (*Cicer arietinum* L.). *Euphytica* 146, 213–222.

Kashiwagi, J., Krishnamurthy, L., Crouch, J.H. and Serraj, R. (2006) Variability of root length density and its contributions to seed yield in chickpea (*Cicer arietinum* L.) under terminal drought stress. *Field Crops Research* 95, 171–181.

Kathiria, K.B., Nayagapara, D.R., Vaddoria, M.A. and Poshiya, V.K. (1997) Screening of chickpea genotypes for salinity tolerance during germination and early seedling growth. *Gujarat Agricultural University Research Journal* 22, 28–32.

Knights, E.J. and Siddique, K.H.M. (2002) Chickpea status and production constraints in Australia. In: Bakr, M.A., Siddique, K.H.M. and Johansen, C. (eds) *Integrated Management of Botrytis Grey Mould of Chickpea in Bangladesh and Australia*. Summary Proceedings of a Project Inception Workshop. Bangladesh Agricultural Research Institute, Joydebpur, Gazipur, Bangladesh, pp. 33–41.

Knights, E.J., Southwell, R.J. and Schwinghamer, M.W. (2003) Evaluation of wild *Cicer* species for resistance to phytophthora root rot. In: Sharma, R.N., Shrivastava, G.K., Rathore, A.L., Sharma, M.L. and Khan, M.A. (eds) *Chickpea Research for the Millennium: Proceedings of International Chickpea Conference*. 20–22 January 2003, Indira Gandhi Agricultural University, Raipur, India, pp. 54–57.

Kumar, J. and Rao, B.V. (1996) Super early chickpea developed at ICRISAT Asia Center. *International Chickpea and Pigeonpea Newsletter* 3, 17–18.

Kumar, J., Yadav, S.S., Singh, B.B., Hedge, V.S. and Yadav, S.K. (2003) Incorporation of multiple resistance in chickpea cultivars. In: *Chickpea Research for the Millennium, Proceedings of the International Chickpea Conference*. 20–22 January 2003, Indira Gandhi Agricultural University, Raipur, India, pp. 28–32.

Kumar, J., Yadav, S.S., Kumar, R., Hegde, V., Singh, A.K. and Singh, N. (2005a) Evaluation of chickpea (*Cicer arietinum* L.) germplasm for stunt virus and phenological traits. *Abstracts of 4th International Food Legumes Research Conference*. 18–22 October 2005, New Delhi, India, p. 372.

Kumar, J., Yadav, S.S., Mishra, S.D. and Kumar, R. (2005b) Inheritance of nematode resistance in chickpea. *Abstract International Conference on Sustainable Crop Production in Stress Environments: Management and Genetic Options*. 9–12 February 2005, Jabalpur, Madhya Pradesh, India, p. 41.

Lateef, S.S. and Pimbert, M.P. (1990) The search for host resistance to *Helicoverpa armigera*. In: *Chickpea and Pigeonpea at ICRISAT: Summary Proceedings of First Consultative Group Meeting on Host Selection Behaviour of Helicoverpa armigera*. March

1990, ICRISAT, Hyderabad, India, pp. 14–18.

Lather, V.S. (2000) Promising chickpea ideotype for higher plant density. *International Chickpea and Pigeonpea Newsletter* 7, 26–27.

Malhotra, R.S., Singh, K.B., van Rheenen, H.A. and Pala, M. (1996) Genetic improvement and agronomic management of chickpea with emphasis on the Mediterranean region. In: Saxena, N.P., Saxena, M.C., Johansen, C., Virmani, S.M. and Harris, H. (eds) *Adaptation of Chickpea in the West Asia and North Africa Region*. ICRISAT, Hyderabad, India, and ICARDA, Aleppo, Syria, pp. 217–232.

Malhotra, R.S., Singh, K.B., Di Vito, M., Greco, N. and Saxena, M.C. (2002) Registration of ILC 10765 and ILC 10766 chickpea germplasm lines resistant to cyst nematode. *Crop Science* 42, 1756.

Malhotra, R.S., Baum, M., Udupa, S.M., Bayaa, B., Kabbabe, S. and Khalaf, G. (2003) Ascochyta blight resistance in chickpea: present status and future prospects. In: Sharma, R.N., Shrivastava, G.K., Rathore, A.L., Sharma, M.L. and Khan, M.A. (eds) *Chickpea Research for the Millennium: Proceedings of International Chickpea Conference*. 20–22 January 2003, Indira Gandhi Agricultural University, Raipur, India, pp. 108–117.

Maliro, M.F.A., McNeil, D., Kollmorgen, J., Pittock, C. and Redden, R. (2004) Screening chickpea (*Cicer arietinum* L.) and wild relatives germplasm from diverse country sources for salt tolerance. *Proceedings of International Crop Science Congress*. 26 September–1 October, Brisbane, Australia. Available at: http://www.cropscience.org.au/icsc2004/poster/3/6/2/1280_malirom.htm

Mhase, N.L., Mote, U.N., Shelke, S.S. and Kadam, D.B. (1999) Evaluation of varieties/lines of pulse crops against root-knot nematode, *Meloidogyne incognita* race-2. *Indian Journal of Nematology* 29, 78–111.

Millan, T., Clarke, H.J., Siddique, K.H.M., Buhariwalla, H.K., Gaur, P.M., Kumar, J., Gil, J., Kahl, G. and Winter, P. (2006) Chickpea molecular breeding: new tools and concepts. *Euphytica* 147, 81–103.

Musa, A.M., Harris, D., Johansen, C. and Kumar, J. (2001) Short-duration chickpea to replace fallow after *aman* rice: the role of on-farm seed priming in the high Barind tract of Bangladesh. *Experimental Agriculture* 37, 509–521.

Mustafa, A.F., Thacker, P.A., McKinnon, J.J., Christensen, D.A. and Racz, V.J. (2000) Nutritional value of feed grade chickpeas for ruminants and pigs. *Journal of the Science of Food and Agriculture* 80, 1581–1588.

Nene, Y.L., Haware, M.P. and Reddy, M.V. (1981) Chickpea diseases: resistance-screening techniques. *Information Bulletin No. 10*. ICRISAT, Hyderabad, India, p. 12.

Nene, Y.L., Haware, M.P., Reddy, M.V., Phillips, J.C., Castro, E.L., Kotasthane, S.R., Gupta, O., Singh, G., Shukla, P. and Sah, R.P. (1989) Identification of broad-based and stable resistance to wilt and root-rots of chickpea. *Indian Phytopathology* 42, 499–505.

Pande, S., Singh, G., Narayana Rao, J., Bakr, M.A., Chaurasia, P.C.P., Joshi, S., Johansen, C., Singh, S.D., Kumar, J., Rahman, M.M. and Gowda, C.L.L. (2002) *Integrated Management of Botrytis Grey Mold of Chickpea. Information Bulletin No. 61*. ICRISAT, Hyderabad, India, p. 26.

Pande, S., Kishore, G.K. and Narayana Rao, J. (2004) Evaluation of chickpea lines for resistance to dry root rot caused by *Rhizoctonia bataticola*. *International Chickpea and Pigeonpea Newsletter* 11, 37–38.

Pande, S., Siddique, K.H.M., Kishore, G.K., Bayaa, B., Gaur, P.M., Gowda, C.L.L., Bretag, T.W. and Crouch, J.H. (2005) Ascochyta blight of chickpea (*Cicer arietinum* L.): a review of biology, pathogenicity and disease management. *Australian Journal of Agricultural Research* 56, 317–332.

Pande, S., Galloway, J., Gaur, P.M., Siddique, K.H.M., Tripathi, H.S., Taylor, P., MacLeod, M.W.J., Basandrai, A.K., Bakr, A., Joshi, S., Krishna Kishore, G., Isenegger, D.A., Narayana Rao, J. and Sharma, M. (2006) Botrytis grey mould of chickpea: a review of biology, epidemiology and disease management. *Australian Journal of Agricultural Research* 57, 1137–1150.

Phillips, J.C. (1988) A distinct race of chickpea wilt in California. *International Chickpea Newsletter* 18, 19–21.

Pundir, R.P.S., Reddy, K.N. and Mangesha, M.H. (1988) *ICRISAT Chickpea Germplasm Catalog: Evaluation and Analysis*. ICRISAT, Hyderabad, India, p. 94.

Rahul, C., Singh, I.S. and Gupta, A.K. (1995) Inheritance of resistance to Botrytis grey mould in chickpea (*Cicer arietinum* L.). *Legume Research* 18, 1–4.

Rao, P.K.A. and Haware, M.P. (1987) Inheritance of dry root rot (*Rhizoctonia bataticola*) resistance in chickpea (*Cicer arietinum*). *Plant Breeding* 98, 349–352.

Reddy, M.V. and Singh, K.B. (1984) Evaluation of a world collection of chickpea germplasm accessions for resistance to Ascochyta blight. *Plant Disease* 68, 900–901.

Rewal, N. and Grewal, J.S. (1989) Inheritance of resistance to *Botrytis cinerea* Pers. in *Cicer arietinum* L. *Euphytica* 44, 61–63.

Rizvi, A.H., Manohar, M., Kumar, J. and Yadav, S.S. (2006) Functional genomics for drought tolerance in chickpea (*Cicer arietinum* L.). *Abstract Proceedings of International meeting on Biotic and Abiotic Stress Responses in Plants held between 11–13 December 2006 at International Centre for Genetic Engineering and Biotechnology*. New Delhi, India, p. 38.

Robertson, L.D., Singh, K.B. and Ocampo, B. (1995) *A Catalog of Annual Wild Cicer Species*. ICARDA, Aleppo, Syria, p. 171.

Ryan, J.G. (1977) A global perspective on pigeonpea and chickpea sustainable production systems: present status and future potential. In: Asthana, A.N. and Ali, M. (eds) *Recent Advances in Pulses Research*. Indian Institute of Pulses Research, Kanpur, India, pp. 1–31.

Sandhu, J.S., Bains, T.S. and Sidhu, P.S. (2002) Evaluation of super early chickpea genotypes for vegetable purpose as a catch crop. *International Chickpea and Pigeonpea Newsletter* 9, 10–12.

Sanyal, I., Singh, A.K., Kaushik, M. and Amla, D.V. (2005) *Agrobacterium*-mediated transformation of chickpea (*Cicer arietinim* L.) with *Bacillus thuringiensis cry1Ac* gene for resistance against pod borer insect *Helicoverpa armigera*. *Plant Science* 168, 1135–1146.

Saxena, N.P. (1984) Chapter 12: Chickpea. In: Goldsworthy, P.R. and Fisher, N.M. (eds) *The physiology of Tropical Field Crops*. John Wiley, Chichester, UK, pp. 419–452.

Saxena, N.P. (2003) Management of drought in chickpea: a holistic approach. In: Saxena, N.P. (ed.) *Management of Agricultural Drought: Agronomic and Genetic Options*. Oxford & IBH Publishing, New Delhi, India, pp. 103–122.

Saxena, N.P. and Johansen, C. (1990) Chickpea ideotypes for genetic enhancement of yield and yield stability in South Asia. In: *Chickpea in the Nineties: Proceedings of the Second International Workshop on Chickpea Improvement*. 4–8 December 1989, ICRISAT, Hyderabad, India, pp. 81–85.

Saxena, N.P., Krishnamurthy, L. and Johansen, C. (1993) Registration of a drought-resistant chickpea germplasm. *Crop Science* 33, 1424.

Sedgley, R.H., Siddique, K.H.M. and Walton, G.H. (1990) Chickpea ideotypes for Mediterranean environments. In: *Chickpea in the Nineties: Proceedings of the Second International Workshop on Chickpea Improvement*. 4–8 December 1989, ICRISAT, Hyderabad, India, pp. 87–91.

Sethi, S.C., Byth, D.E., Gowda, C.L.L. and Green, J.M. (1981) Photoperiodic response and accelerated generation turnover in chickpea. *Field Crops Research* 4, 215–225.

Sharma, H.C., Gowda, C.L.L., Sharma, K.K., Gaur, P.M., Mallikarjuna, N., Buhariwalla, H.K. and Crouch, J.H. (2003) Host plant resistance to pod borer, *Helicoverpa armigera* in chickpea. In: Sharma, R.N., Shrivastava, G.K., Rathore, A.L., Sharma, M.L. and Khan, M.A. (eds) *Chickpea Research for the Millennium: Proceedings of International Chickpea Conference*. 20–22 January 2003, Indira Gandhi Agricultural University, Raipur, India, pp. 118–137.

Sharma, H.C., Pampapathy, G., Lanka, S.K. and Ridsdill-Smith, T.J. (2005) Antibiosis mechanism of resistance to pod borer, *Helicoverpa armigera* in wild relatives of chickpea. *Euphytica* 142, 107–117.

Sharma, H.C., Bhagwat, M., Pampapathy, G., Sharma, J. and Ridsdill-Smith, T. (2006) Perennial wild relatives of chickpea as potential sources of resistance to *Helicoverpa armigera*. *Genetic Resources and Crop Evolution* 53, 131–138.

Sharma, S.B., Sikora, R.A., Greco, N., Di Vito, M. and Caubel, G. (1994) Screening techniques and sources of resistance to nematodes in cool season food legumes. *Euphytica* 73, 59–66.

Singh, K.B. and Ocampo, B. (1997) Exploitation of wild *Cicer* species for yield improvement in chickpea. *Theoretical and Applied Genetics* 95, 418–423.

Singh, K.B. and Reddy, M.V. (1993) Resistance to six races of *Ascochyta rabiei* in the world germplasm collection of chickpea. *Crop Science* 33, 186–189.

Singh, K.B., Reddy, M.V. and Nene, Y.L. (1984) International testing of chickpeas for resistance to ascochyta blight. *Plant Disease* 68, 782–784.

Singh, K.B., Malhotra, R.S., Halila, M.H., Knights, E.J. and Verma, M.M. (1994) Current status and future strategy in breeding chickpea for resistance to biotic and abiotic stresses. *Euphytica* 73, 137–149.

Singh, K.B., Malhotra, R.S. and Saxena, M.C. (1995) Additional sources of tolerance to cold in cultivated and wild *Cicer* species. *Crop Science* 35, 1491–1497.

Singh, N.B. and Singh, R.A. (1984) Screening for alkalinity and salinity tolerance in chickpea. *International Chickpea Newsletter* 11, 25–26.

Singh, S., Gumber, R.K., Joshi, N. and Singh, K. (2005) Introgression from wild *Cicer reticulatum* to cultivated chickpea for productivity and disease resistance. *Plant Breeding* 124, 477–480.

Sinha, S.K. and Swaminathan, M.S. (1984) New parameters and selection criteria in plant breeding. In: Vose, P.B. and Blixt, S.G. (eds) *Crop Breeding: A Contemporary Basis*. Oxford: Pergamon Press, pp. 1–31.

Sinha, S.K., Aggarwal, P.K. and Khanna-Chopra, R. (1985) Phenological and physiological irrigation in India. *Advances in Irrigation* 3, 130–212.

Slinkard, A.E., Soth, M.B. and Vandenberg, A. (2000) Breeding for yield: the direct approach. In: Knight, R. (ed.) *Linking Research and Marketing Opportunities for Pulses in the 21st Century*. Kluwer Academic, Dordrecht, The Netherlands, pp. 1183–1190.

Srinivasan, A., Johansen, C. and Saxena, N.P. (1998) Cold tolerance during early reproductive growth of chickpea (*Cicer arietinum* L.): characterization of stress and genetic variation in pod set. *Field Crops Research* 57,181–193.

Subbarao, G.V., Kumar Rao, J.V.D.K., Kumar, J., Johansen, C., Deb, U.K., Ahmed, I., Krishna Rao, M.V., Venkataratnam, L., Hebbar, K.R., Sai, M.V.S.R. and Harris, D. (2001) *Spatial Distribution and Quantification of Rice-fallows in South Asia: Potential for Legumes*. ICRISAT, Hyderabad, India.

Vadez, V., Krishnamurthy, L., Gaur, P.M., Upadhyaya, H.D., Hoisington, D.A., Varshney, R.K., Turner, N.C. and Siddique, K.H.M. (2006) Tapping the large genetic variability for salinity tolerance in chickpea. In: *Proceedings of the 13th Australian Society of Agronomy Conference*. 10–14 September 2006, Perth, Australia. Available at: http://www.regional.org.au/au/asa/2006/concurrent/environment/ 4561_vadez.htm

Yadav, L.B. and Udit, N. (1993) Sources of resistance in chickpea to Alternaria blight. *Annals of Plant Protection Services* 1, 115–116.

Yadav, S.S. and Kumar, J. (2005) Chickpea improvement for dry farming system utilizing diverse genetic sources. In: Beltagy, A.E. and Saxena, M.C. (eds) *Sustainable Development and Management of Drylands in the Twenty-first Century*. ICARDA, Aleppo, Syria, pp. 280–285.

Yadav, S.S., Kumar, J. and Turner, N.C. (2002a) Breeding for resistant to moisture stress in chickpea (*Cicer arietinum* L.). In: *Plant Breeding for the 11th Millennium: Proceedings of the 12th Australasian Plant Breeding Conference*. 15–20 September 2002, Perth, Australia, pp. 240–242.

Yadav, S.S., Turner, N.C. and Kumar, J. (2002b) Commercialization and utilization of wild genes for higher productivity in chickpea. In: *Plant Breeding for the 11th Millennium: Proceedings of the 12th Australasian Plant Breeding Conference*. 15–20 September 2002, Perth, Australia, pp. 155–160.

Yadav, S.S., Kumar, J., Turner, N.C., Berger, J., Redden, R., McNeil, D., Materne, M., Knights, E.J. and Bahl, P.N. (2004a) Breeding for improved productivity, multiple resistance and wide adaptation in Chickpea *(Cicer arietinum* L.*). Plant Genetic Resources* 4(3), 181–187.

Yadav, S.S., Kumar. J., Turner. N.C. and Muehlbauer, F. (2004b) Chickpea breeding for adaptation to water limited environments. In: *Abstract Proceedings of the International Caucasian Conference on Cereals and Food Legumes*. 14–17 June 2004, Tbilisi, Georgia, pp. 411–412.

Yadav, S.S., Kumar, J., Turner, N.C., Muehlbauer, F., Knights, E.J., Redden, B., McNeil, D. and Berger, J.D. (2004c) Enhancing adaptation of large seeded kabuli chickpea to drought prone environments. In: *Proceedings of the 4th International Crop Science Congress*. 26 September–1 October 2004, Brisbane, Queensland, Australia, p. 117.

Yadav, S.S., Kumar, J. and Kumar, R. (2005) Chickpea breeding for rice-based agrosystem in north India. In: *Chickpea Production and Productivity Constraints*. NCIPM, New Delhi, India, pp. 33–43.

Yadav, S.S., Kumar, J., Yadav, S.K., Singh, S., Yadav, V.S., Turner, N.C. and Robert, R. (2006) Evaluation of Helicoverpa and drought resistance in desi and kabuli chickpea. *Plant Genetic Resources* 4, 1–7.

Yahia, A.A., Ouada, M.A., Hadjarab, K., Belfendes, R. and Sarni, K. (1997) Identification of chickpea stunt viruses in Algeria. *Bulletin-OEPP* 27, 265–268.

20 Chickpea Seed Production

A.J.G. van Gastel,[1] Z. Bishaw,[1] A.A. Niane,[1]
B.R. Gregg[2,*] and Y. Gan[3]

[1]*International Center for Agricultural Research in the Dry Areas (ICARDA), PO Box 5466, Aleppo, Syria;* [2]*Mississippi State University, Starkville, MS 39762, USA;* [3]*Agriculture and Agri-Food Canada, Airport Road East, Box 1030, Swift Current, Saskatchewan, S9H 3X2, Canada*

Introduction

International and national research institutes have developed several varieties of chickpea, which are adapted to local conditions with high and stable grain yield, better grain and nutritional quality and tolerance to biotic and abiotic stresses. The impact of such agricultural research can, however, only be realized if farmers have better access to high-quality seeds of these varieties. Access to high-quality seeds of modern crop varieties is a prerequisite for the transformation and improvement of agriculture. However, the availability of quality seed of cool season food legumes in general and of chickpea in particular does not satisfy the annual requirements.

Lack of improved varieties, mechanization problems in developing nations and high production costs of seed multiplication are major constraints hindering the development of an effective and efficient chickpea seed industry.

In the seed industry, crops are known as either 'high value seed crops' such as hybrids, vegetables and flowers, or 'low value seed crops' such as many of the open-pollinated and self-pollinated crops. Chickpea belongs to the category of 'low value seed crop' because: (i) they are highly self-pollinated, enabling the farmers to store their own seeds; (ii) most of the farmers use only a small fraction of seeds stored on-farm; and (iii) the profit margin from such seed crops is low and does not attract private sector investment. As a result, the formal seed industry pays less-than-needed attention in supplying seeds of the cool season food legumes, except in some areas of more developed economies. For example, in North America, there is a highly developed seed industry to supply high-quality grain legume seeds such as pea seed, lentil and chickpea. There are usually more customers and highly modernized farms produce on large

*Retired Professor.

areas for supply to commercial freezing or canning operations, and are willing to pay more for high-quality seeds of specific varieties.

By contrast, small farmers in developing countries generally do not have access to good seeds of improved food legume varieties, and will not pay a higher price for seed, rather they use their own or seeds from local market for planting. In most of these areas, the development of the seed industry is at a low level, focusing mostly on higher-profit seeds such as hybrids and imported vegetable seeds. Little attention is paid to the supply of higher-yielding seeds of low-profit crops such as cool season food legumes. Unfortunately, these are the areas where food legumes are most important in family nutrition.

Seed Classes

In seed production, a limited generation system is used to minimize the risk of contamination (Table 20.1). Although different nomenclatures are used, their procedures are essentially the same. In this chapter, the Organization for Economic Cooperation and Development (OECD) nomenclature is used. Breeder seed is the initial source of seed, produced by the breeder (his agent) or plant breeding institution. Pre-basic seed is the progeny of breeder seed and is usually produced under the supervision of a breeder or his designated agency. Basic seed is the progeny of breeder or pre-basic seed and is produced under the supervision of a breeder or his designated agency and under the control of a seed quality control agency. Certified seed is the progeny of basic seed and produced on contract by selected seed growers under the supervision of the public or private seed enterprise. Certified seed can be used to produce further generations of certified seed, or can be planted by farmers for production. Commercial seed may be produced according to internal quality guidelines, without formal certification. Seed is usually tested before marketing. Commercial seed is often used in situations where natural disasters have occurred, often referred to as 'emergency seed'.

Table 20.1. Comparative nomenclature in selected countries of West Asia and North Africa (WANA). (From ICARDA, 2002; AOSCA, http://www.aosca.org; OECD, http://www.oecd.org.)

Generation	OECD	AOSCA	Egypt	Ethiopia	Morocco	Syria
First generation	Breeder	Breeder	Breeder	Breeder	Epis-lignes (G_0)	Nucleus
Second generation	Pre-basic	Foundation	Foundation	Pre-basic	Prébase (G_1, G2, G_3)	Foundation
Third generation	Basic	Registered	Registered	Basic	Base (G_4)	Registered
Fourth generation	Certified 1	Certified	Certified	Certified 1	Reproduction 1 (R_1)	Certified
Fifth generation	Certified 2	–	–	Certified 2	Reproduction 2 (R2)	–

The rest of this chapter deals with issues such as: (i) variety evaluation and testing; (ii) variety maintenance; (iii) techniques in seed multiplication; (iv) seed cleaning and treatment; (v) seed storage; and (vi) seed quality assurance.

Variety Evaluation and Testing

New chickpea varieties must pass through a series of evaluation and registration tests before they are released for commercial seed production. Principally, two types of tests are carried out: (i) distinctness, uniformity and stability (DUS); and (ii) value for cultivation and use (VCU) trials.

VCU tests are often referred to as 'variety trials and national performance trials'. They focus on the value (V) of cultivation (C) and use (U) of the variety, i.e. the benefit of the new variety to end-users (farmers and consumers). VCU trials are multilocation trials and are conducted in different agroecological zones to assess the variety's response to different agroecological conditions. The effects of different agronomic management practices are also assessed. In VCU trials, new varieties are compared with existing standard varieties. VCU trials are usually performed for 3 years. In some countries, the variety is further tested in on-farm verification trials under farmer's management conditions before final release.

A DUS test, establishing distinctness (D), uniformity (U) and stability (S), is a descriptive assessment. Morphological, physiological, cytological or chemical characters are used to establish the varietal identity by assessing the distinctness, uniformity and stability. The test is usually conducted for a minimum of 2 years in at least one location, where the variety is going to be released. In these tests, new varieties are compared with a wide range of existing varieties. At the end, the variety description and differences with other varieties are prepared. Since chickpea is a self-fertilizing and thus homozygous crop, only little variation and segregation occurs. The major sources of variation are environmental conditions, but some variations may result from outcrossing, as chickpea is reported to have an outcrossing percentage of 1–2% (Gowda, 1981; Singh and Saxena, 1999). The International Union for the Protection of Plant Varieties (UPOV) has published test guidelines (http://www.upov.int) for morphological description of chickpea (UPOV, 2002).

Harmonization of variety release

Procedures for variety, performance and registration testing and release are essentially similar in most of the countries. This provides an opportunity for developing testing protocols, sharing data and developing flexible and harmonized variety release schemes within regional or international contexts to make available a wider choice of varieties to farmers.

Variety Maintenance

Seed production follows a generation system (Table 20.1), where a small quantity of seeds, received from breeders, is multiplied into larger quantities of certified seeds for marketing to farmers. In the multiplication process, seed is constantly prone to deteriorative influences that could reduce genetic, physical and health quality of the seeds. Because of this potential for deterioration, new lots of breeder seeds must be produced and used at regular intervals in a process called variety maintenance. Laverack (1994) defined variety maintenance as 'the perpetuation of a small stock of parental material through repeated multiplication following a precise procedure'. Legume variety maintenance procedures have been described by Bouwman (1992), Drijfhout (1981), Julen (1983), Sharma (1987) and Singh and Saxena (1999). For chickpea there are several options.

1. *Purification of varieties* – The best available seed field is selected, and obvious contaminants are removed from the field (negative mass selection). Plants not conforming to the variety are rogued out, the crop is bulk-harvested and the seed is then used for further multiplication. This purification aims at producing a very clean starting stock for further multiplication, and is used in situations where there is no organized maintenance procedure.
2. *Mass selection* – In this method the best individual plants are selected from a field. These selected individual plants are bulk-harvested and the rest of the field is discarded. The bulked seed is used for further multiplication.
3. *Plant-to-row* – It is considered to be the best method for variety maintenance of chickpea (Fig. 20.1). Single plants typical of the variety are selected and harvested, and kept separately. These seeds are planted in rows (plant rows); during production, off-type rows and off-type plants within the rows are discarded. Only those rows which conform to the varietal description are maintained and bulk-harvested as breeder seed. When there is doubt regarding varietal purity, plant rows may be individually harvested and planted as small plots for further observation. Within plots, negative mass selection is carried out before plots are bulk-harvested as breeder seed.

Production arrangements

The best method of variety maintenance is the production of enough breeder seeds to satisfy estimated requirements for the lifetime of a variety (Bouwman, 1992), because this method involves less work and minimizes the risk of contamination. However, this breeder seeds must be stored under optimum conditions for medium to long-term storage (upper limits of 4°C and 40% relative humidity [RH]). This method is not appropriate for chickpea, because its multiplication rate is low and the seeds are bulky. A suggested alternative is to multiply breeder seed at regular intervals, after initially producing a sufficiently large quantity to meet seed needs for 4–5 years (Julen, 1983). This breeder seed must be stored under relatively low temperature and RH (20°C and 50% RH or 10°C and 60% RH). The best approach is to produce new lots of breeder seeds whenever stocks are drawn for the production of basic seed.

It is important that breeder seed must be planted in at least two different locations to avoid the loss of the complete generation. Crop management practices and procedures are similar to those used for other seed generations (see the section Seed Production), but for breeder seed, the best possible practices

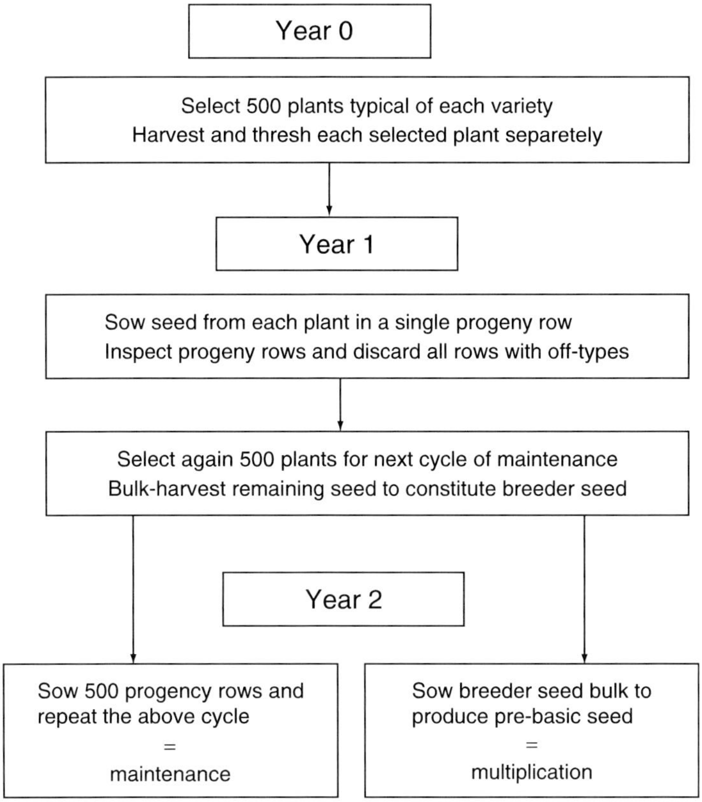

Fig. 20.1. Variety maintenance of chickpea.

must be applied. Breeder seed is the earliest and lowest-volume generation, and any mistake made in this stage will be difficult to correct at a later stage.

Responsibility for variety maintenance

Plant breeders, particularly from the public sector, often lose interest once a variety is officially released. Consequently, in many developing countries neither the plant-breeding institutions nor the seed-producing agencies give the required attention to variety maintenance and breeder seed production. As a result, lack of availability of breeder seeds becomes a bottleneck in the adoption of new varieties by farmers. For example, in Morocco, a separate unit within the Institut National de la Recherche Agronomique (INRA) is responsible for producing breeder seeds of common varieties. In Ethiopia, the Ethiopian Seed Enterprise (ESE) is responsible for the production of pre-basic and basic seeds from its own seed farm. These approaches overcome the problem of seed availability for the major crops. Laverack (1994) described different arrangements and management options for breeder seed production that are useful and adopted in developing countries.

Seed Production

Key components of producing quality seed of chickpea varieties include:

- Site selection to avoid high risk environments;
- Clean field selection to eliminate volunteer plants and noxious weeds from preceding crops;
- Previous cropping to avoid build-up of volunteers and soil-borne pests;
- Isolation from sources of contamination;
- Avoiding outcrossing;
- Inoculation with rhizobia to promote nodulation and nitrogen fixation;
- Roguing to remove contaminants;
- Limited generations of multiplication to reduce contamination;
- Maintaining the cleanliness of farm machinery during planting, harvesting and transportation;
- Maintaining the cleanliness of machinery to avoid admixtures during cleaning, treatment and storage;
- Production arrangements using specialized contract growers;
- Implementation of quality assurance systems.

Seed production practices for legume crops have been described by Agrawal (1985), Doerfler (1976), Erskine *et al.* (1988), Saxena and Singh (1987), Singh (1986), Singh and Saxena (1999) and Wellving (1984). Van Gastel *et al.* (2002a,b) provided a detail description of the wheat seed production practices, which can be used as the basis for chickpea seed production. The most important aspects, as related to chickpea, are discussed below.

Previous cropping

In seed production, requirements for previous cropping on the seed field specify the crops that should not be grown for a specified time preceding the production of the seed crop. Previous cropping differs from rotation, which is usually done to maintain soil fertility and control seed-borne or soil-borne diseases and insect pests. Appropriate previous cropping will avoid volunteer plants which may reduce the varietal purity, and the build-up of diseases and noxious weeds. For chickpea, the land selected to produce seed should be free of any other chickpea variety for at least 2 years for pre-basic and basic seed, unless the previous crop was of the same variety and of the same or higher seed category. For certified seed, only 1 year between two crops of different varieties is required. Fields planted with *Vicia*, *Lathyrus*, pea, small-seeded faba bean and *Phaseolus* bean in the previous year should be avoided, because the seeds of such species are of similar size and difficult to remove during seed cleaning. *Cuscuta*, a parasitic weed that remain viable for more than 5 years in the soil, is a major problem for chickpea seed production and fields with a history of *Cuscuta* infestation should be avoided.

Seedbed preparation, planting methods and seed rates

For chickpea, a 'deep' seedbed preparation is necessary, because of its long main (tap) root system. Good seedbed preparation is also a prerequisite for combine harvesting. Generally for seed production, row planting of the seeds is preferable to broadcasting, as it requires less seeds, and facilitates mechanized weed control, roguing and field inspection (Galanopoulou *et al.*, 1996).

Seed rates differ from variety to variety, depending on seed size. However, for a new variety the multiplication rate (yield per unit of seed planted) is more important than maximum yield. The higher the multiplication rate, the faster the new variety is multiplied and the sooner it is made available to farmers. Using lower seed rate (and having larger initial amounts of breeder seed) is the best management strategy for achieving faster seed multiplication. At International Center for Agricultural Research in the Dry Areas (ICARDA), Aleppo, Syria, unpublished data on the relationship between seed rate and yield showed that decreasing the seed rate increases the multiplication rate in chickpea. Low seed rates also result in a more open crop stand that reduces disease incidence. However, when the seed crop is grown under close supervision only low seed rates can be used, because it is not economical for seed growers (less yield/hectare).

Rhizobium inoculation and fertilization

In legumes, root nodules are highly specialized structures formed as a result of interactions between the host plant and the invading *Rhizobium*. Chickpea seeds must be inoculated with *Rhizobium* at planting time, particularly when they are planted in new areas. Under favourable growing conditions, chickpea fixes a significant portion of their nitrogen requirements (60–80%).

Adequate levels of phosphorus, potassium, calcium and other nutrients are required to ensure proper plant growth and development. Fertilizer application should be based on soil analysis to determine the nutrient requirement of the crop under particular growing conditions.

Weed and pest control

Chickpea is a poor competitor with weeds at all stages of growth, and requires season-long weed management because of its slow growth in the seedling stage and relatively sparse optimum plant population with an open crop canopy. To control weeds, an integrated weed control strategy is required, which includes:

1. Use of a relatively high plant population density to improve the competitive ability of chickpea plants to weeds;
2. Planting in soils with an optimum moisture and temperature to ensure rapid emergence and fast canopy closure;

3. Use of wide row spacing to facilitate mechanical tillage practices and also roguing of unwanted plants;
4. Use of disease-free clean seeds to increase performance and health of emerging seedlings;
5. Use of broad-spectrum herbicides (pre-planting) to allow chickpea plants to have a clear start;
6. Use of herbicides for grassy and broadleaf weeds.

It also involves: (i) selecting clean fields; (ii) planting in rows and cultivating to remove weed seedlings; and (iii) roguing to remove troublesome weeds.

Fields infested with *Cuscuta* should not be used for seed production; if *Cuscuta* is found in seed fields, the affected plants should be uprooted, removed and burned. If a strip of *Cuscuta* is found, the area should be chemically treated to sterilize the soil temporarily.

Seed dressing with effective fungicides minimizes seed-borne and soil-borne diseases. The chemicals used at ICARDA are described in Table 20.2.

Ascochyta

Ascochyta blight has been reported as the most widespread and economically devastating disease of chickpea, causing yield reduction from <10% in resistant varieties, up to 100% in susceptible genotypes (Singh and Saxena, 1999). The production of *Ascochyta*-free seeds involves an integrated approach that involves: (i) using tolerant varieties; (ii) selection of appropriate site; (iii) crop rotation (3 years); (iv) field sanitation; (v) seed treatment; (vi) appropriate sowing time; and (vii) chemical sprays.

Field sanitation – *Ascochyta rabiei* can multiply on chickpea debris after harvest; therefore, crop residues in the field should be burned or buried deeply (10–15 cm) immediately after harvest. Deep ploughing will hasten the decomposition of infested straw and eliminate it as a source of disease inoculum.

Sowing time – Despite the higher yield potential of winter-sown chickpea in West Asia and North Africa, spring-planted varieties may, in certain areas, be the only option to produce disease-free seed. However, spring-sown crops require one or two supplementary furrow irrigations to obtain a good yield.

Table 20.2. Chemical herbicides for weed control in chickpea. (From M. Pala, Aleppo, 2005, personal communication.)

	Chemical control practices			
Growth stages	Trade name	Active ingredient	Rate (kg a.i./ha)	Target weeds
Pre-emergence	Igran	Terbutryne	1.5–2.5	Dicotyledons
	Maloran	Chlorbromuron	1.0–2.0	Dicotyledons
	Kerb	Pronamide	0.5	Monocotyledons
	Stomp	Pendimethalin	1.0	*Cuscuta*
Post-emergence	Fusilade	Fluazifop-P-butyl	0.25–0.5	Monocotyledons
	Focus-Ultra	Cycoxydim	0.25–0.5	Monocotyledons
	Agil	Propaquizafop	0.1–0.2	Monocotyledons
	Cadre-Oroban	Imazapic	3–5 g/ha X 2	*Orobanche*

Chemicals sprays – When field conditions are above 15°C and 80% RH, *Ascochyta* infection of susceptible chickpea varieties can be prevented by spraying with chlorothalonil (Bravo) at an interval of 15 days.

For comprehensive management options for ascochyta blight control, refer to a recent review paper by Gan *et al.* (2006), which offers several agronomic options for the control of *Ascochyta* in chickpea.

Field pests

Specific and broad-spectrum pesticides and insecticides are available for effective control of field and storage pests of chickpea (Table 20.3).

Isolation

Isolation involves growing a seed crop away from any source of contamination whether genetic, mechanical or pathological. Isolation can be achieved by maintaining a distance from the seed field, or through time, by planting so that the seed crop grows and matures on a different time schedule. Minimum isolation distances are usually prescribed based on field size, crop pollination habit, direction as well as speed of wind and presence of natural barriers. Chickpea is self-fertilizing, with a very low percentage (2%) of outcrossing (Singh and Saxena, 1999). Therefore, a physical distance of 1–2 m between two fields is sufficient. However, slightly longer isolation distances are recommended, i.e. 5 m for pre-basic seed and 3 m for basic and certified seeds. Contamination can also be reduced by growing different generations of the same variety side by side. For example, in some North African countries, G_0 is planted in a field surrounded by G_1 and then by G_2, G_3 and G_4 (Table 20.1) to avoid contamination. Thus, the contamination is from the same variety and does not affect the varietal purity.

Since microorganisms and insects (virus vectors) can move over longer distances, isolation will not prevent the problem of viral diseases.

Roguing

Roguing is defined as the systematic examination of seed fields and removal of undesirable plants that may contaminate the seed crop (Gregg *et al.*, 1990a,b;

Table 20.3. Chemicals for control of insects in seed production of chickpea.

Active ingredient	Type	Chemical group	Target pest
Deltamethrin	Insecticide	Pyrethroid	Leaf minor, bruchids, pod borers
Methidathion	Insecticide	Organophosphate	Leaf minor, bruchids, pod borers
Imidacloprid	Insecticide	Chloronicotinyl	Aphids
Pirimiphos-methyl	Insecticide	Organophosphate	Bruchids in seed and grain store

Laverack and Turner, 1995). Roguing maintains varietal purity and, to some extent, protects the crops from seed-borne diseases. The following contaminants are removed: (i) off-type and other variety of crop plants; (ii) other crop species (which have similar seed size); (iii) weed plants whose seeds are not easily separated in cleaning; (iv) parasitic weeds such as *Cuscuta* spp.; and (v) plants infected with seed-borne fungal diseases and viruses.

Harvest

The main risks of seed quality due to mechanical harvesting are: (i) mechanical damage of seeds and (ii) physical admixture with seeds of other varieties.

Legume seeds are more prone to mechanical damage than most agricultural crops, and should be harvested or combined carefully (proper combine speeds, cylinders, sieves and air blast). Legume crops should be harvested at full maturity (approximate moisture content is 12%). Combining should be done early in the morning when seed moisture content is higher, so as to minimize seed cracking and loss. Chickpea normally has low shattering potential, although pod drop and pod shattering may occur when harvest is delayed or is under unusually hot temperatures. Ellis *et al.* (1988) reported that delaying chickpea seed harvest by 1 month reduced germination from 99% to 80%, and a further delay of 1 month reduced germination to 30%.

Combines can be used to harvest chickpea seed multiplication fields. Winter-planted chickpeas are easier to harvest due to its increased height as compared with traditional spring-sown crops. However, chickpea seeds have a characteristic small protruding beak-like structure that can be easily damaged in harvest and handling. Seed damage can be minimized by using conveyor belts (or by keeping the standard augers as full as possible, although augers should not be used in any seed cleaning or handling) and operating them at slower speeds. The cylinder speed should be adjusted to below 500 rpm and the distance between concave and drum adjusted near its maximum. Upper sieves of 10–15 mm and lower sieves of 8–12 mm are used for chickpea, depending on seed size (Erskine *et al.*, 1987). Size of sieves can vary, depending on seed size (Acikgoz and Kutlu, 1987; Saxena *et al.*, 1987).

Seed Processing

Seed processing includes all steps involved in preparing harvested seed for marketing, i.e. drying, pre-cleaning, cleaning, upgrading, treating and packaging. If necessary, seeds are dried immediately after harvest to remove excess moisture. Then the dried seeds are cleaned to remove undesirable contaminants such as: (i) inert matter such as plant parts, soil particles and stones; (ii) weed seed; (iii) other crop seed; (iv) other variety seed; and (v) seeds of the variety which are immature, shriveled, broken, damaged or deteriorated (Thomson, 1979). Upgrading is done by removing poor seed and applying chemical treatment. Cleaning and upgrading is based on physical differences between good seed, poor seed and undesirable contaminants (Boyd *et al.*, 1975). Separation can be done because seeds vary in length, width, thickness, weight, shape, surface texture and colour (Vaughan *et al.*, 1968; Gregg *et al.*, 1970; Boyd *et al.*, 1975; Brandenburg, 1977; Brandenburg and Park, 1977; International Seed Testing Association (ISTA), 1977; Food and Agricultural Organization (FAO), 1981; van der Burg, 1986).

Principles of cleaning

Farmers traditionally use air and/or sieves to clean the seeds. Modern seed plants apply some of the traditional methods like mechanical means in a large-scale operation. Seed is cleaned using screens, cylinders and air blast. For chickpea, the screens and air blasts are most important, because cylinders are rarely used for round seeds. Generally, different machines must be combined in a specific sequence to make a proper separation. The choice of machines depends on the type of contaminants and the quality standard that must be achieved (Boyd *et al.*, 1975). Since chickpea seed is susceptible to mechanical damage during cleaning and handling, it requires the use of smallest possible number of machines, and that certain modifications are made to eliminate long drops that can cause mechanical damage to seed.

Pre-cleaner – Pre-cleaners are high-capacity machines with one or two sieves with large round holes and a powerful air blast to remove larger contaminants. A pre-cleaner is not often required for chickpea.

Fine cleaner – The fine cleaner, also known as air-screen cleaner, is the most important and basic machine of seed processing. It separates seeds according to width, thickness, shape and terminal velocity by using a combination of several screens and air blast. In many cases, the air-screen cleaner is sufficient to clean a chickpea seed, because it separates seeds based on its major physical characteristics.

Indented cylinder – Commercial processing plants have indented cylinders in the standard processing sequence, separating on the basis of differences in seed length. Indented cylinders are not essential for chickpea, which are rounded. However, if unthreshed pods of chickpea have a size (width and thickness) similar to that of the large seed they may escape the scalping screens and must be removed by a cylinder (long grain application).

Gravity separator – The gravity, a density separator (Vaughan *et al.*, 1968), can be used to separate stones and soil particles, which have the same size as the seed. The gravity is also very useful to remove bruchid-infested seed from chickpea seed lots.

Scarifier – Hardseededness, impermeability of the seed coat to water, is common in almost all legumes including chickpea. A scarifier scratches the seed coat to facilitate water intake, and removes hardseededness and improves germination.

Elevators and conveyors – During processing, seeds must flow efficiently from one machine to another and should not be damaged. Elevators move the seeds vertically, whereas conveyor belts move the seeds horizontally or at an angle. In most cases, seeds also slide down in a pipe at an angle. Choosing the right elevators and conveyors for chickpea seed is important, because they are sensitive to mechanical damage. Screw conveyors should not be used, because they damage the seed. Elevating seed by using a high-speed air stream (as in airlift elevators) should be avoided. Pipes feeding seed into bins or machines should be sloped at 45° to ensure that seeds do not get blocked in it. For chickpea seed, fix 'impact-absorbing' materials (rubber) or equipment such as 'bean ladders' at places where seeds fall from more than 1 m, and/or strike hard surfaces. Do not try to reduce the elevator speed for reducing the negative impacts, as this adversely affects discharge of seeds from the elevator buckets, and may increase seed damage.

Cleaning equipment – Cleanliness in seed plants is essential to avoid mechanical mixture. Seed cleaning machines must be thoroughly cleaned between seed lots, using an air compressor and industrial vacuum cleaner. Mechanical admixture can also be reduced by careful planning, such as by cleaning early-generation seed immediately after cleaning a later generation seed of the same variety (or very similar ones) to reduce the chances of contamination by mixing with seed of another variety.

Seed treatment

Seeds may carry fungi, bacteria, viruses, nematodes and insects, either on the seed surface or internally. Protection from pathogens is one of the most important quality attributes. Seed infection may lead to low germination, reduced plant stand, severe yield loss or total crop failure. Chemical treatment of seed lots is a standard procedure for disease control in many crops. Chemicals may be used in different formulations, e.g. dust, wettable powder for slurry treatment or liquid concentrates. In general, dusts are applied at ~2 g/kg seed, and slurries or liquids at 5–10 ml/kg seed. Liquids or slurries are preferred over dust, because they enable easier exact measurement, allow better seed coating, and create less potential hazard for operators. Different admixtures may be used in dust treatment to avoid such problems. A dextrine solution of 0.2% is a cheap additive, added at the rate of 3–5 ml/kg seed. Special 'incrusters' such as Sacrust are more effective, but more expensive also.

In North America, chickpea seeds are usually treated with an effective fungicide such as carbathiin, thiabendazole and metalaxyl before planting (Table 20.4) (Gan *et al.*, 2006).

Equipment for seed treatment – In general, chemical seed treatment is a part of cleaning operation. However, even with excellent chemicals and equipment, it is difficult, if not impossible, to achieve 100% uniform coverage of seeds with

Table 20.4. Fungicides to control pathogenic inoculum on chickpea seed and seedlings.

Fungicides	Active ingredients	Rate/kg	Target disease
Vitavax-200	Thiram + Carboxin	2–3 cc	Wide spectrum
Tecto WP	Thiabendazole	2 gr	Wide spectrum

chemicals. During seed treatment, some seeds do not receive proper treatment, while others receive more than the recommended dosage. These problems can be minimized by choosing the right equipment and calibrating it properly.

On-farm seed treatment – The simplest way of mixing the seeds and chemicals is with a shovel, but this does not meet the requirements of even-mixing and operator safety. Better results are achieved by using a concrete mixer or a hand-driven or motor-driven drum, preferably in a diagonal position. Care must be taken to ensure that these devices are operated long enough to distribute the chemicals evenly. A major disadvantage is that the required amount of chemicals is measured by hand.

Commercial seed treatment – For large-scale treatment, machines have been developed that ensure automatic measuring of chemicals. In Gustafson treaters, for instance, the weight of seed measured in a weighing pan is used to operate the chemical measuring system. By adjusting a counterweight, a fixed quantity of seed is treated with a fixed quantity of chemical, measured in standard cups and operated by tripping (dumping) the weighing pan.

Most commercially available treaters are designed for slurry (suspension of a wettable powder formulation in water) or liquid applications. The solid particles in slurry may settle as sediment; so a stirring device is indispensable in the treater tank. These particles also may clog nozzles, so the mixture is applied from small measuring buckets on an endless chain. Liquid formulations are usually sprayed on seed, as in Mist-O-Matic treaters, to assure uniform coverage of the seed.

There is a general tendency to use chemicals that are safe for both the user and the environment. New chemicals are replacing toxic substances such as organic mercurials (Ceresan and others) and persistent fungicides such as hexachlorobenzene (HCB). Safety instructions must be followed strictly even with new, less toxic chemicals.

Monitoring seed quality

Seed plants should have adequately equipped and well-staffed internal seed quality control facilities. The internal quality control laboratory should carry out simple purity, germination and moisture tests, which are necessary to monitor quality of incoming seed material, the cleaning and treating process and storage activities. The laboratory can ensure that contractual agreements have been observed by the grower to effect payment, the need for drying incoming seed lots; International Plant Genetic Resources Institute (IPGRI) seed bank standards recommend 15% RH drying to reduce seed moisture to 6–7%, selection of appropriate machines for cleaning and to achieve the desired quality standards.

Seed Storage

Seeds attain maximum germination capacity and vigour at physiological maturity. Legume seeds reach physiological maturity at seed moisture contents ranging from 45% to 50% (Ellis *et al.*, 1988). Loss of vigour and viability is a natural phenomenon; all seeds lose viability during storage, and loss of vigour precedes loss in germination. This is manifested in various physiological and biochemical events. This loss appears to be associated with loss of membrane integrity, changes in molecular structure of nucleic acid and reduction in enzyme activity, which results in reduced rate of germination, increase in number of abnormal seedlings, lower vigour and field emergence (Roberts and Osei-Bonsu, 1988). Deterioration moves inexorably towards death without stopping; it cannot be reversed nor eliminated. Delouche *et al.* (1973) and Ellis *et al.* (1988) summarized the requirements for storage of cool season food legumes.

Long dry conditions during seed maturation in the field are very important for seed quality, while unfavourable weather conditions (rain, high humidity and high temperature) or delay in harvest will have a negative impact on seed quality. Once the seed is harvested, it is important to provide proper storage conditions to retard the rate of seed deterioration and minimize losses of physiological quality. McDonald and Nelson (1986) described the physiology of seed deterioration, and the principles and practices of storage are described by Justice and Bass (1978). Some mathematical models have been developed to relate the viability of seeds to their storage environment, and to facilitate prediction of seed longevity (Ellis and Roberts, 1980).

Based on the physiological behaviour, Roberts (1972) classified seeds into two major groups: (i) orthodox and (ii) recalcitrant seeds. Orthodox seeds desiccate on the mother plant and can be dried to low moisture contents without damage, and (when dry) are tolerant to temperatures far below zero. Decrease in seed moisture and temperature increases the longevity of orthodox seeds. Recalcitrant seeds lose viability upon desiccation, and die if their moisture content is reduced below a certain level. Chickpea belongs to the orthodox group, and its seed can be kept viable for many years, depending upon storage conditions. Storability of chickpea seed is considered to be moderate to relatively good (Delouche, 1988). There are, however, differences in storability of chickpea types; the desi types can be stored better than the kabuli types (Saxena, 1987).

Temperature and moisture content

Harrington's classical rules of thumb describe the effects of temperature and moisture on seed deterioration, which can be combined and have geometric effects as follows (Harrington and Douglas, 1970):

- Seed life is doubled for every 5°C decrease in storage temperature when temperatures are between 0°C and 50°C.
- Seed life is doubled with every 1% decrease in seed moisture content when moisture content is between 5% and 14%.

Table 20.5 shows the influence of temperature and RH on chickpea seed germination after different periods of storage (Ellis *et al.*, 1988). This study shows no loss of germination of good-quality chickpea seed lots stored for 18 months at 20°C with seed moisture content in equilibrium with 60% RH. However, the higher the temperature and/or RH, the shorter time the seeds can be stored.

Table 20.5. Influence of different controlled environments on germination (%) of chickpea (accession BG 216) after different storage periods.[a] (From Ellis *et al.*, 1988, cited in Agrawal and Kharlukhi, 1985.)

	Temperature (°C)				
	20			33	
Relative humidity (%)	30	60	80	30	60
Storage period (days)					
62	96	96	94	92	100
100	92	96	87	94	100
165	93	95	46	92	63
255	96	97	10	100	–
373	96	96	–	95	–
531	95	94	–	95	–

[a]The seed lot with initial germination of 95%.

A computer program based on the seed survival equation developed by Ellis and Roberts (1980) is used to predict the period of seed storage of several crops, including cool season food legumes (Kraak, 1992). This programme can be used to calculate: (i) initial viability; (ii) viability after storage; (iii) storage period; (iv) seed moisture content; and (v) temperature during storage, if three of the five parameters are known and one can be made variable.

Fungi, mites and insects

Storage fungi can severely reduce seed quality by: (i) decreasing germination; (ii) heating; (iii) developing mustiness and caking; and (iv) total decay. Such fungi do not cause damage to seed during storage if seed moisture content is in equilibrium with an RH of 65–70% or less. As a general rule, most of the storage fungi, mites and insects do not develop below 0°C, 5°C and 15°C, respectively.

Infestations of storage pests are one of the serious problems in chickpea. Most serious pest is bruchids; when seed is infested by bruchids, complete loss of viability occurs within 2–4 months of storage.

Mechanical damage

The amount of cracking (of testa), which results from a given handling treatment, is dependent upon seed moisture content. Mechanically injured seeds are less storable; they deteriorate faster, and are more susceptible to damage by storage fungi and seed treatment. Hence, seeds have to be harvested, threshed, processed and handled very carefully. Large-seeded chickpea varieties are most susceptible to physical injury, especially when seed moisture content is <11–12% (Delouche, 1988).

Seed storage period

Chickpea seeds can be stored from planting to harvest for a period of few months or as carryover seeds for more than one season. The latter is true particularly in situations where there are recurrent natural disasters to overcome seed shortages. Moreover, early generation seeds can be kept even for longer period, for various reasons. The period of storage can be classified into three categories: (i) short-term; (ii) medium-term; and (iii) long-term.

Short-term storage

For short-term storage (up to a maximum of 9 months), seed storage conditions or facilities that range between an average of ~30°C with 50% RH and 20°C with 60% RH are satisfactory for most of the orthodox seeds, including chickpea (Delouche, 1988).

Medium-term storage

In many formal seed production programmes, ~100% of early generation seeds and 20–25% of certified seeds are carried over through one growing season to the second planting season (up to 18 months) as a guarantee against crop failure or other disasters. According to Ellis *et al.* (1988), the recommended storage conditions of chickpea are 30°C with 40% RH (~9–10% seed moisture content), 20°C with 50% RH (10–12% moisture content), or 10°C with 60% RH (12–14% moisture content).

Long-term storage

Sometimes, breeder seeds may be produced at an interval of several years or only once for the lifetime of the variety, and the seed must be stored for longer periods. For a storage period of 4–6 years, a temperature of 10°C and 45% RH (9.5–10.5% moisture content) is recommended. In general, authentic samples of new varieties, breeding materials and seed for genetic conservation are also stored for longer periods. FAO recommends a storage temperature between −18°C and −20°C, with 5% seed moisture content (in equilibrium with 10–15% RH for starchy seed and 20–25% RH for oily seed). Storage under such conditions is usually not necessary in a market-oriented seed production programme.

Safe seed storage conditions

Storage facilities must protect seeds from damage and deterioration, and maintain seed quality as expressed by vigour, germination, purity, identity and physical condition. If seed moisture does not exceed 10% during storage, seeds may be stored under rather temperate ambient conditions for at least 18 months and still meet the minimum germination requirements for seed certification. Since RH and storage temperature are the two most important factors influencing seed viability during storage, seed stored in places with low RH and low temperature, will lose viability more slowly, i.e. maintain viability longer. The optimum level of RH for safe storage is 70%. Temperature of 20°C is considered safe for seed storage, with an upper limit of 30°C as reasonable (Agrawal, 1976). Encourage lower temperature and RH wherever possible for quality and viability of seed.

To avoid losses and to keep seeds free from insect pests during storage, careful planning and management is essential to maintain clean, cool and dry conditions.

Safe seed storage facilities

Seed storages must be specially designed, equipped and managed to provide clean, cool, dry and safe conditions. Well-constructed, ventilated stores are adequate for short-term storage of chickpea seed in most production areas (Delouche, 1988). A good seed storage structure should have no windows, and must be in a dry, well-drained area. The floor should be 1 m above ground level, or at truck bed height, with a built-in vapour barrier equivalent to 0.25 mm polyethylene sheeting installed with hot-brushed bitumen. There should be only one or two doors in the middle of the building's short sides (ends), to minimize floor space lost to aisles for handling seeds. The storages must be rainproof, moisture–vapourproof and insectproof. There should be no cracks in the walls or floors, to facilitate cleaning and to eliminate cover for insect pests. The roof and walls should be joined without cracks, to prevent the entry of birds and pests. There should be a ratproof lip extending out ~35 cm around the building at about 1 m height.

Seed storages must be designed and constructed so as to minimize the entry of solar radiation and outside heat. The walls and ceiling should provide enough thermal insulation to minimize solar heat gain. Non-insulated wall construction may be adequate, but the ceiling or roof usually requires some thermal insulation. The roof, walls and doors should be painted with a light-coloured reflective paint, to reduce solar heat intake. An extensive roof overhang will provide shade and cool the walls, and protect the ventilation openings from rain. Orientation of the building from east to west will also minimize solar heat on the longer side walls.

An exhaust fan may be used for ventilation when outside temperature is lower than the temperature inside the seed storage, but RH of the outside air should also be kept in mind when planning to ventilate the seed storage. All ventilation openings should be screened with 6 mm wire mesh, and located so as to remove hot air from the upper part of the building and moist air from near the floor level.

Control of Seed Storage Pests

There is a wide range of storage pests, including rodents, birds, insects, mites, fungi and bacteria. Losses due to storage pests vary according to climatic conditions, crop and storage facilities. Quantitative storage losses are estimated to reach up to 30% worldwide. Qualitative losses (in the viability of seed) are more difficult to estimate, and the consequences of these losses may be more drastic. Preventing losses of stored seed deserves special attention. Information on control of storage pests can be found in Bond (1984), Diekmann (1988), Gwinner *et al.* (1990) and Sauer (1992).

Rats and mice

Rodents damage seed, spoil seed with urine and excrement, damage bags, fumigation sheets, electrical wires, buildings and transmit diseases that are dangerous to humans. Moreover, rodents have a high reproductive capacity, and are extremely adaptable and very clever.

Effective control depends on the rodent species, but unfortunately rodents are not seen easily and can be indirectly identified through: (i) shape, size and place of droppings; (ii) type of damage; and (iii) footprints. The brown rat (*Rattus norvegicus*) has round droppings and the roof rat (*Rattus rattus*) has more oblong, banana-shaped droppings; mice (*Mus musculus*) droppings are much smaller. Rodents damage bags also; the presence of spilled seed under a stack (pallet) indicates their presence. The behaviour and areas where they live can also be used to identify the species; rats are suspicious and mistrust all new things, while mice are curious, investigating everything. *R. rattus* lives in roofs of the storage and take different travel routes. *R. norvegicus* lives outside the store and always take the same route to get into the store as well as within the store. Rats generally move along walls, pallets and bags. Mice move around in all directions, but live in a very restricted area (1 m^3) and may never be visible.

It is important to make daily inspections for the presence of rodents and to identify the species based on droppings, damage to bags and footprints. Preventive measures are most important, particularly the cleanliness of the seed storage and its surroundings, which should be kept spotlessly clean and dry, without vegetation, which provides hiding places for brown rat. To avoid entry of rats and mice into the storage, all openings should be screened with wire mesh and storages should, as much as possible, be ratproof. Biological control measures such as keeping cats in seed storages, and traps in and outside the store should be taken.

Rats

If a high population of rats is present, use acute poisons that have a very high toxicity and kill the rats almost on the spot. But rats sometimes show bait shyness and will not touch these baits anymore. To avoid this, pre-baiting is used, whereby rats are first attracted by high-quality baits without poison. After a few days, the poison is added to the baits. But all rats will not be killed at a time. They become cautious after that and hence the baits should be changed after a few days (4–5 days). The most commonly used poison is zinc phosphide.

Chronic poisons are used for long-term control. The 'old' chemicals need several intakes by the rat (up to seven times) to kill, whereas the 'new' generation of chemicals requires only one intake. The effect of the chemical is not rapid, and it takes place after few hours (the rat becomes sleepy and dies) and no bait shyness is developed. These chemicals are anticoagulants and the rat is killed through internal bleeding. A well-known 'several-intake' chemical is Warfarin; and Brodifacoun is a well-known 'one-intake' chemical.

Mice

Mice are more difficult to control because they live in very small areas (1 m^3). Hence, a dense distribution of bait is required. Most widely used chemical is Calciferol.

Storage insects and mites

Mites mainly play a damaging role by transmitting spores of storage fungi and causing skin irritation and allergies in persons handling infested seed. Insects can be very destructive to the seed. Different species feed on food legumes, and their different life cycles determine the appropriate control measures. In general, development of insects is faster with higher temperatures, but there are differences among species. Approximately 35°C is considered the maximum temperature for development, and ~38°C the maximum temperature for survival. Most insects cannot survive temperature below 0°C for more than 2–3 weeks. A certain moisture level also is required for insect development. Seed moisture of 10–11% is, in general, considered to be the minimum for insect development. The number of insects increases with increasing moisture, up to a level where too many microorganisms develop (~15–16% seed moisture).

While in many legumes infestation with storage pests like *Callosobruchus* takes place before harvest, this is not a problem in chickpea (Reed *et al.*, 1987). Typical storage pests of chickpea are *Callosobruchus* spp. and *Bruchidius* spp., which may cause complete loss of viability within 2–4 months of storage if appropriate control action is not taken. There are many sources of infestation; the most important is contamination from infested seed already in the storage. Frequently, the storages have hidden places such as crevices, corners, spilled seed outside the storage and empty sacks with some leftover seed, where insects survive. So control of these areas is extremely important. It is desirable to detect infestation at early stages, before the insect population builds up. There are different methods to detect infestations before they become clearly visible; some of them are simple (flotation of grains), while others are more sophisticated (x-ray). However, insect infestation can be detected mainly by visual inspection.

Infestation with storage pests is a major problem, and *Callosobruchus* spp. has worldwide distribution. Resistance to *Callosobruchus* has been reported by many workers (Bushara, 1988). However, the kabuli chickpea appears most susceptible whereas desi types are most resistant, but not immune (Reed *et al.*, 1987).

The optimum condition for maximum activity of insects is in the range of 30–35°C and 60–80% RH. Insect infestation can be reduced or eliminated by proper drying and storing at low temperatures, or can be controlled by proper sanitation and pesticides.

Spraying protective insecticides

Distinction should be made between insecticides: (i) those which kill the insect population and have a long-lasting effect (pyrethroids and organophosphorus insecticides) and (ii) those which kill insects, but have no residual effect (fumigants). These pesticides should be used as complementary to each other, and in combination with storage sanitation. Storage insects are mostly controlled chemically where a wide range of products are available (e.g. Actellic with Malathion and K-othrine). Some alternative methods have also been tried, e.g. the use of olive oil and salt mixtures, neem extracts (ICARDA, 1990) or traditionally used methods by farmers (Reed *et al.*, 1987).

Insecticides may be applied in different ways, i.e. dusting, spraying, fogging and evaporation. Dusting does not require much equipment; a powder formulation is either (i) mixed with the seed; (ii) applied in layers ('sandwich method'); or (iii) dusted over stacks. The latter can only prevent reinfestation (e.g. after fumigation), because most of the powder insecticides do no penetrate into the stacks to control internal infestation. For spraying, either a suspension (solid insecticide, usually a wettable powder formulation, suspended in water) or a solution (liquid insecticide diluted in water) can be used.

Sprays can be applied with knapsack sprayers. For suspensions, care should be taken that the particles remain suspended and do not sink to the bottom of the sprayer. This requires special stirring devices or frequent shaking of the spray container. Fogging is a technique used specially in storages. Fog droplets are much finer than those in spraying. Special fogging equipment is required for this. Evaporation can be used, in special cases, to control flying insects (moths). This technique requires volatile insecticides and well-closed stores.

Based on efficacy, low toxicity, long-lasting effect and minimum side effects on seed viability, the following spraying scheme is recommended for stored chickpea seed:

- Use, as a preventive measure, spray insecticide once in every 3–4 months;
- Alternate applications of Actellic with Malathion and K-othrine;
- Use fumigation with Phostoxin when insects are detected.

Fungi and bacteria

The fungi that infect seeds in storage are not seed-transmitted, which cause plant diseases in the field and are carried on or in seed. Bacteria normally do not affect stored seed, unless seed moisture content is very high and temperature is raised by fungal infection. Storage fungi require a high moisture content to grow (minimum of ~12–14% seed moisture); so they do not play an important role in dry climates. The most important genera are *Aspergillus* and *Penicillium*. These are mostly saprophytes; i.e. they are unable to attack living tissue. They grow on dead cells of the seed surface, where they produce toxins, cause seed decay and may kill the embryo. The best way to control storage fungi is to maintain low moisture/humidity in seed and storage facilities. Low temperature reduces the growth of fungi. Fungicidal seed treatment often does not give the expected results, because of lack of free water, which is required for many fungicides to become effective.

Fumigation

When insects are found in a seed lot, the entire lot should immediately be fumigated. The main advantage of fumigation is that all stages of the insect including eggs, larvae and pupae are controlled, along with other storage pests including rodents. Mainly two products are used in fumigation: methyl bromide and aluminium phosphide. These two fumigants are active in the gaseous phase, and have good penetration into piles of sacks or seed stored in bins, but are hazardous to human beings.

The only advantage of *methyl bromide* is its quick action, as compared with Phostoxin. Fumigated seed stacks can be aerated after 12–24 h of methyl bromide fumigation. Methyl bromide is often used in seaports because storage in ports is expensive and fumigation must be done as quickly as possible. Methyl bromide has the following properties: (i) extremely toxic and accumulates in the body of human beings; (ii) odourless and colourless and difficult to detect without proper devices; (iii) residues will remain in the seed; (iv) heavier than air, so fans are needed to recirculate the gas; and (v) has an adverse influence on seed germination. Because of its toxicity to humans and other harmful characteristics, methyl bromide is banned in many countries.

Phostoxin releases a gas called phosphine. It has excellent penetration capacity because of the small size of its gas molecules. Phosphine penetrates bags, carton boxes and other containers. It has no influence on germination, so seed can be treated repeatedly. Phostoxin is inflammable at normal temperatures and care should be taken while fumigating.

Conditions for fumigation – Smaller quantities of seeds can be fumigated in airtight fumigation chambers. Complete fumigation of seed storages is not practical because it is impossible to make it sufficiently airtight. Normally, large quantities of seed are fumigated in individual stacks under airtight fumigation sheets. It is most important to hermetically seal the area under fumigation. It is also a good practice to treat the fumigated seed stacks with an insecticide of some persistence, because fumigation does not have a lasting effect and reinfestation may take place immediately after fumigation.

Based on the characteristics of the fumigants, aluminium phosphide is recommended for use in seed, but not methyl bromide.

Seed Quality Assurance

A quality assurance system ensures that the seed quality meets required seed production standards. The national quality assurance system is a combination of technical and administrative procedures and guidelines supported by legislation intended to maintain and ensure that seed offered for sale must meet established standards of genetic, physical, physiological and health quality. It involves varietal certification through field inspection of growing crops, and laboratory testing of seed quality attributes to ensure standards prescribed by legislation including labelling and sealing of seeds offered for sale. It includes labels placed inside and outside on bags of seed for identification of variety, status (basic), year of production, place of production and certification of purity levels – inspections field and processing. It ensures that seed sold to farmers is of the designated variety and have the desired specific quality attributes. All seed quality control and certification programmes have the main features of setting field and seed standards, monitoring, supervising and enforcing seed quality standards during production and marketing. Seed quality control in developing countries has been described by van Gastel *et al.* (2002a,b).

Field standards

The standards suggested here are based on those used in several countries of West Asia and North Africa (ICARDA, 2002) and those given by Doerfler (1976). Field standards for chickpea seed production are usually set for: (i) off-types and other varieties; (ii) other crops; and (iii) parasitic plants (*Cuscuta*), virus-infected plants and *Ascochyta* (Table 20.6).

Table 20.6. Suggested field standards for chickpea seed production.

Factor	Pre-basic seed	Basic seed	Certified seed I	Certified seed II
Off-types and other varieties (maximum %)	0.1 (1:1000)	0.3 (1:333)	0.5 (1:200)	1.0 (1:100)
Other crops (maximum %)	0.1 (1:1000)	0.3 (1:333)	0.5 (1:200)	1.0 (1:100)
Cuscuta (maximum %)	0.05 (1:2000)	0.1 (1:1000)	0.2 (1:500)	0.3 (1:333)
Ascochyta on pods (maximum %)	0.3 (1:333)	0.4 (1:250)	0.5 (1:200)	1.0 (1:100)
	–	–	–	–
Total diseases (maximum %)	1 (1:100)	1.0 (1:100)	1.0 (1:100)	2 (1:50)

Numbers within parentheses represent the ratio of one contaminant plant to the specified number of chickpea crop plants in which one contaminant is allowed.

Inspecting seed fields

Field inspection should be made at the time when potential contamination is likely to happen and contaminants are easy to identify. During the inspection, it is verified that the standards are met as prescribed in the regulations. At least two inspections should be made; one during flowering and another during crop maturity. An additional inspection for ascochyta blight should be made in the vegetative stage.

The Association of Official Seed Certifying Agencies (AOSCA), as described by Revier and Young (1970) and modified by Gregg *et al.* (1990a,b), is suggested to make field inspection. It consists of two steps: 'field overview' and 'field inspection sample'. In the field overview, the field is generally observed to assess the uniformity of crop stand and quality. The field inspection sample involves inspecting randomly selected representative areas of the field and counting the actual number of each contaminant in a statistically determined number of plants. The numbers of contaminants found are compared with the standards; if above the tolerance level, the field is rejected.

Seed standards

To evaluate seed quality, seed standards for quality attributes should be established (Hampton, 1998). Laboratory tests are conducted to ensure that standards have been met. Suggested seed standards for chickpea seeds are given in Table 20.7, based on Moroccan national seed standards (ICARDA, 2002) and those suggested by Doerfler (1976).

Table 20.7. Laboratory quality standards for legume seed production.

	Pre-basic seed	Basic seed	Certified seed 1	Certified seed 2
Physical purity				
Purity (minimum %)	98.0	98.0	98.0	97.0
Other crop seed (maximum %)	0.2	0.2	0.2	0.5
Weed seeds (maximum %)				
Chickpea	0	0	0	0
Number of *Cuscuta* seeds in 100 g	1	2	2	2
Germination				
Germination (minimum %)	85.0	85.0	85.0	85.0
Hard seed expected to germinate (maximum %)				
Chickpea	0	0	0	0
Insects				
Life insects	0	0	0	0
Bruchid-damaged seed (maximum %)	3	3	3	5
Diseases				
Ascochyta (maximum %)	1.0	1.0	1.0	2.0
Seed moisture (maximum %)	12.0	12.0	12.0	12.0

Seed quality tests

Laboratory seed testing procedures are intended to minimize the risk of crop failure, by assessing the quality of seeds before sowing. Seed quality is evaluated by several standardized tests performed on representative samples taken from the seed lot. Seed quality comprises different attributes, including genetic (varietal purity and identity), physical (analytical and size), physiological (germination, vigour and moisture), health (freedom from seed-borne pests and parasitic weeds) and uniformity of seed lots. Although all attributes are important, only a selected few, i.e. physical purity, germination and moisture content, are routinely evaluated in the laboratory (Hampton, 2002). For chickpea, at least the following tests should be conducted: (i) physical purity; (ii) germination; (iii) moisture content; and (iv) seed health. All tests are described in detail by the International Seed Testing Rules (ISTA, 2003) and the Rules for Seed Testing of the Association of Official Seed Analysts (AOSA, 1981); only a few important issues are discussed here.

Seeds of the parasitic weed *Cuscuta* may be mixed with crop seeds or adhere to the surface of crop seed, and be planted with the crop and thus infest clean fields. Since seed of parasitic weeds are very small, they may not be easily detected in the purity analysis. Therefore, a washing test may be carried out, in which the seed sample is submerged in water (with a few drops of household detergent to eliminate surface tension) and poured over a sieve covered with a white filter paper. The number of *Cuscuta* seeds can then be determined under a stereo microscope.

Impermeability of the seed coat to water (hardseededness) is known in all food legumes. The development of seed coat impermeability is associated with the dehydration of seed during later stages of maturation, particularly low RH during ripening (Agrawal, 1985; Ellis *et al.*, 1988). Hardseededness is a substantial problem in drier regions; for example, in India 8–20% hard seeds have been reported for chickpea (Anonymous, 1984). The number of hard seeds is determined during the germination test.

Ascochyta transmission by seed is of major importance, as compared with other sources of inoculum, and special seed health tests are required. Two test methods, agar plate test and blotter test, can be used. The latter is recommended by ISTA and is easy and inexpensive, whereas the former is more sensitive.

Control plot tests

A well developed certification scheme usually includes field plots to provide additional checks on varietal identity and purity and occurrence of weeds and seed-borne diseases. Post-control plots are planted with samples taken from seed lots that were approved (certified) in the previous season. The main benefit of control plots is to ascertain whether or not the certification system works satisfactorily. A sample from each approved basic seed lot and 10–20% of certified seed lots is taken and planted in post-control test plots to confirm that varietal characters remained unchanged during the seed multiplication process. Cereal post-control plots (OECD, 1971, 2000) can be used as the basis for food legumes.

Lot numbering

An effective seed quality assurance programme must use a lot numbering system that can track each lot back to producers, processors or distributors, and to the seeds used for planting. Each seed lot must be given a unique number, which provides information on production year, field or grower, crop, processor and seed class.

Managing seed quality assurance

The seed certification authority should remain independent of seed production so that it can impartially serve both seed producers and seed users. However, with the emergence of a diverse seed industry, certification should be flexible and decentralized with more responsibility given to seed producers than the comprehensive and compulsory certification schemes. Tripp *et al.* (1997) suggested that the certification agency performs its task with greater efficiency by using appropriate criteria in a transparent manner through the participation of stakeholders.

Harmonizing seed certification procedures

Procedures for seed certification are essentially similar in most countries, with the possible exception of seed classes where there is variation in nomenclature. Moreover, most national seed regulations and standards are similar in many ways. Countries should work together to develop a flexible regionally harmonized or internationally acceptable seed certification scheme, for the benefit of the national seed industry development and to ensure adequate supply of high-quality seeds to farmers.

References

Acikgoz, N. and Kutlu, Y.Z. (1987) Mechanization of chickpea production in Turkey. In: Mechanization of Field Experiments in Semi-arid Areas, *Proceedings of the IAMFE/ICARDA Conference*. 23–27 May 1987. ICARDA, Aleppo, Syria, pp. 197–201.

Agarwal, P.K. (1976) Identification of suitable seed storage places in India on the basis of temperature and relative humidity conditions. *Seed Research* 4, 6–11.

Agrawal, P.K. (1985) Seed production technology for chickpea (*Cicer arietinum*) and lentil (*Lens culinaris*). In: Saxena, M.C. and Varma, S. (eds) *Faba Beans, Kabuli Chickpeas, and Lentils in the 1980s*. ICARDA, Aleppo, Syria, pp. 271–279.

Anonymous (1984) Annual progress report. Division of Seed Science and Technology, New Delhi, India.

AOSA (Association of Official Seed Analysts) (1981) Rules for seed testing. *Journal of Seed Technology* 6(2), 1–126.

Bond, E.J. (1984) *Manual of Fumigation for Insect Control*. FAO, Rome, Italy.

Bouwman, A.J. (1992) Maintenance breeding and multiplication of pea and faba bean cultivars. Maintenance of protein peas (*Pisum sativum*) and field beans (*Vicia faba*). *Euphytica* 61, 213–215.

Boyd, A.H., Dougherty, G.M., Matthews, R.K. and Rushing, K.W. (1975) Seed drying and processing. In: Feistritzer, W.P. (ed.) *Cereal Seed Technology*. FAO, Rome, Italy, pp. 60–86.

Brandenburg, N.P. (1977) The principles and practice of seed cleaning: separation with equipment that senses dimensions, shape, density and terminal velocity of seeds. *Seed Science and Technology* 5(2), 173–186.

Brandenburg, N.P. and Park, J.K. (1977) The principles and practice of seed cleaning: separation with equipment that senses surface texture color, resilience and electrical properties of seeds. *Seed Science and Technology* 5(2), 187–197.

Bushara, A.G. (1988) Insect depredation during storage. In: Summerfield, R.J. (ed.) *World Crops: Cool Season Food Legumes*. Kluwer Academic, Dordrecht, The Netherlands, pp. 367–378.

Delouche, J.C. (1988) Seed storage practices and problems for cool season legumes. In: Summerfield, R.J. (ed.) *World Crops: Cool Season Food Legumes*. Kluwer Academic, Dordrecht, The Netherlands, pp. 331–339.

Delouche, J.C., Matthews, R.K., Dougherty, G.M. and Boyd, A.H. (1973) Storage of seed in sub-tropical and tropical regions. *Seed Science and Technology* 1, 671–700.

Diekmann, M. (1988) Control of storage pests. In: van Gastel, A.J.G. and Kerley, J. (eds) *Quality Seed Production*. ICARDA, Aleppo, Syria, pp. 73–81.

Doerfler, Th. (1976) *Seed Production Guide for the Tropics*. GTZ, Eschborn, Germany.

Drijfhout, E. (1981) Maintenance breeding of beans. In: Seeds, FAO Plant Production and Protection Paper 39. *Proceedings of FAO/SIDA Technical Conference on Improved Seed Production*. 2–6 January 1981. Nairobi, Kenya.

Ellis, R.H. and Roberts, E.H. (1980) Improved equations for prediction of seed longevity. *Annals of Botany* 45, 13–30.

Ellis, R.H., Agrawal, P.K. and Roos, E.E. (1988) Harvesting and storage factors that affect seed quality in pea, lentil, faba bean and chickpea. In: Summerfield, R.J. (ed.) *World Crops: Cool Season Food Legumes*. Kluwer Academic, Dordrecht, The Netherlands, pp. 303–329.

Erskine, W., Singh, K.B., Robertson, L.D., Saxena, M.C. and Diekmann, J. (1987) Mechanization of field experimentation in faba bean, kabuli chickpea and lentil. In: Karisson, D. (ed.) *Mechanization of Field Experiments in Semi-Arid Areas: Proceedings of the IAMFE/ICARDA Conference*. 23–27 May 1987. ICARDA, Aleppo, Syria, pp. 169–176.

Erskine, W., Singh, K.B. and Robertson, L.D. (1988) Seed production of food legumes in the Mediterranean area. In: van Gastel, A.J.G. and Hopkins, J.D. (eds) *Seed Production in and for Mediterranean Countries*. ICARDA, Aleppo, Syria, pp. 120–130.

FAO (Food and Agriculture Organization) (1981) Cereal and grain-legume seed processing. Technical guidelines. FAO, Rome, Italy.

Galanopoulou, S., Facinelli, M. and Lorenzetti, F. (1996) General agronomic aspects of seed production. In: van Gastel, A.J.G., Pagnotta, M.A. and Porceddu, E. (eds) *Seed Science and Technology: Proceedings of a Train-the-trainers Workshop Sponsored by Med-campus Programme (EEC)*. 24 April–9 May 1993. Amman, Jordan; ICARDA, Aleppo, Syria.

Gowda, C.L.L. (1981) Natural outcrossing in chickpea. *International chickpea Newsletter* 5, p. 6.

Gregg, B.R., Law, A.G., Virdi, S.S. and Balis, J.S. (1970) *Seed Processing*. Avion Printers, New Delhi, India.

Gregg, B., van Gastel, A.J.G., Homeyer, B., Holm, K., Gomaa, A.S.A., Salah Wanis, M., Ghanem, A.E., Abdel Monem, A., Gouda, A. and Shehata, O. (1990a) *Procedures for Inspecting Wheat Seed Fields*. NARP Publication No. 39. Cairo, Egypt.

Gregg, B., van Gastel, A.J.G., Homeyer, B., Holm, K., Gomaa, A.S.A. and Salah Wanis, M. (1990b) *Roguing Seed Production Fields*. NARP Publication No. 40. Cairo, Egypt.

Gwinner, J., Harnisch, R. and Mück, O. (1990) *Manual on the Prevention of Post-harvest*

Grain Losses. GTZ, Postfach 5180, D-6236 Eschborn, Germany.

Hampton, J.G. (1998) Seed quality standards-their use misuse and the future. *Journal of Applied Seed Production* 16, 31–48.

Hampton, J.G. (2002) What is seed quality? *Seed Science and Technology* 30, 1–10.

Harrington, J.F. and Douglas, J.E. (1970) *Seed Storage and Packaging*. National Seeds Corporation and Rockefeller Foundation, New Delhi, India.

ICARDA (1990) Food Legume Improvement Program Annual Report. Aleppo, Syria.

ICARDA (2002) WANA Catalogue of field and seed standards. Network Publication No. 25/02. ICARDA, Aleppo, Syria.

ISTA (International Seed Testing Association) (1977) Seed cleaning and processing. *Seed Science and Technology* 5(2).

ISTA (2003) *International Rules for Seed Testing and Supplement*. International Seed Testing Association, Bassersdorf, Switzerland pp. 1–330.

Julen, G. (1983) *Maintenance of Varietal Purity for Seed Production. More Food from Better Technology*. FAO, Rome, Italy.

Justice, O.L. and Bass, L.N. (1978) *Principles and Practices of Seed Storage*. USDA Agriculture Handbook 506. US Government Printing Office, Washington, DC.

Kraak, H.L. (1992) A computer program to predict seed storage behavior. *Seed Science and Technology* 20, 337–338.

Laverack, G.K. (1994) Management of breeders seed production. *Seed Science and Technology* 22, 551–563.

Laverack, G.K. and Turner, M.R. (1995) Roguing seed crops for genetic purity: a review. *Plant Varieties and Seeds* 8, 29–45.

McDonald, M.B. Jr and Nelson, C.J. (eds). (1986) *Physiology of Deterioration*, CSSA Special Publication Number 11. Crop Science Society of America, Madison, Wisconsin.

OECD (1971) Guide to the methods used in plot tests and to the methods of field inspection of cereal seed crops. *Proceedings of the International Seed Testing Association* 36, 495–519.

OECD (2000) OECD schemes for varietal certification of seed moving in international trade. OECD, AGR/CA/S(2000)9, Paris, France.

Reed, W., Cardona, C., Sithanantham, S. and Lateef, S.S. (1987) The chickpea insect pests and their control. In: Saxena, M.C. and Singh, K.B. (eds) *The Chcikpea*. CAB International, Wallingford, UK, pp. 283–318.

Revier, P.R. and Young, A.W. (1970) *Field Inspection Techniques in Seed Production*. Department of Agronomy, Texas Technological College, Lubbock, Texas.

Roberts, E.H. (1972) Storage environments and the control of viability. In: Roberts, E.H. (ed.) *Viability of Seeds*. Chapman & Hall, London.

Roberts, E.H. and Osei-bonsu, K. (1988) Seed and seedling vigour. In: Summerfield, R.J. (ed.) *World Crops: Cool Season Food Legumes*. Kluwer Academic, Dordrecht, The Netherlands, pp. 897–910.

Sauer, D.B. (1992) *Storage of Cereal Grains and Their Products*. American Association of Cereal Chemists, St. Paul, Minnesota.

Saxena, M.C. (1987) Agronomy of chickpea, In: Saxena, M.C. and Singh, K.B. (eds) *The Chickpea*. CAB International, Wallingford, UK, pp. 207–232.

Saxena, M.C. and Singh, K.B. (1987) *The Chickpea*. CAB International, Wallingford, UK.

Saxena, M.C., Diekmann, J., Erskine, W. and Singh, K.B. (1987) Mechanization of harvest in lentil and chickpea in semi-arid areas. In: *Mechanization of Field Experiments in Semi-arid Areas: Proceedings of the IAMFE/ICARDA Conference*. 23–27 May 1987. ICARDA, Aleppo, Syria, pp. 211–228.

Sharma, S.P. (1987) Breeder seed production in self-pollinated crops. In: Agrawal, P.K. and Dadlani, M. (eds) *Techniques in Seed Science and Technology*. South Asian Publishers, New Delhi, India.

Singh, K.B. (1986) Principles and techniques for seed production in chickpea. In: Srivastava, J.P. and Simarski, L.T. (eds) *Seed Production Technology*. ICARDA, Aleppo, Syria, pp. 247–256.

Singh, K.B. and Saxena, M.C. (1999) Chickpeas. In: Coste, R. (ed.) *The Tropical Agriculturist*. Macmillan, London.

Thomson, J.R. (1979) *An Introduction to Seed Technology*. Leonard Hill, Glasgow, UK.
Tripp, R., Louwaars, N.P., van der Burg, W.J., Virk, D.S. and Witcombe, J.R. (1997) Alternatives for seed regulatory reform: an analysis of variety testing, variety regulation and seed quality control. Agricultural Research and Extension Network. Network Paper No. 69. ODI, London.
Turner, M. and Bishaw, Z. (2000) Linking participatory plant breeding to seed systems. Paper Presented in International Symposium of Scientific Basis of Participatory Plant Breeding and Conservation of Genetic Resources. 9–13 October 2000. Oaxtepec, Mexico.
van der Burg, W.J. (1986) An introduction to seed cleaning. In: Srivastava, J.P. and Simarski, L.T. (eds) *Seed Production Technology*. ICARDA, Aleppo, Syria, pp. 107–126.
van Gastel, A.J.G., Bishaw, Z. and Gregg, B.R. (2002a) Wheat seed production. In: Curtis, B.C., Rajaram, S. and Gomez Macpherson, H. (eds) *Bread Wheat – Improvement and Production*. FAO, Rome, pp. 463–482.
van Gastel, A.J.G., Gregg, B.R. and Asiedu, E.A. (2002b) Seed quality control in developing countries. *Journal of New Seeds* 4(1–2), 117–130.
UPOV (International Union for the Protection of new Varieties of Plants) (1993) Chickpea: guidelines for the conduct of tests for distinctness, homogeneity and stability. TG/143/3, UPOV, Geneva, Switzerland.
UPOV (International Union for the Protection of new Varieties of Plants) (2002) General introduction to the examination of distinctness, uniformity and stability and the development of harmonized descriptions of new varieties of plants. TG/1/3, UPOV, Geneva, Switzerland.
Vaughan, C.E., Gregg, B.R., Delouche, J.C. (1968) *Seed Processing and Handling*. Handbook No. 1. Seed Technology Laboratory, Mississippi State University, Starkville, Mississippi.
Wellving, A.H.A. (ed.) (1984) *Seed Production Handbook of Zambia*. Department of Agriculture, Lusaka, Zambia.

21 Ciceromics: Advancement in Genomics and Recent Molecular Techniques

P.N. RAJESH,[1] K.E. MCPHEE,[2] R. FORD,[3] C. PITTOCK,[4] J. KUMAR[5] AND F.J. MUEHLBAUER[2]

[1]*Department of Crop and Soil Sciences, Washington State University, Pullman, WA 99164-6434, USA;* [2]*USDA-ARS, Grain Legume Genetics and Physiology Research Unit, Washington State University, Pullman, WA 99164-6434, USA;* [3]*BioMarka, School of Agriculture and Food Systems, Faculty of Land and Food Resources, The University of Melbourne, VIC 3010, Australia;* [4]*The University of Melbourne, PB 260, Horsham, VIC 3401, Australia;* [5]*Hendrick Beans-for-Health Research Foundation, 11791 Sandy Row Inkerman, Ontario, K0E 1J0, Canada*

Introduction

Genomics is a branch of science that decodes the encrypted information from the building block of organisms – DNA. It unveils vital information such as the number of genes present in the genome, genome organization and content that has tremendous application in agriculture, medicine, evolutionary biology and other areas of biology. Although, genomics has allowed revolutionary advancements in plants such as *Arabidopsis*, rice, wheat, barley and the model legumes, *Medicago truncatula* and *Lotus japonicus*, it is an upcoming area of science in chickpea (*Cicer arietinum* L.). Genomic research in this economically important crop has grown rapidly since the beginning of this century with the availability of genome sequence information and various tools such as bacterial artificial chromosome (BAC) libraries, cDNA libraries and targeting induced local lesion in genome (TILLING) mutants. The objective of this chapter is, for the interest of chickpea researchers, to discuss important branches of genomics, where significant progress has been made, and to indicate future needs. We believe that this review will inspire scientists worldwide to advance chickpea genomic research to an unprecedented level in the future.

Molecular Markers and Their Use in Genetic Improvement

The change in the focus of chickpea research from germplasm collection and determination of the genetics of important traits to linkage map development was evidenced by a review of chickpea genetics by Muehlbauer and Singh (1987). This article described genetic linkages of several morphological traits, and subsequent studies by Gaur and Slinkard (1988, 1990) and Simon and Muehlbauer (1997) added relatively few loci to the *Cicer* linkage map.

The first integrated molecular marker map with 354 markers, including 118 sequence tagged microsatellite sites (STMSs), 96 DNA amplification fingerprintings (DAFs), 70 amplified fragment length polymorphisms (AFLPs), 37 intersimple sequence repeats (ISSRs), 17 random amplified polymorphic DNAs (RAPDs), 8 isozymes, 3 cDNAs and 2 sequence characterized amplified regions (SCARs), covering a total distance of 2077.9 centimorgans (cM) was a result of an international collaborative effort and was published by Winter *et al.* (2000). Five additional linkage maps were generated using different mapping populations with different morphological and molecular markers (Santra *et al.*, 2000; Cho *et al.*, 2002, 2004; Collard *et al.*, 2003; Flandez-Galvez *et al.*, 2003). Several agronomically important traits were mapped and flanking markers were identified in these linkage maps. Targeted marker density increase near the genes governing these traits and their selective validation will ultimately have an implication in marker-assisted selection and map-based cloning.

Genes for fusarium wilt resistance *foc1*, *foc4* and more recently *foc3* (Tullu, 1996; Mayer *et al.*, 1997; Tullu *et al.*, 1998, 1999; Winter *et al.*, 2000; Sharma *et al.*, 2004) and the double-podded trait, *s*, have been mapped in the *Cicer* genome (Cho *et al.*, 2002; Rajesh *et al.*, 2002b). In addition, 19 resistance gene analogs (RGAs) have been mapped in different populations (Huttel *et al.*, 2002; Rajesh *et al.*, 2002a; Flandez-Galvez *et al.*, 2003). Furthermore, a number of genes, which were amplified using gene-specific primers from related species, were localized in the chickpea genome (Pfaff and Kahl, 2003).

Inheritance of ascochyta blight resistance (ABR) has been studied extensively by various researchers as it is the major biotic constraint that causes severe damage to chickpea crops worldwide. Santra *et al.* (2000) reported two quantitative trait loci (QTLs), *QTL*1 and *QTL*2, that confer resistance to ascochyta blight in chickpea in the USA. These QTLs accounted for an estimated 34.4% and 14.6% of the total phenotypic variation, respectively (Santra *et al.*, 2000; Tekeoglu *et al.*, 2000, 2002). Subsequently, two QTLs conferring seedling ABR in Australia were identified using an interspecific cross between *C. arietinum* and *C. echinospermum* (Collard *et al.*, 2003). The authors reported that one of the QTL, *QTL*1, is the same as *QTL*2 identified by Santra *et al.* (2000) on the basis of common markers. At the International Center for Agricultural Research in the Dry Areas (ICARDA), Aleppo, Syria, Udupa and Baum (2003) identified a single major gene conferring resistance to pathotype I and two QTLs for pathotype II. Using an interspecific cross, Rakshit *et al.* (2003) from Germany mapped OPS06-1, a DAF marker, between the flanking markers at *QTL*1 of Santra *et al.* (2000) along with several other DAF markers.

Cho *et al.* (2004) used an intraspecific recombinant inbred line (RIL) population to demonstrate that *Ar*19 (or *Ar*21d) on LG2 + 6 appeared to provide the majority of quantitative resistance to pathotype I and partial resistance to pathotype II of *Ascochyta rabiei*, and that a QTL on LG4A, equivalent to *QTL*1, was required for resistance to pathotype II of *A. rabiei*. The environmental effect of ascochyta blight *QTL*1 and *QTL*2 was analysed at Eskisehir, Turkey (Tekeoglu et al., 2004) using CRIL-7 (Santra *et al.*, 2000) developed in the USA from the interspecific cross FLIP84-92C (*C. arietinum*) × PI599072 (*C. reticulatum*). The same two QTLs were identified at both sites indicating their robust nature; however, the effect of *QTL*1 was greater than *QTL*2 at Pullman, Washington, USA, whereas the effect of *QTL*2 was greater than *QTL*1 at Eskisehir indicating possible differences in pathogen populations and environmental interactions (Tekeoglu et al., 2004). Iruela *et al.* (2006) used an intraspecific population developed from the cross ILC3279 × WR315 and identified two QTLs that were the same as that of Santra *et al.* (2000). These reports demonstrate the availability and wealth of information on molecular markers and in-depth studies for several economically important traits, especially ABR in chickpea.

The frequency of single nucleotide polymorphism (SNP) in expressed and genomic sequences of the chickpea genome was determined using an interspecific cross between *C. arietinum* and *C. reticulatum*. It was estimated that one SNP was present for every 74 nucleotides in the genomic regions studied and the frequency of SNPs was greater in expressed sequences. These SNPs were utilized to increase marker density at ABR-*QTL*1 by designing cleavage amplified polymorphic site (CAPS) and derived CAPS (dCAPS) markers (Rajesh *et al.*, 2005). High sequence homology (98–100%) observed within *Cicer* spp. hinders mapping efforts based on intraspecific crosses. However, SNP information is useful to the chickpea research community to genetically map expressed sequence tags (ESTs), BAC end and other sequence-specific markers (F.J. Muehlbauer, Pullman, 2006, personal communication).

Positional Cloning

In order to isolate agronomically important gene(s) by positional or map-based cloning, it is imperative to have a high density map to identify tightly linked flanking markers for genes of interest. Availability of such genetic maps makes this approach easier in small genomes like *Arabidopsis*, rice and in large genomes such as wheat and barley. Large insert libraries such as yeast artificial chromosome (YAC), BAC and P1-derived artificial chromosome (PAC) libraries made a significant impact on gene isolation. Development of large insert libraries has reduced the time and number of steps required for chromosome walking compared to small insert libraries. Chickpea genomics has made significant gains in recent years through the production of genomic and EST libraries. New resources for genomic analysis of chickpea include: the production of BAC genomic libraries (Rajesh *et al.*, 2004; Lichtenzveig *et al.*, 2005), progress towards linking the physical genome to genomic linkage groups (Vláčáilová *et al.*, 2002, Valárik *et al.*, 2004) and demonstration of whole-of-genome or candidate gene analysis for functional purposes (Corum and Pang, 2005a; Gremigni *et al.*, 2005; Jayashree *et al.*, 2005; D.R. Cook, Davis, 2006, personal communication).

Most disease resistance genes that have been isolated in other crops by map-based cloning so far are governed by single genes (Brueggeman *et al.*, 2002; Feuillet *et al.*, 2003). Identification of candidate genes for a trait governed by QTL is a difficult task and, although several QTLs have been characterized in chickpea, no quantitative resistance loci (QRL) have been isolated. In chickpea, some of the agronomically important traits such as seed weight, seed number (Cho *et al.*, 2002) and ABR (Santra *et al.*, 2000) are governed by QTL.

Positional cloning is in progress using a chickpea BAC library to isolate ABR genes from *QTL*1. Screening the library identified new markers from BAC ends and facilitated increased marker density utilizing SNPs at *QTL*1 and also identified additional markers for ascochyta blight. This effort identified an open reading frame (ORF) gene involved in signal transduction, flavonoid 3-*O*-galactosyl transferase and a transcription factor as candidate genes in this genomic region. However, additional work is being performed to correlate the function of the candidate genes with the phenotypic trait (P.N. Rajesh *et al.*, Pullman, 2006, unpublished).

The increasing number of ESTs (more than 500 entries; www.ncbi.nlm.nih.gov/chickpea), BAC sequences (734,456 bp; (http://www.genome.ou.edu/plants_totals.html) and other sequences with potential for new SNP markers are expected to allow generation of a high density consensus linkage map for chickpea. This will facilitate identification of tightly linked flanking markers to perform map-based cloning of genes of interest in chickpea.

Functional Genomics in Chickpea

The aim of functional genomics is to discover the biological function of genes, and to determine how sets of genes and their expressed products interact to result in a particular phenotype. This is achieved using computational and high throughput technologies, often at the whole genome level, such as differential display and microarray analysis. Understanding the functional characteristics and expression of a particular trait may also involve techniques such as EST library construction, mutant identification, RNA interference (RNAi) experiments and overexpression studies. The area of functional genomics research is growing rapidly and now includes protein function studies (proteomics) and, even more broadly, studies on the metabolism of an organism (metabolomics).

Several highly sought chickpea characteristics are currently being targeted for intense scrutiny of their molecular and related physiological mechanisms through functional genomics approaches. For example, module 3 'Seed Composition and Quality' of the Grain Legumes Integrated Project within the 6th European Union Framework Programme aims to understand the determinants of seed composition and quality in chickpea through protein and metabolite analyses. Candidate genes are also sought using reverse genetic tools from related model legume genomes. Indeed, many legume genome information services have been developed for comparative analysis and in determining gene function among related species (see Beavis, 2004 for review).

Meanwhile, preliminary investigations have been carried out in chickpea to determine important functional genes involved in traits such as abiotic tolerances (Mantri *et al.*, 2006), seed quality (Gremigni *et al.*, 2004) and biotic disease resistances (Jaiswal *et al.*, 2004; Corum and Pang, 2005a). These studies were conducted in the hope that the major genes and genetic pathways responsible, and potentially the signaling molecules that trigger and govern them, will be identifiable. Once identified, it is envisaged that progress towards agronomic improvement may be leapfrogged through their future selection and incorporation into elite breeding germplasm, or the pathways they are integral to will be manipulated to provide a beneficial phenotypic response, through direct gene modification.

Jayashree *et al.* (2005) reported the development and availability of a chickpea root-specific EST database comprising an excess of 2800 sequences. This was constructed using subtractive suppressive hybridization (SSH) among root tissue from two closely related chickpea genotypes possessing different sources of drought avoidance and tolerance. Since the database contains many stress-responsive sequences, from broad functional categories, it is predicted to be an invaluable resource in future studies of agronomic responses to overcome many root stresses. Alternatively, serial analysis of gene expression (SAGE; Matsumura *et al.*, 2003) has been employed to quantitatively determine the transcriptome of chickpea plants under drought and pathogen stresses (Winter *et al.*, 2005).

One of the most recent and perhaps most advanced targets for functional genomics research in chickpea has been to understand defense mechanisms to the major fungal pathogen *A. rabiei*. Jaiswal *et al.* (2004) recently reported using a polymerase chain reaction (PCR) subtraction method to identify genes that were upregulated in response to *A. rabiei*. These were subsequently characterized as immediate, early or late-responsive based on the timing of expression after inoculation via reverse northern analysis. Detected sequences were homologous to known signalling and transcription factors (Jaiswal *et al.*, 2004). In another approach, an EST library was constructed from the challenged resistant accession ICC3996 for the isolation and characterization of candidate resistance genes. Subsequently, a cDNA microarray was constructed to include all previously isolated chickpea defense and stress-related gene sequences, to determine the differential regulation of candidate genes among ICC3996 and a susceptible accession, 'Lasseter', with and without exposure to the pathogen. A small subset of ESTs was shown to be significantly upregulated in ICC3996, not Lasseter, when inoculated with an aggressive fungal isolate. This group included a potential regulatory leucine zipper protein, a SNAKIN2 antimicrobial peptide precursor and an elicitor-induced receptor protein (Corum and Pang, in press). Further research, through temporal and spatial RT-PCR analyses and gene silencing and/or overexpression studies, is required to more fully characterize the transcriptomes and validate the observed defence responses.

To correlate the function of the identified genes with the trait, it is essential to have suitable technologies such as *Agrobacterium*-mediated transformation and/or reverse genetic approaches including mutants generated by various methods and RNA interference. In reverse genetics, alteration of a gene sequence or its expression, and comparison of this genetic manipulation to the wild suggests gene function. In chickpea, 9800 targeted TILLING mutants that are generated by point mutation using ethyl methane sulphonate (EMS) mutagenesis are available. TILLING is becoming increasingly popular for studies on reverse genetics. Utilization of this important genetic resource and development of additional tools will reveal the function of several unannotated genes from the chickpea genome (F.J. Muehlbauer, Pullman, 2006, personal communication).

Comparative Genomics

Comparative analysis attempts to answer questions related to function, evolution and organization of genomes. The availability of the genome sequence from model legumes such as *M. truncatula* and *L. japonicus*, and the increase in DNA sequences from other crop legumes in the database have generated resources to perform comparative genomics on a scale similar to that among the cereal crops. A finite number of chromosomal rearrangements have occurred during the evolution of angiosperm plants resulting in significant blocks of syntenic genetic material among genomes of related species (Bennetzen, 2000; Devos and Gale, 2000; Michelmore, 2000).

Soybean when compared with *Arabidopsis* for comparative genome analysis showed substantial microsynteny (Grant *et al.*, 2000). Among legumes, higher levels of genome conservation were observed among soybean, *Vigna radiata, V. uniguiculata, Vicia faba* and *Phaseolus vulgaris* (Lee *et al.*, 2001; Young *et al.*, 2003). Recent analyses show that ~400 soybean BAC contig groups exhibit extensive microsynteny with *M. truncatula* (Yan *et al.*, 2003).

Genome characterization and isolation of genes have been demonstrated in other legumes using *M. truncatula* genomic information as a reference (Yan *et al.*, 2003). For example, synteny analysis was used to better define the *Sym2* locus by fine genetic mapping in pea (Gualteri *et al.*, 2002) and facilitated cloning of a gene for Nod-factor perception in lucerne (Endre *et al.*, 2002). Additional examples for isolation of genes using legume microsynteny include cloning of *NN1* from *M. sativa* and *Sym2* from *L. japonicus* along with their respective orthologues *DM12* in *M. truncatula* and *Sym19* in pea (Endre *et al.*, 2002; Stracke *et al.*, 2002).

Among cool season grain legumes nearly 40% of the arrangements on the lentil genetic linkage map can be found on the pea map with several similarities in gene order (Weeden *et al.*, 1992). Simon and Muehlbauer (1997) showed that at least five genomic regions are similar between pea, lentil and chickpea. The synteny chickpea shares with pea and lentil, has advantages for studies in regions of linkage conservation (Weeden *et al.*, 1992). Many of the observations of synteny were based on linkage maps and should be extended using specific nucleotide sequences of the syntenic regions. Pfaff and Kahl (2003) made use of conserved gene sequences to map several gene-specific markers designed from species such as *Arabidopsis, Phaseolus, Pisum, Lotus* and *Medicago* in an interspecific cross of *Cicer*. Their study demonstrated successful transferability of genetic information between related leguminous species.

Comparative sequence analysis of the chickpea ABR-*QTL1* genomic region and *M. truncatula* identified three orthologous regions present in three different linkage groups (Rajesh *et al.*, unpublished data). Similarly, *Medicago* linkage group V was aligned with chickpea linkage groups II and VIII (D.R. Cook,

Davis, 2006, personal communication). Both observations postulated that segmental reshuffling events between chromosomes might have occurred during the course of speciation. Extensive analysis at the whole genome level will reveal genome conservation across the legumes and the molecular dynamics of different genomic regions.

Genomics to Field

The applied use of genomics for the enhancement of cultivated chickpea requires substantial knowledge of the chickpea genome, molecular and bioinformatic tools for the selection or manipulation of key traits, and a skill base within breeding programmes to bring genomics-based gains to fruition for farmers.

Genomic analysis of chickpea is in transition from an anonymous marker-PCR driven exploration of the genome to a whole-of-genome and expressed-gene exploration. The pairing of these paradigms is led by the application of 'pre-genomics' marker-driven mapping information to whole-of-genome exploration resources to enable BAC walking. The organized arrangement of representative genomes, expressed or otherwise, in library or microarray form will become the basis for selection and functional analysis of gene combinations in the context of future breeding selection processes.

Development of comparative and trans-genomic tools is progressing towards the ability to gain value from the deeper knowledge sets of other species. Such tools include extensive candidate gene analyses for disease resistance (D.R. Cook, Davis, 2006, personal communication), links to available model legume maps such as *M. truncatula* (Gutierrez *et al.*, 2005) and demonstrated transgenesis (Polowick *et al.*, 2004, Senthil *et al.*, 2004; Sanyal *et al.*, 2005;) with follow-up studies on the fate of transgenes within the genome (Polowick *et al.*, 2004). Linking the chickpea genome with model species has been difficult due to low transferability of anonymous DNA markers, but positive results are evident from this approach (Muehlbauer *et al.*, 2005).

Applied genomic analysis of chickpea has taken two main approaches into consideration, the first being reliant upon traditional linkage mapping (via anonymous markers) with subsequent BAC library screening for target gene(s) (Muehlbauer *et al.*, 2005), whilst the second is a candidate gene approach as demonstrated by the cloning of most or all of the RGAs (D.R. Cook, Davis, 2006, personal communication). Alternatively, SAGE (Matsumura *et al.*, 2003) has been employed to quantitatively determine the transcriptome of chickpea plants under drought and pathogen stresses (Winter *et al.*, 2005). The transition from anonymous and untranscribed markers to genic and transcribed markers gives greater ability to find direct markers for the genes controlling the traits, and should raise the ability to transfer such information from the genomes of model species to the cultigen genome.

To date the vast majority of genomics effort in chickpea has focused on biotic stress resistance to two diseases ascochyta blight (Muehlbauer *et al.*, 2005) and fusarium wilt (Benko-Iseppon *et al.*, 2003; Rubio *et al.*, 2003), with some targeted cloning and characterization of developmentally interesting genes (β-galactosidase, Esteban *et al.*, 2005; apyrase, Cannon *et al.*, 2003), seed quality trait-related genes (Carotenoid content, Abbo *et al.*, 2005; seed size, Gremigni *et al.*, 2005) and abiotic stress tolerance (Bhattarai and Fettig, 2005). It will be possible and more cost-effective to apply genomic tools to lower-priority traits when higher throughput platforms permit lower cost-per-locus screening, i.e. to consider opportunity cost.

At present the most readily applicable selection technique for taking genomics to the field is the microarray screening method based on that employed by Corum and Pang (2005b). Key genic measures of the suitability of the transcriptome of breeding lines to resist ascochyta blight challenge may be made via targeted screening of lines with custom-designed gene arrays – in effect putting the work of Cho and Muehlbauer (2004) into a higher throughput and multiplex platform. The component of work by Cho and Muehlbauer (2004) where the expression of a panel of defence-related genes was examined across a sample of RILs when challenged with different pathotypes of *A. rabiei* reflects the basis for conducting comparative expression analysis in context of developing a selection tool for allelic combinations, which confer optimal expression profiles under this biotic stress. The ability to add further elements to arrays at a low per-addition cost reflects the opportunity to be more holistic in the screening of genomes.

Bioinformatic tools for the analysis and interpretation of large-scale multiplexed analyses currently exist; however, the development of applied breeding and selection interfaces is necessary to capitalize on the power of such platforms and their data-sets without suffering 'interpretative bottlenecks'. One initiative in this regard has been proposed (D.R. Cook, Davis, 2006, personal communication), and would seem to be a worthy accompaniment to the development of high throughput molecular tools for breeding and research.

Adoption of genomic tools in applied breeding programmes will be determined primarily on a cost–benefit analysis. Thus, it becomes crucial that the economics of using new genomic tools in various modes of application are assessed and compared with the status quo. A genetic and economic analysis of marker-assisted wheat-breeding strategies (Kuchel *et al.*, 2005) points the way towards gaining an understanding of the reality of molecular breeding in the hands of breeders.

Chickpea as a Model Crop Legume: A Perspective

M. truncatula has been established as the model species for basic genetic inquiry among legumes. Although *M. truncatula* was chosen for a number of valid reasons, it could be argued that chickpea would have been an excellent choice as a model with more direct application to crop production since it is a crop species of worldwide importance and ranks third in world production among legumes after dry bean (*P. vulgaris* L.) and dry pea (*Pisum sativum* L.). In addition, it is a staple in many developing countries where implementation of advancements made from basic studies could have a profound and direct impact on human health and well-being.

Chickpea a self-pollinating, annual, diploid with $2n = 2x = 16$ chromosomes and has an estimated genome size of 740 Mb (Arumuganathan and Earle, 1991), only modestly larger than *M. truncatula* (540 Mb). Its chromosomes are relatively free of cytological aberrations. The growth and development cycle from seed to seed averages 4 months. Significant effort has been placed on locating specific disease-resistance genes on the chickpea genetic map; especially for ascochyta blight and fusarium wilt resistance (Santra *et al.*, 2000; Tekeoglu *et al.*, 2002; Winter *et al.*, 2000). A number of genetic maps have been created through these studies and offer a great opportunity to establish a consensus map once a sufficient number of common markers among the populations are identified. The relatively simple genome, short life cycle and availability of genetic and genomic information make this crop species an ideal model for cool season grain legume genomics.

Conclusions

Genomics of chickpea is currently in initial and formative stages; however, the basis for rapid development of genomic tools is present and progressing rapidly. Genetic reference maps are available and are based on both interspecific and intraspecific crosses. The general lack of polymorphism within the cultivated species initially hindered map development, but the development and use of SSRs and other co-dominant types of markers has generally overcome that obstacle. A number of BAC libraries have been developed from germplasm accessions that have genes for resistance to important diseases, particularly ascochyta blight and fusarium wilt. Progression to map-based cloning of important genes is envisaged and the needed tools for their verification have been or are being developed. Specifically, there are protocols for agrobacterium-mediated gene transfer and the tools for reverse genetics (TILLING) are being established. The rationale for designating chickpea as the 'model' cool season food legume has merit from the standpoint of current genetic knowledge, its importance to world agriculture and use as a food staple in developing countries. The biotic and abiotic stresses that adversely affect chickpea are the subjects of breeding programmes worldwide. Genomic information concerning these stresses provides geneticists and breeders with insights into important genetic mechanisms and avenues for improvement of this important food legume crop.

References

Abbo, S., Molina, C., Jungmann, R., Grusak, M.A., Berkovitch, Z., Refien, R., Kahl, G., Winter, P. and Reifen, R. (2005) Quantitative trait loci governing carotenoid concentration and weight in seeds of chickpea (*Cicer arietinum* L.). *Theoretical and Applied Genetics* 111, 185–195.

Arumuganathan, K. and Earle, E.D. (1991) Nuclear DNA content of some important plant species. *Plant Molecular Biology Reporter* 9, 208–218.

Beavis, W. (2004) Legume Information Resources. Available at:http://catg.ucdavis.edu/Legume% 20information% 20resources-wdb.pdf

Benko-Iseppon, A.M., Winter, P., Heuttel, B., Staginnus, C., Muehlbauer, F.J. (2006) and Kahl, G. (2003) Molecular markers closely linked to fusarium resistance genes in chickpea show significant alignments to pathogenesis-related genes located on *Arabidopsis* chromosomes 1 and 5. *Theoretical and Applied Genetics* 107, 379–386.

Bennetzen, J.L. (2000) Comparative sequence analysis of plant nuclear genomes: microcolinearity and its many exceptions. *Plant Cell* 12, 1021–1029.

Bhattarai, T. and Fettig, S. (2005) Isolation and characterization of a dehydrin gene from *Cicer pinnatifidum*, a drought-resistant wild relative of chickpea. *Physiologia Plantarum* 123, 452–458.

Brueggeman, R., Rostoks, N., Kudrna, D., Kilian, A., Han, F., Chen, J., Druka, A., Steffenson, B. and Kleinhofs, A. (2002) The barley stem rust-resistance gene Rpg1 is a novel disease-resistance gene with homology to receptor kinases. *Proceedings of the National Academy of Sciences of the United States of America* 9, 9328–9333.

Cannon, C.B., McCombie, W.R., Sato, S., Tabata, S., Denny, R., Palmer, L., Katari, M., Young, N.D. and Stacey, G. (2003) Evolution and microsynteny of the apyrase gene family in three legume genomes. *Molecular and General Genetics* 270, 347–361.

Cho, S. and Muehlbauer, F.J. (2004) Genetic effect of differentially regulated fungal response genes on resistance to necrotrophic fungal pathogens in chickpea (*Cicer arietinum* L.). *Physiological and Molecular Plant Pathology* 64, 57–66.

Cho, S., Chen, W. and Muehlbauer, F.J. (2004) Pathotype-specific genetic factors in chickpea (*Cicer arietinum* L.) for quantitative resistance to Ascochyta blight. *Journal of Theoretical and Applied Genetics* 109, 733–739.

Cho, S., Kumar, J., Shultz, J.L., Anupama, K., Tefera, F. and Muehlbauer, F.J. (2002) Mapping genes for double podding and other morphological traits in chickpea. *Euphytica* 128, 285–292.

Collard, B.C.Y., Pang, E.C.K., Ades, P.K. and Taylor, P.W.J. (2003) Preliminary investigation of QTLs associated with seedling resistance to ascochyta blight from *Cicer echinospermum*, a wild relative of chickpea. *Theoretical and Applied Genetics* 107, 719–729.

Corum, T. and Pang, E.C.K. (2005a) Isolation and analysis of candidate ascochyta blight defence genes in chickpea. Part I. Generation and analysis of an expressed sequence tag (EST) library. *Physiological and Molecular Plant Pathology* 66, 192–200.

Corum, T. and Pang, E.C.K. (2005b) Isolation and analysis of candidate ascochyta blight defence genes in chickpea. Part II. Microarray expression analysis of putative defence-related ESTs. *Physiological and Molecular Plant Pathology* 66, 201–210.

Devos, K.M. and Gale, M.D. (2000) Genome relationships: the grass model in current research. *Plant Cell* 12, 637–646.

Endre, G., Kereszt, A., Kevei, Z., Mihacea, S., Kalo, P. and Kiss, G.B. (2002) A receptor kinase gene regulating symbiotic nodule development. *Nature* 417, 962–966.

Esteban, R., Labrdor, E. and Dopico, B. (2005) A family of β-galactosidase cDNAs related to the development of vegetative tissue in *Cicer arietinum. Plant Science* 168, 457–466.

Feuillet, C., Travella, S., Stein, N., Albar, L., Nublat, A. and Keller, B. (2003) Map based isolation of the leaf rust disease resistance gene Lr10 from the hexaploid wheat (*Triticum aestivum* L.) genome. *Proceedings of the National Academy of Sciences of the United States of America* 100, 15253–15258.

Flandez-Galvez, H., Ford, R., Pang, E.C.K. and Taylor, P.W.J. (2003) An intraspecific linkage map of the chickpea (*Cicer arietinum* L.) genome based on sequence tagged microsatellite site and resistance gene analog markers. *Theoretical and Applied Genetics* 106, 1447–1456.

Gaur, P.M. and Slinkard, A.E. (1988) Inheritance and linkage relationships of some isozyme loci in *Cicer. Genome* 30, 243.

Gaur, P.M. and Slinkard, A.E. (1990) Genetic control and linkage relations of isozyme markers in chickpea. *Theoretical and Applied Genetics* 80, 648–656.

Grant, D., Cregan, P. and Shoemaker, R.C. (2000) Genome organization in dicots:

genome duplication in *Arabidopsis* and synteny between soybean and *Arabidopsis*. *Proceedings of the National Academy of Sciences of the United States of America* 97, 4168–4173.

Gremigni, P., Furbank, R.T. and Turner, N.C. (2004) Genetic manipulation of seed size in chickpea. In: *Legumes for the Benefit of Agriculture, Nutrition and the Environment: Their Genomics, Their Products, and Their Improvement*. Conference Handbook of the joint 5th European Conference on Grain Legumes and 2nd International Conference on Legume Genomics and Genetics. 7–11 June 2004. Dijon, France, p. 141.

Gremigni, P., Furbank, R.T. and Turner, N.C. (2005) Genetic manipulations to increase seed size in chickpea (*Cicer arietinum* L.). In: *Proceedings of 2nd Australian Model Legume Workshop. Application to Crop and Pasture Improvement*. 5–8 April 2005. Rottnest Island, Perth, Western Australia, p. 50.

Gualtieri, G., Kulikova, O., Limpens, E., Kim, D.J., Cook, D.R., Bisseling, T. and Geurts, R. (2002) Microsynteny between pea and *Medicago truncatula* in the *SYM2* region. *Plant Molecular Biology* 50, 225–235.

Gutierrez, M.V., Vaz Patto, M.C., Huguet, T., Cubero, J.I., Moreno, M.T. and Torres, A.M. (2005) Cross-species amplification of *Medicago truncatula* microsatellites across three major pulse crops. *Theoretical and Applied Genetics* 110, 1210–1217.

Huttel, B., Santra, D., Muehlbauer, F.J. and Kahl, G. (2002) Resistance gene analogues of chickpea (*Cicer arietinum* L.): isolation, genetic mapping and association with a fusarium resistance gene cluster. *Theoretical and Applied Genetics* 105, 479–490.

Iruela, M., Rubio, J., Barro, F., Cubero, J.I., Millan, T. and Gil, J. (2006) Detection of two quantitative trait loci for resistance to ascochyta blight in an intra-specific cross of chickpea (*Cicer arietinum* L.): development of SCAR markers associated with resistance. *Theoretical and Applied Genetics* 112, 278–287.

Jaiswal, P., Singh, A., Kumar, K., Upadhyaya, K. and Verma, P. (2004) Functional genomics of chickpea in response to ascochyta infection. 12th Plant and Animal Genome Conference. San Diego, California.

Jayashree, B., Buhariwalla, H.K., Shinde, S. and Crouch, J.H. (2005) A legume genomics resource: the chickpea root expressed sequence tag database. *Electronic Journal of Biotechnology* 8, 3.

Kuchel, H., Ye, G., Fox, R. and Jefferies, S. (2005) Genetic and economic analysis of a targeted marker-assisted wheat breeding strategy. *Molecular Breeding* 16, 67–78.

Lee, J.M., Grant, D., Vallejos, C.E. and Shoemaker, R.C. (2001) Genome organization in dicots. II. *Arabidopsis* as a bridging species to resolve genome evolution events among legumes. *Theoretical and Applied Genetics* 103, 765–773.

Lichtenzveig, J., Scheuring, C., Dodge, J., Abbo, S. and Zhang, H.B. (2005) Construction of BAC and BIBAC libraries and their applications for generation of SSR markers for genome analysis of chickpea, *Cicer arietinum* L. *Theoretical and Applied Genetics* 110, 492–510.

Mantri, N., Coram, T., Ford, R., De Vicente, C., Mal, B. and Pang, E.C.K. (2006) Gene expression profiling of chickpea responses to drought, cold and salinity. In: *Proceedings of 13th Australian Plant Breeding Conference*. 18–21 April 2006. Christchurch, New Zealand, pp. 698–707.

Matsumura, H., Reich, S., Ito, A., Saitoh, H., Winter, P., Kahl, G., Reuter, M., Krüger, D. and Terauchi, R. (2003) SuperSAGE: A universal functional genomics tool for any eukaryotes. *Proceedings of the National Academy of Sciences of the United States of America* 100, 15723–15781.

Mayer, M.S., Tullu, A., Simon, C.J., Kumar, J.S., Kraft, J.M., Kaiser, W.J. and Muehlbauer, F.J. (1997) Development of a DNA marker for fusarium wilt resistance in chickpea. *Crop Science* 37, 1625–1629.

Michelmore, R. (2000) Genomic approaches to plant disease resistance. *Current Opinion in Plant Biology* 3, 125–131.

Muehlbauer, F.J. and Singh, K.B. (1987) Genetics of chickpea. In: Saxena, M.C. and Singh, K.B. (eds) *The Chickpea*. CAB International, Wallingford, UK, pp. 99–125.

Muehlbauer, F.J., Chen, W., McPhee, K., Pak, J., Cho, S. and Rajesh, P.N. (2005) Stalking ascochyta blight resistance genes in chickpea with help from model species. In: *Proceedings of 2nd Australian Model Legume Workshop. Application to Crop and Pasture Improvement*. 5–8 April 2005. Rottnest Island, Perth, Western Australia, p. 16.

Pfaff, T. and Kahl, G. (2003) Mapping of gene specific markers on the genetic map of chickpea (*Cicer arietinum* L.). *Molecular and General Genetics* 269, 243–251.

Polowick, P.L., Baliski, D.S. and Mahon, J.D. (2004) *Agrobacterium tumefaciens*-mediated transformation of chickpea (*Cicer arietinum* L.): gene integration, expression and inheritance. *Plant Cell Reports* 23, 485–491.

Rajesh, P.N., Tekeoglu, M., Gupta, V.S., Ranjekar, P.K. and Muehlbauer, F.J. (2002a) Molecular mapping and characterization of RGAPtokin1-2_{171} in chickpea. *Euphytica* 128, 427–433.

Rajesh, P.N., Tullu, A., Gil, J., Gupta, V.S., Ranjekar, P.K. and Muehlbauer, F.J. (2002b) Identification of an STMS marker for the double-podding gene in chickpea. *Theoretical and Applied Genetics* 105, 604–607.

Rajesh, P.N., Coyne, C., Meksem, K., Sharma, K.D., Gupta, V.S. and Muehlbauer, F.J. (2004) Construction of a *Hin*dIII bacterial artificial chromosome library and its use in identification of clones associated with disease resistance in chickpea. *Theoretical and Applied Genetics* 108, 663–669.

Rajesh, P.N., McPhee, K. and Muehlbauer, F.J. (2005) Detection of polymorphism using CAPS and dCAPS markers in two chickpea genotypes. *International Chickpea and Pigeonpea Newsletter* 12, 4–6.

Rakshit, S., Winter, P., Tekeoglu, M., Juarez Muñoz, J., Pfaff, T., Benko-Iseppon, A.M., Muehlbauer, F.J. and Kahl, G. (2003) DAF marker tightly linked to a major locus for ascochyta blight resistance in chickpea (*Cicer arietinum* L.). *Euphytica* 132, 23–30.

Rubio, J., Hajj-Moussa, E., Kharra, M., Moreno, M.T., Millan, T. and Gil, J. (2003) Two genes and linked RAPD markers involved in resistance to *Fusarium oxysporum* f. sp. *Ciceris* race 0 in chickpea. *Plant Breeding* 122, 188–191.

Santra, D.K., Tekeoglu, M., Ratnaparkhe, M.B., Gupta, V.S., Ranjekar, P.K. and Muehlbauer, F.J. (2000) Identification and mapping of QTLs conferring resistance to ascochyta blight in chickpea. *Crop Science* 40, 1606–1612.

Sanyal, I., Singh, A.K., Kaushik, M. and Amla, D.V. (2005) *Agrobacterium*-mediated transformation of chickpea (*Cicer arietinum* L.) with *Bacillus thuringiensis cry1Ac* gene for resistance against pod borer insect *Helicoverpa armigera*. *Plant Science* 168, 1135–1146.

Senthil, G., Williamson, B., Dinkins, R.D. and Ramsay, G. (2004) An efficient transformation system for chickpea *(Cicer arietinum* L.). *Plant Cell Reports* 23, 297–303.

Sharma, K.D., Winter, P. and Muehlbauer, F.J. (2004) Molecular mapping of *Fusarium oxysporum* f. sp. *ciceris* race 3 resistance gene in chickpea. *Theoretical and Applied Genetics* 108, 1243–1248.

Simon, C.J. and Muehlbauer, F.J. (1997) Construction of a chickpea linkage map and its comparison with maps of pea and lentil. *Journal of Heredity* 88, 115–119.

Stracke, S., Kistner, C., Yoshida, S., Mulder, L., Sato, S., Kaneko, T., Tabata, S., Sandal, N., Stougaard, K., Szczyglowski, K. and Parniske, M. (2002) A plant receptor kinase required for both bacterial and fungal symbiosis. *Nature* 417, 959–962.

Tekeoglu, M., Santra, D.K., Kaiser, W.J. and Muehlbauer, F.J. (2000) Ascochyta blight resistance inheritance in three chickpea recombinant inbred line populations. *Crop Science* 40, 1251–1256.

Tekeoglu, M., Rajesh, P.N. and Muehlbauer, F.J. (2002) Integration of sequence tagged microsatellite sites to the chickpea genetic map. *Theoretical and Applied Genetics* 105, 847–854.

Tekeoglu, M., Isuk, M. and Muehlbauer, F.J. (2004) QTL analysis of ascochyta blight resistance in chickpea. *Turkish Journal of Agricultural Forecasting* 28, 183–187.

Tullu, A. (1996) Genetics of fusarium wilt resistance in chickpea. PhD thesis, Crop and Soil

Sciences Department, Washington State University, Pullman, Washington, DC.

Tullu, A., Muehlbauer, F.J., Simon, C.J., Mayer, M.S., Kumar, J., Kaiser, W.J. and Kraft, J.M. (1998) Inheritance and linkage of a gene for resistance to race 4 of fusarium wilt and RAPD markers in chickpea. *Euphytica* 102, 227–232.

Tullu, A., Kaiser, W.J. and Muehlbauer, F.J. (1999) Inheritance of a second gene for resistance to race 4 of fusarium wilt in chickpea and linkage with a RAPD marker. *Euphytica* 109, 43–50.

Udupa, S.M. and Baum, M. (2003) Genetic dissection of pathotype specific resistance to ascochyta blight resistance in chickpea (*Cicer arietinum* L.) using microsatellite markers. *Theoretical and Applied Genetics* 106, 1196–1202.

Valárik, M., Bartoš, J., Kovářová, P., Kubaláková, M., de Jong, J.H. and Doležel, J. (2004) High-resolution FISH on super-stretched flow-sorted plant chromosomes. *The Plant Journal* 37, 940–950.

Vláčáilová, K., Ohri, D., Vrána, J., Čihalíková, J., Kubaláková, M., Kahl, G. and Doležel, J. (2002) Development of flow cytogenetics and physical genome mapping in chickpea (*Cicer arietinum* L.). *Chromosome Research* 10, 695–706.

Weeden, N.F., Muehlbauer, F.J. and Ladizinsky, G. (1992) Extensive conservation of linkage relationships between pea and lentil genetic maps. *Journal of Heredity* 83, 123–129.

Winter, P., Benko-Iseppon, H.B., Ratnaparkhe, M., Tullu, A., Sonnante, G., Ptaff, T., Tekeoglu, M., Santra, D., Sant, V.J., Rajesh, P.N., Kahl, G. and Muehlbauer, F.J. (2000) A linkage map of the chickpea (*Cicer arietinum* L.) genome based on recombinant inbred lines from a *c. arietinum* × *c. reticulatum* cross: localization of resistance genes for *Fusarium* wilt races 4 and 5. *Journal of Theory and Applied Genetics* 101, 1155–1163.

Winter, P., Jungmann, R., Molina, C., Cobos, M.J., Udupa, S., Benko-Iseppon, A.M., Kido, E. and Kahl, G. (2005) Comparative mapping in legumes: poor performance of cross-genome markers in chickpea and possible solutions. In: *Proceedings of 2nd Australian Model Legume Workshop. Application to Crop and Pasture Improvement.* 5–8 April 2005. Rottnest Island, Perth, Western Australia.

Yan, H.H., Mudge, J., Kim, D.J., Larsen, D., Showmaker, R.C., Cook, D.R. and Young, N.D. (2003) Estimates of conserved microsynteny among the genomes of *Glycine max*, *Medicago truncatula* and *Arabidopsis thaliana*. *Theoretical and Applied Genetics* 106, 1256–1265.

Young, N.D., Mudge, J. and Ellis, T.H.N. (2003) Legume genomes: more than peas in a pod. *Current Opinion in Biotechnology* 6, 199–204.

22 Development of Transgenics in Chickpea

K.E. McPhee,[1] J. Croser,[2] B. Sarmah,[3] S.S. Ali,[3] D.V. Amla,[4] P.N. Rajesh,[5] H.-B. Zhang[6] and T.J. Higgins[7]

[1]*USDA-ARS, Grain Legume Genetics and Physiology Research Unit, Washington State University, Pullman, WA 99164-6434, USA;* [2]*Center for Legumes in Mediterranean Agriculture, 35 Stirling Highway, Crawley, WA 6009, Australia;* [3]*Department of Agricultural Biotechnology, Assam Agricultural University, Jorhat, Assam 785013, India;* [4]*National Botanical Research Institute, Rana Pratap Marg, Lucknow 226 001, India;* [5]*Department of Crop and Soil Sciences , Washington State University, Pullman, WA 99164-6434, USA;* [6]*Department of Soil and Crop Sciences, Texas A&M University, College Station, 2474 TAMU, TX 77843-2474, USA;* [7]*CSIRO Plant Industry, Black Mountain Laboratories, Clunies Ross Street, Black Mountain, ACT 2601, Australia*

Application of Genetic Transformation to Chickpea Improvement

Chickpea (*Cicer arietinum* L.) is an important food crop in most of the developing countries in the world and ranks third among food legumes in production. India is the largest producer of chickpea, which is a major protein source in the subcontinent. Chickpea production is limited worldwide by drought, insect damage from *Helicoverpa armigera* and disease pressure, especially fusarium wilt (*Fusarium oxysporum* f. sp. *ciceri*) and ascochyta blight. Efforts to control pod borer (*H. armigera*) through integrated pest management (IPM) have only been partially successful, forcing breeders to explore biotechnological means such as introduction of resistance through expression of the Bt toxin gene, *cry1A(c)* (Kar *et al.*, 1997; Sanyal *et al.*, 2005).

The genetic transformation of chickpea began with the initial transformation of callus (Mohapatra and Sharma, 1991; Islam *et al.*, 1994). Production of transgenic plantlets derived from co-cultivation of embryo axes was reported by Fontana *et al.* (1993) and Kar *et al.* (1996); however, each of these reports did not verify inheritance of the introduced genes, *nptII* and *uidA*. Stable integration of the *cry1A(c)* gene for the production of the Bt toxin was the first report of chickpea transformation with a gene other than a reporter gene (Kar *et al.*, 1997). This was also the first report using biolistic direct DNA transfer to transform immature embryonic axes of chickpea.

The apical meristem and radicle were removed from the embryonic axis prior to bombardment and co-transformation with the *cry1A(c)* and *nptII* genes. Co-transformation allows the separation of two or more introduced genes through selection of natural segregants allowing a single desired gene to be retained in the progeny. Chakrabarty *et al.* (2000) reported transformation of chickpea using both mature and immature decapitated embryo axes. Jayanand *et al.* (2003) and Sanyal *et al.* (2003) reported successful transformation of cotyledonary meristems using *A. tumefaciens*. Polowick *et al.* (2004) reported successful transformation of chickpea using thin slices of embryonic axes similar to the procedure reported for pea by Schroeder *et al.* (1993).

Two examples of gene transfer to chickpea production with agronomic benefit include bruchid resistance and resistance to pod borer (*H. armigera*). Sarmah *et al.* (2004a) reported stable integration and expression of the bean α-amylase inhibitor gene in chickpea through *A. tumefaciens*-mediated transformation. The gene product was accumulated up to 4.2% of seed protein and inhibited development of the bruchid species, *Callosobruchus maculatus* and *C. chinensis*. The Bt toxin gene, *cry1A(c)*, has been introduced into chickpea by two independent laboratories. Kar *et al.* (1997) used biolistic technology, while Sanyal *et al.* (2005) used *A. tumefaciens* to introduce the gene. Both reports demonstrate successful integration and expression of *cry1A(c)*.

Application of Genetic Transformation to Chickpea Genomics

Comprehensive genomics research of chickpea, including cloning-mapped genes and quantitative trait loci (QTLs) by map-based or positional cloning procedures, has been facilitated by construction of large-insert bacterial artificial chromosome (BAC) and binary bacterial artificial chromosome (BIBAC) libraries (Rajesh *et al.*, 2004; Lichtenzveig *et al.*, 2005). The libraries have average insert sizes of 100, 121 and 145 kb, collectively covering 10.8-fold of the chickpea haploid genome. These libraries provide a tool to access mapped genes and QTLs based on flanking DNA markers and chromosome walking. Map-based cloning requires genetic transformation technology to verify whether the candidate clone contains the target gene(s) or QTLs through genetic complementation. Construction of chickpea BIBAC libraries in the *Agrobacterium*-mediated plant transformation competent binary vector (pCLD04541) (Jones *et al.*, 1992; Tao and Zhang, 1998), streamlines the genetic transformation procedure because the candidate clones can be directly transformed without subcloning. While techniques for large DNA fragment transformation remain to be established in chickpea, the successful transfer of DNA fragments of >100 kb cloned in BIBACs by *Agrobacterium* has been reported in tobacco (Hamilton *et al.*, 1996), tomato (Hamilton *et al.*, 1999), *Arabidopsis thaliana* (Liu *et al.*, 1999) and rice (Liu *et al.*, 2002; He *et al.*, 2003). Therefore, the chickpea BIBAC libraries will provide easy-to-use tools for molecular cloning and genetic engineering of genes and QTLs of agronomic importance.

Genetic transformation has also been used to determine the function of genes by T-DNA-based or transposon-based mutagenesis and gene tagging in several models and crop plant species; the most prominent include *A. thaliana* (Krysan *et al.*, 1999), maize (Brutnell, 2002) and rice (An *et al.*, 2003). This method involves generating a large number of transformed plants or plant cells using a T-DNA or transposon construct, followed by selection for a desired phenotype. The T-DNA or transposon integrated into a genome not only induces the gene mutation, but also tags the gene when the T-DNA or transposon sequence resides in the position. The tagged gene sequences are isolated by inverse polymerase chain reaction (PCR) or thermal asymmetric interlaced (TAIL) PCR using T-DNA-specific or transposon-specific primers. A technique capable of transforming plants or plant cells with high efficiency is required for tagging genes and determining their function; therefore, further improvement of genetic transformation of chickpea will be essential to make T-DNA and/or transposon gene tagging viable for gene identification and functional genomics research in chickpea.

In addition to direct identification of genes in chickpea, a large number of genes have been identified in other crop plants and model plant species, especially *A. thaliana* (Arabidopsis Genome Initiative, 2000) and rice (International Rice Genome Sequencing Project, 2005). Comparative analysis of rice genes with *A. thaliana* genes revealed that 70% of rice genes have homologues in *A. thaliana* and 90% of *A. thaliana* genes have homologues in rice (International Rice Genome Sequencing Project, 2005), suggesting that plants, in spite of great divergence (dicot vs monocot), share a large number of common genes. Furthermore, the gene-rich regions of *Medicago truncatula* and *Lotus japonicus* will soon be available (Young *et al.*, 2005) and sequencing of the soybean genome has been scheduled (http://www.energy.gov/news/2979.htm). These genome projects will provide a wealth of genetic resources for chickpea genetic improvement and genetic transformation will be the method of choice to introduce foreign genes into chickpea.

Application of Tissue Culture to Chickpea Improvement

Breeding efforts in chickpea are hampered by a low level of genetic variability within the cultivated germplasm and a lack of biotechnological techniques adapted to the species. There is scope for the use of tissue culture to improve the limited genetic variability in the *Cicer* germplasm, through mutation of single cells prior to regeneration, exploitation of useful genes from wild relatives, targeted transformation as proposed above and also to make breeding more efficient through techniques such as micropropagation and doubled haploidy. At present, the practical use of tissue culture in chickpea improvement has been limited to *in vitro*-assisted interspecific hybridization for the development of breeding lines and the use of *in vitro* screening of callus cultures to identify genotypes with tolerance to NaCl and resistance to ascochyta blight (*Ascochyta rabiei*) and the herbicide atrazine (Rao and Ramulu, 1997; Singh *et al.*, 1999; Jaiswal and Singh, 2001; Ramulu and Ram, 2004). Chickpea callus cultures have also been used to better understand the biochemical basis of defence mechanisms to ascochyta blight and fusarium wilt (Sindhu *et al.*, 1998; Singh *et al.*, 2003).

Regeneration from callus culture

There are several routine methods for chickpea regeneration from callus cultures through somatic embryogenesis and organogenesis from explants of leaves, immature cotyledons, immature embryos, mature embryo axes, mature cotyledons, shoot tips, roots and internodes of chickpea (Rao and Chopra, 1989; Barna and Wakhlu, 1993; Shanker and Ram, 1993; Dineshkumar *et al.*, 1994; Eapen and George, 1994; Dineshkumar *et al.*, 1995; Sharma and Amla, 1998; Arockiasamy *et al.*, 2000; Tikkas *et al.*, 2002; Batra *et al.*, 2004; Arora and Chawla, 2005). A generalized three-step formula has been found to be successful for the induction of embryoids and regeneration of whole plants from callus. First, callus and somatic embryos are induced on a Murashige and Skoog basal salts (MS)-based nutrient medium (Murashige and Skoog, 1962) with B5 vitamins (Gamborg *et al.*, 1968) supplemented with a high level of auxin (usually 2,4-D or naphthalene acetic acid [NAA]) and a relatively low level of cytokinin (usually 6-benzylaminopurine [BAP]) under dark culture. The second step involves the maturation of somatic embryos on a cytokinin-rich medium cultured under light. The third and final step is the germination of embryos and growth to plantlets on a plant growth regulator-free nutrient medium. Standardization of a protocol for callus production and plantlet regeneration has facilitated the use of this technique for practical screening for biotic and abiotic stress resistance.

In vitro interspecific hybridization

One method of addressing the low level of genetic variability in cultivated chickpea germplasm is to introgress genes of interest from the wild *Cicer* gene pool. Recent research has increased our understanding of evolutionary relationships within the *Cicer* genus, identified wild species with potential to contribute advantageous traits and developed novel technologies to assist in the transfer of genes from the wild to the cultivated species. Chickpea has 43 wild relatives, of which 8 are of particular interest to breeders as they share an annual growth habit with chickpea. Of the eight annual species, only two, *C. reticulatum* and *C. echinospermum*, are cross-compatible with the cultivated species using conventional crossing techniques.

Comprehensive screening of wild *Cicer* collections has identified accessions with resistance to abiotic and biotic stresses, but their exploitation in breeding programmes has so far been limited. Introgression of cyst nematode resistance (Singh *et al.*, 1996; Erskine *et al.*, 2001; Malhotra *et al.*, 2002) as well as resistance to fusarium wilt, foot rot (*Operculella padwickii*) and root rot (*Rhizoctonia bataticola*) diseases (Singh *et al.*, 2005) from various germplasm accessions has been reported. In Australia, both *C. echinospermum* and *C. reticulatum* have been used as sources of resistance to phytophthora root rot (*C. sech*) and root-lesion nematode (both species) in crossing programmes; however, these plants are currently at an early stage (BC_2F_1) (E. Knights, New South Wales, 2006, personal communication).

In the case of interspecific hybrids from the remaining six annual *Cicer* species, hybrid embryo development is often arrested at an early post-zygotic stage.

Embryo necrosis is normally due to cytoplasmic incompatibility that leads to the failure of the endosperm to provide the necessary conditions and nutrients for embryo growth and development. In such situations, a technique called 'embryo rescue' can be applied, whereby the hybrid embryos can be removed from the maternal plant and placed into an *in vitro* nutrient medium that fulfils the nutritional role of the endosperm and promotes development of the embryo to maturity and germination. Attempts to develop tissue culture techniques to enable hybridization between chickpea and the more distantly related annual wild species like *C. cuneatum, C. pinnatifidum* and *C. bijugum* (Singh and Singh, 1989; Swamy and Khanna, 1991a,b; Verma *et al.*, 1995; Badami *et al.*, 1997; van Dorrestein *et al.*, 1998; Mallikarjuna, 1999, 2001; Stamigna *et al.*, 2000) (Table 22.1) were not so successful. In an important breakthrough, collaboration between Australian (Centre for Legumes in Mediterranean Agriculture [CLIMA]), Indian (International Crops Research Institute for Semi-Arid Tropics [ICRISAT]) and Canadian (Conservation Data Centre [CDC]) scientists has recently led to the recovery of confirmed *in vitro*-rescued hybrids between *C. arietinum* × *C. bijugum* (Lulsdorf *et al.*, 2005; Clarke *et al.*, 2006).

Table 22.1. Reported *in vitro* ovule culture and embryo rescue in chickpea (*Cicer arietinum*).

Reference	Cross	Rescued at days after planting (days)	Fate of F_1
Singh and Singh (1989)	*C. arietinum* × *C. cuneatum*	5	2 plants – no flowering
Badami *et al.* (1997)	*C. arietinum* × *C. pinnatifidum*	18	2 albino hybrids (confirmed)
van Dorrestein *et al.* (1998)	*C. arietinum* × *C. pinnatifidum*	–	6 putative hybrid plants produced
	C. arietinum x *C. judaicum*	–	–
	C. arietinum × *C. bijugum*	–	–
Mallikarjuna (1999)	*C. arietinum* × *C. pinnatifidum*	18 and 25	4 hybrid plants produced, but infertile
Stamigna *et al.* (2000)	*C. arietinum* × *C. bijugum*	–	Not produced
	C. arietinum × *C. judaicum*	–	–
	C. arietinum × *C. pinnatifidum*	–	–
Mallikarjuna (2001)	*C. arietinum* × *C. canariense*	14	Not produced
Clarke *et al.* (2006)	*C. arietinum* × *C. pinnatifidum*	–	–
	C. arietinum × *C. bijugum*	–	–
Lulsdorf *et al.* (2005)	*C. arietinum* × *C. bijugum*	–	Confirmed hybrids, no seed produced as yet

Doubled haploids and somatic hybridization for chickpea

At present, there are a number of additional tissue culture techniques under development for application in chickpea germplasm enhancement. Doubled haploid production is one of the techniques that are useful in germplasm enhancement programmes for improving the efficiency of breeding, creating populations for molecular mapping and giving access to novel ways to improve genetic variability. Reddy and Reddy (1996) reported the *in vivo* production of 51 haploid plants from the progeny of a cross between a male sterile mutant and kabuli chickpea line ICC 4973. Despite a number of efforts, *in vitro* haploid plant production has yet to be achieved in chickpea (Khan and Ghosh, 1983; Altaf and Ahmad, 1986; Bajaj and Gosal, 1987; Gosal and Bajaj, 1988; Huda *et al.*, 2001; Croser, 2002; Vessal *et al.*, 2002; Croser *et al.*, 2004, 2005; Croser and Lulsdorf, 2004). At present, most attempts to develop *in vitro*-doubled haploids in chickpea have targeted the male gametophyte and used intact anthers as explants, with varied success. The reports by Huda *et al.* (2001) and Croser (2002) indicated that it was possible to regenerate plantlets from anther-derived callus by somatic embryogenesis, but did not confirm an androgenic origin of the regenerants. According to Croser (2002), flow cytometry revealed that all tested regenerants have a diploid chromosome complement, indicating either a somatic origin or a high level of spontaneous diploidy during culture.

Croser (2002) undertook preliminary research into the culture of isolated microspores of chickpea and obtained androgenic structures, but no plant regeneration. Recent collaborative efforts between Australian and Canadian researchers (Croser *et al.*, 2004, 2005) have led to the induction of sporophytic division, multinucleate synctiums, early stage embryogenesis and in some cases caulogenesis from isolated microspores of chickpea. As yet, no haploid plant regeneration has been achieved. In chickpea, as with many other species, microspore culture is most successful when microspores are harvested and cultured at the early to mid-uninucleate stage. The optimum mode and intensity of stress pretreatment appears to vary with genotype. Both cold and heat stress have been used to induce sporophytic development in chickpea, with heat shock appearing to be the most effective method across the widest range of genotypes. It is projected that with ongoing research, this technology will be available to chickpea breeders in the near future.

Another potentially useful but unproven technique is somatic hybridization whereby genomes from different species and genera are combined through protoplast fusion without pollination. Interspecific somatic hybrid plants have been obtained by symmetrical electrofusion of mesophyll protoplasts of the leguminous *Medicago sativa* × *M. arborea* (Nenz *et al.*, 1996). Sagare and Krishnamurthy (1991) have proposed fusion of protoplasts as an alternative means of enabling interspecific hybridization between widely related *Cicer* species. This technique is currently limited by the absence of protocols for the regeneration of whole plants from single cells or protoplasts in any of the *Cicer* species. The progress made towards isolated microspore culture should have important implications for regeneration from other single cells and may lead to a breakthrough in protoplast culture for this species.

Transformation Methods

Targeted transfer of genes from the wild *Cicer* species into the cultivated species would represent a very elegant application of transformation technology. Recent advances in molecular mapping and chickpea transformation have brought the application of genetic transformation much closer to reality (Fontana *et al.*, 1993; Hamblin *et al.*, 1998; Chakrabarty *et al.*, 2000; Krishnamurthy *et al.*, 2000; Sharma and Ortiz, 2000; Jaiwal *et al.*, 2001; Sanyal *et al.*, 2003; Sarmah *et al.*, 2004a). The development of an efficient method for regeneration from single cells would also open opportunities for transformation of cells prior to culture, reducing the development of chimeras and enabling efficient production of large numbers of genetically altered plants. The applicability of transformation will, however, depend on the isolation of key genes, and public acceptance of food from genetically modified cultivars.

Two widely used gene transfer techniques in crop plants are: (i) particle bombardment and (ii) *Agrobacterium*-mediated gene transfer. Prerequisites for successful application of gene technology in crop plants have been recently reviewed (Popelka *et al.*, 2004). One prerequisite is the development of a robust and reproducible *in vitro* regeneration and transformation system. Chickpea plants regenerate *in vitro* through organogenic (shoot buds are organized by meristematic activity of cells) and embryogenic (cells undergo differentiation to produce somatic embryos) pathways. Morphogenesis occurs either directly from the explants or indirectly by the formation of undifferentiated callus. Choice of gene delivery technique is largely determined by the regeneration pathway, i.e. organogenesis or embryogenesis from organized explant tissue or callus.

Several studies have reported *in vitro* regeneration of chickpea using different explants and mostly through organogenic pathways (Table 22.2). With the availability of *in vitro* regeneration protocols attempts were made to transform chickpea with alien genes of agronomic importance. Although microprojectile bombardment and *Agrobacterium*-mediated transformation have been attempted for gene delivery, most recent reports are focused on an *Agrobacterium* system.

The first report of successful chickpea transformation after cocultivation of embryonic axes with *Agrobacterium tumefaciens* came from Fontana *et al.* (1993). By 2000, various researchers published on development of transgenic chickpea using a similar protocol with minor modifications to media composition (Table 22.3). In all these protocols, the antibiotic resistance gene, *nptII* was used as selectable marker with the exception of the *Bar* gene used by Senthil *et al.* (2004). Transformation frequency and reproducibility in these early breakthroughs were low, which limited their practical application. However, recent reports have shown increases in both transformation frequency and reproducibility (Polowick *et al.*, 2004; Sarmah *et al.*, 2004a,b; Senthil *et al.*, 2004; Sanyal *et al.*, 2005). A comparative analysis of these four reports was made by Popelka *et al.* (2004). In all these reports, the embryonic axis was used as the explant. Cotyledonary explants containing half embryogenic axes were used by Sarmah *et al.* (2004a,b); while longitudinally sliced embryonic axes were used by Polowick *et al.* (2004) and Senthil *et al.* (2004), and Sanyal *et al.* (2005) used pre-cultured explants. Pre-cultured explants (L2 layer) were grown for 24 h on

Table 22.2. A selection of *in vitro* regeneration systems for chickpea (*Cicer arietinum* L.).

Explant	Best medium	Means of regeneration	References
Freeze-preserved meristem	MS + 11.4 mg/1 IAA + 2.3 mg/1 Kin	Shoot meristem organogenesis	Bajaj and Dhanju (1979)
Meristem tips	MS + 4.4 μM BA + 1.0 μM NAA	Shoot meristem organogenesis	Kartha *et al.* (1981)
Immature cotyledon	B5 + 2,4-D + 2,4,5-T + NAA + IAA + BA + Kin + Zeatin + ABA	Organogenesis from cotyledon-like structure	Shri and Davis (1992)
Mature seeds	MS + 10 mM TDZ	Direct organogenesis	Malik and Saxena (1992)
Immature cotyledon	MS + 3 mg/1 2,4,5-T	Direct somatic embryogenesis	Sagare *et al.* (1993)
Leaf	MS + 25 μM 2,4-D	Organogenesis from callus	Barna and Wakhlu (1993, 1994)
Meristem tips	DKW + 4.4 μM BA + 0.05 μM IBA	Direct organogenesis	Brandt and Hess (1994)
Embryonic axes without apices	MS + B5 vitamins + 2.0 mg/1 BAP + 0.05 mg/1 NAA	Direct organogenesis	Kar *et al.* (1996)
Epicotyl	Basal medium + BAP + Kin + IAA		Vani and Reddy (1996)
Cotyledonary nodes	MS + 10 μM TDZ + 10 μM L-Proline	Direct somatic embryogenesis	Murthy *et al.* (1996)
Shoot tips; cotyledonary nodes	MS + 2.0 mg/1 BAP + 0.05 mg/1 IBA MS + 1 mg/1 TDZ + 0.05 mg/1 IBA	Direct organogenesis	Sharma and Amla (1998)
Hypocotyl	B5 + BAP	Direct organogenesis	Islam *et al.* (1999)
Internode	MS + B5 vitamins + 2,4-D + BAP + NAA + Kin + IAA	Organogenesis from callus and direct organogenesis	Huda *et al.* (2000)
Embryo axes; germinating seeds	MS + 2iP + TDZ + Kin + GA3 + IBA + NAA	Direct organogenesis	Jayanand *et al.* (2003)

MS: Murashige and Skoog basal salts; B5: Gamborg basal salts; IAA: indole acetic acid; Kin: Kinetin; BA and BAP: 6-benzylaminopurine; DKW: juglans basal salt mixture; NAA: naphthalene acetic acid; 2,4-D: 2,4-dichlorophenoxy acetic acid; 2,4,5-T: 2,4,5-trichlorophenoxyacetic acid; ABA: abscisic acid; TDZ: thidiazuron; IBA: indole-3-butyric acid; 2iP: N6-(2-isopentyl)adenine; GA3: gibberellic acid.

Table 22.3. Reports on genetic transformation of chickpea (*Cicer arietinum* L.).

Explant	Cultivar	Method of gene transfer	Genes transferred	References	Remarks
Embryonic axis without apical meristem	Unknown	AT (LBA 4404)	*nptII*, *uidA*	Fontana *et al.* (1993)	Low reproducibility, expression and integration of transgene in T_0 by western and southern analysis, respectively
Embryonic axis without apical meristem	ICCV-1, ICCV-6, desi	AT (LBA 4404)	*nptII*, *uidA*	Kar *et al.* (1996)	Low reproducibility, transgene integration by southern in T_0
Embryonic axis	ICCV-1, ICCV-6	MPB	*nptII*, *cry1A(c)*	Kar *et al.* (1996)	Low reproducibility, transgene transmitted to T_1 (PCR analysis)
Embryonic axis	PG 1, chafa, Turkey, PG 12	AT (C58, C1, GV2260)	*nptII*, pat, *uidA*	Krishnamurthy *et al.* (2000)	Low reproducibility, *uidA* gene expressed in T_0, transmission into T_1 (PCR)
Cotyledons with half embryonic axis	Vijay	AT (AGL-1)	*nptII*, *Bt-cry1A(c)*	Das and Sarmah (2003)	Transgene transmitted to T_1 (PCR for *nptII*)
Embryonic axis	P 362, P 1042, P 1043	AT (EHA 101)	*nptII*, *uidA*, *bar*	Tewari-Singh *et al.* (2004)	Low reproducibility, Gus gene expressed in T_0, transmission to T_1 (PCR)
Cotyledons with half embryonic axis	Semsen	AT (AGL-1)	*nptII*, α*AI1*	Sarmah *et al.* (2004a)	Good reproducibility, gene integration, transmission up to T_1 and expression in T_1 shown by southern and western analysis,

					respectively. High αAI1 activity in T_1 seeds
Sliced embryonic axis	ICCV-5, H 208, ICCL 87322, K 850	AT (AGL-1)	*bar*, *uidA*, PGIP	Senthil *et al.* (2004)	Good reproducibility, transmission and expression of *uidA* up to T_3
Sliced embryonic axis	CDC Yuma	AT (EHA 105)	*uidA*, *nptII*	Polowick *et al.* (2004)	Good reproducibility, transmission and expression of *uidA* up to T_3
L2 layer of cotyledonary nodes	C 235, BG 256, Pusa 362, Pusa 372	AT (LBA 4404)	*uidA*, *nptII*, *cry1A(c)*	Sanyal *et al.* (2005)	Good reproducibility, transmission and expression of *nptII* and *cry1A(c)* in T_1 was shown. Good *cry1A(c)* activity

AT: *Agrobacterium*-mediated transformation; MPB: microprojectile bombardment.
Values given in parentheses indicate *A. tumefaciens* strain used.

solid MS basal medium supplemented with 1 mg/l BAP. A common element in all four reports was that mature seeds were used as the source of explants, young meristems were the target tissue and explants were submerged in *Agrobacterium* suspensions and co-cultivated for 4 days. Explants were subcultured frequently on fresh selection and regeneration medium, and putatively transformed shoots were eventually induced to root or grafted prior to transfer to the glasshouse. All four protocols required 4–6 months to obtain T_0 plants and had a low but reproducible frequency of transformation (Table 22.3).

Most of the transgenic events were developed using insect resistance genes. Major achievements are introduction of the *Bt-cry1A(c)* gene through particle bombardment (Kar *et al.*, 1997) to confer resistance against the pod borer (*H. armigera*). However, the transgenic lines developed by Kar *et al.* (1997) are not being carried forward in the breeding programme. Recently, the *Bt-cry1A(c)* gene was re-introduced into chickpea using the *Agrobacterium*-mediated system (Sarmah *et al.*, 2004b). Sarmah *et al.* (2004a) reported transgenic chickpeas possessing the α-amylase inhibitor gene to confer resistance against stored grain pests.

To generate a sufficient number of transgenic lines with the desired expression level, a large number of explants need to be co-cultivated because the frequency of transformation is still low (0.1–1.0%). In crops such as, maize and soybean, frequency of transformation has been dramatically enhanced with the use of thiol compounds and L-cystine (Olhoft and Somers, 2001; Olhoft *et al.*, 2003). Similar efforts including the use of supervirulent strains of *A. tumefaciens* may also improve chickpea transformation efficiency.

Conclusions

Significant advance in genetic improvement of chickpea has been accomplished through traditional plant breeding but, like many crop species, genetic resistance to many agronomically important pests, and abiotic stresses are lacking in the primary gene pool. As a result, breeders must look beyond the primary gene pool to unrelated species for the necessary genes to overcome these important yield constraints. Genetic transformation methods using particle bombardment and *A. tumefaciens*-mediated gene transfer technology have been developed for chickpea. Genetic transformation requires an extensive tissue culture period for regeneration of transformed plantlets and the large-seeded grain legume species have been generally recalcitrant to tissue culture. These limitations have been largely overcome in chickpea and tissue culture now has a variety of further applications in chickpea crop improvement, including embryo rescue and micropropagation.

Genetic transformation of chickpea has been applied most widely to introduction of the Bt toxin gene, *cry1A(c)*, for resistance to *H. armigera*, a serious insect pest in India. Introduction of this gene has been shown to significantly reduce crop loss in cotton and could be expected to provide similar protection in chickpea. A less frequently discussed, but very powerful application of genetic transformation is in the area of genomics. Verification of gene function for the increasing number of genes being cloned from all plant species, including chickpea, will be achieved most conveniently through complementation.

Efficient use of transformation for this purpose will require improved efficiency, which will be the focus of future research. In summary, genetic transformation will play a significant role in chickpea variety development and will be indispensable for future gene discovery in chickpea genomics.

References

Altaf, N. and Ahmad, M.S. (1986) Plant regeneration and propogation of chickpea (*Cicer arietinum* L.) through tissue-culture techniques. *Proceedings of the Conference on Nuclear Techniques and in vitro Culture for Plant Improvement.* Food and Agriculture Organisation and International Atomic Energy Agency. Vienna, Austria. p. 407–417.

An, S., Park, S., Jeong, D.-H., Lee, D.-Y., Kang, H.-G., Yu, J.-H., Hur, J., Kim, S.-R., Kim, Y.-H., Lee, M., Han, S., Kim, S.-J., Yang, J., Kim, E., Wi, S.J., Chung, H.S., Hong, J.-P., Choe, V., Lee, H.-K., Choi, J.-H., Nam, J., Kim, S.-R., Park, P.-B., Park, K.Y., Kim, W.T., Choe, S., Lee, C.-B. and An, G. (2003) Generation and analysis of end sequence database for T-DNA tagging lines in rice. *Plant Physiology* 133, 2040–2047.

Arabidopsis Genome Initiative (2000) Analysis of the genome sequence of the flowering plant *Arabidopsis thaliana*. *Nature* 408, 796–815.

Arockiasamy, S., Varghese, G. and Ignacimuthu, S. (2000) High frequency regeneration of chickpea (*Cicer arietinum* L.) plantlets from leaf callus. *Phytomorphology* 50, 297–302.

Arora, A. and Chawla, H.S. (2005) Organogenic plant regeneration via callus induction in chickpea (*Cicer arietinum* L.) – role of genotypes, growth regulators and explants. *Indian Journal of Biotechnology* 4, 251–256.

Badami, P.S., Mallikarjuna, N. and Moss, J.P. (1997) Interspecific hybridization between *Cicer arietinum* and *C. pinnatifidum*. *Plant Breeding* 116, 393–395.

Bajaj, Y.P.S. and Dhanju, M.S. (1979) Regeneration of plants from apical meristem tips of some legumes. *Current Science* 48(20), 906–907.

Bajaj, Y.P.S. and Gosal, S.S. (1987) Pollen embryogenesis and chromosomal variation in cultured anthers of chickpea. *International Chickpea and Pigeonpea Newsletter* 17, 12–13.

Barna, K.S. and Wakhlu, A.K. (1993) Somatic embryogenesis and plant regeneration from callus cultures of chickpea (*Cicer arietinum* L.). *Plant Cell Reports* 12, 521–524.

Barna, K.S. and Wakhlu, A.K. (1994) Whole plant regeneration in *Cicer arietinum* from callus culture via organogenesis. *Plant Cell Reports* 13, 510–513.

Batra, P., Yadav, N.R., Sood, S.K., Choudhary, V.K. and Choudhary, J.B. (2004) Shoot regeneration in callus cultures of chickpea cultivars. *Indian Journal of Plant Physiology* 9, 426–428.

Brandt, E.B. and Hess, D. (1994) *In vitro* regeneration and propagation of chickpea (*Cicer arietinum* L.) from meristem tips and cotyledonary nodes. *In Vitro Cellular and Developmental Biology* 30, 75–80.

Brutnell, T.P. (2002) Transposon tagging in maize. *Functional and Integrative Genomics* 2, 4–12.

Chakrabarty, R., Singh, D.T. and Garg, G.K. (2000) Transformation studies on chickpea embryo axis. *Journal of Plant Biochemistry and Biotechnology* 9, 107–110.

Clarke, H.J., Wilson, J.G., Kuo, I., Kuo, J., Lulsdorf, M., Mallikarjuna, N. and Siddique, K.H.M. (2006) Embryo rescue and plant regeneration *in vitro* of selfed chickpea (*Cicer arietinum* L.) and its wild annual relatives. *Plant Cell Tissue and Organ Culture* 85, 197–204.

Croser, J. (2002) Haploid and zygotic embryogenesis in chickpea (*Cicer arietinum* L.). PhD thesis, University of Melbourne, Melbourne, Australia, p. 189.

Croser, J.S. and Lulsdorf, M.M. (2004) Progress towards haploid division in chickpea (*Cicer arietinum* L.), field pea (*Pisum sativum* L.) and lentil (*Lens culinaris* Medik.) using isolated microspore culture. In: *Proceedings of the 4th European Grain Legume Conference.* Dijon, France, p. 189.

Croser, J.S., Lulsdorf, M., Cheng, B., Allen, K., Wilson, J., Dament, T., Siddique, K., Warkentin, T. and Vandenberg, A. (2004) Embryogenesis from isolated microspores of chickpea and field pea – progress towards a doubled haploid protocol as a tool for crop improvement. In: *Proceedings of the 4th International Crop Science Congress*. Brisbane, Queensland, Australia, pp. 255–256.

Croser, J., Lulsdorf, M., Davies, P., Wilson, J., Sidhu, P., Grewal, R., Allen, K., Dament, T., Warkentin, T., Vandenberg, A. and Siddique, K. (2005) Haploid embryogenesis from chickpea and field pea – progress towards a routine protocol. In: *Proceedings of the Conference of the Australian Branch of the International Association for Plant Tissue Culture and Biotechnology – Contributing to a Sustainable Future*. Perth, Australia, pp. 71–82.

Das, A. and Sarmah, B.K. (2003) An efficient *in vitro* regeneration system for genetic transformation of chickpea (*Cicer arietinum* L.) cultivars. *Research Journal of Contemporary Concerns* 1, 1–5.

Dineshkumar, V., Kirti, P.B., Sachan, J.K.S. and Chopra, V.L. (1994) Plant regeneration via somatic embryogenesis in chickpea (*Cicer arietinum* L.). *Plant Cell Reports* 13, 468–472.

Dineshkumar, V., Kirti, P.B., Sachan, J.K.S. and Chopra, V.L. (1995) Picloram induced somatic embryogenesis in chickpea (*Cicer arietinum* L.). *Plant Cell Reports* 13, 468–472.

Eapen, S. and George, L. (1994) Somatic embryogenesis in *Cicer arietinum* L.: influence of genotype and auxins. *Biologia Plantarum* 36, 343–349.

Erskine, W., Siddique, K., Khan, T. and Cowling, W. (2001) Utilization of grain legume diversity. In: Maxted, N. and Bennett, S.J. (eds) *Plant Genetic Resources of Legumes in the Mediterranean*. Kluwer Academic, Dordrecht, The Netherlands, pp. 311–326.

Fontana, G.S., Santini, L., Caretto, S., Frugis, G. and Mariotii, D. (1993) Genetic transformation in the grain legume *Cicer arietinum* L. (Chickpea). *Plant Cell Reports* 12, 194–198.

Gamborg, O.L., Miller, R.A. and Ojima, K. (1968) Nutrient requirements of suspension cultures of soybean root cells. *Experimental Cell Research* 50, 151–158.

Gosal, S.S. and Bajaj, Y.P.S. (1988) Pollen embryogenesis and chromosomal variation in anther of three food legumes – *Cicer arietinum*, *Pisum sativum* and *Vigna mungo*. *SABRAO Journal* 20, 51–58.

Hamblin, J., Barton, J., Atkins, C., Jones, M., Smith, P., Wylie, S., Schroeder, H., Molvig, L., Tabe, L. and Higgins, T.J. (1998) Interspecific hybridization between cultivated chickpea (*Cicer arietinum* L.) and the wild annual species *C. judaicum* Boiss, *C. pinnatifidum* Jaub. & Sp. and *C. bijugum* K.H. Rech. In: *Proceedings of the 3rd European Grain Legume Conference*. Valladolid, Spain; AEP, Paris, France, pp. 70–71.

Hamilton, C.M., Frary, A., Lewis, C. and Tanksley, S.D. (1996) Stable transfer of intact high molecular weight DNA into plant chromosomes. *Proceedings of the National Academy of Sciences of the United States of America* 93, 9975–9979.

Hamilton, C.M., Frary, A., Xu, Y., Tanksley, S.D. and Zhang, H.-B. (1999) Construction of tomato genomic DNA libraries in a binary-BAC (BIBAC) vector. *Plant Journal* 18, 223–229.

He, R.-F., Wang, Y., Shi, Z., Ren, X., Zhu, L., Weng, Q. and He, G.-C. (2003) Construction of a genomic library of wild rice and *Agrobacterium*-mediated transformation of large-insert DNA linked to BPH resistance locus. *Gene* 321, 113–121.

Huda, S., Islam, R. and Bari, M.A. (2000) Shoot regeneration from internode derived callus of chickpea (*Cicer arietinum* L.). *International Chickpea and Pigeonpea Newsletter* 7, 28–29.

Huda, S., Islam, R., Bari, M.A. and Asaduzzaman, M. (2001) Anther culture of chickpea. *International Chickpea and Pigeonpea Newsletter* 8, 24–26.

International Rice Genome Sequencing Project (2005) The map-based sequence of the rice genome. *Nature* 436, 793–800.

Islam, R., Malik, T., Husnain, T. and Riazuddin, S. (1994) Strain and cultivar specificity in the Agrobacterium-chickpea interaction. *Plant Cell Reports* 13, 561–563.

Islam, R., Farooqui, H. and Riazuddin, S. (1999) Improved efficiency in chickpea tissue culture: effects of presoaking and age of explants on in vitro shoot proliferation.

International Chickpea and Pigeonpea Newsletter 6, 36–37.

Jaiswal, R. and Singh, N.P. (2001) Plant regeneration from NaCl tolerant callus/cell lines of chickpea. *International Chickpea and Pigeonpea Newsletter* 8, 21–23.

Jaiwal, P.K., Upadhyaya, S. and Upadhyaya, K.C. (2001) Chickpea regeneration and transformation. *Current Science* 80, 1368–1369.

Jayanand, B., Sudarsanam, G. and Sharma, K.K. (2003) An efficient protocol for the regeneration of whole plants of chickpea (*Cicer arietinum* L.) by using axillary meristem explants derived from *in vitro* germinated seedling. *In Vitro Cellular and Developmental Biology – Plant* 39, 171–179.

Jones, J.D.G., Shlumukov, L., Carland, F., English, J., Scofield, S.R., Bishop, G.J. and Harrison, K. (1992) Effective vectors for transformation, expression of heterologous genes, and assaying transposon excision in transgenic plants. *Transgenic Research* 1, 285–297.

Kar, S., Johnson, T.M., Nayak, P. and Sen, S.K. (1996) Efficient transgenic plant regeneration through *Agrobacterium* mediated transformation of chickpea (*Cicer arietinum* L.). *Plant Cell Reports* 16, 32–37.

Kar, S., Basu, D., Das, S., Ramkrishnan, N.A., Mukherjee, P., Nayak, P. and Sen, S.K. (1997) Expression of *cryIA(c)* gene of *Bacillus thuringiensis* in transgenic chickpea plants inhibits development of pod-borer (*Heliothis armigera*) larvae. *Transgenic Research* 6, 177–185.

Kartha, K.K., Pahl, K., Leung, N.L. and Mroginski, L.A. (1981) *In vitro* culture of shoot apical meristem in chickpea (*Cicer arietinum* L.). *Canadian Journal of Botany* 59, 1671–1679.

Khan, S.K. and Ghosh, P.D. (1983) *In vitro* induction of androgenesis and organogenesis in *Cicer arietinum* L. *Current Science* 52, 891–893.

Krishnamurthy, K.V., Suhasini, K., Sagare, A.K., Meixner, M., de-Kathen, A., Pickardt, T. and Schieder, O. (2000) Agrobacterium mediated transformation of chickpea (*Cicer arietinum* L.) embryo axes. *Plant Cell Reports* 19, 235–240.

Krysan, P.J., Young, J.C. and Sussman, M.R. (1999) T-DNA as an insertional mutagen in *Arabidopsis*. *Plant Cell* 11, 2283–2290.

Lichtenzveig, J., Scheuring, C., Dodge, J., Abbo, S. and Zhang, H.-B. (2005) Construction of BAC and BIBAC libraries and their applications for generation of SSR markers for genome analysis of chickpea, *Cicer arietinum* L. *Theoretical and Applied Genetics* 110, 492–510.

Liu, Y.-G., Shirano, Y., Fukaki, H., Yanai, Y., Tasaka, M., Tabata, S. and Shibata, D. (1999) Complementation of plant mutants with large genomic DNA fragments by a transformation-competent artificial chromosome vector accelerates positional cloning. *Proceedings of the National Academy of Sciences of the United States of America* 96, 6535–6540.

Liu, Y.-G., Liu, H., Chen, L., Qiu, W., Zhang, Q., Wu, H., Yang, C., Su, J., Wang, Z., Tian, D. and Mei, M. (2002) Development of new transformation-competent artificial chromosome vectors and rice genomic libraries for efficient gene cloning. *Gene* 282, 247–255.

Lulsdorf, M., Mallikarjuna, N., Clarke, H. and Tar'an, B. (2005) Finding solutions for interspecific hybridization problems in chickpea (*Cicer arietinum* L.). In: Kharkwal, M.C. (ed.) *Proceedings of the 4th International Food Legumes Research Conference*. The Indian Society of Genetics and Plant Breeding, New Delhi, India, p. 44.

Malhotra, R.S., Singh, K.B., Di Vito, M., Greco, N. and Saxena, M.C. (2002) Registration of ILC 10765 and ILC 10766 chickpea germplasm lines resistant to cyst nematode. *Crop Science* 42, 1756.

Malik, K.A. and Saxena, P.K. (1992) *In vitro* regeneration of chickpea. *Plant Physiology* 18, 138–142.

Mallikarjuna, N. (1999) Ovule and embryo culture to obtain hybrids from interspecific incompatible pollinations in chickpea. *Euphytica* 110, 1–6.

Mallikarjuna, N. (2001) Prospects of using *Cicer canariense* for chickpea improvement. *International Chickpea and Pigeonpea Newsletter* 8, 23–24.

Mohapatra, S.T. and Sharma, R.P. (1991) Agrobacterium-mediated genetic transformation of chickpea, *Cicer arietinum* L. *Indian Journal of Experimental Botany* 29, 758–761.

Murashige, T. and Skoog, F. (1962) A revised medium for rapid growth and bio assays

with tobacco tissue cultures. *Physiologia Plantarum* 15, 473–496.
Murthy, B.N.S., Victor, J., Singh, R.P., Fletcher, R.A. and Saxena, P.K. (1996) *In vitro* regeneration of chickpea (*Cicer arietinum* L.). Stimulation of direct organogenesis and somatic embryogenesis by thidiazuron. *Plant Growth Regulators* 19, 233–240.
Nenz, E., Pupilli, F., Damiani, F. and Arcioni, S. (1996) Somatic hybrid plants between the forage legumes *Medicago sativa* L. and *Medicago arborea* L. *Theoretical and Applied Genetics* 93, 183–189.
Olhoft, P.M. and Somers, D.A. (2001) L-Cysteine increases *Agrobacterium*-mediated T-DNA delivery into soybean cotyledonary-node cells. *Plant Cell Reports* 20, 706–711.
Olhoft, P.M., Flagel, L.E., Donovan, C.M. and Somers, D.A. (2003) Efficient soybean transformation using hygromycin B selection in the cotyledonary-node method. *Planta* 216, 723–735.
Polowick, P.L., Baliski, D.S. and Mahon, J.D. (2004) *Agrobacterium tumefaciens*-mediated transformation of chickpea (*Cicer arietinum* L): gene integration, expression and inheritance. *Plant Cell Reports* 23, 485–491.
Popelka, J.C., Terryn, N. and Higgins, T.J.V. (2004) Gene technology for grain legumes: can it contribute to the food challenge in developing countries? *Plant Science* 167, 195–206.
Rajesh, P.N., Coyne, C., Meksem, K., DevSharma, K., Gupta, V. and Muehlbauer, F.J. (2004) Construction of a *Hin*dIII bacterial artificial chromosome library and its use in identification of clones associated with disease resistance in chickpea. *Theoretical and Applied Genetics* 108, 663–669.
Ramulu, C.A. and Ram, A.S. (2004) Tissue culture induced genetic variability in certain food legumes for improvement of agronomic traits. *Journal of Phytological Research* 17, 201–203.
Rao, B.G. and Chopra, V.L. (1989) Regeneration in chickpea (*Cicer arietinum* L.) through somatic embryogenesis. *Journal of Plant Physiology* 134, 637–638.
Rao, D. and Ramulu, C.A. (1997) Selection for herbicide resistance in tissue culture and isolation of phenotypic variants in cowpea and chickpea. In: Reddy, S.M. and Srivastava, H.P. (eds) *Microbial Biotechnology: Professor K.S. Bilgrami Commemoration volume*. Vedams Books, New Delhi, India, p. 266.
Reddy, V.D. and Reddy, G.M. (1996) *In vivo* production of haploids in chickpea (*Cicer arietinum* L.). *Journal of Genetics and Breeding* 51, 29–32.
Sagare, A.P. and Krishnamurthy, K.V. (1991) Protoplast regeneration in chickpea *Cicer arietinum* L. *Indian Journal of Experimental Biology* 29, 930–932.
Sagare, A.P., Suhasini, K. and Krishnamurthy, K.V. (1993) Plant-regeneration via somatic embryogenesis in chickpea (*Cicer arietinum* L.). *Plant Cell Reports* 12(11), 652–655.
Sanyal, I., Singh, A.K. and Amla, D.V. (2003) *Agrobacterium tumefaciens* mediated transformation of chickpea (*Cicer arietinum* L.) using mature embryonic axes and cotyledonary nodes. *Indian Journal of Biotechnology* 2, 524–532.
Sanyal, I., Singh, A.K., Kaushik, M. and Amla, D.V. (2005) *Agrobacterium*-mediated transformation of chickpea (*Cicer arietinum* L.) with *Bacillus thuringiensis cryIAc* gene for resistance against pod borer insect *Helicoverpa armigera*. *Plant Science* 168, 1135–1146.
Sarmah, B.K., Moore, A., Tate, W., Molvig, L., Morton, R.L., Rees, D.P., Chiaiese, P., Chrispeels, M.J., Tabe, L.M. and Higgins, T.J.V. (2004a) Transgenic chickpea seeds expressing high levels of a bean α-amylase inhibitor. *Molecular Breeding* 14, 73–82.
Sarmah, B.K., Das, P.K., Acharjee, S., Das, A., Kumar, P.A. and Deka, P.C. (2004b) Expression of *Bt-cry1Ac* gene in chickpea to confer resistance to pod borer (*Helicoverpa armigera*. L.). In: *Proceedings of the National Symposium on Biochemical Approaches for Utilization and Exploitation of Commercially Important Plants*. Assam Agricultural University, Jorhat, India, pp. 193–197.
Schroeder, H.E., Gollasch, S., Moore, A., Tabe, L.M., Craig, S., Hardie, D.C., Chrispeels, M.J., Spencer, D. and Higgins, T.J.V. (1993) Bean alpha-amylase inhibitor confers resistance to the pea weevil (*Bruchus pisorum*) in transgenic peas (*Pisum sativum* L.). *Plant Physiology* 107, 1233–1239.
Senthil, G., Williamson, B., Dinkins, R.D. and Ramsay, G. (2004) An efficient transforma-

tion system for chickpea (*Cicer arietinum* L.). *Plant Cell Reports* 23, 297–303.

Shanker, S. and Ram, H.Y.M. (1993) Induction of somatic embryogenesis in the tissue cultures of chickpea (*Cicer arietinum* L.). *Proceedings of the Indian National Science Academy Part B Biological Sciences* 59(6), 629–632.

Sharma, K.K. and Ortiz, R. (2000) Program for the application of genetic transformation for crop improvement in the semi-arid tropics. *In Vitro Cellular and Developmental Biology–Plant* 36, 83–92.

Sharma, L. and Amla, D.V. (1998) Direct shoot regeneration in chickpea (*Cicer arietinum* L.). *Indian Journal of Experimental Biology* 36, 605–609.

Shri, P.V. and Davis, T.M. (1992) Zeatin induced shoot regeneration from immature chickpea. *Plant Cell Tissue Organ Culture* 28, 45–51.

Sindhu, A., Singh, R., Saini, N., Sangwan, W. and Singal, H.R. (1998) Studies on defense mechanism of chickpea (*Cicer arietinum* L.) callus cultures against *Ascochyta rabiei*. *Legume Research* 21, 105–108.

Singh, A., Singh, N.P., Gurha, S.N. and Asthana, A.N. (1999) *In vitro* selection against Ascochyta blight of chickpea (*Cicer arietinum* L.). *Journal of Plant Biochemistry and Biotechnology* 8, 117–119.

Singh, K.B., DiVito, M., Greco, N. and Saxena, M.C. (1996) Registration of ILWC 292, a chickpea cyst nematode-resistant germplasm of Cicer reticulatum Ladiz. *Crop Science* 36(5), 1421–1422.

Singh, R., Sindhu, A., Singal, H.R. and Singh, R. (2003) Biochemical basis of resistance in chickpea (*Cicer arietinum* L.) against Fusarium wilt. *Acta Phytopathologica et Entomologica Hungarica* 38, 13–19.

Singh, R.P. and Singh, B.D. (1989) Recovery of rare interspecific hybrids of gram *Cicer arietinum* × *C. cuneatum* L. through tissue culture. *Current Science* 58, 874–876.

Singh, S., Gumber, R.K., Joshi, N. and Singh, K. (2005) Introgression from wild *Cicer reticulatum* to cultivated chickpea for productivity and disease resistance. *Plant Breeding* 124(5), 477–480.

Stamigna, C., Crino, P. and Saccardo, F. (2000) Wild relatives of chickpea: multiple disease resistance and problems to introgression in the cultigen. *Journal of Genetics and Breeding* 54, 213–219.

Swamy, A.V.S.R. and Khanna, V.K. (1991a) Pollen grain germination and pollen tube growth following interspecific pollinations in *Cicer*. *SABRAO Journal* 23(2), 137–145.

Swamy, A.V.S.R. and Khanna, V.K. (1991b) Effect of some growth regulators on *in vitro* pollen germination and tube growth in *Cicer*. *International Chickpea and Pigeonpea Newsletter* 25, 32–34.

Tao, Q. and Zhang, H.-B. (1998) Cloning and stable maintenance of DNA fragments over 300 kb in *Escherichia coli* with conventional plasmid-based vectors. *Nucleic Acids Research* 26, 4901–4909.

Tewari-Singh, N., Sen, J., Kiesecher, H., Reddy, V.S., Jacobsen, H.J. and Guha-Mukherjee, S. (2004) Use of a herbicide or lysine plus threonine for non-antibiotic selection of transgenic chickpea. *Plant Cell Reports* 22, 576–583.

Tikkas, P.B., Ghorade, R.B., Shivankar, R.S. and Rathod, T.H. (2002) Studies on callus induction in chickpea (*Cicer arietinum* L.). *Annals of Plant Physiology* 16, 57–63.

van Dorrestein, B., Baum, M. and Malhotra, R.S. (1998) Interspecific hybridization between cultivated chickpea (*Cicer arietinum* L.) and the wild annual species *C. judaicum* Boiss, *C. pinnatifidum* Jaub. & Sp. and *C. bijugum* K.H. Rech. In: *Proceedings of the 3rd European Grain Legume Conference*. Valladolid, Spain; AEP, Paris, France, p. 362.

Vani, A.K.S. and Reddy, V.D. (1996) Morphogenesis from callus culture of chickpea (*Cicer arietinum* L.). *Indian Journal of Experimental Biology* 34, 285–287.

Verma, M.M., Ravi and Sandhu, J.S. (1995) Characterization of the interspecific cross *Cicer arietinum* L. × *C. judaicum* (Boiss). *Plant Breeding* 114, 549–551.

Vessal, S.R., Bagheri, A. and Safarnejad, A. (2002) The possibility of *in vitro* haploid production in chickpea (*Cicer arietinum* L.). *Journal of Science and Technology of Agriculture and Natural Resources* 6, 67–76.

Young, D.Y., Cannon, S.B., Sato, S., Kim, D., Cook, D.R., Town, S.D., Roe, B.A. and Tabata, S. (2005) Sequencing the gene space of *Medicago truncatula* and *Lotus japonicus*. *Plant Physiology* 137, 1174–1181.

23 Abiotic Stresses

C. Toker,[1] C. Lluch,[2] N.A. Tejera,[2] R. Serraj[3] and K.H.M. Siddique[4]

[1]*Department of Field Crops, Faculty of Agriculture, Akdeniz University, TR-07059 Antalya, Turkey;* [2]*Department of Plant Physiology, Faculty of Sciences, University of Granada, Campus de Fuentenueva s/n, 18071 Granada, Spain;* [3]*International Rice Research Institute, Dapo Box 7777, Metro Manila, Philippines;* [4]*Institute of Agriculture, The University of Western Australia, Faculty of Natural and Agricultural Sciences, 35 Stirling Highway, Crawley, WA 6009, Australia*

Introduction

Seed yield of chickpea is generally low, unstable and less than its potential as it is most often grown on marginal lands with minimum inputs. Despite its high yield potential, reportedly more than 6 t/ha (Singh, 1987), low and unstable yields are generally due to the adverse effects of biotic and abiotic stresses. On a global basis, annual yield losses in chickpea were estimated to be 6.4 million tonnes due to abiotic stresses and 4.8 million tonnes due to biotic stresses (Ryan, 1997). The most common abiotic stresses affecting chickpea production, in the order of importance, are drought, heat and cold (Singh *et al.*, 1994; Croser *et al.*, 2003). Drought together with heat causes 3.3 million tonnes yield loss per annum (Ryan, 1997). Other abiotic stresses specific to some regions are salinity, waterlogging, soil alkalinity and acidity, and nutrient deficiencies and toxicities (Ryan, 1997; Siddique *et al.*, 2000).

This chapter summarizes our current knowledge of the main abiotic constraints facing chickpea production in the world. It also reviews selection criteria and available genetic resources for stress resistance, and finally proposes future strategies for chickpea improvement under abiotic stress conditions.

Drought Tolerance

Drought is a meteorological term and environmental event, defined as a water stress due to lack or insufficiency of rainfall and/or inadequate external water supply. Drought stress is a complex syndrome affected by several climatic, edaphic and agronomic factors, and involves three main parameters: timing, duration and intensity (Serraj *et al.*, 2003). Drought stress depends not only on rainfall and its distribution, but also on evaporation, soil water-holding capacity and crop water requirements.

Chickpea target environment and types of drought

The major chickpea-growing areas are in the arid and semiarid zones. In South Asia and northern Australia, chickpea is predominantly grown in the post-rainy season on receding stored soil moisture (Siddique *et al.*, 2000). In the Mediterranean – including West Asia and North Africa (WANA), and Central Asia – chickpea is traditionally grown in spring on stored soil moisture from winter and early spring rainfall. In both environments, chickpea is subjected to declining soil residual moisture, increasing temperatures and vapour pressure deficits during flowering and maturity (Singh *et al.*, 1997; Leport *et al.*, 1998; Siddique *et al.*, 2000).

Chickpea generally faces (i) terminal drought, increasing towards the end of the growing season, due to the depletion of soil moisture; or (ii) intermittent drought, caused by a break-in rainfall followed by insufficient rains during the growing season. Approximately 90% of the world's chickpea is grown under rainfed conditions (Kumar and Abbo, 2001) where terminal drought is one of the major constraints limiting productivity. Terminal drought stress is typical in the post-rainy season in semiarid tropical regions where crops grow and mature on a progressively depleting soil moisture profile. Characterizing drought in post-rainy-season crops such as chickpea is relatively simple compared with intermittent drought experienced by rainy season crops. Factors governing crop growth and water use in the post-rainy season such as radiation, temperature, vapour pressure and potential evaporation are relatively stable and predictable, so simulation modelling of both crop growth and the effects of various crop traits is feasible.

Impact of drought

Drought accompanied by heat stress can affect chickpea growth at any stage of the crop. As a result, the growing season may be shortened, affecting the production of yield components, which are total biomass, pod number, seed number, seed weight and quality, and seed yield per plant.

The critical soil moisture level for germination differs among cool-season food legumes, with chickpea having a higher moisture requirement than field pea (*Pisum sativum* L.), lentil (*Lens culinaris*) and faba bean (*Vicia faba* L.) (Saxena *et al.*, 1993*a*). Maximum seedling emergence in chickpea occurs when soil moisture is at field capacity and declines with increasing drought stress in a curvilinear manner (Sharma, 1985). Siddique and Loss (1999) concluded that the optimum sowing depth for chickpea was between 5 and 8 cm. They suggested that sowing at this depth might improve crop establishment where moisture from rainfall was stored in the subsoil below 5 cm, thus reducing damage from herbicides and improving the survival of specific bacteria inoculated on the seed.

Khanna-Chopra and Sinha (1987) showed that flowering and pod setting in chickpea are the stages of development that are most sensitive to drought stress. A number of yield components including dry matter and harvest index of rainfed chickpea were less in plants exposed to drought stress than irrigated plants (Davies *et al.*, 1999; Leport *et al.*, 1999, 2005). Leport *et al.* (2005) also showed that pod abortion was greater on secondary than on primary branches and in kabuli (white seed coat) types than in desi (coloured seed coat) varieties when drought occurred soon after pod set began. Wild species of chickpea are found to be superior to the cultigen in terms of drought resistance (Nayyar *et al.*, 2005c). Although drought resistance is relatively higher in chickpea than in lentil, field pea and faba bean (Siddique, 2000), seed yield losses due to drought range from 30% to 100% (Saxena *et al.*, 1993a; Leport *et al.*, 1999; Toker and Canci, 2006) depending on genotype and the type of drought experienced in the target environment.

Resistance mechanisms

Mechanisms of drought resistance include escape and tolerance (Ludlow and Muchow, 1990). Drought resistance is classified into three major categories (Turner *et al.*, 2001):

1. Drought escape through early maturity:
(a) *Earliness* (days to first flowering, 50% flowering and maturity) is the most important mechanism to escape onset of drought under terminal conditions. Drought escape can also be managed by early sowing. Short-duration varieties maturing before the onset of severe terminal drought have proved to be successful in increasing yield under drought-prone conditions (Kumar *et al.*, 1996).The optimum crop duration for maximum yield depends upon water availability, so varieties need to be matched with the longest growing period.
(b) *Early vigour* is considered one of the most important drought escape mechanisms in grain legumes (Thomson and Siddique, 1997; Turner *et al.*, 2001). Early vigour should be combined with appropriate phenology for the target environment. Correlation between initial growth vigour and other characteristics in recombinant inbred lines of chickpea showed that high growth vigour had significant negative correlation with days to first flower, flowering, first pod and maturity (Sabaghpour *et al.*, 2003).

2. Dehydration postponement can be achieved by maintaining water uptake and reducing water loss. Two important drought avoidance traits have been characterized and widely used for the genetic enhancement of chickpea: (i) large root system for greater extraction of available soil moisture; and (ii) smaller leaf area reducing transpirational water loss. Chickpea line ICC 4958 has large roots, rapid rates of root development and water extraction. Lines ICC 5680 and ICC 10480 have smaller leaf areas, either due to narrow (ICC 10480) or fewer (ICC 5680) pinnules. Recombinants with traits of ICC 4958 and ICC 5680 showed higher midday leaf-related water content compared to the parents in field trials conducted at International Crops Research Institute for the Semi-Arid Tropics (ICRISAT), Hyderabad, India (Saxena, 2003).

Other mechanisms related to maintaining water uptake and reducing water loss are:
(a) *Turgor* may be maintained by water uptake, reducing water loss and osmotic adjustment (Ludlow and Muchow, 1990; Turner *et al.*, 2001). Stomata play a crucial role in drought tolerance and regulating water loss.

Stomatal conductance differs in response to water potential in the leaf of chickpea (Lawn, 1982; Muchow, 1985).

(b) *Abscisic acid (ABA)* causes stomatal closure and stimulates uptake of water into the roots (Davies, 1995). Nayyar *et al.* (2005c) recently found higher ABA contents in *Cicer reticulatum* Ladiz. than in the cultigen up to the 8th day of water stress.

(c) *Osmotic adjustment (osmoregulation)* maintains stomatal conductance and photosynthesis, and delays leaf senescence and death. Osmotic adjustment also reduces flower abortion and maintains root growth with the accumulation of some solutes in response to water stress (Turner *et al.*, 2001). Osmotic adjustment is positively correlated with yield in drought conditions (Morgan *et al.*, 1991). Moinuddin and Khanna-Chopra (2003) found that high osmotic adjustment group of chickpea had better plant water status, seed yield and yield parameters than the low osmotic adjustment group under water deficit. However, the relationship between osmotic adjustment and grain yield in a range of drought stress conditions is still questionable (Lecoeur *et al.*, 1992; Serraj and Sinclair, 2002; N.C. Turner, Australia, 2005, unpublished data).

(d) *Root density and depth* is an important trait related to drought resistance. Root traits in the available genetic diversity in chickpea were evaluated with some genotypes being selected for drought resistance (Saxena *et al.*, 1993b; Johansen *et al.*, 1994; Krishnamurthy *et al.*, 1998; Kashiwagi *et al.*, 2006). Chickpea genotypes with high root length densities and root dry weights are sources of drought resistance (Kashiwagi *et al.*, 2005).

(e) *Leaf types and movements* include folding, rolling and wilting, as well as diaheliotropic and paraheliotropic movements, in response to water deficit in cereals (Ludlow and Muchow, 1990). Chickpea possesses glandular droplets in its green parts mainly consisting of organic acids such as succinic, malic, oxalic and citric (Toker *et al.*, 2003), which may help lower leaf temperature and protect the crop from heat and drought (Lauter and Munns, 1986; Lecoeur *et al.*, 1992). Chickpea with small leaflets are more likely to be drought-resistant, especially in water-limited terminal stress environments (Saxena, 2003). However, Toker and Canci (2006) did not observe drought resistance in multipinnate or tiny leaf types in the cultigen.

(f) *Phenological plasticity* is the mechanism whereby the duration from germination to maturity varies depending on the extent of the water deficit. Since chickpea has an indeterminate growth habit, flowers and pods borne on the upper part of the canopy are more affected by terminal drought than early-formed pods. Short-duration chickpea does not respond to additional rainfall towards the end of the life cycle in long-season environments.

3. Dehydration tolerance is the ability of cells to metabolize at low leaf water status:

(a) *Membrane stability* is associated with leakage of solutes from the cell. Drought stress injury in chickpea has recently been shown using electrolyte leakage (EL) and 2,3,5-triphenyl tetrazolium chloride (TTC) reduction assays (Nayyar *et al.*, 2005c).

(b) *Lethal water potential* is the degree to which plant organs withstand desiccation, expressed as the relative water content or water potential at which leaves die (Nayyar *et al.*, 2005c).

(c) *Chlorophyll content* is another trait used for evaluating drought resistance in water-stressed chickpea (Nayyar *et al.*, 2005c).
(d) *Proline accumulation* is a response to environmental stress but it has not been found to be a useful selection tool for drought resistance improvement in chickpea (Turner *et al.*, 2001).
(e) *Polyamines* are synthesized in plant cells when plants are subjected to high temperatures and drought. The most commonly induced polyamines – thermopolyamines – are essential for continued protein synthesis at high temperature both *in vitro* and *in vivo* (Davies, 1995; Nayyar and Chander, 2004).
(f) *Brassinosteroids, jasmonates, phosphatidic* and *salicylic acids* are widespread in plant cells and can exert regulatory control over growth and development at micromolar concentrations (see Davies, 1995), and may play a role in drought tolerance (Testerink and Munnik, 2005).

Large seed size (100-seed weight) and fast groundcover growth have been associated with drought resistance (Singh *et al.*, 1994). Basu and Singh (2003) found seed size to be highly correlated with root length and root dry weight. It is still controversial to use seed size alone as a selection criterion for drought resistance (Toker and Canci, 2006).

Drought-related yield components

Turner *et al.* (2001) recently reviewed yield component frameworks with an emphasis on grain legumes: crop growth models, water use and transpiration efficiency, combinations of soil water balance and crop phenological models, and harvest index.

Fischer and Maurer (1978) proposed a drought susceptibility index, defined as $S = (1 - Y/Y_P)/D$, where Y is the yield under drought stress and Y_P is the yield under non-stress conditions. D is drought intensity; $D = 1 - X/X_P$, where X and X_P represent mean yield of all varieties under stressed and non-stressed conditions, respectively; and D ranges from zero to 1. Toker and Cagirgan (1998) found a significant correlation in chickpea between the drought susceptibility index and seed yield, biological yield, harvest index and mean productivity in drought-stressed environments. Silim and Saxena (1993) used a drought response index (DRI) to describe the response of individual chickpea genotypes to drought conditions and fitted a multiple regression of stressed grain yield on unstressed grain yield and days to flower: $Y_0 = a - bF + cY_i$ and DRI $= (Y_0 - \hat{Y}_0)$/standard error of $\hat{Y}_0$, where Y_0 is yield under drought, $\hat{Y}_0$ is regression estimate of yield under drought, Y_i represents yield potential and F is the days to flowering. Silim and Saxena (1993) determined the relative contributions of drought escape, yield potential and DRI to yield in chickpea. Deep roots, high leaf water potential at dawn and large number of seeds were related to the DRI. Anbessa and Bejiga (2002) identified two drought-tolerant genotypes, ACC 41235 and ACC 209025, based on the DRI. They concluded that reduced water loss from the plant and extensive extraction of soil moisture are factors involved in the adaptation of chickpea to drought conditions. Baker (1994) chose to approximate the response to stress as a linear function of increasing stress. Productivity (Y_j) at any level of stress (S_j) can be represented by a linear regression equation such as $Y_j = M - TS_j$, where

M is the maximum productivity in the absence of stress, *T* represents a measure of tolerance and S_j is a quantitative measure of the level of stress. High tolerance, represented by low values of *T*, will result in low-level changes in productivity with changes in stress level. A low-level stress (S_l) and high-level stress (S_h), $T(S_l - S_h)$, will decrease productivity.

Soltani *et al.* (2000) reported that leaf area expansion and transpiration in chickpea did not change until the fraction of transpirable soil water reached 0.48 and 0.34, respectively. Siddique *et al.* (2001) studied several grain legumes at two locations in Australia and stated that the total transpiration (E_t) ranged from 180 mm at Mullewa to 270 mm at Merredin, and water use efficiency (WUE) in terms of dry matter ranged from 11.1 kg/ha/mm at Merredin to 18.3 kg/ha/mm at Mullewa. They concluded that early phenology with rapid groundcover and dry matter production allowed high E_t post flowering.

Inheritance of drought resistance traits

Morphological and physiological characters related to drought resistance in chickpea have different types of inheritance patterns (monogenic and polygenic) and gene actions (additive and non-additive). Sabaghpour *et al.* (2003) showed that growth vigour in chickpea is controlled by two genes with duplicate dominant epistasis. Root length density and root dry weight among recombinant inbred lines derived from ICC 4958 and Annigeri appeared to be under polygenic control. The broad-sense heritability for root length and root dry weight was estimated to be 0.23 and 0.27, respectively (Serraj *et al.*, 2004).

Despite the availability of many screening and selection techniques for drought tolerance in chickpea, only three have been successful under field conditions: (i) line-source sprinkler irrigation systems (Saxena *et al.*, 1993a); (ii) root trait characteristics (root length, root density, root biomass, root length density) at ICRISAT (Serraj *et al.*, 2004); and (iii) delayed sowing strategy (Singh *et al.*, 1994) at the International Center for Agricultural Research in the Dry Areas (ICARDA), Aleppo, Syria. The first two methods are restricted to large-scale breeding programmes. Singh *et al.* (1997) rated 4165 chickpea genotypes on a scale of 1–9 to eliminate susceptible lines. Drought-resistant lines performed well under both conditions with 19 lines producing more than 1 t/ha seed under drought conditions and more than 2 t/ha under non-stressed conditions. Recent screening of the chickpea mini core collection at ICRISAT characterized the genetic variability of drought-avoidance root traits and selected suitable parents to develop populations for molecular mapping. The accessions identified in the mini core collection could be used as valuable alternative sources for diversification of mapping populations with varying growth duration and to obtain the required polymorphism for successfully mapping root traits in chickpea (Kashiwagi *et al.*, 2005). ICC 4958, ICC 8261 and FLIP 87–59C are drought-tolerant germplasm lines (Saxena *et al.*, 1993b; Singh *et al.*, 1994; Kashiwagi *et al.*, 2005). ICCV 96029 has greater early vigour, super-early flowering and double pods (Kumar and Rao, 2001).

Some characters such as seed size (Singh *et al.*, 1994), fast groundcover (Siddique *et al.*, 2001), early vigour (Sabaghpour *et al.*, 2003), biological yield and harvest index (Toker and Canci, 2006) can be evaluated as putative traits related to yield under drought conditions. These traits seem to be effective and easier to adopt in large-scale breeding programmes, especially under terminal drought conditions.

Heat Tolerance

Types of heat

Blum (1988) and Wery *et al.* (1993) defined two types of heat stress according to the interaction of time and temperature: (i) heat shock (lethal temperatures from a few minutes to a few hours); and (ii) moderate heat (higher than optimum temperatures during the growing season). Among cool-season food legumes, critical temperatures vary between genera, species within a genus and genotypes within a species. Critical temperature appears to be higher in chickpea than in lentil, field pea and faba bean (Saxena *et al.*, 1988). Optimal growth and development of chickpea occurs at temperatures between 15°C and 30°C (van der Maesen, 1972). Heat or high temperature stress is common in major chickpea-growing areas in the world. Chickpea is sensitive to high temperatures, especially in the reproductive phase (Malhotra and Saxena, 1993; Singh *et al.*, 1994).

Impact of heat

In general, heat stress has negative effects on production, especially during gamete development, flowering and podding. Yield reductions have been observed at a temperaure of 30°C at 50% flowering and >30°C for 3–4 days at 100% flowering (Summerfield *et al.*, 1984). van Rheenen *et al.* (1997) reported that such conditions reduced the duration of flowering and pod filling, resulting in large yield losses. Saxena *et al.* (1988) stated that heat caused withering, burning of lower leaves, desiccation of poorly developed plants, stunting flower and pod abortion, and reduced root nodulation, nitrogen (N) fixation and seed yield (Saxena *et al.*, 1988; Kurdali, 1996). The optimal temperature for germination of chickpea pollen is 25°C with drastic reductions at cooler and warmer temperatures (Jaiwal and Mehta, 1983). Flowers are the most sensitive organs to heat (Wery *et al.*, 1993; Toker and Canci, 2006).

Resistance mechanisms

Heat stress resistance is defined as the ability of a genotype to perform better when subjected to the stress. Mechanisms of resistance to heat stress vary (Wery *et al.*, 1993):

1. Heat escape occurs when plant escapes heat shock with accelerated phenology. Earlier flowering and maturity would be an effective resistance mechanism to heat shock.
2. Heat avoidance is the ability of a genotype to reflect radiation energy, usually from the leaves.

3. Heat tolerance involves accumulation in solutes (proline, glycine betaine) and heat shock proteins. Accumulation of solutes could play a protecting role for pollen (Blum, 1988).

Laurie and Stewart (1993) observed that chickpea responds to heat stress differently at different growth stages. Heat tolerance in cool-season food legumes has not attracted much attention from researchers due to the difficulty in distinguishing heat stress from drought stress in the field (Malhotra and Saxena, 1993). However, progress made in other crop species can be easily adapted for chickpea. Screening for heat tolerance should be assessed with drought as both stresses generally occur together in chickpea-growing regions.

Cold Tolerance

Types of cold

Cold-related stress can be defined in terms of either chilling (between 0°C and 12°C) or freezing or below 0°C without snow cover (Wery *et al.*, 1993). Mean daily temperatures below 15°C have caused flower and pod abortion in parts of India and Australia (Savithri *et al.*, 1980; Srinivasan *et al.*, 1999; Clarke and Siddique, 2004; Nayyar *et al.*, 2005a). Chickpea is susceptible to chilling temperatures below 10°C, especially in the reproductive phase (Srinivasan *et al.*, 1999; Nayyar *et al.*, 2005b). Freezing stress is common during vegetative growth in WANA, Europe and Central Asia, especially when sown in autumn or in early spring (Singh *et al.*, 1994).

Impact of cold

Effects of chilling at the cellular level are on membrane stability decreasing with modifications of lipids and proteins, loss of integrity of membrane increasing in some solutes such as K^+, amino acids and carbohydrates, and membrane altering chloroplast function (Wery *et al.*, 1993; Croser *et al.*, 2003). These effects reduce growth rate and increase chlorosis and necrosis of older leaves at the whole-plant level. Meiosis is adversely affected by cold (Blum, 1988). Reproductive organs are very sensitive to cold, resulting in sterile flowers (Savithri *et al.*, 1980; Srinivasan *et al.*, 1999; Clarke and Siddique, 2004; Nayyar *et al.*, 2005a,b). Germination and emergence may be delayed by low temperatures (Auld *et al.*, 1988; C. Toker, Turkey, 1999, unpublished data).

Cold resistance mechanisms

Wery *et al.* (1993) reviewed mechanisms of resistance to freezing stress:

1. Freezing escape is the resistance of autumn-sown chickpea if emergence occurs after the main freezing period. Susceptibility increases with rapid growth before the onset of a severe winter.

2. Freezing avoidance is the avoidance of intracellular ice formation by two mechanisms: super cooling and osmotic adjustment (Wery *et al.*, 1993). Super cooling (ice formation at termperatures above freezing point) is induced by some bacteria such as *Pseudomonas syringae*, *Erwinia herbicola* and *P. fluorescens* (Lindlow, 1980). Genetic resistance to these bacteria can serve as a selection tool for freezing avoidance. Super cooling is also restricted by low water contents (Blum, 1988). Osmotic adjustment retains water in the cell with limited effects (Wery *et al.*, 1993).
3. Freezing tolerance is the ability of the plant to avoid cell dehydration or increase membrane stability. Hardening, an adaptation of the plant to successive low temperatures, is a freezing avoidance and tolerance mechanism.

Chickpea seedlings subjected to temperatures ranging from 1°C to 7°C (cold-acclimated) had higher ABA contents than non-acclimated controls (Nayyar *et al.*, 2005a). Hydrogen peroxide (H_2O_2) and malondialdehyde contents in roots were greater in non-acclimated seedlings than in acclimated ones. Carbohydrates including sucrose, glucose, fructose and trehalose, as well as proline, were also higher in cold-acclimated seedlings. They concluded that a reduction in chilling damage in chickpea by cold acclimation possibly involves ABA and calcium by modulating the content of cryoprotective and reactive oxygen species. Nayyar *et al.* (2005b) concluded that there was a relationship between glycine betaine as a cryoprotectant and cold tolerance in chickpea. Application of glycine betaine at the bud stage improved pollen germination and viability, pollen tube growth, stigma receptivity and ovule viability, and increased yield per plant (Nayyar *et al.*, 2005b).

Winter sowing in the Mediterranean environments

The yield advantage of autumn or winter sowing over traditional spring sowing in Mediterranean environments has been well documented (Malhotra and Saxena, 1993). Other advantages of winter sowing include longer vegetative and reproductive periods, higher fixation of atmospheric nitrogen (N_2), less soil erosion during winter, earlier harvests and increased plant height suitable for machine harvesting. Disadvantages include susceptibility to ascochyta blight and fusarium wilt, poor weed control, cold damage and a lack of large-seeded cold-tolerant genotypes.

Since cold tolerance variability in the cultigen is not sufficient for use as a winter crop at high altitudes of the Mediterranean, wild *Cicer* spp. have been evaluated for cold tolerance (Robertson *et al.*, 1995). Toker (2005) reported that *C. bijugum* had the highest level of cold tolerance followed by *C. reticulatum* and *C. echinospermum*. *C. pinnatifidum* Jaub. and Spach had tolerant and susceptible accessions. Wild *Cicer* spp., especially annuals, are becoming important for the cultigen as sources of resistance to both abiotic and biotic stresses. Nevertheless, the world collection is limited with 572 accessions held in nine gene banks, many of which are duplicates, misidentified or mislabelled

(Shan *et al.*, 2005). Therefore, additional expeditions and exploration of wild chickpea species are warranted.

Inheritance of cold tolerance

Malhotra and Singh (1991) determined that cold tolerance was dominant over susceptibility, and controlled by at least five genes in the cultigen with both additive and non-additive gene effects. They suggested that selection for cold tolerance would be more effective if dominance and epistatic effects were reduced after selfing generations. Narrow-sense heritability for cold tolerance was 87.9% (Malhotra and Saxena, 1993).

To introduce winter chickpea in WANA, a screening and selection method at ICARDA has been developed to incorporate resistance to ascochyta blight and cold tolerance (Singh *et al.*, 1994). The technique depends on: (i) early sowing (in October) of test materials; (ii) using at least one known susceptible and tolerant accession – ILC 8262 and ILC 8617 are the best sources of cold tolerance at seedling stage (Singh *et al.*, 1994); and (iii) inoculation of test materials with *Ascochyta*-infected crop debris prior to flowering and frequent sprinkler irrigation of test materials if natural rainfall is inadequate, which would encourage an epidemic. Field-screening tests should be repeated for at least 2 years, otherwise screening with selected lines must be confirmed under controlled conditions. Pre-emergence herbicide application and high dose of N fertilization increased cold susceptibility (Malhotra and Saxena, 1993; Malhotra and Singh, 1995), whereas different sowing depths (5, 10, 15 and 20 cm) had no effect (Malhotra *et al.*, 1990). Additionally, Wery *et al.* (1993) reported that screening for cold tolerance in controlled environments developed for field pea and faba bean could easily be adapted for chickpea. The temperature can be progressively increased when ~50% of plants show frost damage.

Prevalence of chilling temperature during flowering is one of the main causes of low seed yield in subtropical South Asia (Savithri *et al.*, 1980; Srinivasan *et al.*, 1999) and Australia (Clarke and Siddique, 2004; Clarke *et al.*, 2004) due to floral abortion and low pod and seed set. Srinivasan *et al.* (1999) found that the greater pod-setting ability of the cold-tolerant genotypes ICCV 88502 and ICCV 88503 over the cold-susceptible Annigeri was related to higher pollen germination, tube growth and ovule ability at low temperature. Clarke and Siddique (2004) showed that pollen function in chilling-sensitive plants was clearly affected by low temperature, particularly pollen tube growth down the style before fertilization. Using pollen selection techniques, Clarke *et al.* (2004) developed two chilling-tolerant genotypes, which have been commercially released in Australia as Rupali and Sonali, respectively. This is an excellent example where basic physiological studies and screening procedures have been used in a breeding programme to develop improved crop varieties. Croser *et al.* (2003) reviewed controlled environment and laboratory-based screening techniques for cold tolerance (chilling and freezing) in chickpea. In conclusion, genotypes should be screened for chilling tolerance in the field over successive generations before confirming selection under controlled environments to improve chilling tolerance in chickpea for target environments.

Salinity Tolerance

There are ~955 million hectares of salt-affected soil in the world (Szabolcs, 1994) and ~323 million hectares are considered saline or sodic (Saxena *et al.*, 1993a). These types of soils are particularly common in arid and semiarid regions of West and Central Asia, and Australia, where evaporation exceeds precipitation leading to salt accumulation on the soil surface. The most widely accepted definitions of saline and sodic soils are: a soil is saline when the electrical conductivity (EC) of saturated soil extract is >2 dS/m and a soil is sodic when the sodium adsorption ratio (SAR) is >15 Ds/m (see Saxena *et al.*, 1993a).

Farmers generally do not consider growing cool-season food legumes in salt-affected soils since they are relatively salt-sensitive compared to cereal crops (Saxena *et al.*, 1993a). Yet, Ryan (1997) reported annual yield losses in chickpea of 0.9 million tonnes worldwide due to salinity.

Types of salinity

Salt-affected soils can be divided into the following groups: saline (dominantly Na_2SO_4 and NaCl, seldom $NaNO_3$), alkaline (mainly $NaCO_3$ and $NaHCO_3$, seldom Na_2SiO_3 and $NaHSiO_3$), gypsifer (mainly $CaSO_4$ and seldom $CaCl_2$), magnesium (magnesium ions) and acid sulphate ($Al_2(SO_4)_3$ and $Fe_2(SO_4)_3$) (Szabolcs, 1994).

Impact of salinity

The deleterious effects of salinity on plant growth are associated with (i) water stress (low osmotic potential of soil solution); (ii) nutrient ion imbalance; (iii) salt stress (specific ion effects); and (iv) a combination of all these factors (Ashraf and Harris, 2004). All these factors cause adverse pleiotropic effects on plant growth and development at physiological, biochemical, molecular and whole-plant levels.

Germination, flower and pod production, and gamete development of chickpea are adversely affected by salinity, leading to severe yield loss. Genetic variations for salinity in chickpea have been demonstrated and evaluated both in the cultigen and wild species (Jaiwal *et al.*, 1983; Lauter and Munns, 1986; Dhingra and Varghese, 1993; Dua and Sharma, 1997; Katerji *et al.*, 2001). Progressive delays in germination of chickpea occur with increasing levels of salinity (Jaiwal *et al.*, 1983). After germination, the first signs of salinity damage are usually necrosis of the outer margins of leaves followed by yellowing of the older leaves. Leaves will eventually abscise and die due to excess ion accumulation (Saxena *et al.*, 1993a). Salinity increases anthocyanin pigmentation in leaves and stems in desi chickpea while leaves and stems of the kabuli variety appear yellow.

Lauter and Munns (1986) found that only one genotype (L-550) survived at 50 mM chloride treatment after screening 160 chickpea genotypes. Sodium concentration in shoots of the least sensitive genotype was less than in the most sensitive genotypes. In contrast, potassium levels were highest in shoots of the least sensitive genotype. Ashraf and Waheed (1993) observed reductions in fresh and dry weights of leaves and roots in a range of genotypes treated with 40 mol/m^3 NaCl. Dua and Sharma (1997) reported that genotypes CSG 8977, CSG 8962 and CSG 8943 had the best yield under saline conditions. Genotypic variation for salinity tolerance was attributed to differences in uptake and distribution of sodium and chloride ions (Dua and Sharma, 1997). Katerji *et al.* (2001) found that FLIP 89–57C (drought-tolerant) senesced and flowered earlier, whereas ILC 3279 (drought-sensitive) responded differently with new leaf growth and flowers, and a delay in senescence. Salinity reduces plant height, leaf number, leaf, stem and root dry weights, and seed emergence (Esechie *et al.*, 2002; Welfare *et al.*, 2002).

Resistance mechanisms

The principal features of tolerance at the cellular level have been accepted (Gorham *et al.*, 1985):

1. Large quantities of salts (mainly sodium chloride) absorbed into leaves contributing to osmotic adjustment are mainly accumulated in vacuoles under saline conditions when tissue concentrations exceed 200 mol/m^3 (0.9 MPa).
2. Concentrations of inorganic ions in the cytoplasm, especially in meristematic cells, range from 100 to 200 mol/m^3, and the cytoplasm has strong selectivity for potassium over sodium, magnesium over calcium, and phosphate over chloride or nitrate.
3. Maintenance of osmotic adjustment across the tonoplast requires accumulation in the cytoplasm of non-toxic organic solutes (compatible solutes) under hyperosmotic conditions (0.9 MPa).

The ability of plants to regulate the influx of ions is one of the major factors determining salt tolerance. The most salt-tolerant species have high internal salt concentrations (Gorham *et al.*, 1985), which suggests that this is at least as important as the ability to restrict accumulation. The expression of ion concentrations on a tissue water basis appears to be more useful than on a dry weight basis to screen for salinity tolerance in chickpea (Sharma and Kumar, 1992).

Although screening methods in the field have been reported for selection of salt-tolerant chickpea (Saxena *et al.*, 1993a), its routine use in breeding programmes seems to be very limited. This is due to the complex nature of salinity. Moreover, field screening for salinity tolerance requires considerable time, labour and other resources, and it is difficult to separate environmental effects from genetic variations. A number of other non-field-screening methods are available for selection of salt-tolerant chickpea (Epitalawage *et al.*, 2003). Several criteria have been used to assess salinity tolerance including cell survival, germination, dry matter accumulation, leaf death and senescence, ion concentrations (ratio Na^+/K^+ or K^+/Na^+), leaf necrosis, osmoregulation and yield. In conclusion, no single selection criterion has been found for salinity tolerance.

Nitrogen Fixation

N fixation is recognized as a process of great agronomic importance, and a variety of leguminous plants and some non-leguminous plants can obtain their N from the atmosphere by association with symbiotic bacteria (members of the *Rhizobiaceae* family). Chickpea in symbiosis with an efficient strain of *Mesorhizobium ciceri* constitutes an important component of cropping systems, and is capable of supplying between 80 and 120 kg N/ha to the soil (Herridge *et al.*, 1995). The selection and characterization of salt-tolerant strains with efficient symbiotic performance may be a strategy to improve *C. arietinum–M. ciceri* symbiosis in adverse environments.

Soil factors affecting legume production and survival of *Rhizobium* spp. include extremes of soil pH: acidity, alkalinity and salinity (Slattery *et al.*, 2004). Grain legumes are particularly vulnerable because of their low salinity tolerance and high sensitivity of symbiotic N fixation to stress, e.g. infection of root hairs by rhizobia and subsequent nodule development.

Mesorhizobium ciceri and salinity

There are ten *Mesorhizobium* spp. (Bacterial nomenclature up-to-date from November, 2005). Most of these species have a wide host range, capable of nodulating more than one legume. Cross-inoculation studies of Gaur and Sen (1979) revealed that chickpea rhizobia are highly host-specific (only *M. ciceri* and *M. mediterraneum* nodulate chickpea). *M. ciceri* actively respond to variations in osmolality including transient adjustments in ionic balance and pronounced changes in the metabolism of cytoplasmic low-molecular-weight compounds. Two methods of adaptation to high osmolality are: (i) accumulation of very high intracellular concentration of ions – a strategy enlisted by bacteria whose entire physiology is adapted to a high-saline environment; and (ii) intracellular amassing of osmotically active solutes that are highly congruous with cellular functions (Soussi *et al.*, 2001a). Osmoprotectants are defined as organic solutes that enhance bacterial growth in media of high osmolality. These substances may themselves be compatible solutes with metabolic functions or they may act as precursor molecules that can be enzymatically converted into these compounds to help in osmoregulation and prevention of cytoplasmic dehydration in saline environments. In this regard, Soussi *et al.* (2001b) found a proline accumulation, even higher than glutamate, in *M. ciceri* under salt stress.

Nitrogen fixation–salinity interaction

Salt tolerance of symbiotic N fixation depends on both the plant and the rhizobial genotype. It is necessary to identify host cultivars, strains of *Mesorhizobium* and combinations of both that have superior N fixation capacity.

Studies on cultivar–strain interactions show that the plant is the determining factor for symbiosis tolerance (Soussi *et al.*, 1999). Chickpea, an amide producer, is relatively tolerant to salt stress in terms of N fixation compared with ureide producers; nevertheless, salt affects its growth, nodulation and nitrogenase activity, photosynthesis, and carbon metabolism in root nodules (Soussi *et al.*, 1999). Salinity levels inhibiting symbiosis between chickpea and *M. ciceri* differ from those inhibiting growth of the individual symbiont. Salt stress influences all aspects of nodulation and symbiotic N fixation, including reducing rhizobial survival, nodulation and nitrogenase activity. It is often difficult to isolate the effects of salinity on successful inoculation from symbiotic functioning and N fixation. Several studies have shown that N fixation of nodules is more extensively affected by salinization than plant growth (Soussi *et al.*, 1999; Tejera *et al.*, 2006). In Tunisia, chickpea grown in saline soil had fewer nodules per plant, which were smaller and apparently ineffective compared with the controls (Abdelly *et al.*, 2005).

Although salinity effects on N fixation have been extensively studied in several legume species, the physiological mechanisms involved are poorly understood. One hypothesis to explain the inhibition of nitrogenase activity in chickpea nodules under salt stress is related to the lack of photosynthate supply to nodules. Most comparative studies on chickpea cultivars differing in salinity tolerance reveal a higher inhibition of plant growth, photosynthesis, nodulation and N fixation in the salt-sensitive cultivar compared with salt-tolerant cultivars. Photosynthesis is inhibited by salt stress since leaf chlorophyll and Rubisco activity are reduced; however, the limitation in photosynthate supply to nodules does not appear to inhibit nitrogenase activity, since the stress promotes an accumulation of soluble sugars (Soussi *et al.*, 1998). In addition, a reduction in nitrogenase activity under salt stress has been associated with a decrease in sucrose synthase expression (Gordon *et al.*, 1997) and with the enzymatic inhibition of sucrose breakdown activities (Soussi *et al.*, 1999). At the same time, alkaline invertase activity increases in salt-tolerant cultivars, which presumably compensates for the lack of the sucrose synthase hydrolytic activity (Soussi *et al.*, 1999).

It is possible that the supply of malate to bacteroids is limited under saline conditions. The production of malate in legume nodules is mediated by the phosphoenolpyruvate carboxylase (PEPC) and malate dehydrogenase activities. In general, malate concentration declines under salt stress; however, it has increased in salt-tolerant chickpea cultivars (Soussi *et al.*, 1999). On the other hand, PEPC activity always increases in nodules of different legume species (pea, lucerne and chickpea) in such conditions.

Chickpea grown in soils with high salt concentrations has reduced tissue osmotic potential with the accumulation of inorganic and organic solutes. Such a response is considered an adaptation of plants against osmotic stress (Tejera *et al.*, 2006); therefore, accumulation of proline, amino acids and carbohydrates in leaves has been related to salt, and in nodules, protein content was boosted by salt (Soussi *et al.*, 1998). Thus, accumulation of some compatible solutes could be considered a consequence of damage produced by salt stress rather than a protective strategy (Soussi *et al.*, 1998). Exposure to NaCl increases Na^+ and Cl^- content of all chickpea tissues and cultivars, but it is higher in roots and nodules than in shoots (Soussi *et al.*, 1998; Tejera *et al.*, 2006). Baalbaki *et al.* (2000) postulated the involvement of two physiological mechanisms to reduce the damage associated with excessive sodium levels in soils: (i) sodium compartmentalization in roots; and (ii) K/Na selectivity. Chickpea excluded and/or accumulated sodium in roots, resulting in a decrease of root osmotic potential and a limitation of sodium toxicity in shoots (Tejera *et al.*, 2006).

Screening of tolerant N-fixing chickpea genotypes may help to improve the fertility of saline soils. Selection of indicators of salinity tolerance under symbiotic conditions as well as screening of available germplasm to identify tolerant varieties is useful to enhance the productivity of chickpea in areas adversely affected by salt stress (Abdelly *et al.*, 2005). In a recent review, Sessitsch *et al.* (2002) summarized technologies and strategies for selecting rhizobial strains by more closely examining the complex interaction between the edaphic environment with genotypes of both the legume and its microsymbiont. Different physiological and nutritional indicators have been proposed for selection of chickpea salt-tolerant genotypes including nodulation capacity (Garg and Singla, 2004), higher root/shoot ratio, normalized nodule weight, shoot K/Na ratio and reduced foliar accumulation of Na (Tejera *et al.*, 2006).

Waterlogging Tolerance

Waterlogging is considered to be a major cause of reduced yields on fine-textured and duplex soils in the major chickpea-producing regions in Australia (Siddique, 2000). Among cool-season food legumes, faba bean is more tolerant to waterlogging than lentil, field pea and chickpea (Siddique, 2000). Waterlogging reduces germination, seedling emergence, root and shoot growth, and plant density by up to 80%, in addition to being conducive to seedling diseases. Yield losses vary up to almost 100%. Management practices to reduce the effects of waterlogging include paddock selection, sowing time, seeding rate and drainage. Genetic variation to waterlogging tolerance in chickpea has not been reported and deserves attention.

Nutrient Deficiency and Toxicity

Micronutrient deficiencies and toxicities in chickpea result in yield losses estimated to be ~360,000 t/year worldwide (Ryan, 1997). Micronutrients of interest in chickpea are iron (Fe), molybdenum (Mo), and zinc (Zn) deficiencies and boron (B) toxicity. Nutrient deficiencies or toxicities have been reported to

Table 23.1. The effects of micronutrient deficiencies and toxicities in chickpea.

Micronutrient	Symptom	Reference
Fe	Interveinal chlorosis in younger leaves, uniform yellow-green colour	Saxena *et al*., 1990; Srinivasarao *et al*., 2003; Smith and Pieters, 2005
Mo	Chlorotic interveinal mottling of older leaves, leaf wilting	Srinivasarao *et al*., 2003; Nautiyal and Chatterjee, 2004
Zn	Pale green followed by red-brown pigmentation, reduced internode, rosette type growth	Srinivasarao *et al*., 2003; Smith and Pieters, 2005
B	Yellowing of tips and serrated margins of leaflets	Srinivasarao *et al*., 2003; Smith and Pieters, 2005

cause yield losses of varying magnitude in chickpea (Table 23.1), e.g. 22–50% due to Fe deficiency and up to 100% due to B toxicity (Ali *et al*., 2002).

Ideotypes for Stress Conditions

Many chickpea ideotypes have been proposed (Siddique and Sedgley, 1985; Khanna-Chopra and Sinha, 1987; Singh, 1987; Saxena and Johansen, 1990; Saxena *et al*., 1997). Ideotypes have been defined in terms of morphological and physiological traits, sowing time, and low-input as well as high-input conditions. Farmers' and consumers' requirements were of little consideration. We have defined ideotypes of chickpea in relation to major abiotic stresses in target environments (Table 23.2).

Future Strategies

Worldwide there have been limited breeding efforts in chickpea compared with major cereal and oilseed crops. Primitive landraces still account for much of the crop area in developing countries despite new varieties being released by national programmes and major international centres (ICRISAT and ICARDA). Negative effects of abiotic stresses can be minimized using both genetic and agronomic approaches. The world's largest chickpea collections in ICRISAT, ICARDA and other national gene banks have been screened for resistance to various abiotic and biotic stresses. Apart from a few stress-resistant characteristics in the cultigen, desirable sources of resistance have not been found in the collections. The wild *Cicer* spp., especially those easily crossable (*C. reticulatum* and *C. echinospermum*) with the cultigen, appear to be important gene sources for resistance to abiotic stresses.

Recent advances in biotechnology, linkage mapping, marker-assisted breeding, embryo rescue techniques and transformation technology hold much promise for enhancing chickpea yield under various abiotic stresses.

Table 23.2. Chickpea ideotypes for tolerance to major abiotic stresses in target environments.

Target stresses	Morphologic traits	Physiologic traits	Farmer's requirements
1. Drought (a) Terminal (b) Intermittent	1. Very early flowering and early maturing (in terminal drought) 2. Early seedling vigour 3. Fast groundcover 4. High root length density 5. Development plasticity (in intermittent drought) 6. Tiny leaflets and small leaves 7. At least two flowers per node	1. High water use efficiency 2. High stomatal frequency 3. High membrane stability 4. High chlorophyll content 5. High photosynthetic capacity 6. High water potential 7. High transpiration efficiency 8. High turgor capacity 9. Remobilization of assimilates 10. High malic acid content 11. High pollen viability 12. High fertilization capacity	1. High and stable seed yield 2. Resistance to local diseases and pests 3. High N_2 fixation capacity 4. Extra large-seeded kabuli types (for some Mediterranean countries and Americas) 5. Medium-seeded kabuli types (for some Mediterranean countries) 6. Desi types (generally Asia, Australia and WANA) 7. Iron deficiency-resistant 8. Suitable plant height or combine harvest 9. High yield potential 10. Herbicide-resistant
2. Heat (a) Heat shock (b) Moderate heat	1. Short-duration phenology (to escape heat shock) 2. Developmental plasticity (to avoid moderate heat) 3. Tiny leaflets and small leaves	1. Same as above	
3. Cold (a) Freezing (b) Chilling	1. Rosette type at seedling stage for freezing 2. Dark green foliage colour for freezing 3. Freezing-tolerant at vegetative stage (up to –15°C without snow) 4. Extra early to escape chilling 5. Chilling-tolerant at flowering stage (up to –1.5°C) 6. Two or more flowers per node and multiseeds per pod	1. Low lethal leaf water potential 2. High chlorophyll content 3. Suitable osmoregulation 4. High pollen viability 5. High fertilization capacity 6. Remobilization of assimilates 7. High malic acid content	

References

Abdelly, C., Krouma, A. and Drevon, J.J. (2005) Nitrogen fixation and yield of chickpea in saline Mediterranean zones. *Grain Legumes* 42, 16–17.

Ali, M.Y., Krishnamurthy, L., Saxena, N.P., Rupela, O.P., Kumar, J. and Johansen, C. (2002) Scope for genetic manipulation of mineral acquisition in chickpea. *Plant and Soil* 245, 123–134.

Anbessa, Y. and Bejiga, G. (2002) Evaluation of Ethiopian chickpea landraces for tolerance to drought. *Genetic Resources and Crop Evolution* 49, 557–564.

Ashraf, M. and Waheed, A. (1993) Responses of some genetically diverse lines of chickpea (*Cicer arietinum* L.) to salt. *Plant and Soil* 154, 257–266.

Ashraf, M. and Harris, P.J.C. (2004) Potential biochemical indicators of salinity tolerance in plants. *Plant Science* 166, 3–16.

Auld, D.L., Bettis, B.L., Crock, J.L. and Kephart, K.D. (1988) Planting date and temperature effects on germination, emergence, and seed yield of chickpea. *Agronomy Journal* 80, 909–914.

Baalbaki, R.Z., Zurayak, R.A., Adlan, M.A.M. and Saxena, C.M. (2000) Effect of nitrogen source and salinity levels on salt accumulation of two chickpea genotypes. *Journal of Plant Nutrition* 23, 805–814.

Bacterial nomenclature up-to-date (2005) Genus *Mesorhizobium*. Available at: http://www.dsmz.de/bactnom/nam5731.htm

Baker, R.J. (1994) Breeding methods and selection indices for improved tolerance to biotic and abiotic stresses in cool season food legumes. In: Muehlbauer, F.J. and Kaiser, W.J. (eds) *Expanding the Production and Use of Cool Season Food Legumes*. Kluwer Academic, Dordrecht, The Netherlands, pp. 429–438.

Basu, P.S. and Singh, D.N. (2003) Physiology and abiotic stresses in chickpea. In: Ali, M., Kumar, S. and Singh, N.B. (eds) *Chickpea Research in India*. Indian Institute of Pulses Research, Kanpur, India, pp. 137–166.

Blum, A. (1988) *Plant Breeding for Stress Environments*. CRC Press, Boca Raton, Florida.

Clarke, H.J. and Siddique, K.H.M. (2004) Response of chickpea genotypes to low temperature stress during reproductive development. *Field Crops Research* 90, 323–334.

Clarke, H.J., Khan, T.N. and Siddique, K.H.M. (2004) Pollen selection for chilling tolerance at hybridization leads to improved chickpea cultivars. *Euphytica* 139, 65–74.

Croser, J.S., Clarke, H.J., Siddique, K.H.M. and Khan, T.N. (2003) Low-temperature stress: implications for chickpea (*Cicer arietinum* L.) improvement. *Critical Review in Plant Science* 22, 185–219.

Davies, J. (1995) *Plant Hormones: Physiology, Biochemistry and Molecular Biology*. Kluwer Academic, Dordrecht, The Netherlands.

Davies, S.L., Turner, N.C., Siddique, K.H.M., Plummer, J.A. and Leport, L. (1999) Seed growth of desi and kabuli chickpea (*Cicer arietinum* L.) in a short-season Mediterranean-type environment. *Australian Journal of Experimental Agriculture* 39, 181–188.

Dhingra, H.R. and Varghese, T.M. (1993) Flowering and male reproductive functions of chickpea (*Cicer arietinum* L.) genotypes as affected by salinity. *Biologia Plantarum* 35, 447–452.

Dua, R.P. and Sharma, S.K. (1997) Suitable genotypes of gram (*Cicer arietinum* L.) and mechanism of their tolerance to salinity. *Indian Journal of Agricultural Sciences* 67, 440–443.

Epitalawage, N., Eggenberg, P. and Strasser, R.J. (2003) Use of fast chlorophyll a fluorescence technique in detecting drought and salinity tolerant chickpea (*Cicer arietinum* L.) varieties. *Archives des Sciences* 56, 79–93.

Esechie, H.A., Al-Saidi, A. and Al-Khanjari, S. (2002) Effect of sodium chloride salinity on seedling emergence in chickpea. *Journal of Agronomy and Crop Science* 188, 155–160.

Fischer, R.A. and Maurer, R. (1978) Drought resistance in spring wheat cultivars. I. Grain yield responses. *Australian Journal of Agricultural Research* 29, 897–912.

Garg, N. and Singla, R. (2004) Growth, photosynthesis, nodule nitrogen and carbon fixation in the chickpea cultivars under salt stress. *Brazilian Journal of Plant Physiology* 16, 137–146.

Gaur, Y.D. and Sen, A.N. (1979) Cross inoculation group specificity in *Cicer–Rhizobium* symbiosis. *New Phytologist* 83, 745–754.

Gordon, A.J., Minchin, F.R., Skot, L. and James, C.L. (1997) Stress-induced declines in soybean N_2 fixation are related to nodule sucrose synthase activity. *Plant Physiology* 114, 937–946.

Gorham, J., Wyn Jones, R.G. and McDonnel, E. (1985) Some mechanisms of salt tolerance in crop plants. *Plant and Soil* 89, 15–40.

Herridge, D.F., Marcellos, H., Felton, W.L., Turner, G.L. and Peoples, M.B. (1995) Chickpea increases soil-N fertility in cereal systems through nitrate sparing and N-fixation. *Soil Biology and Biochemistry* 27, 545–551.

Jaiwal, P.K. and Mehta, K. (1983) Thermosensitivity of pollen behaviour in *Cicer arietinum* L. *International Chickpea Newsletter* 9, 15.

Jaiwal, P.K., Bhambie, S. and Mehta, K. (1983) Effects of salinity on seed germination and seedling growth of chickpea. *International Chickpea Newsletter* 9, 15–16.

Johansen, C., Krishnamurthy, L., Saxena, N.P. and Sethi, S.C. (1994) Genotypic variation in moisture response of chickpea grown under line-source sprinklers in a semi-arid tropical environment. *Field Crops Research* 37, 103–112.

Kashiwagi, J., Krishnamurthy, L., Upadhyaya, H.D., Krishna, H.S., Vandez, C.V. and Serraj, R. (2005) Genetic variability of drought-avoidance root traits in the mini-core germplasm collection of chickpea (*Cicer arietinum* L.). *Euphytica* 146, 213–222.

Kashiwagi, J., Krishnamurthy, L., Crouch, J.H. and Serraj, R. (2006) Variability of root length density and its contributions to seed yield in chickpea (*Cicer arietinum* L.) under terminal drought stress. *Field Crops Research* 95, 171–181.

Katerji, N., van Hoorn, J.W., Hamdy, A., Mastrorilli, M., Oweis, T. and Malhotra, R.S. (2001) Response to soil salinity of two chickpea varieties differing in drought tolerance. *Agricultural Water Management* 50, 83–96.

Khanna-Chopra, R. and Sinha, S.K. (1987) Chickpea: physiological aspects of growth and yield. In: Saxena, M.C. and Singh, K.B. (eds) *The Chickpea*. CAB International, Wallingford, UK, pp. 163–189.

Krishnamurthy, L., Ito, O., Johansen, C. and Saxena, N.P. (1998) Length to weight ratio of chickpea roots under progressively receding soil moisture conditions in a vertisol. *Field Crops Research* 58, 177–185.

Kumar, J. and Abbo, S. (2001) Genetics of flowering time in chickpea and its bearing on productivity in semiarid environments. *Advances in Agronomy* 72, 107–138.

Kumar, J. and Rao, B.V. (2001) Registration of ICCV 96029, a super early and double podded chickpea germplasm. *Crop Science* 41, 606–606.

Kumar, J., Sethi, S.C., Johansen, C., Kelly, T.G., Rahman, M.M. and van Rheenen, H.A. (1996) Potential of short-duration chickpea varieties. *Journal of Dryland Agricultural Research and Development* 11, 28–32.

Kurdali, F. (1996) Nitrogen and phosphorus assimilation, mobilization and partitioning in rainfed chickpea (*Cicer arietinum* L.). *Field Crops Research* 47, 81–92.

Laurie, S. and Stewart, G.R. (1993) Effects of nitrogen supply and high-temperature on the growth and physiology of the chickpea. *Plant Cell and Environment* 16, 609–621.

Lauter, D.J. and Munns, D.N. (1986) Salt resistance in chickpea genotypes in solutions salinized with NaCl or Na_2SO_4. *Plant and Soil* 95, 271–279.

Lawn, R.J. (1982) Responses of four grain legumes to water stress in south eastern Australia. I. Physiological response mechanisms. *Australian Journal of Agricultural Research* 33, 481–496.

Lecoeur, J., Wery, J. and Turc, O. (1992) Osmotic adjustment as a mechanism of dehydration postponement in chickpea (*Cicer arietinum* L.) leaves. *Plant and Soil* 144, 177–189.

Leport, L., Turner, N.C., French, R.J., Tennant, D., Thomson, B.D. and Siddique, K.H.M.

(1998) Water relation, gas-exchange, and growth of cool-season grain legumes in a Mediterranean-type environment. *European Journal of Agronomy* 9, 295–303.

Leport, L., Turner, N.C., French, R.J., Barr, M.D., Duda, R., Davies, S.L., Tennant, D. and Siddique, K.H.M. (1999) Physiological response of chickpea genotypes to terminal drought in a Mediterranean-type environment. *European Journal of Agronomy* 11, 279–291.

Leport, L., Turner, N.C., Davies, S.L. and Siddique, K.H.M. (2005) Variation in pod production and abortion among chickpea cultivars under terminal drought. *European Journal of Agronomy* 24, 236–246.

Lindlow, S.E. (1980) Population dynamics of epiphytic ice nucleation active bacteria on frost sensitive plants and frost control by means of antagonistic bacteria. In: Li, P.H. and Sakai, A. (eds) *Plant Cold Hardiness and Freezing Stress. Mechanisms and Crop Implications*, Vol. 2. Academic Press, New York, pp. 395–416.

Ludlow, M.M. and Muchow, R.C. (1990) A critical evaluation of traits for improving crop yields in water-limited environments. *Advances in Agronomy* 43, 107–153.

Malhotra, R.S. and Saxena, M.C. (1993) Screening for cold and heat tolerance in cool-season food legumes. In: Singh, K.B. and Saxena, M.C. (eds) *Breeding for Tolerance in Cool Season Food Legumes*. Wiley, Chichester, UK, pp. 227–244.

Malhotra, R.S. and Singh, K.B. (1991) Gene action for cold tolerance in chickpea. *Theoretical and Applied Genetics* 82, 598–601.

Malhotra, R.S. and Singh, K.B. (1995) Effects of nitrogen fertilizer application on cold tolerance in chickpea. *International Chickpea and Pigeonpea Newsletter* 2, 24–25.

Malhotra, R.S., Singh, K.B. and Saxena, M.C. (1990) Effect of depth of sowing on cold tolerance in chickpea. *International Chickpea Newsletter* 22, 19–21.

Moinuddin, and Khanna-Chopra, R. (2003) Osmotic adjustment in chickpea in relation to seed yield and yield parameters. *Crop Science* 44, 449–455.

Morgan, J.M., Rodriguez-Maribona, B. and Knights, E.J. (1991) Adaptation to water-deficit in chickpea breeding lines by osmoregulation: relationship to grain yields in the field. *Field Crops Research* 27, 61–70.

Muchow, R.C. (1985) Stomatal behaviour in grain legumes grown under different soil water regimes in a semi-arid tropical environment. *Field Crops Research* 11, 291–307.

Nautiyal, N. and Chatterjee, C. (2004) Molybdenum stress-induced changes in growth and yield of chickpea. *Journal of Plant Nutrition* 27, 173–181.

Nayyar, H. and Chander, K. (2004) Protective effects of polyamines against oxidative stress induced by water and cold in chickpea. *Journal of Agronomy and Crop Science* 190, 355–365.

Nayyar, H., Bains, T. and Kumar, S. (2005a) Chilling stressed chickpea seedlings: effect of cold acclimation, calcium and abscisic acid on cryoprotective solutes and oxidative damage. *Environmental and Experimental Botany* 54, 275–285.

Nayyar, H., Chander, K., Kumar, S. and Bains, T. (2005b) Glycine betaine mitigates cold stress damage in chickpea. *Agronomy for Sustainable Development* 25, 381–388.

Nayyar, H., Kaur, S., Smita, Singh, K.J., Dhir, K.K. and Bains, T. (2005c) Water stress-induced injury to reproductive phase in chickpea: evaluation of stress sensitivity in wild and cultivated species in relation to abscisic acid and polyamines. *Journal of Agronomy and Crop Science* 191, 450–457.

Robertson, L.D., Singh, K.B. and Ocampo, B. (1995) *A Catalog of Annual Cicer Species*. ICARDA, Aleppo, Syria.

Ryan, J.G. (1997) A global perspective on pigeonpea and chickpea sustainable production systems: present status and future potential. In: Asthana, A.N. and Ali, M. (eds) *Recent Advantages in Pulses Research*. Indian Society of Pulses Research and Development, IIPR, Kanpur, India, pp. 1–31.

Sabaghpour, S.H., Kumar, J. and Rao, T.N. (2003) Inheritance of growth vigour and its association with other characters in chickpea. *Plant Breeding* 122, 542–544.

Savithri, K.S., Ganapathy, P.S. and Sinha, S.K. (1980) Sensitivity to low temperature in pollen germination and fruit-set in *Cicer arietinum* L. *Journal of Experimental Botany* 31, 475–481.

Saxena, M.C., Saxena, N.P. and Mohamed, A.K. (1988) High temperature stress. In: Summerfield, R.J. (ed.) *World Crops: Cool Season Food Legumes*. Kluwer Academic, Dordrecht, The Netherlands, pp. 845–856.

Saxena, M.C., Malhotra, R.S. and Singh, K.B. (1990) Iron-deficiency in chickpea in the Mediterranean region and its control through resistant genotypes and nutrient application. In: El-Bassam, N., Dambroth, M. and Loughman, B.C. (eds) *Genetic Aspects of Mineral Nutrition*. Kluwer Academic, Dordrecht, The Netherlands, pp. 381–384.

Saxena, N.P. (2003) Management of drought in chickpea: a holistic approach. In: Saxena, N.P. (ed.) *Management of Agricultural Drought: Agronomic and Genetic Options*. Oxford and IBH Publishing, New Delhi, pp. 103–122.

Saxena, N.P. and Johansen, C. (1990) Chickpea ideotypes for genetic enhancement of yield and yield stability in South Asia. In: van Rheenen, H.A. and Saxena, M.C. (eds) *Chickpea in the Nineties: Proceedings of the 2nd International Workshop on Chickpea Improvement*. ICRISAT, Hyderabad, India, pp. 81–85.

Saxena, N.P., Johansen, C., Saxena, M.C. and Silim, S.N. (1993a) Selection for drought and salinity tolerance in cool-season food legumes. In: Singh, K.B. and Saxena, M.C. (eds) *Breeding for Tolerance in Cool Season Food Legumes*. Wiley, Chichester, UK, pp. 245–270.

Saxena, N.P., Krishnamurthy, L. and Johansen, C. (1993b) Registration of drought-resistant chickpea germplasm. *Crop Science* 33, 1424.

Saxena, N.P., Saxena, M.C. and Johansen, C. (1997) Chickpea ideotypes for low and high-input conditions. In: Asthana, A.N. and Ali, M. (eds) *Recent Advantages in Pulses Research*. Indian Society of Pulses Research and Development, IIPR, Kanpur, India, pp. 217–231.

Serraj, R. and Sinclair, T.R. (2002) Osmolyte accumulation: can it really help increase crop yield under drought conditions? *Plant Cell and Environment* 25, 335–341.

Serraj, R., Bidinger, F.R., Cauhan, Y.S., Seetharama, N., Nigam, S.N. and Saxena, N.P. (2003) Management of drought in ICRISAT cereal and legume mandate crops. In: Kijne, J.W., Barker, R. and Molden, D. (eds) *Water Productivity in Agriculture: Limits and Opportunities for Improvement*. CAB International, Wallingford, pp. 127–144.

Serraj, R., Krishnamurthy, L., Kashiwagi, J., Kumar, J., Chandra, S. and Crouch, J.H. (2004) Variation in root traits of chickpea (*Cicer arietinum* L.) grown under terminal drought. *Field Crops Research* 88, 115–127.

Sessitsch, A., Howieson, J.G., Perret, X., Antoun, H. and Martinez-Romero, E. (2002) Advances in *Rhizobium* research. *Critical Review in Plant Science* 21, 323–378.

Shan, F., Clarke, H.C., Plummer, J.A., Yan, G. and Siddique, K.H.M. (2005) Geographical patterns of genetic variation in the world collections of wild annual *Cicer* characterized by amplified fragment length polymorphisms. *Theoretical and Applied Genetics* 110, 381–391.

Sharma, R.A. (1985) Influence of drought stress on emergence and growth of chickpea seedlings. *International Chickpea Newsletter* 12, 15–16.

Sharma, S.K. and Kumar, S. (1992) Effect of salinity on Na^+, K^+ and Cl^- content in different organs of chickpea and the basis of ion expression. *Biologia Plantarum* 34, 311–317.

Siddique, K.H.M. (2000) Understanding growth stresses of cool season pulses. In: O'Connell, L. (ed.) *Shared Solutions Cropping Manual*, Australian Grain. Berekua, Brisbane, pp. 12–16.

Siddique, K.H.M. and Loss, S.P. (1999) Studies on sowing depth for chickpea (*Cicer arietinum* L.), faba bean (*Vicia faba* L.) and lentil (*Lens culinaris* Medik.) in a Mediterranean-type environment of south-western Australia. *Journal of Agronomy and Crop Science* 182, 105–112.

Siddique, K.H.M. and Sedgley, R.H. (1985) The effect of reduced branching on yield and water use of chickpea (*Cicer arietinum* L.) in a Mediterranean-type environment. *Field Crops Research* 12, 251–269.

Siddique, K.H.M., Brinsmead, R.B., Knight, R., Knights, E.J., Paull, J.G. and Rose, I.A. (2000) Adaptation of chickpea (*Cicer arietinum* L.) and faba bean (*Vicia faba* L.) to Australia. In: Knight, R. (ed.) *Linking Research and Marketing Opportunities for Pulses in the 21st Century*. Kluwer Academic, Dordrecht, The Netherlands, pp. 289–303.

Siddique, K.H.M., Regan, K.L., Tennant, D. and Thomson, B.D. (2001) Water use and water use efficiency of cool season grain legumes in low rainfall Mediterranean-type environment. *European Journal of Agronomy* 15, 267–280.

Silim, S.N. and Saxena, M.C. (1993) Adaptation of spring-sown chickpea to the Mediterranean basin. I. Response to moisture supply. *Field Crops Research* 34, 121–136.

Singh, K.B. (1987) Chickpea breeding. In: Saxena, M.C. and Singh, K.B. (eds) *The Chickpea*. CAB International, Wallingford, UK, pp. 127–162.

Singh, K.B., Malhotra, R.S., Halila, M.H., Knights, E.J. and Verma, M.M. (1994) Current status and future strategy in breeding chickpea for resistance to biotic and abiotic stresses. *Euphytica* 73, 137–149.

Singh, K.B., Omar, M., Saxena, M.C. and Johansen, C. (1997) Screening for drought resistance in spring chickpea in the Mediterranean region. *Journal of Agronomy and Crop Science* 178, 227–235.

Slattery, J., Siddique, K.H.M. and Howieson, J. (2004) Cool-season grain legumes production and rhizboial interactions in Australian dryland agriculture. In: Rao, S.C. and Ryan, J. (eds) *Challenges and Strategies for Dryland Agriculture*. Crop Science Society of America, Madison, Wisconsin, pp. 1–14.

Smith, F.W. and Pieters, W.H.J. (2005) *Foliar Symptoms of Nutrient Disorders in Chickpea* (Cicer arietinum *L.*). Available at: http://www2.icrisat.org

Soltani, A., Khooie, F.R., Ghassemi-Golezani, K. and Moghaddam, M. (2000) Thresholds for chickpea leaf expansion and transpiration response to soil water deficit. *Field Crops Research* 68, 205–210.

Soussi, M., Ocaña, A. and Lluch, C. (1998) Effects of salt stress on growth, photosynthesis and nitrogen fixation in chickpea (*Cicer arietinum* L.). *Journal of Experimental Botany* 49, 1329–1337.

Soussi, M., Lluch, C. and Ocaña, A. (1999) Comparative study of nitrogen fixation and carbon metabolism in two chickpea (*Cicer arietinum* L.) cultivars under salt stress *Journal of Experimental Botany* 50, 1701–1708.

Soussi, M., Khadri, M., Lluch, C. and Ocaña, A. (2001a) Effects of salinity on protein and lipopolysaccharide pattern in a salt-tolerant strain of *Mesorhizobium ciceri*. *Journal of Applied Microbiology* 90, 476–481.

Soussi, M., Khadri M., Lluch, C. and Ocaña, A. (2001b) Carbon metabolism and bacteroid respiration in nodules of chickpea (*Cicer arietinum* L.) plants grown under saline conditions. *Plant Biosystems* 135, 157–164.

Srinivasan, A., Saxena, N.P. and Johansen, C. (1999) Cold tolerance during early reproductive growth of chickpea (*Cicer arietinum* L.): genetic variation in gamete development and function. *Field Crops Research* 60, 209–222.

Srinivasarao, C.H., Ganeshamurthy, A.N. and Ali, M. (2003) *Nutritional Constraints in Pulse Production*. Indian Institute of Pulses Research, IIPR, Kanpur, India.

Summerfield, R.J., Hadley, P., Roberts, E.H., Minchin, F.R. and Rawsthorne, S. (1984) Sensitivity of chickpeas (*Cicer arietinum* L.) to hot temperatures during the reproductive period. *Experimental Agriculture* 20, 77–93.

Szabolcs, I. (1994) Soils and salinization. In: Pessarakli, M. (ed.) *Handbook of Plant and Crop Stress*. Marcel Dekker, New York, pp. 3–11.

Tejera, N., Soussi, M. and Lluch, C. (2006) Physiological and nutritional indicators of tolerance to salinity in chickpea plants growing under symbiotic conditions. *Environmental and Experimental Botany* 58, 17–24.

Testerink, C. and Munnik, T. (2005) Phosphatidic acid: a multifunctional stress signaling lipid in plants. *Trends in Plant Science* 10, 368–375.

Thomson, B.D. and Siddique, K.H.M. (1997) Grain legume species in low rainfall Mediterranean-type environments. II. Canopy development, radiation interception, and dry-matter production. *Field Crops Research* 54, 189–199.

Toker, C. (2005) Preliminary screening and selection for cold tolerance in annual wild *Cicer* species. *Genetic Resources and Crop Evolution* 52, 1–5.

Toker, C. and Cagirgan, M.I. (1998) Assessment of response to drought stress of chickpea (*Cicer arietinum* L.) lines under rainfed conditions. *Turkish Journal of Agriculture and Forestry* 22, 615–621.

Toker, C. and Canci, H. (2006) Selection for drought and heat resistance in chickpea under terminal drought conditions. *4th International Food Legumes Research Conference: Food Legumes for Nutritional Security and Sustainable Agriculture*. 18–22 October 2005. New Delhi, India, (in press).

Toker, C., Karhan, M. and Ulger, S. (2003) Endogenous organic acid variations in different chickpeas. *Acta Agriculturae Scandinavica Section B. Plant and Soil* 54, 42–44.

Turner, N.C., Wright, G.C. and Siddique, K.H.M. (2001) Adaptation of grain legumes (pulses) to water-limited environments. *Advances in Agronomy* 71, 193–231.

van der Maesen, L.J.V. (1972) Cicer *L.: A Monograph of the Genus, with Special Reference to the Chickpea (*Cicer arietinum *L.), Its Ecology and Cultivation.* Mededelingen Landbouwhooge School, Wageningen, The Netherlands.

van Rheenen, H.H., Singh, O. and Saxena, N.P. (1997) Using evaluation techniques for photoperiod and thermo-insensitivity in pulses improvement. In: Asthana, A.N. and Ali, M. (eds) *Recent Advantages in Pulses Research*. Indian Society of Pulses Research and Development, IIPR, Kanpur, India, pp. 443–458.

Welfare, K., Yeo, A.R. and Flowers, T.J. (2002) Effects of salinity and ozone, individually and in combination, on the growth and ion contents of two chickpea (*Cicer arietinum* L.) varieties. *Environmental Pollution* 120, 397–403.

Wery, J., Turc, O. and Lecoeur, J. (1993) Mechanism of resistance to cold, heat and drought in cool-season legumes, with special reference to chickpea and pea. In: Singh, K.B. and Saxena, M.C. (eds) *Breeding for Tolerance in Cool Season Food Legumes*. Wiley, Chichester, UK, pp. 271–291.

24 Diseases and Their Management

G. Singh,[1] W. Chen,[2] D. Rubiales,[3] K. Moore,[4]
Y.R. Sharma[1] and Y. Gan[5]

[1]*Department of Plant Breeding, Genetics and Biotechnology, Punjab Agricultural University, Ludhiana 141004, Punjab, India;* [2]*United States Department of Agriculture, Agricultural Research Service, Grain Legume Genetics and Physiology Research, Washington State University, Pullman, WA 99164, USA;* [3]*Institute for Sustainable Agriculture, Alameda del Obispo s/n, Apdo. 4084, 14080 Cordoba, Spain;* [4]*New South Wales Department of Primary Industries, Tamworth Agricultural Institute, 4 Marsden Park Rd, Calala, New South Wales 2340, Australia;* [5]*Agriculture and Agri-Food Canada, Airport Road East, Swift Current, Saskatehewan, S9H 3X2, Canada*

Introduction

Chickpea (*Cicer arietinum* L.) is a major food legume in the Indian subcontinent and several other countries. Although the yield potential of chickpea varieties in the current scenario exceeds 4 t/ha, the average yield realized is <0.8 t/ha. The gap between the potential and average yield is mainly due to diseases and poor management practices. The major diseases of chickpea in the order of their global importance are ascochyta blight, fusarium wilt, botrytis grey mould (BGM), dry root rot, collar rot, foot rot and stem rot (Nene and Reddy, 1987; Gaur and Singh, 1996; Singh and Sharma, 2002). This chapter focuses on the most economically important diseases and their management.

Ascochyta Blight

Distribution and economic importance

Ascochyta blight was first identified in the then North-western Province of India, presently in Pakistan. It has been reported in 37 countries: Bangladesh, China, India, Iran, Iraq, Israel, Jordan, Lebanon, Libya, Pakistan, Syria, Turkey, Algeria, Cyprus, Egypt, Ethiopia, Morocco, Sudan, Tanzania, Tunisia, Bulgaria, France, Greece, Hungary, Italy, Portugal, Romania, Spain, Volga, Northern, Caucasia, Canada, USA, Columbia, Mexico and Australia.

The economic importance of the disease is evident from the frequent occurrence of epidemics causing serious losses in many countries. In 1981–1983, the chickpea crop was completely destroyed in the north-western states of India and Pakistan, and the latter was compelled to import pulses worth $7.43 million (Verma *et al.*, 1981; Singh *et al.*, 1982b). Substantial losses were also reported from the Mediterranean and up to 100% in Italy and Spain (Hawtin and Singh, 1984). In the Pacific North-west, a severe epidemic of ascochyta blight caused financial loss of more than $1 million in 1987 (Kaiser and Muehlbauer, 1988). The 1998 Australian epidemic (Galloway and MacLeod, 2003) devastated the industry and caused a shift in production from 105,000 ha in 1998 to 15,000 ha in 1999 in a single state (Moore *et al.*, 2004).

Causal organism and symptoms

Ascochyta blight of chickpea, which is caused by the fungus *Ascochyta rabiei* (Pass.) Lab. (Teleomorph: *Didymella rabiei* (Kovachevski) vs Ark. The pathogen produces its asexual stage as minute dot-like black fruiting bodies (pycnidia) in concentric rings. The pycnidia contain numerous hyaline conidia. The sexual stage was first reported by Kovachevski (1936) from Bulgaria, and has subsequently been found in most countries where the disease occurs (Trapero-Casas and Kaiser, 1992b), the most recent being Australia (Galloway and MacLeod, 2003; Galloway *et al.*, 2005).

The disease attacks all above-ground parts of chickpea plants at all growth stages, causing necrotic lesions on leaves, stems and pods. Depending on varietal susceptibility, temperature and humidity, lesions appear 4–11 days after infection, and are usually circular on leaves and pods, and oval on stems, and brown or dark brown in colour. A characteristic feature of ascochyta blight is the formation of many black specks (pycnidia) in the lesions, usually in concentric circles. Infected stems often collapse and break off. Pod infection may result in seed infection or contamination.

Disease cycle and epidemiology

Infection and disease progression occur from 5°C to 25°C with an optimum of ~16–20°C, and a minimum of 6 h leaf wetness. As the duration of leaf wetness and relative humidity increase, so does disease severity (Trapero-Casas and Kaiser, 1992a). Cloudiness and prolonged intermittent rains favour rapid development and spread of the disease (Singh *et al.*, 1995). The pathogen survives on infected or contaminated seeds, infected chickpea debris left in the field and volunteer chickpea plants. Where both mating types exist, and cool (5–15°C), moist conditions occur, the pathogen undergoes sexual reproduction forming pseudothecia and ascospores (Trapero-Casas and Kaiser, 1992b). The ascospores are discharged during storms (Kaiser, 1992), and carried by wind to adjacent fields, serving as primary inoculum. Infected seed is the main source of introducing the pathogen into countries that were earlier known to be disease-free (Kaiser, 1992; Dey and Singh, 1994). Secondary spread of the disease occurs by conidia. Conidia are disseminated by winds, hail and rains (Kaiser, 1992; Dey and Singh, 1994).

Pathogen variability

Virulence variability was detected in isolates of *A. rabiei* in several countries including India, Pakistan, Iran, Italy, USA, Syria and Lebanon, and a variety of classification systems have been proposed (Vir and Grewal, 1974; Reddy and Kabbabeh, 1985; Singh, 1990; Porta-Puglia *et al.*, 1996; Jamil *et al.*, 2000; Chongo *et al.*, 2004; Maden *et al.*, 2004; Pande *et al.*, 2005a,b). Classification of isolates into either a two- or three-pathotype system (I, II and III) according to their levels of virulence has recently become the prevailing view (Udupa *et al.*, 1998; Chen *et al.*, 2004; Jayakumar *et al.*, 2005).

The presence of races of *A. rabiei* was contested by Strange *et al.* (2004). Their argument was based on the absence of a true hypersensitive response and gene-for-gene interaction between *A. rabiei* and chickpea (implicit in the concept of races). Pathotype is a more appropriate term than race in the case of ascochyta blight.

Screening for resistance

A number of screening techniques have been developed including field, growth room, cut-twig and detached leaf screening techniques (Singh, 1989; Bashir *et al.*, 1995; Sharma *et al.*, 1995). These have been recently revised by Tivoli *et al.* (2006).

In field screening, test material and susceptible chickpea lines grown in short single rows are inoculated with conidia or diseased residue (Singh *et al.*, 1982b). Maintaining free moisture and high humidity (>85%) with sprinkler or mist systems encourages disease but is not essential under frequent rainfall conditions.

In greenhouse or growth chamber screening, test entries along with susceptible checks are inoculated with conidia and kept in moist *dasuti* cloth chambers (Singh *et al.*, 1982b) for 24–48 h. An inexpensive and flexible mini-dome technique is also effective in screening chickpea for ascochyta blight resistance (Chen *et al.*, 2005). Detached tissue (cut-twig and leaf) screening techniques have also been developed for assaying the disease (Sharma *et al.*, 1995). The reliability of these detached tissue assays remains to be shown.

Due to the scarcity of sources of resistance to ascochyta blight in cultivated chickpea, a small number of resistant accessions have been widely used in the crossing programmes most notably ILC482, ILC3279, FLIP84-92C and FLIP84-79C (Singh, 1989). In search for additional resistance genes, wild species have received increased attention. Resistance against ascochyta blight has been identified not only in *C. judaicum*, *C. pinnatifidum* (Singh *et al.*, 1982a,b), but also in the species crossable with *C. arietinum*, *C. echinospermum* and *C. reticulatum*. Resistance is already available in segregating populations originating from interspecific crosses (Singh, 2004). With knowledge of resistance sources and molecular markers, current research efforts are using gene pyramiding to enhance disease resistance.

Host plant resistance

Resistance of chickpea plant to ascochyta blight is associated with a number of morphological and biochemical factors (Pieters and Tahiri, 1986; Dey and Singh, 1993). However, none of the biochemical factors could be detected by pathogenicity assay using reciprocal grafting between resistant and susceptible chickpea genotypes (Chen *et al.*, 2005). The genetics of chickpea resistance to ascochyta blight is complex depending on chickpea genotypes and pathogen pathotypes. Several genetic theories have been advanced including single dominant and recessive genes (Hafiz and Ashraf, 1953; Vir *et al.*, 1975; Singh and Reddy, 1983; Iqbal and Ghafoor, 2005), and two complimentary genes (Dey and Singh, 1993). At present, it is not clear whether the reported resistance genes represent the same or different loci because allelic tests were not performed (for a recent review see Winter *et al.*, 2003). Modifier genes may also be involved in this process.

A number of quantitative trait loci (QTLs) have been identified in different segregating populations against different pathotypes (Santra *et al.*, 2000; Tekeoglu *et al.*, 2000; Flandez-Galvez *et al.*, 2003; Rakshit *et al.*, 2003; Udupa and Baum, 2003; Cho *et al.*, 2004; Iruela *et al.*, 2006). A recent literature review is presented by Millán *et al.* (2006). *A. rabiei* pathotype I resistance is governed by a major gene on linkage group two (LG2) close to marker GA16. This or an adjacent locus is also partly responsible for resistance to pathotype II. A QTL flanked by sequence tagged microsatellite site (STMS) 11 and TR20 on LG4, confirmed by all researchers, is responsible for resistance to pathotype II and eventually also for resistance during the seedling stage. Since apparently all major blight resistance QTLs are tagged with STMS markers, pyramiding of resistance genes via marker-assisted selection (MAS) should now be feasible and awaits proof-of-principle.

Disease management

A number of cultural practices are available in managing ascochyta blight. These include planting resistant cultivars, production and use of disease-free seed, crop rotation with cereal or other non-host crops, deep ploughing, roguing volunteer chickpea plants, sanitation and intercropping with wheat, barley and mustard to reduce the disease spread. Delayed planting, where practical, also helps escape or reduce the initial primary inoculum source from forcibly discharged air-borne ascospores, as does isolating current season crops from paddocks sown previously to chickpea. Planting resistant cultivars should be the first choice in managing the disease.

Chemical control of ascochyta blight is through both seed treatment and foliar spray. The primary seed-borne infection can easily be controlled by treating the seed with calixin-M (11%) tridemorph (+ 36% Thiram), Bavistin + Thiram (1:1), Hexacap, captan, captafol, thiabendazole + Thiram and Rovral. Seed treatment also enhances seed germination (Singh and Singh, 1990). The secondary spread of the disease can be controlled by timely application of foliar fungicides. In India, complete control of the disease can be achieved by treating the seed with any of the above fungicides and applying 2–3 sprays of Hexacap, captan, captaf, Indofil M-45 or Kavach. In Australia, chlorothalonil was more effective than mancozeb both as a preventative and salvage treatment, providing application takes place before the rains (Moore *et al.*, 2006). Fungicide spray is applied based on a disease forecasting model, or alternatively, if the weather remains cloudy with intermittent rainfall, they can be sprayed immediately after each rain (Singh and Singh, 1990).

Botrytis Grey Mould

Distribution and economic importance

BGM is a devastating disease of chickpea in South Asian countries like India, Bangladesh, Nepal and Pakistan, and seriously damages crops in other countries like Australia, Argentina, Myanmar, Canada, Columbia, Hungary, Mexico, Spain, Turkey, USA and Vietnam. In India, BGM appeared as an epidemic in 1968 on the chickpea crop in Tarai area extending to the submontane region of Uttar Pradesh (Joshi and Singh, 1969) and later on, during 1981–1983, the disease occurred along with ascochyta blight in the north-western states of India, such as Punjab, Haryana, Himachal Pradesh, Rajasthan, parts of Bihar and West Bengal causing 70–100% yield loss (Singh *et al.*, 1982a; Grewal and Laha, 1983). The disease frequently causes total yield loss in the Indo-Gangetic Plains (IGP) of India, Nepal and Bangladesh (Singh and Sharma, 2002; Pande *et al.*, 2005b). Losses up to 96% have also been reported from Argentina (Carranza *et al.*, 1965).

Causal organism and disease symptoms

BGM is caused by *Botrytis cinerea* Pers. Ex Fr. (Teleomorph: *Botryotinia fuckeliana* Grover and Loveland). Initial fungal growth is white and cottony and turns grey with age. The conidia are smooth, hyaline to pale brown, ellipsoidal, one-celled 6–18 × 4–11 ∝m. Conidiophores are tall and dark bearing short branches with terminal ampulla on their apex on which clusters of conidia are formed on short denticles. The teleomorphic state of this fungus has been reported from sclerotia of *B. cinerea* on chickpea in India.

The disease can appear at any growth stage, but the maximum development of the disease takes place at reproductive stage. It attacks flowers, pods, leaves and stems, of which the flowers are more vulnerable. Initial symptoms under artificial conditions are water soaking and softening of affected plant parts. On these parts, brown spots are produced, which are rapidly covered with dense sporophore mass of conidia and mycelium. Under field conditions such grey fungal growth can be seen on flowers, pods and stems hidden under a dense canopy in wet conditions. On stem, BGM symptoms are gradually replaced by dark grey to black sporodochia. When relative humidity is low, irregular brown spots without any fungal growth appear on leaves. Sometimes, small, tiny black sclerotia are produced on dead tissues and water-soaked lesions on pods. The affected pods either do not produce any seed or produce small, shriveled seeds (Singh and Sharma, 2002).

Disease cycle and epidemiology

B. cinerea can survive through infected seeds, crop debris and other host plants either parasitically or saprophytically. The infected seeds carry the pathogen from season to season, to new areas, which were earlier known to be disease-free. It can survive on seeds externally as well as internally up to 5 years at a temperature of 18°C (Pande *et al.*, 2005b).

The pathogen can also survive in the soil in the form of mycelium and sclerotia. Small infected plant debris mixed in the seed also plays an important role in the survival of the pathogen. *B. cinerea* is a facultative parasite having a wide host range. It infects several plant species such as vegetables, fruits, ornamental plants, legumes and several weed hosts (Farr *et al.*, 2006).

Severe epidemics of BGM occur frequently in many South Asian countries. Free moisture, high relative humidity and temperature ~20–25°C are congenial for the infection and development of the disease. *B. cinerea* requires a wet period of 6 h, incubation period of 36 h and latent period of 72 h (Singh and Kapoor, 1985). The pathogen can complete the entire disease cycle in 7 days under favourable conditions. The pathogen can attack the basal parts of the plant early in the season and move from the seed to epicotyl portion. A closed chickpea canopy with reduced light penetration and air movement creates ideal conditions for fast spread of the disease. Under such conditions, there is abundant sporulation of the fungus on dead plant parts, particularly, on flowers and pods (Pande *et al.*, 2005b).

Host plant resistance

Growth room and cut-twig screening techniques are available (Pande *et al.*, 2005b). More than 13,000 chickpea germplasm and advance breeding lines have been screened against BGM, resulting in the identification of a reasonable level of resistance in the cultigen *C. arietinum* and higher resistance in wild *Cicer* spp. (Singh *et al.*, 1982a,b, 1991). Attempts have been made to enhance BGM resistance through interspecific and intraspecific hybridization, and gene pyramiding.

Several lines having reasonably good resistance with rating 3.0–5.0 were identified and extensively used in a disease-resistance breeding programme (Verma *et al.*, 1981; Singh *et al.*, 1991). Materials (GL no. 84133, 90168, 91040, 96165, 97125, GNG 734, FG 575 and FLIP 82–150 C) derived through intraspecific crossing have higher levels of resistance to BGM and have been extensively used in breeding programmes. Good resistance was found in a number of wild *Cicer* spp. including *C. judaicum* (ILWC 19-2), *C. bijugum* (ILWC 7/S-1), *C. echinospermum* (ILWC 35/S 1) and *C. pinnatifidum* (189) (Singh *et al.*, 1991). Coincidentally all these accessions also had a higher degree of resistance to ascochyta blight. However, many of these wild species are difficult to cross with cultivated genotypes. Successful crosses have been made with *C. judaicum* 182, *C. pinnatifidum* 188 and *C. bijugum* ILWC 7/S 1 in Punjab Agricultural University (PAU), Ludhiana, India. Breeding materials derived from these crosses are in various stages of testing and are promising to make breakthroughs to enhance BGM resistance (G. Singh, 2002, unpublished data).

Management

Late planting, lower seed rate and increased plant spacing often help avoid or reduce the disease. New chickpea varieties with moderate type of resistance to BGM are also available. Chemical control has been effective either as seed treatment or as foliar spray. Repeated spray is necessary if disease-conducive conditions persist. The seed-borne infection can be completely eliminated by treating the seed with a combination of Bavistin + Thiram (1:1). Effective fungicides include Indofil M 45, Bavistin, Thiabendazole, Rovral and Ronilan.

Fusarium Wilt

Distribution and economic importance

Chickpea wilt is a very important disease and occurs in 32 countries falling in the six continents (Nene *et al.*, 1991; Singh and Sharma, 2002). The pathogen in association with other soil-borne pathogens like root rots and foot rot also causes extensive damage to chickpea crop. It causes around 10% yield loss in India but under certain conditions and specific locations, the losses may go up to 60%. The disease is becoming a major constraint in chickpea production in California, USA, and the Mediterranean (Haware, 1990).

Causal organism and symptoms

Fusarium oxysporum f. sp. *ciceris* (Padwick) synd. and Hans causes fusarium wilt of chickpea. It produces microconidia, macroconidia and chlamydospores. Microconidia (2.5–4.5 $\times$ 5–11 $\propto$) are oval or cylindrical, straight or curved. Macroconidia (3.5–4.5 $\times$ 25–65 $\propto$) are 3–5 $\propto$ septate or fusoid. Both microconidia and macroconidia are generally sparse on solid media and are formed abundantly in potato dextrose broth. Chlamydospores are formed in 15-day-old cultures singly, in pairs or in a chain, and are smooth or rough-walled. Best growth of the fungus can be obtained at 25°C and pH 6.0.

The disease appears 20 days after sowing and even earlier on susceptible cultivars. The characteristic symptoms of wilt are drooping of petioles, rachis and leaflets. The lower leaves are chlorotic, gradually turn yellow and then light brown or straw-coloured, and finally dry up. Discoloration of xylem vessels extends towards stem and branches and can be seen when split open vertically. Sometimes only a few branches are affected, resulting in partial wilt. Affected plants do not show external root discoloration. However, co-infection with other soil-borne pathogens may cause external root discoloration.

Disease cycle and epidemiology

The pathogen is both soil- and seed-borne, and its infection is systemic. It can be isolated from the aerial plant parts, including seeds. The pathogen can survive on infected crop residues, roots and stem buried in the soil for up to 6 years. The infected seeds play an important role in spread of the disease to uninfected areas. Other leguminous host plants such as lentil, pea, pigeon pea, bean and faba bean have been identified as symptomless carriers. The pathogen can survive through mycelium and chlamydospores in seed and soil (Haware and Nene, 1982).

The epidemiology of root-infecting fungi in the soil is complex and involves several factors. Environmental and physical factors such as soil moisture, soil temperature, soil nutrients, pH, inoculum density, race, plant age and host genotype play an important role in the development of the disease (Singh and Sharma, 2002). Light and sandy soils, alkaline soils, soil moisture and temperature around 25°C favour development of the disease (Chauhan, 1963).

Pathogenic variability

Development of resistant cultivars is hindered by the pathogenic variability of the fungus. To date, eight races of *F. oxysporum* f. sp. *ciceris* have been reported from India, Spain and the USA (0, 1A, 1B/C, 2, 3, 4, 5 and 6; Haware and Nene, 1982; Jiménez-Díaz *et al.*, 1989; Jiménez-Gascó *et al.*, 2004). Each of these races forms a monophyletic lineage, which acquires its virulence on different chickpea lines in simple, stepwise patterns (Jiménez-Gascó *et al.*, 2004). Recently, Sharma *et al.* (2005) proposed a new set of chickpea differential lines to differentiate the six races of *F. oxysporum* f. sp. *ciceris* unambiguously.

Screening for resistance

Screening for resistance to fusarium wilt can be carried out either in the field or in the greenhouse (Nene, 1988; Nene *et al.*, 1989; Infantino *et al.*, 2006). Field screening can be performed in infected wilt-sick plots. A susceptible check, JG 62 (ICC 4951) is planted after every two test rows. Seed germination is recorded 20 days after sowing and wilt incidence is recorded at monthly intervals. The detailed procedure for the development of wilt-sick plot and identification of resistant donors is given by Nene *et al.* (1989).

Greenhouse screening is useful for confirming resistance identified in field screening, and for genetic or other critical studies. Inoculum can be prepared from either a sand–maize medium (9:1) or in liquid V8 (Sharma *et al.*, 2005) for 15–21 days at 25°C. The inoculum in sand–maize medium can be mixed with sterilized soil. After incubating for 4 days, when the fungus establishes itself, chickpea seeds are planted and kept at 25°C. Plant mortality as disease incidence is recorded 15 and 45 days after planting. Alternatively, conidia at 2×20^5 spores/ml harvested from the liquid medium can be used as inoculum. Two-week-old seedlings grown in perlite are uprooted, their roots dipped in the conidia suspension for 5 min, transplanted to 1:1 perlite:potting soil mix and then incubated at 25°C /22°C day/night temperatures. Disease incidence is recorded starting 2 weeks after inoculation at weekly intervals for up to 8 weeks (Sharma *et al.*, 2005).

Host plant resistance

In chickpea, resistance to fusarium wilt is governed by major resistance genes. In particular, resistance to races 1A, 2 and 4 is either under the control of two or three genes, whereas resistance to races 3 and 5 is monogenic (Sharma *et al.*, 2005). The resistance to races 1 and 2 of the pathogen in CPS1 and WR 315 is governed by a single recessive allele at the same locus in both lines. Further studies indicated that resistance to race 1 appears to be controlled by at least three independent loci designated H-1, H-2 and H-3. Any two of these alleles together confer complete resistance (Sindhu *et al.*, 1983; Singh *et al.*, 1988). Rubio *et al.* (2003) found two genes that were responsible for resistance to race 0 in a cross between two resistant chickpea cultivars, CA1938 and JG-62, with resistance that can be conferred by the presence of one of them.

More than 200 wilt-resistant lines have been identified (Singh *et al.*, 1986; Nene *et al.*, 1989; Sharma *et al.*, 2005). It is believed that currently sufficient material with high level of resistance to wilt is available. In addition, stable and multiple disease resistance (MDR) lines have also been identified (López-García, 1974; Singh *et al.*, 1984, 1986; Nene, 1988; Haware *et al.*, 1990; Singh and Sharma, 2002). Several chickpea lines (GL 84170884200, 84254, 85058, 66059, 86071, 86072, 90134, 90145, PPL 41-1, 57, 146, 155, GG 762 and GG 774) have shown resistance to wilt, dry root rot and foot rot (Haware *et al.*, 1980; Singh *et al.*, 1991). Wilt resistance has also been found in some wild *Cicer* spp. Several national and international institutions have also developed wilt-resistant varieties for controlling the disease.

Management

Collection of diseased plant debris and deep ploughing after harvest can reduce the inoculum and consequently reduce disease incidence. Soil solarization by covering it with a transparent polythene sheet for 6–8 weeks during summer effectively reduces the pathogen population (Katan, 1980). Crop rotation and delayed planting are also very effective in controlling the disease. Use of healthy seed produced in a disease-free area helps in reducing seed-borne inoculum.

Planting resistant cultivars is the most efficient measure in controlling the disease. A number of resistant varieties are available. It is important to know which race is present in the soil and select varieties that are resistant to the targeted race.

Seed-borne inoculum can be controlled by treating the seed with fungicides such as Benlate-T and Bavistin (Haware *et al.*, 1978). Some biocontro agents may also be effective but additional research is needed for practical applications (Bhan and Chand, 1998).

Dry Root Rot

Distribution and economic importance

Dry root rot is a serious problem and has been reported from Australia, Ethiopia, Iran, Pakistan, Bangladesh, Nepal and several other countries (Nene *et al.*, 1991; Singh and Sharma, 2002). Although it is found in all chickpea growing areas of India, it is most severe in Central and South India, where the crop is grown under rainfed conditions. High day temperature ~30°C and dry soil conditions at flowering and podding stage rapidly increase the severity of the disease (Gurha *et al.*, 2003).

Causal organism and disease symptoms

Dry root rot is caused by *Macrophomina phaesolina* (Maub.) Ashby. The disease is more severe in sandy poor soils and generally appears at flowering and podding stage. The petioles and leaflets droop only at the top of plant. Tap roots turn black, show sign of rotting and are devoid of lateral roots. The lower portion of the tap root usually remains in the soil when the plant is uprooted. Sometimes a greyish mycelium can be seen on the tap roots. Dead roots are brittle and show shredding of bark. The tip of the root is easily broken when touched. Minute sclerotia can be seen with the arid of handles on the exposed wood of the root and inner side of the bark or whenever split open at the collar region vertically.

Disease cycle and epidemiology

The disease is both seed- and soil-borne. It also has a wide range of host plant species particularly belonging to family *Leguminosae*. It is most severe at a temperature of 30°C or above in sandy and poor soils. Deficient soil moisture is favourable for disease development (Singh and Sharma, 2002).

Disease management

Deep ploughing and removal of infected host debris from the soil reduce disease severity. Moisture-stress conditions should be avoided. Timely sowing of early maturing varieties is a good option to escape the hot weather conditions during maturity of the disease (Gurha *et al.*, 2003).

Both pot culture and field screening techniques have been developed by using multiple disease sick-plots for wilt, dry root rot, collar rot and foot rot. More than 10,000 chickpea germplasm accessions and cultivated genotypes were screened for resistance to dry root rot along with wilt, and several resistant sources have been identified. These resistant sources have been utilized in disease-resistance breeding programmes and consequently varieties resistant or tolerant to multiple diseases have been developed (Gurha *et al.*, 2003)

Treating seeds with fungicides such as Captan or Thiram is helpful in reducing the disease (Gurha *et al.*, 2003). Seed treatment with biocontrol agents *Trichoderma viride* has shown some benefits in managing the disease (Gurha *et al.*, 2003).

Sclerotinia Stem Rot

Distribution and economic importance

Sclerotinia stem rot (or white mould) occurs in most of the chickpea growing regions of the world. It is widely prevalent in Australia, Bangladesh, Chile, India, Iran, Morocco, Nepal, Pakistan, Syria, USA and Tunisia. It often appears in its severe form in the north-western states of India. It appeared as an epidemic and caused heavy losses in eight growing regions and destroyed the crop completely (Singh *et al.*, 1989; Chen *et al.*, 2006).

Causal organisms and disease symptoms

Sclerotinia stem rot is caused by *Sclerotinia sclerotiorum* (Lib.) de Bary. The disease can appear at any stage. The leaves of the affected plants turn yellow, dry up and become straw-coloured. In the initial stage chlorotic and dried plants can be seen scattered in the field. The characteristic symptoms of the disease include production of a web of white mycelial strands at the collar region and above, or on branches. Black irregular sclerotial bodies are seen mingled with mycelial strands; sometimes, shredded straw-coloured lesions are also seen on the aerial parts of the plant due to infection of the pathogen. Under wet and high humid conditions, white fluffy fungal growth is seen on the soil surface (Singh and Sharma, 2002).

Disease cycle and epidemiology

The pathogens survive in soil primarily as sclerotia. It has also a wide host range. Sclerotia mixed with seed also play an important role in perpetuation of the pathogen from one crop to another. Under high humid conditions sclerotia are germinated and produce ascospores, which are forcibly discharged causing aerial infections. Sclerotia may also germinate in soil and directly infect adjacent plant stems.

Stem rot almost needs similar environmental conditions as that of ascochyta blight and BGM, i.e. free moisture, high relative humidity, temperature ~20°C and dense plant canopy, particularly at flowering, podding and near maturity stages of the crop.

Disease management

Cultural practices aimed at reducing soil-borne sclerotia and increasing airflow under crop canopy are recommended for managing sclerotinia stem rot. Deep ploughing to bury sclerotia is helpful in preventing sclerotia germination and promoting decomposition of sclerotia. Flooding the field in summer also helps destroy sclerotia. Late planting in certain production regions and wide row-spacing can also be used to reduce the plant growth and consequently the disease.

No complete resistance to stem rot is known in chickpea. However, some chickpea lines (GL 84012, GL 88223, GLK 8814, GF 89-75) do show moderate tolerance to the disease. Some accessions in wild *Cicer* spp. (*C. judaicum, C. reticulatum, C. pinnatifidum* and *C. yamashitae)* were found to have good resistance to stem rot.

The disease can be managed to some extent by treating the seed with chemicals (Bavistin, Derosal, or Bavistin + Thiram (1:1)). Foliar application of the chemicals immediately after rain was found to be effective in controlling the disease and increasing yield.

Foot Rot

Distribution and economic importance

Foot rot of chickpea was first reported by Kheswalla in 1941 from India. It is quite severe in the north-western states of India (Singh *et al.*, 1991). It has also been reported from Bangladesh and Nepal (Singh and Sharma, 2002).

Causal organism and disease symptoms

Foot rot is caused by *Operculae padwickii* Kheswalla. Symptoms include brown lesions on the collar region of the plant. These lesions increase in size, become black in colour and involve epicotyl and basal taproot of the plant. In the advanced stage of disease development, a complete girdling of the plant at the collar region takes place, resulting in death of the plant. The affected plants start drying from lower leaves and ultimately the whole plant dries up. The phloem of the diseased plant becomes brown in color and there is no blackening of xylem vessels as in the case of wilt (Nene, 1979).

Disease cycle and epidemiology

The disease is soil-borne and the pathogen is carried from one season to the next through infected plant debris left over after the harvest of the crop. The diseased debris mixed with farmyard manure also plays a role in the perpetuation of the disease.

The moisture content of the soil has significant effect on the development of the disease. The disease is severe when soil moisture is somewhat below field capacity. Acidic soil favours disease development. Optimum temperature for disease development is 25°C.

Disease management

Late planting is shown to help escape the disease, and crop rotation with wheat, barley or raya is helpful in reducing soil-borne inoculum and the disease. Many chickpea accessions were found to be resistant to the disease and several resistant varieties have been released. Particularly interesting is the variety PBG 5, which is resistant to foot rot and several diseases and was recently released in Punjab, India (Singh and Sharma, 2002). Seed treatment with Dodine, Triforine, Tecto-60, Delan and Benlate–T, and especially BASF 3302 is very effective in reducing the disease incidence.

Rust

Distribution and economic importance

Chickpea rust is a disease of local importance but it is present in almost every region of the world where chickpea is grown. The disease is widespread in the Mediterranean, South-east Europe, South Asia, East Africa and Mexico (Reddy *et al.*, 1980; Ragazzi, 1982; Jones, 1983; Venette and Stack, 1987; Rubiales *et al.*, 2001). Normally, chickpea rust epidemic begins late in the season so yield components are usually less affected by the infection. However, early infections can incite important losses. Severe outbreaks have been reported in India, Central Mexico and Italy (Ragazzi, 1982; Jones, 1983).

Causal organism and disease symptoms

Chickpea rust is caused by *Uromyces ciceris-arietini* (Grogon) Jacz. & Boyer (Asthana, 1957). The rust cause large pustules (up to 1 mm) on leaves that appears first as small, round, brown spots, which may coalesce and turn dark later. Masses of brown uredospores develop under the epidermis at the centre of the spots and are released from the mature pustule when the epidermis ruptures.

Host plant resistance

Partial resistance has been identified only recently (Rubiales *et al.*, 2001); earlier, no sources of resistance to rust had been identified and only escape by sowing dates and chemical control had been proposed (Gómez and Paredes, 1985; Díaz-Franco and Pérez-García, 1995). The resistance to chickpea rust is mainly of incomplete non-hypersensitive nature. This type of resistance is characterized by a prolonged latency period, reduced number of pustules and decrease of pustule size, which implied a reduction of disease severity in the field. Resistance was stronger on wild *Cicer* spp. than on cultivated chickpea (Rubiales *et al.*, 2001).

Management

When necessary, a number of fungicides (propiconazol, pyraclostrobin and chlorothalonil) could be applied to control chickpea rust.

Parasitic Weeds

Distribution and economic importance

More than 4000 species of angiosperms are parasitic plants, but only a few of them parasitize cultivated plants, becoming agricultural weeds. Nevertheless, these weedy parasites pose a tremendous threat to farmers' economy, mainly because they are at present almost uncontrollable (Rubiales, 2003). Most important are weedy root parasites like broomrapes (*Orobanche* spp.) that connect to host roots, and climbers like doders (*Cuscuta* spp.) that parasitize shoots.

Orobanche crenata Forsk. (crenate broomrape) is widely distributed in the Mediterranean and West Asia where it is a major constraint in the production of grain and forage legumes. Chickpea is a host of *O. crenata* that does not suffer important levels of infection in traditional spring sowings (ICARDA, 1989); however, with the continuous increase of winter sowing, crenate broomrape might become a problem in chickpea (ICARDA, 1989; Linke *et al.*, 1991). Delayed sowing reduces the broomrape incidence (Rubiales *et al.*, 2003).

This species is highly adapted to agricultural conditions. It has large erect plants, branching only from their underground tubercle. The spikes may reach a height of up to 1 m, bearing many flowers of diverse pigmentation, from yellow, through white to pink and violet.

O. aegyptiaca Pers. is an important pest of many crops in the eastern parts of the Mediterranean, in the Middle East and parts of Asia (Joel *et al.*, 2006). It has branched stems with flowers that may significantly differ in colour, from white to dark blue.

O. foetida Poir. is known as a weed only in North Africa (Tunisia and Morocco) (Kharrat *et al.*, 1992), but the species is common in native habitats also in Spain and France. The plant has unbranched stems that bear red or purple flowers, which release an unpleasant smell.

Host plant resistance

Resistance to *O. crenata* is scarce and of complex nature in faba beans, lentils, vetches and peas, making breeding for broomrape resistance a difficult task in these crops (Cubero, 1991; Rubiales *et al.*, 2006). However, resistance is common in chickpea germplam (Rubiales *et al.*, 2003) and in wild species of *Cicer* (Rubiales *et al.*, 2004) and may be used in breeding programmes. Low induction of germination of the seeds is a major component of resistance to *O. crenata* in chickpea germplasm (Rubiales *et al.*, 2003, 2004). Hypersensitive reactions blocking haustorial penetration is common in chickpea infected by *O. crenata*, but there is evidence that the darkening is due to blocking and death of the parasite intrusive tissue, rather than to death of the surrounding host tissue (Rubiales *et al.*, 2003). The total number of broomrape tubercles might be relatively high in some accessions, but most of those shoots would neither emerge nor develop further.

Although there are no clear reports on the existence of races of *O. crenata* (Román *et al.*, 2001), we cannot exclude the possibility of selection of more aggressive populations, better adapted to germination by chickpea root exudates and/or able to bypass subsequent mechanisms of resistance. In addition to new races of *O. crenata*, attention should also be paid to other broomrape species such as *O. foetida* that has recently been reported on faba bean and chickpea in Tunisia (Kharrat *et al.*, 1992). The same applies for *O. aegyptiaca* that is widespread in the eastern side of the Mediterranean (Joel *et al.*, 2006).

Disease management

The main obstacle in the long-term management of *Orobanche*-infested fields is the durable seedbank, which may remain viable for decades, and gives rise to only a very low annual germination percentage, if at all. As long as the seed bank is not controlled, the need to apply means to control the parasite will persist. Three main approaches are possible for the control of seed bank demise: (i) fumigation, (ii) solarization and (iii) biological control. Although soil solarization is effective under optimal conditions to the upper soil layer (15–20 cm), it is expensive and may be considered for use only at localized geographical regions with a long and sunny summer.

Several strategies have been employed to control broomrape. Chickpea is very sensitive to the standard glyphosate treatment recommended for broomrape control in faba bean (García-Torres *et al.*, 1999). Knowledge of the developmental stage of broomrape is crucial to achieve an effective control with herbicides. It has been shown that broomrape can be controlled in faba bean by applying low rates of glyphosate during the early stages of attachments (Mesa-García and García-Torres, 1985). Adjustment of the time and rate of application is crucial as chickpea is more sensitive to glyphosate than faba bean. However, it seems that chickpea tolerates pre-emergence treatment of other herbicides suitable for broomrape control such as imazethapyr (75–100g active ingredient/ha) (García-Torres *et al.*, 1999). Satisfactory control can be achieved by use of resistant cultivars complemented with other control strategies such as an intermediate sowing date like December or pre-emergence treatment with imazethapyr.

Cuscuta

Cuscuta may cause 100% yield loss in chickpea crops. *Cuscuta campestris* is selectively controlled by pre-emergence application of pronamide with chlorthaldimethyl (Graf *et al.*, 1982). Soil solarization effectively reduces cuscuta seed bank in the soil (Haidar *et al.*, 1999).

Other Diseases

Chickpea plants are also affected by a number of other diseases with local or limited importance. Information about those diseases is presented in Table 24.1.

Table 24.1. Diseases of chickpea with local or limited importance.

Disease	Causal organism	Distribution	Importance	Reference
Fungal diseases				
Phoma blight	*Phoma medicaginis* Mablbr. and Roum.	Australia, India, Bangladesh, USA	Minor	Boerema *et al.*, 2004
Alternaria blight	*Alternaria alternata* (Fr.) Keisser)	Bangladesh, Hungary, India, Nepal, Sri Lanka	Minor	Nene *et al.*, 1991
Stemphylium blight	*Stemphylium sarciniforme* (Cav.) Wilts	Bangladesh, India, Iran, Syria	Minor	Nene *et al.*, 1991
Powdery mildew	*Leveillula taurica* (Lev.) Arn.	Ethiopia, India, Mexico, Sudan	Minor	
Stem anthracnose	*Colletotrichum dematium* (Pers. Ex Fr.) Grove	India	Minor	Nene *et al.*, 1991
Black root rot	*Fusarium solani* (Mart.) Sacc.	Bangladesh, India, Spain	Minor	Nene *et al.*, 1991

Table 24.1. *Continued*

Disease	Causal organism	Distribution	Importance	Reference
Wet root rot	*Rhizoctonia solani* Kuhn	Widespread	Minor	Nene, 1979
Phytophthora root rot	*Phytophthora medicaginis* Hansen and Maxwell	Australia	Major	Irwin and Dale, 1982
Black root rot	*Thielaviopsis basicola* (Berk. and Br.) Ferr.	USA	Minor	Bowden *et al.*, 1985
Verticillicum wilt	*Verticillium albo-atrum* Reinke and Berth, and *V. dahliae* Kleb.	India, USA	Minor	Nene *et al.*, 1991
Viral diseases				
Stunt	Pea leaf roll virus	Widespread	Important in Iran, India, Pakistan and several other countries	Nene *et al.*, 1991
Narrow leaf	Bean yellow mosaic potyvirus	Australia, USA, Iran, India	Locally important, more important in Iran	Thomas *et al.*, 2004
Yellow mosaic	Beet western yellows luteovirus	Australia, India, Spain, Syria, USA	Minor, locally	Nene *et al.*, 1991; Makkouk *et al.*, 2003
Mosaic, bud necrosis, wilt	Alfalfa mosaic virus	Widespread	Locally important in Australia, Iran	Thomas *et al.*, 2004
Cucumber mosaic virus	Cucumber mosaic cucumovirus	Widespread	Locally important in Iran	Nene *et al.*, 1991
Enation mosaic	Pea enation mosaic virus	USA, Italy	Minor	Makkouk *et al.*, 2003
Necrotic yellows	Lettuce necrotic yellow rhabdovirus	Australia	Important in northern New South Wales and Southern Queensland	Behncken, 1983
Tomato spotted wilt	Tomato spotted wilt virus	Australia	Minor	Thomas *et al.*, 2004
Pea seed-borne mosaic	Pea seed-borne mosaic virus	Australia	Minor	Thomas *et al.*, 2004

Continued

Table 24.1. *Continued*

Disease	Causal organism	Distribution	Importance	Reference
Bean leaf roll	Bean leaf roll	Australia	Minor	Makkouk *et al.*, 2003; Thomas *et al.*, 2004
Sub clover stunt	Subterranean clover stunt virus	Australia	Minor	Thomas *et al.*, 2004
Sub clover red leaf	Subterranean clover red leaf virus	Australia	Minor	Thomas *et al.*, 2004

References

Asthana, R.D. (1957) Some observations on the incidence of *Uromyces ciceris-arietini* (Grognon) Jakz. and Boyer on *Cicer arietinum* L. *Nagpur Agricultural College Magazine* 31, 1–4.

Bashir, A., Alam, S.S. and Qureshi, S.H. (1995) Chickpea germplasm evaluation for resistance to ascochyta blight under artificial conditions. *International Chickpea Newsletter* 12, 24–26.

Behncken, G.M. (1983) A disease of chickpea caused by lettuce necrotic yellows virus. *Australasian Plant Pathology* 12, 64–65.

Bhan, U. and Chand, H. (1998) Fusarial wilt of chickpea, pea and lentil with special reference to integrated management. In: Upadhyay, R.K., Mukerji, K.G. and Rajak, R.L. (eds) *IPM system in Agriculture*, Vol. 4: *Pulses*. Aditya Books, New Delhi, India, pp. 477–500.

Boerema, G.H., de Gruyter, J., Noordeloos, M.E. and Hamers, M.E.C. (2004) *Phoma Identification Manual. Differentiation of Specific and Infra-specific Taxa in Culture*. CAB International, Wallingford, UK.

Bowden, R.L., Wiese, M.V., Crock, J.E. and Auld, D.L. (1985) Root rot of chickpeas *Cicer arietinum* and lentils *Lens culinaris* caused by *Thielaviopsis basicola*. *Plant Disease* 69, 1089–1091.

Carranza, J.M. (1965) Wilt of chickpea (*Cicer arietinum*) caused by *B. cinerea* (In Es.). *Revista de la Facultad de Agronomia, Universidad Nacional de la Plata* 41, 135–138.

Chauhan, S.K. (1963) Influence of different soil temperatures on the incidence of fusarium wilt of gram (*Cicer arietinum* L.). *Proceedings of the National Academy of Sciences of the United States of America* 33, 552–554.

Chen, W., Coyne, C., Peever, T. and Muehlbauer, F.J. (2004) Characterization of chickpea differentials for ascochyta blight and identification of resistance sources for *Didymella rabiei*. *Plant Pathology* 53, 759–769.

Chen, W., McPhee, K.E. and Muehlbauer, F.J. (2005) Use of a mini-dome bioassay and grafting to study chickpea resistance to ascochyta blight. *Journal of Phytopathology* 153, 579–587.

Chen, W., Schatz, B., Henson, B., McPhee, K.E. and Muehlbauer, F.J. (2006) First report of sclerotinia stem rot of chickpea caused by *Sclerotinia sclerotiorum* in North Dakota and Washington. *Plant Disease* 90, 114.

Cho, S., Chen, W. and Muehlbauer, F.J. (2004) Pathotype-specific genetic factors in chickpea (*Cicer arietinum* L.) for quantitative resistance to ascochyta blight. *Theoretical and Applied Genetics* 109, 733–739.

Chongo, G., Gossen, B.D., Buckwaldt, L., Adhikari, T. and Rimmer, S.R. (2004) Genetic diversity of *Ascochyta rabiei* in Canada. *Plant Disease* 88, 4–10.

Cubero, J.I. (1991) Breeding for resistance to *Orobanche* species: a review. In: Wegmann, K. and Musselman, L.J. (eds) *Progress in Orobanche Research*. Eberhard-Karls Universität, Germany, pp. 257–277.

Dey, S.K. and Singh, G. (1993) Resistance to ascochyta blight in chickpea – genetic basis. *Euphytica* 68, 147–153.

Dey, S.K. and Singh, G. (1994) Dissemination and development of ascochyta blight in chickpea. *Plant Disease Research* 9, 105–111.

Díaz-Franco, A. and Pérez-García, P. (1995) Control químico de la roya y la rabia del garbanzo y su influencia en el rendimiento de grano. *Revista Mexicana de Fitopatología* 13, 123–125.

Farr, D.F., Rossman, A.Y., Palm, M.E. and McCray, E.B. (2006) Fungal Databases, Systematic Botany and Mycology Laboratory, Agriculture Research Service, United States Department of Agriculture. Available at: http://nt.ars-grin.gov/fungaldatabases/

Flandez-Galvez, H., Ford, R., Pang, E.C.K. and Taylor, P.W.J. (2003) An intraspecific linkage map of the chickpea (*Cicer arietinum* L.) genome based on sequence-tagged microsatellite site and resistance gene analog markers. *Theoretical and Applied Genetics* 106, 1447–1456.

Galloway, J. and MacLeod, W. (2003) *Didymella rabiei*, the teleomorph of *Ascochyta rabiei*, found on chickpea stubble in Western Australia. *Australasian Plant Pathology* 32, 127–128.

Galloway, J., MacLeod, W., Fuhlbohm, M. and Moore, K. (2005) Asexual and sexual spore production by *Ascochyta rabiei* on previous seasons chickpea stubble in QLD, NSW and WA. In: *Proceedings of the 15th Australasian Plant Pathology Society Conference*. 26–29 September 2005. Geelong, Victoria, Australia.

García-Torres, L., López-Granados, F., Jurado-Expósito, M. and Díaz-Sánchez, J. (1999) Chemical control of *Orobanche* in legumes: achievements and constraints. In: Kroschel, J., Abderabihi, M. and Betz, H (eds) *Workshop on Joint Actions to Control Orobanche in the WANA-Region: Experiences from Morocco*, Vol. II. Margraft Verlag, Weikersheim, Germany, pp. 239–250.

Gaur, R.B. and Singh, R.D. (1996) Evaluation of chickpea cultivars for resistance to ascochyta blight. *Indian Journal of Mycology and Plant Pathology* 26, 50–55.

Gómez, G.R. and Paredes, P. (1985) El cultivo del garbanzo blanco en el centro de Sinaloa. Campo experimental Valle de Culiacán, INIFAP, Folleto técnico. No. 4.

Graf, S., Kleifeelds, Y., Berguiiti, A. and Retig, B. (1982) Dodder control in chickpeas. *Hassadeh* 62, 1388–1389.

Grewal, J.S. and Laha, S.K. (1983) Chemical control of botrytis blight of chickpea. *Indian Phytopathology* 36, 516–520.

Gurha, S.N., Singh, G. and Sharma, Y.R. (2003) Diseases of chickpea and their management. In: Masood Ali, Kumar, S. and Singh, N.B. (eds) *Chickpea Research in India*. Army Printing Press, Lucknow, India, pp. 195–227.

Hafiz, A. and Ashraf, M. (1953) Studies on the inheritance of resistance to mycosphaerella blight in gram. *Phytopathology* 43, 580–581.

Haidar, M.A., Iskandarani, N., Sidahmad, M. and Baalbaki, R. (1999) Response of dodder (*Cuscuta* spp.) seeds to soil solarization and chicken manure. *Journal of Crop Protection* 18, 253–258.

Haware, M.P. (1990) Fusarium wilt and other important diseases in the Mediterranean areas. *Options Méditerranéennes, Série Séminaires* 9, 61–64.

Haware, M.P. and Nene, Y.L. (1982) Symptomless carriers of the chickpea wilt Fusarium. *Plant Disease* 66, 250–251.

Haware, M.P., Nene, Y.L. and Rajeswari, R. (1978) Eradication of *Fusarium oxysporum* f. sp. *ciceris* in chickpea seed. *Phytopathology* 68, 1364–1367.

Haware, M.P., Kumar, J. and Reddy, M.V. (1980) Disease resistance in *kabuli-desi* introgression. In: *Proceedings of International Workshop on Chickpea Improvement*. February–March, 1979. ICRISAT, Hyderabad, India.

Haware, M.P., Jiménez-Díaz, R.M., Amin, K.S., Phillips, J.S. and Halila, H. (1990) Integrated management of wilt and root rot of chickpea. In: *Chickpea in the Nineties. Proceedings of the 2nd International Workshop on Chickpea Improvement.* 4–8 December, 1989. ICRISAT, Hyderabad, India, pp. 129–133.

Hawtin, G.C. and Singh, K.B. (1984) Prospectus and potential of winter sowing of chickpea in the Mediteerranean region. In: Saxena, M.C. and Singh, K.B. (eds) *Proceedings of the Workshop on Ascochyta Blight and Winter Sowing of Chickpea.* Martinus Nijhoff/Junk, The Hague, The Netherlands, pp. 7–16.

ICARDA (1989) Food Legume Improvement Program: Annual Report 1988, ICARDA, Aleppo, Syria.

Infantino, A., Kharrat, M., Riccioni, L., Coyne, C.J., McPhee, K.E. and Grünwald, N.J. (2006) Screening techniques and sources of resistance to root diseases in cool season food legumes. *Euphytica* 147, 201–221.

Iqbal, S.M. and Ghafoor, A. (2005) Identification of blight resistant genotypes from local and exotic chickpea genetic resources. *Pakistan Journal of Botany* 37(1), 79–86.

Iruela, M., Rubio, J., Barro, F., Cubero, J.I., Millán, T. and Gil, J. (2006) Detection of two QTL for resistance to ascochyta blight in an intraspecific cross of chickpea (*Cicer arietinum* L.): Development of SCAR markers associated to resistance. *Theoretical and Applied Genetics* 112, 278–287.

Irwin, J.A.G. and Dale, J.L. (1982) Relationships between *Phytophthora megasperma* isolates from chickpea, lucerne and soybean. *Australian Journal of Botany* 30, 199–210.

Jamil, F.F., Sarwar, N., Sarwar, M., Khan, J.A., Geistlinger, J. and Khal, G. (2000) Genetic and pathogenic diversity within *Ascochyta rabiei* (Pass.) Lab. populations in Pakistan causing blight of chickpea (*Cicer arietinum* L.). *Physiological and Molecular Plant Pathology* 57, 243–254.

Jayakumar, P., Gossen, B.D., Gan, Y.T., Warkentin, T.D. and Banniza, S. (2005) Ascochyta blight of chickpea: infection and host resistance mechanisms. *Canadian Journal of Plant Pathology* 27, 499–509.

Jiménez-Díaz, R.M., Trapero-Casas, A. and Cabrera-de-la-Colina, J. (1989) Races of *Fusarium oxysporum* f. sp. *ciceris* infecting chickpea in southern Spain. In: Jamos, E.C. and Beckman, C.H. (eds) *Vascular Wilt Diseases of Plants* NATO ASI Series, H 28. Springer, Berlin, pp. 515–520.

Jiménez-Gascó, M.M., Milgroom, M.G. and Jiménez-Díaz, R.M. (2004) Stepwise evolution of races in *Fusarium oxysporum* f. sp. *ciceris* inferred from fingerprinting with repetitive DNA sequences. *Phytopathology* 94, 228–235.

Joel, D.M., Hershenhorn, Y., Eizenberg, H., Aly, R., Ejeta, G., Rich, P.J., Ransom, J.K., Sauerborn, J. and Rubiales, D. (2006) Biology and Management of Weedy Root Parasites. Horticultural Reviews, Vol. 38, Wiley, New York.

Jones, D.R. (1983) Chickpea rust – a disease new to Australia. Queensland Department of Primary Industry Farm Note AGDEX 168/673.

Joshi, M.M. and Singh, R.S. (1969) A botrytis gray mold of gram. *Indian Phytopathology* 22, 125–128.

Kaiser, W.J. (1992) Epidemiology of *Ascochyta rabiei*. In: Singh, K.B. and Saxena, M.C. (eds) *Proceedings of the Disease Resistance Breeding in Chickpea.* ICARDA, Aleppo, Syria, pp. 117–134.

Kaiser, W.J. and Muehlbauer, F.J. (1988) An outbreak of ascochyta blight of chickpea in the Pacific NorthWest USA in 1987. *International Chickpea Newsletter* 18, 16–17.

Katan, J. (1980) Solar pasteurization of soils for disease control: status and prospects. *Plant Disease* 64, 450–454.

Kharrat M., Halila, M.H., Linke, K.H. and Haddar, T. (1992) First report of *Orobanche foetida* Poiret on faba bean in Tunisia. *FABIS Newsletter* 30, 46–47.

Linke, K.H., Singh, K.B. and Saxena, M.C. (1991) Screening technique for resistance to *Orobanche crenata* Forks. in chickpea. *International Chickpea Newsletter* 24, 32–34.

López-García, H. (1974) Inheritance of resistance to wilt (*Fusarium* sp.) in chickpea (*Cicer arietinum*) under field conditions. *Agricultura Técnica en Mexico* 3, 286–289.

Maden, S., Dolar, F.S., Babaliogullu, I., Bayraktar, H. and Demirci, F. (2004) Determination of pathogenic variability of *Ascochyta rabiei* and reactions of chickpea cultivars against the pathotypes in Turkey. In: *5th European Conference on Grain Legumes and 2nd International Conference on Legume Genomics and Genetics*. Academy of Engineering in Poland, Poland, pp. 307 (book of abstracts).

Makkouk, K.M., Kumari, S.G., Shahraeen, N., Fazlali, Y., Farzadfar, S., Ghotbi, T. and Mansouri, A. (2003) Identification and seasonal variation of viral diseases of chickpea and lentil in Iran. *Zeitschrift fuer Pflanzenkrankheiten und Pflanzenschutz* 110, 157–169.

Mesa-García, J. and García-Torres, L. (1985) *Orobanche crenata* Forsk control in *Vicia faba* L. with glyphosate as affected by herbicide rates and parasite growth stages. *Weed Research* 25, 129–134.

Millán, T., Clarke, H.J., Siddique, K.H.M., Buhariwalla, H.K., Gaur, P.M., Kumar, J., Gil, J., Kahl, G. and Winter, P. (2006) Chickpea molecular breeding: new tools and concepts. *Euphytica* 147, 81–103.

Moore, K., Knights, E., Verrell, A. and Nash, P. (2004) Learning from the Australian experience. In: *Proceedings of the Saskatchewan Pulse Growers Pulse Days 2004*, 12–13 January 2004, Saskatoon, Saskatchewan, Canada, pp. 12–24.

Moore, K., Knights, E., Nash, P. and Verrell, A. (2006) Update on *Ascochyta* and disease management in chickpeas for 2006. GRDC Grains Research Updates, 21–22 February, 2006, Dubbo, New South Wales, Australia. Available at: http://www2.rangemedia.com.au/library/programs/grdcupdates/fullcdrom/start.htm

Nene, Y.L. (1979) *Proceedings of the Consultants' Group Discussion on the Resistance to Soil-borne Diseases of Legumes*. 8–11 January 1979. ICRISAT, Hyderabad, India.

Nene, Y.L. (1988) Multiple disease resistance in grain legumes. *Annual Review of Phytopathology* 26, 203–217.

Nene, Y.L. and Reddy, M.V. (1987) Chickpea disease and their control. In: Saxena, M.C. and Singh, K.B. (eds) *The Chickpea*. CAB International, Wallingford, UK, pp. 233–270.

Nene, Y.L., Haware, M.P. and Reddy, M.V. (1981) Chickpea disease resistance screening techniques. *Information Bulletin* No. 10, ICRISAT, Hyderabad, India.

Nene, Y.L., Haware, M.P., Reddy, M.V., Phillips, J.C., Castro, E.L., Kotasthane, S.R., Gupta, O.M., Singh, G., Shukla, P. and Saha, R.P. (1989) Identification of broad based and stable resistance to wilt and root rots of chickpea. *Indian Phytopathology* 42, 499–505.

Nene, Y.L., Reddy, M.V., Haware, M.P., Ghaneka, A.M. and Amin, K.S. (1991) Field diagnosis of chickpea diseases and their control. *Information Bulletin* No. 28, ICRISAT, Hyderabad, India.

Pande, S., Siddique, K.H.M., Kishore, G.K., Bayaa, B., Gaur, P.M., Gowda, C.L.L., Bretag, T.W. and Crouch, J.H. (2005a) Ascochyta blight of chickpea (*Cicer arietinum* L.): a review of biology, pathogenicity, and disease management. *Australian Journal of Agricultural Research* 56(4), 317–332.

Pande, S., Stevenson, P., Rao, J.N., Neupane, R.K., Chaudhary, R.N., Grzywacz, D., Bourai, V.A. and Kishore, G.K. (2005b) Reviving chickpea production in Nepal through integrated crop management, with emphasis on botrytis gray mold. *Plant Disease* 89, 1252–1262.

Pieters, R. and Tahiri, A. (1986) Breeding chickpea for horizontal resistance to ascochyta blight in Morocco. *FAO Plant Protection Bulletin* 34, 99–105.

Porta-Puglia, A., Crino, P. and Mosconi, C. (1996) Variability in virulence to chickpea of an Italian population of *Ascochyta rabiei*. *Plant Disease* 80, 39–41.

Ragazzi, A. (1982) Un grave attacco di ruggine su foglie di cece. (A serious attack of rust on *Cicer arietinum* leaves). *Informatore Fitopatologico* 32, 41–43.

Rakshit, S., Winter, P., Tekeoglu, M., Juarez-Muñoz, M., Pfaff, T., Benko-Iseppon, A.M., Muehlbauer, F.J. and Kahl, G. (2003) DAF marker tightly linked to a major locus for ascochyta blight resistance in chickpea (*Cicer arietinum* L.). *Euphytica* 132, 23–30.

Reddy, M.V. and Kabbabeh, S. (1985) Pathogenic variability in *Ascochyta rabiei* (Pass.) Lab.

in Syria and Lebanon. *Phytopatholgia Mediterranea* 24, 265–266.

Reddy, M.V., Gridley, H.E. and Kaack, H.J. (1980) Major disease problems of chickpea in North Africa. *International Chikpea Newsletter* 3, 13–14.

Román, B., Rubiales, D., Torres, A.M., Cubero, J.I. and Satovic, Z. (2001) Genetic diversity in *Orobanche crenata* populations from southern Spain. *Theoretical and Applied Genetics* 103, 1108–1114.

Rubiales, D. (2003) Parasitic plants, wild relatives and the nature of resistance. *New Phytologist* 160, 459–461.

Rubiales, D., Moreno, I., Moreno, M.T. and Sillero, J.C. (2001) Identification of partial resistance to chickpea rust (*Uromyces ciceris-arietini*). In: *Proceedings of 4th European Conference on Grain Legume*, pp. 194–195.

Rubiales, D., Alcántara, C., Pérez-de-Luque, A., Gil, J. and Sillero, J.C. (2003) Infection by crenate broomrape (*Orobanche crenata*) in chickpea *(Cicer arietinum*) as influenced by sowing date, weather conditions and genetic resistance. *Agronomie* 23, 359–362.

Rubiales, D., Alcántara, C. and Sillero, J.C. (2004) Variation in resistance to crenate broomrape (*Orobanche crenata*) in species of *Cicer. Weed Research* 44, 27–32.

Rubiales, D., Pérez-de-Luque, A., Fernández-Aparicio, M., Sillero, J.C., Román, B., Kharrat, M., Khalil, S., Joel, D.M. and Riches, C.H. (2006) Screening techniques and sources of resistance against parasitic weeds in grain legumes. *Euphytica* 147, 187–199.

Rubio, J., Hajj-Moussa, E., Kharrat, M., Moreno, M.T., Millán, T. and Gil, J. (2003) Two genes and linked RAPD markers involved in resistance to *Fusarium oxysporum* f. sp. *ciceris* race 0 in chickpea. *Plant Breeding* 122, 188–191.

Santra, D.K., Tekeoglu, M., Ratnaparkhe, M., Kaiser, W.J. and Muehlbauer, F.J. (2000) Identification and mapping of QTLs conferring resistance to ascochyta blight in chickpea. *Crop Science* 40, 1606–1612.

Sharma, K.D., Chen, W. and Muehlbauer, F.J. (2005) Genetics of chickpea resistance to five races of fusarium wilt and a concise set of race differentials for *Fusarium oxysporum* f. sp. *ciceris. Plant Disease* 89, 385–390.

Sharma, Y.R., Singh, G. and Kaur, L. (1995) A rapid technique for ascochyta blight resistance in chickpea. *International Chickpea and Pigeonpea Newsletter* 2, 34–35.

Sindhu, J.S., Singh, K.P. and Slinkard, A.E. (1983) Inheritance of resistance to fusarium wilt in chickpea. *Journal of Heredity* 74, 68.

Singh, G. (1989) Identification of chickpea lines resistant to ascochyta blight. *Plant Disease Research* 4, 128–132.

Singh, G. (1990) Identification and designation of physiologic races of *Ascochyta rabiei* in India. *Indian Phytopathology* 43, 48–52.

Singh, G. (2004) My experiences of Ascochyta blight of chickpea: Lead the way towards eradication. *Plant Disease Research* 19(1), 1–9.

Singh, G. and Kapoor, S. (1985) Screening for combined resistance to Botrytis gray mold and ascochyta blight of chickpea. *International Chickpea Newsletter* 12, 21–22.

Singh, G. and Sharma, Y.R. (2002) Fungal diseases of pulses. In: Gupta, V.K. and Paul, Y.S. (eds) *Diseases of field crops*. Indus Publishing, New Delhi, India, pp. 155–192.

Singh, G. and Singh, M. (1990) Chemical control of ascochyta blight of chickpea. *Indian Phytopathology* 43, 59–63.

Singh, G., Kapoor, S. and Singh, K. (1982a) Screening for gray mold resistance in chickpea. *International Chickpea Newsletter* 7, 13–14.

Singh, G., Singh, K. and Kapoor, S. (1982b) Screening for sources of resistance to ascochyta blight of chickpea. *International Chickpea Newsletter* 6, 15–17.

Singh, G., Kapoor, S., Singh, K. and Gill, A.S. (1984) Screening for resistance to gram wilt. *Indian Phytopathology* 37, 393–394.

Singh, G., Kapoor, S., Gill, A.S. and Singh, K. (1986) Chickpea varieties resistant to Fusarium wilt and foot rot. *Indian Journal of Agricultural Science* 56, 344–346.

Singh, G., Gill, A.S., Verma, M.M. and Kaur, L. (1989) High susceptibility of chickpea to stem rot in Punjab, India. *International Chickpea Newsletter* 20, 16.

Singh, G., Kaur, L. and Sharma, Y.R. (1991) Ascochyta blight and gray mold resistance in wild species of *Cicer*. *Crop Improvement* 18, 150–151.

Singh, G., Kaur, P., Kumar, A., Verma, M.M., Kaur, L. and Sharma, Y.R. (1995) Primary and secondary spread of ascochyta blight of grain. In: Gupta, G.K. and Sharma, R.C. (eds) *Integrated Disease Management*. Scientific Publishers, Jodhpur, India, pp. 65–69.

Singh, H., Kumar, J., Smithson, J.B. and Haware, M.P. (1988) Association among fusarium wilt resistance flower colour and number of flowers per fruiting node in chickpea (*Cicer arietinum*). *Journal of Agricultural Sciences* 110, 407–409.

Singh, K.B. and Reddy, M.V. (1983) Inheritance of resistance to ascochyta blight in chickpea. *Crop Science* 23, 9–10.

Strange, R.N., Gewiss, E., Gil, J., Millán, T., Rubio, J., Daly, K., Kharrat, M., Cherif, M., Rhaiem, A., Maden, S., Dólar, S. and Dusunceli, F. (2004) Integrated control of blight of chickpea, *Cicer arietinum*, caused by the fungus *Ascochyta rabiei*: an overview. In: *The 5th European Conference on Grain Legumes and 2nd International Conference on Legume Genomics and Genetics*. Academy of Engineering in Poland, Poland, pp. 71–76.

Tekeoglu, M., Santra, D.K., Kaiser, W.J. and Muehlbauer, F.J. (2000) Ascochyta blight resistance inheritance in three chickpea recombinant inbred line populations. *Crop Science* 40, 1251–1256.

Thomas, J.E., Schwinghamer, M.W., Parry, J.N., Sharman, M., Schilg, M. and Dann, E.K. (2004) First report of *Tomato spotted wilt virus* in chickpea (*Cicer arietinum* L.) in Australia. *Australasian Plant Pathology* 33, 597–599.

Tivoli, B., Baranger, A., Avila, C.M., Banniza, S., Barbetti, M., Chen, W., Davidson, J., Lindeck, K., Kharrat, M., Rubiales, D., Sadiki, M., Sillero, J.C., Sweetingham, M. and Muehlbauer, F.J. (2006) Screening techniques and sources of resistance to foliar diseases caused by major necrotrophic fungi in grain legumes. *Euphytica* 147, 223–253.

Trapero-Casas, A. and Kaiser, W.J. (1992a) Influence of temperature, wetness period, plant age and inoculum concentration on infection and development of ascochyta blight of chickpea. *Phytopathology* 82, 589–596.

Trapero-Casas, A. and Kaiser, W.J. (1992b) Development of *Didymella rabiei*, the teleomorph of *Ascochyta rabiei* on chickpea straw. *Phytopathology* 82, 1261–1266.

Udupa, S.M. and Baum, M. (2003) Genetic dissection of pathotype specific resistance to ascochyta blight resistance in chickpea (*Cicer arietinum* L.) using microsatellite markers. *Theoretical and Applied Genetics* 106, 1196–1202.

Udupa, S.M., Weigand, F., Saxena, M.C. and Kahl, G. (1998) Genotyping with RAPD and microsatellite markers resolves pathotype diversity in the ascochyta blight pathogen of chickpea. *Theoretical and Applied Genetics* 97, 299–307.

Venette, J.R. and Stack, R.W. (1987) First report of rust (*Uromyces ciceris-arietini*) on garbanzo (chickpea) in the United States. *Plant Disease* 71, 101.

Verma, M.M., Singh, G., Sandhu, T.S., Brar, H.S., Singh, K. and Bhullar, B.S. (1981) Sources of resistance to gram blight and mold. *International Chickpea Newsletter* 4, 14–16.

Vir, S. and Grewal, J.S. (1974) Physiologic specialization in *Ascochyta rabiei*, the causal organism of gram blight. *Indian Phytopathology* 27, 355–360.

Vir, S., Grewal, J.S. and Gupta, V.P. (1975) Inheritance of resistance to ascochyta blight in chickpea. *Euphytica* 24, 209–211.

Winter, P., Staginnus, C., Sharma, P.C. and Kahl, G. (2003) Organisation and genetic mapping of the chickpea genome. In: Jaiwal, P.K. and Singh, R.P. (eds) *Improvement Strategies of Leguminosae Biotechnology*. Kluwer Academic, Dordrecht, The Netherlands, pp. 303–351.

25 Host Plant Resistance and Insect Pest Management in Chickpea

H.C. Sharma,[1] C.L.L. Gowda,[1] P.C. Stevenson,[2] T.J. Ridsdill-Smith,[3] S.L. Clement,[4] G.V. Ranga Rao,[1] J. Romeis,[5] M. Miles[6] and M. El Bouhssini[7]

[1]Genetics Resources Divisions, International Crops Research Institute for the Semi-Arid Tropics (ICRISAT), Asia Center, Hyderabad 502324, India; [2]Natural Resources Institute, University of Greenwich, Chatham, ME4 4TB, UK and Royal Botanic Gardens, Kew, Surrey, TW9 3AB, UK; [3]CSIRO Entomology, Private Bag No 5, Wembley, WA 6913, Australia; [4]USDA-ARS, 59 Johnson Hall, Washington State University, Pullman, WA 99164-6402, USA; [5]Agroscope ART Reckenholz Tänikon, Reckenholzstr, 191, 8046 Zurich, Switzerland; [6]Department of Primary Industries and Fisheries, 203 Tor Street, PO Box 102, Toowoomba, Queensland 4350, Australia; [7]International Center for Agricultural Research in the Dry Areas (ICARDA), Aleppo, Syria

Insect Pest Problems in Chickpea

Chickpea (*C. arietinum* L.) is the third most important legume crop in the world, after dry beans and peas (FAO, 2003). It is cultivated in 42 countries in South Asia, North and Central America, the Mediterranean region, West Asia and North and East Africa. In recent years, it has become an important crop in Australia, Canada and the USA. Nearly 60 insect species are known to feed on chickpea (Reed *et al.*, 1987) (Table 25.1). The important insect pests damaging chickpea in different regions are:

- Wireworms: false wireworm – *Gonocephalum* spp.;
- Cutworm: black cutworm – A. ipsilon (Hfn.) and turnip moth – A. segetum Schiff.;
- Termite: *Microtermes obesi* (Holm.) and *Odontotermes* sp.;
- Leaf-feeding caterpillars: cabbage looper – Trichoplusia ni (Hub.), leaf caterpillar – S. exigua (Hub.) and hairy caterpillar – S. oblique Walker;
- Semilooper: *Autographa nigrisignia* Walker;
- Leaf miners: *L. cicerina* (Rondani) and *L. congesta* (Becker);

Table 25.1. *Insect pests feeding on chickpea.*

Common name	Scientific name	Family	Distribution	Nature of damage
Order: *Orthoptera*				
Surface grasshopper	*Chrotogonus trychypterus* Blanch.	Acridiidae	India	Feeds on tender leaves, flowers and tender pods
Grasshopper	*Ailopus simulatrix* Wlk.	Acridiidae	India, Africa	Feeds on tender leaves and flowers
Field cricket	*Liogryllus bimaculatus* De Geer.	Gryllidae	India	Feeds on developing pods and seeds
Order: *Isoptera*				
Termites	*Microtermes obesi* (Holm.) *Odontotermes* sp.	Termitidae	Asia	Damages tap root
Order: *Hemiptera*				
Black aphid	*Aphis craccivora* Koch	Aphididae	Worldwide	Sucks sap from tender leaves, flower stalks and pods
Pea aphid	*Acrythosiphon pisum* (Harris)	Aphididae	Worldwide	Sucks sap from growing tips, flowers and pods
Cow bug	*Tricentrus bicolor* Dist.	Membracidae	India	Sucks sap
Order: *Lepidoptera*				
Cutworms	*Agrotis ipsilon* (Hfn.)	Noctuidae	Worldwide	Cuts the whole plant or growing tips and feeds on the leaves
	A. flammatra Schiff.	Noctuidae	Asia	Cuts the stem and growing tips
	Euxoa spinifera (Hub.) [=*A. spinifera* Hub.]	Noctuidae	Asia	Cuts the plant at ground level
	E. segetum Schiff (=*A. segetum* Dennis and Schiff.)	Noctuidae	Asia	Cuts the plant at ground
Semiloopers	*Autographa nigrisigna* Walker	Noctuidae	Asia	Feeds on leaves and pods
	Plusia orichalcea F.	Noctuidae	Asia	Feeds on leaves pods
	P. signata F.	Noctuidae	Asia	Feeds on leaves pods
	Chrysodeixis chalcites (Esp.)	Noctuidae	Asia	Feeds on leaves flowers
Cabbage looper	*Trichoplusia ni* (Hub.)	Noctuidae	America	Feeds on leaves
Western yellow striped armyworm	*Spodoptera praefica* (Grote)	Noctuidae	America	Feeds on leaves
Tobacco caterpillar	*S. litura* F.	Noctuidae	Asia	Feeds on leaves

Continued

Table 25.1. *Continued*

Common name	Scientific name	Family	Distribution	Nature of damage
Leaf caterpillar	*S. exigua* (Hub.)	Noctuidae	Asia, America	Feeds on leaves
Pod borers	*Helicoverpa armigera* (Hub.)	Noctuidae	Asia, Africa, Australia,	Feeds on leaves flowers and bores holes on the pod and eat away the seeds
	H. punctigera (Wallengren)	Noctuidae	Australia	Feeds on leaves, flowers and pods
	H. zea (Boddie.)			
	Heliothis virescens (Fab.)			
	H. assulta Cn.	Noctuidae	Asia	Feeds on leaves, flowers and pods
Noctuid caterpillar	*Rhyacia herwlea* C&D	Noctuidae	Asia	Feeds on leaves
Green leaf caterpillar	*Anticarisisa irrorata* (F.)	Noctuidae	Asia	Feeds on leaves
Fig moth	*Caudra cautella* (Wlk.)	Phycitidae	Asia	Feeds on stored grain
Bihar hairy caterpillar	*Diacrisia obliqua* (L.)	Arctiidae	Asia	Feeds on leaves
Order: *Diptera*				
Gram stem miner	*Ophiomyia cicerivora* Spencer	Agromyzidae	Asia	Feeds on the stem
Leaf miner	*Chromatomyia horticola* (Goureau)	Agromyzidae	Asia	Larvae mine leaves and feed on green matter
Pea leaf miner	*Phytomyza articornis* (Meig.)	Agromyzidae	Asia	Larvae mine leaves and feed on mesophyll
Chickpea leaf miner	*Liriomyza cicerina* (Rondani)	Agromyzidae	North Africa Asia	Larvae mine leaves and feed on mesophyll
Order: *Coleoptera*				
False wireworms	*Gonocephalum* spp.	Tenebrionidae	Asia	Damages the seedlings
Gujhia weevil	*Tanymecus indicus* F.	Curculionidae	Asia	Damages the seedlings
Pea leaf weevil	*Sitona lineatus* (L.)	Curculionidae	America	Adults feed on seedlings
Pumpkin beetle	*Aulacophora foveicolis* (Lucas)	Chrysomelidae	Asia	Feeds on leaves
Bruchids	*Callosobruchus chinensis* L.	Bruchidae	Worldwide	Feeds on stored seed
	C. maculatus (F.)	Bruchidae	Worldwide	Feeds on stored seed
	C. phaseolli (Gylh.)	Bruchidae	Worldwide	Feeds on stored seed
	C. analis (F.)	Bruchidae	Worldwide	Feeds on stored seed
	Acanthoscelides obtectus (Say)	Bruchidae	Worldwide	Feeds on stored seed

- Aphids: *A. craccivora* Koch and *Acyrthosiphon pisum* (Harris);
- Nodule-damaging fly: *Metopina ciceri* Disney;
- Pod borers: cotton bollworm – H. armigera (Hub.), native budworm – H. punctigera (Wallengren) and corn earworm – H. zea (Boddie.);
- Bruchids: Chinese bruchid – Callosobruchus chinensis L., bean bruchid – Acanthoscelides obtectus (Say), pulse weevil – C. analis F. and pulse bruchid – C. phaseoli (Gylh.).

The pod borer, *H. armigera* and the aphid, *A. craccivora* are the major pests of chickpea in the Indian Subcontinent. In the Mediterranean region, the most important pest is the leaf miner, *L. cicerina*. The black aphid, *A. craccivora* is important as a vector of the chickpea stunt disease, while *C. chinensis* is the most dominant species in storage.

In Australia, the major pests of chickpea are the two pod borers, *H. armigera* and *H. punctigera* (Knights and Siddique, 2002). Chickpea has a few pest problems in the USA (Miller *et al.*, 2002; Margheim *et al.*, 2004; Glogoza, 2005). Occasional pests in the Pacific Northwest are the western yellow striped armyworm, *S. praefica* (Grote) (Clement, 1999), pea leaf weevil, *Sitona lineatus* (L.) (Williams *et al.*, 1991), pea aphid, *A. pisum* and cowpea aphid, *A. craccivora* (Clement *et al.*, 2000). The potential pests are early season cutworms, loopers, corn earworm (*H. zea*), wireworms, aphids, grasshoppers and an agromyzid leafminer. Larvae of the agromyzid fly mine the chickpea leaves, but the impact of damage has not been established (Miller *et al.*, 2002; Margheim *et al.*, 2004). The major pest problems in chickpea and their management options are discussed below.

Pod Borers: *Helicoverpa armigera* and *Helicoverpa punctigera*

Chickpea production is severely threatened by increasing difficulties in controlling the pod borers, *H. armigera* and *H. punctigera* (Matthews, 1999). The extent of losses due to *H. armigera* in chickpea have been estimated to be over $328 million in the semi-arid tropics (ICRISAT, 1992). Worldwide, losses due to *Heliothis/Helicoverpa* in cotton, legumes, vegetables, cereals, etc. may exceed $2 billion, and the cost of insecticides used to control these pests may be over $1 billion annually (Sharma, 2005). Field surveys in the early 1980s indicated that less than 10% of the farmers used pesticides to control *H. armigera* in chickpea in India (Reed *et al.*, 1987). However, the shift from subsistence to commercial production and the resulting increase in prices have provided the farmers an opportunity to consider application of pest management options for increasing chickpea production (Shanower *et al.*, 1998).

Population monitoring and forecasting

Efforts have been made to develop a forewarning system for *H. armigera* on cotton, pigeonpea and chickpea in India (Das *et al.*, 1997; Puri *et al.*, 1999).

A thumb rule has been developed to predict *H. armigera* population using surplus/deficit rainfall in different months in South India (Das *et al.*, 2001). A combination of surplus rains during the monsoon and deficit rainfall during November indicated low incidence, while deficit rains during the monsoon and surplus rains during November (A–, B+) indicated severe attack. Additional information on November rainfall gives precise information on the level of attack (low, moderate or severe). In Australia, population monitoring with sex pheromone-baited traps is used to detect the onset of immigration or emergence from local diapause. Abundance of *H. armigera* and *H. punctigera* as measured by light traps showed that seasonal rainfall and local crop abundance gave a reasonable prediction of the timing of population events and the size of subsequent generations (Maelzer and Zalucki, 1999; Zalucki and Furlong, 2005). Timing of control is determined by field monitoring of larval densities in crops through the period of crop susceptibility. Control is only recommended when larval populations in post flowering crops exceed the threshold of 2–4 larvae per metre row (Lucy and Slatter, 2004).

Host-plant resistance

The development of crop cultivars resistant or tolerant to *H. armigera* has a major potential for use in integrated pest management, particularly under subsistence farming conditions in the developing countries (Fitt, 1989; Sharma and Ortiz, 2002). More than 14,000 chickpea germplasm accessions have been screened for resistance towards *H. armigera* at ICRISAT, Hyderabad, India, under field conditions (Lateef and Sachan, 1990). Several germplasm accessions (ICC 506EB, ICC 10667, ICC 10619, ICC 4935, ICC 10243, ICCV 95992 and ICC 10817) with resistance to *H. armigera* have been identified, and varieties such as ICCV 7, ICCV 10 and ICCL 86103 with moderate levels of resistance have been released for cultivation (Gowda *et al.*, 1983; Lateef, 1985; Lateef and Pimbert, 1990) (Table 25.2). Pedigree selection appears to be effective in selecting lines with resistance to *Helicoverpa*. However, most of these lines are highly susceptible to fusarium wilt. Therefore, concerted efforts are being made to break the linkage by raising a large population of crosses between *Helicoverpa* and wilt resistant parents.

Wild relatives of chickpea are an important source of resistance to leaf miner, *L. cicerina* and the bruchid, *C. chinensis* (Singh *et al.*, 1997). Based on leaf feeding, larval survival and larval weights, accessions belonging to *C. bijugum* (ICC 17206, IG 70002, IG 70003, IG 70006, IG 70012, IG 70016 and IG 70016), *C. judaicum* (IG 69980, IG 70032 and IG 70033), *C. pinnatifidum* (IG 69948) (Sharma *et al.*, 2005a) and *C. reticulatum* (IG 70020, IG 72940, IG 72948 and IG 72949, and IG 72964) (Sharma *et al.*, 2005b) showed resistance to *H. armigera*. With the use of interspecific hybridization, it would be possible to transfer resistance genes from the wild relatives to cultivated chickpea. Some of the wild relatives of chickpea may have different mechanisms of resistance than those in the cultivated types, which can be used in crop improvement to diversify the bases of resistance to this pest.

Table 25.2. Identification and utilization of host plant resistance to *Helicoverpa armigera*.

Genotypes	Remarks	Reference
Desi: short-duration		
ICC 506, ICCV 7 (ICCX 730041-1-1P-BP), ICC 10667, ICC 6663, ICC 10619, ICC 10817, ICCL 861992, ICCL 86103, ICCX 73008-8-1-IP-BP-EB, ICCX 730162-2-IP-B-EB, ICCX 730213-9-1-3HB, C 10, PDE 2, PDE 5, DPR/CE 72, DPR/CE 1-2, DPR/CE 3-1 and DPR/CE 2–3	DR < 3.8 compared to 6.0 in Annigeri	Lateef and Sachan (1990)
Desi: medium-duration		
ICC 4935-E-2793, ICCX 730094-18-2-IP-BP-EB, BDN 9-3, ICCX 730185-2-4- H1-EB, ICCX 730190-12-1H-B-EB, ICCX 730025-11-3-IH-EB, ICC 3474-4EB, ICC 5800, S 76, N 37 and PDE 1	DR < 4.6 compared to 8.5 in ICC 3137	
ICCL 86101, ICCL 86102, ICCL 86103 and ICCL 86104		
Desi: long-duration		
ICC 10243, ICCX 730020-11-1-1H-B-EB, GL 1002, Pant G 114 and PDE 7	DR 4.3 compared to 6.0 in H 208	
Kabuli: - medium-duration		
ICC 10870, ICC 5264-E10, ICC 8835, ICC 4856, ICC 7966, ICC 2553-3EB, ICC 2695-3EB, ICC 10243 and ICCX 730244-17-2-2H-EB	DR < 5.4 compared to 6.0 in L 550	
GL 645, Dhulia, 6–28, GGP Chaffa, P 1324-11, P 1697, P 6292 and selection 418	Suffered <5% pod damage compared to 16.1 to 36% damage in G 130 and L 550	Chhabra *et al.* (1990)
ICC 506EB, ICC 2397, ICC 6341, ICC 4958 and ICC 8304	Suffered <12% pod damage compared to 42% in ICC 14665	Bhagwat *et al.* (1995)
PDE 2-1, IC 16, Annigeri, BGM 42 and C 21–79	These lines had 6–9 larvae per meter row compared to 32 larvae in H 86-18	Chauhan and Dahiya (1994)
BG 372, B 390, GNG 469, PDE 2-1 and PDE 3-2	Performed better than H 82-2 based on pod damage and grain yield	
DHG 84-11, P 240, DHG 88-20, ICP 29, DHG 86-38, SG 90-55, KBG 1, H 83-83, NP 37, DHG 87-54, GNG 669 and SG 89-11	These varieties were better or on par with the commercial cultivars 240, P 256, C 235 and BR 77	Singh and Yadav (1999 a,b)
ICC 12475, ICC 12477, ICC 12478, ICC 12479, ICC 14876, ICCV 96782, ICCL 87316, ICCL 87317 and ICCV 95992	Stable resistance to pod borer across seasons	Sreelatha (2003); Lakshmi-narayanamma (2005)

Molecular marker-assisted selection (MAS) can be used to accelerate the introgression of desirable genes into improved cultivars (Sharma *et al.*, 2002). Preliminary results on development of molecular markers for resistance to *H. armigera* have been reported in chickpea (Lawlor *et al.*, 1998) based on bulk segregant analysis with amplified fragment length polymorphism (AFLP) analysis of F_2 and F_4 generations. Recombinant inbred lines (RILs) derived from ICCV2 × JG 62 cross have shown considerable variation for susceptibility to *H. armigera*. A skeletal molecular map is already available from this mapping population (Cho *et al.*, 2002). Another mapping population derived from ICC 506EB × Vijay is being currently evaluated for resistance to pod borer (H.C. Sharma, 2005, India, unpublished data). A susceptible *C. arietinum* variety (ICC 3137) has been crossed with a *C. reticulatum* accession (IG 72934) resistant to *H. armigera*, and the F_2 plants have been screened for resistance to *H. armigera*. Significant progress has been made over the last decade in introducing foreign genes into plants, providing opportunities to modify crops to increase yields, impart resistance to biotic and abiotic stresses and improve nutritional quality (Sharma *et al.*, 2004). Kar *et al.* (1997) developed transgenic chickpea plants with *cry1Ac* gene. Efforts are underway at ICRISAT and elsewhere to develop transgenic plants of chickpea with *Bacillus thuringiensis* (Bt) and soybean trypsin inhibitor (SBTI) genes for resistance to *H. armigera* (Sharma *et al.*, 2004). Efficient tissue culture and transformation methods by using *Agrobacterium tumefaciens* have been standardized at ICRISAT (Jayanand *et al.*, 2003).

Cultural manipulation of the crop and its environment

A number of cultural practices such as time of sowing, spacing, fertilizer application, deep ploughing, interculture and flooding have been reported to reduce the survival of and damage by *Helicoverpa* spp. (Lal *et al.*, 1980, 1985; Reed *et al.*, 1987; Murray and Zalucki, 1990; Shanower *et al.*, 1998; Romeis *et al.*, 2004). Intercropping or strip-cropping with marigold, sunflower, linseed, mustard and coriander can minimize the extent of damage to the main crop. Strip-cropping also increases the efficiency of chemical control. Hand-picking of large-sized larvae can also be practised to reduce *Helicoverpa* damage. However, the adoption of cultural practices depends on the crop husbandry practices in a particular agro-ecosystem. Rotations do not help manage these polyphagous and very mobile insects, although it has been noted that some crops (e.g. lucerne) are more attractive to the moths, and susceptible crops should not be planted too close to the main crop. Habitat diversification to enhance pest control has been attempted in Australia. An area-wide population management strategy has been implemented in regions of Queensland and New South Wales to contain the size of the local *H. armigera* population, and chickpea trap crops have played an important role in this strategy (Ferguson and Miles, 2002; Murray *et al.*, 2005b). Chickpea trap crops are planted after the commercial crops to attract *H. armigera* as they emerge from winter diapause. The emergence from diapause typically occurs when commercial chickpea has senesced, and before summer crops (sorghum, cotton and mung bean) are attractive to moths (October to November). However, moths are diverted to weeds for oviposition (including wheat, *Triticum aestivum*) when they grow above the chickpea crop canopy (Sequeira *et al.*, 2001). Trap crops are managed in the same way as commercial crops, but destroyed by cultivation before larvae begin to pupate. The trap crops reduce the size of the local *H. armigera* population before it can infest summer crops and start to increase in size. As a result, the overall *H. armigera* pressure on summer crops is reduced, resulting in greater opportunity for the implementation of soft control options, reduced insecticide use and greater natural enemy activity.

Biological control

The importance of both biotic and abiotic factors on the seasonal abundance of *H. armigera* is poorly understood. Low activity of parasitoids has been reported from chickpea because of dense layer of trichomes and their acidic exudates (Jalali *et al.*, 1988; Murray and Rynne, 1994; Romeis *et al.*, 1999). The ichneumonid, *Campoletis chlorideae* (Uchida), is probably the most important larval parasitoid on *H. armigera* in chickpea in India. *Carcelia illota* (Curran), *Goniophthalmus halli* Mesnil and *Palexorista laxa* (Curran) have also been reported to parasitize up to 54% larvae on chickpea (Yadava *et al.*, 1991; King, 1994; Romeis and Shanower, 1996), although Bhatnagar *et al.* (1983) recorded only 3% parasitism on chickpea. Predators such as *Chrysopa* spp., *Chrysoperla* spp., *Nabis* spp., *Geocoris* spp., *Orius* spp. and *Polistes* spp. are the most common in India. Provision of bird perches or planting of tall crops that serve as resting sites for insectivorous birds such as myna and drongo helps reduce the numbers of caterpillars.

The use of microbial pathogens including *H. armigera* nuclear polyhedrosis virus (HaNPV), entomopathogenic fungi, Bt, nematodes and natural plant products such as neem, custard apple and karanj (*Pongamia*) kernel extracts have shown some potential to control *H. armigera* (Sharma, 2001). HaNPV has been reported to be a viable option to control *H. armigera* in chickpea (Rabindra and Jayaraj, 1988; Cowgill and Bhagwat, 1996; Butani *et al.*, 1997; Ahmad *et al.*, 1999; Cherry *et al.*, 2000). Jaggery (0.5%), sucrose (0.5%), egg white (3%) and chickpea flour (1%) are effective in increasing the activity of HaNPV (Sonalkar *et al.*, 1998). In Australia, the efficacy of HaNPV in chickpea has been increased by the addition of milk powder, and more recently the additive Aminofeed® (Anonymous, 2005). Spraying Bt formulations in the evening results in better control than spraying at other times of the day (Mahapatro and Gupta, 1999). Entomopathogenic fungus, *Nomuraea rileyi* (10^6 spores per ml), results in 90–100% larval mortality, while *Beauveria bassiana* (2.68×10^7 spores per ml) resulted in 6% damage in chickpea compared to 16.3% damage in the untreated control plots (Saxena and Ahmad, 1997). In Australia, specific control of *H. armigera* and *H. punctigera* on chickpea is being achieved using the commercially available HaNPV, with an additive that increases the level of control. Bt formulations are also used as a spray to control *Helicoverpa*.

Chemical control

Management of *Helicoverpa* in India and Australia in chickpea and other high-value crops relies heavily on insecticides. There is substantial literature on the comparative efficacy of different insecticides against *Helicoverpa*. Endosulfan, cypermethrin, fenvalerate, thiodicarb, profenophos, spinosad and indoxacarb have been found to be effective for *H. armigera* control on chickpea in Australia (Murray *et al.*, 2005a). Spray initiation at 50% flowering has been found to be most effective (Sharma, 2001). The appearance of insecticide resistance in *H. armigera,* but not in *H. punctigera* is considered to be related to the greater mobility of the later species (Maelzer and Zalucki, 1999, 2000). However, *H. armigera* populations in the northern Australia are largely resistant to pyrethroids, carbamates and organophosphates. Introduction of new chemistry, notably indoxacarb and spinosad, is being managed to minimize the development of resistance in *H. armigera* through a strategy that takes into account its use in all crops throughout the year (Murray *et al.*, 2005a). Consequently, the use of indoxacarb in chickpea is limited to one application with a cut-off date for application to ensure that one generation of *H. armigera* is not exposed to the product in any crop before the commencement of its use in summer crops (cotton and mung bean).

Leaf Miner: *Liriomyza cicerina*

The leaf miner, *L. cicerina,* is an important pest of chickpea in the Mediterranean region and eastern Europe (Weigand *et al.*, 1994). It has also been reported from North India (Naresh and Malik, 1986). Efforts are currently underway at ICARDA, Aleppo, Syria, to breed lines that combine leaf miner resistance and high yield. Spraying with neem seed kernel extract is relatively effective, but the persistence is limited (Weigand *et al.*, 1994). Studies in Syria have also identified a parasitic wasp (*Opius* sp.) that feeds on the leaf miner larvae, but further research is required before this insect can be used for biological control in the field. *Opius monilicornis* Fischer parasitizes the larvae of *L. cicerina* in May and June in chickpea fields (M. El Bouhssini, 2006, Syria, unpublished data). It was observed that *L. cicerina* parasitism was 0–23.91%. Early-sown crops usually escape leaf miner damage.

Black Cutworm: *Agrotis ipsilon*

The black cutworm is a pest of chickpea, pea, lentil, potato and other crops in North India (Ahmad, 2003). It cuts the plants and drags them into cracks between soil clods. Dry weather during April–May affects the cutworms adversely. *A. flammatra* Schiff and *A. spinifera* (Hub.) are of minor importance. Heavy damage by cutworms occur in areas that remain flooded during the rainy season. *A. ipsilon* has four generations in North India. Chaudhary and Malik (1983) reported up to 9.5% plant damage at 40 days after crop emergence. Eggs are laid on earth clods, and on the chickpea plants. The pre-oviposition and oviposition periods vary from 3.9 to 5.5 and 5.8 to 8.3 days, respectively. A female may lay as many as 639–2252 eggs, and the egg incubation, larval and pupal periods vary from 2.7 to 5.1, 18.2 to 39.5 and 31.4 to 69.8 days, respectively. Larval mortality is as high as 70% during the early instars. In summer, it survives on the weeds in wastelands. It has been suggested that it may migrate to hills during the summer. Ploughing the fields before planting and after crop harvest reduces cutworm damage. The plants at times are able to recover from cutworm damage. Endosulfan dusts or sprays (Chaudhary and Malik, 1981; Kumar *et al.*, 1983) and endosulfan bait have been found to be effective for cutworm control. In India, the braconids such as *Microgaster* sp., *Bracon kitcheneri* (Will.) and *Fileanta ruficanda* (Cam.) parasitize the cutworm larvae, while *Broscus punctatus* (Klug.) and *Liogryllus bimaculatus* (DeGeer) are common predators (Nair, 1975).

Aphid: *Aphis craccivora*

The black aphid, *A. craccivora* causes substantial damage to chickpea in North India. This aphid is capable of transmitting a number of viral diseases in chickpea (Kaiser *et al.*, 1990). The most important is a strain of the bean leaf roll luteovirus, the incitant of chickpea stunt disease, which has assumed economic importance (Nene and Reddy, 1976). It is transmitted in a persistent manner. Chickpea chlorotic dwarf, a monogeminivirus (Horn *et al.*, 1995), is transmitted in a persistent, non-propagative and circulative manner by the leafhopper, *Orosius orientalis* (Matsumura) (Horn *et al.*, 1994). In Australia, lucerne mosaic virus, subterranean clover red leaf virus, beet western yellow virus, cucumber mosaic virus and bean leaf roll virus infect chickpea (Knights and Siddique, 2002). Aphids transmit many of these viruses, and may require chemical sprays (Loss *et al.*, 1998). Both nymphs and adults suck the sap from the leaves and the pods, causing depletion of photosynthates. In case of severe infestation, the leaves and shoots are deformed, and the plants become stunted. Life cycle from nymph to adult stage is completed in 8–10 days. Varieties with low trichome density or devoid of trichomes are highly susceptible to aphid damage. The aphids are active throughout the year (Bakhetia and Sidhu, 1977). A gravid female can produce over 100 nymphs in 15 days (Talati and Bhutani, 1980). The aphid incidence is greater under drought conditions. Nymphs undergo four molts. There are three population peaks on chickpea at Hisar, Haryana, India (Sithanantham *et al.*, 1984). Early sowing leads to early canopy closure, which also helps reduce virus spread in chickpea. Aphid infestation is greater under wider spacing. The genotypes, H 75-35 and H 2184, are less susceptible (Lal *et al.*, 1989). Additionally, Mushtaque (1977) observed that the lines H 6560, H 6576 and H 424 were less susceptible to aphid damage. *Coccinella septempunctata* L., *C. transversalis* (F.), *C. nigritis* (F.), *Cheilomenes sexmaculatus* (F.), *Brumus suturalis* (F.), *Chrysoperla* spp. and *Ischiodan javana* (Weid.) are common predators, while *Trixys indicus* (Subbarao & Sharma) and *Lipolexix scutellaris* Mackaur are important parasitoids (Singh and Tripathi, 1987). Generally, there is no need for aphid control on chickpea in India, but chemical control may become necessary to prevent secondary spread of the chickpea viruses (Reed *et al.*, 1987). A number of insecticides such as methomyl, oxy-demeton methyl and monocrotophos are effective for aphid control. *Aphis craccivora* has also developed resistance to some commonly used insecticides (Dhingra, 1994).

Semilooper: *Autographa nigrisigna*

The semilooper, *A. nigrisigna* occasionally damages chickpea in North India in January. The larvae feed on leaf buds, flowers and the young pods. The young larvae scratch the leaf, which becomes whitish. The larvae of *A. nigrisigna* feed on the whole pod, leaving the peduncle behind. The egg, larval and pupal stages last for 3–6, 8–80 and 5–13 days, respectively (Mahmood *et al.*, 1984). Males live for 4–5 days, while the females survive for 7–9 days. One generation is completed in 18–52 days. Its populations increase under high humidity. Endosulfan, phosalone, dichlorvos and malathion have been recommended for controlling this pest (Rizvi and Singh, 1983; Mahmood *et al.*, 1984; Chhabra and Kooner, 1985). Bt formulations have also been found to be effective against this pest (Saxena and Ahmad, 1997).

Bruchids: *Callosobruchus* spp.

In India, bruchid infestation levels approaching 13% have been reported (Mookherjee *et al.*, 1970; Dias and Yadav, 1988). Total losses have been reported from the Near East (Weigand and Tahhan, 1990). Extensive screening of kabuli type chickpea has not shown any acceptable level of resistance (Weigand and Pimbert, 1993). However, high levels of resistance have been observed in desi type chickpea (Raina, 1971; Schalk *et al.*, 1973; Weigand and Tahhan, 1990). Lines showing resistance to bruchids usually have small seeds with a rough seed coat. Such grain is not acceptable to the consumers (Reed *et al.*, 1987). Chickpea seed that is split for *dhal* is unattractive to ovipositing bruchid females (Reed *et al.*, 1987). Chemical insecticides are little used in chickpea storage in India (Srinivasu and Naik, 2002). Biological control of bruchids has not really been exploited in India. For more information, see Van Huis (1991) for a comprehensive review of biological control of bruchids in the tropics.

Need for Future Research

Insect-resistant chickpea cultivars will form the backbone of integrated pest management in future. The development and deployment of chickpea plants with resistance to insects would offer the advantage of allowing some degree of selection for specificity effects, so that pests, but not the beneficial organisms, are targeted.

Deployment of insect-resistant chickpea will result in decreased use of chemical pesticides and increased activity of natural enemies, and thus, higher yields. For pest management programmes to be effective in future, there is a need for:

- In-depth understanding of the population dynamics of insect pests in chickpea growing areas to develop appropriate control strategies;
- Combined resistance to insects with resistance to important diseases and cold tolerance;
- Utilization of wild relatives of chickpea to diversify the genetic basis, and thus increase the levels of resistance to the target insect pests;
- Identification of quantitative trait loci (QTLs) associated with resistance to insects to increase the levels of resistance through gene pyramiding;
- Development of insect-resistant varieties through genetic transformation using genes with diverse modes of action;
- Insecticide resistance management, development of biopesticides with stable formulations and strategies for conservation of natural enemies for integrated pest management.

Conclusion

Nearly 60 insect species are known to feed on chickpea, of which cutworms (black cutworm – *Agrotis ipsilon* and turnip moth – *A. segetum*), leaf-feeding caterpillars (leaf caterpillar – *Spodoptera exigua* and hairy caterpillar – *Spilarctia oblique*), leaf miners (*Liriomyza cicerina*), aphids (*Aphis craccivora*), pod borers (cotton bollworm – *Helicoverpa armigera* and native budworm – *H. punctigera*) and the bruchids (*Callosobruchus* spp.) are the major pests worldwide. The pod borer, *H. armigera* and aphids, *A. craccivora* (as a vector of chickpea stunt virus) are the major pests in the Indian subcontinent; while the leaf miner, *L. cicerina* is an important pest in the Mediterranean region. Bruchids, *Callosobruchus* spp. cause extensive losses in storage all over the world. Low to moderate levels of resistance have been identified in the germplasm, and a few improved varieties with resistance to pod borer and high grain yield have been developed. Germplasm accessions of the wild relatives of chickpea (*Cicer bijugum*, *C. judaicum* and *C. reticulatum*) can be used to increase the levels and diversify the bases of resistance to *H. armigera*. Efforts are also underway to utilize molecular techniques to increase the levels of resistance to pod borer. Synthetic insecticides, agronomic practices, nuclear polyhedrosis virus (NPV), entomopathogenic fungi, bacteria and natural plant products have been evaluated as components of pest management in chickpea.

References

Ahmad, R. (2003) Insect pests of chickpea and their management. In: *Chickpea Research in India*. Indian Institute of Pulses Research, Kanpur, Uttar Pradesh, India, pp. 229–260.

Ahmad, R., Yadava, C.P. and Lal, S.S. (1999) Efficacy of nuclear polyhedrosis virus for the management of *Helicoverpa armigera* (Hubner) infesting chickpea. *Indian Journal of Pulses Research* 12, 92–96.

Anonymous (2005) Using NPV to manage *Helicoverpa* in field crops. Queensland Department of Primary Industries and Fisheries, Brisbane, Queensland, Australia. Publication ISSN 0727-6273 QI04080.

Bakhetia, D.R.C. and Sidhu, A.S. (1977) Biology and seasonal activity of the groundnut aphids, *Aphis craccivora* Koch. *Journal of Research, Punjab Agriculture University* 14, 299–309.

Bhagwat, V.R., Aherker, S.K., Satpute, V.S. and Thakre, H.S. (1995) Screening of chickpea (*Cicer arietinum* L.) genotypes for resistance to *Helicoverpa armigera* (Hb.) and its relationship with malic acid in leaf exudates. *Journal of Entomological Research* 19, 249–253.

Bhatnagar, V.S., Sithanantham, S., Pawar, C.S., Jadhav, D.S., Rao, V.K. and Reed, W. (1983) Conservation and augmentation of natural enemies with reference to IPM in chickpea and pigeonpea. In: *Proceedings of the International Workshop on Integrated Pest Control in Grain Legumes*, 4–9 April 1983. Empresa Braseleira Pesquisa Agropocuria (EMRAPA), Goiania, Brazil, pp. 157–180.

Butani, P.G., Kapadia, M.N. and Parsana, G.J. (1997) Comparative efficacy and economics of nuclear polyhedrosis virus (NPV) for the control of *Helicoverpa armigera* (Hubner) on groundnut. *Journal of Oilseeds Research* 14, 85–87.

Chaudhary, J.P. and Malik, V.S. (1981) Extent of damage and screening of dust formulations of insecticides against gram cutworm *Agroris ipsilon* Hufn. on gram, *Cicer arietinum* L. *Indian Journal of Entomology* 43, 153–157.

Chaudhary, J.P. and Malik, V.S. (1983) Some observations on the reproductive biology of gram cut worm, *Agrotis ipsilon* Hufn. *Bulletin of Entomology* 24, 31–34.

Chauhan, R. and Dahiya, B. (1994) Response of different chickpea genotypes to *Helicoverpa armigera* at Hissar. *Indian Journal of Plant Protection* 22, 170–172.

Cherry, A.J., Rabindra, R.J., Parnell, M.A., Geetha, N., Kennedy, J.S. and Grzywacz, D. (2000) Field evaluation of *Helicoverpa armigera* nucleopolyhedrovirus formulations for control of the chickpea podborer, *H. armigera* (Hubn.), on chickpea (*Cicer arietinum* var. Shoba) in southern India. *Crop Protection* 19, 51–60.

Chhabra, K.S. and Kooner, B.S. (1985) Synthetic pyrethroids for the control of gram pod borer, *Heliothis armigera* (Hub.) on gram (*Cicer arietinum* L.). *Pesticides* 19, 44–65.

Chhabra, K.S., Kooner, B.S., Sharma, A.K. and Saxena, A.K. (1990) Sources of resistance in chickpea, role of biochemical components on the incidence of gram podborer *Helicoverpa* (*Heliothis*) *armigera* (Hubner). *Indian Journal of Entomology* 52, 423–430.

Cho, S., Kumar, J., Shultz, J., Anupama, K., Tefera, F. and Muehlbauer, F.J. (2002) Mapping genes for double podding and other morphological traits in chickpea. *Euphytica* 128, 285–292.

Clement, S.L. (1999) Ex situ genebank practices: protecting germplasm nurseries from insect pests. *Plant Genetic Resources Newsletter* 119, 46–50.

Clement, S.L., Wightman, J.A., Hardie, D.C., Bailey, P., Baker, G. and McDonald, G. (2000) Opportunities for integrated management of insect pests of grain legumes. In: Knight, R. (ed.) *Linking Research and Marketing Opportunities for Pulses in the 21st Century*. Kluwer Academic, Dordrecht, The Netherlands, pp. 467–480.

Cowgill, S.E. and Bhagwat, V.R. (1996) Comparison of the efficacy of chemical control and *Helicoverpa* NPV for the management of *Helicoverpa armigera* (Hubner) on resistant and susceptible chickpea. *Crop Protection* 15, 241–246.

Das, D.K., Sharma, O.P., Trivedi, T.P., Puri, S.N., Srivastava, C.P., Wightman, J.A., Shanower, T.G., Bilapate, G.G. and Nathan, K.K. (1997) Forecasting of *Helicoverpa armigera*. In: *Symposium on Integrated Pest Management for Sustainable Crop Production*. Indian Agricultural Research Institute, New Delhi, India.

Das, D.K., Trivedi, T.P. and Srivastava, C.P. (2001) Simple rule to predict attack of *Helicoverpa armigera* on crops growing in Andhra Pradesh. *The Indian Journal of Agricultural Sciences* 71, 421–423.

Dhingra, S. (1994) Development of resistance in the bean aphid, *Aphis craccivora* Koch.

to various insecticides used for nearly a quarter century. *Journal of Entomological Research* 18, 105–108.

Dias, C.A.R. and Yadav, T.D. (1988) Incidence of pulse beetles in different legume seeds. *Indian Journal of Entomology* 50, 457–461.

FAO (2003) FAOSTAT, Agriculture. Food and Agriculture Organization of the United Nations, Rome, Italy. Available at: http://apps.fao.org

Ferguson, J.M. and Miles, M.M. (2002) Area-wide pest management on the Darling Downs: has it worked? In: *Proceedings of the 11th Australian Cotton Conference*. ACGRA, Brisbane, Australia, pp. 711–719.

Fitt, G.P. (1989) The ecology of *Heliothis* species in relation to agroecosystems. *Annual Review of Entomology* 34, 17–52.

Glogoza, P. (2005) *Field Crop Insect Management Recommendations*. North Dakota State University Extension Bulletin E-1143, Fargo, North Dakota.

Gowda, C.L.L., Lateef, S.S., Smithson, J.B. and Reed, W. (1983) Breeding for resistance to *Heliothis armigera* in chickpea. In: *Proceedings of the National Seminar on Breeding Crop Plants for Resistance to Pests and Diseases*, 25–27 May 1983. School of Genetics, Tamil Nadu Agricultural University, Coimbatore, Tamil Nadu, India, pp. 36–39.

Horn, N.M., Reddy, S.V. and Reddy, D.V.R. (1994). Virus–vector relationships of chickpea chlorotic dwarf geminivirus and the leafhopper *Orosius orientalis* (Hemiptera, Cicadellidae). *Annals of Applied Biology* 124, 441–450.

Horn, N.M., Reddy, S.V. and Reddy, D.V.R. (1995). Assessment of yield losses caused by chickpea chlorotic dwarf geminivirus in chickpea (*Cicer arietinum*) in India. *European Journal of Plant Pathology* 101, 221–224.

ICRISAT (International Crops Research Institute for the Semi-Arid Tropics). (1992) *Mid Term Plan*. ICRISAT, Hyderabad, India.

Jalali, S.K., Singh, S.P., Kumar, P. and Ballal, C.R. (1988) Influence of the food plants on the degree of parasitism of larvae of *Heliothis armigera* by *Cotesia kazak*. *Entomophaga* 33, 65–71.

Jayanand, B., Sudarsanam, G. and Sharma, K.K. (2003) An efficient protocol for regeneration of whole plant of chickpea (*Cicer arietinum* L.) by using axillary meristem explant derived from in vitro germinated seedlings. *In Vitro Cellular & Developmental Biology – Plant* 39, 171–179.

Kaiser, W.J., Ghanekar, A.M., Nene, Y.L., Rao B.S. and Anjaiah, V. (1990) Viral diseases of chickpea. In: *Chickpea in the Nineties: Proceedings of the Second International Workshop on Chickpea Improvement*, 4–8 December 1989. ICRISAT, Hyderabad, India, pp. 139–142.

Kar, S., Basu, D., Das, S., Ramkrishnan, N.A., Mukherjee, P., Nayak, P. and Sen, S.K. (1997) Expression of cryIA(c) gene of *Bacillus thuringiensis* in transgenic chickpea plants inhibits development of pod borer (*Heliothis armigera*) larvae. *Transgenic Research* 6, 177–185.

King, A.B.S. (1994) *Heliothis/Helicoverpa* (Lepidoptera: Noctuidae). In: Matthews, G.A. and Tunstall, J.P. (eds) *Insect Pests of Cotton*. CAB International, Wallingford, UK, pp. 39–106.

Knights, E.J. and Siddique, K.H.M. (2002) Chickpea status and production constraints in Australia. In: Bakr, M.A., Siddique, K.H.M. and Johansen, C. (eds) *Integrated Management of Botrytis Grey Mould of Chickpea in Bangladesh and Australia*. Australian Centre for International Agricultural Research, ACT, Canberra, Australia, pp. 33–41.

Kumar, K., Sinha, R.B.P. and Sinha, P.K. (1983) Testing of different insecticidal formulations against greasy surface caterpillars, *Agroris ipsilon* Rutt. *Journal of Soil Biology and Ecology* 3, 35–38.

Lal, S.S., Dias, C.A.R., Yadava, C.P. and Singh, D.N. (1980) Effect of sowing dates on the infestation of *Heliothis armigera* (Hub.) and yield. *International Chickpea Newsletter* 3, 14–15.

Lal, S.S., Sachan, J.N. and Chandra, S. (1985) Cultural and varietal tools for integrated pest management for increasing pulse production. *Plant Protection Bulletin* 37, 1–10.

Lal, S.S., Yadava, C.P., Dias, C.A.R. and Nigam, R. (1989) Effect of planting density and chickpea cultivars on the infestation of black aphid, *Aphid craccivora* Koch. *Madras Agricultural Journal* 76, 461–462.

Lateef, S.S. (1985). Gram pod borer (*Heliothis armigera* (Hub.) resistance in chickpea. *Agriculture, Ecosystem, and Environment* 14, 95–102.

Lateef, S.S. and Pimbert, M.P. (1990) The search for host plant resistance of *Helicoverpa armigera* in chickpea and pigeonpea at ICRISAT. In: *Proceedings of the Consultative Group Meeting on the Host Selection Behavior of Helicoverpa armigera*, 5–7 March 1990. ICRISAT, Hyderabad, India, pp. 185–192.

Lateef, S.S. and Sachan, J.N. (1990) Host plant resistance to *Helicoverpa armigera* (Hub.) in different agro-economical conditions. In: *Chickpea in Nineties: Proceedings of the Second International Workshop on Chickpea*, 4–8 December 1989. ICRISAT, Hyderabad, India, pp. 181–189.

Lawlor, H.J., Siddique, K.H.M., Sedgley, R.H. and Thurling, N. (1998) Improving cold tolerance and insect resistance in chickpea and the use of AFLPs for the identification of molecular markers for these traits. *Acta Horticulturae* 461, 185–192.

Lakshminarayanamma, V. (2005) Genetics of resistance to pod borer, *Helicoverpa armigera* in chickpea (*Cicer arietinum*). PhD thesis. ANGR Agricultural University, Hyderabad, India.

Loss, S., Brandon, N. and Siddique, K.H.M. (eds) (1998). *The Chickpea Book*. Bulletin 1326. Western Australian Department of Agriculture, Perth, Australia.

Lucy, M. and Slatter, J. (2004) Chickpeas. In: Lucy. M., French, V. and Bullen, K. (eds) *Crop Management Notes*, Winter Edition. Queensland Department of Primary Industries, Brisbane, Queensland, Australia.

Maelzer, D.A. and Zalucki, M.P. (1999) Analysis of long-term light-trap data for *Helicoverpa* spp. (Lepidoptera: Noctuidae) in Australia; the effect of climate and crop host plants. *Bulletin of Entomological Research* 89, 455–463.

Maelzer, D.A. and Zalucki, M.P. (2000) Long-range forecasts of the numbers of *Helicoverpa punctigera* and *H. armigera* (Lepidoptera: Noctuidae) in Australia using the Southern Oscillation Index and the Sea Surface Temperature. *Bulletin of Entomological Research* 90, 133–146.

Mahmood, I., Shah, T. and Shah, H.A. (1984) Biology and chemical control of *Autographa* (=*Plusia*) *nigrisigna* Wlk. (Noctuidae: Lepidoptera), an unusual insect pest of gram, *Cicer arietinum* L. *Pakistan Journal of Zoology* 16, 159–163.

Mahapatro, G.K. and Gupta, G.P. (1999) Evenings suitable for spraying Bt formulations. *Insect Environment* 5, 126–127.

Margheim, J., Baltensperger, D.D., Wilson, R.G., Lyon, D.J., Hein, G.L., Harveson, R.M., Burgener, P., Krall, J.M., Cecil, J.T., Rickertsen, J.R., Merrigan, A.P., Watson, M.H. and Hansen, B.J. (2004) *Chickpea Production in the High Plains*. University of Nebraska Cooperative Extension EC04-183, Lincoln, Nebraska.

Matthews, M. (1999) *Heliothine Moths of Australia. A Guide to Pest Bollworms and Related Noctuid Groups*. Monograph on Australian Lepidoptera, Volume 7. CSIRO Publishing, Callingford, Victoria, Australia.

Mookherjee, P.B., Jotwani, M.G., Yadav, T.D. and Sircar, P. (1970) Studies on incidence and extent of damage due to insect pests in stored seeds. II. Leguminous and vegetable seeds. *Indian Journal of Entomology* 32, 350–355.

Miller, P., McKay, K., Jenks, B., Riesselman, J., Neill, K., Buschena, D. and Bussan, A.J. (2002) Growing chickpea in the Northern Great Plains. Montana State University Extension Service Bulletin 2002, Bozeman, Montana.

Murray, D.A.H. and Rynne, K.P. (1994) Effect of host plant on parasitism of *Helicoverpa armigera* (Lep.: Noctuidae) by *Microplitis demolitor* (Hym.: Braconidae). *Entomophaga* 39, 251–255.

Murray, D.A.H. and Zalucki, M.P. (1990) Survival of *Helicoverpa punctigera* (Wallengren) and *H. armigera* (Hubner) (Lepidoptera: Noctuidae) pupae submerged in water. *Journal of Australian Entomological Society* 29, 191–192.

Murray, D.A.H., Lloyd, R.J. and Hopkinson, J.E. (2005a) Efficacy of new insecticides for management of *Helicoverpa* spp.

(Lepidoptera: Noctuidae) in Australian grain crops. *Australian Journal of Entomology* 44, 62–67.

Murray, D.A.H., Miles, M.M., Mclennan, A.J., Lloyd, R.J. and Hopkinson, J.E. (2005b) Area-wide management of *Helicoverpa* spp. in an Australian mixed cropping agroecosystem. In: *Proceedings of the 2005 Beltwide Cotton Conference*. New Orleans, Louisiana.

Mushtaque, A. (1977) Preliminary studies on the infestation of *Aphis craccivora* Koch to chickpea and lentil. *Journal of Agricultural Research* 15, 31–35.

Nair, M.R.G.K. (1975) *Insects and Mites of Crops in India*. Indian Council of Agricultural Research, New Delhi, India.

Naresh, J.S. and Malik, V.S. (1986) Observations on the insect pests of chickpea (*Cicer arietinum* L.) in Haryana. *Bulletin of Entomology* 27, 75–77.

Nene, Y.L. and Reddy, M.V. (1976) Preliminary information on chickpea stunt. *Tropical Grain Legume Bulletin* 5, 31–32.

Puri, S.N., Das, D.K., Trivedi, T.P., Sharma O.P., Srivastava, C.P., Wightman, J.A., Shanower, T.G., Bilapate, G.G. and Nathan, K.K. (1999) Modeling to predict *Helicoverpa armigera* in Deccan region (abstract). In: *National Workshop on Dynamic Crop Simulation Modeling for Agrometeorological Advisory Services*. National Centre for Median Range Weather Forecasting, Department of Science and Technology, New Delhi, India.

Rabindra, R.J. and Jayaraj, S. (1988) Efficacy of nuclear polyhedrosis virus with adjuvants as high volume and ultra low volume applications against *Helicoverpa armigera* on chickpea. *Tropical Pest Management* 34, 441–444.

Raina, A.K. (1971) Comparative resistance to three species of *Callosobruchus* in a strain of chickpea (*Cicer arietinum* L.). *Journal of Stored Product Research* 7, 213–216.

Reed, W., Cardona, C., Sithanantham, S. and Lateef, S.S. (1987) The chickpea insect pests and their control. In: Saxena, M.C. and Singh, K.B. (eds) *The Chickpea*. CAB International, Wallingford, UK, pp. 283–318.

Rizvi, S.M.A. and Singh, H.M. (1983) Chemical control of *Autographa nigrisigna* Wlk. and *Heliothis armigera* Hub. in chickpea. *Pesticides* 17, 27.

Romeis, J. and Shanower, T.G. (1996) Arthropod natural enemies of *Helicoverpa armigera* (Hübner) (Lepidoptera: Noctuidae) in India. *Biocontrol Science and Technology* 6, 481–508.

Romeis, J., Shanower, T.G. and Zebitz, C.P.W. (1999) Why *Trichogramma* (Hymenoptera: Trichogrammatidae) egg parasitoids of *Helicoverpa armigera* (Lepidoptera: Noctuidae) fail on chickpea. *Bulletin of Entomological Research* 89, 89–95.

Romeis, J., Sharma, H.C., Sharma, K.K., Das, S. and Sharmah, B.K. (2004) The potential of transgenic chickpeas for pest control and possible effects on non-target arthropods. *Crop Protection* 23, 923–938.

Saxena, H. and Ahmad, R. (1997) Field evaluation of *Beauveria bassiana* (Balsamo) Vuillemin against *Helicoverpa armigera* (Hubner) infecting chickpea. *Journal of Biological Control* 11, 93–96.

Schalk, J.M., Evans, K.H. and Kaiser, W.J. (1973) Resistance in lines of chickpea to attack by *Callosobruchus maculatus* in Iran. *FAO Plant Protection Bulletin* 21, 126–131.

Sequeira, R.V., McDonald, J.L., Moore, A.D., Wright, G.A. and Wright, L.C. (2001) Host plant selection by *Helicoverpa* spp. in chickpea – companion cropping systems. *Entomologia Experimentalis et Applicata* 101, 1–7.

Shanower, T.G., Kelley, T.G. and Cowgill, S.E. (1998) Development of effective and environmentally sound strategies to control *Helicoverpa armigera* in pigeonpea and chickpea production systems. In: Saini, R.K. (ed.) *Tropical Entomology: Proceedings of the Third International Conference on Tropical Entomology*. International Center for Insect Physiology and Ecology (ICIPE) Science Press, Nairobi, Kenya, pp. 239–260.

Sharma, H.C. (2001) *Crop Protection Compendium:* Helicoverpa armigera. *Electronic Compendium for Crop Protection*. CAB International, Wallingford, UK.

Sharma, H.C. (ed.) (2005) Heliothis/Helicoverpa *Management: Emerging Trends and Strategies for Future Research*. Oxford and IBH, New Delhi, India.
Sharma, H.C. and Ortiz, R. (2002) Host plant resistance to insects: an eco-friendly approach for pest management and environment conservation. *Journal of Environmental Biology* 23, 11–35.
Sharma, H.C., Crouch, J.H., Sharma, K.K., Seetharama, N. and Hash, C.T. (2002) Applications of biotechnology for crop improvement: prospects and constraints. *Plant Science* 163, 381–395.
Sharma, H.C., Pampapathy, G., Lanka, S.K. and Ridsdill-Smith, T.J. (2005a) Antibiosis mechanism of resistance to pod borer, *Helicoverpa armigera* in wild relatives of chickpea. *Euphytica* 142, 107–117.
Sharma, H.C., Pampapathy, G. and Ridsdill-Smith, T.J. (2005b) Exploitation of wild *Cicer reticulatum* germplasm for resistance to *Helicoverpa armigera*. *Journal of Economic Entomology* 98, 2246–2253.
Sharma, H.C., Sharma, K.K. and Crouch, J.H. (2004) Genetic transformation of crops for insect resistance: potential and limitations. *CRC Critical Reviews in Plant Sciences* 23, 47–72.
Singh, B. and Yadav, R.P. (1999a) Location of sources of resistance amongst chickpea (*Cicer arietinum* L.) genotypes against gram pod borer (*Heliothis armigera* Hub.) under normal sown conditions using new parameters. *Journal of Entomological Research* 23, 19–26.
Singh, B. and Yadav, R.P. (1999b) Field screening of chickpea (*Cicer arietinum* L.) genotypes against gram pod borer (*Heliothis armigera* Hub.) under late sown conditions. *Journal of Entomological Research* 23, 133–140.
Singh, K.B., Weigand, S. and Saxena, M.C. (1997) Registration of ILWC 39 and ILWC 181: *Cicer echinospermum* germplasm lines with resistance to *Callosobruchus chinensis* (L.). *Crop Science* 37, 634.
Singh, R. and Tripathi, N. (1987) Record of parasitoids from Tarai belt of eastern Uttar Pradesh. *Journal of Aphidology* 1, 89–92.
Sithanantham, S., Sethi, S.C. and Benitwal, S.P.S. (1984) A preliminary study of incidence of *Aphis craccivora* in chickpea at Hisar, India. *International Chickpea Newsletter* 10, 19–20.
Sonalkar, V.U., Deshmukh, S.D., Satpute, U.S. and Ingle, S.T. (1998) Efficacy of nuclear polyhedrosis virus in combination with adjuvants against *Helicoverpa armigera* (HBN). *Journal of Soils and Crops* 81, 67–69.
Sreelatha, E. (2003) *Stability, Inheritance and Mechanisms of Resistance to Helicoverpa armigera (Hub). in Chickpea*. PhD thesis. ANGR Agricultural University, Hyderabad, India.
Srinivasu, C.S. and Naik, L.K. (2002) Survey for adoptable indigenous methods for the control of stored grain pests. *Karnataka Journal of Agricultural Sciences* 15, 715–716.
Talati, G.M. and Bhutani, P.G. (1980) Reproduction and population dynamics of groundnut aphids. *Gujarat Agricultural University, Journal of Research* 5, 54–56.
Van Huis, A. (1991) Biological control methods of bruchid control in the tropics: a review. *Insect Science and its Application* 12, 87–102.
Weigand, S. and Tahhan, O. (1990) Chickpea insect pests in the Mediterranean zones and new approaches to their management. In: *Chickpea in the Nineties: Proceedings of the Second International Workshop on Chickpea Improvement*, 4–8 December 1989. ICRISAT, Hyderabad, India, pp. 169–175.
Weigand, S. and Pimbert, M.P. (1993) Screening and selection criteria for insect resistance in cool-season food legumes. In: Singh, K.B. and Saxena, M.C. (eds) *Breeding for Stress Tolerance in Cool Season Food Legumes*. Wiley, Chichester, UK, pp. 145–156.
Weigand, S., Lateef, S.S., El Din Sharaf, N., Mahmoud, S.F., Ahmed, K. and Ali, K. (1994) Integrated control of insect pests of cool season food legumes. In: Muehlbauer, F.J. and Kaiser, W.J. (eds) *Expanding the Production and Use of Cool Season Food Legumes*. Kluwer Academic, Dordrecht, The Netherlands, pp. 679–694.

Williams, L., O'Keeffe, L.E. and Schotzko, D.J. (1991) Chickpea, *Cicer arietinum*: a new host for the pea leaf weevil, *Sitona lineatus*. *Field Crops Research* 27, 377–380.

Yadava, C.P., Sachan, J.N., Ahmad, R. and Lal, S.S. (1991) Record of natural enemies of pests of chickpea and pigeonpea. *Journal of Biological Control* 5, 52–54.

Zalucki, M.P. and Furlong, M.J. (2005) Forecasting *Helicoverpa* populations in Australia: a comparison of regression based models and a bioclimatic based modeling approach. *Insect Science* 12, 45–56.

26 Storage of Chickpea

C.J. Demianyk,[1] N.D.G. White[1] and D.S. Jayas[2]

[1]*Cereal Research Centre, Agriculture and Agri-Food Canada, 195 Dafoe Road, Winnipeg, MB R3T 2M9, Canada;* [2]*University of Manitoba, Winnipeg, MB R3T 2N2, Canada*

Introduction

In today's world where prices for cereal crops have been severely depressed, chickpea has been seen as an economically beneficial crop in some traditional wheat-growing areas of Canada, the USA and Australia. With its increase of importance in these new markets information is required for safe handling and storage. Modern technology can provide knowledge and equipment to handle and store grains under ideal conditions such that damage and spoilage are kept to a minimum; however, the majority of chickpea production occurs in areas of the world where this technology may be lacking or is too expensive for grass roots production and storage. India produces 68% of the world's chickpea (Food and Agriculture Organization, 2006), most being grown for personal consumption or for local market. Therefore, much literature is available reporting on traditional techniques and developing methods for chickpea storage under tropical and semi-tropical conditions aimed at helping family-sized rural farmers protect their stocks. With a shift towards larger production in developed countries with more temperate climates, storage information and technology developed for cereals and oilseeds is being adapted and improved upon to provide optimal storage conditions for chickpea under these climatic conditions.

Harvest and Handling

Storage guidelines for chickpea recommend following those of pulses in general (Hnatowich, 2000). Susceptibility of pulse crops to mechanical damage from harvest is extremely high. Crops should be handled gently to reduce seed damage including splitting, cracking and reduced germination since the seedling root of chickpea projects from the seed; kabuli seed is more sensitive to damage also because of its thin seed coat (Hnatowich, 2000). It is recommended that chickpea be handled no more than six times before delivery to minimize damage (Department of Primary Industries and Fisheries, 2005). Even non-visible damage to the seed may reduce germination by allowing infection by fungi. The use of belt conveyers is more efficient at preventing damage than using spiral augers; however, several techniques can reduce the impact of augering the seed. Using a large-diameter auger at low incline, operating slowly and fully with an optimal flight clearance of typically 50% of grain size will minimize damage (Department of Primary Industries and Fisheries, 2005).

Physical properties of different seed types are important and required in the design of equipment for postharvest cleaning, conveying and storage, as well as planting and harvesting. With chickpea this is somewhat complicated by the existence of two distinct desi and kabuli varieties. Masoumi and Tabil (2003) reported on bulk density, porosity, static coefficient of friction, angle of repose, terminal velocity and cohesion of desi, large kabuli and small kabuli (chico) chickpea. In India, traditional manual threshing by bamboo sticks produced seed with the lowest damage and highest germination when compared with cattle or tractor treading (Verma *et al.*, 1999). Afterwards, the seed required separation from all other material by hand, which was a tedious and time-consuming process. Tabatabaeefar *et al.* (2002) developed a sieving machine that was capable of cleaning chickpea with a separation efficiency of 93%, and also separating grades of seed for quality and end use. This was important for the local marketing of the crop, as well as introducing a simple, low priced and easily maintained machine technologically appropriate for Iranian farmers.

Many recommendations for the handling and storing of chickpea suggest using those for field peas, *Pisum sativum* var. *arvense* (L.) Poir. as the characteristics of both these legumes are thought to be similar. However, chickpea has a greater loss in germination and is likely to spoil more rapidly than field peas under similar conditions (Cassells and Caddick, 2000). The threshold relative humidity (RH) limit for fungal growth occurs at lower levels for chickpea, making it more susceptible to quality loss. Table 26.1 provides moisture contents (MCs) (wet weight basis) for several oilseeds, grains and pulses, including chickpea, used in the Canadian grading system (Canadian Grain Commission, 2005). Straight grade for chickpea is similar to lentil up to 14.0% MC under Canadian conditions; field peas and beans can be stored safely at considerably highe MCs. It is important to realize that preharvest, harvest and postharvest conditions all influence the storability of the seed. Weathered seed typically has poor uniformity in seed size, and darkened or poor colour in the seed coat (Cassells and Caddick, 2000). Producers must be aware that chickpea continues to age over time; desi chickpea will darken considerably, lose germination and vigour, and this rate of decline is enhanced by elevated MC, temperature, RH and seed condition. Vigilance is required as pulse crops are often harvested at MCs above the safe level for storage to prevent excess damage (Hnatowich, 2000).

Table 26.1. Conversion table for several oilseeds, cereals, and pulses including chickpea, for use with the Model 9191 3.5" moisture meter. (From Canadian Grain Commission, Winnipeg, Canada.)

Grain	Weight (g)	Conversion table number	Tough (%)	Damp (%)
Flaxseed and solin	225	6	10.1–13.5	Over 13.5
Canola and rapeseed	250	5	10.1–12.5	Over 12.5
Mustard seed, all classes	250	8 – Brown mustard 7 – Oriental mustard 6 – Yellow mustard	9.6–12.5	Over 12.5
Peas, green and yellow	250	2	16.1–18.0	Over 18.0
Split peas, green and yellow	250	1	16.1–18.0	Over 18.0
Chickpea	250	1	14.1–16.0	Over 16.0
Pea beans	250	2	No tough	Over 18.0
Lentils	250	1	14.1–16.0	Over 16.0
Peas, green and yellow	250	2	16.1–18.0	Over 18.0
Beans				
Black	250	1	No tough	Over 18.0
Cranberry	225	1	No tough	Over 18.0
Faba	250	2	16.1–18.0	Over 18.0
Dark red kidney	250	1	No tough	Over 18.0
Pinto	250	1	No tough	Over 18.0
Buckwheat	225	3	16.1–18.0	Over 18.0
Triticale	250	1	14.1–17.0	Over 17.0
Mixed grain	Use the conversion table and tough and damp ranges for the predominant grain.			

Safe Storage Time

A range of safe storage times for chickpea stored at several temperature and MC combinations is presented in Table 26.2 (Department of Primary Industries and Fisheries, 2005). A combination of cool, dry storage conditions provide an environment where seed can be stored almost indefinitely; both temperature and seed moisture seem to be the most critical variables. Seed at 15% MC and 20°C can be stored for a much longer time than seed at 12% MC and 40°C (Table 26.2). Storage tests at the Australian Temperate Field Crops Collection indicated that storage of high quality seed at 2°C and 7% MC will maintain a high level of seed viability for more than 25 years (Redden, personal communication). High ambient temperatures and RH accelerate darkening of the seed coat and loss of colour which can produce a less marketable product (Cassells and Caddick, 2000). Storage at high temperature (>25°C) and RH (>65%) also affects end-use properties in chickpea with the development of hard-to-cook (HTC) phenomena caused by a decrease in seed water absorption capacity (Reyes-Moreno *et al.*, 2000). A simple test for examining quality loss in chickpea is by surface sterilizing seeds, washing with distilled water and plating for germination on Petri dishes. Hayat and Ahmad (1996) found that seed stored for 1 year had a germination loss of 23%, and greater leakage of

Table 26.2. Effect of moisture content and temperature on storage life of chickpea seed in Australia. (From Department of Primary Industries and Fisheries, Queensland, Australia.)

Storage moisture (%)	Storage temperature (°C)	Longevity of seed (days)
12	20	More than 2000
12	30	500–650
12	40	110–130
15	20	700–850
15	30	180–210
15	40	30–50

nitrogen, phosphorus, potassium, sugars and proteins than fresh seeds. Seed pre-treatment soaking with sodium dikegulac (1000 to 4000 μg/ml) solution for 6 h, then drying to original MC resulted in decreased leakage of sugars and amino acids, and slowed losses in viability and concentrations of insoluble compounds (Rai *et al.*, 1992). Modern grain storage structures can be fitted with aeration fans and ducts designed to maintain optimum storage conditions (White, 2001).

Fungi

Seeds entering storage carry microflora composed of field and storage fungi. A survey of chickpea pods during development found primarily *Alternaria alternata* (Fr.) Keissl., *Cladosporium herbarum* (Persoon) and *Fusarium* spp. in the standing crop (Singh and Ahmed, 1989). *Alternaria* sp., *Cladosporium* and *Curvularia* sp. were also predominant fungi in 24 pathogens isolated from stored chickpea (Aminuzzaman and Chudhury, 1989). Field fungi are gradually displaced by storage fungi under poor storage conditions. Storage fungi, especially *Aspergillus flavus* Link, *A. niger* van Tieghem, *A. nidulans* (Eidam) Winter, *A. ochraceus* K. Wilh. and *Penicillium* spp., can grow vigorously under suitable conditions to initiate grain spoilage and produce mycotoxins hazardous to both animals and humans. Dwivedi and Shukla (1990) noted *Rhyzopus arrhizus* R., *A. niger* and *A. flavus* were the most commonly occurring fungi in storages of chickpea irrespective of treatments and containers tested. Of 36 species of fungi isolated from pulse crops, *Aspergillus* spp. represented 60% of the total with *A. flavus* being predominant with 24% occurrence (Husain and Ahmed, 1971). A survey of the toxigenic potential of *A. flavus* isolates found over 59% positive to aflatoxin production; 10–15% of samples collected from harvested plant piles and dehusked matured seeds prepared after threshing and winnowing had aflatoxin contamination ranging from 30 to 110 μg/kg (Singh and Ahmed, 1989). Seeds stored in semi-dry pods with a MC of 60% contained medium to high levels of aflatoxins associated with infection by *A. flavus* (Habish, 1972). Testing common Egyptian foods including nuts and seeds, spices, herbs, cereal grains and dried vegetables found the lowest prevalence and concentration of B1 aflatoxin in chickpea and highest occurrences in watermelon seeds, in-shell and shelled groundnuts (Selim *et al.*, 1996).

Ahmad and Singh (1991) noted that fungal species became prevalent more rapidly in chickpea stored in jute bags than in seed stored in metal bins. MC and aflatoxin contamination of seeds in jute bags were maximum during September–October influenced by external environmental conditions. The internal environment of metal bins is comparatively less influenced by external conditions and initially restricts fungal growth and aflatoxin production. Prolonged storage resulted in increased seed MC and high aflatoxin levels after 6 months storage in metal bins (Ahmad and Singh, 1991). High rainfall and humidity during the storage season also made the seed vulnerable to fungal attack; germination was reduced by 30–90% when seed MC exceeded the safe level of 8% under rural tropical conditions (Aminuzzaman and Chudhury, 1989). One note of interest was that chickpea stored at 40°C showed no survival of the pathogenic grey mould fungi *Botrytis cinerea* Pers. Fr., common on seed stored at lower temperatures (Singh and Tripathi, 1993) and which can cause seed emergence problems in the following crop year.

Aeration and Drying

Aeration is the practice of forcing unheated air, by means of a fan, through grain to maintain its condition and reduce the chances of spoilage and heating. This system lowers the temperature of the bulk if the grain is above ambient air temperature, maintains a uniform temperature throughout the grain mass, reduces or eliminates moisture migration and hot spots, reduces mould and insect growth and removes storage odours (Mills, 1989). Field peas stored near 15% MC may develop a surface crust during winter as a result of moisture migration and snow seepage, particularly when they are stored warm without aeration (White, 2001). Seeds tend to clump, and if left undisturbed become blackened as a result of fungal activity. In smaller bins, a periodical walk across the top of the bin or moving the top 30 cm of stocks with a shovel will prevent clumping (White, 2001). Crusting in larger bins can be prevented by using a front-end loader to divide stocks and disturb the surface layer; the use of the large bucket on the loader also results in less handling damage. Using a low aeration rate of 0.15 (l/s)/t prevented crusting in chickpea in a conventional Bulk Handling Company bunker in Murtoa, Victoria, Australia (Darby, 1999). Generally, aeration at a rate of 1–2 (l/s)/m^3 is sufficient to prevent seed spoilage. Aeration principles and the proper operation of aeration systems are described by Friesen and Huminicki (1986).

Unfortunately, the majority of chickpea is grown in semi-tropical and tropical regions of the world where the climate is hot and humid, and advanced technology may be limited; therefore, much attention should be given to the conditions the seed is placed into storage. It is essential for seed MC to be monitored to ensure that the seed dries thoroughly. Large kabuli seeds can often test-dry; however, the interior of the seed may still remain moist resulting in spoilage and quality loss (Hnatowich, 2000). For bulk field peas, the maximum drying temperatures are 45°C for seeding purposes, 70°C for commercial seed and 80–100°C for feed; temperatures higher than 45°C will harm germination

(White, 2001). Jain and Tiwari (2003) noted open sun drying (OSD) as the most common method of crop drying in developing countries because of its simplicity and economics. To aid local Indian farmers, a solar crop dryer designed to dry 1 tonne of grain was developed using a single glass-covered flat plate solar air heater with a 50 m^2 collector area (10 m × 5 m), and a 2.2 kW electrical motor-operated blower. Chickpea dried from 29% to 12% MC during a total drying period of 10 h (2 days with overnight tempering) at a hot-air temperature of 50 ± 5°C when ambient air temperature was 33 ± 1°C (Singhal and Thierstein, 1984). A two-stage method of drying provided best results for reducing seed splitting of pea-shaped seeds compared to chickpea varieties with semi-spherical or owl-head-shaped seeds (Rao *et al.*, 1990). Initial slow drying at 15°C and 30–40% RH until MC reached 9% or less, followed by drying at 15% RH until seeds reached 7% MC was best for long-term germplasm storage which is required to maintain the seed in its highest quality (Rao *et al.*, 1990). Technically it is also feasible to use the sun for a solar-powered, cold storage plant to maintain a grain temperature of 4–5°C. A 0.5-tonne capacity solar-powered, cold storage unit was developed at the Indian Institute of Technology, Bombay; however, this system had too high of an initial cost for farmers with limited resources (Sukhatme *et al.*, 1979).

Traditional Storage

Proper drying followed by storage of seed under 1 inch of ash or sand or by treatment with vegetable oils and leaves of neem and tobacco provided better storage than traditional storage systems using gunny sack, bamboo doles, metal tins and earthen pitchers in Bangladesh (Aminuzzaman and Chudhury, 1989). Pushpamma *et al.* (1985) reported favourable results where chickpea stored in pots, tins and bags showed only a 0.5% weight loss after 9 months of storage. Seed in kangi (local bin) and kangi with a plastic lining maintained lower MC, temperature, percentage weight loss, insect damage and higher germination rates than seed in jute bags after 7 months of storage (Kenghe and Kanawade, 1992). Seed stored under ambient conditions for 1 year in polythene bags had higher germination and lowest fungal infections compared with grain stored by other methods (Dwivedi and Shukla, 1990). This may be in part due to reduced insect infestation and damage which can allow invasion by fungi. Three indigenous north-eastern Indian storage structures, *duli*, *mer* and *bukari*, were modified for storing pulse grain for a period of 10 months. Small modifications to improve fit and sealing provided better storage as indicated by higher seed germination counts, and a lower rise in MC and insect damage when compared with unmodified traditional structures (Kalita *et al.*, 2002).

Pests of Chickpea

Open storages of chickpea are susceptible to losses by rodents, birds and insects. Losses are quantitative due to pests directly consuming the seeds (many more seeds may be partially eaten), and qualitative due to contamination with pest hairs or feathers, and faeces and excreta which can render grains unpalatable for human or animal consumption. Damaged seed is also prone to insect infestation and secondary infection by fungi which can cause further nutritional losses, present health risks and reduce germination affecting the success of subsequent crop plantings. The simplest method to prevent such losses is to have a physical barrier to exclude the large vertebral pests; the size of storage container whether bin, barrel or jar is less important than ensuring the exclusion of the pest. Traps, predators and chemical poisons are also available for their control.

Insects present more of a challenge because their small size makes them difficult to detect, and in many instances heavy infestations do not become apparent until offspring emerge from seeds as adults. Stored-product insects develop at optimal temperatures between 25°C and 35°C making them a serious problem in chickpea-producing areas of the world with high postharvest humidity, where yearly temperatures rarely fall below 18°C, the point at which these insects cease laying eggs (Sinha and Watters, 1985). In the Mediterranean region, the most dominant bruchid is the Chinese bean weevil, *Callosobruchus chinensis* (L.) (Cardona, 1985) with infestation rates ranging up to 79% (Weigand and Tahhan, 1990). Chickpea provides suitable nutrition for *C. chinensis* as demonstrated by a 76% survival rate with an egg-to-adult development of <38 days at 28°C and 70% RH (Pandey and Singh, 1997). A survey of farmers' storage conditions in Bangladesh during 1981–1984 found that bruchids infested 64% of chickpea, 35% of lentil and 24% of mung bean at 3 months of storage (Aminuzzaman and Chudhury, 1989). *C. chinensis* was declared a major pest of chickpea in Pakistan based on infestation rates which caused more than 10% damage in samples collected from farmers, villagers and general public (Aslam, 2004). More important was the conclusion that damaged grains were unfit for human and animal consumption because of undesirable odour. Insect species other than bruchids generally are of lesser importance in chickpea. Lesser grain borer, *Rhyzopertha dominica* (F.) and red flour beetle, *Tribolium castaneum* Herbst adults survived a 14-day exposure to chickpea and also produced viable progeny, whereas the rice weevil, *Sitophilus oryzae* (L.) could only survive on split-decorticated seed (Chander *et al.*, 1992).

Detection and Control of Pests

The simplest method of insect detection is by visual inspection as well as by sieving of grain samples. These techniques are simple at the farm level where stocks tend to be in small allotments and may be stored within the home or in close locality. The presence of adult insects and emergence holes is a clue for immediate attention to prevent further damage. Farmers can also rely on a simple machine to determine if seeds are infested with storage pests such as the cowpea weevil, *C. maculatus* (F.). The Central Warehouse Corporation of the Food Corporation of India has regularly used a four-part device consisting of an upper chamber, top lid fitted with a magnifying lens, with detachable circular sieves of varying sizes over an insect collecting chamber (Bhagwat and Thakur,

2003). Sieves can be changed according to the commodity being tested and magnification helps locate any extracted insects. Where modern technology is not available or is in short supply farmers and villagers must make use of local products and remedies. By simply choosing an appropriate storage container which is less attractive to insect pests, product quality can be retained without much effort. *C. maculatus* was found to lay up to four times as many eggs on cloth bags stored at room temperature for 8 months compared with polythene bags when both contained pulse grain; this resulted in a tenfold less weight loss of seed stored in polythene bags (Singh, 1995).

Locally prepared storage agents can be effective in reducing infestation by storage insects. Rhizome powder from calamus, *Acorus calamus* L., used at 0.1–0.5% by weight effectively prevented breeding by *C. chinensis* (Khan and Borle, 1985). Pulses stored in gunny sacks treated with 10% aqueous extracts from tubers of nutgram, *Cyperus rotundus* L., leaves of neem, *Melia azadirachta* Linn., or pignut, *Hyptis suaveolens* (L.) Poit., reduced egg production and adult emergence of *C. maculatus* for 1–2 months (Ragu *et al.*, 2001). Srinivasu and Naik (2002) reported protection against *C. chinensis* over a period of 6 months in bengal gram treated with 1.0% and 2.0% boric acid. *C. maculatus*, the maize weevil, *S. zeamais* (Motsch.) and the rusty grain beetle, *Cryptolestes ferrugineus* (Stephens), were repelled from chickpea seeds and holding bags treated with acetone and ethanol extracts from the fruit of ackee, *Blighia sapida* Koenig with high mortality (>60%) achieved against *C. maculatus* exposed for over 24 h (Khan and Gumbs, 2003). A traditional storage method of seed protection using olive oil with salt provided 90% protection during 4 months of storage (Weigand and Tahhan, 1990). Freshly harvested seeds treated with such common oils as coconut, sesame, rapeseed and cottonseed applied at 1, 2 or 4 ml/kg offered protection against infestation by *C. chinensis* for up to 1 year (groundnut and mustard oil protected up to 8 months); seed quality and seedling vigour were unaffected (Sharma and Singh, 1993). Khaire *et al.* (1992) found that neem, palm and karanj oils offered storage protection for 100 days; these oils were cheaper to use in comparison to other oils. Refined neem seed oil, soybean and crude caster oils applied at 10 ml/kg on chickpea proved effective surface protectants against *C. chinensis* for up to 6 months (Das, 1989), but mustard oil treatments caused 100% inhibition of seed germination on several seed lines (Khalique *et al.*, 1988). Caster oil at 5 and 10 ml/kg prevented infestation by *C. maculatus* for 2 and 5 months, respectively, and by *C. phaseoli* (Gyll.) for 2 and 3 months, respectively (Pacheco *et al.*, 1995). Soybean oil was not an effective protectant, but did inhibit both these bruchids for a reduced period of time. One major drawback was the presence of off-odours which compromised product acceptability with the 10 ml/kg oil treatments (Pacheco *et al.*, 1995).

Insect control practices by farmers in Dwarwad, Karnataka, India, include sun drying, using a mixture of neem leaves and ashes, boric acid and the application of insecticides and fumigants; sun drying was the most practical with 75–100% of surveyed households reporting its use (Srinivasu and Naik, 2002). The sun's energy dries crops relatively quickly to both preserve its quality and also kill insects which may be present within the seed; low MC also prevents the growth of fungi which can deteriorate the nutritional value of the crop or produce harmful toxins. Commercial freezers have been used to control

C. maculatus in organic garbanzo bean (Johnson and Valero, 2003). No insect survival occurred in small bulks of seed exposed to −18°C for 14 days; portions of the bulks had only reached −10°C which appears to be adequate for control (Johnson and Valero, 2003).

Some protection from insect infestation may be offered as new chickpea cultivars are developed. The difference in oviposition by *C. maculatus* between most preferred and least preferred cultivar in a laboratory screening test of 22 varieties was 824.7 eggs/50 g of seeds to 1 egg/50 g of seed, respectively (Salunkle and Jadhav, 1982). Seeds with a smooth surface were preferred sustaining ~50% infestation with 20.5% weight loss, while a resistant variety with a rough and spiny surface suffered only 0.76% infestation and 0.28% weight loss (Salunkle and Jadhav, 1982). Such a dramatic reduction in insect pest losses due to seed coat texture would surely benefit farmers if confirmed in further studies, and providing the new seed type is accepted by local populations.

Commercial Chemical Control

Commercial fungicidal seed treatments all improved germination of chickpea (83.2–97.3%) over a storage period of 1 year when compared to untreated seed (Dwivedi and Shukla, 1989). The obvious concerns with such treatments are that seed be restricted for seeding and new crop production, labelled as hazardous for human and animal consumption, and be secured with necessary controls in place for safe distribution when required.

Bags treated with the insecticide dichlorvos remained free from infestation by *C. chinensis* for 2 months while those treated with fenitrothion and malathion were free for 1 month (Sultana *et al.*, 1980). Dilute dust formulations of pirimiphos-methyl and permethrin were found to control adult *C. chinensis* on several pulses with a recommended application rate of 5 mg/kg (Taylor and Evans, 1980). Although treatment had little effect on reducing the numbers of emerging adults, their death prevented breeding and further infestation, thus reducing overall damage. Chlorpyrifos-methyl (50% EC, 50 mg/m^2) and malathion (25% WP, 100 mg/m^2) persisted for up to 90 and 110 days, respectively, on jute bags against *C. maculatus* (Yadav and Singh, 1994). Deltamethrin (25% flowable powder, 20 mg/m^2) remained persistent up to 6 months and protected chickpea seed for up to 3 months as did fluvalinate (20% EC, 50 mg/m^2) (Yadav and Singh, 1994). Deltamethrin (2.8 EC) sprayed at a rate of 15 and 25 mg/m^2 on jute sacks filled with chickpea was effective in controlling *C. chinensis* with 76–78% mortality for up to 6 months (Lal and Dikshit, 2001). Cypermethrin treated directly on grain at 3 and 5 mg/kg was found to have a high persistence (62–71% of original application) after 6 months. Decontamination processes by washing and steaming only removed 37–49% and 63–74% of the residues, respectively; however, the majority of the residue was located on the seed coat only (Dikshit, 2001).

Seed viability was found to be highest (79.2–90.3%) in pulses fumigated with ethylene dibromide before being placed into metal bins or jute bags and stored for 1 year (Vimala and Pushpamma, 1990). Phosphine gave effective control against *S. oryzae* which can cause significant damage to stored chickpea when harvest time coincides with high RH. Effective control was achieved at

3 mg/l with as little as 4 days exposure; however, with seeds exposed to 6 mg/l for 12 days, seed viability was adversely affected (Mazzuferi *et al.*, 2000). A phosphine and methyl bromide fumigation study on chickpea seed germination and vigour during storage for 3.5 years at temperatures and seed MCs ranging from ambient to –20°C and 12–6% MC, respectively, found the effect of moisture, temperature, fumigant, and storage period to be significant (Singh *et al.*, 2003). A decline in germination and vigour responded to fumigation at high seed MC and temperature. Phosphine was found to be more detrimental to seed viability than methyl bromide in chickpea (Singh *et al.*, 2003). *C. maculatus* infestations in postharvest legumes have been routinely controlled with methyl bromide and phosphine fumigants within the USA (Johnson and Valero, 2003). Regulatory loss of methyl bromide (UNEP, 1992) and proposed restrictions on the use of phosphine (USEPA, 1998) are beginning to generate interest in non-chemical disinfestation treatment methods. Great care is necessary in the choice of fumigant in combination with storage conditions before fumigating seed destined for germplasm conservation.

Irradiation

The use of irradiation for the disinfestation of legumes was first recommended in 1980 (Potter *et al.*, 1980). The cost for food irradiation is significantly higher than for application of contact insecticides and fumigants. Major capital costs include the radiation source, irradiator, conveyance and control systems, radiation shield and land; additional costs include staff salaries, utilities, taxes and insurance and miscellaneous other operating costs (Forsythe and Evangelou, 1993). However, with tighter restrictions on methyl bromide fumigations and its associated increasing costs, it is likely irradiation will become more economical and a viable control strategy.

Irradiation studies on three bruchids, the pulse beetle, *C. analis* F., *C. chinensis* and *C. maculatus* showed that a dose rate of 0.08 kilogray (kGy) was sufficient to suppress populations in stored products by rendering the insects sterile (Gill and Pajni, 1991). With irradiation of chickpea, lentil, mung and *masha kaloi*, *Phaseolus roxburghii* stored for 5 months at ambient temperatures, chickpea showed the greatest reduction in fungal microflora, specifically *A. flavus*, with a 90.6% decrease at a 1 kGy dose rate (Begum *et al.*, 1995). Doses used to disinfest foods and agricultural products such as seeds can range between 0.1 and 2 kGy (Snyder and Poland, 1995). Graham *et al.* (2002) found that chickpea subjected to irradiation of up to 50 kGy had increased shelf storage life owing to a reduction in insect infestation; physical and chemical changes within the seeds also improved cooking quality.

Diatomaceous Earth

Inert dusts (diatomaceous earth) have been used to protect stored products from insect infestation as a residual protective deposit in empty storage structures or by being directly applied to bulk grain. Mixing with grain can cause significant changes to its physical properties by reducing bulk density and flow rate, and increasing angle of repose; it can also cause accelerated wear on handling equipment. Changes in handling properties lead to its prohibition for use on grain delivered to the bulk-handling systems in many countries. Chickpea flow rate and bulk density decreased by 25% and 5.9%, respectively, when inert dust was mixed with grain at a rate of 500 g/t. These changes can be of great concern and consequence where large bulks of seed are routinely handled by mechanical means (Jackson and Webley, 1994). Potential problems may include slower loading rates, the plugging of legs and spouts and excessive wear to mechanical equipment. The resulting delays add costs to grain-handling companies. However, due to its tremendous benefits of effective insect control, being non-residual and safe for workers to handle, recommended concentrations were established while noting changes to quality criteria such as grain test weight. The negative concerns are of minimal consequence for chickpea produced in developing countries at the family farm level where the majority of harvesting and handling is done by hand and simple equipment. The benefits of applying inert dusts to small bulks of seed at the local farm level are effectiveness, ease of use and safeness of use; however, costs may be prohibitive for the majority of subsistence farmers who are poor.

Grading System

Grain in a grade is similar, but not identical and it represents a range of quality. It varies within the limits set by the standards for the grade for colour, absence of disease, presence of foreign matter, etc. (Canadian Grain Commission, 2003). Grading systems may differ in the various grain producing countries of the world. Some aspects of each may have specific advantages or disadvantages when compared to others, but the purpose of each is the same. Grain grades must reflect the end-use quality of grain accurately and consistently so producers will be adequately compensated for the grain they produce and consumers can be reassured that they can purchase a product for the use they desire. Table 26.3 shows Canadian grading factors for both kabuli and desi chickpea (Canadian Grain Commission, 2005). A grade is assigned based on assessing several visual criteria from a representative sample.

The following descriptions are excerpts from the *Official Grain Grading Guide* (Canadian Grain Commission, 2005):

Colour – Sound, well-matured seeds having a uniform normal colour grade *good*; immature, but not green, having moderate amounts of adhered soil, are lightly stained but otherwise moderately discoloured from natural causes grade *fair*; and seeds that do not meet the definition of *fair* colour grade *poor*. This only applies to kabuli chickpea.
Damage – Damaged chickpea include whole or broken seeds that are sprouted, frost damaged, heated, damaged by insects, distinctly deteriorated or discoloured

Table 26.3. Primary and export grade determinant tables for Canadian chickpeas: chickpeas, Canada Western (CW) Kabuli and Canada Western (CW) Desi. (From Canadian Grain Commission, Winnipeg, Canada.)

Standard of quality					Foreign material	
Grade name	Colour	Damage (%)	Mechanical damage including splits (%)	Green (%)	Insect parts (%)	Total (%)
			Chickpeas, Canada Western Kabuli			
No. 1 CW	Good, natural colour	0.5	1	0.5	0.02	0.10
No. 2 CW	Fair	1	2	1.0	0.02	0.2
No. 3 CW	Poor	2	3	2	0.02	0.2
Grade, if no. 3 specs not met	Chickpeas, sample CW kabuli account colour	Chickpeas, sample CW kabuli account damaged	Chickpeas, sample CW kabuli account mechanical damage and splits	Chickpeas, sample CW kabuli account green	Chickpeas, sample CW kabuli account foreign material	Chickpeas, sample CW kabuli account foreign material
			Chickpeas, Canada Western Desi			
No. 1 CW		1	2	1.0	0.02	0.10
No. 2 CW		2	3.5	2.0	0.02	0.2
No. 3 CW		3	5	3	0.02	0.2
Grade, if no. 3 specs not met	Chickpeas, sample CW desi account colour	Chickpeas, sample CW desi account mechanical damage and splits	Chickpeas, sample CW desi account green		Chickpeas, sample CW desi account foreign material	Chickpeas, sample CW desi account foreign material

by weather or by disease, or that are otherwise damaged in a way that seriously affects their quality. In kabuli chickpea, white and shriveled seeds and yellow or water-stained seeds should be cut and examined for damage. If the cotyledons show any signs of visible damage, they are considered damaged; if no signs of visible damage, they are considered in the evaluation of colour.

Mechanical damage – Mechanical damage including splits includes whole chickpea with more than 10% of the seed broken off, and/or split seeds. It is important to note that seeds with hairline cracks and chipped seed coats are not considered to have mechanical damage.

Green – Chickpea may be considered green regardless of the cause. Frost-damaged seeds which are green are considered under the grade determinant for *green* while frost-damaged seeds with no green colour are considered under the grade determinant for *damage*. Kabuli varieties are considered green if they show any green colour of any size area anywhere on the seeds or seed coats. Desi chickpea is considered green if it shows distinctly green colour throughout the seed when cut to expose the cotyledons.

Insect parts – Insect parts refer to pieces of insects such as grasshoppers and lady-bird beetles that remain in the sample after cleaning or processing. Samples are analysed for the percentage of insect fragments and graded according to established tolerances. If pulse crops come into contact with insects during the harvesting process, it may result in seed staining and earth adhering to the seed and may result in samples having an objectionable odour. Samples containing staining of this nature will be considered to be earth tagged and graded according to colour definitions. Samples having a distinct objectionable odour not associated with the quality of the grain will be graded (*Type of Grain*) Sample Account Odour. Additional to this may be further degrading factors such as earth or fertilizer pellets, ergot, excreta, fireburnt, heated, foreign material, stones, contaminated grain, treated grain and other chemical substances. Chickpea of excellent quality, harvested, handled and stored under optimum conditions, will garner the highest grades, bring the highest economical return and be most suitable for human consumption, processing and seeding. Lower grades reflect decreasing quality based on the established criteria, have less economic value and are suitable for fewer end uses; such seed may be blended, processed into food or non-food products or used as animal feed. The poorest grades may have factors rendering them suitable only for industrial purposes, may be a total loss or even hazardous, requiring disposal.

Conclusions

Chickpea can generally be stored in a manner similar to other pulses but they are somewhat more difficult to manage due to specific pests and diseases. Chickpea continues to age in storage; hence, reduction of both temperature and seed moisture are key control measures. Quality in stored chickpea can be maintained by good management practices, whether by large commercial operations or by subsistence farmers. As more research is conducted specifically on chickpea, generalized recommendations will need to be updated appropriately.

Acknowledgements

The authors wish to thank the Department of Primary Industries and Fisheries, Queensland, Australia, and the Canadian Grain Commission, Winnipeg, Manitoba, Canada, for permission to use information provided by their departments; and the Canada Research Chairs Program for partial funding.

References

Ahmad, S.K. and Singh, P.L. (1991) Mycofloral changes and aflatoxin contamination in stored chickpea seeds. *Food Additives and Contaminants* 8(6), 723–730.

Aminuzzaman, M. and Chudhury, J.C.S. (1989) Present status and future of postharvest technological research on pulses. In: *Proceedings of the Second National Workshop on Pulses; Advances in Pulses Research in Bangladesh*. 6–8 June 1989. Bangladesh Agricultural Research Institute (BARI), Joydebpur, Gazipur, Bangladesh, pp. 147–154.

Aslam, M. (2004) Pest status of stored chickpea beetle, *Callosobruchus chinensis* Linnaeus on chickpeas. *Journal of Entomology (Pakistan)* 1(1), 28–33.

Bhagwat, V.R. and Thakur, A.K. (2003) Insect detection device from infested stored grains. *Journal of Applied Zoological Researches* 14(2), 175–177.

Begum, F., Joarder, G.K., Hossain, M.M., Bhuyan, A.D. and Islam, M.S.C. (1995) Effect of radiation on fungal flora of pulses during storage. *Bangladesh Journal of Science and Industrial Research* 30(2/3), 131–146.

Canadian Grain Commission (2003) Grain grades and standards, fact sheet. Canadian Grain Commission, Winnipeg, Canada. Available at: http://www.grainscanada.gc.ca/Pubs/factsheets/grades/grades-e.htm

Canadian Grain Commission (2005) *Official Grain Grading Guide*. Canadian Grain Commission, Winnipeg, Canada.

Cardona, C. (1985). Insect pests of faba beans, lentils and chickpeas in North Africa and West Asia: a review of their economic importance. In: Saxena, M.C. and Varma, S. (eds) *Proceedings of the International Workshop on Faba Beans, Kabuli Chickpeas and Lentils in the 1980s: An International Workshop*. 16–20 May 1983. ICARDA, Aleppo, Syria, pp. 159–167.

Cassells, J. and Caddick, L.P. (2000) Storage of desi type chickpeas. In: Wright, E.J., Banks, H.J. and Highley, E. (eds) *Stored grain in Australia: Proceedings of the Australian Postharvest Technical Conference*. 1–4 August 2000. Commonwealth Scientific and Industrial Research Organization (CSIRO) Stored Grain Research Laboratory, Adelaide, pp. 206–214.

Chander, H., Kulkarni, S.G. and Berry, S.K. (1992) Breeding potential of some cereal pests in pulses. *Journal of Insect Science* 5(2), 193–195.

Darby, J. (1999) Bunker aeration-maintenance trail, 1996/7. Bunker 701, chickpeas, Murta, Victoria. In: Banks, H.J., Wright, E.J. and Damcevski, K.A. (eds) *Stored grain in Australia. Proceedings of the Australian Postharvest Technical Conference*. 26–29 May 1998. CSIRO Stored Grain Research Laboratory, Canberra, pp. 245–251.

Das, G.P. (1989) Effect of the duration of storing chickpea seeds treated with neem oil on the oviposition of the bruchid *Callosobruchus chinensis* L. (Bruchidae: Coleoptera). *Bangladesh Journal of Zoology* 17(2), 199–201.

Department of Primary Industries and Fisheries, Queensland, Australia (2005) Available at: http:/www2.dpi.qld.gov.au/fieldcrops/9404.html

Dikshit, A.K. (2001) Persistence of cypermethrin on stored pulses and its decontamination. *Pesticide Research Journal* 13(2), 141–146.

Dwivedi, S.N. and Shukla, T.N. (1989) Effect of seed treatment after storage upon germinability of gram (*Cicer arietinum* L.) seeds. *Farm Science Journal* 4(1/2), 125–127.

Dwivedi, S.N. and Shukla, T.N. (1990) Effect of methods of storage on germinability and mycoflora of gram (*Cicer arietinum* L.) seed. *Seed Research* 18(1), 82–85.

Food and Agriculture Organization (2006) World Chickpeas Production Data, FAOSTAT Database, Food and Agriculture Organization of the United Nations. Available at: http://faostat.fao.org/faostat/form?collection=Production.Crops.Primary&Domain=Production&servlet=1&hasbulk=0&version=ext&language=EN

Forsythe, K.W., Jr and Evangelou, P. (1993) Costs and benefits of irradiation and other selected quarantine treatments for fruit and vegetable imports to the United States of America. *Proceedings of an International Symposium on Cost–Benefit Aspects of Food Irradiation Processing*, Aix-en-Provence, France. 1–5 Mar 1993. International Atomic Energy Agency, Vienna, Austria.

Friesen, O.H. and Huminicki, D.N. (1986) *Grain aeration and unheated air drying*. Agdex 732-1, Manitoba Agriculture, Winnipeg, Canada.

Gill, M. and Pajni, H.R. (1991) Disinfestation of stored pulses with gamma radiations. *Entomology for Defense Services,* pp. 239–243.

Graham, J.A., Panozzo, J.F., Lim, P.C. and Brouwer, J.-B. (2002) Effects of gamma irradiation on physical and chemical properties of chickpeas (*Cicer arietinum*). *Journal of the Science of Food and Agriculture* 82, 1599–1605.

Habish, H.A. (1972) Aflatoxin in haricot bean and other pulses. *Experimental Agriculture* 8(2), 135–137.

Hayat, S. and Ahmad, A. (1996) The level of proteins and leakage of solutes in germinating fresh and stored seeds of *Cicer arietinum* L. *Biologia Plantraum* 38(4), 597–599.

Hnatowich, G. (2000) Chickpea. In: Slinkard, A. (ed.) *Pulse Production Manual 2000, Saskatchewan Pulse Growers*. Food Focus Saskatoon Inc. and Colorshape Communications, Saskatoon, Canada, pp. 8.8–8.12.

Husain, S.S. and Ahmed, M.A. (1971) Studies on stored food fungi. Part IV. Fungi from pulses. *Pakistan Journal of Scientific and Industrial Research* 14(6), 507–511.

Jackson, K. and Webley, D. (1994) Effects of Dryacide^R on the physical properties of grains, pulses and oilseeds. In: Highley, E., Wright, E.J., Banks, H.J. and Champ, B.R. (eds) *Stored Product Protection. Proceedings of the 6th International Working Conference on Stored-products Protection*, Vol. 2. Canberra, Australia. 17–23 April 1994. CAB International, Wallingford, UK, pp. 635–637.

Jain, D. and Tiwari, G.N. (2003) Thermal aspects of open sun drying of various crops. *Energy* 28(1), 37–54.

Johnson, J.A. and Valero, K.A. (2003) Use of commercial freezers to control cowpea weevil, *Callosobruchus maculatus* (Coleoptera: Bruchidae), in organic garbanzo beans. *Journal of Economic Entomology* 96(6), 1952–1957.

Kalita, J., Gogoi, P., Bhattacharyya, P.R. and Handique, R. (2002) Development and testing of some modified indigenous structures of north-east India for safe storage of pulse grains. *Journal of Applied Zoological Researches* 113(2/3), 251–254.

Kenghe, R.N. and Kanawade, L.R. (1992) Storage studies on chickpea (*Cicer arietinum* L.) seed. In: Salokhe, V.M. and Ilangantileke, S.G. (eds) *Proceedings of the International Agricultural Engineering Conference*, Vol. 2. 7–10 December 1992. Asian Institute of Technology, Bangkok, pp. 541–547.

Khaire, V.M., Kachare, B.V. and Mote, U.N. (1992) Efficacy of different vegetable oils as grain protectants against pulse beetle, *Callosobruchus chinensis* L. increasing storability of pigeonpea. *Journal of Stored Products Research* 28(3), 153–156.

Khalique, F., Ahmed, K., Afzal, M., Malik, B.A. and Malik, M.R. (1988) Protection of stored chickpea, *Cicer arietinum* L. from attack of *Callosobruchus chinensis* L. (Coleoptera: Bruchidae). *Tropical Pest Management* 34(3), 333–334, 366, 370.

Khan, M.I. and Borle, M.N. (1985) Efficacy of some safer grain protectants against the pulse beetle, *Callosobruchus chinensis* L. infesting stored Bengal gram (*Cicer arietinum* L.). *PKV Research Journal* 9(1), 53–55.

Khan, A. and Gumbs, F.A. (2003) Repellent effect of ackee (*Blignia sapida* Koenig) component fruit parts against stored product insects pest. *Tropical Agriculture* 80(1), 19–27.

Lal, A.K. and Dikshit, A.K. (2001) The protection of chickpea (*Cicer arietinum* L.) during storage using deltamethrin on sacks. *Pesticide Research Journal* 13(1), 27–31.

Masoumi, A.A. and Tabil, L.G. (2003) Physical properties of chickpeas (*C. arietinum*) cultivars. Paper No. 036058, *2003 American Society of Agricultural Engineers (ASAE) Annual International Meeting,* Las Vegas, Nevada, 27–30 July 2003.

Mazzuferi, V., Carreras, J. and Casanoves, F. (2000) Uso de fosfamina para el control de *Sitophilus oryzae* (L.) (Coleoptera: Curculionidae) en semillas de garbanzo (*Cicer arietinum* L.) y efectos sobre su viabilidad. *Agriscientia* 17, 65–68.

Mills, J.T. (1989) *Spoilage and Heating of Stored Agricultural Products: Prevention, Detection, and Control.* Publication 1823E, Agriculture and Agri-Food Canada, Research Branch, Canadian Government Publishing Centre, Ottawa.

Pacheco, I.A., De Castro, M.F.P.P.M., De Paula, D.C., Lourencao, A.L., Bolonhezi, S. and Barbieri, M.K. (1995) Efficacy of soybean and caster oil in the control of *Callosobruchus maculatus* (F.) and *Callosobruchus phaseoli* (Gyllenhal) in stored chick-peas (*Cicer arietinum* L.). *Journal of Stored Products Research* 31(3), 221–228.

Pandey, N.K. and Singh, S.C. (1997) Observations on the biology of the pulse beetle *Callosobruchus chinensis* (Linn.) infesting stored pulses. *Uttar Pradesh Journal of Zoology* 17(1), 38–42.

Potter, W.T., Elias, P.S. and Gottschalk, H.M. (1980) International project in the field of food irradiation news. *Food Irradiation Information* 10, 46.

Pushpamma, P., Rao, K.C., Reddy, K.S. and Rao, K.K. (1985) Home level storage of legumes in Andhra Pradesh, India. *Legume Research* 8(1), 17–24.

Ragu, N., Babu, A., Dorn, S. and Ignacimuthu, S. (2001) Potential of plants for protecting stored pulses from *Callosobruchus maculatus* (Coleoptera: Bruchidae) infestation. *Biological Agriculture and Horticulture* 19(1), 19–27.

Rai, A.S., Chhetri, D.R. and Bhattacharjee, A. (1992) The effect of sodium-dikegulac on maintenance of viability of a few crop seeds under adverse storage conditions. *Environment and Ecology* 10(4), 814–824.

Rao, N.K., Mengesha, M.H. and Pundir, R.P.S. (1990) Cleavage damage due to rapid drying in pea-shaped seeds of chickpeas (*Cicer arietinum*). *Indian Journal of Agricultural Sciences* 60(4), 255–258.

Reyes-Moreno, C., Okamura-Esparza, J., Armienta-Rodelo, E., Gomez-Garza, R.M. and Milan-Carrillo, J. (2000) Hard-to-cook phenomenon in chickpeas (*Cicer arientinum* L.): Effect of accelerated storage on quality. *Plant Foods for Human Nutrition* 55(3), 229–241.

Salunkle,V.S. and Jadhav, L.D. (1982) Relative susceptibility of some gram (*Cicer arietinum* L.) varieties to pulse beetle (*Callosobruchus maculatus* Fab.) in storage. *Legume Research* 5(1), 45–48.

Selim, M.I., Popendorf, W., Ibrahim, M.S., El-Sharkawy, S. and El-Kashory, E.S. (1996) Aflatoxin B1 in common Egyptian foods. *Journal of the AOAC International* 79(5), 1124–1129.

Sharma, J.K. and Singh, H.B. (1993) Aftereffects of some oils on the seedling trials in stored chickpea seeds. *Crop Research Hisar* 6(2), 317–322.

Singh, G. (1995) Use of polythene bags to check the infestation of pulse beetle, *Callosobruchus maculatus* F. in stored moong. *Journal of Insect Science* 8(2), 212–214.

Singh, M.P. and Tripathi, H.S. (1993) Effect of storage temperatures on the survival of *Botrytis cinerea* in chickpea seeds. *Indian Journal of Mycology and Plant Pathology* 23(2), 177–179.

Singh, N., Kapur, M.L. and Mahajan, R.K. (2003) Effect of fumigation on germination and vigour in chickpeas and green gram during prolonged storage. *Seed Science and Technology* 31(1), 161–168.

Singh, P. and Ahmed, S.K. (1989) Pre-storage aflatoxin contamination and *Aspergillus flavus* association with various growth stages of chickpeas (*Cicer arietinum*). *Annals of Agricultural Science* 34(2), 805–812.

Singhal, O.P. and Thierstein, G.E. (1984) The use of solar energy in post-harvest technology. In: Achaya, K.T. (ed.) *Interfaces between Agriculture, Nutrition, and Food Science: Proceedings of a Workshop Held at Hyderabad, India*. 10–12 November 1981. United Nations University Press, Tokyo, Japan, pp. 142–159.

Sinha, R.N. and Watters, F.L. (1985) *Insect Pests of Flour Mills, Grain Elevators, and Feed Mills and Their Control*. Publication 1776E, Agriculture Canada, Canadian Government Publishing Centre, Ottawa.

Snyder, O.P. and Poland, D.M. (1995) *Food Irradiation Today*. Hospitality Institute of Technology and Management (HITM). Rubbright-Brody, St. Paul, Minnesota.

Srinivasu, C.S. and Naik, L.K. (2002) Survey of adoptable indigenous methods of control of storage grain pests. *Karnataka Journal of Agricultural Sciences* 15(4), 715–716.

Sukhatme, S.P., Khandilikar, S.G., Sharma, G.K. and Arun Kumar, S. (1979) Performance studies of a flat plate collector along with booster mirrors and associated hot water thermal storage tank. *Proceedings of the National Solar Energy Convention*. 13–15 December 1979. Indian Institute of Technology, Bombay.

Sultana, P., Ahmed, A. and Ahmed, A. (1980) The effect of impregnation of gunny sacks with some insecticides on *Callosobruchus chinensis* (L.) attacking stored pulse. *Bangladesh Journal of Zoology* 6(1), 71–72.

Tabatabaeefar, A., Aghagoolzadeh, H. and Mobli, H. (2002) The design and development of a chickpea second sieving and grading machine. Paper No. 021175, *2002 American Society of Agricultural Engineers (ASAE) Annual International Meeting*, Chicago, USA, 28–31 July 2002.

Taylor, W.D. and Evans, N.J. (1980) Laboratory evaluation of pirimiphos-methyl and permethrin dilute dust for control of bruchid beetles attaching stored pulses. *International Pest Control* 22(5), 108–110.

UNEP (United Nations Environmental Programme) (1992) *Fourth Meeting of the Parties to the Montreal Protocol on Substances that Deplete the Ozone Layer*, 23–25 November 1992. UNEP, Copenhagen, Denmark.

USEPA (United States Environmental Protection Agency) (1998) *Reregistration Eligibility Decision: Aluminum and Magnesium Phosphide – Cases 0025 and 0645*. USEPA, Office of Pesticide Programs, Special Reviews and Reregistration Division, Washington, USA.

Verma, O.P., Singh, P.V. and Kushwaha, G.D. (1999) Effect of threshing methods on seed quality and storability in gram (*Cicer arietinum*). *Seed Research* 27(1), 119–121.

Vimala, V. and Pushpamma, P. (1990) The effect of three prestorage treatments on the viability of pulses stored in different containers. *Indian Journal of Nutrition and Dietetics* 27(1), 7–17.

Weigand, S. and Tahhan, O. (1990) Chickpea insect pests in the Mediterranean zones and new approaches to their management. In: van Rheenen, H.A. and Saxena, M.C. (eds) *Proceedings of the Second International Workshop on Chickpea Improvement: Chickpea in the Nineties*. 4–8 December 1989. ICRISAT, Hyderabad, India, pp. 169–175.

White, N.D.G. (2001) *Protection of Farm-stored Grains, Oilseeds and Pulses from Insects, Mites and Moulds*. Publication 1851E (revised), Agriculture and Agri-Food Canada, Research Branch, Winnipeg.

Yadav, T.D. and Singh, S. (1994) Persistence toxicity and efficacy of four insecticides as jute fabric treatment to protect cereal and legume seeds. *Indian Journal of Entomology* 56(2), 146–155.

27 International Trade

F. Dusunceli,[1] J.A. Wood,[2] A. Gupta,[3] M. Yadav[4] and S.S. Yadav[5]

[1]*The Central Research Institute for Field Crops, PO Box 226, Ulus, Ankara, Turkey;* [2]*Tamworth Agricultural Institute, NSW Department of Primary Industries, Tamworth, NSW 2340, Australia;* [3]*Canny Overseas Private Limited, B-170, Priyadarshini Vihar, New Delhi 110092, India;* [4]*College of Business Administration, Tarleton State University, (Texas A&M University System), Stephenville, TX 76402, USA;* [5]*Pulse Research Laboratory, Division of Genetics, Indian Agricultural Research Institute, New Delhi 110012, India*

Introduction

Chickpea is an important crop for farming communities in many developing countries, especially in Asia, the Middle East and North Africa. Its nutritional properties and affordable prices compared to animal protein mean chickpea is one of the major staple foods, especially for people with limited income. In addition, consumer preference and religious festivals also play a role in the consumption of chickpea in many countries. Despite the great demand for chickpea, many developing countries have difficulties meeting their own needs. This creates a significant marketplace for other countries to fill this gap by producing and exporting chickpea.

India is the major chickpea producer in the world, followed by Turkey and Pakistan (Fig. 27.1). These three countries produced 5,770,000 t, 620,000 t and 611,100 t respectively in 2004, accounting for ~70% of global chickpea production. Compared to other agricultural commodities, the size of global chickpea trade is relatively small. In 2004, the total recorded chickpea exports were 0.49% of wheat and <0.07% of agricultural products exported globally (Table 27.1). However, chickpea exports (680,951 t with a value of $333 million) rank fourth among the pulses after dry peas, dry beans and lentils. Considering chickpea is rarely used for animal feed (as peas and beans certainly are) chickpea is an important human food traded in the international arena.

Exports

The majority of chickpea was internally marketed and consumed within the countries where they were produced. Whilst ~15% of total pulse production was traded internationally over recent years, only 8% of chickpea produced

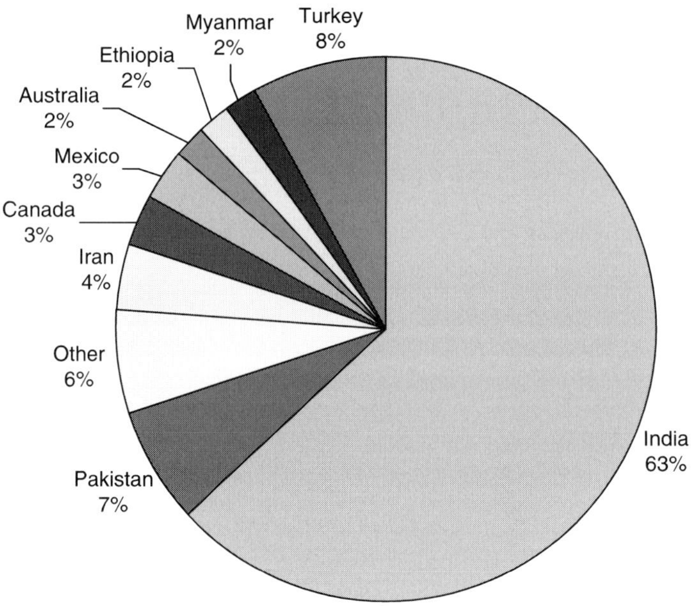

Fig. 27.1. Annual chickpea production (averaged over 5 years: 2000–2004). (From FAOSTAT, 2006.)

Table 27.1. Global chickpea exports compared to other pulses and commodities in 2004. (From FAOSTAT, 2006.)

	Quantity exported		Value exported	
Commodity	Quantity ('000 t)	% of total agricultural commodities	Value (million $)	% of total agricultural commodities
Chickpea	681	0.07	333	0.06
Broad bean	553	0.06	153	0.03
Lentils	1,127	0.12	497	0.08
Bean	3,011	0.32	1,299	0.21
Pea	3,103	0.33	677	0.11
Other pulses	910	0.10	89	0.01
Wheat	139,120	14.66	40,154	6.64
Barley	328,961	34.67	83,690	13.85
Maize	85,967	9.06	12,803	2.12
Rice	30,046	3.17	9,118	1.51
Other small grain cereals	18,390	1.94	9,065	1.50
Potato	13,571	1.43	5,719	0.95
Soybean	116,789	12.31	32,542	5.38
Others	206,545	22.00	408,190	68.00
Total agricultural commodities	948,774	100.00	604,329	100.00

globally were exported in 2004 (Boubaker, 2005). Data taken from the Food and Agriculture Organisation (FAO, 2006) show that a total of 71 countries exported chickpea in 2004. However, most of these countries export very small amounts. Currently the majority of world chickpea exports are dominated by five major producers: Australia, Turkey, Iran, Mexico and Canada accounting for 76% of total chickpea exports in 2004 (Table 27.2). The next ten chickpea exporting countries have a share of 0.5–4.3% in world chickpea exports. The total annual global chickpea exports gradually increased 3.6-fold from 1980 to 2004 with sharp variations in some years (Fig. 27.2).

The major traders of annual chickpea exports between 1980 and 2004 are shown in Fig. 27.3. World chickpea exports were dominated by Turkey until the early 1990s. This was the result of the very successful integrated fallow replacement project which included price support (Dusunceli and Aydoğan, 2004). However, following the abandonment of this support system in Turkey, Australia became the largest competitor in the world market. Australia started exporting significant amount of chickpea in the late 1980s and passed Turkey for the first time in 1993. Australia generally dominated in the 1990s, with a peak of 379,735 t in 1997. In 1998, ascochyta blight (*Ascochyta rabiei*) devastated many Australian chickpea crops and took time to rebuild with increasingly resistant varieties. Canada has also felt the effects of ascochyta, responsible for much of their export decline over the last 5 years. Ascochyta was initially spread to Australia, Canada and the USA by poor quarantine, whereas it had always been present in most other countries, with year-to-year disease levels influenced by local climate conditions. Mexico improved its exports gradually from 83,250 t in 1980 to a peak of 207,093 t in 2001. Canada and Iran have only become important chickpea exporters since 2000. The annual exports by each of these five countries have narrowed since 2002, the minimum and maximum exports

Table 27.2. Chickpea exports (quantity, value and cost/t) in 2004. (From FAOSTAT, 2006.)

Country	Quantity (t)	% in total	Value ($1000)	% in total	Value ($/t)
Australia	149,500	21.9	46,500	14.0	312
Turkey	133,000	19.5	69,000	20.8	520
Iran	85,000	12.5	47,000	14.1	552
Mexico	83,000	12.2	67,500	20.3	812
Canada	68,500	10.0	32,500	9.7	474
Syria	29,500	4.3	25,500	7.7	871
Tanzania	25,000	3.6	7,500	2.2	298
Pakistan	17,500	2.6	5,000	1.6	298
Morocco	13,500	2.0	5,000	1.5	374
USA	12,500	1.8	7,500	2.3	610
India	12,000	1.8	8,000	2.4	647
Russia	9,000	1.4	2,000	0.6	214
UAE	8,500	1.3	2,500	0.8	317
Spain	5,000	0.7	3,500	1.1	765
Others	29,500	4.3	3,500	1.0	118
Total – world	681,000	100.0	332,500	100.0	489

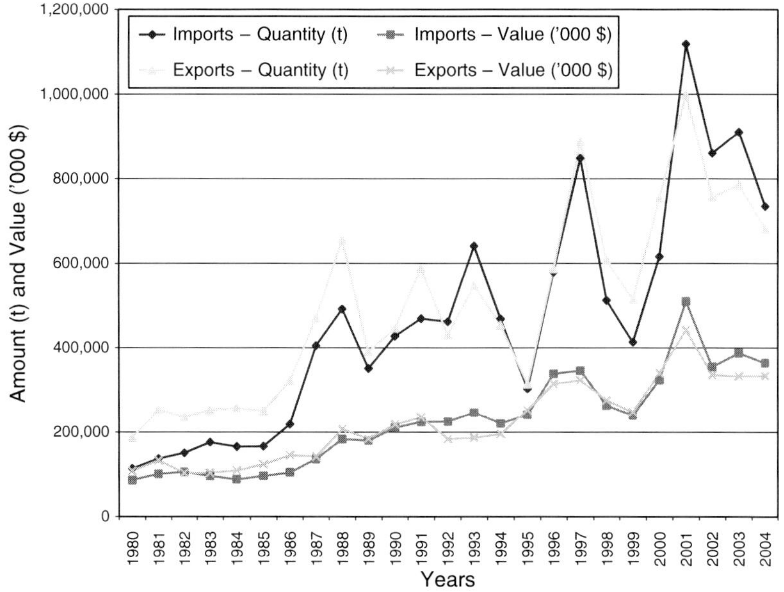

Fig. 27.2. Changes in global chickpea imports and exports. (From FAOSTAT, 2006.)

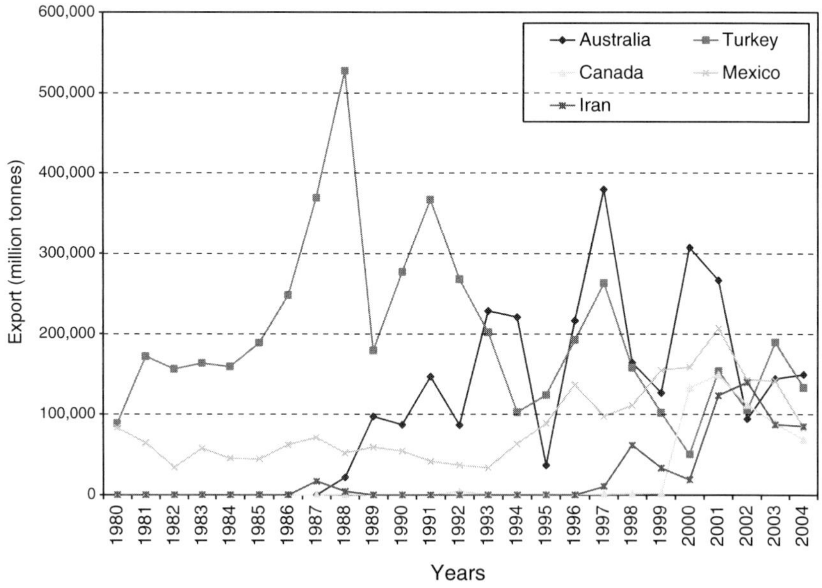

Fig. 27.3. Trends in chickpea exports from the most active countries. (From FAOSTAT, 2006.)

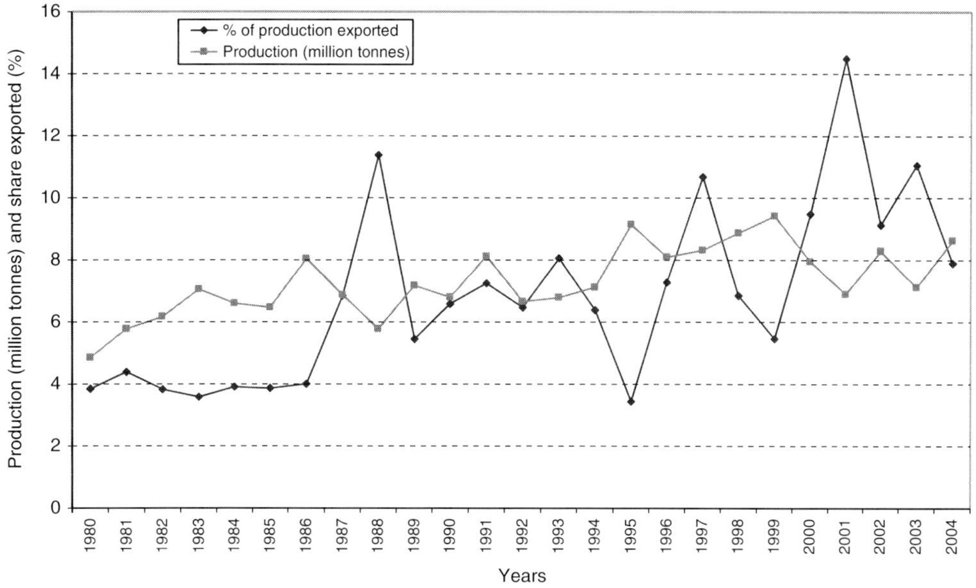

Fig. 27.4. Changes in world chickpea production and percentage exported. (From FAOSTAT, 2006.)

being 68,265 and 149,321 t respectively in 2004. These five countries remain competitive in the firmest chickpea market ever recorded.

The data of last 20 years indicate that global chickpea trade has been affected significantly by production. In general, the share of traded chickpea increased in low production years and decreased in high production years (Fig. 27.4). The largest export years were recorded when there was relatively low production globally, most evident in 1988, 1993, 1997 and 2001. In contrast, chickpea trade decreased in years with high chickpea production, most obviously in 1989, 1995, 1999 and 2002. In parallel to this trend, global chickpea trade declined to 680,951 t with a reduction of 13% from 2003, while production increased by 21% to 8.6 million t.

Imports

The Asian countries of India, Pakistan and Bangladesh not only have the highest chickpea consumption in the world but they are also the largest importers accounting for 41.6% of the trade in 2004 (FAO, 2006). These are followed by European and African countries which made up 21.1% and 12.7% of the total world chickpea imports respectively. In total 138 countries import chickpea supporting the wide consumption of chickpea across the globe (Table 27.3).

In the last 20 years annual chickpea imports by Bangladesh, Spain, Algeria and Italy were relatively stable, remaining <75,000 t, with the exception of Bangladesh importing 104,000 t in 2004 (Fig. 27.5). For India and Pakistan,

Table 27.3. Chickpea imports (quantity, value and cost/t) in 2004. (From FAOSTAT, 2006.)

Country	Quantity (t)	% in total	Value ('000 $)	% in total	Value ($/t)
India	132,500	18.0	51,000	14.1	387
Bangladesh	104,500	14.2	31,500	8.6	299
Pakistan	69,500	9.4	24,000	6.5	343
Spain	58,000	7.9	54,000	14.8	926
Algeria	48,500	6.6	38,500	10.6	791
Italy	28,500	3.9	18,500	5.1	657
Jordan	24,500	3.3	11,000	3.0	443
Sri Lanka	22,500	3.1	12,000	3.3	524
UAE	21,000	2.8	9,500	2.6	457
Tunisia	20,500	2.8	3,500	0.9	163
UK	20,500	2.8	11,500	3.2	566
Saudi Arabia	17,500	2.4	9,000	2.4	503
USA	13,500	1.9	9,500	2.6	681
Portugal	11,500	1.6	8,500	2.3	743
Colombia	9,500	1.3	4,500	1.3	505
Others	132,500	18.0	68,000	18.7	515
Total – world	735,000	100.0	364,500	100.0	495

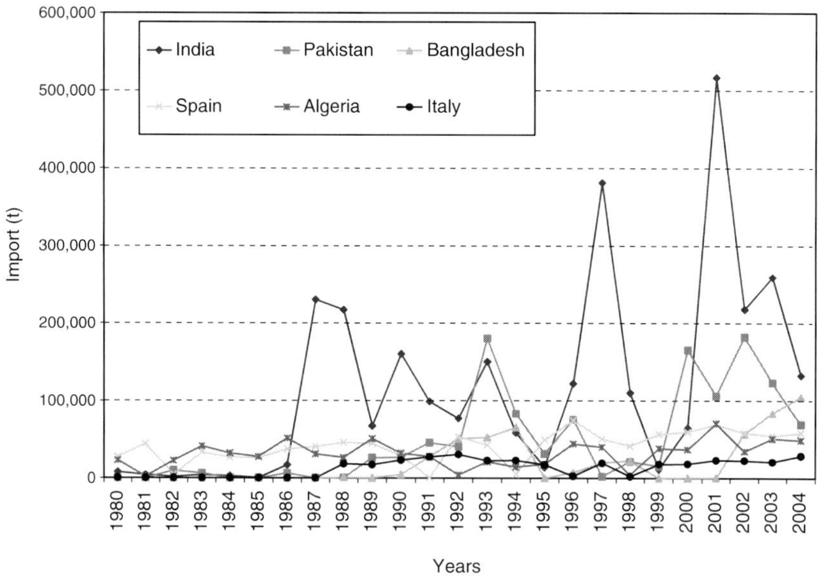

Fig. 27.5. Trends in chickpea imports from the most active countries. (From FAOSTAT, 2006.)

the major chickpea importing countries, FAO statistics recorded imports of <200,000 t and 80,000 t respectively in most years. However, large fluctuations in the amount imported can occur. The import demands of these countries, especially of India, are generally attributed to the status of local production as affected by the favourable or adverse growing conditions.

The Value of Traded Chickpea and Trade Balance

The FAO statistics (2006) showed a gradual increase in the value of traded chickpea from $108 million in 1980 to $333 million in 2004, with some fluctuations. In contrast the cost per tonne showed a decreasing trend and more significant variation over the years (Fig. 27.6). The cost per tonne of exported and imported chickpea were $579 and $762 respectively in 1980, decreasing to just above $300/t until 1987. The entrance of Australia, Canada and Iran as chickpea exporters played an important role in market stabilization. In the 1990s the cost was relatively stable between $350 and $550 per tonne. Despite the overall stability in recent years, the cost of exported chickpea can significantly change due to supply and demand. For example, in 1995 the cost of chickpea reached an all-time high of $800/t, in a year when very little chickpea was traded. India produced a record amount of chickpea in 1995. As a result India reduced its imports just to 13,662 t, the second lowest in the last 20 years. Consequently, a much higher proportion of exported chickpea were kabuli types and therefore had higher prices.

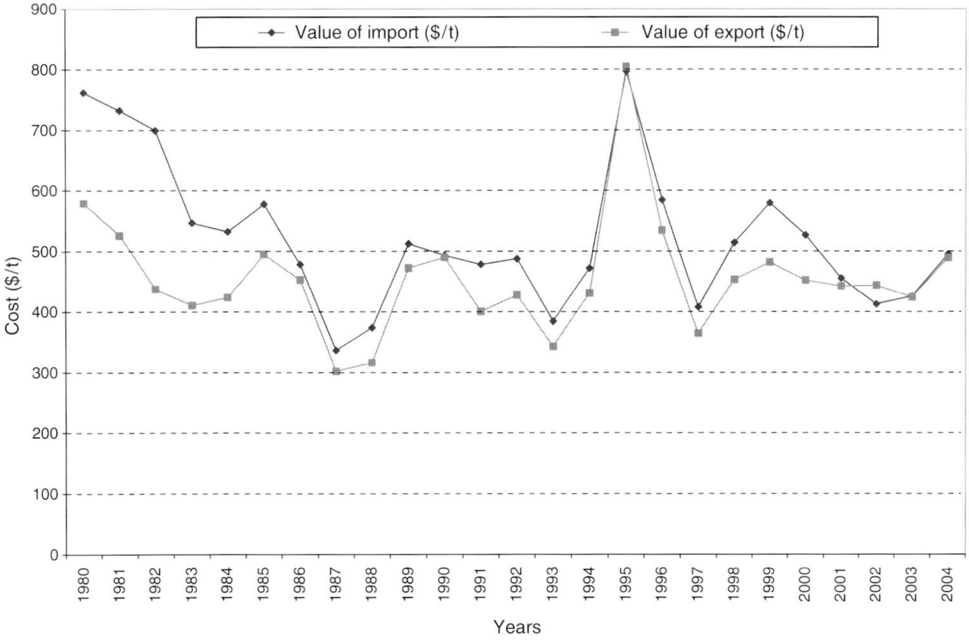

Fig. 27.6. Changes in value of globally exported and imported chickpeas. (From FAOSTAT, 2006.)

Whilst India, Bangladesh and Pakistan collectively imported 42% of total chickpea in 2004, their value was only 29% of the global chickpea imports. In comparison, the next three (Spain, Algeria and Italy) accounted for 19% of the total chickpea imported and 31% of its value. The main difference in the imported quantities and values of these two groups is that the former group of countries import mostly desi chickpea while the latter group import mostly kabuli chickpea. Kabuli types fetch higher prices than desi types, as do larger sizes. These differences are also reflected in the cost per tonne for each country (see Table 27.3). The rest of the countries import much smaller quantities (<3% each), together accounting for 39.3% of total global imports and 40.3% of its value.

In terms of trade balance, Asia, Europe, Africa and South America are the net importing continents, with net import figures for 2004 of −144, 100, −127, 100, −53,800 and −16,4t costing −21, −99, −39 and −46 million dollars respectively. In contrast, the trade balance of Australia and North–Central America were +149,000 and +138,000t with values of +92 and +46 million dollars respectively.

Regulations in Trade

Government policies

Chickpea production is supported in many countries through various farmer subsidy tools. These support systems may include direct payments as in the European Union (EU), minimum support price systems and certain levies in India (Agbola *et al.*, 2002a), and marketing loans/schemes (Grains and Oilseeds Payment Program and Agricultural Income Stabilisation scheme) in Canada (Agriculture and Agri-Food Canada, 2006a). Similarly in the USA, farm bills facilitate a marketing loan programme which serves as price support for domestic pulse producers (Margheim *et al.*, 2004). The US Farm Bill only recently added chickpea to the 2006 eligibility list for collecting loan deficiency payments (LDPs). In Turkey, the Grain Board ceased directly supporting their producers with base rates for chickpea in 1994. Now support is provided only for inputs such as certified seeds and fuel (Dusunceli *et al.*, 2006; Anonymous, 2006a). Australia is one of the few countries with no farmer subsidies or schemes promoting chickpea production.

Most countries involved in the international chickpea market liberalized their trade arrangements in the 1990s (Agbola *et al.*, 2002b; Dusunceli *et al.*, 2006). This had little effect on chickpea production in some countries, such as India. In contrast, it reduced chickpea production in Turkey from 860,000t in 1990 to 610,000t in 2005 as farmers shifted production to other crops (Çiftçi, 2004). In Turkey this was beneficial to the consumers as it facilitated imports of wider chickpea qualities at competitive prices.

Many countries try to restrict or regulate imports/exports for many reasons: to protect domestic markets against over- or undersupply, to protect a country against disease/pest introduction or to protect consumers against pesticide/chemical contamination. One method is to impose trade barriers. Another is to impose high import duties. A combination of the two is where a country allows imports duty free up to a specific tonnage (or at a reduced tariff), then charges a high tariff on any imports above the quota. For example, pulse production in India has not kept pace with population growth. At the same time globalization has allowed local Indian producers to grow more profitable crops for export in preference to pulses for the domestic market. India has, therefore, restricted its pulse exports (by imposing high duties and quantitative restrictions) and liberalized imports (low import duties) to maintain an adequate supply for local consumption (Agbola, 2004).

The World Trade Organization is trying to establish fair and competitive market conditions by encouraging countries to reduce/remove farmer support and import tariffs in order to free up international trade. The least developed countries are granted special exceptions within this framework.

Permits and quarantine

Most countries require an import permit that matches the standards and quarantine criteria set by the importing country. On arrival the shipment of chickpea will be checked for defects against this permit. Failure to meet the permit may result in refusal of the shipment or require special treatments like fumigation.

Chickpea shipments have the potential, like any other commodity, to spread pests, weeds and diseases to other countries. Therefore, many countries imply quarantine requirements to prevent the introduction of diseases, pests, weed seeds and other unwanted foreign material. The importance given by an importing country to the respective diseases and pests will depend upon what diseases/pests they already have. One of the most important diseases undergoing quarantine restriction is ascochyta blight (*A. rabiei*) which may have different pathotypes in different countries. Similarly, the *Bruchid* sp. of beetle, a stored grain pest, is given a high priority in many quarantine procedures. Ideally quarantine aims to keep a country's environment safe from any potentially damaging organisms. This can sometimes mean particular commodities from certain countries may not be allowed by some importing countries.

Pricing

The price paid for chickpea is primarily determined, like any other commodity, by supply and demand. However, other factors such as the time of harvest, seed quality, exchange rates and government policies will also affect chickpea prices. A recent price-elasticity study by Agbola and Damoense (2005) shows that economic growth and urbanization in India should stimulate increased demand for chickpea. A 10% increase in real incomes in India should increase the quantity of import demand by 3.4% for chickpea. Conversely, a 10% increase in import price could reduce the quantity of import demand for chickpea by 17%. These results show chickpea to be a preferred, but substitutable, food in Indian diets.

Supply and demand

The market price of chickpea, like any other commodity, is determined by both the supply and the demand for it, and has been recognized since 1890 (Wagner, 1891). Today the supply–demand model is one of the fundamental concepts of economics. The price (*P*) is determined by the point at which the quantity supplied (*S*) equals quantity demanded (*D*), graphically represented as the point where the two lines cross (Fig. 27.7). For example:

1. If the actual price (P_2) is higher than the price at which the curves cross (P_1), then there is a surplus, and sellers would lower their price in order to sell the surplus.
2. If the actual price (P_1) is below the price where the curves cross (P_2), then the quantity demanded would increase beyond what is supplied, creating a shortage, then the seller would increase the price to increase profits.

Similarly, if there is a shift in *S* or *D*, a new price equilibrium point will be created. For example, at the beginning of 1997 desi chickpea prices were quite low. However, prices increased considerably ($P_1 \rightarrow P_2$) due to supply shortage (decreased pulse production in Turkey and a dry season in eastern Australia) and increased demand ($D_1 \rightarrow D_2$) from the Indian subcontinent.

Factors affecting supply and demand include:

- Population growth in the major consuming countries;
- Consumption rates in the major consuming countries (influenced by income and changes in taste preferences);
- Consumption rates of health-conscious consumers in affluent countries;
- Growing and harvest conditions in major producing regions;
- Chickpea production within each producing/importing and exporting country (see Chapter 7, this volume);
- Timing of chickpea harvest and timing of export arrivals;
- Amount of chickpea seed in storage;

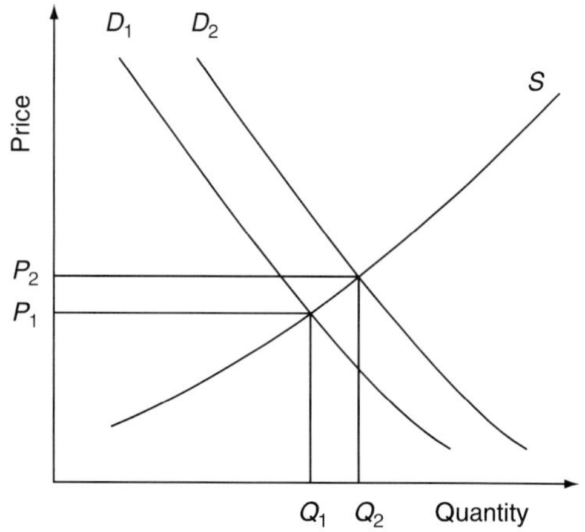

Fig. 27.7. The supply–demand model.

- Quality of chickpea seed available;
- Relative price and availability of other pulses that could be substituted for chickpea (yellow peas and pigeon pea);
- Relative price and availability of alternative protein sources (cereal and animal products);
- Sentiments of traders;
- Exchange rates of importing and exporting countries and the US dollar;
- Shipping rates.

Timing of harvests and transit times

Chickpea growing season and harvest period differ from country to country. Storage and transit times of exported chickpea also influence the time of arrival in the importing country. The harvest times of the major producing/exporting countries and approximate times of arrival in the Indian subcontinent are shown in Table 27.4.

The time taken from harvest to export can vary due to a number of factors including:

- On-farm storage time;
- Internal trading, storage and transport from grain receival sites to processors and exporters;
- Processing and/or storage by processors and/or exporters;
- Storage at docks and transit time during export;
- Prevailing market situation.

Australia, Canada and Mexico are disadvantaged by their distance from the major importer, the Indian subcontinent. They deal with longer transit times during export (15–45 days) compared to almost immediate availability within producing countries and shorter transit times (5–15 days) for the other main exporting countries, Turkey and Iran.

Table 27.4. Timing of chickpea harvests and export arrivals in the Indian subcontinent.

Country	Status	Harvest time	Arrival in Indian subcontinent
India	Producer/importer	February–April	All year
Bangladesh	Producer/importer	February–March	February–April
Pakistan	Producer/importer/ exporter	March–April	March–June
Mexico	Exporter	March–May	April–June
Iran	Exporter	April–May	May–September
Turkey	Exporter	June–August	July–October
Canada	Exporter	August–October	September–December
Australia	Exporter	October–January	November–July

The timing of harvests and export arrivals impacts on the price of chickpea in several ways:

1. Supply and demand between countries. Countries that try to export chickpea around the harvest time of producing/importing countries may affect prices if supply outstrips demand. For example, Mexico and Iran (April–September export) and late shipments from Australia (February–July) compete with the major producing/importing countries' own harvests (Bangladesh, India and Pakistan – February–June supply). In contrast, the major importers will be most active when their own production of chickpea is exhausted (July–January). Hence, some exporters, such as Iran and Australia, may choose to store chickpea until then to capture the increased demand when producing countries' supplies start to run out.
2. Direct competition between exporting countries (when their chickpea types for export are similar), e.g. Mexico and Iran, Turkey and Canada, Canada and Australia.
3. Ramadan occurs in the ninth month of the Islamic (lunar) calendar and differs annually by 10–12 days, falling on an earlier date in the mainstream (solar) calendar each year. During this month Muslims observe the Fast of Ramadan which lasts for 30 days. They fast in the daylight hours and traditionally break the fast in the evening with *Harira* (a soup based on chickpea). Ramadan concludes with festivities and feasting; hence, there is increased demand for chickpea in preparation for, during and immediately after Ramadan. Different exporting countries benefit as the timing of Ramadan changes. For example, Australia benefited in the 1990s and up to 2002 by the increased demand for chickpea due to the Ramadan season; thereafter it could no longer supply chickpea early enough and Canada, Iran and Turkey became beneficiaries of the timing of this religious event.

Currency exchange rates

Chickpea is internationally traded in US currency. However, the exchange rate at the time of export/import can affect the price obtained/paid in the respective countries. Currency values change on a daily basis, and their worth fluctuates from country to country. However, if the US dollar falls relative to the importing and exporting countries' currencies, then the exporting countries would lose profit, whilst importing countries pay less of their own currency. In this case the exporter may try to pay less to the farmer and/or ask for higher prices internationally (in US dollars). Caution must be exercised as this may make chickpea more attractive from competing producing/exporting countries, pushing their prices up and increasing their local production, potentially reducing their imports.

The inverse is also true: gains in the US dollar relative to other currencies would increase profits of exporting countries and increase the purchase cost to importing countries.

Whilst currency exchange rates affect the actual prices paid and received, it is very difficult to forecast future currency changes. Some traders may exercise day trading of currency to try to optimize currency rates, although this is a very speculative strategy.

Quality

The main quality parameters of importance will ultimately depend on the end-use and the market for which the seed is destined. The main determinant of end-use is the type of chickpea: desi or kabuli. Desi types are traditionally dehulled and split to produce dhal, or further ground into flour (*besan*). Some particular desi types, sometimes called *gulabi*, are preferred for roasting. Kabuli types are traditionally cooked as a whole vegetable, roasted, canned or mashed into hommus. Although prices for kabuli chickpea are traditionally higher than for desi types, they are also more volatile.

Two new chickpea markets have emerged more recently. The first is for larger, lighter coloured desi chickpea that is consumed as whole seed, predominantly in Bangladesh and the UAE. This class of chickpea commands a premium of $20/t (range $0–50) over traditional desi types. However, when India and to a lesser extent Pakistan are dominating the market there may be no premium as they predominantly use desi chickpea for dhal and *besan*, and are therefore less concerned with appearance.

The second new market is for small (6–7 mm) kabuli chickpea. Used as a cheaper vegetable or ground into *besan*, they trade at similar prices to (and compete directly with) the desi types. The Canadian varieties 'B90' and 'Chico' have exceptional water imbibing properties and trade at premium prices between those of 7 and 8 mm kabuli chickpea (Agriculture and Agri-Food Canada, 2001).

Seed quality also impacts on the prices received along the distribution chain. The most obvious quality traits can be visually observed and include seed size, colour, shape and defects. Moisture content, imbibition and ease of dhal milling are less obvious but not necessarily less important.

Seed size

The most economically important, and commonly traded, quality parameter in both desi and kabuli types is seed size. Kabuli types are commonly graded into particular size ranges before sale as the larger seeds will bring premium prices. Whether grading is performed on a batch of chickpea will depend on whether the extra profit is greater than the cost of grading. As a guide, a 1 mm increment in seed size can equate to around $80–190/t for kabuli chickpea (Tables 27.5 and 27.6). Size is also important for desi types, with larger seeds gaining premiums in recent times. Premiums for larger-sized desi chickpea are driven by who is the most aggressive buyer in the market at any given time.

Table 27.5. Prices (CAN$/t) to Canadian growers. (From STAT Communications Ltd, Canada: www.statpub.com.au.)

Chickpea classification	Average of previous 3 years	2004/05
Desi	264.30	309.58
Kabuli B90	371.80	360.84
Kabuli 7 mm	303.53	309.21
Kabuli 8 mm	585.91	518.15
Kabuli 9 mm	810.59	687.58
Kabuli 10 mm	909.40	777.48

Table 27.6. Turkish kabuli chickpea export prices in 2005. (From Sadikoglu Paketleme Paz. San. Tic. Ltd. Sti.: www.alibaba.com/catalog/11108347/Chickpea_And_Lentils.html.)

Chickpea classification	2005 Price $/t (FOB Turkey)
Kabuli 8.0 mm	831
Kabuli 8.5 mm	913
Kabuli 9.0 mm	1025
Kabuli 10.0 mm	1107

Uniformity of size is also very important. A batch of chickpea with uniform size looks much more presentable than one containing seed of variable sizes. Size uniformity is also very important to ensure cooking uniformity and milling efficiencies. Seed size is determined primarily by genetic factors with some influence of environment.

Seed colour

Colour is the next most important parameter and is the single most important factor in determining whether chickpea is marketable. Seed colour is determined by a strong genetic component and some environmental influence. Kabuli types are preferably a very light cream/pinkish to white colour, depending on the market. Some markets, e.g. Spain, prefer a brilliant white colour, whilst others, e.g. India, view this as undesirable as it looks 'bleached' (bleaching can result from weathering or chemical treatment). Desi types have traditionally been a dark brown colour; however, some new varieties can be much lighter in colour with yellow or pinkish hues. These lighter coloured desis are now generally favoured by most markets as they look fresher, despite the majority of desis still ending up dehulled. Large, lighter coloured desis have been attracting premiums in recent years.

EU countries import mainly large seeded kabuli types (Agriculture and Agri-Food Canada, 2006b) while Bangladesh, India and Pakistan import both desi and kabuli types (McVicar *et al.*, 2000).

Seed shape

Seed shape is also an important parameter when trading, and is believed to be a characteristic of genotype with little change due to the environment. Kabulis should have a traditional rounded ram's head shape for most markets and not be too elongated. In comparison, desi types were traditionally very angular, with a prominent beak. Some markets still prefer this shape; however, slightly more rounded desis with less prominent beaks have been attracting attention recently. The gulabi type of desi chickpea is also much rounder in shape and is used for roasting and puffing.

Moisture content

Seed moisture is also an important parameter and is a result of harvest time and environmental conditions. Excessive seed moisture can indicate immature seeds and/or green cotyledons and promote mould growth during storage.

Since chickpea is traded on a weight basis, and as buyers prefer lower moisture contents, the percent moisture of a shipment will affect the price. Dhal millers also prefer seed with low moisture contents as they are easier to dehull and split. However, seed that is too dry will also split during harvest, handling and transport, resulting in losses and possible dockage.

Most countries accept seed with 12–14% moisture content. Australia now accepts a maximum of 14% moisture, recognizing the loss of seed that has been occurring at lower moisture levels in the Australian dry climate. Canada has the opposite problem: forced early harvests mean seed is often harvested at 18% moisture and must be subjected to artificial drying to reduce seed moisture (Agriculture and Agri-Food Canada, 2001). This can sometimes cause bleaching or burning of the cotyledons.

Channa dhal

Chickpea is most commonly traded as whole seeds; however, the split product, channa dhal, is also traded within countries and internationally on a smaller scale. India is the major producer of channa dhal. Western countries do export some but find it hard to compete with India which has a cheaper labour force. India also produces very good quality dhal: its prolonged dry season enables seed to be sun-dried on roof tops, a process conducive to more efficient dehulling and splitting. Good quality dhal is free of husk and chips and the cotyledons show no or minimal abrasion and have a bright yellow/orange colour.

Grading and standards

Unfortunately there is no universally accepted grading system or nomenclature for chickpea. Grading systems vary from country to country, although all are size-based. In general, the larger the seed size, the higher the value.

The USA does not have specific chickpea grades, they are covered under the General Provisions section of the Grain Standards Document produced by the US Department of Agriculture (USDA, 1999). The local Indian market only specifies one grade to farmers: Fair-Average Quality (FAQ). Any seed not meeting this grade is termed non-FAQ. Approximately 80% of the Indian chickpea is classified as FAQ grade (Choudhary *et al.*, 2004). However, such classification is not practical for international trading and more standardized grades are required. Turkey has a slightly more detailed grading approach, classifying chickpea types as red chickpea (desi) for roasting to be used as snacks (*leblebi*) and graded as 7, 8 and 9 mm seed sizes. White chickpea (kabuli) for cooking and for export purposes is graded as 7, 8, 9 and 10 mm (Akova, 2002).

Quality standards are much more important in the international trading arena. Major trading countries have set up their own standards for export and import purposes. These establish criteria for the minimum requirements acceptable for each different class of chickpea. Australia, for example, has six chickpea classifications for export: 3 desi, 2 kabuli and 1 for channa dhal (Pulse Australia, 2005). Similar standards are also applied in Turkey (Anonymous, 2001), in India (Government of India, 2002, 2004a,b,c) and Canada (Canadian Grain Commission, 2005). India has 1 kabuli, 1 desi, 1 channa dhal and 1 roasted desi (split) classification. Failure to meet the requirements of a particular standard at local receival sites will result in the seed being downgraded to a lower classification, end up as stock feed or rejected altogether. Seed to be exported is usually sub-sampled and sent to the importer with specifications of the sample to minimize the potential for disputes over quality and price.

Defects

A chickpea batch may be downgraded or rejected for many reasons. Depending on the quality standards of importing countries, most defects are acceptable at low levels, although a few have nil tolerance. Examples are:

1. Defective seeds – seeds not of type; seeds that are broken, split, damaged (by insects, hail, heat, shrivelled, sprouted, frosted, caked, bin burnt, mouldy), whole pods and screenings;
2. Small seeds that pass through a sieve of a particular size (size specified in standards);
3. Poor coloured seeds – seeds that are blemished, green or of a different colour from the predominant class;
4. Obviously diseased seeds;
5. Foreign material – unmillable material (soil, stones, metals and non-vegetable matter) and all vegetable matter other than the specified seed;
6. Presence of field insects, snails and/or foreign seeds;
7. Seeds with objectionable odours.

Prices are discounted if batches fail to meet the relative standards. This applies to both desi and kabuli types, despite most desi types being dehulled prior to consumption (seed coat removal will eliminate any superficial seed markings – black seeds, blotching and speckling). The levels at which these defects become significant are uncertain and depend on the importing country, but more than 5% would almost certainly result in some sort of discount unless demand is greater than supply.

Premiums are sometimes paid for superior quality and appearance. Other important quality parameters, not used directly in trading, may also be used to differentiate preferred suppliers of particular classifications of chickpea.

Water imbibition and cooking quality

Kabuli chickpea for canning and cooking should hydrate quickly, absorb a high amount of water (at least doubling their weight) and have no unhydratable seeds (Wood and Harden, 2006). A shorter cooking time is also preferred by consumers and processors. These same parameters are equally important for desi chickpea destined for whole seed use.

Ease of dhal milling

Desi chickpea destined for dehulling should preferably be uniform in size and have a thin seed coat that is easily removed from the cotyledons to maximize dhal yield. Australian desi types are often preferred for splitting due to their high dhal yields, uniform size/shape and fewer impurities (Agbola *et al.*, 2002a). These attributes should also apply to small kabuli types used for dehulling.

Flour quality and rheology of chickpea flour have received little attention to date (Meares *et al.*, 2004; Kaur and Singh, 2005) but may assume more importance with increased processing competitiveness and consumer awareness.

Reputation of seed quality

Different countries generally produce different quality chickpeas and develop a reputation for doing so. Where this is favourable, it is important that the quality of new varieties be at least maintained to protect the national reputation. Once developed, reputations tend to be maintained irrespective of their factual underpinning. Some reputations in the current marketplace are:

- India and Pakistan produce chickpea with traditional quality.
- Indian gulabi (desi) types (also supplied by Myanmar and Tanzania) are preferred for roasting/puffing as they taste the best (Agbola *et al.*, 2002a).
- Canadian small seeded kabulis have superior hydration properties (cvs. Chico and B90).
- Australia produces desi chickpea for whole seed consumption with large size and exceptional quality (cvs. Gully, Jimbour, Howzat).
- Australian desi types are preferred for splitting due to their high dhal yields, uniform size/shape and fewer impurities (Agbola *et al.*, 2002a).
- Mexico currently has the 'preferred' kabuli (large and sweet tasting), closely followed by Australian kabulis (Agbola *et al.*, 2002a).
- Spain has a brilliant white and very large kabuli (cv. Spanish white).
- Turkey has large, cream coloured kabulis as well as white kabulis.

Effect of seed quality on prices

The price of chickpea in the national and international markets varies greatly according to the product type (desi or kabuli), grain size and quality status. The Indian surveys of Agbola *et al.* (2002a) indicate that the price of kabuli chickpea can be 2.6 times higher than desi types. Similarly, the best quality desi and kabuli types can have prices of 1.1 and 2.6 times more, respectively, than those of low quality chickpea in the market. Moreover, the price differences between the different grades of kabuli types are generally more than the difference between the desi grades. These differences are likely to be more significant in international trading.

Forward contracts, futures and options

In many countries, farmers are able to locally pre-sell chickpea they produce to guarantee a certain price by taking out a *Forward* contract. Forward contracts, in combination with crop insurance, enable farmers some management of price and production risks. Some farmers trade Futures and Options markets: but these can be risky, especially for the inexperienced.

Similar contracts are available for grain traders via *Futures* contracts (for bank bills, Treasury bonds and other commodities, sometimes including chickpea). Futures contracts are also commitments to make or accept delivery of a specific quantity and quality at a specific date and price. However, traders generally use Futures contracts as a tool for managing price risk, a strategy known as *hedging*. Hedging allows a trader to lock in a position in advance that is equal and opposite to the physical market position, thereby reducing the potential for unanticipated loss or competitive disadvantage. In fact, very few chickpea Futures contracts result in the delivery of chickpea as traders generally offset their futures positions before the contracts mature. The difference between the initial purchase/sale price and the price of the offsetting transaction represents the realized profit or loss.

A very active chickpea futures market operates within India, but Indian law prohibits non-Indian companies from participating. This market is actively utilized by most of the chickpea importers and 'multinationals' that have India-based trading companies.

Where there is no futures market for chickpea (Canada, Australia), prices are negotiated directly between traders and customers based on supply–demand, chickpea type and quality. The prices negotiated could be for immediate delivery or delivery at a future date.

Marketing Organizations

There are many grower and marketing organizations in chickpea trading countries that provide information to their members and some information may be available to the general public on their websites (Table 27.7).

The pulse traders have established a global confederation, Confederation Internationale du Commerce et des Industries des Legumes/The International

Table 27.7. Agencies providing information related to marketing of chickpea.

Agency	Website
Agriculture and Agri-Food Canada	http://www.agr.gc.ca/index_e.phtml
Alberta Pulse Growers, Canada	http://www.pulse.ab.ca
Canadian Grain Commission	http://www.grainscanada.gc.ca/main-e.htm
Department of Agriculture and Co-operation, Government of India	http://agmarknet.nic.in
International Pulse Trade and Industry Confederation (CICILS/IPTIC)	http://www.cicilsiptic.org
Mediterranean Grain-Pulse Exporters Union, Turkey	http://www.akib.org.tr
MediumGrain Rice.com	http://www.mediumgrainrice.com/
Multicommodity Exchange of India	http://www.mcxindia.com
National Agency for Export Development (NAFED), Indonesia	http://www.nafed.go.id/
Pulse Australia	http://www.pulseaus.com.au
Saskatchewan Pulse Growers Association, Canada	http://www.saskpulse.com
STAT Communications Ltd, Canada	http://www.statpub.com
Turkish Exporters Association	http://www.turkishtime.org/22/106_en.asp
USA Dry Peas, Lentils and Chickpeas	http://www.pea-lentil.com/nwdev/

Pulse Trade and Industry Confederation (CICILS/IPTIC). It represents 12 national pulse organizations and more than 250 leading companies in 32 countries. CICILS/IPTIC represents the international pulse trade, notably in the International Bodies of the United Nations (Organization for Economic Cooperation and Development, Food and Agriculture Organisation, World Health Organization, Codex Alimentarius), Regional Governmental Organisations (EU) and International Trade Organizations. It maintains regular contacts with these bodies in order to promote the viewpoint of the pulse industry. This essentially involves defending the general interests of its members, notably in improving the conditions of international pulse trade, in resisting trade barriers and regulations of protectionist nature.

Conclusion and Future Outlook

World production of chickpea in 2004 was 8.6 million tonnes and increased to 9.2 million tonnes in 2005 (FAO, 2006). Agriculture and Agri-Food Canada (2006c) forecasts that production will remain stable; however, world supply is expected to decrease marginally due to lower carry-in stocks. Whilst there are no official statistics for the ratio of desi and kabuli chickpea types produced globally, it was believed to be in the region of 85% desi and 15% kabuli in 2001 (Agriculture and Agri-Food Canada, 2001). Estimates calculated from inferring the production statistics and estimates of desi/kabuli ratios in each country over the last decade suggest the balance may be closer to 70% desi and 30% kabuli. Agriculture and Agri-Food Canada (2006b) have forecast a continuation of the increasing ratio of kabuli to desi chickpea for the 2006/07 season. Population growth continues in the major consuming countries thereby increasing demand for chickpea.

Acknowledgement

The authors wish to thank Mr R. Brealey for his contribution to this chapter.

References

Agbola, F.W. (2004) Agricultural policy reform in India: implications for pulse trade, prices and production, 1970–1999. *Agribusiness Perspectives Papers 2004*, Paper No. 63. Available at: www.agrifood.info/perspectives/2004/Agbola.html.

Agbola, F.W. and Damoense, M.Y. (2005) Time-series estimation of import demand functions for pulses in India. *Journal of Economic Studies* 32, 146–157.

Agbola, F.W., Kelley, T.G., Bent, M.J. and Rao, P.P. (2002a) Marketing channels in India: Challenges and opportunities. *Agribusiness Perspectives Papers 2002*, Paper No. 53.

Agbola, F.W., Kelley, T.G., Bent, M.J. and Rao, P.P. (2002b) Eliciting and valuing market preferences with traditional food crops: the case of chickpea in India. *International Food and Agribusiness Management Review* 5, 7–21.

Agriculture and Agri-Food Canada (2001) Chickpeas situation and outlook. *Bi-weekly Bulletin* 14(3), 1–4.

Agriculture and Agri-Food Canada (2006a) Canada: pulse and special crops outlook. *Bi-weekly Bulletin* (April 28), 1–2.

Agriculture and Agri-Food Canada (2006b) Canadian outlook for grains, oilseeds, pulses and special crops. *Bi-weekly Bulletin* 18(12), 1–4.

Agriculture and Agri-Food Canada (2006c) Canadian outlook for grains, oilseeds, pulses and special crops. *Bi-weekly Bulletin* 19(1), 1–8.

Akova, Y. (2002) Bakliyat Sektörü Dis Pazar arastirmasi, IGEME, Ihracati Gelistirme Etüd Merkezi, Ankara.

Anonymous (2001) Turkish standard: Nohut. Institute of Turkish Standards, Ankara, pp. 1–8.

Anonymous (2006a) Republic of Turkey, Ministry of Agriculture. Available at: http://www.tarim.gov.tr/arayuz/9/icerik.asp?fl=mevzuat/mevzuat.htm.

Boubaker, B. (2005) Global pulse markets: situation and outlook. In: *Proceedings of the CICILS-IPTIC World Convention*. 1–2 June 2005. Cairo. Available at: http://www.fao.org/documents/show_cdr.asp?url_file=/docrep/008/J6801e/J6801e00.htm.

Canadian Grain Commission (2005) Chickpeas. *Official Grain Grading Guide*. Canadian Grain Commission, Winnipeg, Canada, pp. 22/1–22/11.

Choudhary, M.K., Sherkar, B.D., Singh, H.P. and Bhatia, G.R. (2004). *Post Harvest Profile of Bengal Gram*. Directorate of Marketing, Government of India.

Çiftçi, C.Y. (2004) Dünyada ve Türkiye'de Yemeklik Tane Baklagiller Tarimi, TMMOB Ziraat Mühendisleri Odasi Teknik Yayinlar Dizisi No.5, 2004, 200 sayfa, Ankara.

Dusunceli, F. and Aydoğan, A. (2004) Turkey: a prominent grain legumes trading country. *Grain Legumes Magazine*, Vol. 41. January 2005.

Dusunceli, F., Kilinç, H., Dölekoğlu, C.Ö., Eken, S., Yalçin, H., Serifoğlu, S., Tüzün, S., Yücer, A., Polat, S., Perkin, A.Y., Günlü, H., Aydoğan, A. and Karahan, S. (2006) *Grain Legumes Common Market Regimes Sub Committee Study Report*, Ministry of Agriculture, Ankara, Turkey (In Turkish), February 2006.

FAO (Food and Agriculture Organisation) (2006) *FAOSTAT Database*. Food and Agriculture Organisation of the United Nations, Rome, Italy.

Government of India (2002) Schedule II – Grade designation and definition of quality of Roasted Bengal Gram (split). *The Gazette of India*, Part II, Section 3, Subsection (i), Department of Agriculture and Co-operation, New Delhi, India, pp. 2686–2694.

Government of India (2004a) Schedule XII – Grade designation and definition of quality of Kabuli Chana. *The Gazette of India*, Part II, Section 3, Subsection (i), Department of Agriculture and Co-operation, New Delhi, India, pp. 680–714.

Government of India (2004b) Schedule XIII – Grade designation and definition of quality of Chana whole (Bengal gram). *The Gazette of India*, Part II, Section 3, Subsection (i), Department of Agriculture and Co-operation, New Delhi, India, pp. 680–714.

Government of India (2004c) Schedule XIV – Grade designation and definition of quality of Chana split (husked)/Dal Chana. *The Gazette of India*, Part II, Section 3, Subsection (i), Department of Agriculture and Co-operation, New Delhi, India, pp. 680–714.

Kaur, M. and Singh, N. (2005) Studies on functional, thermal and pasting properties of flours from different chickpea (*Cicer arietinum* L.) cultivars. *Food Chemistry* 91, 403–411.

McVicar, R., Pearse, P., Panchuk, K., Brenzil, C., Hartley, S., Harris, C., Yasinowski, J., Goodwillie, D., Warkentin, T. and Banniza, S. (2000) *Chickpea in Saskatchewan*. Available at: http://www.agr.gov.sk.ca/docs/crops/pulses/production_information/chickpea2002.asp?firstPick=Crops&secondpick=Pulses&thirdpick=Production%20Information.

Margheim, J.F., Baltensperger, D.D., Wilson, R.G., Lyon, D. J., Hein, G.L., Harveson, R.M., Burgener, P., Krall, J.M., Cecil, J.T., Rickertson, J.R., Merrigan, A.P., Watson, M.H., and Hansen, B.J. (2004) *Chickpea Production*

in the High Plains. University of Nebraska, Cooperative Extension, EC03–183.
Meares, C.A., Bogracheva, T.Y., Hill, S.E. and Hedley, C.L. (2004) Development and testing of methods to screen chickpea flour for starch characteristics. *Starch/Starke* 56, 215–224.
Pulse Australia. (2005) Australian Pulse Trading Standards 2005/2006. *National Agricultural Commodities Standards Manual*, National Agricultural Commodities Marketing Association (NACMA), Sydney, Australia, pp. 1–61.
USDA (United States Department of Agriculture) (1999) General Provisions. In: *Official United States Standards for Grain*. The Grain Inspection, Packers and Stockyards Administration (GIPSA), Washington, USA, pp. A/1–A/5.
Wagner, A. (1891) Marshall's Principles of Economics. *The Quarterly Journal of Economics* 5, 319–338.
Wood, J.A. and Harden, S. (2006) A method to estimate the hydration and swelling properties of chickpeas (*Cicer arietinum* L.). *Journal of Food Science* 71, E190–E195.

28 Crop Simulation Models for Yield Prediction

M.R. Anwar,[1] G. O'Leary,[1] J. Brand[1] and R.J. Redden[2]

[1]*Primary Industries Research Victoria – Horsham 110 Natimuk Road, Horsham, VIC 3400, Australia;* [2]*Australian Temperate Field Crops Collection, Department of Primary Industries, Victorian Institute for Dryland Agriculture, Horsham, VIC 3401, Australia*

Abstract

The integration of seasonal climate forecasts with crop simulation models offers ways to reduce the impact of climate variability on chickpea production. We applied various methods to evaluate the forecasting quality of the five Southern Oscillation Index (SOI) phase system and observed significant changes in the probability distribution functions of grain yields and spring rainfall by using June/July SOI phase. We observed significant changes in shift and dispersion of the probability distribution functions of chickpea yield and spring rainfall by using June/July of the SOI phase system. These changes in shift and dispersion can be useful to select appropriate farming management options. March/June is known as the 'predictability barrier' and model skill is at its lowest across this span of months in Australia. We observed significant El Niño Southern Oscillation (ENSO) signals on the climate and grain yield in an example region (Birchip, an emerging chickpea area of south-eastern Australia) that can be used to produce outlooks of seasonal conditions and chickpea yield as early as the end of July. This knowledge can be used operationally to forecast rainfall distributions and season types for many parts of the world. When combined with crop simulation models that account for the major environmental effects on crop growth, producers are able to strategically adjust their crop management options, reducing climatic risks.

Introduction

Climate influences almost every agronomic decision growers make in their desire to produce a profitable chickpea crop each season. However, the climate in many chickpea cropping regions is highly variable from year to year and within the year (e.g. extreme weather events) and is difficult to predict. High rainfall variability is the major source of dryland chickpea yield fluctuations in south-eastern Australia (see Fig. 28.2) and elsewhere in the semiarid tropics and subtropics (Anwar *et al.*, 2003). Currently, growers in south Australia implement agronomic practices such as disease resistant cultivars, pesticide strategies and soil-nutrient management that are best suited to long-term average climatic conditions for the region. Our understanding of the influence of ocean and atmospheric dynamics on climate variability in specific regions has greatly increased over the last decade, enabling significant improvements in predictability (Australian Bureau of Meteorology: http://www.bom.gov.au). The requirement for a certain lead-time between deciding on a course of action and realizing its results is a characteristic of managing crop productivity. This chapter outlines a range of methods that address how crop management of chickpea can be improved to reduce climate risk in highly variable climatic regions by integrating seasonal climate forecasting tools with a cropping systems model to help growers make informed decisions. In particular, we focus on the dryland cropping regions of South Australia.

Chickpea Cropping in Australia

Rainfed chickpea cropping in Australia occurs on about 0.2 million hectares distributed from the summer-dominated rainfall region of the central highlands of the north-east of Queensland in an arc around South Australia to the winter-dominated rainfall areas around Geraldton in West Australia (Fig. 28.1). In the

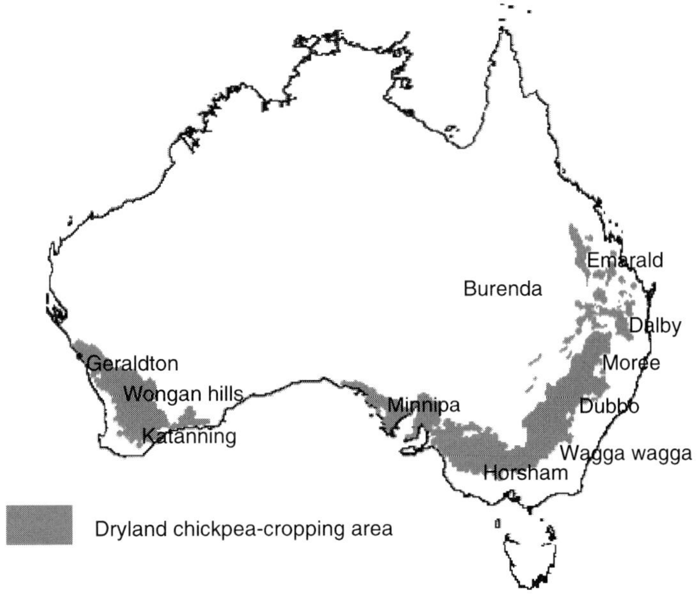

Fig. 28.1. Chickpea production areas in Australia. (From Loss *et al.*, 1998.)

western and southern regions, the predominantly cool season rainfall (i.e. autumn to spring) allows chickpea crops to take place on a variety of soil types – from the sands of West Australia to the heavy clay soils of the Wimmera in Victoria. In contrast, in the northern regions, cropping is largely restricted to heavier soils with high water-holding capacity that can store the predominantly summer rainfall so that it is available for the cool season crops. This diversity of climatic patterns in Australia makes chickpea production very risky. Chickpea production, in particular, can be greatly affected by wet springs (e.g. in south-eastern Australia). Even new ascochyta resistant varieties will still be vulnerable to this disease during the pod filling phase, and will be vulnerable to botrytis disease if conditions are warm and wet (Hobson *et al.*, 2004). Planning for these contingencies may be assisted by seasonal climate forecasting.

Climate Forecasting

Climate varies at a range of scales

Climate can be defined as the ensemble of weather (http://www.bom.gov.au). As such, it is inherent to the atmosphere, but is affected by interactions among the atmosphere and the ocean, the biosphere, the land surface and the cryosphere. Because climate varies on all timescales (e.g. Parry *et al.*, 1988; Power *et al.*, 1999; IPCC, 2001), an appropriate mean can serve as a reference state for the study of variability on shorter time scales, while itself changing on longer timescales. In practice, whatever the definition used for the mean, an anomaly is the difference between the instantaneous state of the climate system and that mean. Climate variability and change are characterized in terms of these anomalies. As the study of climate progresses, it is becoming increasingly apparent that the variations are not randomly distributed in space and time, but often appear to be organized into relatively coherent spatial structures that tend to preserve their shape – or assume a limited number of related shapes – while their amplitude, phase and sometimes geographic position change through time.

Climate, and in particular rainfall variability and its interaction with land management, has shaped Australian agriculture since the beginning of European settlement over 200 years ago. In south-eastern Australia the primary climatic influence on the productivity of rainfed farming systems is from the interannual climatic system, El Niño Southern Oscillation (ENSO) (Power *et al.*, 1999; Potgieter *et al.*, 2002). The ENSO phenomenon strongly influences rainfall distribution around the world (Stone *et al.*, 1996a). El Niño refers to the warming of the oceans in the equatorial eastern and central Pacific; Southern Oscillation is the changes in atmospheric pressure (and climate systems) associated with this warming (hence 'Southern Oscillation Index' to measure these changes). ENSO is used colloquially to describe the whole suite of changes associated with El Niño and La Niña events – to rainfall, oceans, atmospheric pressure, etc. The understanding of the influence of the ENSO, as well as other important couplings of ocean currents and atmospheric dynamics, on climate variability in specific regions has enhanced our ability to forecast events such

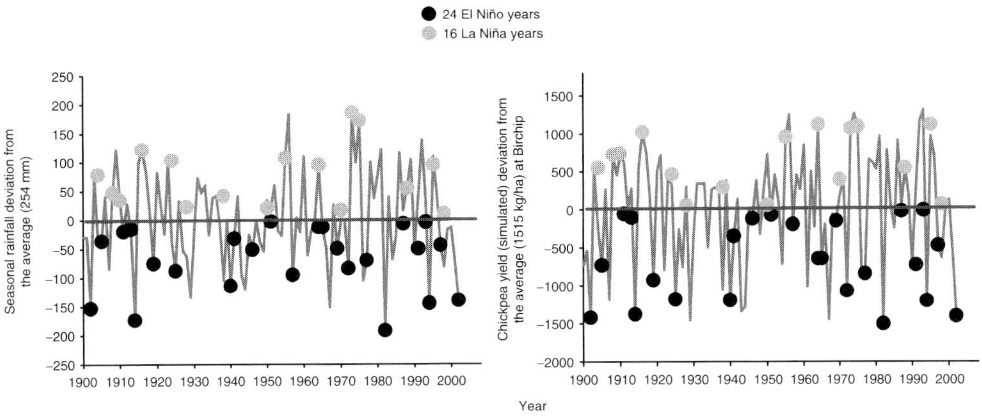

Fig. 28.2. Effect of typical 24 El Niño years and 16 La Niña years after 1990 at Birchip, south-eastern Australia, on seasonal rainfall and simulated chickpea yield.

as El Niño and La Niña years on a regional basis (http://www.bom.gov.au). The general impact on crop yield of the climate regimes associated with the El Niño phenomenon is reasonably well understood and effectively captured in a number of different crop simulation models (e.g. Power *et al.*, 1998; Meinke and Hochman, 2000; Stephens *et al.*, 2000; Anwar *et al.*, 2006b). The ENSO affects chickpea yields (Fig. 28.2) through its influences on seasonal conditions – i.e. rainfall, temperatures, etc. (Rimmington and Nicholls, 1993). Figure 28.2 illustrates the simulated association of chickpea yields and ENSO at a location in south Australia, upwards in La Niña years and downwards in El Niño years. Chickpea, like wheat, is a long-day plant sown in winter in eastern and southern Australia, and winter rainfall is influenced by ENSO patterns. Relatively dry winters and wet springs can still result in a favourable yield potential in chickpea, compared to a cereal crop (Whish and Castor, 2005).

Impact of El Niño and La Niña on rainfall

The five phases of the Southern Oscillation Index (SOI) as defined by Stone and Auliciems (1992) can be used as a climate indicator of the state of ENSO (Stone *et al.*, 1996a), and its influence on seasonal conditions (McBride and Nicholls, 1983; Lythgoe *et al.*, 2004, Anwar *et al.*, 2004). In El Niño conditions, the impact on Australian rainfall generally results in below-average rainfall in winter and spring, particularly in inland eastern cropping regions (Fig. 28.3a). The effect is reduced in coastal regions. The La Niña phase is the opposite (Fig. 28.3b) and was often the prevailing phase during the record rainfall periods that occurred in the 1950s and 1970s in south-eastern Australia (http://www.nrm.qld.gov.au/silo). ENSO is not a regular cycle (Figs 28.2 and 28.4) but it does have some clear, now well understood, and very useful patterns (http://www.longpaddock.qld.gov.au). Figure 28.4 shows both persistent El Niño and La Niña events. In 1997, consistently negative (CN) SOI values were recorded through the winter

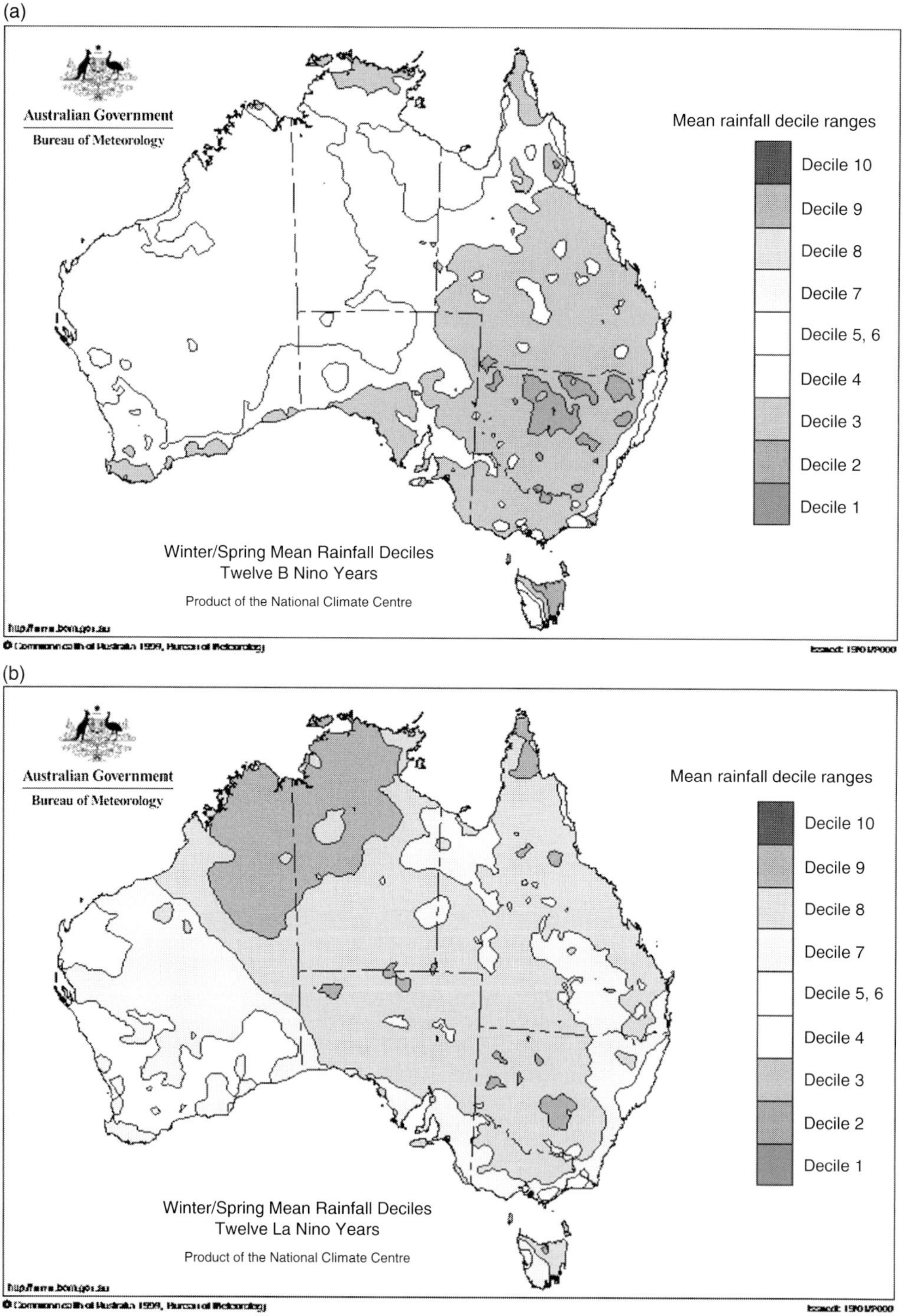

Fig. 28.3. Impact of ENSO on Australian winter/spring mean rainfall deciles for 12 (a) El Niño and (b) La Niña events. (From http://www.bom.gov.au)

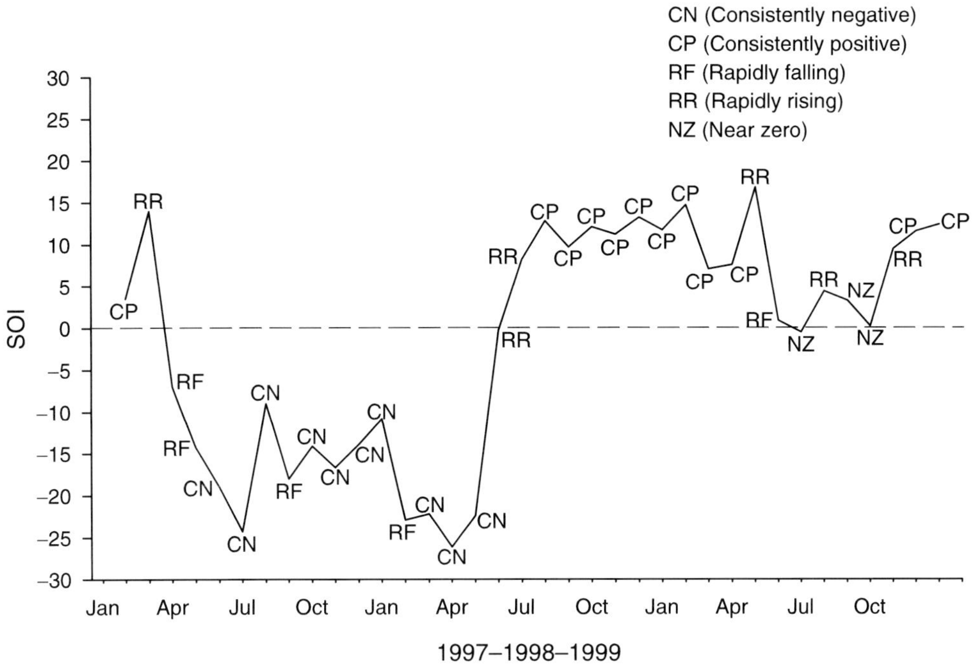

Fig. 28.4. The Southern Oscillation Index (SOI) value and the associated SOI phases across 1997 through 1999. (From www.longpaddock.qld.gov.au)

and spring forming a similar pattern to previous El Niño events. In winter and spring 1998 through autumn 1999 consistently positive (CP) index values were recorded, indicative of a La Niña phase. The rainfall patterns in previous events give the basis for a forecast of the likely impact on rainfall probabilities.

Application of seasonal climate forecasts

Seasonal climate forecasting (SCF) is the use of meteorological or climate information to produce estimates of the future (3–12 months) condition of the climate, e.g. seasonal rainfall, risk of frost, etc. Information from SCF can be used to make informed changes in a decision that might generate improved outcomes from the farming system. The SCF deals with climatic mechanisms that behave predictably over season-to-season timescales. The ENSO is currently the most important climate phenomenon providing a level of interseasonal predictability. The SCF can increase preparedness for dealing with climate variability and extreme climate events. However, it is vitally important to distinguish between application of a seasonal climate forecast and the impact of a particular climatic event. There is a rich literature on impacts of ENSO on agriculture and other sectors (see Glantz, 1996; White, 2000; Hansen *et al.*, 2001) and a number of specific reports of impacts on crop production in various countries (Nicholls, 1985; Rimmington and Nicholls, 1993; Khandekar, 1996; Kirono and Khakhim, 1998; Ferris, 1999; Ferreyra *et al.*, 2001; Park *et al.*, 2002; Anwar *et al.*, 2006). Application of SCF differs considerably from impact or anticipated impact. While impact refers to the extent of effect, application of a forecast refers to the pre-emptive management response aimed at changing the anticipated impact. In this way, the forecast can acquire value. An SCF has no value unless it provides sufficient information in a timely manner that allows producers to make informed management decisions that enhance on-farm profitability and viability. A considerable body of effort in various parts of the world is now being focused on this issue of applying climate predictions to improve agricultural systems. Some studies have examined tactical changes (such as fertilizer management, sowing date and cultivar selection) in decisions associated with a seasonal forecast (Keating *et al.*, 1993; Marshall *et al.*, 1996; Meinke and Stone, 1997; Mjelde *et al.*, 1997, 2000). In Australia, benefits from SCF include extra profit resulting from selecting management options according to particular phases of SOI (Hammer *et al.*, 1996, Meinke and Hochman, 2000, White, 2000, Robinson and Butler, 2002). For Australian chickpea growers, the benefit of seasonal climate forecasts is particularly dependent on the timing of information. For example, a forecast given in June and July, post-sowing, enables growers to predict potential grain yields and forecast that amount of grain that they will have available for marketing. It is unlikely that a forecast at this stage will change in-crop management decisions. However, forecasts prior to sowing (April) enable crucial changes in the cropping system. A grower with knowledge of the climate in the season ahead would be able to make crop, variety and management choices that reduce production risks and enhance profitability.

The inherent probabilistic nature of SCF presents particular challenges as financial loss can be incurred even from the appropriate use of climate information and SCF. Matters get more complex when probabilistic approaches need to be tested. Potgieter *et al.* (2003) indicated that in order to measure the skill of probabilistic forecasting schemes, any skill-scoring system must measure the two basic components of the system: (i) the shift in the forecast mean and median from the reference distribution and (ii) the change in the dispersion of forecast vs reference distribution. Based on case studies of forecasts of regional wheat and sugar yields, they concluded that no single skill-scoring scheme is able to measure both attributes well, and consequently there is no single statistical procedure available that measures 'skill' adequately. This must be taken into account when developing acceptable verification schemes for climate forecasts.

Cropping system models and probability forecasts

Timing and frequency of rainfall events strongly influence crop growth and yield. This is because of variation in other factors like initial soil water content at sowing time, rate of phenological development and soil nutritional status as well as fertilization strategies. The general impact on crop yield of the climate regimes associated with the El Niño phenomenon is reasonably well understood and effectively captured in a number of different crop simulation models (e.g. Power *et al.*, 1998; Meinke and Hochman, 2000; Stephens *et al.*, 2000; Anwar *et al.*, 2006). Cropping systems models that attempt to mimic the biophysical state through time are useful tools because of the interconnection between the key factors that determine crop yield, e.g. Agricultural Production Systems sIMulator (APSIM) (McCown *et al.*, 1996), Decision Support System for Agrotechnology Transfer (DSSAT) (Jones *et al.*, 2003) and CropSyst (Stockle *et al.*, 1994).

A cropping field and its management consists of the crops, the soil and its physical and chemical attributes, and the manager undertaking a range of management practices, such as sowing, tillage, weed control and harvesting (Fig. 28.5). Inputs to this system include daily weather, fuel, fertilizer and pesticide. Outputs include harvested products, runoff and soil loss. One could also look at balances for elemental C and N in compounds entering and leaving the system. The dynamics involve crop growth and development processes, and soil chemical and physical processes given the specific conditions of the system. Management of a system is the manipulation of that system to achieve specific objectives. Hence, managers must be considered an element of the system, or at least their actions as inputs to the system. The consequences of management manipulations may be predicted using system models, although the adequacy of such models for this task must be examined critically.

In this chapter we used the APSIM modelling system (Robertson *et al.*, 2002; Keating *et al.*, 2003). Hammer *et al.* (1999) have provided a general overview on predictive capability and application of crop simulation in APSIM. Additionally, a major advantage of the modelling approach is that a large number of simulated yields would allow the exploration of the influence of ENSO on extreme (much above or below normal) outcomes, a difficult approach with limited historical yields series. The five-phase SOI forecasting system (Stone *et al.*, 1996a) allows historical records of rainfall to be grouped into analogue years and applying this system to simulate yields (Hammer *et al.*, 1996; Anwar *et al.*, 2006). This systems analysis approach not only integrates the amount of rainfall but also other

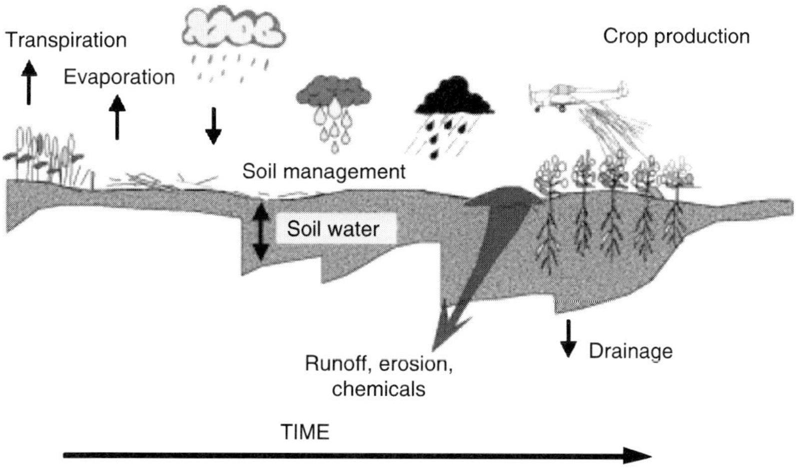

Fig. 28.5. Schematic representation of a cropping system showing transitions of the system state through time in response to management activities. Some system inputs and outputs are indicated.

climate variables of impact on growth and production – i.e. rainfall distribution, maximum temperature, minimum temperature and solar radiation.

As an example (Fig. 28.6), at Birchip (south-eastern Australia) the long-term (1889–2005) simulated chickpea yield indicates that below average, average and above average chickpea yields are seasons having <1197 kg/ha, between 1197 and 1901 kg/ha and > 1902 kg/ha, respectively. The Pie chart (Fig. 28.6), however, indicates how the odds of achieving below average, average and above average chickpea yield shifts from 33% when the SOI during the previous March/April period is consistently negative (CN), consistently positive (CP), rapidly falling (RF), rapidly rising (RR) and near zero (NZ). Although this Pie chart will never be fully green, yellow or red, any shift in the odds from the long-term values can assist in decision making, to increase the area sown to chickpea, or to plan seeding rates and levels of inputs of fertilizer, herbicides and crop protection chemicals. Seasonal forecasting tools are not perfect and never will be. Therefore, all seasonal outlooks should also include a measure of their likelihoods. Simulation techniques can be used to produce incremental forecasts of crop yield that are of great value for farmers (Fig. 28.6).

Probabilistic Forecast Quality

Statistical measures

Probabilistic forecasts are distributions, and the complete nature and information content of the distributions can be retained if their quality is assessed relative to

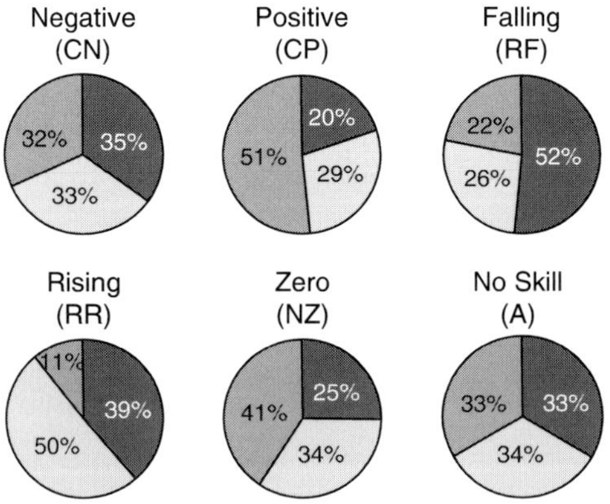

Fig. 28.6. Probabilities of simulated chickpea yields for Birchip, south-eastern Australia, from analogue years derived for each of the five Southern Oscillation Index (SOI) phases in March/April. Low yields are defined as the lower 33% of the data (low •, <1197 kg/ha); average yields as the middle tercile, i.e. 33–66% (average •, 1197–1901 kg/ha); and high yields as the upper tercile, i.e. 66% (high •, 1901–2837 kg/ha), of the probability distribution function. CN: consistently negative; CP: consistently positive; RF: rapidly falling; RR: rapidly rising; NZ: near zero; A: with no predictable knowledge of the coming season.

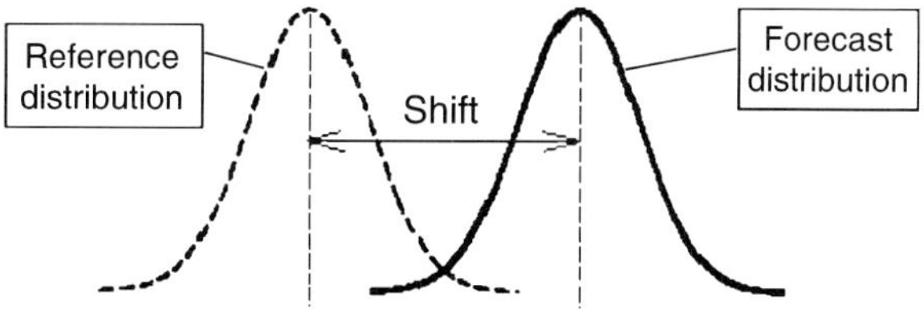

Fig. 28.7. Representation of a shift in median between two distributions with the same variance.

a standard reference forecast. There are many appropriate attributes of forecast quality (see Wilks, 1995). In this chapter, we use measurements of distribution discrimination (shift and dispersion). The distribution shift (i.e. shift in the central location, e.g. mean or median) of the forecast distribution from the reference distribution (Fig. 28.7) can be quantified as the absolute median difference (AMD):

$$\text{Absolute median difference (AMD)}=\left|\overline{X}_R - \overline{X}_F\right| \qquad (28.1)$$

where $\overline{X}_F$ and $\overline{X}_R$ are the sample medians of the forecasted distribution (*F*) for a given variable and reference distribution (*R*) – i.e. the all-years distributions.

The changes in the dispersion (i.e. variability) of the forecast distribution relative to the reference distribution (Fig. 28.8) can be quantified using inter-quartile ratios (IQRs) (Wilks, 1995):

$$\text{Inter-quartile ratio (IQR)}=\frac{(X_{75} - X_{25})_F}{(X_{75} - X_{25})_R} \qquad (28.2)$$

where X_p indicates the specific percentile value (p = 75% and 25%) for distribution *F* and *R*. Values of the IQR higher (smaller) than 1 indicate a higher (smaller) dispersion (spread) in the forecasted values with respect to the reference distribution.

The overall significance of shifts in medians for all the examined variables (e.g. chickpea yields, spring rainfall, in-crop rainfall, etc.) for each SOI phase (forecasts) with respect to the rest of the years can be tested using a Kruskall–Wallis (KW) test (Kruskall and Wallis, 1952). Whenever significant differences were detected a Kolmogorov–Smirnov (KS) test (Conover, 1971) can be used

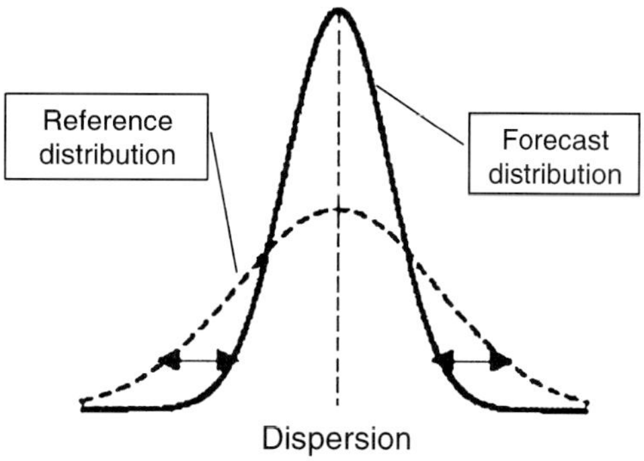

Fig. 28.8. Change in dispersion or variance between two distributions with the same median.

to test differences in cumulative distributions between each of the individual phases and the rest of the years (Wilks, 1995; Meinke *et al.*, 1996).

To independently verify the hindcast skill of the forecasting systems, a cross-validation technique can be applied to several climate variables and simulated yields. Cross-validation can be a useful method of estimating the predictive skill of a forecasting system by comparing outlooks with observations from historical data. Cross-validation involves the subdivision of a limited number of observations into two data sets, a training data-set, and a validation data-set (Wilks, 1995). This methodology results in a measure of skill expressed as percent consistent, that is the number of consistent forecasts for a category (e.g. above median), divided by the total number of forecasts made for that category (for further details, see http://www.bom.gov.au/silo/products/verif/).

Analogue years

Analogue years are past years that resemble the current season in key attributes. Stone *et al.* (1996a) and Drosdowsky (2002) have used analogue years in their forecasting systems. Such systems allow historical records of climate variables to be used to simulate yields using a cropping systems simulator. In this paper we tested the capacity of the five SOI phase system, as defined by Stone and Auliciems (1992), to forecast climate variables and chickpea yields by the end of July. Analogue years (Table 28.1) with the same SOI phase over the June/July months and SOI phase over the March/April months (see Fig. 28.6) were obtained from the long paddock website (http://www.longpaddock.qld.gov.au/). We analysed the following climatic variables: annual rainfall, seasonal rainfall, spring rainfall, date of last frost and break of the season. 'Seasonal rainfall' was defined as rainfall between 1 April and 30 November; 'spring rainfall' as rainfall between 1 September and 30 November; 'break of the season' as the day of the year when 10 mm of accumulated rainfall had fallen within three consecutive days, allowing the sowing of a chickpea crop; and the 'date of last frost' as the last day of the year when the minimum temperature was lower than or equal to 0°C.

Case Study in the Southern Mallee Region of South-eastern Australia

In the southern Mallee region, chickpea is sown between May and July and is reliant on in-crop rainfall. The availability of soil water around the time of flowering and podding during the spring months is particularly critical to achieving profitable grain yields. As indicated previously high rainfall variability is the major source of dryland chickpea yield fluctuations in south-eastern Australia. These year-to-year yield fluctuations lead to high uncertainty of income. To remain economically viable, growers devise management options that can produce long-term, sustainable profits in such a variable environment. This requires some knowledge of climatic conditions for the season ahead. In this case study, we have considered a rainfed chickpea growing location 'Birchip (35.98°S,

Table 28.1. The analogue years for the five phase SOI system in June/July and March/April. (From http://www.longpaddock.qld.gov.au/)

SOI phase in June/July	Years
Consistently negative (CN)	1896, 1899, 1905, 1911, 1914, 1919, 1940, 1941, 1946, 1972, 1977, 1982, 1987, 1992, 1993, 1994, 1997, 2004
Consistently positive (CP)	1892, 1893, 1900, 1901, 1909, 1910, 1917, 1920, 1921, 1924, 1931, 1934, 1938, 1945, 1950, 1952, 1955, 1956, 1964, 1968, 1973, 1975, 1981, 1986, 1989, 1996, 1998
Rapidly falling (RF)	1889, 1918, 1923, 1925, 1937, 1951, 1965, 1970, 1976
Rapidly rising (RR)	1898, 1903, 1906, 1912, 1916, 1926, 1928, 1933, 1936, 1939, 1943, 1947, 1948, 1949, 1954, 1960, 1963, 1974, 1979, 1984, 1985, 1988, 1995, 1999, 2003
Near zero (NZ)	1890, 1891, 1894, 1895, 1897, 1902, 1904, 1907, 1908, 1913, 1915, 1922, 1927, 1929, 1930, 1932, 1935, 1942, 1944, 1953, 1957, 1958, 1959, 1961, 1962, 1966, 1967, 1969, 1971, 1978, 1980, 1983, 1990, 1991, 2000, 2001, 2002, 2005
SOI phase in March/April	**Years**
Consistently negative (CN)	1897, 1900, 1905, 1915, 1926, 1940, 1941, 1966, 1977, 1980, 1981, 1983, 1987, 1991, 1992, 1998
Consistently positive (CP)	1890, 1892, 1898, 1899, 1901, 1902, 1903, 1910, 1917, 1921, 1923, 1925, 1927, 1928, 1931, 1935, 1939, 1950, 1956, 1959, 1960, 1963, 1971, 1974, 1975, 2000
Rapidly falling (RF)	1909, 1912, 1914, 1922, 1924, 1944, 1945, 1947, 1952, 1965, 1967, 1969, 1976, 1993, 1994, 1995, 1997, 2004, 2005
Rapidly rising (RR)	1889, 1891, 1904, 1916, 1918, 1919, 1933, 1934, 1936, 1938, 1943, 1948, 1954, 1961, 1984, 1985, 1989, 1990, 1999
Near zero (NZ)	1893, 1894, 1895, 1896, 1906, 1907, 1908, 1911, 1913, 1920, 1929, 1930, 1932, 1937, 1942, 1946, 1949, 1951, 1953, 1955, 1957, 1958, 1962, 1964, 1968, 1970, 1972, 1973, 1978, 1979, 1982, 1986, 1988, 1996, 2001, 2002, 2003

142.92°E)′ in the southern Mallee region of Victoria, Australia. Probabilistic rainfall forecasts for the months of August, September and October could be used in Birchip to primarily help farmers make tactical changes to their marketing decisions, based on grain yields that are likely to be achieved. The major goal of this modelling effort was to produce a time series of simulated chickpea yields (in this case 117 yield outcomes for the 1889–2005 period) to test whether simulation models can be used to capture signals from ENSO apart from rainfall amount and also investigate the vulnerability of current chickpea production systems to ENSO-related climate variability in south-eastern Australia.

To explore the impact of climate variability on chickpea yield, we applied the APSIM systems model to simulate a long-term time series of chickpea yield based on common agronomic practices (Table 28.2) at Birchip. Key parameters used for the simulation by APSIM are listed in Table 28.3. Starting conditions (soil water, soil N and residues) for each simulation were set on 1 January of each simulated year. In-crop rainfall (from sowing to maturity) for all 117 years was taken from the model outputs. Long-term historical climate data (1 January

Table 28.2. Typical management strategy for chickpea grown in Birchip, south-eastern Australia, and the basic assumptions for chickpea crop simulation with APSIM.

Management choice	Typical practice
Previous crop	Cereal (e.g. wheat, barley)
Rainfall (annual)	370 mm
Soil type and pH	Sands – loamy clays. pH 5.5–9.0. Avoid areas with high salinity and/ or subsoil nutrient toxicities.
Variety	Genesis 090
Sowing window	Early May/early June
Sowing rate	30–45 plants/m^2
Row spacing	20–40 cm
Fertilizer	50 kg/ha Grain legume superphosphate + 1% Zn
Tillage	Minimum tillage or no-till
Inoculant	Group N
Herbicides	Pre-sowing: glyphosate and trifluralin
	Post sow pre-emergent: simazine or simazine + isoxaflutole
	Post emergent: clethodim
Insecticides	At emergence: endosulfan
	During flowering and podding: alpha-cypermethrin
Fungicides	Seed: thiram + thiabendazole
	In crop: chlorothalonil applied during podding

Table 28.3. Chickpea cultivar parameters used to simulate grain yield with APSIM.

Description	Units	Value
Leaf growth		
Thermal time required for node appearance on main stem	Degree-days	46.0
Number of leaves per plant per main stem node	Leaf/node	13.4
Number of nodes at start of branching		7.0
Rate of death of nodes on the main stem	Degree-days	47.0
Fraction of total leaf number senescing for each node that senesces		0.02
Radiation interception, biomass accumulation		
Extinction coefficient (at default row spacing)		0.70
Radiation-use efficiency (RUE)	g/MJ	0.72
Cardinal temperatures for relationship between average temp and 0–1 stress factor on RUE		0, 15, 30, 40
Biomass partitioning		
Fraction allocated to leaves pre-flowering		0.48
Fraction allocated to leaves post-flowering		0.40
Fraction allocated to leaves in grain filling		0.00
Fraction allocated to pod before grain fill		0.10
Fraction allocated to pod relative to grain during grain fill		0.28
Potential rate of increase in harvest index	Per/day	0.010

1889–31 December 2005) for Birchip in the southern Mallee, obtained from the SILO patch-point data-set (http://plum.nre.vic.gov.au) were used to parametize the model. In this study, all crop simulations were undertaken assuming year 2005 technology levels. Grain yield results from the simulation and in-crop rainfall were then segregated based on the five SOI phases during June/July resulting in cumulative distribution functions (Fig. 28.9). The KW test (Kruskall and Wallis, 1952) was used to compare median values in grain yields, while KS tests (Conover, 1971) were used to examine the significance of the shift in the medians for each SOI phase with respect to the rest of the years.

The long-term median simulated chickpea yield and in-crop rainfall at Birchip was 1657 kg/ha and 241 mm (Fig. 28.9) with a coefficient of variation of 49% and 33%, respectively. This simulated chickpea yield corresponds well with actual long-term farm yield of the region. Figure 28.9 shows important shifts in the median values of the cumulative probability distribution function for simulated chickpea yields and in-crop rainfall during CN, CP, RF, RR and NZ SOI phases at the end of July, compared with the rest of the years. During CN and RF phases in June/July, that are frequently associated with El Niño-like conditions and below-median rainfall for much of Australia (Stone *et al.*, 1996a), the probability distribution function shifts towards lower yields. Under CP and RR SOI phase, higher chickpea yields are associated with higher in-crop rainfall (Fig. 28.9). The KW test showed significant (p = <0.001) shifts in the median values of simulated yields among all SOI phases (Fig. 28.9). Whilst the chickpea model does not explicitly simulate crop disease, the analysis of the effects of disease is possible by considering the likelihood of favourable weather conditions for such disease outbreaks. For example, if a wet spring is forecast, farmers need to be alert for the risk of botrytis disease.

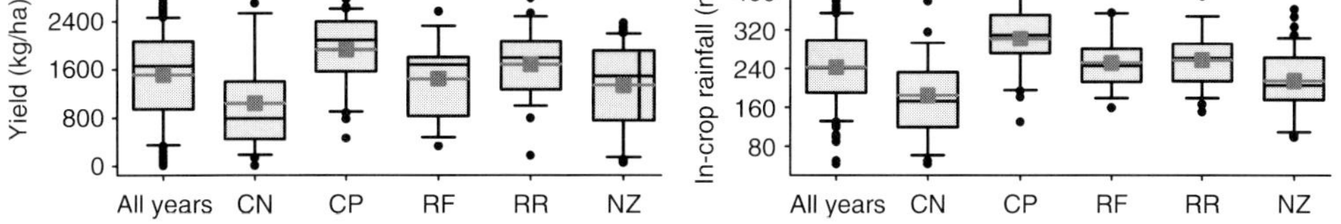

Fig. 28.9. Distribution of (a) simulated chickpea yields and (b) in-crop rainfall at Birchip, south-eastern Australia, for each Southern Oscillation Index (SOI) phase in June/July. The significance of the shift in the medians in each phase away from the all-years value was measured using the Kruskal–Wallis (KW) test and the similarity of distributions via the Kolmogorov–Smirnov (KS) p value test. The shift was measured using the absolute median difference (AMD), while changes in the distribution dispersion were measured using the inter-quartile ratio (IQR). All years: actual years used – 1889–2005; CN: consistently negative; CP: consistently positive; RF: rapidly falling; RR: rapidly rising; and NZ: near zero.

The changes in the median and in the dispersion of the probability distribution functions of simulated chickpea yield were derived from the analogue years for each of the five SOI phases in June/July. Indices quantifying the shift were computed as the AMD, and the dispersion was evaluated by calculating IQRs (Fig. 28.9). For instance, at Birchip (Fig. 28.9a), the long-term simulated median chickpea yield was 1657 kg/ha, and under CN SOI phase, the median of the simulated chickpea yields has shifted downward compared with the long-term median by 989 kg/ha with highly variable cropping conditions (i.e. higher dispersion of IQR = 1.10). Under CP SOI conditions higher yield (AMD = 584 kg/ha) can be expected with smaller IQR (0.82), implying lower risk levels. This information on the shift of the median and the dispersion is available at the end of July before crop protection on new disease-resistant varieties and gives farmers a reasonable lead time to prepare management options depending on SOI phases and associated long-term yield trends as illustrated for Birchip.

Capacity of the June/July SOI phase system to discriminate spring rainfall at Birchip

Differences were quantified in the median values and dispersion among phases for spring rainfall at Birchip (Fig. 28.10). There was a clear signal of the impact of the June/July SOI phases over spring rainfall. Significant (p = <0.001) differences in the median values of spring rainfall for the different five SOI phases compared to the rest of the years were detected (Fig. 28.10). The CN and RF SOI phases in June/July indicated a greater probability of lower amounts of spring rainfall. CP and RR SOI phases in June/July indicated greater probabilities for higher spring

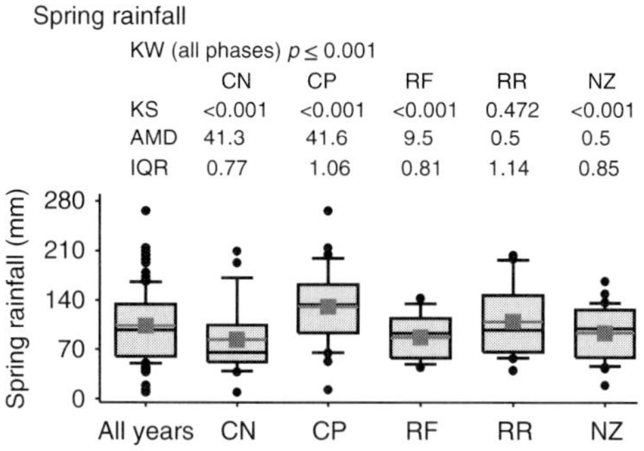

Fig. 28.10. Distribution of spring rainfall at Birchip, south-eastern Australia, for each Southern Oscillation Index (SOI) phase in June/July. The significance of the shift in the medians in each phase away from the all-years value was measured using the Kruskal–Wallis (KW) test and the similarity of distributions via the Kolmogorov–Smirnov (KS) p value test. The shift was measured using the absolute median difference (AMD), while changes in the distribution dispersion are measured using the inter-quartile ratio (IQR). All years: actual years used – 1889–2005; CN: consistently negative; CP: consistently positive; RF: rapidly falling; RR: rapidly rising; and NZ: near zero.

rainfall. Additionally, different SOI phases in June/July had contrasting discrimination capacities (KS tests) (Fig. 28.10) as all five SOI phases, except RR were able to significantly discriminate spring rainfall.

We further quantified the changes in the shift (AMD) and dispersion (IQR) among the probability distribution functions derived from analogue years for each of the five SOI phases in June/July. These AMD and IQR values are listed in Figure 28.10. In general, spring rainfall was higher for CP and RR phases than for CN and RF phases, with NZ phases falling in between. From these results, it can be summarized that different SOI phases at the end of July have different capacities to forecast spring rainfall. These results confirm that the ENSO is an important component of the variability (Figs 28.9 and 28.10) in spring rainfall and simulated chickpea yields in the Birchip region of south-eastern Australia. The use of the five phase SOI forecasting system (Stone *et al.*, 1996a) allows historical records of rainfall to be grouped into analogue years and crop models to be applied to produce crop yield outlooks/forecasts similar to those for wheat crops (Hammer *et al.*, 1996). The additional value of forecasting simulated yields instead of climatic variables probably rests in the fact that crop yield results from the integral of many interacting processes involving not only the amount of rainfall but also the impact of ENSO on rainfall distribution during the cropping season, temperatures and other important climatological variable such as solar radiation and vapour pressure deficits. Simulated chickpea yields agreed fairly well with actual farm yields across south-eastern Australia.

Seasonal climate outlook assessment

The skill of the five SOI phase forecasting system in south-eastern Australia (Birchip) was tested by cross-validation techniques, and some skill of predictive accuracy was evident (Fig. 28.11a–f). The overall skill of the five SOI phase system for simulated chickpea yields and spring rainfall at the end of July varied from 60% to 67% consistent (Fig. 28.11a and b). Cross-validation techniques were also applied to evaluate the early season predictive skill for seasonal rainfall (Fig. 28.11c), date of last frost (Fig. 28.11d), break of the season (Fig. 28.11e) and annual rainfall (Fig. 28.11f). These estimates are made at the end of April for seasonal rainfall, date of last frost, break of the season and at the end of May for annual rainfall. As an example, for CN years at the end of July, the predictive accuracy for simulated chickpea yield was 86% consistent, 74% consistent for the CP years and 65% consistent for the RF years (Fig. 28.11a). The level of skill for predicting spring rainfall at the end of July in the years of CN showed 73% consistency and 76% consistency in the years of CP SOI phase (Fig. 28.11b). The predictive accuracy of spring rainfall was low for the RR years (58% consistent), and NZ years (51% consistent). Values lower than 50% consistent indicate the absence of skill.

The skill for prediction of seasonal rainfall (Fig. 28.11c) at the end of April ranged between 60% and 76% consistent among SOI phases and the CN phase showed no skill in predicting seasonal rainfall at the end of April. Similarly, the predictive accuracy of SOI phases as RF and RR was higher than 90% consistent

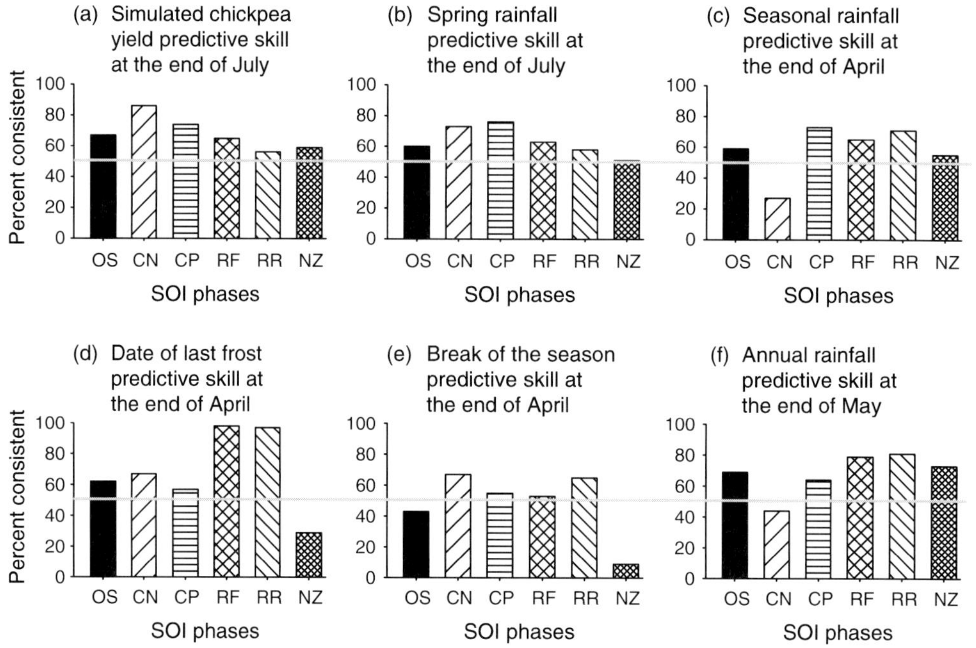

Fig. 28.11. Accuracy of the Southern Oscillation Index (SOI) phase forecast systems (via cross-validation of percent consistent) at Birchip, south-eastern Australia: (a) simulated chickpea yields; (b) spring rainfall; (c) seasonal rainfall; (d) date of last frost; (e) break of the season; and (f) annual rainfall. OS: overall skill; CN: consistently negative; CP: consistently positive; RF: rapidly falling; RR: rapidly rising; NZ: near zero; and horizontal line at 50% consistent.

for date of last frost at the end of April (Fig. 28.11d) and no skill was evident for years under the NZ phase. There was no skill to forecast the break of the season at the end of April (Fig. 28.11e) using the five SOI phase system and for annual rainfall at the end of May (Fig. 28.11f), the skill of the five SOI phase system, with the exception of the CN phase, ranged between 64% and 81% consistent. We observed important variations in the predictive skill among the five SOI phases at different months of the year. Nevertheless, some skill of predictive accuracy indicates that there is some potential for farmers to benefit from the adoption of more flexible management approaches to take full advantage of the climate information. Different phases of the SOI showed different contributions to the overall predictive accuracy of the forecasts. Consequently, some phases might be more useful than others although this warrants further investigation.

Yield fluctuations caused by a highly variable rainfall environment are of concern to chickpea producers in Birchip (south-eastern Australia) (Hobson *et al.*, 2004). The approach using a climate forecasting system (here we have used SOI phases) and a dynamic simulation crop model in conjunction with long-term historical data is shown to be useful and efficient for seasonal forecast of chickpea yields, seasonal conditions and associated climatic risks. However, other forecasting systems like the Pacific Ocean sea-surface temperature phase system (Drosdowsky, 2002) and very sophisticated climate-forecasting models called General Circulation Models (GCMs) are also being used to provide climatic and forward-estimation of crop production (Hansen and Indeje, 2004). Stone *et al*. (1996a) showed how median rainfall values following a certain SOI phase can differ strongly from the climatological median in many parts of the world. Our finding of receiving higher probability of in-crop rainfall and spring rainfall at Birchip (south-eastern Australia) following a CP and RR SOI phase in June/July is consistent with findings by Stone *et al*. (1996a). Higher probability of rainfall during CP and RR SOI phases in June/July, in turn, influences crop growth and realization of a higher chickpea grain yield through appropriate crop protection strategies. Additionally, this information is available 4–5 months before harvest and thus gives the chickpea growers a reasonable lead-time to adjust their crop management operations.

This methodology can be extended to include other chickpea growing regions and crops. Stone *et al*. (1996b) have shown that a similar system can also be applied to estimate frequency, timing and severity of damaging frost around anthesis in northern Australia. Similar types of analyses need to be conducted for other regions/parts of the world. While these findings are scientifically interesting, the question that needs to be addressed is: How can this knowledge be applied to increase yield stability and ultimately farm profits? We applied various methods to evaluate the forecasting quality of the five SOI phase system (operational forecasting system from http://www.bom.gov.au; and http://www.longpaddock.qld.gov.au). Their forecasting quality was examined for predictions of seasonal rainfall, spring rainfall, date of last frost, break of the season and simulated chickpea yields. Spring rainfall could be forecasted by the end of July based on SOI phases which would allow farmers to make tactical crop management plans and forecasts to make potentially useful seasonal outlooks for crop production. We observed significant changes in the probability distribution functions of grain yield and spring rainfall, when using the SOI phase systems in June/July. These changes (in shift and dispersion) can be useful to select appropriate farming management options.

The 'forecasting' method applied in this study is based on using historic climate patterns as analogues of future climate. It assumes that the variability of future climate will be similar to that of the historic record and that the existing historic climate records are an adequate representation of the true climatic variability. Although both assumptions are frequently challenged by climatologists (e.g. Meehl and Washington, 1996; Nicholls, 1996), potential errors are unlikely to be substantial considering the time frame of the forecast (up to 5 months) and that of the historic record (starting in 1889). Excellent correspondence of predictive skills with realized rainfall and chickpea yields, e.g. the June/July NZ phase in Figs 28.9 and 28.10, provides confidence in the use of climate prediction tools.

Conclusions

- Integrating a climate forecasting system (five SOI phase system) of Stone and Auliciems (1992) and a dynamic simulation crop model in conjunction with long-term historical data is shown to be useful for seasonal forecasts of chickpea yield and seasonal conditions. This methodology can be applied to other chickpea areas and other crops.

- From this work we concluded that there are significant ENSO signals on the climate and simulated chickpea yield in south-eastern Australia that can be used to produce forecasts of seasonal conditions and yield outlooks as early as the end of July.
- Spring rainfall condition can be forecasted in July based on the SOI phase system and this allows farmers to make tactical crop management decisions.
- The overall predictive skill (cross-validation analysis) for simulated chickpea yield and spring rainfall ranged from 60% to 67% consistent. Different phases of SOI had different contributions to the overall predictive skills; consequently, some phases might be more useful than others.

Acknowledgements

The authors thank Dr Kim Lowell of Primary Industries Research – Victoria (Knoxfield Centre) for helpful comments on the manuscript. This study was supported by Primary Industries Research – Victoria and the Grains Research and Development Corporation, Australia, through the grant DAV00006.

References

Anwar, M.R., McKenzie, B.A. and Hill, G.D. (2003) Phenology and growth response to irrigation and sowing date of Kabuli chickpea (*Cicer arietinum* L.) in a cool-temperate sub-humid climate. *Journal of Agricultural Science, Cambridge* 141, 273–284.

Anwar, M.R., Rodriguez, D., De Li, Liu and Power, S. (2004) Signals from ENSO on climate and grain yield in South-eastern Australia. *16th Australia–New Zealand Climate Forum*. 8–10 November 2004, Lorne, Victoria, Australia. Available at: http://www.bom.gov.au/events/anzcf2004/

Anwar, M., O'Leary, G., McNeil, D., Hossain, H. and Nelson, R. (2006a) Climate change impact on field crops and daptation options in south eastern Australia. In: Turner, N.C., Acuna, T. and Johnson, R.C. (2006). Ground-breaking stuff. *Proceedings of the 13th Australian Agronomy Conference*, 10–14 September 2006, Perth, Western Australia. Australian Society of Agronomy. Available at: http://www.regional.org.au/au/asa/2006/plenary/water/4515_anwar.htm.

Anwar, M.R., Rodriguez, D., De Li, Liu, O'Leary, G. and Power, S. (2006b) Accuracy of probabilistic spring rainfall and wheat yield predictions for South-eastern Australia. *The 13th National Conference on Climate, Water and Sustainability*. The Australian Meteorological and Oceanographic Society (AMOS). 6–8 February 2006, Newcastle, NSW, Australia. Available at: http://www.amos.org.au/AMOS06_progr_ABSTRACTS.pdf

Conover, W.J. (1971) *Practical Nonparametric Statistics*. Wiley, New York.

Drosdowsky, W. (2002) SST phases and Australian rainfall. *Australian Meteorological Magazine* 51, 1–12.

Ferreyra, R.A., Podesta, G.P., Messina, C.D., Letson, D., Dardanelli, J., Guevara, E. and Meira, S. (2001) A linked-modeling framework to estimate maize production risk associated with ENSO-related climate variability in Argentina. *Agricultural and Forest Meteorology* 107, 177–192.

Ferris, J. (1999) An analysis of the impact of ENSO (El Nino/Southern Oscillation) on global crop yields. *American Journal of Agricultural Economics* 81, 1309–1309.

Glantz, M.H. (1996) The ENSO Signal. *Science in the Service of Society* 5, 3–5.

Hammer, G.L., Holzworth, D.P. and Stone, R. (1996) The value of skill in seasonal climate forecasting to wheat crop management in a region with climatic variability. *Australian Journal of Agricultural Research* 47, 717–37.

Hammer, G.L., Keating, B.A., Meinke, H., Carberry, P.S., Freebairn, D.M., Probert, M. and Holzworth, D.P. (1999) An integrated systems approach to improving agricultural systems using the agricultural production systems simulator, APSIM. In: *International Symposium, Modelling Cropping Systems: Proceedings of the European Society for Agronomy, Division of Agroclimatology and Agronomic Modelling*. 21–23 June 1999, Lleida, Spain, pp. 31–37.

Hansen, J.W. and Indeje, M. (2004) Linking dynamic seasonal climate forecasts with crop simulation for maize yield prediction in semi-arid Kenya. *Agricultural and Forest Meteorology* 125, 143–157.

Hansen, J.W., Jones, J.W., Irmak, A. and Royce, F.S. (2001) ENSO impacts on crop production in the Southeast US. In: Rosenzweig, C., Boote, K.J., Hollinger, S., Iglesias, A. and Phillips, J. (eds) *Impacts of El Niño and Climate Variability on Agriculture*. ASA Special Publication No. 63, American Society of Agronomy, Madison, Wisconsin, pp. 55–76.

Hobson, K., Materne, M., Bretag, T., Ivess, B. and Holding, B. (2004) Chickpea variety development. *Birchip Cropping Group – Crop and Pasture Production Manual 2004–2005*. The Southern Mallee and Northern Wimmera, pp. 119–123.

IPCC (2001) Summary for policymakers. In: Houghton, J.T., Ding, Y., Griggs, D.J., Noguer, M., van Der Linden, P.J. and Xioaosu, D (eds) *Climate Change 2001: The Scientific Basis*. Contribution of Working Group I to the Third Assessment Report of the Intergovernmental Panel on Climate Change, Cambridge University Press, Cambridge.

Jones, J.W., Hoogenboom, G., Porter, C.H., Boote, K.J., Batchelor, W.D., Hunt, L.A., Wilkens, P.W., Singh, U., Gijsman, A.J. and Ritchie, J.T. (2003) The DSSAT cropping system model. *European Journal of Agronomy* 18, 235–265.

Keating, B.A., McCown, R.L. and Wafula, B.M. (1993) Adjustment of nitrogen inputs in response to a seasonal forecast in a region of high climatic risk. In: Penning de Vries, F.W.T., Peng, P.S. and Metselaar, K. (eds) *Systems Approaches for Agricultural Development*, Kluwer Academic, Dordrecht, The Netherlands, pp. 233–252.

Keating, B.A., Carberry, P.S., Hammer, G.L., Probert, M.E., Robertson, M.J., Holzworth, D., Huth, N.I., Hargreaves, J.N.G., Meinke, H., Hochman, Z., McLean, G., Verburg, K., Snow, V., Dimes, J.P., Silburn, M., Wang, E., Brown, S., Bristow, K.L., Asseng, S., Chapman, S., McCown, R.L., Freebairn, D.M. and Smith, C.J. (2003) An overview of APSIM, a model designed for farming systems simulation. *European Journal of Agronomy* 18, 267–288.

Khandekar, M.L. (1996) El Niño/Southern oscillation, Indian monsoon and world grain yields – a synthesis. *Advances in Natural and Technological Hazards Research* 7, 79–95.

Kirono, D.G.C. and Khakhim, N. (1998) Rainfall and El Nino southern oscillation: Links and its impact on crop production (A case study of Yogyakarta Special Province of Indonesia). *The Indonesian Journal of Geography* 30, 21.

Kruskall, W.H. and Wallis, W.A. (1952) Use of ranks on one-criterion variance analysis. *Journal of the American Statistical Association* 48, 907–911.

Loss, S., Brandon, N. and Siddique, K.H.M. (1998) The Chickpea Book: A technical guide to chickpea production. Bulletin 1326, ISSN 1326 – 415X, Chief Executive Officer, Agriculture, Western Australia.

Lythgoe, B., Rodriguez, D., De Li, Liu, Brennan, J., Scott, B., Murry, G. and Hayman, P. (2004) Seasonal climate forecasting has economic value for farmers in south eastern Australia. In: *Proceedings of the 4th International Crop Science Congress, Brisbane, Australia*. 26 September–1 October 2004. Available at: http://www.cropscience.org.au/icsc2004/poster/2/6/524_rodriguezd.htm

Marshall, G.R., Parton, K.A. and Hammer, G.L. (1996) Risk attitude, planting conditions and the value of seasonal forecasts to a dryland wheat grower. *Australian Journal of Agricultural Economics* 40, 211–233.

McBride, J.L. and Nicholls, N. (1983) Seasonal relationships between Australian rainfall and the Southern Oscillation. *Monthly Weather Review* 111, 1998–2004.

McCown, R.L., Hammer, G.L., Hargreaves, J.N.G., Holzworth, D.P. and Freebairn, D.M. (1996) APSIM: a novel software system for model development, model testing, and simulation in agricultural systems research. *Agricultural Systems* 50, 255–271.

Meehl, G.A. and Washington, W.M. (1996) El-Niño-like climate change in a model with increased atmospheric CO_2 concentrations. *Nature* 382, 56–60.

Meinke, H. and Hochman, Z. (2000) Using seasonal climate forecasts to manage dryland crops in northern Australia. In: Hammer, G.L., Nicholls, N. and Mitchell, C. (eds) *Applications of Seasonal Climate Forecasting in Agricultural and Natural Ecosystems – The Australian Experience*, Kluwer Academic, Dordrecht, The Netherlands, pp. 149–165.

Meinke, H. and Stone, R.C. (1997) On tactical crop management using seasonal climate forecasts and simulation modelling – a case study for wheat. *Scientia Agricola* 54, 121–129.

Meinke, H., Stone, R.C. and Hammer, G.L. (1996) SOI phases and climatic risk to peanut production: a case study for northern Australia. *International Journal of Climatology* 16, 783–789.

Mjelde, J.W., Thompson, T.N., Hons, F.M., Cothren, J.T. and Coffman, C.G. (1997) Using southern Oscillation information for determining corn and sorghum profit-maximizing input levels in east-central Texas. *Journal of Production Agriculture* 10, 168–175.

Mjelde, J.W., Penson, J.B. and Nixon, C.J. Jr. (2000) Dynamic aspects of the impact of the use of perfect climate forecasts in the Corn Belt region. *Journal of Applied Meteorology* 39, 67–79.

Nicholls, N. (1985) Impact of the Southern Oscillation on Australian crops. *Journal of Climatology* 5, 553–560.

Nicholls, N. (1996) Climate variability|climate change. In: *Proceedings of the Second Australian Conference on Agricultural Meteorology*. 1–4 October 1996. University of Queensland, pp. 49–53.

Park, J., Mjelde, J.W., Fuller, S.W., Malaga, J.E., Rosson, C.P. and Dainello, F.J. (2002) An assessment of the effects of ENSO events on fresh vegetable and melon supplies. *Hortscience* 37, 287–291.

Parry, M.L., Carter, T.R. and Konijn, N.T. (1988) *The Impact of Climatic Variations on Agriculture*, Vol. 2: *Assessments in Semi-Arid Regions*. Kluwer Academic, Dordrecht, The Netherlands.

Potgieter, A.B., Hammer, G.L. and Butler, D. (2002) Spatial and temporal patterns in Australian wheat yield and their relationship with ENSO. *Australian Journal of Agricultural Research* 53, 77–89.

Potgieter, A.B., Everingham, Y.L. and Hammer, G.L. (2003) On measuring quality of a probabilistic commodity forecast for a system that incorporates seasonal climate forecasts. *International Journal of Climatology* 23, 1195–1210.

Power, S., Tseitkin, F., Torok, S., Lavery, B., Dahni, R. and McAvaney, B. (1998) Australian temperature, Australian rainfall and the Southern Oscillation, 1910–1992: coherent variability and recent changes. *Australian Meteorological Magazine* 47, 85–101.

Power, S., Casey, C., Folland, C. and Mehta, V. (1999) Inter-decadel modulation of the impact of ENSO on Australia. *Climate Dynamics* 15, 319–324.

Rimmington, G.M. and Nicholls, N. (1993) Forecasting wheat yields in Australia with the Southern oscillation Index. *Australian Journal of Agricultural Research* 44, 625–632.

Robertson, M.J., Carberry, P.S., Huth, N.I., Turpin, J.E., Probert, M.E., Poulton, P.L., Bell, M., Wright, G.C., Yeates, S.J. and Brinsmead, R.B. (2002) Simulation of growth and development of diverse legume species in APSIM. *Australian Journal of Agricultural Research* 53, 429–446.

Robinson, J.B. and Butler, D.G. (2002) An alternative method for assessing the value of the Southern Oscillation Index (SOI), including case studies of its value for crop management in the northern grainbelt of Australia. *Australian Journal of Agricultural Research* 53, 423–428.

Stephens, D.J., Butler, D.G. and Hammer, G.L. (2000) Using seasonal climate forecasts in forecasting the Australian wheat crop. In: Hammer, G.L., Nicholls, N. and Mitchell, C. (eds) *Applications of Seasonal Climate Forecasting in Agricultural and Natural Ecosystems – The Australian Experience*, Kluwer Academic, Dordrecht, The Netherlands, pp. 351–366.

Stockle, C.O., Martin, S.A. and Campbell, G.S. (1994) CropSyst, a cropping systems simulation model: Water/nitrogen budgets and crop yield. *Agricultural Systems* 46, 335–359.

Stone, R.C. and Auliciems, A. (1992) SOI phase relationships with rainfall in eastern Australia. *International Journal of Climatology* 12, 625–636.

Stone, R.C., Hammer, G.L. and Marcussen, T. (1996a) Prediction of global rainfall probabilities using phases of the Southern Oscillation Index. *Nature* 384, 252–255.

Stone, R.C., Nicholls, N. and Hammer, G.L. (1996b) Frost in NE Australia: trends and influences of the Southern Oscillation Index. *Journal of Climate* 9, 1896–1909.

Whish, J. and Castor, P. (2005) How much water do chickpeas require? In: Focus 2005: Chickpeas in Farming Systems, *Proceedings National Conference on Chickpeas in Farming Systems*. Ed Pulse Australia, pp. 56–60.

White, B. (2000) The importance of climate variability and seasonal forecasting to the Australian economy. In: Hammer, G.L., Nicholls, N. and Mitchell, C. (eds) *Applications of Seasonal Climate Forecasting in Agricultural and Natural Ecosystems – The Australian Experience*. Kluwer Academic, Dordrecht, The Netherlands, pp. 1–22.

Wilks, D.S. (1995) *Statistical Methods in the Atmospheric Sciences – An Introduction*. Academic Press, London.

Appendix

Glossary and Definitions

Analogue years: A past situation, which resembles the current situation, is called an analogue. Analogue can be used as a guide for what might come. For example, by finding similar SOI patterns to the current year, we can then examine rainfall patterns from that time to use as a guide as to what will happen this year. The Bureau of Meteorology's Seasonal Climate Outlook includes a listing of several analogue years in which the SOI behaved similarly to the current year.

Box plot: Box plots show the range of data falling between the 25th and 75th percentiles, the horizontal line inside the box shows the median value, and the whiskers show the complete range of the data.

Break of season: The first time sufficient rain (i.e. when rain exceeds evaporation) has fallen within a 'window' of sowing opportunity, e.g. between 1 April and 31 July for southern Mallee in south-eastern Australia.

Climatic time series: A historical record of atmospheric data is termed a climatic time series.

Cross-validation: Another method of estimating the predictive skill of a forecasting system compares outlooks with observations from historical data.

Date of last frost: The date of last frost is defined as the last day of the year when the minimum temperature is lower or equal to a particular threshold, usually 2.2°C at screen height.

Decile: Deciles are used to rank data. The terms tercile, quartile, quintile and decile should refer to the percentiles that divide the distribution into 3, 4, 5 or 10 equal parts, respectively. These terms are, however, often used to signify a specified fraction of the total number of ranked values. Thus the value of a particular element (e.g. a single rainfall observation, or a student's height in a classroom of students) may be predicted to be in a specified tercile or quintile of the appropriate distribution.

El Niño: The term 'El Niño' refers to the extensive warming of the central and eastern Pacific that leads to a major shift in weather patterns across the Pacific. In Australia (particularly eastern Australia), El Niño events are associated with an increased probability of drier conditions.

ENSO: Stands for El Niño Southern Oscillation. 'El Niño' used here refers to the warming of the oceans in the equatorial eastern and central Pacific; Southern Oscillation is the changes in atmospheric pressure (and climate systems) associated with this warming (hence 'Southern Oscillation Index' to measure these changes). 'ENSO' is used colloquially to describe the whole suite of changes associated with an 'El Niño' and 'La Niña' events – to rainfall, oceans, atmospheric pressure, etc. More information available at: http://www.bom.gov.au; and http://www.longpaddock.qld.gov.au.

Forecast: A prediction of the future state of the climate, production or any other variable of interest. Real-time forecasts are made for future times, for which the verifying observations are not yet available.

Forecast verification: Forecast verification is the process of assessing the quality of a forecast, and it can be compared, or verified, against a corresponding observation of what actually occurred, or some good estimate of the true outcome. The verification can be qualitative ('does it look right?') or quantitative ('how accurate was it?'). In either case it should give information about the nature of the forecast errors.

Hindcasts: Hindcasts are forecasts made for past data using even earlier data, for which the verifying observations are already available.

La Niña: The extensive cooling of the central and eastern Pacific Ocean. In Australia (particularly eastern Australia), La Niña events are associated with increased probability of wetter conditions.

Median: The median for the month is the value of an element that exceeds half the occurrences for that element for that month over the period on record. However, there is a 50% probability of the element being below the median value. With many meteorological quantities the mean and median values are close and the use of either value is acceptable. Although this may often be the case with annual rainfall, for shorter periods the mean can differ significantly from the median, as the mean can be influenced by an extremely heavy or light value, while the median is not. Hence the median is usually taken as giving a better description of the characteristics of rainfall.

Neutral conditions: The middle situation, when there is neither an El Niño event nor a La Niña. The SOI usually has small values around zero.

Percent consistent: Percent consistent refers to the percentage of forecasts that were consistent with the category later observed. In other words, it is the number of consistent forecasts for a category (e.g. above median), divided by the total number of forecasts made for that category. More information available at: http://www.bom.gov.au.

Percentile: The term for denoting thresholds or boundary values in frequency distributions. Thus the 5th percentile is that value that marks the lowest 5% of the observations from the rest, the 50th percentile is the same as the median, and the 95th percentile exceeds all but 5% of the values. When percentiles are estimated by ranking the items of a finite sample, the percentile generally falls between two of the observed values, and the midway value is often taken.

Persistence forecast: A forecast that the future seasonal mean will be the same as the most recently observed seasonal mean; often used as a standard of comparison in measuring the degree of skill of forecasts prepared by other methods.

Pie chart: Pie charts are used to represent the chances of three possible outcomes: below average (lowest 33% of the years in red), average (middle 33% of the years in yellow) and above average (highest 33% of the years in green).

Probabilistic forecasts: An attempt to convey the uncertainty in a forecast by expressing its likelihood of occurrence as a probability, or percentage. High probabilities do not guarantee an outcome – they merely indicate that that outcome is highly likely. Probabilities are usually based on the frequency of occurrence in the historical record. For instance, if the chance of receiving above-median rainfall in a particular climate scenario is 60%, then 60% of past years when that scenario occurred had above median rainfall, and 40% had below-median rainfall.

Probability: The chance of an event happening expressed as a percentage. A probability of 70% means the event can be expected to occur in 7 out of 10 years.

Seasonal climate forecasting: Seasonal climate forecasting (SCF) is the use of meteorological or climate information to produce estimates of the future (3–12 months) condition of the climate, e.g. seasonal rainfall. Information from SCF can be used to make informed changes in a decision that might generate improved outcomes from the farming system. The SCF deals with climatic mechanisms that actually behave predictability over season-to-season timescales. The SCF can increase preparedness for dealing with climate variability and extreme climate events. In an SCF, the lead-time is the length of time between issuing of the forecast and the commencement of the period of time that is the subject of the forecast (usually expressed as a number of months).

Seasonal rainfall: Defined as rainfall during the season, e.g. between 1 April and 31 October for southern Mallee, Victoria.

Seasons: In Australia, the seasons are defined by grouping the calendar months in the following way:

Spring – 3 transition months: September, October and November.
Summer – 3 hottest months: December, January and February.
Autumn – 3 transition months: March, April and May.
Winter – 3 coldest months: June, July and August.

In south-eastern Australia, winter crops in most of the years are sown in May and June, but due to variable rainfall, crops can be sown at an early seasonal-break year as early as April and as late as late July.

Skill: The relative accuracy of the forecast over some reference forecast. The reference forecast is generally an unskilled forecast such as random chance, persistence (defined, as the most recent set of observations, 'persistence' implies no change in condition) or climatology. Skill refers to the increase in accuracy due purely to the 'smarts' of the forecast system. Weather forecasts may be more accurate simply because the weather is easier to forecast – skill takes this into account.

SOI: The Southern Oscillation Index (SOI) is calculated from the monthly or seasonal fluctuations in the air pressure difference between Tahiti and Darwin. Sustained negative values of the SOI often indicate El Niño episodes. These negative values are usually accompanied by sustained warming of the central and eastern tropical Pacific Ocean, a decrease in the strength of the Pacific Trade Winds, and a reduction in rainfall over eastern and northern Australia. The most recent strong El Niño was in 1997/98. Positive values of the SOI are associated with stronger Pacific Trade Winds and warmer sea temperatures to the north of Australia, popularly known as a La Niña episode. Waters in the central and eastern tropical Pacific Ocean become cooler during this time. Together these give an increased probability that eastern and northern Australia will be wetter than normal. The most recent strong La Niña was in 1988/89; a moderate La Niña event occurred in 1998/99, which weakened back to neutral conditions before reforming for a shorter period in 1999/2000.

SOI phases (Stone and Auliciems, 1992): In this scheme the future state of the climate system is defined in terms of the past 2 months' SOI values. The phases are defined through a principal component analysis of these two values and their difference. The first two components, which represent the mean value and the month to month trend, are used to place each month into one of five possible clusters or phases describing the state of the SOI over the current and previous months. These phases are:

Phase 1: consistently negative SOI (CN)
Phase 2: consistently positive SOI (CP)
Phase 3: rapidly falling SOI (RF)
Phase 4: rapidly rising SOI (RR)
Phase 5: near zero SOI (NZ)

Spring rainfall: Rainfall between 1 September and 30 November.

Weather forecast: Short-term weather forecasts cover immediate weather conditions out to a few days. Medium-term weather forecasts focus on the period from a few days ahead to a few weeks ahead. Current forecasts/outlooks extend up to 7–10 days ahead. More information available at: http://www.bom.gov.au; http://www.longpaddock.qld.gov.au.

Weather or climate: Climate is the meteorological conditions that characteristically prevail in a particular region. It is important to understand the difference between 'weather' and 'climate'. Climate is what you expect; weather, what you get!

29 Chickpea Farmers

J. Kumar,[1] F. Dusunceli,[2] E.J. Knights,[3] M. Materne,[4] T. Warkentin,[5] W. Chen,[6] P.M. Gaur,[7] G. Bejiga,[8] S.S. Yadav,[9] A. Satyanarayana,[10] M.M. Rahman[11] and M. Yadav[12]

[1]*Hendrick Beans-for-Health Research Foundation, 11791 Sandy Row Inkerman, Ontario, K0E 1J0, Canada;* [2]*The Central Research Institute for Field Crops, PO Box 226, Ulus, Ankara, Turkey;* [3]*New South Wales Department of Primary Industries, Tamworth Agricultural Institute, Calala, NSW 2340, Australia;* [4]*Victoria Institute of Dryland Agriculture, Horsham, VIC 3401, Australia;* [5]*Crop Development Centre, University of Saskatchewan, Saskatoon, SK S7N 5A8, Canada;* [6]*USDA-ARS, Grain Legume Genetics and Physiology Research Unit, Washington State University, Pullman, WA 99164-6402, USA;* [7]*International Crops Research Institute for the Semi-Arid Tropics (ICRISAT), Hyderabad 502324 India;* [8]*Green Focus Ethiopia, PO Box 802, Code No.1110, Addis Ababa, Ethiopia;* [9]*Pulse Research Laboratory, Division of Genetics, Indian Agricultural Research Institute, New Delhi 110012, India;* [10]*Nuziveedu Seeds Ltd, Hyderabad, Andhra Pradesh, India;* [11]*Bangladesh Agricultural Research Institute, Joydebpur, Gazipur, Bangladesh;* [12]*College of Business Administration, Tarleton State University (Texas A&M University System) Stephenville, TX 76402, USA*

Introduction

Legumes are the most important source of food for humans after cereals. These enhance soil health and thus play a significant role in sustainability of agriculture. Recent nutrition research has shown that legumes are important not only as a source of protein but also for their complex starch, fibre and phytoestrogens that benefit human health. Therefore, these are being increasingly associated with prevention of diseases linked with lifestyles in developed countries, e.g. some forms of cancer, cardiovascular diseases, obesity, diabetes, menopausal symptoms and osteoporosis (Scarafoni *et al.*, 2005). Therefore, promotion of legumes as health foods will add value and make these more competitive with cereals.

Chickpea (*Cicer arietinum* L.) is the fourth largest grain-legume crop in the world following soybean, dry pea and dry bean (FAO, 2005). In 2005 their productivity stood at 2234, 1067, 698 and 781 kg/ha compared with 1965 figures of 1129, 1043, 493 and 649 kg/ha, respectively. Thus over the last four decades, soybean productivity doubled, with no change for dry pea, 42% increase for dry bean and 20% increase for chickpea. There are several reasons for the improvement in soybean productivity (Orf *et al.*, 2004). These include the cultivation environment, research inputs and the farmers' market choices that define local, national and international trade. Soybean has been developed as a commercial crop, and more recently as a nutraceutical crop. It received strong research support especially in the USA. The scientific advances in soybean have led to the development of varieties that are responsive to particular photoperiods and high input cultivation. These varieties do not lodge and lend themselves to large-scale mechanical crop management operations. Soybean is grown in environments that are generally free of extreme moisture stresses, faced so often by chickpea. However, when cultivated under high input conditions chickpea tends to put up excessive vegetative growth and lodges. Dense canopy leads to high humidity and increased risk of foliar diseases. Chickpea lacks scientific breakthroughs like determinate growth habit and resistances to major foliar diseases like ascochyta blight and to the pest helicoverpa pod borer. Thus, adaptation of the present chickpea cultivars to favourable conditions is poor (Berger *et al.*, 2004, 2006).

During the last 40 years the global chickpea area decreased by 10%. However, the productivity increased by 35% and production by 23%. Potential experimental chickpea seed yields of up to 5 t/ha have been reported. Why does such a large gap exist between the potential and realized yields? This paper will deal with major developments in chickpea production. We shall analyse the recent success stories, and the reasons why such breakthroughs have not occurred in other regions, and suggest future research needs.

Chickpea is grown in nearly 50 countries around the world. However, the bulk of the crop is produced in some 15 countries (Table 29.1; FAO, 2005). In 2004 Asia contributed 89.5% (7.67 t), Africa 4.4% (0.37 t), North and Central America 3.4% (0.29 t), Australia 1.6% (0.14 t) and Europe 1.0% (0.09 t). Although North and Central America and Australia together contributed about 5% of the world chickpea production, these regions have the highest chickpea productivity (1.32–1.47 t/ha). In contrast, Asia and Africa show the lowest productivity (0.80 t/ha). The countries with largest chickpea production include India, Turkey, Pakistan and Iran in Asia; Ethiopia in Africa; and Mexico in North America. Large variation in chickpea yield from 0.35 t/ha in Iran to 1.6 t/ha in Mexico is reported. The crop in Iran is grown under extremely dry conditions and in Mexico it is generally irrigated (Tay *et al.*, 2000). Drought is thus the common major constraint to higher productivity.

Cropping Seasons

Agro-climatic conditions and cropping systems in which chickpea is cultivated vary across growing regions. In the subtropics like the Indian subcontinent the crop is planted at the end of the rainy season under relatively high temperatures. The vegetative growth is on the residual soil moisture in the winter season. Flowering occurs in the cooler temperatures. There is little rainfall during

Table 29.1. Area (3-year mean), production and productivity for major chickpea producing countries during 1963–1965, 1983–1985 and 2003–2005.

	1963–1965			1983–1985			2003–2005		
	Area			Area			Area		
Region/ country	Million hectares	Production tonnes	Yield kg/ha	Million hectares	Production tonnes	Yield kg/ha	Million hectares	Production tonnes	Yield kg/ha
World	11.75	6.77	577	9.83	6.7	682	10.61	8.3	781
Indian subcontinent									
India	9.14	5.21	572	7.15	4.87	680	6.72	5.30	784
Pakistan	1.183	0.65	553	0.94	0.51	542	1.01	0.72	706
Myanmar	0.123	0.680	558	0.172	0.145	842	0.208	0.229	1104
Bangladesh	0.053	0.035	666	0.113	0.850	750	0.140	0.010	749
West Asia, North Africa and southern Europe									
Iran	0.012	0.063	523	0.197	0.114	581	0.755	0.31	411
Turkey	0.084	0.089	654	0.359	0.342	957	0.640	0.61	953
Syria	0.046	0.118	521	0.076	0.038	701	0.093	0.063	671
Morocco	0.135	0.077	567	0.067	0.043	645	0.072	0.042	591
Spain	0.224	0.118	521	0.091	0.047	521	0.075	0.042	539
Italy	0.063	0.042	663	0.011	0.013	1224	0.005	0.006	1158
North America									
Mexico	0.132	119	901	0.135	0.155	1154	0.150	0.240	1600
Canada	0	0	0	0	0	0	0.055	0.072	1306
Eastern Africa									
Ethiopia	0.169	0.118	698	0.158	0.11	759	0.159	0.128	806
Tanzania	0.010	0.003	298	0.057	0.02	350	0.070	0.032	457
Malawi	0.016	0.01	654	0.032	0.2	621	0.088	0.035	398
Oceania									
Australia	0	0	0	0.012	0.014	1154	0.154	0.169	1154

this period. Podding begins as the temperatures start rising and the crop faces increasing drought stress at this critical stage. In some parts of India winter rains may alleviate the moisture stress. Occasional excessive precipitation may induce excessive vegetative growth and lodging. The crop is harvested in the dry season at the beginning of summer.

In the temperate climates in West Asia, North Africa, southern Europe, USA and Canada chickpea is planted in the spring under low temperatures and the vegetative growth occurs in the summer. There is some rainfall during flowering and podding. The crop matures in the fall when the temperatures begin to fall (Anbessa *et al.*, 2006). Harvest time temperatures are down also in some regions in West Asia and North Africa. Early frost may discolour and damage the maturing seed leading to lower quality produce.

In West Asia and North Africa some chickpea is seeded as a winter crop before snowfall. As soon as the snow melts in the spring, the seeds emerge, plants grow, flower and pod in much better moisture regime than the summer crop. Seed yields are often much higher under such cultivation. However, higher moisture conditions and better crop canopy favour ascochyta blight development and the crop is often damaged. Thus variations in cropping system and crop season differ region to region and influence the crop productivity drastically. Under such seasonal variations in cropping system the chickpea growers are compelled to harvest this crop with great variations in productivity.

Farmers' Holdings

Chickpea farmers in the developing world with >90% chickpea area have smallholdings generally 1–10 ha. The land is often double-cropped. Most operations are manual. The farmers in the developed world (USA, Canada, Australia, Spain) grow chickpea as a commercial crop on large farms often larger than 200 ha. The harvest is mainly exported. All the field, seed handling and cleaning operations are mechanical. Thus the varietal requirements are different not only for machine operations but also for marketing. Kabuli types are preferred over desi types because of their higher market price. On the other hand, much of the crop in the Indian subcontinent is for consumption as dhal and chickpea flour for which desi types are preferred. Of course larger farmers in India also prefer kabuli types for marketing. Therefore, the landholdings of farmers in different continents differ greatly which affect both the crop cultivation system and crop production. With these variations in landholdings the situation of chickpea growers differs which influences its cultivation globally.

Crop Losses

Chickpea crops face biotic and abiotic stresses during their growing seasons. The major constraints to increased productivity include drought, fusarium wilt, ascochyta blight and helicoverpa pod borer. There is a lack of research that addresses these problems. Fortunately all of these problems may not occur at one time and in the same region. The rough estimates suggest than in India alone damages due to biotic and abiotic stresses exceeded more tha 40% annually (Yadav *et al.*, 2004, 2006). In other countries the damages due to these stresses are also high; in Australia due to drought >30% loss occurs annually (Leport *et al.*, 1999). Likewise losses due to fusarium wilt and ascochyta blight are very high ranging between 40% and 60% in Turkey (Dusunceli *et al.*, 2004) and Australia (Knights *et al.*, 2005). Ascochyta blight in the USA and Canada also causes huge losses to chickpea crops. Stable multiple resistant high-yielding cultivars are needed by farmers to increase chickpea production. Early maturity, increased seed yield, improved harvestability and disease resistance are required to stabilize and expand the area under this crop. Development of extra early and cold tolerant chickpea ICCV 96029 has helped stabilize chickpea yields by reducing crop duration (Kumar and Rao, 2001). This variety is also shown to allow chickpea cultivation as a catch crop between rice and wheat in northern India (Sandhu *et al.*, 2002).

Considering the published statistics, we feel that the success stories of chickpea expansion in Andhra Pradesh in India, Myanmar, Turkey, Australia

Table 29.2. Area, production and productivity of pulses and chickpea in India, 1950–2005.

	Pulses			Chickpea		
Area/production	1950–1951	1970–1971	2000–2005	1950–1951	1970–1971	2000–2005
Area (million hectares)	19.09	22.54	23.90	7.57	8.00	6.30
Production (tonnes)	8.41	11.82	14.41	3.65	5.50	5.20
Productivity (kg/ha)	441	524	635	482	660	830

and Canada deserve detailed discussion, as well as a major decline in chickpea cultivation in the Indo-Gangetic plains in the Indian subcontinent and in southern Europe.

India

The green revolution in cereals in the 1970s progressively relegated chickpea to more marginal lands in India (Kumar *et al.*, 1996). Until the 1970s it was a premier crop of northern region. With the adoption of irrigated wheat-cropping system, especially in Punjab, Haryana, northern Rajasthan and western Uttar Pradesh, the farmers shifted land away from the cultivation of chickpea (Anonymous, 2000). The other pulses did not suffer as much (Table 29.2). Consequently, the chickpea cultivation moved southwards into warmer and more drought-prone environments. Nearly 3 million hectares were lost in the North. However, expansion took place in the central and peninsular India especially in Madhya Pradesh, Maharashtra, Karnataka and Andhra Pradesh (Deshmukh, 2005). Chickpea being a cool season crop suffered drought and heat stress in the warmer regions (Kumar and Abbo, 2001). This negated some of the varietal improvements that had been made for the longer season environments of North India.

In the hot and dry areas of central and peninsular India, the crop is grown on about 4 million hectares (more than half of the total chickpea area in the country). However, these areas are traditionally characterized by low-input farming systems under rainfed conditions, poor chickpea yields and a stagnant market. To address chickpea production in this region, three major important constraints are drought, fusarium wilt and helicoverpa pod borer. These were addressed by the national chickpea breeders working under national chickpea improvement programme and chickpea team lead by the International Crops Research Institute for the Semi-Arid Tropics (ICRISAT), Hyderabad, India, and major area, productivity and production improvements took place in the 1990s (Kumar and Abbo, 2001). We discuss the case of Andhra Pradesh where this crop expanded exponentially during the last two decades.

Silent chickpea revolution in Andhra Pradesh

Chickpea was not even a minor crop in Andhra Pradesh until 1985. Chickpea varieties were susceptible to fusarium wilt and root rots, and suffered losses

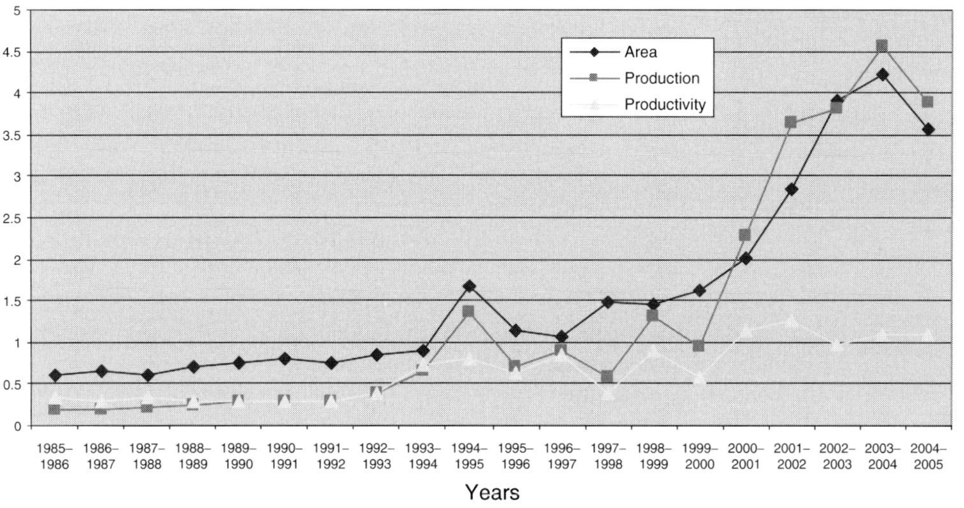

Fig. 29.1. Area ('000 ha), production ('000 t) and productivity (t/ha) of chickpea in Andhra Pradesh, India: 1985–2004.

from heat and drought stresses because they matured too late. Over the last two decades chickpea has become a popular pulse crop in Andhra Pradesh. Good market prospects combined with the introduction of high-yielding short-duration varieties that escape drought and are resistant to wilt and to some extent pod borer have helped trigger a silent pulse revolution (ICRISAT, 2002; Figure 29.1). Short- and extra-short-duration varieties were developed to avoid moisture stress during podding. Notable among these are ICCV 2, ICCC 37, ICCV 10, BG 372, BG 391, BGD 72, BG 1053, BG 1108 and JKG 1. ICCV 2 is resistant to wilt and escapes drought stress because of early vigour and extra-short-duration maturity. It was the first kabuli cultivar developed for tropical conditions. It was released as Swetha in Andhra Pradesh in 1989, as ICCV 2 in Maharashtra in 1991 and as Yezin 3 in Myanmar. It fits well in the existing production systems following a cereal crop. This and new large-seeded kabuli varieties have spread rapidly and are replacing crops such as cotton, tobacco and chillies. ICCV 37 is resistant to wilt and tolerant of root rot. It was released as Kranthi in Andhra Pradesh. It has become very popular in Maharashtra and Madhya Pradesh. ICCV 10 is resistant to wilt and root rot and tolerant of drought. It was released as Bharati in central India. Because of the availability of the seed of superior varieties chickpea production in Andhra Pradesh increased 22-fold in the last two decades. The area increased from 60,000 ha in 1985 to 425,000 ha in 2004, and production rose sharply from 20,000 to 450,000 t during the same period (Figure 29.1; Agricultural Production Statistics, 2005). The chickpea productivity of 1.1 t/ha in Andhra Pradesh in 2004 was the highest for any state in India. The crop duration here is <3 months, much shorter than that in the more favourable growth environment in North India.

North-western India

Urban Indian consumers and middle-class families have a preference for extra-large-seeded types. The market price for the large-seeded types is very high; ranging between $700/t and $900/t. Progressive chickpea growers want to cultivate such economically attractive cultivars. However, varieties that possess large seed size of 10–12 mm (diameter) have not been developed yet for India. Varieties having 8–9 mm seed size are available to the farmers. The most popular kabuli cultivars among chickpea growers in this part of India are BG 1053, BG 1088 and BG 1108 and of desi types are BG 256, BG 362, BG 372 and BG 1103. For example, Rana manages his 300 acres farm in Sirsa district, Haryana, the wheat/rice belt of North India. He is a pioneer in promoting chickpea after rice in this belt. He grows both desi and new kabuli cultivars, BG 1103 and BG 1088, on >40 ha after the rice harvesting. The winter rains are very scanty and at maturity terminal drought in March prevails. Being a progressive chickpea grower he is adopting integrated crop production and management technologies. During the last few years he harvested 3.0–3.5 t/ha. The cost of cultivation and sowing is approximately $300–350/ha. Due to a high demand for chickpea seed he is able to sell his produce for use as seed to the neighbouring farmers at the rate of $700/t. Likewise the chickpea farmers in other northern states of India are following this example and adopting other improved varieties like BG 1053, BG 1105, BG 1108 for cultivation in a big way.

The Rice Fallows – Barind, Bangladesh

About 14 million hectares land is left fallow after the paddy harvest in South Asia (Subbarao *et al.*, 2001). By the time farmers harvest and handle their paddy crop and return for planting the post-rainy season crops, the upper soil layers loose the moisture. In the absence of irrigation, no crops can be produced in the post-rainy season in those lands.

In the late 1980s experiments were conducted in the Barind region of north-western Bangladesh. Chickpea was planted soon after the paddy harvest when the topsoil was still moist and soft. Chickpea emerged and established. The deep root system allowed the crop to produce good canopy. Relatively dry climate helped chickpea avoid foliar diseases that affect the crop in other parts of the country. The seed yields of about 1 t/ha are harvested from the land that would otherwise remain fallow (Musa *et al.*, 2001), exceed chickpea productivity in the rest of the country. From hardly 100 ha in 1981 the crop expanded to about 14,000 ha by 2000. This development has a promise of expanding chickpea cultivation to the rice fallows elsewhere in the region.

Kabuli Chickpea Revolution in Myanmar

Chickpea is the third most important food legume in Myanmar after black gram and green gram. Production of chickpea after rice in Myanmar increased 336% over the last 40 years (FAO, 2005; ICRISAT, 2003). This includes doubling the seed yield from 558 to 1104 kg/ha and increasing the area from 123,000 to 208,000 ha. Chickpea is mainly grown in the northern dry zone comprising Sagaing, Mandalay and Magway divisions. ICCV 2 has made significant impact in the Sagaing division that accounts for 85% of the total chickpea area in Myanmar. During the last 3 years, the average chickpea yield in Sagaing division increased from 626 to 1154 kg/ha. In this region the climatic conditions are similar to those of central and peninsular India. The mean annual rainfall is 700 mm and winter temperatures are mild. Farmers prefer the extra-short-duration kabuli chickpea ICCV 2 (Yezin 3 in Myanmar), because it fetches a higher price than the desi types and it escapes terminal drought because of its early maturity (ICRISAT, 2003).

Other varieties such as ICCV 10 and ICCC 37 performed well and were released. Expansion of kabuli ICCV 2 has transformed the country as a significant exporter of kabuli chickpea. Latest figures indicate that kabuli type cultivation in the country accounts for nearly half the area under the crop. In 2004 ICCV 2 accounted for nearly 90% area under kabuli type and 44% of the 208,000 ha under chickpea in Myanmar.

A farmer U Ba Than of Aleban village is happy that he switched from wheat to ICCV 2 cultivation. The ICCV 2 seed yields are 25% higher and price 50% more than those for wheat. Thus chickpea is definitely more remunerative in short-growing environments (ICRISAT, 2003).

Ethiopia

Ethiopia is a leading country in Africa for chickpea production. In the 1970s, the production of chickpea in Ethiopia was stable and higher than that in the 1980s and early 1990s. In 2002, chickpea accounted for 58% of the total revenue earned from the export of pulses in Ethiopia. The incentives helped researchers promote improved new varieties and other agronomic packages to the farmers. Some desi chickpea varieties released since 1991 were Worku and Akaki. In addition, kabuli chickpea varieties released were Shasho, Arerti, Habru, Chefe, Ejere and Teji. A farmer-to-farmer seed multiplication model was developed by chickpea scientists in Bichena, Adaa and Chefe Donsa areas using Mariye (my honey), a large-seeded variety from ICRISAT. This variety is popular both for dry and green seeds. The farmers who are involved in the seed production of these kabuli varieties obtained higher prices. They have higher economic growth and improved living conditions compared to those who grow chickpea or other crops for consumption. Currently, the high domestic prices for chickpea do not allow large exports. The relative yield levels and size and colour of seeds of these varieties determine their popularity. The release of new varieties has generated stiff competition among the chickpea farmers to multiply and get high price for the foundation seeds they produce.

Successful Chickpea Production Turkey

Kadinhani is a small town in the province of Konya in Central Anatolia, Turkey. Being in Central Anatolia, annual rainfall is limited to 300 mm and agriculture is based on rainfed systems. Chickpea is a major component of the production in the locality. It is grown mostly in rotation with wheat or barley, but other crops such as sunflower or sugarbeet can also be included in the system. Chickpea cultivation dropped to 1000 ha in 1994 as a result of susceptibility of the local cultivars to ascochyta blight. However, the cultivar Gokce improved by the Central Research Institute for Field Crops, Ankara, has made a real change in the chickpea production system of the town. Registered in 1997, it is a high-yielding genotype with good adaptation to Central Anatolia. Its 44–46 g/100 seed weight is a bit smaller than the largest types, but its tolerance to ascochyta blight impressed the farmers.

With Gokce introduction in the region, chickpea production gradually increased to 6000 ha in 2006 (Provincial Directorate of Agricultural Extension of Konya, Turkey). The mean yield doubled from 500 kg/ha in 1991 to 1000 kg/ha in 2005 (Kadinhani Agricultural Extension Directorate, Turky; available at: http://www.kadinhanitarim.gov.tr/tarimsalyapi.asp). Currently, local chickpea varieties have been abandoned completely. Because of Gokce's tolerance to ascochyta blight, the farmers are able to sow it as early as 15 March, representing a 4–6 weeks advancement of sowing date resulting in better utilization of the available moisture. Irrigation to chickpea has also been introduced in the region.

The sowing date trials conducted as part of an EU-supported project (ASCORAB) demonstrated the advantages of early sowing in the area (Dusunceli *et al.*, 2004). The current production system is based completely on mechanized sowing and harvest. The winter sowing attempts also demonstrated very promising results. Following the introduction of the cultivar its production increased rapidly in contrast to the decreasing trends in total chickpea production in the country. In this, the contract-farming programme of the Exporters Union Seed Research Company has played an important role. Thus chickpea production can be improved significantly if appropriate technologies are developed and transferred.

Australia

Chickpea was first grown in Australia in the 1880s but the industry developed only after research and breeding programmes were established during the 1970s and 1980s (Knights *et al.*, 2005). Desi variety Tyson (C 235) from India stimulated chickpea production in all the states of the country. The area expanded from 6000 ha in 1984 to 159,000 ha in 1990. Desi chickpea was predominantly grown on alkaline, fertile clay soils in rotations with cereals in southern Queensland, northern New South Wales (NSW) and in the Wimmera region of Victoria. Kabuli chickpea were also grown in Victoria using the variety Kaniva and a small kabuli industry was established in the Ord River region of Western Australia. In the 1990s new varieties with improved adaptation and quality were released by the national chickpea-breeding programme. Based on these varieties and good management strategies, the chickpea industry was entering a phase of increased profitability through increased yields and quality. Area in Australia increased to a record 265,000 ha in 1998 and chickpea became established as a significant crop in Western Australia after a concerted extension programme by DAWA.

In 1998, the previously exotic disease ascochyta blight devastated chickpea crops in Victoria and South Australia. The area in Victoria crashed from a peak of 144,000 ha in 1996 to <10,000 ha by 2000. Ascochyta blight similarly impacted on chickpea in Western Australia 2 years later. Desi production became unprofitable in these regions due to the high cost of applying fungicides and risk of crop failure. However, because of its higher value, susceptible Kabuli chickpea varieties continued to be grown in Victoria with up to eight fungicide applications to control ascochyta blight. In the north-eastern Australia, ascochyta blight is less severe and high quality moderately susceptible or susceptible varieties (Howzat) have continued to be grown successfully using good disease management strategies.

Ascochyta blight necessitated a dramatic shift in breeding objectives and essentially meant a new start to chickpea breeding in Australia (Knights *et al.*, 2005). Varieties were fast tracked into production to re-establish chickpea in southern and western Australia and to stabilize the industry in north-eastern Australia. The first ascochyta blight-resistant desi variety Genesis 508 and kabuli Genesis 090 were released in 2004 and 2005, respectively. The moderately resistant larger-seeded kabulis Almaz and Nafice have also been released for more favourable long season areas. All these varieties were introductions from the International Center for Agricultural Research in the Dry Areas (ICARDA), Aleppo, Syria. Desi varieties with moderate resistance to ascochyta blight have been released for Western Australia (Genesis 836, Sonali) and north-eastern Australia (Flipper, Yorker) to stabilize production in these areas. In the next 5 years ascochyta blight-resistant desi varieties with high yield and quality will be released for all areas of Australia with variety-specific agronomic packages and further impact on chickpea area. In the longer term research will need to address the threat of exotic diseases, pathogen evolution and weeds (Knights *et al.*, 2005). In the next 5 years chickpea production in southern Australia is expected to expand rapidly based primarily on Genesis 090.

Chickpea farmer in New South Wales

Michael O'Brien manages his 14,500 ha family farm 'Kincora' near Walgett, a small town on the dry, western edge of the New South Wales cropping belt. He was a pioneer of the chickpea industry in the Walgett district, growing his first crop in 1991. Chickpea proved not only to be ideally suited to this environment, characterized by low and erratic rainfall (460 mm annually; <200 mm during the winter growing season), deep clay soils with a high moisture holding capacity and high temperatures (24–28°C) at flowering, but also to a farming system based on direct seeding into retained cereal stubble. Initially adopted to counter the worsening problems associated with cereal monoculture (low soil nitrogen and the build-up of weeds and cereal diseases), chickpea is now an integral part of O'Brien's farming system. It usually follows winter crops of wheat or barley, although opportunistic double cropping after early sown sorghum can occur if there is favourable rainfall during March–May. Production is on a large scale, with 2500 ha, amounting to 20% of the farm's cropped area, normally sown to desi type chickpea. All production stages are highly mechanized and integrated into a satellite-guided, controlled traffic system. The large areas farmed require high capacity machinery and usually more than one planter or harvester are used simultaneously. 'No-till' planters, fitted with parallelograms to maintain constant sowing depth, and covering up to 12 m with a single pass, are used to sow directly into standing cereal stubble.

O'Brien's production techniques are broadly characteristic of those used for rainfed chickpea in the drier areas of north-eastern Australia. They include a reliance on chemical fallow and in-crop herbicides (pre-emergent broadleaf and post-emergent grass) to control weeds, rhizobial inoculation and fungicidal protection of seed, deep sowing ('moisture seeking') if there has been inadequate sowing rain, fungicides to control ascochyta blight and insecticides, applied by aeroplane, to control *Helicoverpa* spp. In this low rainfall environment, average yields of 1.8 t/ha, rising to 2.1 t/ha in favourable years, make chickpea the most profitable cropping enterprise on O'Brien's farm. O'Brien's success as a chickpea grower derives both from an appreciation of the benefits of chickpea in farm rotations and an awareness of the crop's vulnerabilities. The technical expertise required to successfully grow chickpea is attested to by the choice of improved variety, optimal timing of operations (especially the application of foliar fungicides before a predicted rain event), modification to machinery (e.g. use of belt conveyors to minimize seed damage during transport) and a willingness to radically change established practices (e.g. harvesting at night to reduce the risks of cracked grain and harvesters catching fire). The farmer's own skills are supplemented by advice from a professional agronomist employed by a local farmer group, of which O'Brien is the founding member. The profitability of chickpea has also been underpinned by a successful marketing strategy: variability in the export price of chickpea has been exploited by a capability to store on-farm all grain produced and then to sell at times of peak prices; moreover, O'Brien markets his seed through a local cooperative which is able to leverage higher prices (ranging from $225–300/t on-farm) on his behalf.

Canada

Chickpea cultivation in Canada increased dramatically from 770 ha in 1995 to 500,000 ha in 2001 (FAO, 2005). The brown and dark brown soil zones of Saskatchewan and to some extent Alberta accounted for this development. This expansion was made possible with availability of seed for chickpea varieties developed by the Crop Development Centre of the University of Saskatchewan and Myles, Sanford and Dwelley from the US Department of Agriculture, Pullman, Washington. Further expansion of the crop was expected in the following years. However, damage by ascochyta blight, drought, untimely rains and early frost resulted in heavy losses to the crop and chickpea cultivation was much reduced in 2002. The recent development of early maturing, and to some extent ascochyta blight-resistant varieties, has helped increase chickpea production in the country (Anbessa *et al.*, 2006). Increased research input will help further area expansion.

Chickpea farmer in Saskatchewan

Jim Moen operates a grain farm near the town of Cabri, Saskatchewan, Canada. His farm is located in the Brown soil zone of western Canada. This is the area of the country best suited to chickpea production. The growing season typically runs from mid-April to late September. Average annual precipitation is 350 mm, most of which occurs during the growing season. The soil is frozen from November to March, thus little evaporative losses occur over the winter. Moen's farm includes 1200 ha of cropped land. The crops he produces include durum wheat, winter wheat, red and green lentil, field pea and chickpea. Moen has been growing chickpea for the last 8 years, and has averaged 160 ha of kabuli chickpea production per year over the last 5 years. He has grown the large-calibre kabuli cultivars Dwelley (from USDA) and CDC Xena (from the University of Saskatchewan) and the medium-calibre cultivar CDC Frontier (from the University of Saskatchewan). His 5-year average yield has been 1700 kg/ha. Market prices over this period have averaged $0.88 CDN/kg for 10 mm and $0.79 CDN/kg for 9 mm seed size kabuli chickpea.

The primary challenges Moen has faced with chickpea production are drought and ascochyta blight (in wetter seasons). In order to reduce the effects of drought, he uses a zero-tillage cropping system. In this system, soil disturbance is minimized to reduce moisture loss. Cereal stubble is left standing to trap snow during the winter which supplements cropping season precipitation. His cropping system is essentially continuous cropping, with minimal fallow utilized. For ascochyta blight control, Moen uses seed that has been tested for near zero levels of seed-borne ascochyta. He uses a 4-year rotation between chickpea crops. During the season, he scouts his chickpea crops for any signs of ascochyta blight, and applies fungicide as needed. On average, four fungicide applications are utilized per season on Dwelley and CDC Xena, with fewer applications required on CDC Frontier that has a higher level of resistance.

Moen has found chickpea to be a risky but profitable crop in his rotation. He expects to maintain a similar area of chickpea production over the next few years.

United States of America

Production of chickpea in the USA has been mainly in two regions: central California and the Pacific Northwest. It is estimated that chickpea yields range from 500 to 1500 kg/ha depending on the production region. In California, large white kabuli type chickpea has been grown for several decades mainly for export to European countries. In central coastal areas of California, chickpea has been grown as a spring-sown crop, often in yearly rotation with winter wheat or barley. In these areas of Mediterranean climate, summer rainfall is essentially absent and chickpea is grown utilizing residual soil moisture derived from winter rains. During the mid-1980s, attempts were made to switch from spring-sown to winter-sown production. The switch not only resulted in higher yield, but also allowed expansion of chickpea production to the central San

Joaquin Valley where high temperatures in the summer preclude chickpea as a spring-sown crop. In addition to drought, other constraints in California are fusarium wilt, ascochyta blight and sclerotinia stem rot.

In the US Pacific Northwest, chickpea has been a rotational crop in the wheat-based production system as a spring-sown crop in Idaho, Oregon and Washington for more than 20 years. The kabuli type has been the dominant type because of its price premium. Farmers have enjoyed the profits of growing chickpea despite the inconvenience of late maturity.

Currently, chickpea production is expanding to the other northern states like Montana, North Dakota and South Dakota where summer rainfall is often abundant. Consequently, late season diseases associated with high moisture, like ascochyta blight and Botrytis grey mould, become problematic. New breeding materials and germplasm lines have shown promise of improved resistance to ascochyta blight in North Dakota.

Conclusion

Among pulse crops chickpea has reasonably good yield potential, excellent food value and high market value among pulses. However, the risks associated with the cultivation of this crop do not encourage farmers to increase their area. The crop area and productivity can increase substantially in both the developing and the developed world if constraints to higher productivity and stability of production are addressed. Drought resistance or tolerance through early maturity, increased seed yield and disease and helicoverpa pest resistance are required to stabilize and expand the area under this crop in the developing countries. In the developed world and for larger farmers in the developing countries mechanical harvestibility is absolutely necessary. Thus development of determinate growth habit is a priority research area. Research on safe defoliants may also be undertaken. A greater input in breeding for ascochyta blight and other foliar disease resistance, larger-seeded desi and kabuli cultivars and better weed control mechanisms will help to increase and stabilize production levels. Promotion of chickpea as a nutraceutical may help attract funds for research and add value to the produce. Development of marker assisted selection techniques can greatly accelerate the breeding process. With these positive developments, chickpea production can increase dramatically in the near future.

References

Anbessa,Y., Warkentin, T., Vandenberg, A. and Ball, R. (2006) Inheritance of time to flowering in chickpea in a short-season temperate environment. *Journal of Heredity* 97,55–61.

Anonymous (2000) Expert committee report on pulses. Technology mission on oil seed and pulses. Department of Agriculture and Cooperation, Ministry of Agriculture, Government of India, New Delhi, pp. 20–21, 43–44, 68.

Agricultural Production Statistics (2005) Department of Agriculture, Government of Andhra Pradesh, Hyderabad, India.

Berger, J.D., Turner, N.C., Siddique, K.H.M., Knights, E.J., Brinsmead, R.B., Mock, I., Edmonson, C., Khan, T.N. (2004) Genotype by environment studies across Australia reveal the importance of phenology for chickpea (*Cicer arietinum* L.) improvement. *Australian Journal of Agricultural Research* 55, 1071–1084.

Berger, J.D., Ali, M., Basu, P.S., Chaudhary, B.D., Chaturvedi, S.K., Deshmukh, P.S., Dwivedi, S.K., Gangadhar, G.C., Gaur, P.M., Kumar, J., Pannu, R.K., Siddique, K.H.M., Singh, D.N., Singh, D.P., Singh, S.J., Turner, N.C., Yadava, H.S. and Yadav, S.S. (2006) Genotype by environment studies demonstrate the critical role of phenology in adaptation of chickpea (*Cicer arietinum* L.) to high and low yielding environments of India. *Field Crops Research* 98, 230–244.

Deshmukh, R.B. (2005) *The Souvenir – The IFLRC IV*. 18–22 October 2005. IARI, New Delhi, India, pp. 25–36.

Dusunceli, F., Meyveci, K., Cetin, L., Avci, M., Surek, D., Karabay, S., Albustan, S., Mert, Z., Akan, K. and Strange R.N. (2004) Management of ascochyta blight of chickpea through utilization of genetic resistance and adjustment of sowing date. *Legumes for the Benefit of Agriculture, Nutrition and the Environment: Their Genomics, Their Products, and Their Environment*, No. 8217. Conference handbook of the 5th European Conference on Grain Legumes, 7–11 June 2004, Dijon, France.

FAO (2005) http://FAOSTAT. FAO.org. Agricultural Production Statistics.

ICRISAT (2002) *Annual Report*. ICRISAT, Hyderabad, India.

ICRISAT (2003). *Annual Report*. ICRISAT, Hyderabad, India.

Knights, T., Ryan, M., Thompson, J., Hobson, K., Materne, M., Khan, T., Clarke, H., Moore, K. and Wood, K. (2005) Review of chickpea breeding in Australia. In: Knights, T. and Ryan, M. (eds) *Proceedings of Focus 2005. Chickpeas in the Farming Systems*. 21–23 September 2005. Goondiwindi, Queensland, Australia, pp. 20–31.

Kumar, J. and Abbo, S. (2001) Genetics of flowering time in chickpea and its bearing on productivity in semiarid environments. *Advances in Agronomy* 72, 107–138.

Kumar, J. and Rao, B.V. (2001) Registration of ICCV 96029, a super early and double podded chickpea germplasm. *Crop Science* 41, 605–606.

Kumar, J. Sethi, S.C., Johansen, C., Kelley, T.G., Rahman, M.M. and van Rheenen, H.A. (1996) The potential of short-duration chickpea varieties. *Indian Journal of Dryland Agriculture Research and Development* 11, 28–32.

Leport, L., Turner, N.C., French, R.J., Barr, M.D., Duda, R., Davies, S.L., Tennant, D. and Siddique, K.H.M. (1999) Physiological responses of chickpea genotypes to terminal drought in a Mediterranean-type environment. *European Journal of Agronomy* 11, 279–291.

Musa, A.M., Harris, D., Johansen, C. and Kumar, J. (2001) Short duration chickpea to replace fallow after aman rice: the role of on-farm seed priming in high Barind tract of Bangladesh. *Experimental Agriculture* 37, 509–521.

Orf, J.H., Diers, B.W. and Boerma, H.R. (2004) Genetic improvement: conventional, and molecular-based strategies. In: Boerma, H.R. and Specht, J.E. (eds) *Soybean: Improvement, Production and Uses*. CSSA monograph 16. Madison, Wisconsin.

Sandhu, J.S., Bains, T.S. and Sidhu, P.S. (2002) Evaluation of super early chickpea genotypes for vegetable purposes as a catch crop. *International Chickpea and Pigeonpea Newsletter* 9, 9–12.

Scarafoni, A., Kumar, J., Magni, C., Sironi, E. and Duranti, M. (2005) Biologically active molecules and nutraceutical properties of legume seeds. Paper presented at the IFLRC IV. 18–22 October 2005. New Delhi, India.

Subbarao, G.V., Kumar Rao, J.V.D.K., Kumar, J., Johansen, C., Deb, U.K., Ahmed, I., Krishna Rao, M.V., Venkatratnam, L., Hebbar, K.R., Sai, M.V.S.R. and Harris, D. (2001) *Spatial Distribution and Quantification of Rice Fallows in South Asia: Potential for Legumes*. International Crops Research Institute for the Semi-Arid Tropics, Hyderabad, India.

Tay, J.U., Penaloza, E. and Moralez, J.A. (2000) Regional reviews – Region 2:

South America. In: Knight, R. (ed.) *Linking Research and Marketing Opportunities for Pulses in the 21st Century*. Kluwer Academic, Dordrecht, The Netherlands, pp. 71–78.

Yadav, S.S., Kumar, J., Turner, N.C., Berger, J., Redden, R., McNeil, D., Materne, M., Knights, E.J. and Bahl, P.N. (2004) Breeding for improved productivity, multiple resistance and wide adaptation in chickpea (*Cicer arietinum* L.). *Plant Genetic Resources* 2, 181–187.

Yadav, S.S., Kumar, J., Yadav, S.K., Shoraj Singh, Yadav, V.S., Turner, Neil, C. and Redden, R. (2006) Evaluation of helicoverpa and drought resistance in desi and kabuli chickpea. *Plant Genetic Resources NIAB, UK* 4(3), 1–7.

30 Genotype by Environment Interaction and Chickpea Improvement

J.D. Berger,[1,2] J. Speijers,[3] R.L. Sapra[4] and U.C. Sood[5]

[1]*CSIRO Plant Industry, Private Bag 5, Wembley, Western Australia 6913, Australia;* [2]*Centre for Legumes in Mediterranean Agriculture, M080, The University of Western Australia, 35 Stirling Highway, Crawley, WA 6009, Australia;* [3]*Western Australia Department of Agriculture and Food, 3 Baron–Hay Court, South Perth, WA 6151, Australia* [4]*Division of Genetics, Indian Agricultural Research Institute, New Delhi 110012, India;* [5]*Indian Agricultural Statistics Research Institute, New Delhi 110012, India*

Introduction

> Ultimately, a plant breeder needs to assess and understand how G × E interactions influence breeding objectives, rather than simply determining their existence or magnitude. There are three ways of dealing with G × E interactions in a breeding program . . . to ignore, to avoid, or to exploit . . .
>
> (Eisemann *et al.*, 1990)

This chapter will briefly review key concepts and methods in G × E analysis and indicate how these have been applied to chickpea improvement, addressing their strengths and weaknesses to show how the state of the art can be improved in this crop.

Multi-environment trials (METs) are typically used in plant-breeding programmes to evaluate material across a range of sites representing target environments for the crop. Genotype by environment (G × E) interaction is the change in relative performance of genotypes across sites (DeLacy *et al.*, 1996a), and is a function of the range of both genotypes and environments sampled (Comstock and Moll, 1963). G × E interaction complicates genetic advance in breeding programmes because of the uncertainty of identifying the superior genotype, particularly when the interaction effect is large relative to the genotype effect (Cooper *et al.*, 1993). In well-conducted G × E trials, which sample both a wide range of environments and genotypes, main effects and interactions are invariably statistically significant, and variance components tend to be ranked in the following order: E >> G × E > G. Under these circumstances G × E interaction cannot be ignored, but should be investigated so that the breeder can decide to either restructure the programme to minimize the interaction effect, or exploit it to produce cultivars with specific adaptation to particular environments (Eisemann *et al.*, 1990). G × E analysis should be viewed as the start of the investigation of genotype behaviour across environments, rather than an end in itself because on its own it cannot provide the biological insights required to make these types of decision (Eisemann *et al.*, 1990; Lawn and Imrie, 1993). To this end Eisemann *et al.* (1990) argue that genotypes and environments must be thoroughly characterized so that the underlying cause of G × E interaction can be elucidated. This then also allows the interpretation of results from the standpoint of repeatability: how typical were the conditions experienced at any given trial, and how do these relate to the target population of environments the trial sites were chosen to represent (Cooper *et al.*, 1993).

A multitude of methods are available for the analysis and visualization of G × E, and these have been extensively reviewed in the last 50 years (Comstock and Moll, 1963; Knight, 1970; Lin *et al.*, 1986; Becker and Leon, 1988; Huehn, 1990a,b; Nyquist, 1991; Cooper *et al.*, 1993; Cooper and Delacy, 1994; Mathur and Horst, 1994; DeLacy *et al.*, 1996a,b; Kang, 1998; Piepho, 1998; Huehn and Truberg, 2002). A key concept in G × E analysis is genotype stability. By definition, genotypes exhibiting a high degree of G × E interaction are unstable across sites. Genotype stability is commonly approached from two different perspectives: the static or dynamic concept of stability (Becker and Leon, 1988). According to the static concept, stable genotypes are defined by an unchanging performance across environments, corresponding to low variance between sites. In chickpea the static concept of stability has merit for analysing characters which ideally should be invariant across environments, such as seed size or disease resistance. However, G × E analysis is most commonly directed at yield, where unchanging performance across sites is not desirable (see Lin *et al.*, 1986 for more discussion). Therefore the dynamic concept, whereby stability is defined by a consistent, undeviating response to environment, parallel to the mean response of all genotypes (Lin *et al.*, 1986; Becker and Leon, 1988) is more commonly applied.

G × E analysis using regression

The most common techniques for the analysis of G × E interaction in chickpea are calculated by regressing genotype means against a site index, usually site mean yield (Finlay and Wilkinson, 1963) or the mean site effect (Perkins and Jinks, 1968) (Table 30.1).

The Finlay–Wilkinson coefficient (b_i) is the slope of the genotype-site mean regression, and is equivalent to the Perkins–Jinks coefficient ($\beta_i = b_i - 1$) (Lin *et al.*, 1986; Becker and Leon, 1988), while its alter ego, the deviation variance (s^2d_i) is a measure of the goodness of fit (Eberhart and Russell, 1966) (Fig. 30.1). These coefficients can be defined as part of a linear mixed model:

$$Y_{ijk} = \mu + G_i + E_j + R_{jk} + \beta_i \bar{Y}_j + \lambda_{ij} + \varepsilon_{ijk} \tag{30.1}$$

where Y_{ijk} is the yield of the kth replicate of genotype i in environment j, μ is the overall mean, G_i is the effect of genotype i, E_j is the effect of environment j, R_{jk} is the effect of the kth replicate at site j, β_i is the Perkins–Jinks regression

Table 30.1. Published chickpea G × E studies categorized by numbers of genotypes and sites, and the methodology employed for analysing G × E interaction.

Genotypes	Sites	Reference	b_i	s^2d_i	W_i^2	ANOVA	AMMI/PCA	HC
<16	≤8	Tomer *et al.* (1973)	Ref64√	√				
		Yadavendra and Dixit (1987)	Ref68√	√				
		Khan *et al.* (1988)	Ref29√	√				
		Banik *et al.* (1997)	Ref4√	√				
		Deshmukh *et al.* (1998)	Ref16√	√				
≥16	≤8	Chandra *et al.* (1971)	Ref10			√		
		Ramanujam and Gupta (1974)	Ref51√	√				
		Singh *et al.* (1974)	Ref57					
		Jain *et al.* (1984)	Ref26√	√				
		Singh and Bejiga (1990)	Ref55√					
		Singh *et al.* (1991)	Ref54√	√				
		Katiyar *et al.* (1992)	Ref28√	√				
		Singh *et al.* (1993)	Ref59√	√				√
		Singh and Kumar (1994)	Ref58√	√				
		Singh *et al.* (1995)	Ref56√	√				
		Aher *et al.* (1998)	Ref1√	√				
		Mhase *et al.* (1998)	Ref42√	√				
		Popalghat *et al.* (1999)	Ref50√	√				
		Khorgade *et al.* (2000)	Ref30√	√				
		Yadava *et al.* (2000)	Ref67√	√				
		Sood *et al.* (2001)	Ref63√	√				
<16	>8	Mehra *et al.* (1980)	Ref41√	√				
		Rubio *et al.* (2004)	Ref52			√	√	
≥16	>8	Malhotra and Singh (1991)	Ref38			√		√
		Singh and Singh (1991)	Ref60√	√				
		Kumar *et al.* (1996)	Ref33√	√		√		
		Kumar *et al.* (1998)	Ref34√		√			
		Arshad *et al.* (2003)	Ref3√	√				
		Berger *et al.* (2004)	Ref9				√	√
		Berger *et al.* (2006)	√				√	√
Total *n*		30	25	22	1	4	3	4

b_i: slope of genotype vs site mean regression.
s^2d_i: deviation mean squares for genotype vs site mean regression.
W_i^2: Wricke's ecovalence.
ANOVA: main effects are further subdivided (i.e. environments into years and locations, genotypes into desi and kabuli), and used to partition G × E interaction.
AMMI: additive main effects and multiplicative interaction.
HC: hierarchical clustering.
PCA: principal components analysis.

coefficient showing the response of genotype i to site mean yield (note that in the Finlay–Wilkinson model environment (E_j) is not included as a main effect), λ_{ij} is the deviation from regression of genotype i in environment j and is normally and independently distributed (NID) with mean = 0 and variance s^2d_i (the Eberhart–Russell parameter) and ε_{ijk} are residual errors which are NID $(0,s^2e_j)$ with the following constraints:

$$\sum_{i=1}^{G} G_i = \sum_{k=1}^{R} R_{jk} = \sum_{i=1}^{G} \lambda_{ij} = \sum_{j=1}^{E} \lambda_{ij} = \sum_{i=1}^{G}\sum_{k=1}^{R} \varepsilon_{ijk} = \sum_{j=1}^{E}\sum_{k=1}^{R} \varepsilon_{ijk} = 0 \qquad (30.2)$$

This model incorporates different experimental variance at each environment and can be fitted using a number of commonly available statistical packages such as Genstat.

The regression slopes describe genotype responses to environmental effects, while the deviation variances capture the predictability of this response. Accordingly, these parameters correspond exactly to the concepts of static and dynamic stability. Stable genotypes are defined by $b_i = 0$ using the static concept, and $b_i = 1$ under the dynamic definition, with $s^2d_i = 0$ in both cases. Although this close correspondence to the common definitions of stability appears to make interpretation of these parameters straightforward, the individual genotype responses should be graphed to facilitate a genuine understanding of interaction behaviour. Figure 30.1 shows two genotypes with similar responsiveness and stability, but different mean yields. In genotype A the instability associated with the Finlay–Wilkinson coefficient is a positive trait because this variety becomes increasingly more productive in favourable environments, and only yields less than the mean at a single low-yielding environment. In genotype B the crossover point is central; therefore, this variety is more productive than the mean in high-yielding environments, but relatively

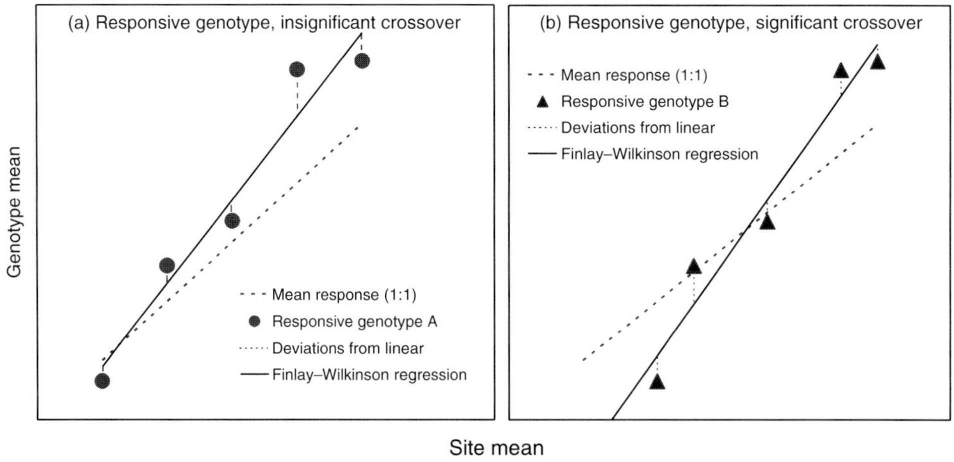

Fig. 30.1. Schematic representation of Finlay–Wilkinson response and Eberhart–Russell stability parameters demonstrating how the crossover point with the average genotype response influences the interpretation of stability.

less productive in low-yielding environments. This type of Finlay–Wilkinson instability is normally regarded as a negative trait, unless the environmental circumstances that define the crossover point are well understood, and can be exploited by the breeder.

Interpretation of Eberhart–Russell stability variances should also be approached graphically. If the reason why a given genotype diverges from the mean in particular environments is well understood (e.g. lack of resistance to a disease encountered in specific environments), then the stability variance can be meaningfully interpreted. The regression-based stability and response parameters have been criticized because the site indices are not independent of the genotype responses, or measured without error, and error variances may not be homogeneous between sites (see discussion in Becker and Leon, 1988). However, their use is valid when analyses are based on large numbers of genotypes, environments and replication, and the analysis of variance (ANOVA) assumptions are met (Skroppa, 1984).

Other stability parameters have rarely been used for the analysis of G × E in chickpea (Table 30.1), but include: genotype variance across environments (s_x^2) (Lin *et al.*, 1986; Becker and Leon, 1988), Plaisted and Peterson's (1959) mean θ_i, Plaisted's (1960) θ_i, Wricke's (1962) ecovalence (W_i^2), Shukla's (1972) stability variance (σ_i^2), Pinthus' (1973) coefficient of determination (r_i^2) (see also DeLacy *et al.* (1996a) for model descriptions). In practice these various parameters tend to behave similarly. Kumar *et al.* (1998) report very high correlations (mean $r = 0.9$) between W_i^2, s^2d_i, σ_i^2, r_i^2, θ_i and mean θ_i, calculated using a range of chickpea METs. In maize, spring barley, oat, winter wheat and winter rye trials W_i^2 and s^2d_i, and b_i and s_x^2 were very closely correlated (r = 0.84–1.0, 0.81–0.99, respectively) (Becker, 1981), suggesting that while the genotype interaction variance parameters were modelling genotype stability in terms of predictability, genotype variance across environments was a measure of responsiveness.

G × E analysis using multivariate techniques

More recently multivariate techniques such as principal components analysis (PCA), cluster analysis and additive main effects and multiplicative interaction (AMMI) have come to the fore in the analysis of G × E interaction in chickpea (Table 30.1). These methods are usually applied to a matrix of G × E means for the trait of interest (typically seed yield) with genotypes as rows and sites as columns. Trait values at each site are therefore treated as different variables and multivariate analysis used to capture the pattern in the matrix. Classification techniques such as hierarchical clustering are normally used to separate genotypes into homogeneous groups (Westcott, 1986; Huehn and Truberg, 2002; Truberg and Huehn, 2002), ideally with no significant within group G × E interactions, using incremental sums of squares clustering strategies (Ward's method) (DeLacy *et al.*, 1996a). Ordination techniques like PCA and AMMI allow relationships within and between genotypes and sites to be readily visualized graphically using biplots (Kroonenberg, 1997).

The outcome of pattern analysis is strongly affected by the type and quality of data presented in the G × E matrix. Because of the large effects of environment in METs, analyses of raw mean matrices are usually dominated by the effect of environment. Therefore G × E matrices are usually centred (column mean = 0) or standardized (column mean = 0, standard deviation = 1) (McLaren, 1996; Kroonenberg, 1997). Centring is analogous to subtracting the main effect of environment from each genotype (equivalent to PCA using a covariance matrix), and therefore the resulting patterns may still be dominated by genotype main effects. Standardization makes the variance between genotypes identical at each site, and therefore highlights rank order within sites (equivalent to PCA using a correlation matrix). As a result environments with similar genotype rankings will tend to group together, even if their productivity is very different (Fox and Rosielle, 1982). While this is effective for breeding purposes, important biological insights can be lost as genotype scale effects are discarded. For example, standardization will ignore Haldane's type 2 G × E interactions where genotypes respond completely differently as site mean yield increases, but do not cross over and change rank order (Fig. 30.2). Pattern analysis of a log-transformed, centred G × E matrix is often an effective compromise which minimizes the influence of high experimental variance associated with high-yielding sites, but still allows genotypic variance to play a role (Berger

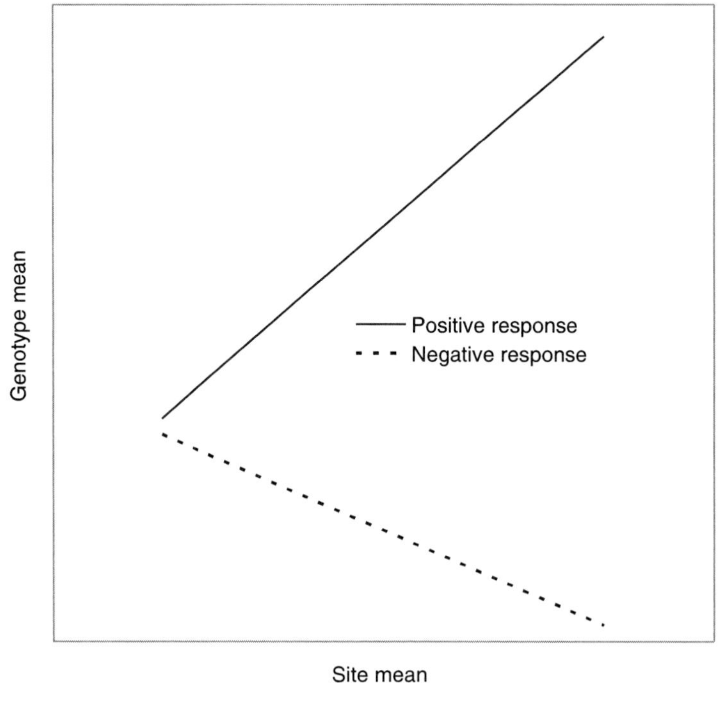

Fig. 30.2. Schematic representation of Haldane's type 2 G × E interactions demonstrating positive and negative genotypic responses to higher-yielding environments without rank changes. (Adapted from Allard and Bradshaw, 1964.)

et al., 2004; Berger *et al.*, 2006). AMMI is PCA-performed on a G × E matrix of interaction residuals alone (i.e. main effects of genotype and environment subtracted) (Gauch and Furnas, 1991). However, when presenting the results as a biplot, AMMI ordinations typically present genotype and environmental main effects on the *x*-axis, and the first interaction PC as the *y*-axis (Gauch and Furnas, 1991). The AMMI model is given below:

$$Y_{ijk} = \mu + G_i + E_j + R_{jk} + \sum_{l=1}^{P} \kappa_l u_{il} v_{jl} + \rho_{ij} + \varepsilon_{ijk} \qquad (30.3)$$

where κ_l is a constant associated with the *l*th of *P* principal components (PCs), u_{il} are genotype loadings (or effects) for the *l*th PC, v_{jl} are site loadings for the *l*th PC and ρ_{ij} are interaction effects not explained by the first *P* principal components.

The AMMI model can be viewed as an extension of the linear regression model since if $P = 1$ and $v_{jl} = (Y_j - Y)$ then model (30.3) becomes model (30.1). It follows that the AMMI and other PCA models must fit the data at least as well as the Finlay–Wilkinson model even with a single PC since the site effects are chosen to maximize the variance explained. Therefore the proponents of G × E pattern analysis argue that these techniques capture far more information than the traditional stability and response parameters (Cooper *et al.*, 1993; DeLacy *et al.*, 1996a). However, others suggest that interpretation can be difficult, because there is no clear correspondence between the traditional stability concepts and the multivariate output, which may or may not bear an obvious relationship to environmental conditions (Westcott, 1986; Becker and Leon, 1988). This is because in PCA the site parameters are a weighted mean of the genotype yields at each site which may have no particular meaning to the plant breeder. We suggest that ordination techniques which present data for both genotypes and sites stimulate the researcher to address adaptation from the point of view of environment because they pose the obvious question: why are particular groups of genotypes performing well (or poorly) in certain trial site subsets? This is precisely the challenge posed by Eisemann *et al.* (1990) and will be addressed in the next section.

Impact of G × E Analysis on Chickpea Improvement

The literature on G × E interaction in chickpea is dominated by experiments based on low numbers of genotypes and/or trial sites, often limited to the home state of the chief investigator (Table 30.1). Nevertheless, all 30 publications listed in Table 30.1 report highly significant G × E interaction for yield, which suggests that the issue is important in chickpea.

Twenty-five of the 30 publications listed in Table 30.1 present Finlay–Wilkinson coefficients, while 22 present Eberhart–Russell stability parameters. These are generally tabulated alongside the genotype means, and individual genotypes recommended on the basis of high productivity, high stability (i.e. non-significant s^2d_i) and high or low environmental responsiveness (b_i). In many cases b_i is positively correlated with mean yield (Ramanujam and Gupta, 1974; Jain *et al.*, 1984; Singh and Bejiga, 1990; Singh and Singh, 1991; Kumar *et al.*, 1996; Arshad *et al.*, 2003), suggesting that high-yielding genotypes are more responsive to favourable environments. Similar results have been reported in other legumes (Berger *et al.*, 2002) and oats (Eagles *et al.*, 1977). A correlation between b_i and genotype mean yield is a consequence of defining the site index by the site mean yield. With few exceptions environments are uncharacterized in the chickpea G × E literature, and no attempt has been made to explain high responsiveness or deviations from linearity among the genotypes from the point of view of environment or genotype biology. As a result, it remains unclear to what the genotypes are responding, and why some genotypes fail to respond to these stimuli. Nor does it allow the reader to judge the value of the experimental approach, because it is not clear to what degree the trials represent real environments encountered by the crop in farmers' fields. This criticism particularly applies to some of the work based on low numbers of trial sites ($n < 4$) (Chandra *et al.*, 1971; Tomer *et al.*, 1973; Jain *et al.*, 1984; Singh and Bejiga, 1990; Katiyar *et al.*, 1992; Singh *et al.*, 1995; Aher *et al.*, 1998; Yadava *et al.*, 2000), or where different environments were created by varying sowing date or agronomy (Ramanujam and Gupta, 1974; Singh *et al.*, 1974, 1991, 1993, 1995; Singh and Singh, 1991; Singh and Kumar, 1994; Banik *et al.*, 1997; Popalghat *et al.*, 1999; Khorgade *et al.*, 2000; Sood *et al.*, 2001). While this approach allows breeders to choose the best material in the particular subset of genotypes evaluated, it does little to further understanding of adaptation in chickpea, makes extrapolation of the results very difficult and therefore fails to meet the challenge set by Eisemann *et al.* (1990).

This is clearly demonstrated by studies which calculate stability parameters across years or experimental subsets (Kumar *et al.*, 1996, 1998). Kumar *et al.* (1996) grew a diverse set of desi and kabuli genotypes ($n = 16$) in 17–31 sites located in Mediterranean-type and subtropical environments across the world over 3 consecutive years. While genotype means averaged over all sites were well correlated across years ($r = 0.75$–0.84), genotype responsiveness (b_i) and stability (s^2d_i) were far less predictable ($r = 0.41$–0.46; $r = 0.40$–0.49, respectively). Similar findings have been made in soybean (Sneller *et al.*, 1997), wheat, barley and oats (Leon and Becker, 1988; Nurminiemi and Rognli, 1996). When sites were divided into low- and high-yielding environments and parameters calculated for each subset, the correlations between b_i, s^2d_i and W_i^2 in chickpea became even weaker (Kumar *et al.*, 1998). These results demonstrate that responsiveness and stability are not absolute properties of genotypes, but instead represent specific responses to particular environmental stimuli. Unless these are investigated as argued by Eisemann *et al.* (1990), most of the valuable information generated by these trials is lost. This point cannot be overemphasized because while genotype interaction behaviour may or may not be repeatable, the biological insights gained by their investigation can be reapplied whenever similar environments are encountered.

The few chickpea G × E interaction studies which incorporate genotypic or environmental information suggest that this approach will be profitable. Kumar *et al.* (1996) demonstrated that site effects could be well explained by contrasting spring and winter sown trials, and genotypic differences explained by comparing desis with kabulis. Their analysis showed that desi and kabuli differences were particularly strong when sites were partitioned into spring and winter sowing. Unfortunately these differences were not presented, nor were the environments characterized. Rubio *et al.* (2004) used a similar approach to demonstrate the consistent superiority of early flowering and a bushy habit, and the negative impact of double-poddedness in autumn sown desi–kabuli crosses evaluated in southern Spain.

Genotypes and environments have been well characterized in the few chickpea G × E publications which have used a multivariate approach as the core of the analysis (Berger *et al.*, 2004; Berger *et al.*, 2006). This work is based on 72 and 39 genotypes evaluated over 9 and 15 site/year combinations in Australia (Berger *et al.*, 2004) and India (Berger *et al.*, 2006), respectively. A similar analytical approach was employed in both cases: (i) a log-transformed G × E matrix was prepared by ANOVA; (ii) genotypes were classified on the basis of yield using Ward's hierarchical clustering; and (iii) an environment-centred PCA was run based on the covariance matrix, and the genotype clusters were superimposed on the biplot of site factor loadings and genotype scores in order to visualize G × E interaction among the clusters. Site climate and biology was well described, with a particular emphasis on sites contributing highly to G × E interaction. The Ward's genotype yield clusters were compared within each site by ANOVA to define cluster behaviour in terms of plant stand, early vigour, phenology, productivity and yield components.

The result of this in-depth analysis was an improved understanding of the important adaptive traits for chickpea in both countries. In Australia, contrasting genotype interaction behaviour was identified in dry, low-yielding, southern Mediterranean-type environments and wetter, longer growing-season environments with dominant summer rainfall in northern NSW (Berger *et al.*, 2004). Genotypes performing well under stress tended to yield well at all sites except northern NSW, and were characterized by early phenology and high harvest index, but were not different than other genotypes in terms of biomass or early vigour (Berger *et al.*, 2004). In India, latitude and its effect on site climate and biology played a pivotal role in G × E interaction. Low-yielding sites were confined to Central and South India and associated with low pre-season rainfall, high temperature, early phenology, short crop duration, low biomass and low fecundity (Berger *et al.*, 2006). Genotypes specifically adapted to low-yielding sites were characterized by low yield responsiveness, low biomass, early phenology and high harvest index, confirming the role of drought escape in South India. Genotypes that were well adapted to higher-yielding northern sites were characterized by later, highly responsive phenology, whereby flowering was delayed significantly longer at late flowering northern sites. Extending the vegetative phase under long season conditions resulted in a greater biomass accumulation prior to reproduction, and delayed flowering until temperatures became sufficiently warm to support pod set. As a result, both source and sink potential were increased in material specifically adapted to productive northern sites (Berger *et al.*, 2006). Widely adapted genotypes were characterized by intermediate flowering and relatively early, responsive maturity, a phenological compromise which provided sufficient drought escape in the south, and enough biomass in the north to produce above average yields in these contrasting environments (Berger *et al.*, 2006).

Conclusions

In the farming community in many of the environments in which it is grown chickpea has a reputation for high risk because of its unpredictable yield. All the literature based on METs suggests that G × E interaction is invariably highly significant. Therefore, it is incumbent on researchers working with the crop to develop a good understanding of adaptation. When analysing METs it is not enough to quantify G × E interaction, regardless of whether traditional stability parameters, or the newer multivariate approaches are being employed. In either case scientists need to rise to the challenge set by Eisemann *et al.* (1990) and characterize both genotypes and environments so that the underlying causes of G × E interaction are understood. Ultimately it will be this understanding, rather than the genotypes themselves, that can be transferred again and again to tackle breeding challenges faced in the ongoing development of the chickpea.

Acknowledgements

The authors would like to acknowledge generous research funding support from CSIRO, DAFWA, CLIMA, ACIAR and GRDC, and the Biometrics Group at DAFWA for stimulating discussion on statistical approaches to visualizing G × E interaction.

References

Aher, R.P., Mhase, L.B., Manjare, M.R. and Kolte, T.B. (1998) Phenotypic stability for grain yield in chickpea under irrigated condition. *Indian Journal of Pulses Research* 11, 115–116.

Allard, R.W. and Bradshaw, A.D. (1964) Implications of genotype–environment interactions in applied plant breeding. *Crop Science* 4, 503–508.

Arshad, M., Bakhsh, A., Haqqani, A.M. and Bashir, M. (2003) Genotype–environment interaction for grain yield in chickpea (*Cicer arietinum* L.). *Pakistan Journal of Botany* 35, 181–186.

Banik, B.R., Sarker, R. and Pandit, D.B. (1997) Genotype environment interaction of seed yield and seed weight in chickpea. *Legume Research* 20, 230–232.

Becker, H.C. (1981) Correlations among some statistical measures of phenotypic stability. *Euphytica* 30, 835–840.

Becker, H.C. and Leon, J. (1988) Stability analysis in plant breeding. *Plant Breeding* 101, 1–23.

Berger, J.D., Robertson, L.D. and Cocks, P.S. (2002) Genotype × environment interaction for yield and other plant attributes among undomesticated Mediterranean *Vicia* species. *Euphytica* 126, 421–435.

Berger, J.D., Turner, N.C., Siddique, K.H.M., Knights, E.J., Brinsmead, R.B., Mock, I., Edmondson, C. and Khan, T.N. (2004) Genotype by environment studies across Australia reveal the importance of phenology for chickpea (*Cicer arietinum* L.) improvement. *Australian Journal of Agricultural Research* 55, 1–14.

Berger, J.D., Ali, M., Basu, P.S., Chaudhary, B.D., Chaturbedi, S.K., Desmuk, P.S., Dharmaraj, P.S., Dwivedi, S.K., Gangadhar, G.C., Gaur, P.M., Kumar, J., Pannu, R.K., Siddique, K.H.M., Singh, D.N., Singh, D.P., Singh, S.J., Turner, N.C., Yadava, H.S. and Yadav, S.S. (2006) Genotype by environment studies demonstrate the critical role of phenology in adaptation of chickpea (*Cicer arietinum* L.) to high and low yielding environments of India. *Field Crops Research*. Submitted.

Chandra, S., Sohoo, M.S. and Singh, K.B. (1971) Genotype × environment interaction for yield in gram. *Journal of Research Punjab Agricultural University* 8, 165–168.

Comstock, R.E. and Moll, R.H. (1963) Genotype-environment interactions. In: Hanson, W.D. and Robinson, H.F. (eds) *Statistical Genetics and Plant Breeding*. National Academy of Sciences, Washington, DC, pp. 164–196.

Cooper, M. and Delacy, I.H. (1994) Relationships among analytical methods used to study genotypic variation and genotype-by-environment interaction in plant breeding multi-environment experiments. *Theoretical and Applied Genetics* 88, 561–572.

Cooper, M., Delacy, I.H. and Eisemann, R.L. (1993) Recent advances in the study of genotype × environment interactions and their application to plant breeding. In: Imrie, B.C. and Hacker, J.B. (eds) *Focused Plant Improvement: Towards Responsible and Sustainable Agriculture*. Australian Plant Breeding Conference Gold Coast, Queensland, pp. 116–131

DeLacy, I.H., Basford, K.E., Cooper, M., Bull, J.K. and McLaren, C.G. (1996a) Analysis of multi-environment trials – An historical perspective. In: Cooper, M. and Hammer, G.L. (eds) *Plant Adaptation and Crop Improvement*. CAB International, Wallingford, UK, pp. 39–124.

DeLacy, I.H., Cooper, M. and Basford, K.E. (1996b) Relationships among analytical methods used to study genoytpe-by-environment interactions and evaluation of their impact on response to selection. In: Kang, M.S. and Gauch, H.G. (eds) *Genotype by Environment Interaction*. CRC Press, Boca Raton, Florida, pp. 51–84.

Deshmukh, R.B., Kute, N.S., Aher, R.P. and Mhase, L.B. (1998) Stability for grain yield in chickpea. *Indian Journal of Pulses Research* 11, 79–81.

Eagles, H.A., Hinz, P.N. and Frey, K.J. (1977) Selection of superior cultivars of oats by using regression coefficients. *Crop Science* 17, 101–105.

Eberhart, S.A. and Russell, W.A. (1966) Stability parameters for comparing varieties. *Crop Science* 6, 36–40.

Eisemann, R.L., Cooper, M. and Woodruff, D.R. (1990) Beyond the analytical methodology – Better interpretation of genotype-by-environment interaction. In: Kang, M.S. (ed.) *Genotype-by-environment Interaction and Plant Breeding*. Louisiana State University, Baton Rouge, Louisiana, pp. 108–117.

Finlay, K.W. and Wilkinson, G.N. (1963) The analysis of adaptation in a plant breeding programme. *Australian Journal of Agricultural Research* 14, 742–754.

Fox, P.N. and Rosielle, A.A. (1982) Reducing the influence of environmental main-effects on pattern analysis of plant breeding environments. *Euphytica* 31, 645–656.

Gauch, H.G. and Furnas, R.E. (1991) Statistical analysis of yield trials with MATMODEL. *Agronomy Journal* 83, 916–20.

Huehn, M. (1990a) Nonparametric measures of phenotypic stability. Part 1. Theory. *Euphytica* 47, 189–194.

Huehn, M. (1990b) Nonparametric measures of phenotypic stability. Part 2. Applications. *Euphytica* 47, 195–2001.

Huehn, M. and Truberg, B. (2002) Contributions to the analysis of genotype × environment interactions: Theoretical results of the application and comparison of clustering techniques for the stratification of field test sites. *Journal of Agronomy and Crop Science* 188, 65–72.

Jain, K.C., Pandya, B.P. and Pande, K. (1984) Stability of yield components of chickpea genotypes. *Indian Journal of Genetics* 44, 159–163.

Kang, M.S. (1998) Using genotype-by-environment interaction for crop cultivar development. *Advances in Agronomy* 62, 199–252.

Katiyar, P.K., Ram, L. and Katiyar, R.P. (1992) Phenotypic stability of advance generation populations of chickpea for yield and its components. *Indian Journal of Genetics and Plant Breeding* 52, 341–346.

Khan, I.A., Malik, B.A. and Bashir, M. (1988) Investigation of genotype × environment interaction for seed yield in chickpea (*Cicer arietinum* L.). *Pakistan Journal of Botany* 20, 201–204.

Khorgade, P.W., Rathi, V.B., Narkhede, M.N. and Sakhare, B.A. (2000) Genotype × environment interaction and stability parameters in chickpea. *Annals of Plant Physiology* 14, 32–35.

Knight, R. (1970) The measurement and interpretation of genoytype–environment interactions. *Euphytica* 19, 225–235.

Kroonenberg, P.M. (1997) *Introduction to Biplots for G × E Tables.* Report no. 51, Department of Mathematics, University of Queensland, Brisbane.

Kumar, J., Singh, K.B., Malhotra, R.S., Miranda, J.H. and Das, G.T. (1996) Genotype × environment interaction for seed yield in chickpea. *Indian Journal of Genetics and Plant Breeding* 56, 69–78.

Kumar, S., Singh, O., Van Rheenen, H.A. and Rao, K.V.S. (1998) Repeatability of different stability parameters for grain yield in chickpea. *Plant Breeding* 117, 143–146.

Lawn, R.J. and Imrie, B.C. (1993) Exploiting physiological understanding in crop improvement. In: Imrie, B.C. and Hacker, J.B. (eds) *Tenth Australian Plant Breeding Conference–Focused Plant Improvement: Towards Responsible and Sustainable Agriculture*. Gold Coast, Queensland, pp. 136–146

Leon, J. and Becker, H.C. (1988) Repeatability of some statistical measures of phenotypic stability-correlation between single year results and multi year results. *Plant Breeding* 100, 137–142.

Lin, C.S., Binns, M.R. and Leftkovitch, L.P. (1986) Stability analysis: Where do we stand? *Crop Science* 26, 894–900.

Malhotra, R.S. and Singh, K.B. (1991) Classification of chickpea growing environments to control genotype by environment interaction. *Euphytica* 58, 5–12.

Mathur, P.K. and Horst, P. (1994) Methods for evaluating genotype–environment interactions illustrated by laying hens. *Journal of Animal Breeding and Genetics* 111, 265–288.

McLaren, C.G. (1996) Method of data standardization used in pattern analysis and AMMI models for the analysis of international multi-environment trials. In: Cooper, M. and Hammer, G.L. (eds) *Plant Adaptation and Crop Improvement*. CAB International, Wallingford, UK, pp. 225–242.

Mehra, R.B., Bahl, P.N. and Pahuja, A.N. (1980) Phenotypic stability and adaptability of newly developed varieties of chickpea in Northern India. *Indian Journal of Agricultural Science* 50, 218–222.

Mhase, L.B., Manjare, M.R., Aher, R.P. and Kolte, T.B. (1998) Stability for grain yield in chickpea under rainfed condition. *Indian Journal of Pulses Research* 11, 120–122.

Nurminiemi, M. and Rognli, O.A. (1996) Regression analysis of yield stability is strongly affected by companion test varieties and locations–Examples from a study of Nordic barley lines. *Theoretical And Applied Genetics* 93, 468–476.

Nyquist, W.E. (1991) Estimation of heritability and prediction of selection response in plant populations. *Critical Reviews in Plant Sciences* 10, 235–322.

Perkins, J.N. and Jinks, J.L. (1968) Environmental and genotype–environmental components of variability. III. Multiple lines and crosses. *Heredity* 23, 339–356.

Piepho, H.P. (1998) Methods for comparing the yield stability of cropping systems–A review. *Journal of Agronomy and Crop Science* 180, 193–213.

Pinthus, M.J. (1973) Estimate of genotypic value: A proposed method. *Euphytica* 22, 121–123.

Plaisted, R.L. (1960) A shorter method for evaluating the ability of selections to yield consistently over locations. *American Potato Journal* 37, 166–172.

Plaisted, R.L. and Peterson, L.C. (1959) A technique for evaluating the ability of selections to yield consistently in different locations or seasons. *American Potato Journal* 36, 381–385.

Popalghat, G.R., Patil, J.V., Deshmukh, R.B. and Mhase, L.B. (1999) Stability for yield and yield components in chickpea (*Cicer arietinum* L.). *Legume Research* 22, 254–258.

Ramanujam, S. and Gupta, V.P. (1974) Stability of yield and its components in Bengal gram and its bearing on plant type. *Indian Journal of Genetics* 34, 757–763.

Rubio, J., Flores, F., Moreno, M.T., Cubero, J.I. and Gil, J. (2004) Effects of the erect/bushy habit, single/double pod and late/early flowering genes on yield and seed size and their stability in chickpea. *Field Crops Research* 90, 255–262.

Shukla, G.K. (1972) Some statistical aspects of partitioning genotype–environmental components of variablity. *Heredity* 29, 237–245.

Singh, G., Brar, H.S. and Verma, M.M. (1991) Genotype environment interaction and phenotypic stability in chickpea. *Crop Improvement* 18, 135–140.

Singh, K.P. and Bejiga, G. (1990) Analysis of stability for some characters in kabuli chickpea. *Euphytica* 49, 223–227.

Singh, K.P., Singh, V.P. and Sareen, P.K. (1995) Stability of yield and its components in chickpea (*Cicer arietinum* L.). *International Journal of Tropical Agriculture* 13, 191–197.

Singh, L., Sharma, D., Baghel, S.S., Tomer, G.S. and Mishra, P.K. (1974) Estimation of genetic and environmental variability in Bengal gram. *SABRAO Journal* 6, 207–211.

Singh, O. and Kumar, S. (1994) Phenotypic stability of yield and related characters in desi gram (Cicer arietinum). *Indian Journal of Agricultural Science* 64, 815–820.

Singh, P.K., Singh, N.B. and Bharti, R.C. (1993) Divergence and stability analysis in chickpea. *Crop Improvement* 20, 59–62.

Singh, V. and Singh, F. (1991) Stability of yield and yield component characters in chickpea. *Indian Journal of Genetics and Plant Breeding* 51, 183–189.

Skroppa, T. (1984) A critical evaluation of methods available to estimate the genotype × environment interaction. *Studia Forestalia Sueccia* 166, 3–14.

Sneller, C.H., KilgoreNorquest, L. and Dombek, D. (1997) Repeatability of yield stability statistics in soybean. *Crop Science* 37, 383–390.

Sood, B.C., Bhambota, S.K. and Gartan, S.L. (2001) Studies on genotype × environment interaction and stability in chickpea. *Indian Journal of Pulses Research* 14, 31–33.

Tomer, G.S., Singh, L., Sharma, D. and Deodhar, A.D. (1973) Phenotypic stability of yield and some seed characteristics in Bengal gram (*Cicer arietinum* L.) varieties. *Jawaharlal Nehru Krishi Vishwa Vidyalaya Journal* 7, 35–39.

Truberg, B. and Huehn, M. (2002) Contributions to the analysis of genotype × environment interactions: Experimental results of the application and comparison of clustering techniques for the stratification of field test sites. *Journal of Agronomy and Crop Science* 188, 113–122.

Westcott, B. (1986) Some methods of analyzing genotype–environment interaction. *Heredity* 56, 243–253.

Wricke, G. (1962) On a method of understanding the biological diversity in the field research. *Zeitschrift fuer Pflanzenzuec-htung.* 47, 92–96.

Yadava, H.S., Prasad, K.V.V. and Agrawal, S.C. (2000) Stability in resistance against wilt in chickpea (*Cicer arietinum* L.). *Indian Journal of Agricultural Sciences* 70, 345–348.

Yadavendra, J.P. and Dixit, S.K. (1987) Stability for seed protein in chickpea. *Indian Journal of Agricultural Sciences* 57, 876–879.

Index